AF357328

10th Anniversary of *Inorganics*: Inorganic Materials

10th Anniversary of *Inorganics*: Inorganic Materials

Editors

Roberto Nisticò
Torben R. Jensen
Luciano Carlos
Hicham Idriss
Eleonora Aneggi

Basel • Beijing • Wuhan • Barcelona • Belgrade • Novi Sad • Cluj • Manchester

Editors

Roberto Nisticò
Department of Materials
Science
University of Milano-Bicocca
Milano
Italy

Torben R. Jensen
Department of Chemistry and
Interdisciplinary Nanoscience
Center
Aarhus University
Aarhus
Denmark

Luciano Carlos
Instituto de Investigación y
Desarrollo en Ingeniería de
Procesos, Biotecnología y
Energías Alternativas,
PROBIEN (CONICET-UNCo)
Universidad Nacional Del
Comahue
Neuquén
Argentina

Hicham Idriss
Institute of Functional
Interfaces (IFG)
Karlsruhe Institute of
Technology (KIT)
Karlsruhe
Germany

Eleonora Aneggi
Department of Agricultural,
Food, Environmental and
Animal Sciences (Di4A)
University of Udine
Udine
Italy

Editorial Office
MDPI
St. Alban-Anlage 66
4052 Basel, Switzerland

This is a reprint of articles from the Special Issue published online in the open access journal *Inorganics* (ISSN 2304-6740) (available at: www.mdpi.com/journal/inorganics/special_issues/5V4QCV6608).

For citation purposes, cite each article independently as indicated on the article page online and as indicated below:

Lastname, A.A.; Lastname, B.B. Article Title. *Journal Name* **Year**, *Volume Number*, Page Range.

ISBN 978-3-7258-0408-5 (Hbk)
ISBN 978-3-7258-0407-8 (PDF)
doi.org/10.3390/books978-3-7258-0407-8

Contents

About the Editors

Roberto Nisticò

Roberto Nistico (RN) is Fixed-Time Assistant Professor (RTD-b), i.e., tenure track Associate Professor, in General and Inorganic Chemistry at the University of Milano-Bicocca (Department of Materials Science, Italy). Up until today, RN published more than 80 papers on renowned international journals (mostly as first and/or corresponding author) and two book chapters. Additionally, RN is member of the Editorial Board of the journal Inorganics (Inorganic Materials Section). The research activity of RN mainly focuses on several aspects at the interface between inorganic chemistry, nanotechnology and materials science, always looking for novel and appealing solutions for a sustainable future. The principal fields of interest are the development of magnet-responsive nanomaterials (*i.e.,* iron oxides, ferrites) and other inorganic systems for the environmental remediation of contaminated wastewater, (photo)catalysis, and for energetic applications. Other fields of interest are functional coatings and the surface functionalization of nanoparticles, mesoporous oxides, metallic nanomaterials, and inorganic fillers for new generations of nanocomposites.

Torben R. Jensen

Torben R. Jensen (TRJ) obtained a PhD degree in Materials Chemistry at the University of Southern Denmark, Odense, in 1999. In October 1998, he joined RisøNatl. Lab as a post-doc, where he changed research topic to Biophysics and conducted surface diffraction and reflectivity of lipids and enzymes. Much of his research was conducted at the synchrotron Hasylab, Hamburg. After 2 years, Oct. 2000, he became an Assistant Prof. at the Department of Chemistry, Aarhus University, and became associated with the Interdisciplinary Nanoscience Center (iNANO). Since then, he has assembled a unique and productive independent research group within energy materials science. His research achievements led to a Doctor of Science (D.Sc.) degree in 2014 and a promotion to Professor in 2016. Over the years, T.R. Jensen developed several new courses in chemistry and nanoscience and educated 16 post-docs, 19 PhDs, 44 MScs, 61 BSc students and hosted about 45 guest professors and researchers at all levels. His research group synthesised a multitude of new metal hydrides by combining solvent-based methods, mechanochemistry, and solid–gas reactions. They also infiltrated materials in nanoporous scaffolds and conducted systematic studies of properties as a function of pore size and surface area to investigate their nanoeffects. This group now focuses on cationic conductivity in the solid state and on the development of novel battery materials. This group is a frequent user of large synchrotron X-ray and neutron facilities and has developed knowhow and new versatile in situ powder X-ray diffraction (PXD) sample environments and gas control systems for the investigation of solid–gas and solid–liquid reactions, also at extreme p,T conditions of up to $p(H_2) = 700$ bar. T. R. Jensen has published >310 peer-reviewed papers.

Luciano Carlos

Luciano Carlos was born in 1978, in Neuquén, Argentina. He received his Chemical degree in 2003 and he obtained his PhD in Chemistry from the National University of La Plata in 2008. From 2009 to 2010, he obtained a postdoctoral position at Institute of Theoretical and Applied Research on Physical Chemistry (INIFTA, La Plata, Argentina). Since, 2011 he has been a member of the CONICET research career. Currently, he is an Independent Researcher and he leads the research group "Applied Nanomaterials" at the Institute of Research and Development in Process Engineering, Biotechnology and Alternative Energies (PROBIEN, Neuquén, Argentina). His research interests encompass the design and functionalization of nanostructured surfaces, the development of nanomaterials with environmental applications, including photocatalysts, catalysts and adsorbents, and nanomaterials of interest in medicine for controlled drug delivery and alternative therapies.

Hicham Idriss

Hicham Idriss received his BSc, MSc, PhD, and Habilitation from the U. of Strasbourg (France). He started his academic career at the Department of Chemistry, U. of Auckland in New Zealand (1995 to 2008), then moved to the U. of Aberdeen and Robert Gordon U. (UK) to hold the Aberdeen Energy Futures Chair and Prof. of Chemistry position. He moved to corporate research at SABIC/KAUST (Saudi Arabia) in 2011. After retiring from SABIC in 2021, Hicham moved to the Institute of Functional Interfaces at KIT (Germany) and is also a visiting scientist at the Joint Research Center of the European Commission (Karlsruhe). His main research area is in catalysis and surface reactions of oxides. Hicham is also Professor (Hon.) at the Department of Chemistry, University College London, UK, since 2013.

Eleonora Aneggi

Dr. Eleonora Aneggi is a temporary researcher (RTD B) in General and Inorganic Chemistry at the University of Udine. She graduated in Chemistry with full marks and honours in 2001 at University of Trieste, and in 2007, she obtained her PhD in Chemical and Energy Technologies at the Department of Chemical Sciences and Technologies of the University of Udine. Since 2009, he has carried out regular teaching activities at the University of Udine. Her main scientific interests are catalytic processes, mainly related to environmental. Her research activity involves the characterization and development of advanced catalytic materials for the abatement of atmospheric pollutants and of biorecalcitrant organic compounds in liquid waste. Her research is also focused on the development of innovative oxidation catalysts based on transition metals with traditional and advanced techniques with lower environmental impact, for sustainable chemistry. She worked on various national and international projects and she collaborates with various public and private research institutions in the study and development of catalysts. She is regularly called upon as reviewer in the field of catalysis for international journals. She is author or co-author of more than 70 scientific publications in top international journals.

Editorial

10th Anniversary of *Inorganics*: Inorganic Materials

Roberto Nisticò [1,*], Hicham Idriss [2], Luciano Carlos [3], Eleonora Aneggi [4] and Torben R. Jensen [5,*]

1 Department of Materials Science, University of Milano-Bicocca, INSTM, Via R. Cozzi 55, 20125 Milano, Italy
2 Institute of Functional Interfaces (IFG), Karlsruhe Institute of Technology (KIT), 76131 Karlsruhe, Germany; hicham.idriss@kit.edu
3 Instituto de Investigación y Desarrollo en Ingeniería de Procesos, Biotecnología y Energías Alternativas, PROBIEN (CONICET-UNCo), Universidad Nacional Del Comahue, Neuquén 8300, Argentina; luciano.carlos@probien.gob.ar
4 Department of Agricultural, Food, Environmental and Animal Sciences (Di4A), University of Udine, 33100 Udine, Italy; eleonora.aneggi@uniud.it
5 Department of Chemistry and Interdisciplinary Nanoscience Center, Aarhus University, DK-8000 Aarhus, Denmark
* Correspondence: roberto.nistico@unimib.it (R.N.); trj@chem.au.dk (T.R.J.)

Citation: Nisticò, R.; Idriss, H.; Carlos, L.; Aneggi, E.; Jensen, T.R. 10th Anniversary of *Inorganics*: Inorganic Materials. *Inorganics* **2024**, *12*, 62. https://doi.org/10.3390/inorganics12030062

Received: 5 February 2024
Revised: 12 February 2024
Accepted: 17 February 2024
Published: 20 February 2024

1. Introduction

To celebrate the 10th anniversary of the journal *Inorganics*, the "Inorganic Materials" section launched this Special Issue entitled "10th Anniversary of *Inorganics*: Inorganic Materials", which collected 25 interesting papers (i.e., 20 articles, 1 communication, and 4 reviews), thus becoming the second most valued Special Issue in terms of the number of published contributions to the journal *Inorganics* (the first Special Issue in the "Inorganic Materials" section,). Moreover, despite the fact that the "Inorganic Materials" section in *Inorganics* is very recent (i.e., the beginning of 2022), it has rapidly grown, becoming an important cornerstone of the journal. The reason for this positive result is the continuously growing demand for advanced functional inorganic materials in a large variety of technological fields and applications.

Furthermore, the recent and numerous public demonstrations in support of climate and ecological justice, and the energy crisis have revealed the actual importance of technological sustainability. The "Inorganic Materials" section in *Inorganics* strongly supports a transition towards a 'green' and sustainable future based on renewable energy and with closed life-cycles for all used materials. The aim of the "Inorganic Materials" section is to serve as a medium for ground-breaking research that forms the fundamentals for new technologies still unknown today.

Therefore, this Special Issue is composed of a collection of several contributions mainly focused on the sustainable production of inorganic materials following alternative eco-friendly methods, new protocols and strategies for the reuse of materials (saving minerals and raw materials), the reduction in waste production, and environmental clean-up approaches.

The final aims of this Special Issue are to increase the knowledge of the latest advances, highlight challenges, address unresolved issues, and present newly emerging areas of interest involving the sustainable use of inorganic materials.

Prior to proceeding with the overview of the contributions, the Guest Editors would like to thank all the reviewers who spent their valuable time thoroughly reviewing and improving the articles published in this volume. We also sincerely thank all the authors for choosing "Inorganics Materials" as the section in which to publish their excellent science.

2. An Overview of Published Articles

As expressed above, the scope of this collection covers the entire focus area where inorganic materials can play a key role in order to reach a sustainable future, and this is exemplified by the various topics covered by the 25 articles published in this Special Issue. This section provides a brief overview of the contributions, organizing them into discreet subsections that include: sustainable synthesis, surface modifications, structural properties and reactivity, environmental remediation and catalysis, energy, composites, and nanomaterials against COVID-19 (Table 1).

Table 1. Correlation between subsections and contributions collected in the present Special Issue.

Subsections	Contribution No.	Title
Sustainable synthesis	1	Preparation of environmentally friendly BiVO4@SiO$_2$ encapsulated yellow pigment with remarkable thermal and chemical stability
	2	Microwave-mediated synthesis and characterization of Ca(OH)$_2$ nanoparticles destined for geraniol encapsulation
	3	Alternative synthesis of MCM-41 using inexpensive precursors for CO$_2$ capture
	4	Continuous and intermittent planetary ball milling effects on the alloying of a bismuth antimony telluride powder mixture
	5	Tetragonal nanosized zirconia: hydrothermal synthesis and its performance as a promising ceramic reinforcement
Surface modifications	6	Modification of graphite/SiO$_2$ film electrodes with hybrid organic–inorganic perovskites for the detection of vasoconstrictor Bisartan 4-butyl-N,N-bis{[2-(2H-tetrazol-5-yl)biphenyl-4-yl]methyl}imidazolium bromide
	7	Enhanced thermal stability of sputtered TiN thin films for their applications as diffusion barriers against copper interconnect
Structural properties and reactivity	8	Combination of multiple operando and in-situ characterization techniques in a single cluster system for atomic layer deposition: unraveling the early stages of growth of ultrathin Al$_2$O$_3$ films on metallic Ti substrates
	9	Gd$_2$O$_3$ doped UO$_2$(s) corrosion in the presence of silicate and calcium under alkaline conditions
	10	Incorporation of antimony ions in heptaisobutyl polyhedral oligomeric silsesquioxanes
	11	Hydrazine oxidation in aqueous solutions I: N$_4$H$_6$ decomposition
Environmental remediation and catalysis	12	Photocatalytic degradation of emerging contaminants with N-doped TiO$_2$ using simulated sunlight in real water matrices
	13	Synthesis of polystyrene@TiO$_2$ core–shell particles and their photocatalytic activity for the decomposition of methylene blue
	14	Method for decontamination of toxic aluminochrome catalyst sludge by reduction of hexavalent chromium
	15	Hydrogen incorporation in Ru$_x$Ti$_{1-x}$O$_2$ mixed oxides promotes total oxidation of propane
	16	Superoxide radical formed on the TiO$_2$ surface produced from Ti(OiPr)$_4$ exposed to H$_2$O$_2$/KOH
Energy	17	Binder-free CoMn$_2$O$_4$ nanoflower particles/graphene/carbon nanotube composite film for a high-performance lithium-ion battery
	18	Hydrogen storage properties of economical graphene materials modified by non-precious metal nickel and low-content palladium
	19	Integration of CO$_2$ capture and conversion by employing metal oxides as dual function materials: recent development and future outlook
	20	ZnO-doped CaO binary core-shell catalysts for biodiesel production via Mexican palm oil transesterification

Table 1. *Cont.*

Subsections	Contribution No.	Title
Composites	21	Basalt-fiber-reinforced phosphorus building gypsum composite materials (BRPGCs): an analysis on their working performance and mechanical properties
	22	The mechanical properties of geopolymers from different raw materials and the effect of recycled gypsum
	23	How to address flame-retardant technology on cotton fabrics by using functional inorganic sol–gel precursors and nanofillers: flammability insights, research advances, and sustainability challenges
Nanomaterials against COVID-19	24	Nanomaterials used in the preparation of personal protective equipment (PPE) in the fight against SARS-CoV-2
	25	Emerging nanomaterials biosensors in breathalyzers for detection of COVID-19: future prospects

2.1. Sustainable Synthesis

The encapsulation of chemicals is an area of research undergoing rapid growth, as it can be used for improving the efficiency, stability, compatibility, safety, and the delivery of end products, thus, allowing their use into a multitude of applications [1,2]. Chen et al. (Contribution 1) reported the encapsulation of $BiVO_4$, a brilliant yellow pigment characterised by having a poor thermo-chemical stability, in a SiO_2 matrix with the formation of a core-shell structure following a sol–gel process, evaluating the effects due to both pre-treatment conditions of the $BiVO_4$ dispersion, and the calcination temperature (350–800 °C) on the phase composition, morphology, and colour-rendering properties. The resulting encapsulated pigment showed comparable chromatic parameters to pure $BiVO_4$, coupled with an enhanced thermal stability (up to 700 °C), resistance against acid corrosion, and good compatibility in PP matrix. Tryfon et al. (Contribution 2) evaluated the possibility of encapsulating geraniol in hydrophobic $Ca(OH)_2$ nanoparticles coated with oleylamine following a one-step microwave-assisted synthesis. The results indicated an efficient encapsulation and loading of geraniol, a release rate of geraniol which can be modulated by varying the pH and temperature, and an antifungal inhibition activity against *B. cinerea*.

Another interesting field of research related to inorganic materials is the development of substrates for the capture and storage of CO_2 [3]. In this context, the use of porous materials as CO_2 sorbents is gaining substantial attraction due to many advantages, such as the low energy requirements for the adsorbent regeneration, high adsorption capacity, and selectivity for CO_2. Among the different systems proposed for this purpose, mesoporous SiO_2 has emerged as particularly noteworthy candidate, due to its well-defined pore structures, high surface area, and chemical surface composition suitable for functionalisation with reactive species selective for the CO_2 chemisorption. Aquino et al. (Contribution 3) reported the synthesis of an MCM-41 SiO_2 starting from low-cost industrial sources of silicon and surfactants. Templates were removed by either oxidant thermal treatments or simple washing steps. Finally, the surface of porous SiO_2 particles was modified by grafting with 3-aminopropyltrimethoxysilane to enhance the CO_2 adsorption capability. The results indicated that: (i) silanol groups are retained under mild conditions, (ii) all amino-functionalised materials showed performances as CO_2 adsorbents comparable to those reported in the literature, with better performance for materials with higher concentration of silanol groups, and (iii) chemisorbed gas is retained, thus suggesting the potential of these inorganic materials for the CO_2 storage.

Samourgkanidis et al. (Contribution 4) investigated the effects of continuous and in-steps mechanical alloying of a bismuth antimony telluride ($Bi_{0.4}Sb_{1.6}Te_{3.0}$) powder mixture via the mechanical planetary ball milling process as a function of milling time and powder mixture amount, and the thermoelectric properties of the alloys obtained at optimised conditions. Because of the mixture's high agglomeration tendency, the results revealed a significant difference in the alloying process in terms of milling time and powder mixture quantity, varying the ball milling process. Regarding the thermoelectric properties of the produced powders, the results revealed a very good thermoelectric profile, consistent with the literature.

ZrO_2 is one of the most studied and industrially applied metal oxides extensively exploited mostly in gas sensing and in biomedicine. Most of the properties of ZrO_2 are strongly influenced by its crystalline organization and the consequent interphase transformation [4]. In this context, Liu et al. (Contribution 5) reported the synthesis of a pure tetragonal phase ZrO_2 using a hydrothermal route in the presence of propanetriol as additive, finding that an appropriate amount of propanetriol was favourable for the generation of the pure tetragonal-phase, whereas the increment of the hydrothermal temperature gradually transformed the monoclinic phase into the tetragonal phase. Interestingly, ultrasonication and mechanical stirring influenced the particle size and distribution, whereas the addition of dispersant effectively alleviated the occurrence of agglomeration phenomena. The resulting nanoparticles were introduced into a 3Y-ZrO_2 (3:1) matrix to produce high-performance ZrO_2 nanopowders with improved flexural strength.

2.2. Surface Modifications

The majority of the electrochemical methods for the modification of electrodes needs the growth of nanostructured porous materials to enhance and increase the electroactive surface area and to improve their mechanical strength and chemical stability [5]. In fact, by using a modifier, it is possible to dramatically improve both the sensitivity and selectivity of an electrode. Papathanidis et al. (Contribution 6) reported the surface functionalisation of a graphite paste/SiO_2 film electrode with a hybrid organic–inorganic perovskite semiconductor to fabricate a promising, sensitive sensor for the vasoconstrictor, bisartan BV6 (an angiotensin II receptor blocker). The perovskite-modified film electrode exhibits an enhanced electrocatalytic activity towards the oxidation of BV6 with a relatively low LOD, high reproducibility, and stability towards BV6 determination.

Interconnect materials (e.g., Cu) are favourable substrates for integration circuits, owing to their high resistance to electro-migration and low electrical resistivity. However, the electrical performance of these circuits is severely damaged/degraded since Cu diffuses into the Si/SiO_2 substrates, thus refractory metal nitrides have been used as diffusion barriers in Cu metallization [6]. Aljaafari et al. (Contribution 7) investigated the deposition of TiN thin film on Si/SiO_2 substrates using a direct current sputtering technique and its application as a diffusion barrier film against Cu interconnect material. The results showed that TiN film can be successfully used as the diffusion barrier for metallization in Cu up to a temperature of 700 °C, whereas at higher temperatures, the diffusion of Cu through the TiN film barrier occurred, with formation of various copper silicide phases.

2.3. Structural Properties and Reactivity

Fundamental research focused on the structure–reactivity relationship in inorganic materials is a very important activity for favouring the integration of this class of materials into different technological fields of emerging interest. For instance, Morales et al. (Contribution 8) reported a new ultra-high vacuum cluster tool to perform systematic studies of the early growth stages of atomic layer-deposited ultrathin Al_2O_3 films on metallic Ti substrates following a surface science approach combining both operando and in situ characterisation techniques. The role of the metallic substrate at both low- and high-temperature conditions has been discussed, considering the hetero- and homo-deposition growth regimes, thus favouring the optimisation of the deposition process, and decreasing the amount of wasted precursor and associated costs. Interestingly, Garcia-Gomez et al. (Contribution 9) investigated the anodic reactivity of UO_2 (i.e., radionuclides are usually located within a UO_2 matrix) and Gd_2O_3-doped UO_2 by electrochemical methods in slightly alkaline conditions in the presence of both silicate and calcium ions. The results showed that both Gd_2O_3 doping and the presence of silicate and calcium in solution are factors that decrease the reactivity of UO_2, which is beneficial for the assessment of repository safety of the spent nuclear fuel, as an important fraction of the radionuclide release would be inhibited. Lastly, Marchesi et al. (Contribution 10) reported the direct incorporation of Sb(V) ions into a polycondensed silsesquioxane (POSS) network based on heptaisobutyl POSS units through a corner-capping reaction carried out under mild experimental conditions, potentially exploitable as novel flame-retardant for polymeric composites.

Regarding the reactivity of inorganic compounds, Breza et al. (Contribution 11) investigated the oxidation mechanism involving a mixture of non-labelled and ^{15}N-labelled hydrazine in an aqueous solution to N_2 molecules, identifying the formation of N_4H_6 intermediate molecules, which decompose following a complex pathway involving different energetically favoured reactions.

2.4. Environmental Remediation and Catalysis

One of the strategies largely studied and adopted to effectively degrade organic pollutants in contaminated water is the employment of photocatalysts [7]. Gaggero et al. (Contribution 12) investigated the photodegradation performances of N-doped TiO_2 photocatalysts synthesized through both the sol–gel and hydrothermal methods with enhanced absorption of visible light for the abatement of some representative emerging contaminants (i.e., benzotriazole, bisphenol A, diclofenac, and sulfamethoxazole). Experiments conducted solely with the visible portion of sunlight exhibited a noticeable advantage in the use of N-doped materials and a sensitive efficiency of the photocatalysts produced through the sol–gel method, both in ultrapure and real water matrices.

Another important parameter which can dramatically affect the photocatalytic efficiency of TiO_2 is the tendency of nanoparticles to agglomerate. To solve this issue, TiO_2 particles have been immobilized on polymer substrates, finding a further difficulty in obtaining uniformly sized particles due to rapid sol–gel reaction rates of TiO_2 precursors. In their study, Toyama et al. (Contribution 13) explored the growth of homogeneous PS@TiO_2 core-shell particles following a sol–gel synthesis. The results showed that the PS@TiO_2 core–shell particles exhibited superior photo-activity against methylene blue than that of a comparable commercial reference sample.

Interestingly, Pyagay et al. (Contribution 14) investigated the neutralization of the harmful effects of aluminochrome catalyst sludge (for the presence of hexavalent chromium), a waste product from the petrochemical process of dehydrogenation of lower paraffins to C3–C5 olefins. In the present study, the reduction of hexavalent chromium was carried out with different reagents: Na_2SO_3, $FeSO_4$, $Na_2S_2O_3$, and $Na_2S_2O_5$, finding that sodium metabisulfite ($Na_2S_2O_5$) is the most preferred reagent working at a neutral pH, precipitating chromium in the form of hydroxide, thus saving the use of concentrated sulfuric acid and the additional precipitant reagents more commonly adopted.

Furthermore, Wang et al. (Contribution 15) reported a synthetic approach for the hydrogenation of mixed RuO_2/TiO_2 metal oxides, varying the oxide compositions. The effectiveness of the hydrogen insertion in the mixed oxide lattice has been evaluated by measuring the catalytic activity against the total oxidation of propane, reporting that after treating metal oxides with hydrogen the highest activity is encountered for compositions where both Ru and Ti have similar concentrations, whereas in the absence of the hydrogen treatment, the higher the Ru concentration the higher the activity. These results demonstrated that hydrogen insertion affects the electronic structure of the mixed RuO_2/TiO_2 metal oxides that is expected to be responsible for the improved catalytic activity. Samoilova et al. (Contribution 16) contributed a study dealing with the formation of superoxide radicals by treating $Ti(OR)_4$ alkoxides with H_2O_2 in the presence of KOH. The use of EPR spectroscopic techniques revealed the stabilization of superoxide radicals near the potassium ion and its involvement in the formation of a strong hydrogen bond with the surface.

2.5. Energy

The considerable progress of human societies since the late part of the 19th century has strongly depended on fossil fuels, with consequently serious environmental pollution problems and the persistent risk of energy depletion. To overcome this issue, novel energy sources and devices for efficient energy storage and conversion are under investigation [8,9]. In their contribution, Tong et al. (Contribution 17) reported on the design of flexible films made with $CoMn_2O_4$ nanoflowers supported on graphene/carbon nanotube 3D networks by means of a filtration strategy followed by a thermal treatment process. These films were deposited onto Ni foams without using binders and were successfully tested as electrodes in Li-batteries. Instead, Chen et al. (Contribution 18) investigated the hydrogen storage capacities of a Ni/Pd co-modified graphene material obtained via the solvothermal route, showing an excellent hydrogen storage performance due to the synergistic hydrogen spill-over effect of the Pd–Ni bimetal.

Tan et al. (Contribution 19) analysed the scientific literature describing the use of mixed metal oxides as dual function materials able to perform both CO_2 capture and CO_2 conversion. The authors recognised that the effectiveness of CO_2 capture and conversion depends on different factors, such as the basicity of the materials, surface area and porosity, crystallite size and dispersion of active sites, reducibility of the active metals, and availability of oxygen vacancies.

Arenas-Quevedo et al. (Contribution 20) investigated the catalytic production of biodiesel (i.e., a sustainable fuel) from Mexican palm oil in the presence of methanol using binary core-shell catalysts made by CaO and ZnO. During the optimisation process, the authors found that several factors (e.g., surface basicity, ZnO content, phase compositions, thermal treatment conditions) might influence the quality of the biodiesel produced.

2.6. Composites

Even in the field of composites, inorganic materials can play a crucial role, enhancing the mechanical properties of the composites or introducing specific properties and abilities [10,11]. For example, Wu et al. (Contribution 21) reported the preparation of a phosphorus-building gypsum composite reinforced with basalt fibres, registering a mechanical improvement in the composite due to the addition of the basalt fibres. However, even if the addition of fibres in the matrix promotes the mechanical properties, there is an adverse effect in terms of working performance (i.e., decrease in fluidity and setting time), thus, the authors evidenced that there is a trade-off between these two important aspects that should be considered for choosing the amount and length of the basalt fibres. Korhonen et al. (Contribution 22) investigated the mechanical response of geopolymers obtained from different raw materials (thus, with a different composition), evaluating the effect of the addition of recycled gypsum on these. The results indicated that the chemical composition of the raw materials had a significant impact on the properties of geopolymers. Furthermore, recycled gypsum affected the setting time of the geopolymers, depending on the calcium content, whereas it had no effect on the total shrinkage of the geopolymers.

Trovato et al. (Contribution 23) analysed the scientific literature describing the latest advances in the use of inorganic fillers and sol–gel-based flame-retardant technologies for textile treatments, with a clear focus on the cotton matrix. In this review, the enhanced safety of inorganic functional fillers with respect to their organic counterpart was evidenced, also providing some perspectives on the use of inorganic flame-retardant solutions coupled with environmentally suitable molecules, in line with the 'green chemistry' principles.

2.7. Nanomaterials against COVID-19

The coronavirus (COVID-19) pandemic arrived like a storm in our society, becoming one of the greatest global crises in recent history, touching many different aspects of our daily living. Apart from the health risks due to COVID-19 disease, what was really shocking was the lack of comprehensive strategies, plans, and toolboxes to manage COVID-19 by the governments of the entire world [12]. To hopefully avoid further serious situations like the ones that happened during the last pandemic, everyone has to try to contribute to this issue. Hence, even the field of inorganic chemistry can positively contribute against the COVID-19 disease, proposing interesting technological solutions involving inorganic materials. This is, for instance, the case of the review paper written by De Luca et al. (Contribution 24), where the key role played by nanomaterials to support the fight against COVID-19 was analysed. In particular, in this work, the authors reported the more representative examples of available personal protective equipment (e.g., masks and fabrics, sensors) involving the use of nanomaterials, explaining their mechanisms of action. Additionally, Rajendrasozhan et al. (Contribution 25), in their review paper, analysed the scientific literature describing the recent advancements in the potential use of integrating nanomaterial-based biosensors within breathalysers to favour the rapid detection of COVID-19, highlighting their principles, applications, and implications.

3. Conclusions

With this Special Issue "10th Anniversary of *Inorganics*" published in "Inorganics Materials" section, and also published as a book, the Editors hope that the high quality of the contributions collected here will receive the visibility and attention they deserve. These would help readers to increase their knowledge in the field of inorganic materials, and be a new source of inspiration for novel, focused investigations.

Acknowledgments: The Editors would like to thank all authors, reviewers, and the entire editorial staff of *Inorganics* who provided their new science, constructive recommendations, and assisted in the realisation of the present Special Issue.

Conflicts of Interest: The authors declare no conflicts of interest.

List of Contributions

1. Chen, R.; Zhang, X.; Tao, R.; Jiang, Y.; Liu, H.; Cheng, L. Preparation of environmentally friendly $BiVO_4$@SiO_2 encapsulated yellow pigment with remarkable thermal and chemical stability. *Inorganics* **2024**, *12*, 17. https://doi.org/10.3390/inorganics12010017.
2. Tryfon, P.; Kamou, N.N.; Mourdikoudis, S.; Vourlias, G.; Menkissoglu-Spiroudi, U.; Dendrinou-Samara, C. Microwave-mediated synthesis and characterization of $Ca(OH)_2$ nanoparticles destined for geraniol encapsulation. *Inorganics* **2023**, *11*, 470. https://doi.org/10.3390/inorganics11120470.
3. Aquino, G.D.; Moreno, M.S.; Piqueras, C.M.; Benedictto, G.P.; Pereyra, A.M. Alternative synthesis of MCM-41 using inexpensive precursors for CO_2 capture. *Inorganics* **2023**, *11*, 480. https://doi.org/10.3390/inorganics11120480.
4. Samourgkanidis, G.; Kyratsi, T. Continuous and intermittent planetary ball milling effects on the alloying of a bismuth antimony telluride powder mixture. *Inorganics* **2023**, *11*, 221. https://doi.org/10.3390/inorganics11050221.
5. Liu, S.; Wang, J.; Chen, Y.; Song, Z.; Han, B.; Wu, H.; Zhang, T.; Liu, M. Tetragonal nanosized zirconia: hydrothermal synthesis and its performance as a promising ceramic reinforcement. *Inorganics* **2023**, *11*, 217. https://doi.org/10.3390/inorganics11050217.
6. Papathanidis, G.; Ioannou, A.; Spyrou, A.; Madrapylia, A.; Kelaidonis, K.; Matsoukas, J.; Koutselas, I.; Topoglidis, E. Modification of graphite/SiO_2 film electrodes with hybrid organic–inorganic perovskites for the detection of vasoconstrictor Bisartan 4-butyl-N,N-bis{[2-(2H-tetrazol-5-yl)biphenyl-4-yl]methyl}imidazolium bromide. *Inorganics* **2023**, *11*, 485. https://doi.org/10.3390/inorganics11120485.
7. Aljaafari, A.; Ahmed, F.; Shaalan, N.M.; Kumar, S.; Alsulami, A. Enhanced thermal stability of sputtered TiN thin films for their applications as diffusion barriers against copper interconnect. *Inorganics* **2023**, *11*, 204. https://doi.org/10.3390/inorganics11050204.
8. Morales, C.; Mahmoodinezhad, A.; Tschammer, R.; Kosto, J.; Chavarin, C.A.; Schubert, M.A.; Wenger, C.; Henkel, K.; Flege, J.I. Combination of multiple operando and in-situ characterization techniques in a single cluster system for atomic layer deposition: unraveling the early stages of growth of ultrathin Al_2O_3 films on metallic Ti substrates. *Inorganics* **2023**, *11*, 477. https://doi.org/10.3390/inorganics11120477.

9. Garcia-Gomez, S.; Gimenez, J.; Casas, I.; Llorca, J.; De Pablo, J. Gd_2O_3 doped UO_2(s) corrosion in the presence of silicate and calcium under alkaline conditions. *Inorganics* **2023**, *11*, 469. https://doi.org/10.3390/inorganics11120469.

10. Marchesi, S.; Bisio, C.; Carniato, F.; Boccaleri, E. Incorporation of antimony ions in heptaisobutyl polyhedral oligomeric silsesquioxanes. *Inorganics* **2023**, *11*, 426. https://doi.org/10.3390/inorganics11110426.

11. Breza, M.; Manova, A. Hydrazine oxidation in aqueous solutions I: N_4H_6 decomposition. *Inorganics* **2023**, *11*, 413. https://doi.org/10.3390/inorganics11100413.

12. Gaggero, E.; Giovagnoni, A.; Zollo, A.; Calza, P.; Paganini, M.C. Photocatalytic degradation of emerging contaminants with N-doped TiO_2 using simulated sunlight in real water matrices. *Inorganics* **2023**, *11*, 439. https://doi.org/10.3390/inorganics11110439.

13. Toyama, N.; Takahashi, T.; Terui, N.; Furukawa, S. Synthesis of polystyrene@TiO_2 core–shell particles and their photocatalytic activity for the decomposition of methylene blue. *Inorganics* **2023**, *11*, 343. https://doi.org/10.3390/inorganics11080343.

14. Pyagay, I.; Zubkova, O.; Zubakina, M.; Sizyakov, V. Method for decontamination of toxic aluminochrome catalyst sludge by reduction of hexavalent chromium. *Inorganics* **2023**, *11*, 284. https://doi.org/10.3390/inorganics11070284

15. Wang, W.; Wang, Y.; Timmer, P.; Spriewald-Luciano, A.; Weber, T.; Glatthaar, L.; Guo, Y.; Smarsly, B.M.; Over, H. Hydrogen incorporation in $Ru_xTi_{1-x}O_2$ mixed oxides promotes total oxidation of propane. *Inorganics* **2023**, *11*, 330. https://doi.org/10.3390/inorganics11080330.

16. Samoilova, R.I.; Dikanov, S.A. Superoxide radical formed on the TiO_2 surface produced from $Ti(OiPr)_4$ exposed to H_2O_2/KOH. *Inorganics* **2023**, *11*, 274. https://doi.org/10.3390/inorganics11070274.

17. Tong, X.; Yang, B.; Li, F.; Gu, M., Zhan, X.; Tian, J.; Huang, S.; Wang, G. Binder-free $CoMn_2O_4$ nanoflower particles/graphene/carbon nanotube composite film for a high-performance lithium-ion battery. *Inorganics* **2023**, *11*, 314. https://doi.org/10.3390/inorganics11080314.

18. Chen, Y.; Habibullah; Xia, G.; Jin, C.; Wang, Y.; Yan, Y.; Chen, Y.; Gong, X.; Lai, Y.; Wu, C. Hydrogen storage properties of economical graphene materials modified by non-precious metal nickel and low-content palladium. *Inorganics* **2023**, *11*, 251. https://doi.org/10.3390/inorganics11060251.

19. Tan, W.J.; Gunawan, P. Integration of CO_2 capture and conversion by employing metal oxides as dual function materials: recent development and future outlook. *Inorganics* **2023**, *11*, 464. https://doi.org/10.3390/inorganics11120464.

20. Arenas-Quevedo, M.G.; Manriquez, M.E.; Wang, J.A.; Elizalde-Solis, O.; Gonzalez-Garcia, J.; Zuniga-Moreno, A.; Chen, L.F. ZnO-doped CaO binary core-shell catalysts for biodiesel production via Mexican palm oil transesterification. *Inorganics* **2024**, *12*, 51. https://doi.org/10.3390/inorganics12020051.

21. Wu, L.; Tao, Z.; Huang, R.; Zhang, Z.; Shen, J.; Xu, W. Basalt-fiber-reinforced phosphorus building gypsum composite materials (BRPGCs): an analysis on their working performance and mechanical properties. *Inorganics* **2023**, *11*, 254. https://doi.org/10.3390/inorganics11060254.

22. Korhonen, H.; Timonen, J.; Suvanto, S.; Hirva, P.; Mononen, K.; Jaaskelainen, S. The mechanical properties of geopolymers from different raw materials and the effect of recycled gypsum. *Inorganics* **2023**, *11*, 298. https://doi.org/10.3390/inorganics11070298.

23. Trovato, V.; Sfameni, S.; Debabis, R.B.; Rando, G.; Rosace, G.; Malucelli, G.; Plutino, M.R. How to address flame-retardant technology on cotton fabrics by using functional inorganic sol–gel precursors and nanofillers: flammability insights, research advances, and sustainability challenges. *Inorganics* **2023**, *11*, 306. https://doi.org/10.3390/inorganics11070306.

24. De Luca, P.; Nagy, J.B.; Macario, A. Nanomaterials used in the preparation of personal protective equipment (PPE) in the fight against SARS-CoV-2. *Inorganics* **2023**, *11*, 294. https://doi.org/10.3390/inorganics11070294.

25. Rajendrasozhan, S.; Sherwani, S.; Ahmed, F.; Shaalan, N.; Alsukaibi, A.; Al-Motair, K.; Khan, M.W.A. Emerging nanomaterials biosensors in breathalyzers for detection of COVID-19: future prospects. *Inorganics* **2023**, *11*, 483. https://doi.org/10.3390/inorganics11120483.

References

1. Andrade, B.; Song, Z.; Li, J.; Zimmerman, S.C.; Cheng, J.; Moore, J.S.; Harris, K.; Katz, J.S. New Frontiers for Encapsulation in the Chemical Industry. *ACS Appl. Mater. Interfaces* **2015**, *7*, 6359–6368. [CrossRef]

2. De Jongh, P.E.; Adelhelm, P. Nanosizing and Nanoconfinement: New Strategies Towards Meeting Hydrogen Storage Goals. *ChemSusChem* **2010**, *3*, 1332–1348. [CrossRef]

3. Zhang, S.; Chen, C.; Ahn, W.-S. Recent progress on CO_2 capture using amine-functionalized silica. *Curr. Opin. Green Sustain. Chem.* **2019**, *16*, 26–32. [CrossRef]

4. Nisticò, R. Zirconium oxide and the crystallinity hallows. *J. Austr. Ceram. Soc.* **2021**, *57*, 225–236. [CrossRef]

5. Nikolaou, P.; Vareli, I.; Deskoulidis, E.; Matsoukas, J.; Vassilakopoulou, A.; Koutselas, I.; Topoglidis, E. Graphite/SiO_2 film electrode modified with hybrid organic-inorganic perovskites: Synthesis, optical, electrochemical properties and application in electrochemical sensing of losartan. *J. Solid State Chem.* **2019**, *273*, 17–24. [CrossRef]

6. Petrov, I.; Hultman, L.; Helmersson, U.; Sundgren, J.E.; Greene, J.E. Microstructure modification of TiN by ion bombardment during reactive sputter deposition. *Thin Solid Films* **1989**, *169*, 299–314. [CrossRef]

7. Varma, R.; Yadav, M.; Tiwari, K.; Makani, N.; Gupta, S.; Kothari, D.C.; Miotello, A.; Patel, N. Roles of Vanadium and Nitrogen in Photocatalytic Activity of VN-Codoped TiO_2 Photocatalyst. *Photochem. Photobiol.* **2018**, *94*, 955–964. [CrossRef] [PubMed]

8. Reveles-Miranda, M.; Ramirez-Rivera, V.; Pacheco-Catalan, D. Hybrid energy storage: Features, applications, and ancillary benefits. *Renew. Sustain. Energy Rev.* **2024**, *192*, 114196. [CrossRef]

9. Guo, T.; Hu, P.; Li, L.; Wang, Z.; Guo, L. Amorphous materials emerging as prospective electrodes for electrochemical energy storage and conversion. *Chem.* **2023**, *9*, 1080–1093. [CrossRef]

10. Yadav, R.; Singh, M.; Shekhawat, D.; Lee, S.-Y.; Park, S.-J. The role of fillers to enhance the mechanical, thermal, and wear characteristics of polymer composite materials: A review. *Compos. Part A Appl. Sci. Manuf.* **2023**, *175*, 107775. [CrossRef]

11. Markandan, K.; Lai, C.Q. Fabrication, properties and applications of polymer composites additively manufactured with filler alignment control: A review. *Compos. Part B Eng.* **2023**, *256*, 110661. [CrossRef]

12. Mondino, E.; Scolobig, A.; Di Baldassarre, G.; Stoffel, M. Living in a pandemic: A review of COVID-19 integrated risk management. *Int. J. Disaster Risk Reduct.* **2023**, *98*, 104081. [CrossRef]

Article

Preparation of Environmentally Friendly BiVO$_4$@SiO$_2$ Encapsulated Yellow Pigment with Remarkable Thermal and Chemical Stability

Renhua Chen [1], Xiaozhen Zhang [1,2,*], Rui Tao [1,2,*], Yuhua Jiang [2], Huafeng Liu [1,2] and Lanlan Cheng [1]

[1] Jiangxi Jinhuan Pigments Co., Ltd., Yichun 336000, China
[2] School of Materials Science and Engineering, Jingdezhen Ceramic University, Jingdezhen 333403, China
* Correspondence: zhangxiaozhen@jcu.edu.cn (X.Z.); ponny2001@163.com (R.T.)

Abstract: The preparation of environmentally friendly inorganic encapsulated pigments with a bright color and sufficient stability provides an effective strategy for expanding their applications in plastic, paint, glass, and ceramic decoration. The challenges facing the use of such pigments include the formation of a dense protective coating with the required endurance, the relatively weak color of the encapsulated pigments, and the preferable inclusion particle size. Environmentally friendly BiVO$_4$ is regarded as a very promising pigment for multiple coloring applications due to its brilliant yellow color with high saturation. However, its poor thermal and chemical stability greatly limit the application of BiVO$_4$. Herein, we report a sol–gel method to synthesize inorganic BiVO$_4$@SiO$_2$ yellow pigment with a core–shell structure. By controlling the synthesis conditions, including the particle size and dispersion of BiVO$_4$ and the calcination temperature, a BiVO$_4$@SiO$_2$ encapsulated pigment with excellent chromatic properties was achieved. The obtained environmentally friendly BiVO$_4$@SiO$_2$ pigment with encapsulation modification has a comparable color-rendering performance to BiVO$_4$, and it has a high thermal stability at 700 °C, excellent acid resistance, and good compatibility in plastics. The present research is expected to expand the application of yellow BiVO$_4$ pigment in harsh environments.

Keywords: encapsulated pigment; BiVO$_4$@SiO$_2$; yellow color; thermal stability; resistance to chemical ero

Citation: Chen, R.; Zhang, X.; Tao, R.; Jiang, Y.; Liu, H.; Cheng, L. Preparation of Environmentally Friendly BiVO$_4$@SiO$_2$ Encapsulated Yellow Pigment with Remarkable Thermal and Chemical Stability. *Inorganics* **2024**, *12*, 17. https://doi.org/10.3390/inorganics12010017

Academic Editors: Torben R. Jensen, Eleonora Aneggi, Hicham Idriss, Roberto Nisticò and Luciano Carlos

Received: 4 December 2023
Revised: 22 December 2023
Accepted: 28 December 2023
Published: 30 December 2023

1. Introduction

Inorganic pigments have been applied in various products such as ceramics, enamels, paints/coatings, plastics, and glasses due to their high hiding power, weather resistance, UV stability, and thermal stability [1–7]. Among them, yellow pigments are particularly interesting because of their great applicability in many industries, such as the building, automotive, decorative paints, and plastics industries [7–10]. In particular, inorganic yellow pigments, with their striking color, are in great demand due to their usefulness in many applications. However, conventional industrial pigments such as chrome yellow (PbCrO$_4$) and cadmium yellow (CdS$_{1-x}$Se$_x$) contain toxic elements (Pb, Cr, Cd, Se, etc.), which have adverse effects on animals, plants, and the environment. Therefore, the development of novel environmentally friendly yellow pigments is especially important. Bismuth vanadate (BiVO$_4$), an eco-friendly type of yellow pigment, has attracted much attention due to its bright color, nontoxicity, and high hiding power [10–14]. This environmentally friendly inorganic yellow pigment has the potential to replace iron yellow, chrome yellow, cadmium yellow, nickel titanium yellow, and chromium titanium yellow pigments [7,15]. Simultaneously, because green or yellow is an eye-catching color, greenish-yellow BiVO$_4$ pigments can be used as traffic marks and building coatings, indicating that greenish-yellow BiVO$_4$ pigments seem to be promising candidates in terms of pigment safety and environmental friendliness [10,14,16].

Many studies on the doping modification of yellow $BiVO_4$ pigment via different synthesis methods have been carried out to tailor its phase structure, particle size, chromatic performance, and spectral characteristics [1,11–14,16–21]. For example, Sameera et al. synthesized V-site Nb-doped $BiVO_4$ via the citrate complexation route [19]. The resulting pigment exhibits excellent color characteristics that are comparable to those of commercial $BiVO_4$ and has a high NIR reflectivity. Wendusu et al. synthesized Bi-site La-doped $BiVO_4$ via a hydrothermal method [20]. The yellowness of the doped pigments (b* > 80) was higher than that of a commercially available $BiVO_4$ pigment. Ca/Zn codoped $BiVO_4$ was also synthesized via the evaporation-to-dryness method by Toshiyuki et al. [1], and the most vivid greenish-yellow hue was obtained for $BiVO_4$ with a high yellowness. It can be concluded from the above-mentioned studies that the doping process is effective in tailoring the properties of $BiVO_4$ pigments, such as the NIR reflectance and color, but it needs to be emphasized that this has no obvious positive effect on improving the thermal and chemical stability of $BiVO_4$. Furthermore, most of the doped pigments show a reddish-yellow color with a depressed b* value instead of the more striking greenish-yellow color for pure $BiVO_4$. Even when applied in plastic coloring, both pure and doped $BiVO_4$ pigments often suffer from discoloration due to the erosion of the plastic melt above 150 °C.

Interestingly, the formation of a core–shell structure can improve the thermal stability of these pigments through the modification of the encapsulation with more stable materials, such as $ZrSiO_4$, SiO_2, and Al_2O_3 [22–25]. For instance, the working temperature of $ZrSiO_4$-encapsulated carbon black via sol–gel-spraying could be effectively increased to 900 °C [22]. Although $ZrSiO_4$ is usually selected as a coating layer to protect the colorant, the synthesis temperature of the dense $ZrSiO_4$ layer is not less than 950 °C [24]. The too-high temperature required for the formation of a dense protective coating is extremely unfavorable for $BiVO_4$ due to its poor thermal stability; at these temperatures, the volatilization of V and thus the alteration of the phase structure and color of the pigment would be unavoidable. Furthermore, the V ions in $BiVO_4$ can react easily with zirconium hydroxide, one of the precursors for synthesizing $ZrSiO_4$ or ZrO_2, which brings about a significant color change in the pigment [26,27]. Thus, in the present work, SiO_2 was selected to encapsulate $BiVO_4$ pigments due to the lower temperature to form a dense surface protective coating. The thermal stability of yellow $BiVO_4$ pigments is similar to that of red γ-Ce_2S_3 pigment. Li and co-workers prepared core–shell γ-Ce_2S_3@SiO_2 red pigment via the Stober method, and the thickness of the dense SiO_2 layer was as thick as 140 nm after multiple coating [25,28]. The oxidization-resistant temperature was enhanced from 300 to 550 °C after coating with a dense SiO_2 layer for γ-Ce_2S_3 pigments. Similarly, the synthesized α-Fe_2O_3@SiO_2 composite revealed greatly incremented thermal stability, less color variation, and strong acid–alkali resistance compared with Fe_2O_3 [29].

In this study, we developed a sol–gel method to construct an environmentally friendly $BiVO_4$@SiO_2 encapsulated pigment. The pretreatment and dispersion of $BiVO_4$ powder, and the effect of calcination temperature on the phase composition, morphology, and color-rendering properties, were systematically studied to achieve bright-color and high-thermal-stability yellow $BiVO_4$@SiO_2 pigment. The spectroscopic properties and resistance to acidic solution and high-temperature plastic melt for the obtained $BiVO_4$@SiO_2 pigment were also investigated. It was found that the encapsulation did not weaken the chromatic performance of $BiVO_4$, while the thermal and chemical stability could be significantly enhanced. This present work opens up a vast space for expanding the application of brilliant yellow $BiVO_4$-based pigments in paints, coatings, plastics, and even ceramic, enamel, and glass decorations.

2. Results and Discussion

2.1. Pretreatment and Dispersion of BiVO₄ Pigment

Typically, a suitable particle size is preferred to improve the coating effect of a pigment. Therefore, a ball-milling process was used to optimize the particle size of the as-received $BiVO_4$ pigment powder. As displayed in Figure 1, the median diameter (D_{50}) of $BiVO_4$

decreased from 0.51 to 0.17 µm as the milling time increased from 0 to 4 h. When a pigment powder has a larger particle size, its specific surface area is smaller, and the wrapping rate is lower. In contrast, if the pigment particles are too small, which is equivalent to the particle size of SiO_2 synthesized via the sol–gel method, it causes electrostatic repulsion, resulting in a decrease in the encapsulation rate. In this study, after the milling time was increased up to 4 h, the particle size did not obviously reduce. As such, the $BiVO_4$ ball-milled for 4 h was selected for the following coating.

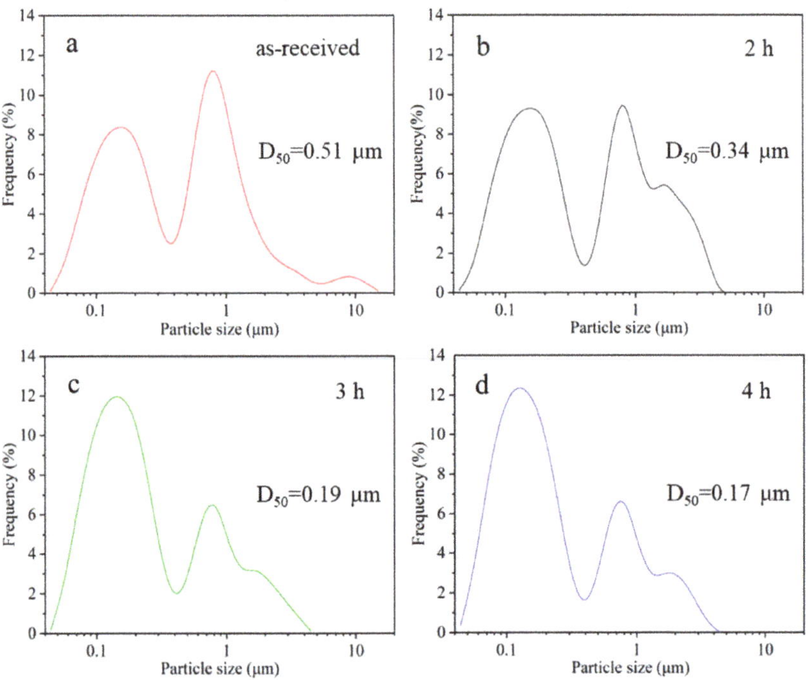

Figure 1. Particle size distributions of $BiVO_4$ (**a**) as-received or treated with ball milling for (**b**) 2 h, (**c**) 3 h, or (**d**) 4 h.

On the other hand, the dispersion of $BiVO_4$ in the reaction system could directly affect the encapsulation effect. Therefore, we used the dilute HCl and NaOH aqueous solutions to adjust the pH value of the $BiVO_4$ suspension and tested the corresponding zeta potentials. The isoelectric point of $BiVO_4$ particles in an aqueous solution was around pH = 6.5 (Figure 2). Under acidic conditions, the surface of $BiVO_4$ is positive, and the zeta potential is relatively low (<20 mV), which indicate the poor stability of the particles in aqueous solutions. Under alkaline conditions, the surface of $BiVO_4$ is negative; especially when the pH value is above 8, the zeta potential is around −40 mV. The much higher zeta potential under alkaline conditions is beneficial for achieving the stable dispersion of $BiVO_4$ particles to improve the encapsulation rate. In addition, OH^- nucleophilic catalysis under alkaline conditions is conducive to the condensation of hydrolytic TEOS, contributing to a dense SiO_2 layer [30]. In contrast, H^+ catalysis under acidic conditions tend to form a linear connection of SiO_2 precipitate with many hydroxyl groups on its surface, which can be easily combined with H_2O and other solvents in the system via intermolecular H bonds, resulting in a loose and porous SiO_2 layer. In addition, the resistance of $BiVO_4$ powder to alkaline conditions is much better than to acid. Therefore, this work carried out the SiO_2 coating process in an alkaline environment (pH ≈ 9.5).

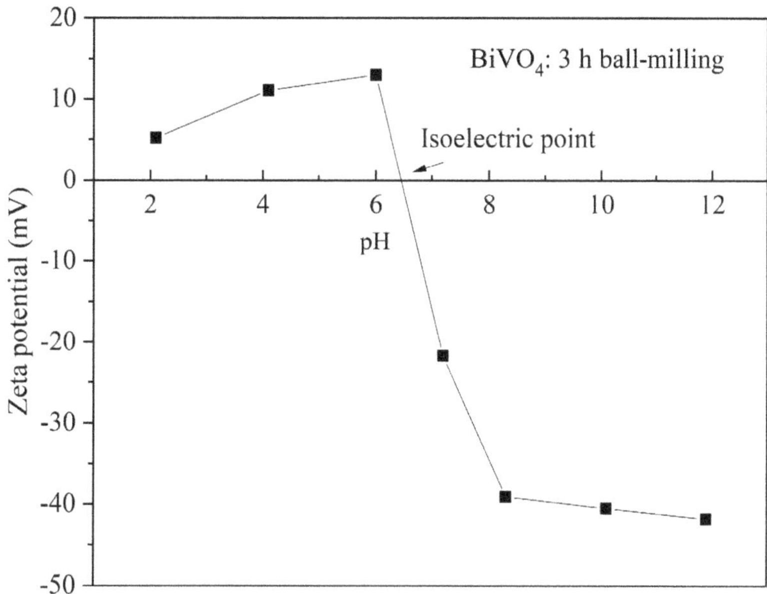

Figure 2. Zeta potentials of $BiVO_4$ at different pH values.

2.2. Phase Composition and Morphology of $BiVO_4@SiO_2$ Pigment

Figure 3a,b present the XRD results of the synthesized $BiVO_4@SiO_2$ pigments prepared at different calcination temperatures with a TEOS concentration of 0.5 mol/L. Only the monoclinic $BiVO_4$ (m-$BiVO_4$, PDF#14-0688) with high crystallinity could be observed for the samples calcinated at 350–750 °C. It is thought that the synthesized SiO_2 was in an amorphous state and could not be detected with XRD. Thus, the process to synthesize SiO_2 from TEOS in alkaline solutions had a negligible effect on the phase structure of $BiVO_4$. As the calcination temperature was further increased to above 800 °C, minor tetragonal SiO_2 (t-SiO_2, $2\theta = 21.94°$, PDF#71-0785) was detected (Figure 3b). Meanwhile, an additional weak diffraction peak at $2\theta = 18.7°$ appeared, which could be ascribed to tetragonal $BiVO_4$ (t-$BiVO_4$, PDF#14-0133) [18]. It is proposed that the scheelite m-$BiVO_4$ consists of isolated VO_4 tetrahedra, which are corner-connected by BiO_8 dodecahedra. At higher temperatures, m-$BiVO_4$ tends to transform to t-$BiVO_4$ due to the volatilization of some V ions, resulting from the defective SiO_2 coating of the encapsulated pigment calcined at 800 °C, as shown in the TEM image (Figure 4i). In t-$BiVO_4$, Bi^{3+} cations are located at the centrosymmetric site of the BiO_8 polyhedra, whereas m-$BiVO_4$ exhibits a distorted BiO_8 dodecahedron due to Bi^{3+} off-centering [31]. Table 1 lists the corresponding lattice parameters of the m-$BiVO_4$ phase for the as-received $BiVO_4$ and $BiVO_4@SiO_2$ prepared at 800 °C, which were calculated from the XRD results using MDI jade software (v6.5). It can be seen that although the monoclinic structure was maintained, tiny lattice distortion occurred after high-temperature calcination, along with a small increase in the lattice size, probably due to the loss of trace amounts of V ions. Particularly, a small reduction in the β angle was also observed, indicating the cell symmetry was slightly improved. These results also suggest that m-$BiVO_4$ undergoes a greenish-to-reddish tone transition at high temperatures due to the generation of cation vacancy defects and thus lattice distortion and/or phase transition. Figure 3c,d show the SEM morphology of the encapsulated $BiVO_4$ pigment. The SiO_2-coated primary particles had a near-spherical or elliptic structure, and a certain degree of particle agglomeration was unavoidable for the encapsulated pigment due to the bonding effect of the amorphous SiO_2.

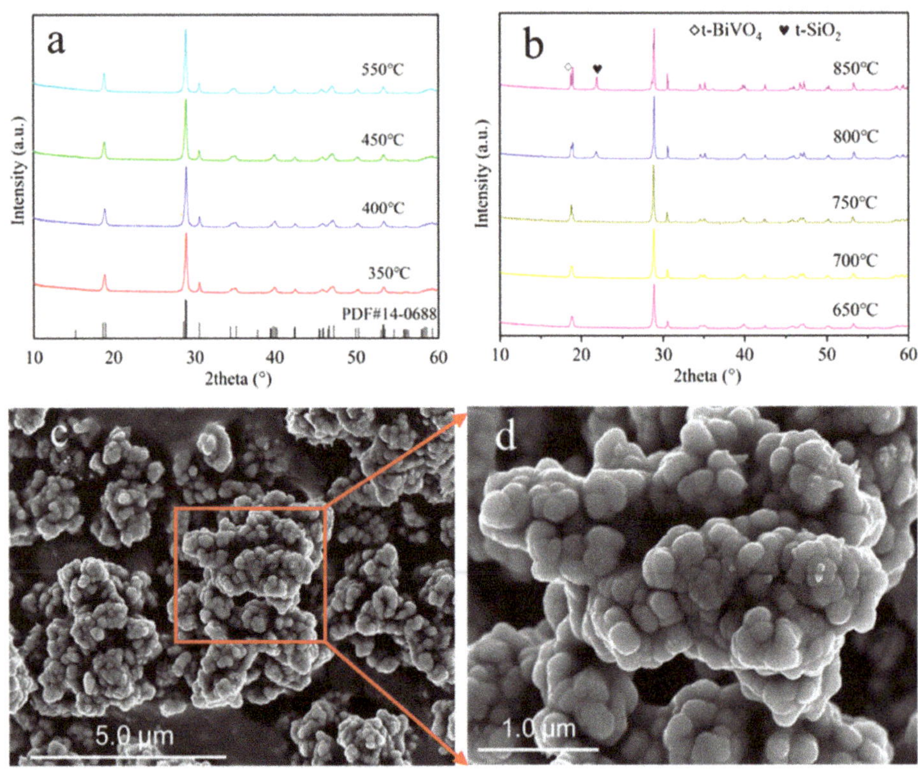

Figure 3. XRD patterns of BiVO$_4$@SiO$_2$ (**a**,**b**) calcined at different temperatures for 6 h. (**c**) SEM images of BiVO$_4$@SiO$_2$ encapsulated pigment calcined at 500 °C and (**d**) corresponding enlarged view of the selected area.

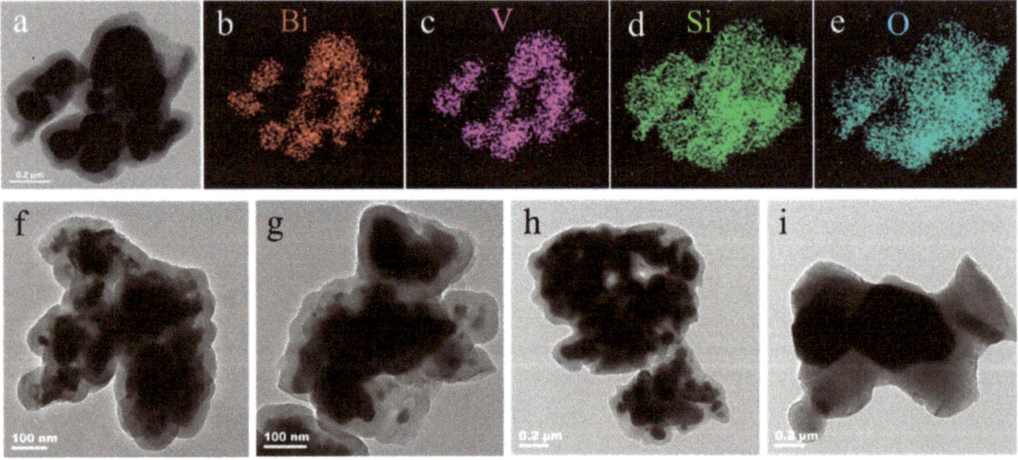

Figure 4. (**a**) TEM image and (**b–e**) corresponding EDS elemental mapping of BiVO$_4$@SiO$_2$ with 20 g of TEOS added and calcination at 500 °C. TEM images of BiVO$_4$@SiO$_2$ at different calcination temperatures of (**f**) 500 °C, (**g**) 600 °C, (**h**) 700 °C, and (**i**) 800 °C.

Table 1. Lattice parameters of m-BiVO$_4$ for the as-received BiVO$_4$ and BiVO$_4$@SiO$_2$ prepared at 800 °C.

Sample	Lattice Parameter					
	a/Å	b/Å	c/Å	α/°	β/°	γ/°
BiVO$_4$	5.186	11.698	5.101	90.000	90.287	90.000
BiVO$_4$@SiO$_2$	5.192	11.717	5.116	90.000	90.184	90.000

TEM and EDS analyses were further conducted to characterize in detail the morphology and size of the BiVO$_4$@SiO$_2$ pigments prepared with a TEOS content of 20 g and calcined at different temperatures (Figure 4). As shown in Figure 4a–e, a dense and uniform SiO$_2$ layer around the BiVO$_4$ particles was successfully achieved, forming a core–shell structure with a coating thickness of about 60~100 nm. Bi and V were located in the center, and Si and O were dispersed throughout the coating layer. The inner BiVO$_4$ particles of the encapsulated pigment exhibited a diameter of 100~200 nm when calcined at 500 °C, as shown in Figure 4f, and their size gradually increased with increasing calcination temperature. In particular, significant grain growth happened as the temperature increased from 700 to 800 °C (300~400 nm, in Figure 4i). This could mostly be ascribed to the coalescence of small BiVO$_4$ particles, since their fusing point was close to 800 °C. On the whole, the agglomerated particle size of the BiVO$_4$@SiO$_2$ pigments was 1~2 μm, which is much smaller than those of the traditional encapsulated ceramic pigments, such as red or yellow CdS$_{1-x}$Se$_x$@ZrSiO$_4$ (>10 μm) [32]. This ensures the colorimetric properties of the pigment and its wide applications, especially in inkjet print decoration for ceramics and glasses. However, when calcined at 800 °C, the SiO$_2$ coating surrounding the BiVO$_4$ particles became discontinuous and tended to separate from the particle surface, resulting in partial exposure of the pigment (Figure 4i). This could be ascribed to the shrinkage of the coating resulting from the crystallization and structural rearrangement of SiO$_2$, which is consistent with the XRD results in Figure 3b. The observed phenomenon suggested that the thickness of the SiO$_2$ layer was insufficient to effectively improve the thermal stability and chromatic performance of BiVO$_4$ at temperatures higher than 800 °C. To solve this problem, it is necessary to increase the coating thickness with a higher TEOS content and optimize the preparation process in future work.

2.3. Chromatic and Spectroscopic Properties of BiVO$_4$@SiO$_2$ Pigments

The CIE-L*a*b* color parameters (L*, a*, b*, C*, h°, ΔE*) of the BiVO$_4$@SiO$_2$ pigments calcined at different temperatures are listed in Table 2. For comparison, the corresponding parameters for the as-received BiVO$_4$ pigment are also provided. It was found that the yellowness (b*) and color saturation (C*) of BiVO$_4$@SiO$_2$ pigment improved to some extent when calcined at 500–700 °C. In particular, the encapsulated sample at 600 °C exhibited the highest b* and C* values, which were comparable to those of Ca/Zn-co-doped BiVO$_4$ and higher than those of most of BiVO$_4$ pigments when doped with other transition metals or rare earth metal ions [1,11,12,20,21]. These obtained results indicate that the introduction of SiO$_2$ (37.5% in mass content) does not affect the chromatic properties of BiVO$_4$ pigment. A similar phenomenon was also found for the synthesis of hybrid SiO$_2$/BiVO$_4$ pigments, with SiO$_2$ as the core, by He et al. [33]. The addition of SiO$_2$ is expected to decrease the use cost of yellow pigment greatly since BiVO$_4$ is much more costly than SiO$_2$. However, when the calcination temperature was further increased to 800 °C, the L*, b*, and C* values of the encapsulated pigment became slightly lower than those of BiVO$_4$. Importantly, as shown in Table 2, the a* value changed from negative to positive, indicating that BiVO$_4$@SiO$_2$ pigment underwent a transition from a greenish to a reddish-yellow color. Corresponding to this, the resulting pigment presented a relatively large chromatic aberration (ΔE* = 8.73) relative to BiVO$_4$. Generally, a greenish-yellow color is preferable to reddish since the former looks brighter and more striking. The degradation of the chromatic properties of BiVO$_4$@SiO$_2$ prepared at 800 °C was probably caused by the incomplete encapsulation of the SiO_2 coating and thus the generated defects of the cation vacancies and lattice distortion, and even the formation of a small amount of the t-$BiVO_4$ phase, as determined via TEM observation (Figure 4i) and the XRD results (Figure 3b and Table 1). In this case, the volatilization of some V ions at high temperatures would be inevitable in consideration of the poor thermal stability of $BiVO_4$, which led to the formation of lattice defects, the changes in structure, and thus the spectral characteristics. Despite this, the hue angle (h°) values for all $BiVO_4@SiO_2$ encapsulated pigments were around 87, indicating an intense yellow color.

Table 2. CIE-L*a*b* chromatic parameters of $BiVO_4$ and $BiVO_4@SiO_2$ pigments calcined at different temperatures.

Calcination Temperature	Color Parameter					
	L*	a*	b*	C*	h° (°)	ΔE* [1]
$BiVO_4$	88.94	−3.93	80.09	80.19	87.19	/
500 °C	90.29	−4.24	80.38	80.49	86.98	2.00
600 °C	88.69	−4.42	83.83	83.95	86.99	3.78
700 °C	88.26	−2.57	81.52	81.56	88.20	2.06
800 °C	87.60	4.63	79.03	79.17	86.65	8.73

[1] Chromatic aberration $\Delta E^* = [(\Delta L)^2 + (\Delta a)^2 + (\Delta b)^2]^{0.5}$.

To display the colors of all the samples more intuitively, the corresponding chromatic coordinates of the $BiVO_4@SiO_2$ pigments are plotted in the L*a*b* color space (Figure 5). The colors of $BiVO_4@SiO_2$ calcined at 500 °C, 600 °C, and 700 °C were quite similar. In addition, the coordinate location of $BiVO_4@SiO_2$ pigment calcined at 500–700 °C was closer to the green region than that prepared at 800 °C, which could be ascribed to the decreased thermal stability resulting from the defective SiO_2 coating at 800 °C.

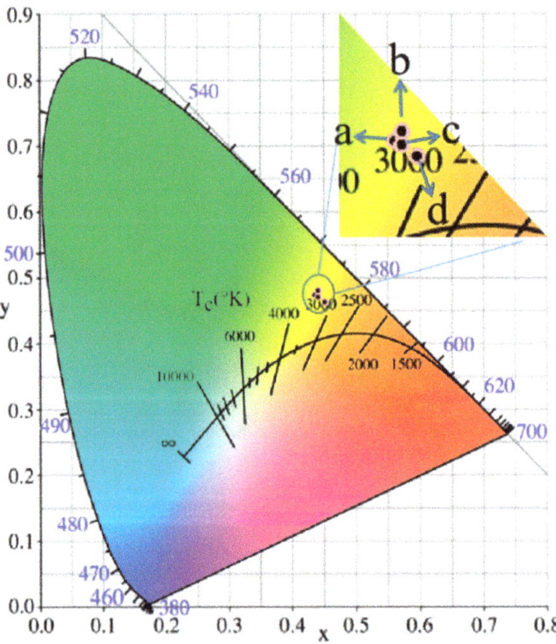

Figure 5. Chromatic coordinates of $BiVO_4@SiO_2$ at different calcination temperatures: (a) 500 °C, (b) 600 °C, (c) 700 °C, and (d) 800 °C.

To further elucidate the effect of SiO_2 encapsulation on the color-rendering performance of the $BiVO_4$ pigment, the UV–Vis diffuse reflectance properties were also investigated, as shown in Figure 6a. The $BiVO_4$ and $BiVO_4@SiO_2$ pigments show a very close high reflectance (>90%) from the yellow to red bands in the wavelength range of 550–780 nm but a relatively low reflectance from blue to purple light in the wavelength range of 380–490 nm. Furthermore, the reflectance of $BiVO_4@SiO_2$ in the 380–490 nm wavelength range is slightly lower than that of $BiVO_4$, which contributes to a little higher yellowness (b*) for the former [19]. In addition, the two samples had the same reflectance/absorption edge of about 500 nm in the green band, since $BiVO_4$ is the main color-rendering substance, and thus they were endowed with a greenish tone (negative a* value). On the whole, the measured visible spectrum characteristics are consistent with the chromatic performance shown in Table 2 and Figure 5. The effect of SiO_2 coating on the NIR reflectance performance of the $BiVO_4$ pigment was also investigated, with results displayed in Figure 6b. Only a minor reduction in the NIR reflectance within the 780–2500 nm wavelengths could be observed when SiO_2 coating with a high mass percentage of about 37.5% was introduced. Both samples possessed an overall sunlight NIR reflectance of more than 90%. The retained high NIR reflectivity of the $BiVO_4@SiO_2$ encapsulated pigment means that it has good prospects for being applied as a cold pigment for energy-saving applications such as architectural coatings, vehicle paints, exterior wall tiles, and other building decoration materials, since NIR radiation accounts for 52% of the energy from sunlight [14,16,34–36]. Furthermore, the preparation cost of $BiVO_4@SiO_2$ is much lower than that of $BiVO_4$ considering the ~37.5% SiO_2 content. Therefore, from the perspectives of colorimetric performance, thermal stability, near-infrared reflectance, and cost, the obtained $BiVO_4@SiO_2$ pigment is an excellent candidate for "cool" roofs and energy-saving coating materials, having excellent weather resistance and thermal stability.

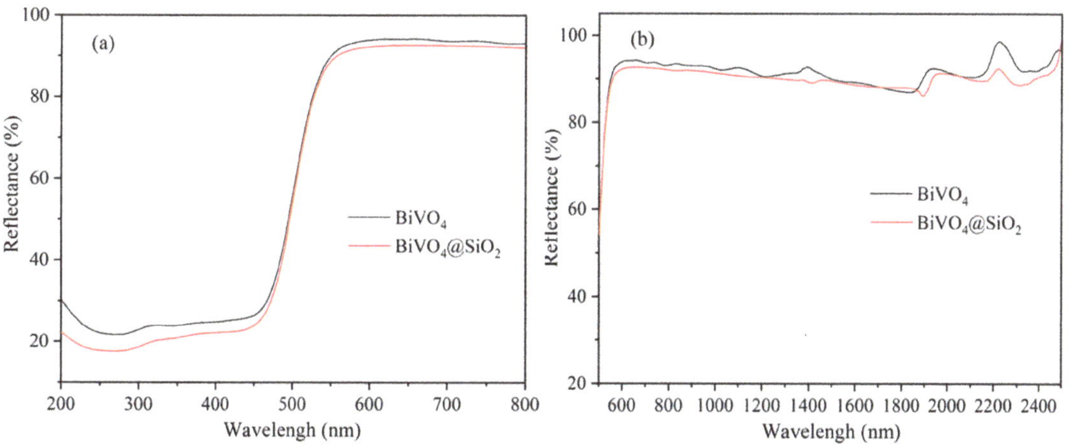

Figure 6. (**a**) UV–Vis and (**b**) NIR reflectance spectra of $BiVO_4$ and $BiVO_4@SiO_2$ calcined at 700 °C.

2.4. Evaluation of Thermal and Chemical Stability of $BiVO_4@SiO_2$ Pigment

In order to verify the effect of coating modification on the thermal stability of $BiVO_4$, the 600 °C-encapsulated and as-received $BiVO_4$ pigments were heat-treated at 500 and 700 °C for 24 h, respectively, and then their color-rendering performance were evaluated. The colors of post-test $BiVO_4$ and $BiVO_4@SiO_2$ are shown in Figure 7, and the corresponding L*, a*, and b* values are listed in Table 3. At 500 °C, $BiVO_4$ showed a more reddish-yellow color compared with $BiVO_4@SiO_2$. Compared with the as-received $BiVO_4$ pigment (Table 2), the brightness (L*) and yellowness (b*) showed no significant change, while the a* value changed from positive to negative, suggesting that the pigment transformed from greenish to reddish-yellow. This clearly shows that even at 500 °C, the BiVO$_4$ pigment does not have a sufficient thermal stability due to the loss of V ions. In sharp contrast, no obvious change in chromatic performance occurred for the BiVO$_4$@SiO$_2$ pigment at 500 °C compared to the as-prepared counterpart (Table 2). It is noted that even with treatment at 500 °C for 24 h, BiVO$_4$@SiO$_2$ still exhibited color-rendering performance superior to that of uncoated BiVO$_4$. When treated at 700 °C, the color of BiVO$_4$ turned brown due to the greatly increased a* and decreased b* values. However, the BiVO$_4$@SiO$_2$ pigment still possessed a brilliant yellow color at 700 °C. Compared with the BiVO$_4$@SiO$_2$ pigments, the L* (from 87.72 to 78.48) and b* (from 80.89 to 62.31) values of BiVO$_4$ were greatly reduced, while the a* value (from 2.73 to 11.28) was greatly enhanced. It is worth noting that despite heat treatment at 700 °C for 24 h, the color-rendering performance of BiVO$_4$@SiO$_2$ was still better than those of some ion-doped BiVO$_4$, such as (BiV)$_{1-x}$(YNb)$_x$O$_4$- and (LiLa)$_{1/2}$MoO$_4$-doped BiVO$_4$ with a yellowness b* of <77 [11,12]. As a result, the temperature resistance of BiVO$_4$@SiO$_2$ pigment is remarkably better than that of BiVO$_4$ pigment, and the silica layer can significantly improve the thermal stability of BiVO$_4$ pigment.

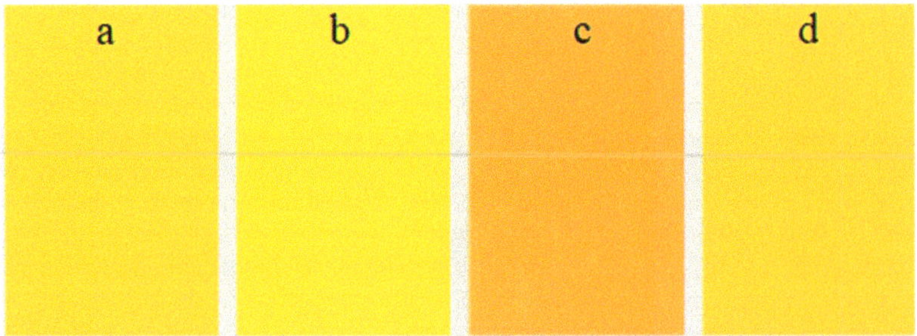

Figure 7. Photographs of BiVO$_4$ at (**a**) 500 °C and (**c**) 700 °C for 24 h; BiVO$_4$@SiO$_2$ at (**b**) 500 °C and (**d**) 700 °C for 24 h.

Table 3. CIE L*, a*, b* values of BiVO$_4$ and BiVO$_4$@SiO$_2$ after heat treatment at 500 °C and 700 °C.

Temperature (°C)	Sample	Chromatic Parameter				
		L*	a*	b*	C*	h° (°)
500	BiVO$_4$	88.21	0.21	79.91	79.91	89.85
	BiVO$_4$@SiO$_2$	89.71	−4.28	82.43	82.54	87.03
700	BiVO$_4$	78.48	11.28	62.31	63.32	79.74
	BiVO$_4$@SiO$_2$	87.72	2.73	80.89	80.93	88.07

Generally, BiVO$_4$ has a poor resistance to acid corrosion. Thus, to test the effectiveness of SiO$_2$ encapsulation in improving its acid corrosion resistance, SiO$_2$-coated and uncoated BiVO$_4$ pigments were treated with a 5 wt.% HCl solution for different durations. As shown in Figure 8, the as-received BiVO$_4$ dissolved almost completely and lost its yellow color after placement in the HCl solution for only 1 min, while BiVO$_4$@SiO$_2$ showed no obvious change after 90 min under the same condition, retaining its vivid yellow color. Further, the UV–Vis reflectance spectra of the BiVO$_4$@SiO$_2$ pigment from before and after the acidic corrosion test are compared in Figure 9. No obvious change in reflectance spectral characteristics was found for the two samples. These findings clearly indicate that excellent acid corrosion resistance was achieved successfully for BiVO$_4$ pigment through the modification with a silica coating. The obtained result also further suggests that the prepared amorphous SiO$_2$ coating is sufficiently dense to prevent the penetration of acidic solutions. Considering that BiVO$_4$ has good alkaline resistance, the prepared BiVO$_4$@SiO$_2$ pigment would have good application prospects in both acidic and alkaline coatings or paints.

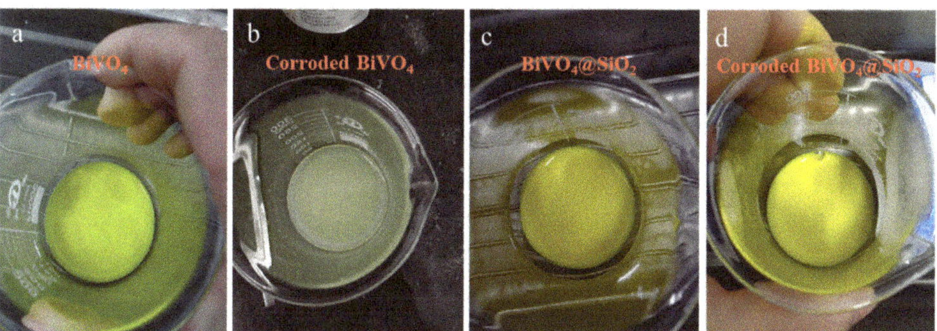

Figure 8. Photographs displaying the acid resistance of the pigments: $BiVO_4$ (**a**) in water and (**b**) treated for 1 min in HCl solution; $BiVO_4@ SiO_2$ (**c**) in water and (**d**) treated for 90 min in HCl solution.

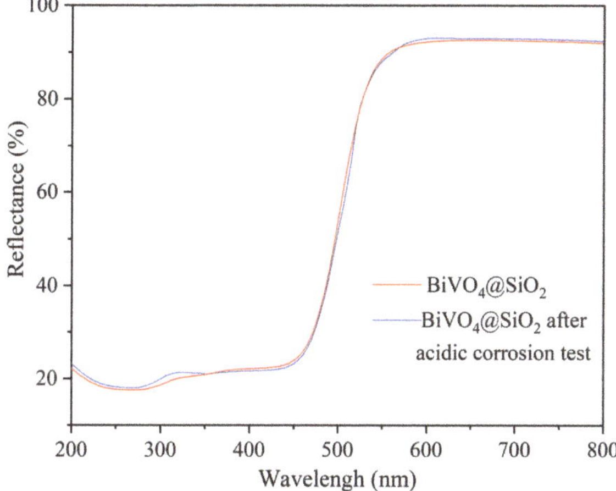

Figure 9. UV–Vis reflectance spectra of $BiVO_4@SiO_2$ before and after acidic corrosion test.

The environmentally friendly and nontoxic $BiVO_4$ pigment is considered to be one of the most promising materials in plastic coloring applications [1,13]. However, in practical applications, it was found that pure $BiVO_4$ pigment is vulnerable to the erosion by some plastic melts, resulting in degraded color-rendering performance. To investigate the influence of SiO_2 encapsulation on the resistance to plastic melt of yellow $BiVO_4$ pigment, the prepared $BiVO_4@SiO_2$ was well mixed into white polypropylene (PP) in a mass ratio of 5:1000, and then injection-molded into plastic plates at 240 °C. For comparison, the $BiVO_4$-colored plastic sample was also prepared under the same conditions. It can be seen from Figure 10 that the $BiVO_4@SiO_2$-colored plastic plate exhibited a bright greenish-yellow color, while the color of the sample with $BiVO_4$ was dull and pale yellow. Accordingly, the former possessed much higher L* and b* values. These results indicate that the SiO_2 coating successfully enhanced the erosion resistance of $BiVO_4$ pigment to high-temperature plastic melt, and the obtained $BiVO_4@SiO_2$ encapsulated pigment had a very strong tinting strength, even though the actual content of $BiVO_4$ was only 0.32 wt.%. In short, the thermal and chemical stability, as well as the color performances, of the $BiVO_4$ pigment can be effectively enhanced via SiO_2 encapsulation modification, which is expected to greatly promote the application of this pigment in plastics, coatings, or paints, and even ceramics and glass surface decorations.

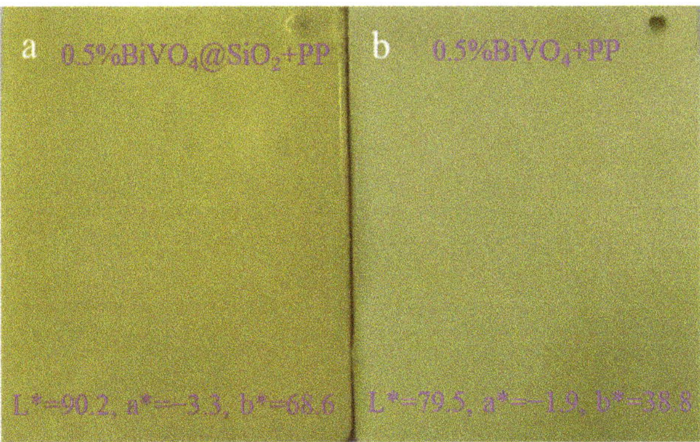

Figure 10. Photographs of PP plastics colored with (**a**) 0.5% BiVO$_4$@SiO$_2$ and (**b**) 0.5% BiVO$_4$ pigment.

3. Experimental Section

3.1. Preparation of Encapsulated Pigments

BiVO$_4$@SiO$_2$ pigment was prepared via a sol–gel method (Figure 11). The bismuth vanadate (BiVO$_4$) pigment with a primary grain size of about 100 nm was synthesized at 350~400 °C and provided by Jiangxi Jinhuan Pigments Co., Ltd., Yichun, China. Before use, the pigment powder was treated with ball milling to ensure good dispersity. The reagents of AR-grade tetraethoxysilane (TEOS) and sodium dodecyl benzene sulfonate (SDBS) were purchased from Sinopharm Chemical Reagent Co., Ltd., Shanghai, China. In a typical synthesis, 10 g of BiVO$_4$, 0.2 g of SDBS, and a certain amount of ethanol were mixed and ball-milled for 3 h to obtain the pigment suspension. Then, an ethanol solution of 20 g TEOS was added, and the mixture with a TEOS content of ~0.5 mol/L was mechanically stirred for 2 h. Subsequently, ammonium hydroxide (NH$_3$·H$_2$O, 28 wt.%, 20 g) was slowly added and stirred for 10 h. Finally, the mixture was dried at 75 °C for 12 h and then annealed at a certain temperature (400~850 °C) for 6 h under an air atmosphere to obtain BiVO$_4$@SiO$_2$ pigment.

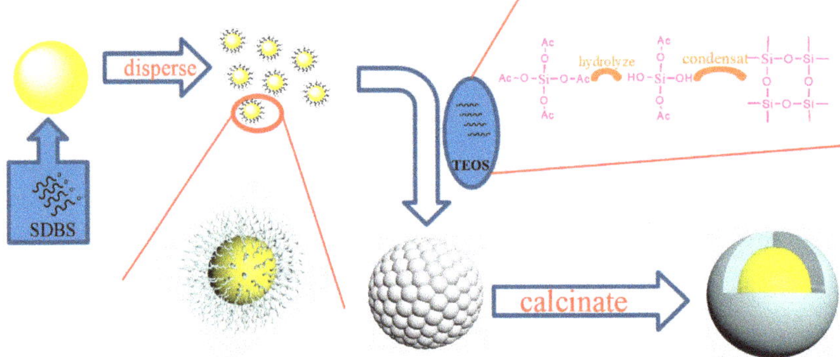

Figure 11. Schematic diagram for the synthesis of BiVO$_4$@SiO$_2$ pigment via the sol–gel method.

3.2. Materials' Characterization

Particle size distribution of the BiVO$_4$ pigment was analyzed with a laser diffraction particle analyzer (Bettersize2600, Bettersize Instruments, Dandong, China). Zeta potential measurements at different pH values were conducted using a Nanosizer (Nano ZS90, Malven, Worcestershire, UK). Phase composition of the encapsulated pigments was recognized with an X-ray diffractometer (XRD, D8-Advance, Bruker, Mannheim, Germany), using Cu Kα radiation (40 kV, 40 mA) at a step width of 0.02° min^{-1}. Field-emission scanning electron microscopy (FE–SEM, SU–8010, Hitachi, Tokyo, Japan) analysis was performed to obtain the morphology of pigment powder. Transmission electron microscope (TEM) and energy-dispersive spectroscopy (EDS) investigations were carried out on a JEOL JEM–2010 (Tokyo, Japan). Ultraviolet–visible (UV–vis) and near-infrared (NIR) spectra were obtained with an ultraviolet–visible–near infrared (UV–Vis–NIR) spectrophotometer (UV–3600, Shimadzu, Kyoto, Japan) using $BaSO_4$ as a reference.

3.3. Colorimetric Measurements

The color-rendering performance of the pigment was quantified using the CIE-L*a*b* chromaticity coordinate value of the International Commission on Illumination, where L* is the brightness of the color (L* = 0 means black, L* = 100 means white), a* represents green (−)/red (+) value, and b* represents blue (−)/yellow (+) value. The values of a* and b* range from −100 to 100. The color density, C* = $[(a^*)^2 + (b^*)^2]^{1/2}$, represents the color saturation. h° (h° = arctan (b*/a*)) is the color angle (for yellow, h° ranges from 70~105°). The L*a*b* chromaticity parameter and the tristimulus values of X, Y, and Z for the pigment were determined using a desktop colorimetric spectrophotometer (Ci7600, X-rite, Grand Rapids, MI, USA).

3.4. Stability Tests

To evaluate the high-temperature stability of $BiVO_4@SiO_2$ pigment, the pigments before and after the SiO_2 encapsulation were incubated at a constant temperature (500 °C, 700 °C) for 24 h, and then the chromaticity properties of the pigments were analyzed. To test the acid corrosion resistance, the SiO_2-coated and as-received $BiVO_4$ pigments were ultrasonically dispersed in a 30 mL of diluted hydrochloric acid solution (5 wt.% HCl), separately, and the dissolutive state of pigments over time was observed. The stability of pigments subjected to the erosion of plastic melt was also investigated in order to evaluate the application feasibility of $BiVO_4@SiO_2$ pigment in plastic coloring. For this purpose, the pigment (0.5 wt.%) was mixed with white polypropylene (PP) granules, followed by shaping into plastic plates via the injection molding technique at 240 °C, and the appearance of the $BiVO_4@SiO_2$-and $BiVO_4$-colored plastics was compared.

4. Conclusions

$BiVO_4@SiO_2$ encapsulated pigment, with its brilliant yellow color and strong tinting strength, was developed via the sol–gel method. The prepared $BiVO_4@SiO_2$ pigments were mostly composed of monoclinic $BiVO_4$ and amorphous SiO_2 calcined 350–800 °C. A core–shell-structured $BiVO_4@SiO_2$ encapsulated pigment was successfully obtained when calcined at 500–700 °C, with a dense SiO_2 coating thickness of 60~100 nm and an inclusion particle size of 1~2 μm. The $BiVO_4@SiO_2$ encapsulated pigment showed comparable chromatic parameters to pure $BiVO_4$, and the encapsulation of SiO_2 coating did not weaken the color-rendering performance of $BiVO_4$. In particular, chromatic parameters of L* = 88.69, a* = −4.42, and b* = 83.83 were achieved for $BiVO_4@SiO_2$ prepared at 600 °C, which are superior to those of pure $BiVO_4$ in yellowness. The obtained $BiVO_4@SiO_2$ pigments also possessed a high NIR reflectivity of more than 90%, though a high content of amorphous SiO_2 phase was introduced.

The thermal stability was effectively increased from ≤350 °C for $BiVO_4$ to at least 700 °C for yellow $BiVO_4@SiO_2$ pigment. Furthermore, the chemical corrosion resistance of the $BiVO_4$ pigment could be remarkably enhanced by the encapsulation of dense SiO_2 coating. The polypropylene plastic colored with only 0.5% $BiVO_4@SiO_2$ pigment at 240 °C displayed an obvious brilliant yellow color, indicating the good erosion resistance to plastic melt and the strong tinting strength of the encapsulated pigment.

In summary, the thermal and chemical stability, as well as the color performance, of $BiVO_4$ pigment can be effectively enhanced via SiO_2 encapsulation modification. Thus, this research provides a cost-effective strategy for realizing applications of environmentally friendly $BiVO_4$ pigment under high temperatures or corrosive environments, such as in plastics, energy-saving coatings, or paints, and even in ceramics and glass surface decoration.

Author Contributions: Conceptualization, X.Z. and R.T.; methodology, R.C., H.L. and R.T.; software, R.C., X.Z. and R.T.; validation, Y.J. and L.C.; formal analysis, R.C., X.Z. and R.T.; investigation, R.C. and R.T.; resources, X.Z.; data curation, Y.J.; writing—original draft preparation, R.C. and R.T.; writing—review and editing, X.Z.; visualization, H.L. and R.T.; supervision, X.Z.; project administration, Y.J.; funding acquisition, R.C. and X.Z. All authors have read and agreed to the published version of the manuscript.

Funding: This research was funded by Jiangxi Provincial Department of Science and Technology of China (No. 20181ACE50017, No. 20181BCD40044) and Jiangxi Provincial Department of Education of China (No. GJJ211303).

Data Availability Statement: The data presented in this study are available on request from the corresponding author. The data are not publicly available due to privacy.

Conflicts of Interest: Yuhua Jiang declares no conflict of interest. Authors Renhua Chen, Xiaozhen Zhang, Rui Tao, Huafeng Liu, and Lanlan Cheng were employed by the company Jiangxi Jinhuan Pigments Co., Ltd. The remaining authors declare that the research was conducted in the absence of any commercial or financial relationships that could be construed as a potential conflict of interest.

References

1. Masui, T.; Honda, T.; Wendusu; Imanaka, N. Novel and environmentally friendly $(Bi,Ca,Zn)VO_4$ yellow pigments. *Dye. Pigment.* **2013**, *99*, 636–641. [CrossRef]
2. Pailhé, N.; Gaudon, M.; Demourgues, A. (Ca^{2+}, V^{5+}) co-doped $Y_2Ti_2O_7$ yellow pigment. *Mater. Res. Bull.* **2009**, *44*, 1771–1777. [CrossRef]
3. Wang, Q.; Seto, T.; Du, Y.; Liu, W.; Wang, Y. Design and preparation of high color rendering red inorganic pigment via red light emission phosphor. *Opt. Mater.* **2023**, *140*, 113886. [CrossRef]
4. Blasco-Zarzoso, S.; Beltrán-Mir, H.; Cordoncillo, E. Su stainable inorganic pigments with high near-infra-red reflectance based on Fe^{3+} doped $YAlO_3$ for high temperature applications. *J. Alloys Compd.* **2023**, *960*, 170695. [CrossRef]
5. Zou, J.; Chen, Y.; Zhang, P. Influence of crystallite size on color properties and NIR reflectance of $TiO_2@NiTiO_3$ inorganic pigments. *Ceram. Int.* **2021**, *47*, 12661–12666. [CrossRef]
6. Liu, H.; Wang, Q.; Chang, Q.; Wang, C.; Wang, Y.; Wang, Y.; Zhang, X. Relationship between the colour and particle size of the ultrafine $V-ZrSiO_4$ and $Pr-ZrSiO_4$ pigments and their mixture. *Mater. Res. Express* **2019**, *6*, 075214. [CrossRef]
7. Zhou, X.; Zhang, X.; Zou, C.; Chen, R.; Cheng, L.; Han, B.; Liu, H. Insight into the Effect of Counterions on the Chromatic Properties of Cr-Doped Rutile TiO_2-Based Pigments. *Materials* **2022**, *15*, 2049. [CrossRef]
8. Kumar, B.; Sharma, R.; Bhoi, H.; Punia, K.; Kumar, S.; Barbar, S.K. Synthesis, structural, morphological and optical properties of environment friendly yellow inorganic pigment $Bi_4Zr_3O_{12}$. *Opt. Mater.* **2023**, *142*, 114040. [CrossRef]
9. Wang, Y.; Jiang, P.; Subramanian, M.A.; Cao, W. Synthesis, properties and applications of novel inorganic yellow pigments based on Ni-doped Al_2TiO_5. *Solid State Sci.* **2023**, *135*, 107088. [CrossRef]
10. Thejusa, P.K.; Nishantha, K.G. Rational approach to synthesis low-cost $BiVO_4$–ZnO complex inorganic pigment for energy efficient buildings. *Sol. Energy Mater. Sol. Cells* **2019**, *200*, 109999. [CrossRef]
11. Sameera, S.; Rao, P.P.; James, V.; Divya, S.; Raj, A.K.V. Influence of $(LiLa)_{1/2}MoO_4$ substitution on the pigmentary properties of $BiVO_4$. *Dye. Pigment.* **2014**, *104*, 41–47. [CrossRef]
12. Sameera, S.F.; Rao, P.P.; Kumari, L.S.; James, V.; Divya, S. Potential NIR Reflecting Yellow Pigments in $(BiV)_{1-x}(YNb)_xO_4$ Solid Solutions. *Chem. Lett.* **2013**, *42*, 521523. [CrossRef]
13. Zhao, G.; Liu, W.; Dong, M.; Li, W.; Chang, L. Synthesis of monoclinic sheet-like $BiVO_4$ with preferentially exposed (040) facets as a new yellow-green pigment. *Dye. Pigment.* **2016**, *134*, 91–98. [CrossRef]
14. Kumari, L.S.; Rao, P.P.; Radhakrishnan, A.N.P.; James, V.; Sameera, S.; Koshy, P. Brilliant yellow color and enhanced NIR reflectance of monoclinic $BiVO_4$ through distortion in VO_4^{3-} tetrahedra. *Sol. Energy Mater. Sol. Cells* **2013**, *112*, 134–143. [CrossRef]
15. Zhang, S.; Pan, Z.; Wang, Y. Synthesis and characterization of (Ni, Sb)-co-doped rutile ceramic pigment via mechanical activation-assisted solid-state reaction. *Particuology* **2018**, *41*, 20–29. [CrossRef]
16. Sameera, S.; Rao, P.P.; Divya, S.; Raj, A.K.V. Brilliant IR reflecting yellow colorants in rare earth double molybdate substituted $BiVO_4$ solid solutions for energy saving applications. *ACS Sustain. Chem. Eng.* **2015**, *3*, 1227–1233. [CrossRef]
17. Tao, R.; Jiang, Y.; Liu, H.; Zhang, X.; Chen, R.; Chang, Q.; Wang, Y. Effects of metal ions doping on the chromatic performance of $BiVO_4$ yellow pigments. *J. Ceram.* **2019**, *40*, 371–376.
18. Tao, R.; Zhang, X.; Liu, H.; Jiang, Y.; Pan, T.; Chang, Q.; Wang, Y.; Zhou, J. Effect of Zn doping on the properties of nano $BiVO_4$ yellow pigments. *J. Synth. Cryst.* **2019**, *6*, 1144–1149.

19. Sameera, S.F.; Rao, P.P.; James, V.; Raj, A.K.V.; Chitradevi, G.R.; Leela, S. Probing structural variation and multifunctionality in niobium doped bismuth vanadate materials. *Dalton Trans.* **2014**, *43*, 15851–15860.
20. Wendusu; Ken-ichi, I.; Toshiyuki, M.; Nobuhito, I. Novel Environment-friendly Yellow Pigments Based on (Bi, La)VO$_4$. *Chem. Lett.* **2011**, *40*, 792–794. [CrossRef]
21. Sameera, S.F.; Rao, P.P.; Kumari, L.S.; Koshy, P. New Scheelite-based Environmentally Friendly Yellow Pigments: (BiV)$_x$(CaW$_{)1-x}$O$_4$. *Chem. Lett.* **2009**, *38*, 1088–1089. [CrossRef]
22. Chang, Q.; Wang, X.; Wang, Y.; Bao, Q.; Zhou, J.; Zhu, Q. Encapsulated carbon black prepared by sol-gel-spraying: A new black ceramic pigment. *J. Eur. Ceram. Soc.* **2014**, *34*, 3151–3157. [CrossRef]
23. Chen, T.; Jiang, W.; Zhang, X.; Liu, J.; Jiang, W.; Xie, Z. Ionic liquid-assisted synthesis of C@ZrSiO$_4$ ceramic inclusion pigment. *Mater. Sci. Forum* **2016**, *848*, 256–261. [CrossRef]
24. Tang, H.; Hu, Q.; Jiang, F.; Jiang, W.; Liu, J.; Chen, T.; Feng, G.; Wang, T.; Luo, W. Size control of C@ZrSiO$_4$ pigments via soft mechano-chemistry assisted non-aqueous sol-gel method and their application in ceramic glaze. *Ceram. Int.* **2019**, *45*, 10756–10764. [CrossRef]
25. Liu, S.; Li, Y.; Wang, Z.; Shen, Z.; Xie, Z. Enhanced high temperature oxidization resistance of silica coated γ-Ce$_2$S$_3$ red pigments. *Appl. Surf. Sci.* **2016**, *387*, 1147–1153. [CrossRef]
26. Zhang, W.; Wang, Q.; Chen, X.; Li, X.; Long, Q.; Wang, C.; Liu, K.; Wang, Y.; Chang, Q. Synthesis of high color performance V-ZrSiO$_4$ blue pigment with low doping amount via inorganic sol–gel route. *Adv. Powder Technol.* **2021**, *32*, 3355–3363. [CrossRef]
27. Hu, Z.; Wang, Q.; Liu, J.; Liu, K.; Zhang, W.; Yang, Y.; Wang, Y.; Chang, Q. Synthesis and chromatic properties of V-doped and V/Y-codoped ZrO$_2$ yellow pigments. *J. Alloys Compd.* **2021**, *856*, 157397. [CrossRef]
28. Li, Y.; Liu, S.; Song, F.; Wang, Z.; Shen, Z.; Xie, Z. Preparation and thermal stability of silica layer multicoated γ-Ce$_2$S$_3$ red pigment microparticles. *Surf. Coat. Technol.* **2018**, *345*, 70–75. [CrossRef]
29. Sultan, S.; Kareem, K.; He, L. Synthesis, characterization and resistant performance of α-Fe$_2$O$_3$@SiO$_2$ composite as pigment protective coatings. *Surf. Coat. Technol.* **2016**, *300*, 42–49. [CrossRef]
30. Huang, J.; Feng, L.; Cao, L. *Sol-Gel Technology and Applications*; Higher Education Press: Beijing, China, 2021; pp. 79–91. (In Chinese)
31. Kweon, K.E.; Hwang, G.S. Structural phase-dependent hole localization and transport in bismuth vanadate. *Phys. Rev. B* **2013**, *87*, 205202. [CrossRef]
32. Qin, W.; Wang, K.; Zhang, Y. Preparation of submicron CdS$_x$Se$_{1-x}$@ZrSiO$_4$ inclusion pigment and its application in ink-jet printing. *J. Eur. Ceram. Soc.* **2021**, *41*, 7878–7885. [CrossRef]
33. He, X.; Wang, F.; Liu, H.; Wang, X.; Gao, J. Synthesis and coloration of highly dispersive SiO$_2$/BiVO$_4$ hybrid pigments with low cost and high NIR reflectance. *Nanotechnology* **2019**, *30*, 295701. [CrossRef] [PubMed]
34. Guan, L.; Fan, J.; Zhang, Y.; Guo, Y.; Duan, H.; Chen, Y.; Li, H.; Liu, H. Facile preparation of highly cost-effective BaSO$_4$@BiVO$_4$ core-shell structured brilliant yellow pigment. *Dye. Pigment.* **2016**, *128*, 49–53. [CrossRef]
35. Kalarikkal, T.P.; Veedu, K.K.; Gopalan, N.K. Inorganic ferrite brown; transformation towards a versatile pigment for energy efficient constructions. *J. Ind. Eng. Chem.* **2023**, *122*, 349–358. [CrossRef]
36. He, X.; Wang, F.; Liu, H.; Li, J.; Niu, L. Fabrication of highly dispersed NiTiO$_3$@TiO$_2$ yellow pigments with enhanced NIR reflectance. *Mater. Lett.* **2017**, *208*, 82–85. [CrossRef]

Article

Microwave-Mediated Synthesis and Characterization of Ca(OH)$_2$ Nanoparticles Destined for Geraniol Encapsulation

Panagiota Tryfon [1], Nathalie N. Kamou [2], Stefanos Mourdikoudis [3,4,5], George Vourlias [6], Urania Menkissoglu-Spiroudi [2] and Catherine Dendrinou-Samara [1,*]

[1] Laboratory of Inorganic Chemistry, Department of Chemistry, Aristotle University of Thessaloniki, 54124 Thessaloniki, Greece

[2] Pesticide Science Laboratory, Faculty of Agriculture Forestry and Natural Environment, School of Agriculture, Aristotle University of Thessaloniki, 54124 Thessaloniki, Greece; nnkamou@gmail.com (N.N.K.); rmenkis@agro.auth.gr (U.M.-S.)

[3] Biophysics Group, Department of Physics and Astronomy, University College London, London WC1E 6BT, UK; s.mourdikoudis@ucl.ac.uk

[4] UCL Healthcare Biomagnetics and Nanomaterials Laboratories, 21 Albemarle Street, London W1S 4BS, UK

[5] Separation and Conversion Technology, Flemish Institute for Technological Research (VITO), Boeretang 200, 2400 Mol, Belgium

[6] Laboratory of Advanced Materials and Devices, Physics Department, Aristotle University of Thessaloniki, 54124 Thessaloniki, Greece; gvourlia@auth.gr

* Correspondence: samkat@chem.auth.gr

Citation: Tryfon, P.; Kamou, N.N.; Mourdikoudis, S.; Vourlias, G.; Menkissoglu-Spiroudi, U.; Dendrinou-Samara, C. Microwave-Mediated Synthesis and Characterization of Ca(OH)$_2$ Nanoparticles Destined for Geraniol Encapsulation. *Inorganics* **2023**, *11*, 470. https://doi.org/10.3390/inorganics11120470

Academic Editors: Roberto Nisticò, Torben R. Jensen, Luciano Carlos, Hicham Idriss and Eleonora Aneggi

Received: 23 October 2023
Revised: 28 November 2023
Accepted: 29 November 2023
Published: 2 December 2023

Abstract: Nanotechnology presents promising opportunities for enhancing pest management strategies, particularly in protecting active ingredients to prolong their shelf life and effectiveness. Among different approaches, the combination of inorganic nanoparticles with active ingredients such as the main constituents of natural essential oils in one nanoarchitecture is challenging. In this study, hydrophobic calcium hydroxide nanoparticles coated with oleylamime [Ca(OH)$_2$@OAm NPs] were synthesized using microwave-assisted synthesis. These primary NPs were physicochemically characterized and subsequently utilized to prepare nanocapsules (NCs) either alone (Ca NCs) and/or in combination with geraniol at different ratios of Ca(OH)$_2$@OAm NPs and geraniol, i.e. 1:1 (CaGer1 NCs), 1:2 (CaGer2 NCs), and 1:3 (CaGer3 NCs), respectively. Among the formulations, the CaGer2 NCs demonstrated higher encapsulation efficiency (EE) and loading capacity (LC) of 95% and 20%, correspondingly. They exhibited a hydrodynamic size of 306 nm, a ζ-potential of −35 mV, and a monodisperse distribution. Release kinetics of geraniol from CaGer2 NCs indicated a pH-dependent slow release over 96 h at both 25 °C and 35 °C. In vitro antifungal assay against *B. cinerea* revealed a concentration-dependent activity, and the EC$_{50}$ values for Ca(OH)$_2$@OAm NPs, Ca NCs, and CaGer2 NCs were estimated to be 654 µg/mL, 395 µg/mL, and 507 µg/mL, respectively. These results underscore the potential of Ca-based nanoformulations to control plant pathogens, suggesting that while Ca NCs showcase potent antifungal attributes, the different architectures/structures play a critical role in the antifungal effectiveness of the nanoformulations that have to be explored further.

Keywords: calcium hydroxide nanoparticles; inorganic nanoparticles; nano-delivery systems; antifungal efficacy; *Botrytis cinerea*; pH-responsive delivery

1. Introduction

The European Commission has set plant health very high on the political agenda since it is fundamental to our well-being and the environment. Plant pests can impact plant health causing serious economic and environmental effects. Therefore, plant protection from pests and diseases is of high priority in the European Union (EU). Since the 1940s, synthetic plant protection chemicals have substantially improved food production [1]. Synthetic fungicides, especially those targeting notorious pathogens like *Botrytis cinerea* (*B. cinerea*), have been the frontline defense against fungal diseases [2]. However, the vast majority of these pesticides end up affecting non-target plants while culminating in significant environmental degradation and elevated health risks [1,3]. The resulting environmental and health implications highlight an urgent need for alternative, eco-friendly crop protection strategies.

Nanomaterials, with their versatile applications, offer novel solutions to this alarming situation [4]. Engineered inorganic nanoparticles (EINPs) provide substantial potential in plant protection. Their unique physicochemical characteristics facilitate effective fungal control and improve soil nutrient availability, while under specific nanoarchitectures, they enable the controlled release of antifungal agents [5,6]. EINPs have been observed to disrupt fungal cellular structures, inhibiting growth and ensuring the targeted delivery of antifungal agents [6]. However, with the progress of nanotechnology, advanced inorganic-based nanostructures have been developed with potential multifunctional properties that depend on their single counterparts. In that vein, and among different strategies for alternative plant protection agents, a promising pathway is the up-regulation of EINPs with natural active ingredients such as essential oils (EOs) or their constituents. Among them, geraniol, a monoterpene alcohol, is prominent. This highly hydrophobic natural product, derived from aromatic plants, is prevalent in various EOs. Due to its potent antimicrobial activity, geraniol has garnered research interest [7,8] and commercial products have already emerged in the market as plant protection products [9,10]. However, agents like geraniol are inherently unstable and volatile, which can hinder their efficacy [11]. An innovative approach within this domain is encapsulating EOs with EINPs. Encapsulation protects the active components of EOs, ensures their controlled release, and enhances delivery precision [11,12].

Thus, recent studies emphasize the synergistic potential of inorganic-based nanocapsules (NCs) and EOs in fighting plant pathogens. NCs formulated with *Zataria multiflora* and zinc oxide nanoparticles (ZnO NPs) demonstrated up to 66.33% increased efficacy against various *Fusarium* isolates and *A. solani* [13,14]. In studies with *B. cinerea*, pure geraniol and nanoemulsions containing geraniol showed EC_{50} values of 235 µg/mL and 105 µg/mL, respectively [15]. Furthermore, NCs integrating ZnO nanorods and geraniol effectively controlled *B. cinerea* without inducing phytotoxic effects in tomato and cucumber plants [9]. These results underline the significant role of NCs in enhancing essential oil efficiency against plant diseases.

Calcium (Ca) is an essential secondary nutrient that plays a crucial role in plant vitality [16]. Beyond its direct nutritional aspects, Ca is commonly applied in agriculture as a fertilizer and soil amendment [17,18]. This dual function of Ca-based compounds opens avenues for their potential fungicidal use while simultaneously enhancing cell wall fortification protection [19]. Notably, the European Food Safety Authority (EFSA) recently approved the use of calcium hydroxide [$Ca(OH)_2$] as a fungicide across various crops [20]. Calcium hydroxide nanoparticles [$Ca(OH)_2$ NPs] distinguish among the array of EINPs through their unique physical and chemical attributes, such as biocompatibility, non-toxicity, synthesis simplicity, and environmental compatibility [21]. The importance of $Ca(OH)_2$ NPs has been recognized in the past two decades, particularly in cultural heritage preservation [21] and dentistry [22]. Calcium-based nanoparticles (Ca-based NPs) are emerging as antibacterial candidates against a range of human pathogens, both Gram-negative (e.g., *Escherichia coli*, *Pseudomonas aeruginosa*) and Gram-positive (e.g., *Bacillus subtilis*, *Streptococcus aureus*) [23]. In the agricultural field, lime-based materials such as limestone ($CaCO_3$) and hydrated lime [$Ca(OH)_2$] are generally considered benign and beneficial when used to improve the soil pH, while each material exhibits diverse pH values and demonstrates a variety of functions. Also, Ca-based NPs have shown efficacy as nematicides against *Meloidogyne incognita* and *Meloidogyne javanica* as pH adjusters [24]. However, plant protection utilizing Ca-based nanoparticles (NPs) has received relatively less attention, despite their potential to be regarded as time-honored agents for phytoprotection.

Based on our previous efforts [9,25], in the present study, different nanoarchitectures were synthesized, characterized, and in vitro tested against *B. cinerea*. Initially, a microwave-assisted synthesis was applied to form relatively small hydrophobic nanoparticles of calcium hydroxide coated with oleylamine [Ca(OH)$_2$@OAm NPs]. These primary NPs were up-regulated to calcium-based nanocapsules (Ca-based NCs); (i) solely (Ca NCs) and (ii) in the presence of geraniol at varying ratios (1:1 for CaGer1, 1:2 for CaGer2, and 1:3 for CaGer3 NCs, respectively) in an attempt to optimize the most efficient formulation. The biodegradable and biocompatible surfactant, sodium dodecyl sulfate (SDS) was used as an emulsifier in all NCs. Comprehensive physicochemical analysis of both primary as-synthesized Ca(OH)$_2$@OAm NPs and the subsequent Ca-based NCs was undertaken using a range of techniques. Moreover, the influence of pH and kinetic analysis was explored for the case of CaGer2 NCs based on different models (first-order, Higuchi, and Korsmeyer–Peppas) at two different temperatures (25 °C and 35 °C) to assess the stability and release behavior of geraniol. An in vitro antifungal evaluation against *B. cinerea* was followed for the primary NPs and the secondary structures of NCs to test their efficacy against the well-known phytopathogen.

2. Results and Discussion

2.1. Physicochemical Characterization of Ca(OH)$_2$@OAm NPs

An X-ray diffractogram of the Ca(OH)$_2$@OAm NPs is presented in Figure 1, revealing a hexagonal Ca(OH)$_2$ in the portlandite phase (JCPDS pdf card #72-0156) [26]. Prominent peaks in the diffraction pattern are evident at angles (2θ) of 18.2°, 28.9°, and 34.4°, corresponding to (001), (100), and (011) reflections, respectively. These well-defined and pronounced diffraction peaks indicate a highly crystalline nature of the material. Utilizing the Scherrer formula based on the (011) plane, the crystalline size was determined to be 27 nm. This size is notably smaller than that observed for tannic acid-coated and PEGylated Ca(OH)$_2$@PEG NPs, which measured 50 nm and 40 nm, respectively [24,27]. The smaller crystalline size of the Ca(OH)$_2$@OAm NPs could be substantial in applications due to the increased surface area, enhanced diffusion, improved stability, and greater reactivity that it can provide. The crystallinity of the Ca(OH)$_2$@OAm NPs was established at 81.7%. Such distinct crystallinity suggests the potential suitability of the current NPs for applications that demand superior mechanical properties, particularly those requiring increased strength and density [28].

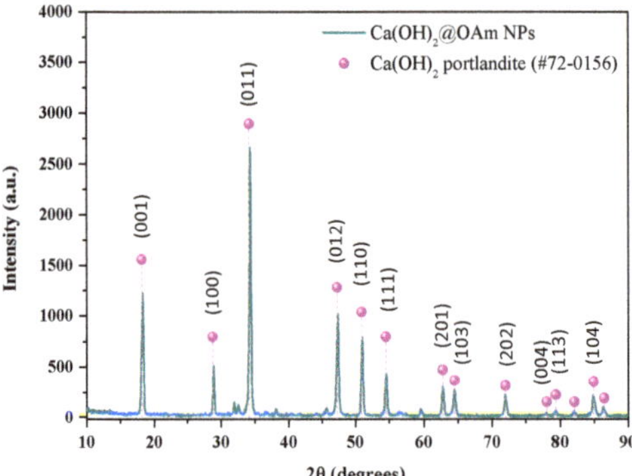

Figure 1. X-ray diffraction (XRD) of Ca(OH)$_2$@OAm NPs synthesized with the microwave-assisted process.

Functional groups present in the Ca(OH)$_2$@OAm NPs were evaluated with FT-IR spectroscopy, as depicted in Figure S1. The FT-IR spectrum exhibits prominent bands that correspond to diverse functional groups in the sample. A prominent and intense peak at 3642 cm^{-1} signifies the stretching mode of the –OH group, corroborating the presence of Ca(OH)$_2$ [24,29]. The observed peaks at 2922 cm^{-1} and 2853 cm^{-1} can be assigned to the asymmetric and symmetric stretching vibrations of CH$_2$ groups, respectively, indicative of the characteristic OAm surfactant [9]. Additionally, the peak at 3347 cm^{-1} corresponds to the stretching vibration of N–H bonds, while the peaks at 1608 cm^{-1} and 1452 cm^{-1} are attributed to NH$_2$ bending and C–H bending vibrations, respectively [30,31]. The peak observed at 430 cm^{-1} corresponds to the characteristic Ca–O stretching vibration band, confirming the formation of Ca(OH)$_2$ particles [24,32].

The TEM analysis of Ca(OH)$_2$@OAm NPs helped to visualize their morphology. Figure 2 presents a TEM image of the NPs, showcasing particles with a hexagonal shape and an edge size in the range of a few hundred nanometers. Despite their thorough washing, the structures appear to be surrounded by an organic coating on their surface, which is associated with the use of OAm as a ligand. The observed structures exhibit a platelet-like morphology. The small thickness is indicated by the light contrast in the TEM images. OAm as a ligand favored this particle morphology by acting as a growth modifier. A partial tendency for agglomeration between the distinct nanoplates was noticed for the depicted Ca(OH)$_2$@OAm NPs.

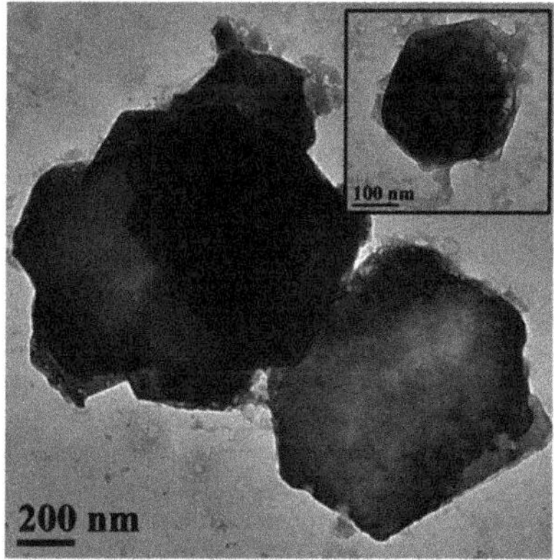

Figure 2. Transmission electron microscope (TEM) image of hexagonal Ca(OH)$_2$@OAm NPs at a scale of 200 nm (inset: scale at 100 nm).

A thermal stability study of Ca(OH)$_2$@OAm NPs was performed using TGA over the temperature range of 25 to 850 °C. An examination of the Ca(OH)$_2$@OAm NPs curve delineates three distinct stages of weight loss, as illustrated in Figure 3. The first weight loss step, accounting for approximately 8% *w/w*, occurs up to 120 °C and corresponds to the evaporation of physically adsorbed water molecules and/or loosely bound hydroxyl groups from the particle surface [33,34]. A subsequent 8% *w/w* decrement, materializing near 300 °C, results from the thermal decomposition of the OAm ligand [35]. OAm, commonly used as a surface-capping agent in metal oxide NP synthesis, decomposes upon heating, leading to the generation of volatile products such as ammonia, alkenes, and alkanes. This OAm detachment reflects the interplay of non-covalent interactions (hydrogen bonds and/or *van der Waals* forces) onto the particle surface, whereas the attrition at escalated temperatures signifies the disruption of covalent bonds [36]. The final weight loss step (15% *w/w*) takes place around 500 °C and is concurrent with the emergence of calcium oxide (CaO) within a 400–460 °C interval [32], notwithstanding the potential ongoing OAm ligand decomposition. Additionally, the DTG plot indicates a positive peak, validating the occurrence of the endothermic reaction in the conversion of Ca(OH)$_2$ into CaO, which is concomitant with the removal of water.

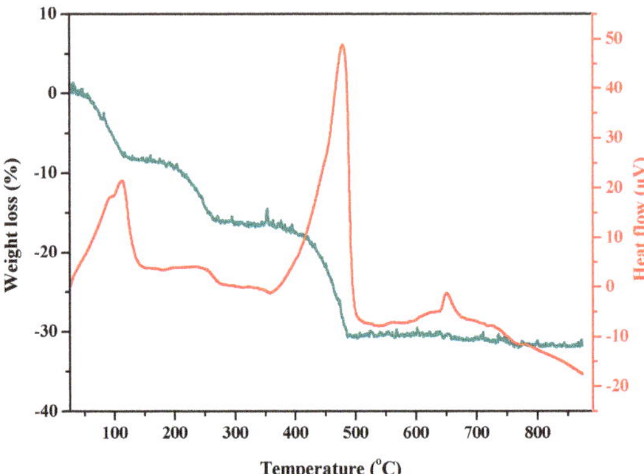

Figure 3. Thermogravimetric analysis (TGA) and derivative thermogravimetric (DTG) of the Ca(OH)$_2$@OAm NPs are presented with blue and red line, respectively.

UV-Vis spectroscopy was used to investigate the optical properties of Ca(OH)$_2$@OAm NPs. A solution containing dispersed Ca(OH)$_2$@OAm NPs in an ethanol/water (1:2) mixture was utilized to explore their absorption properties within the wavelength range of 210−600 nm. This choice of solvent offered high colloidal ability. The absorption spectrum (Figure S2A) of the NPs displayed two distinct absorbance bands observed at approximately 234 and 270 nm, corresponding to electronic transitions within the material. Additionally, a peak at 225 nm was observed, which is attributed to the presence of crystallized OAm. It is important to note that the cutoff wavelengths for water and ethyl alcohol are 190 and 210 nm, respectively. The absence of a shoulder in the absorption spectra indicates that the Ca(OH)$_2$@OAm NPs possess a direct band gap energy, signifying transitions from the valence band to the conduction band. Furthermore, the calculated band gap of the Ca(OH)$_2$@OAm NPs, determined from the Tauc plot (Figure S2B), was approximately 4 eV. This value is lower than the band gap of the naked Ca(OH)$_2$ NPs with a similar crystallite size (25 nm) at 5.20 eV, indicating a blue shift, in agreement with previous reports [29,32]. Generally, a larger band gap energy corresponds to higher electrical resistance and lower optical absorption, whereas a smaller band gap energy indicates an opposite trend.

Investigation into the physical properties of Ca(OH)$_2$@OAm NPs focused on determining the average particle size and ζ-potential value. The results revealed a mean particle size of 221 ± 0.4 nm (Figure S3A), indicating the size distribution of these NPs. Furthermore, ζ-potential assessments showed a value of +6.45 ± 0.85 mV (Figure S3B), suggesting a slight positive charge on the surface of particles. It is pertinent to highlight that agglomeration of NPs might transpire, owing to multiple factors such as weak inter-particle interactions, electrostatic attraction, and *van der Waals* forces, potentially culminating in particle clustering.

2.2. Physicochemical Characterization of Calcium-Based Nanocapsules

The emulsifier SDS was used in all NC formations at critical micelle concentration (CMC, 19.5 mM). SDS, which is an anionic surfactant commonly used in a variety of industrial and laboratory applications, plays a pivotal role in NC preparation by stabilizing emulsions, preventing particle aggregation, and controlling the release of encapsulated substances [37,38]. The evaluation of NC formulations for the smart delivery of bioactive agents in various bio-applications relies on parameters such as EE% and LC%. In the current study, the EE% and LC% of the three Ca-based NCs (CaGer1 NCs, CaGer2 NCs, and CaGer3 NCs) exhibited variations. EE% values were 43%, 95%, and 80% for CaGer1 NCs, CaGer2 NCs, and CaGer3 NCs, respectively. Conversely, the LC% values were 5%, 20%, and 24% for CaGer1 NCs, CaGer2 NCs, and CaGer3 NCs, correspondingly. The successful encapsulation and higher geraniol content compared with NCs combining ZnO@OAm nanorods and geraniol (ZnOGer NCs) with the SDS stabilizer [9] may be attributed to the hydroxyl donors from $Ca(OH)_2$ that interact with the geraniol molecule, which contains a free –OH group [39]. This interaction allows for enhanced loading and encapsulation efficiency of geraniol in the CaGer NCs.

DLS measurements were conducted to assess the hydrodynamic size and ζ-potential of the NCs. The DLS analysis demonstrated that the hydrodynamic sizes of the NCs were 280 ± 2.3 nm (PDI = 0.328), 351 ± 1.1 nm (PDI = 0.064), and 306 ± 2.3 nm (PDI = 0.161) for Ca NCs (Figure S4A), CaGer1 NCs (Figure S4B), and CaGer2 NCs (Figure S4C), respectively. The hydrodynamic sizes are close, and size monodispersity is observed except for CaGer3 NCs, which displayed a polydisperse nature characterized by a multiple-size distribution by intensity, as shown by the presence of three distinct peaks and a relatively higher PDI rate of 0.423 (Figure S3D). This observation can be attributed to geraniol diffusion from the core of the NCs. The increase in the molecular weight and volume of geraniol necessitates the reorientation of polymeric chains, resulting in a less uniform particle size distribution in the CaGer3 NCs formulation. Regarding the ζ-potential, all formulations displayed negative values, with ζ-potential rates of -35.25 ± 0.57 mV for CaGer1 NCs, -35.40 ± 0.74 mV for CaGer2 NCs, and -41.3 ± 0.52 mV for CaGer3 NCs, respectively. The physicochemical characteristics of CaGer NCs are summarized in Table 1.

Table 1. Physicochemical characteristics of the calcium-based nanocapsules (Ca-based NCs).

Nano-Formulations	Mass Ratio NPs/Geraniol	Encapsulation Efficacy (%)	Loading Capacity (%)	DLS (d.nm)	PDI	ζ-Potential (mV)
CaGer1 NCs	1:1	43	5	351 ± 1.1	0.064	-35.2 ± 0.57
CaGer2 NCs	1:2	95	20	306 ± 2.3	0.161	-35.4 ± 0.74
CaGer3 NCs	1:3	80	24	382 ± 1.1	0.423	-41.3 ± 0.52

2.3. pH-Dependent Release Profiles and Kinetic Analysis of CaGer2 Nanocapsules

Plant growth is influenced by various environmental factors, including temperature and pH conditions, which play a crucial role in plant physiology and agricultural applications. Generally, pesticides can be degraded in soil, and the rate and type of chemical degradation are influenced by soil temperature, pH levels, moisture, and the binding of insecticides to the soil [40]. The pH of the surrounding environment can affect the release of entrapped ingredients [41] for several reasons: (i) the pH-dependent release behavior provides valuable insights into the potential of the NCs as pH-responsive delivery systems, (ii) changes in pH can alter the electrostatic interactions between the components of the NCs, leading to structural changes or disintegration of the capsules, and (iii) pH can affect the chemical properties of the environment surrounding the NCs. Additionally, temperature can modulate the release of geraniol from the NCs, with high temperatures generally resulting in a higher release rate compared with lower temperatures [9,15]. This can be attributed to the effects of temperature on the molecular mobility and diffusion properties of the encapsulated geraniol and the NC system. Higher temperatures increase the kinetic energy of the molecules, promoting faster diffusion and release of geraniol from the NCs. Temperature can also affect the solubility of geraniol and the interactions between geraniol and the NC matrix. Furthermore, it can impact the stability and integrity of the NCs, as higher temperatures may induce changes in the structure and properties of the NCs, leading to increased porosity or destabilization.

We investigated the effect of pH (values at 9.2, 7.2, and 5.2) on the release kinetics of geraniol from CaGer2 NCs at different temperatures of 25 °C and 35 °C due to their higher colloidal stability, lower hydrodynamic size, and EE% value than CaGer1 NCs and CaGer3 NCs. For the release mechanism of geraniol, the release data were fitted to various kinetic models, and the regression coefficients were analyzed (Tables S1 and S2). The geraniol release profile of CaGer2 NCs at 25 °C exhibited cumulative release percentages ($Q_{4h\%}$) of 7%, 16%, and 20% for pH values of 9.2, 7.2, and 5.2, respectively (Figure 4A). The first-order model demonstrated a slow-release trend (Figure 4B) with an R^2 value of 0.965 at pH 9.2. Interestingly, the burst release rate at pH 7.2 in PBS media ($Q_{4h\%}$ = 36%) was lower compared with ZnOGer2 NCs [9] and native geraniol ($Q_{4h\%}$ = 55%) [15], resulting in an overall $Q_{96h\%}$ of 36% lower than that of ZnO-based NCs. The Higuchi model confirmed a diffusion mechanism release (Figure 4C), as evidenced by decreasing R^2 values (0.965, 0.949, and 0.914) and K_H values (6.43, 5.69, and 5.54) as pH decreased from 9.2 to 5.2. Moreover, the evaluation of the Korsmeyer–Peppas model revealed a normal Fickian diffusion mechanism for geraniol release from CaGer2 NCs in solutions with pH 7.2 and 5.2 (Figure 4D). However, a non-Fickian mechanism (anomalous diffusion) was observed at the highest pH (pH = 9.2), characterized by an exponent (n) greater than 0.45, which exhibited the best fit with the experimental data and system dimension [42].

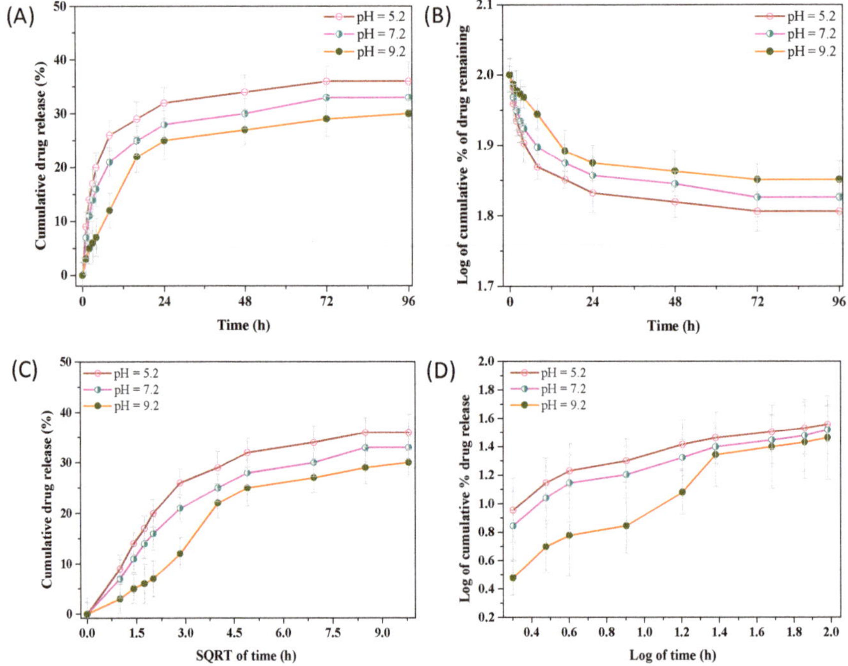

Figure 4. The pH-dependent release profiles of CaGer2 NCs were investigated under various pH conditions (pH = 9.2, 7.2, and 5.2) at a temperature of **25** ± 0.5 °C to simulate environmental conditions. The release profiles of geraniol were analyzed using four kinetic models: zero-order (**A**), first-order (**B**), Higuchi (**C**), and Korsmeyer–Peppas (**D**).

Both diffusion processes and macromolecule relaxation influenced the release process, with solvent diffusion and polymer relaxation transpiring at analogous rates. This triggered/delayed release phenomenon was particularly noticeable under alkaline conditions, resulting in consecutive release profiles over an extended period. Solvent diffusion proceeded at a markedly slower pace compared to micellar relaxation (evidenced by swelling and/or core erosion). The samples displayed an initial burst release followed by a slower continuous drug release over 96 h, indicating an immediate release trend [43]. Also, the pH values in the media affected the size distribution of CaGer2 NCs, as presented in Table S3, with 312 nm (PDI = 0.106) at pH 9.2 and 287 nm (PDI = 0.183) at pH 5.2.

The observed variations in release profiles can be attributed to the architectural differences among the nanosystems, as reported in previous studies where NCs with ZnO nanorods and geraniol [9] and geraniol-loaded nanoparticles prepared with the polymer Pluronic® F-127 [44] demonstrated sustained release over 24 h.

The kinetic profiles of geraniol release from CaGer2 NCs at a higher temperature (35 ± 0.5 °C) revealed cumulative release percentages ($Q_{4h\%}$) of 24%, 30%, and 22% at pH values of 9.2, 7.2, and 5.2, respectively (Figure 5A). A noticeable initial burst release of geraniol from the micelles was observed at pH 5.2 and a temperature of 35 °C. The release rate of geraniol was slightly slower ($Q_{4h\%}$ = 30%) at neutral pH compared with ZnOGer2 NCs [9]. The observed burst release of geraniol (at pH 5.2 and 35 °C) can lead to several potential impacts including faster drug delivery. Burst release can result in rapid delivery of the encapsulated drug/geraniol, which could be beneficial in certain applications where rapid drug action is required. Using the first-order model for NCs at pH 5.2 (Figure 5B) revealed a higher R^2 value (R^2 = 0.953), suggesting a more robust correlation among the variables and a superior data fit to the linear model than was observed at other pH levels (Table S2). Subsequent analysis of geraniol release over time utilized the Higuchi model (Figure 5C), where elevated K_H values signified a more rapid release rate (K_H = 7.24 at pH 5.2) compared with other conditions. Therefore, based on the obtained K_H values, it can be inferred that the release of geraniol from the NCs followed the Higuchi model for all samples. The Korsmeyer–Peppas model (Figure 5D) revealed a non-Fickian diffusion release of geraniol from CaGer2 NCs at pH 5.2 ($n > 0.45$), while a normal Fickian diffusion behavior was observed for the other two cases. Non-Fickian diffusion generally involves a combination of diffusion and other release mechanisms such as swelling, erosion, or polymer relaxation [45]. Hence, the primary mechanism of release from the micelles is considered to be drug diffusion. These measurements highlight the influence of pH on the stability and release behavior of geraniol, providing valuable insights for optimizing the formulation for effective plant protection.

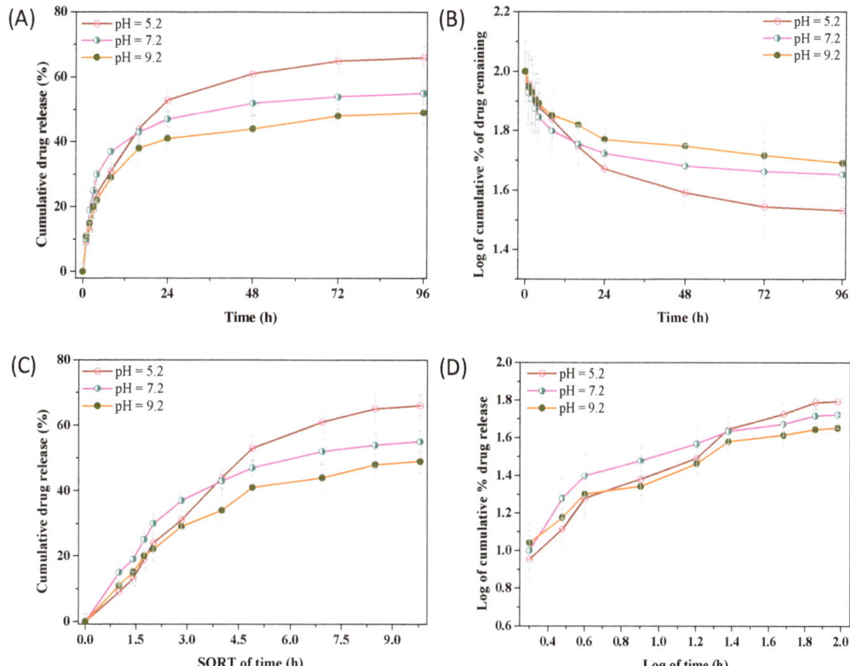

Figure 5. The pH-dependent release profiles of CaGer2 NCs were studied under various pH conditions (pH = 9.2, 7.2, and 5.2) at a temperature of 35 ± 0.5 °C to simulate environmental conditions. The release profiles of geraniol were processed using four kinetic models: zero-order (**A**), first-order (**B**), Higuchi (**C**), and Korsmeyer–Peppas (**D**).

After all, differentiations of the release profile of geraniol from pH and temperature may be considered as a limitation for further application. However, it is important to note that when acute treatment is needed, burst release can be potentially beneficial as rapid depletion of the drug occurs, while sustained release could be favorable in the case of early stages of affection and/or for preventive purposes.

2.4. In vitro Evaluation of Antifungal Activity of Ca-Based NPs and NCs

A preliminary estimation of the antifungal activity against *B. cinerea* of the Ca-based nanoformulations was undertaken. The efficacy of the primary Ca(OH)$_2$@OAm NPs and the Ca-based NCs: Ca NCs and CaGer2 NCs, were evaluated in vitro against *B. cinerea* by measuring their respective EC$_{50}$ values. The determined EC$_{50}$ values for Ca(OH)$_2$@OAm NPs, Ca NCs, and CaGer2 NCs against this pathogen were 654 μg/mL, 395 μg/mL, and 507 μg/mL, respectively, after 96 h (Table 2). The dose–response curves (Figure S5) and mycelium growth in Petri discs (Figure 6) for each case are presented. Interestingly, Ca NCs exhibited higher antifungal efficacy compared with CaGer2 NCs and Ca(OH)$_2$@OAm NPs. Nevertheless, lower EC$_{50}$ values were found before by us in the case of pure geraniol and nanoemulsions containing geraniol against *B. cinerea* [15]. Moreover, NCs composed of ZnO and geraniol in molar ratios of 1:1 and 1:2 exhibited EC$_{50}$ values of 176 μg/mL and 150 μg/mL *against B. cinerea*, respectively [9]. However, a straightforward correlation is not accurate due to the distinct metal base, indicating also that the derived nanostructures can exhibit properties not observed for any of the individual components.

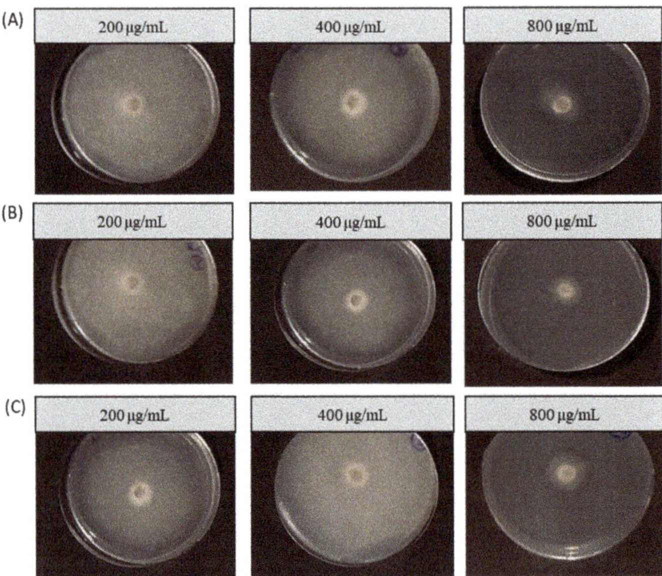

Figure 6. In vitro evaluation of the antifungal activity of Ca(OH)$_2$@OAm NPs (**A**), Ca NCs (**B**), and CaGer2 NCs (**C**) against *B. cinerea* was conducted on Petri plates containing PDA media 96 h after treatment.

Table 2. Half-maximal effective concentration (EC$_{50}$ values) after the application of the Ca-based nanoformulations to *Botrytis cinerea*.

Treatments	EC$_{50}$ Values (μg/mL) $\pm$ SDV
Ca(OH)$_2$@OAm NPs	654 $\pm$ 0.54
Ca NCs	395 $\pm$ 0.73
CaGer2 NCs	507 $\pm$ 0.75

3. Materials and Methods

3.1. Materials

Chemicals and Reagents. All chemicals and reagents were of analytical grade and were used without any further purification: calcium chloride (ACS, VWR Chemicals BDH®, Darmstadt, Germany, M = 110.9 g/mol, $CaCl_2$), oleylamine (Merck, Darmstadt, Germany, M = 267.493 g/mol, OAm), sodium hydroxide (Merck, M = 39.997 g/mol, NaOH), sodium dodecyl sulfate (Sigma-Aldrich, Darmstadt, Germany, ≥99%, M = 288.38 g/mol, SDS), chloroform (VWR Chemicals BDH®, ≥99.8%, $CHCl_3$), diethyl ether (Merck, ≥99%, M = 74.12 g/mol), geraniol (Sigma-Aldrich, 98%, M = 154.25 g/mol, trans-3,7-Dimethyl-2,6-octadien-1-ol), phosphate-buffered saline (Gibco by Life Technologies, Waltham, MA, USA, pH 7.2, 10X, PBS), and potato dextrose agar (BD Difco, Franklin Lakes, NJ, USA, PDA).

3.2. Synthesis of Ca(OH)₂@OAm NPs

A microwave-assisted (MW) route was used for the one-pot synthesis of $Ca(OH)_2$@OAm NPs: 10 mL of a 1.5 M NaOH aqueous solution was added dropwise to 0.8 g of anhydrous $CaCl_2$, which was then dissolved in 30 mL of OAm under vigorous stirring. This mixture was stirred continuously at 35 °C for 15 min before being transferred to a Teflon vessel. The reaction was conducted at 190 °C with a hold time of 30 min and a 15 min ramp time heating step in a MARS 6-240/50-CEM commercial microwave accelerated reaction system, operating at a maximum frequency of 2450 MHz and power of 1800 W. After the microwave treatment, the autoclave was cooled naturally to room temperature. The resulting mixture was centrifuged at 5000 rpm for 20 min and washed three times with chloroform ($CHCl_3$), to remove unnecessary impurities and precursors. The reaction yield was calculated to be 76% based on the initial amount of metal precursor used in the synthesis process and the amount of metal actually incorporated into the NPs.

3.3. Preparation of CaGer Nanocapsules

Calcium@geraniol nanocapsules (CaGer NCs) were synthesized following a previously established method [9]. The synthesis process involved several steps: first, 10 mg of hydrophobic $Ca(OH)_2$@OAm NPs were dissolved in 2 mL of diethyl ether and sonicated for 15 min. Subsequently, pure commercial geraniol was dissolved in 1 mL of diethyl ether and mixed with the NP solution. The resulting mixture was then added to a 30 mL solution of deionized water (diH_2O) containing sodium dodecyl sulfate (SDS) at a critical micelle concentration (CMC) of 19.5 mM. The emulsion was formed with sonication in a closed vial under controlled conditions (<30 °C). After five hours, the vial was opened, and the diethyl ether was slowly evaporated to isolate the respective formulations of NCs: (i) CaGer1 NCs (NPs: geraniol mass ratio of 1:1), (ii) CaGer2 NCs (NPs: geraniol mass ratio of 1:2), and (iii) CaGer3 NCs (NPs: geraniol mass ratio of 1:3). A similar procedure was used for the preparation of calcium hydroxide nanocapsules (Ca NCs), except that this was carried out without adding geraniol. In all procedures, the CMC was maintained at a stable value.

3.4. Characterization Techniques

The crystalline size and structure of $Ca(OH)_2$@OAm NPs were determined using X-ray diffraction (XRD) analysis. XRD pattern was obtained using a two-cycle Rigaku Ultima+ X-ray diffractometer (Rigaku Corporation, Shibuya-Ku, Tokyo, Japan) with a Cu-Kα radiation source (1.541 Å) operating at 40 kV/30 mA. Estimation of the crystallite size was conducted using the Scherrer formula [46]. Furthermore, the crystallinity of the NPs was calculated following the method outlined by Khan et al. (2019) [47]:

$$Crystallinity\ (\%) = \frac{Area\ of\ crystalline\ peaks}{Area\ of\ all\ peaks} \tag{1}$$

Fourier-transform infrared (FT-IR) spectrum of the NPs was performed with a Nicolet iS20 FT-IR spectrometer (Thermo Fisher Scientific, Waltham, MA, USA), featuring a monolithic diamond attenuated total reflection (ATR) crystal, in the wavenumber range of 4000–450 cm^{-1}. TEM imaging was employed to study the morphology of the NPs using a JEOL JEM 1200-EX (Tokyo, Japan) microscope operating at an acceleration voltage of 120 kV. Samples were prepared by placing a drop of diluted NP suspension on a carbon-coated copper grid and air-drying at room temperature. Thermal studies were conducted using a SETA-RAM SetSys-1200 (KEP Technologies, Caluire, France) instrument. Differential thermogravimetric (DTG) and thermogravimetric analysis (TGA) measurements were performed under a nitrogen (N_2) atmosphere with a heating rate of 10 °C/min. Optical property evaluation of the NPs was conducted with a UV-Vis spectrophotometer (V-750, Jasco, Tokyo, Japan), and the band gap energy was calculated based on Tauc's formula $\alpha h\nu = A(h\nu - E_g)^2$ [48]. Assessments of hydrodynamic size (nm), polydispersity index (PDI), and ζ-potential (mV) for the freshly synthesized NCs and NPs were carried out using dynamic light scattering (DLS) analysis at 25 °C, utilizing a Zetasizer (Nano ZS Malvern apparatus VASCO Flex™ Particle Size Analyzer NanoQ V2.5.4.0, Malvern, UK).

3.5. Encapsulation Efficiency and Loading Capacity

Encapsulation efficiency (EE, %) and loading capacity (LC, %) for the nanosystems were calculated to evaluate the compatibility of geraniol with, and the effectiveness of its integration into, CaGer NCs. Evaluation of the EE and LC of geraniol was conducted based on the methodology delineated in prior research [9]. These essential parameters indicate the quantity of geraniol effectively incorporated into the NCs and the overall efficiency of the encapsulation procedure.

3.6. pH-Dependent Release Profiles of Geraniol

The in vitro release profile of geraniol from CaGer NCs was assessed using a dialysis membrane (Pur-A-Lyzer Midi 1000) at two different temperatures: (i) 25 ± 0.5 °C and (ii) 35 ± 0.5 °C. Briefly, 1 mL of the nanoformulation was placed inside a dialysis bag, which was then immersed in 40 mL of PBS solution with varying pH values (9.2, 7.2, and 5.2). The entire system was kept at a constant temperature and stirring rate for 96 h. At predetermined time intervals, 2 mL of the release media was collected and replaced with an equal volume of fresh PBS. The concentration of released geraniol was measured using a UV-Vis spectrophotometer. The obtained data were subjected to mathematical modeling using various models, including zero-order, first-order, Higuchi, and Korsmeyer-Peppas, to determine the transport mechanism, release type, and kinetics of geraniol release [49–51]. Each experiment was performed in triplicate (n = 3) to ensure the reliability and reproducibility of the results.

3.7. Fungal Strain and Growth Conditions

The *Botrytis cinerea* (*B. cinerea*) strain B05 used in this study was obtained from the culture collection of the Plant Pathology Laboratory at the Aristotle University of Thessaloniki. The strain was maintained on potato dextrose agar (PDA) plates at 25 °C in a controlled environment.

3.8. Antifungal Activity Assay

The antifungal activity of Ca(OH)$_2$@OAm NPs, Ca NCs, and CaGer2 NCs was evaluated by mixing the nanoformulations with PDA media and measuring the impact on the growth of *B. cinerea*. To prevent the degradation of the nanoformulations, the media were maintained at approximately 40 °C during the process. Various concentrations of Ca(OH)$_2$@OAm NPs (100, 200, 400, 600, 800, and 1000 µg/mL) and NCs (20, 80, 200, 400, 600, 800, and 1000 µg/mL) were tested. The inhibitory effect of these concentrations on mycelial growth was then assessed. The half-maximal effective concentration

(EC_{50}) value, which is the concentration required to inhibit 50% of mycelial growth, was determined [52,53].

3.9. Statistical Evaluation

The EC_{50} values in the in vitro bioassays were determined according to a non-linear dose-response curve and ten replicates per sample concentration using Origin Pro 8 (Data Analysis and Graphing Software 8.0). Data in kinetic release models were calculated based on the mean of three independent measurements with standard deviations.

4. Conclusions

The development of innovative nanoformulations for the enhanced and targeted control of phytopathogens is challenging. Advanced, second-generation structures amplify advantages primarily due to their unique material properties. In the pursuit of sustainable and eco-friendly solutions for crop protection, $Ca(OH)_2$@OAm NPs were synthesized, which emerged as potent platforms for the encapsulation of geraniol. Remarkably, the $Ca(OH)_2$@OAm NPs were produced with high yields and crystallinity using a one-step and reproducible, microwave-assisted synthesis. The primary NPs were advanced to secondary nanostructures, Ca NPs, CaGer1 NCs, CaGer2 NCs, and CaGer3 NCs, using the emulsifier SDS. Among these, CaGer2 NCs exhibited efficient encapsulation and loading of geraniol, boasting an encapsulation efficiency (EE) of 95% and a loading capacity (LC) of 20%. The release profile evaluation of CaGer2 NCs revealed that variations in pH and temperature influence the release rate of geraniol from the NCs. Antifungal assays against *B. cinerea* showed that Ca NCs demonstrated the highest inhibition in the fungal growth, with an EC_{50} value of 395 µg/mL. Encapsulated $Ca(OH)_2$@OAm NPs demonstrated enhanced efficacy against *B. cinerea* compared with individual NPs. Nonetheless, the newly developed CaGer2 NCs did not exhibit synergistic effects. Further studies are required to test these nanoformulations on a broader spectrum of microorganisms and their potential impact on non-target organisms in order to be deployed for in planta applications.

Supplementary Materials: The following supporting information can be downloaded at: https://www. mdpi.com/article/10.3390/inorganics11120470/s1, Figure S1: FT-IR spectrum of $Ca(OH)_2$@OAm NPs; Figure S2: Absorption spectra as a function of wavelength (210–600 nm) (A) and Tauc plots of $(\alpha h\nu)^2$ vs. $h\nu$ with linear relationship (B) for $Ca(OH)_2$@OAm NPs in ethanol/water solution; Figure S3: Hydrodynamic diameter distribution measured with DLS and ζ-potential for $Ca(OH)_2$@OAm NPs suspension; Figure S4: Hydrodynamic diameter distribution was measured with DLS for the supernatant of CaGer1 NCs (A), CaGer2 NCs (B), and CaGer3 NCs (C) at 25 °C; Table S1: The values (R^2, K_H, n) for the pH-dependent release profiles of CaGer2 NCs are determined using mathematical models: zero-order, first-order, Higuchi, and Korsmeyer-Peppas. The measurements were performed under 25 °C and different pH values (9.2, 7.2, and 5.2) in triplicate; Table S2: The values (R^2, K_H, n) for the pH-dependent release profiles of CaGer2 NCs are derived using mathematical models: zero-order, first-order, Higuchi, and Korsmeyer–Peppas. The measurements were performed under 35 °C and different pH values (9.2, 7.2, and 5.2) in triplicate; Table S3: Dynamic light scattering analysis and ζ-potential values of CaGer2 NCs at pH 5.2 and 9.2; Figure S5: Dose-response growth inhibition curves of *B. cinerea* in response to varying concentrations of $Ca(OH)_2$@OAm nanoparticles (A), Ca nanocapsules (B), and CaGer2 nanocapsules (C). Each data point represents the mean of ten replicates per nanoformulation concentration (2 experiments, 5 replications) 96 h after application.

Author Contributions: Conceptualization, C.D.-S.; methodology, P.T. and N.N.K.; software, P.T., S.M. and G.V.; validation, C.D.-S. and U.M.-S.; formal analysis, P.T.; investigation, P.T., S.M., N.N.K. and G.V.; resources, C.D.-S. and U.M.-S.; data curation, P.T.; writing—original draft preparation, P.T.; writing—review and editing, C.D.-S. and U.M.-S.; visualization, P.T. and S.M.; supervision, C.D.-S. and U.M.-S.; project administration, C.D.-S. and U.M.-S. All authors have read and agreed to the published version of the manuscript.

Funding: Tryfon Panagiota acknowledges the financial support from Greece and the European Union (European Social Fund-ESF) through the Operational Programme "Human Resources Development, Education and Lifelong Learning" in the context of the Act "Enhancing Human Resources Research Potential by undertaking a Doctoral Research" Sub-action 2: IKY Scholarship Programme for PhD candidates in Greek Universities (MIS-5113934).

Data Availability Statement: The data presented in this study are available in this article.

Conflicts of Interest: The authors declare no conflict of interest.

References

1. Bernardes, M.F.F.; Pazin, M.; Pereira, L.C.; Dorta, D.J. Impact of Pesticides on Environmental and Human Health. In *Toxicology Studies-Cells, Drugs, and Environment*; IntechOpen: London, UK, 2015; pp. 195–233.
2. Kim, J.-O.; Shin, J.-H.; Gumilang, A.; Chung, K.; Kim, K.S. Effectiveness of different classes of fungicides on *Botrytis cinerea* causing gray mold on fruit and vegetables. *Plant Pathol. J.* **2016**, *32*, 570–574. [CrossRef] [PubMed]
3. Hernández, A.F.; Gil, F.; Lacasaña, M.; Rodríguez-Barranco, M.; Tsatsakis, A.M.; Requena, M.; Alarcón, R. Pesticide exposure and genetic variation in xenobiotic-metabolizing enzymes interact to induce biochemical liver damage. *Food Chem. Toxicol.* **2013**, *61*, 144–151. [CrossRef] [PubMed]
4. Singh, J.; Yadav, A.N. Natural products as fungicide and their role in crop protection. In *Natural Bioactive Products in Sustainable Agriculture*; Springer: Singapore, 2020; Chapter 9; pp. 131–219.
5. Xu, J.-L.; Luo, Y.-X.; Yuan, S.-H.; Li, L.-W.; Liu, N.-N. Antifungal nanomaterials: Current progress and future directions. *Innov. Digit. Health Diagn. Biomark.* **2021**, *1*, 3–7. [CrossRef]
6. Huang, T.; Li, X.; Maier, M.; O'Brien-Simpson, N.M.; Heath, D.E.; O'Connor, A.J. Using inorganic nanoparticles to fight fungal infections in the antimicrobial resistant era. *Acta Biomater.* **2023**, *158*, 56–79. [CrossRef] [PubMed]
7. Chen, W.; Viljoen, A.M. Geraniol-A review of a commercially important fragrance material. *S. Afr. J. Bot.* **2010**, *76*, 643–651. [CrossRef]
8. Lira, M.H.P.; de Andrade Júnior, F.P.; Moraes, G.F.Q.; Macena, G.S.; Pereira, F.O.; Lima, I.O. Antimicrobial activity of geraniol: An integrative review. *J. Essent. Oil Res.* **2020**, *32*, 187–197. [CrossRef]
9. Tryfon, P.; Kamou, N.N.; Pavlou, A.; Mourdikoudis, S.; Menkissoglu-Spiroudi, U.; Dendrinou-Samara, C. Nanocapsules of ZnO nanorods and geraniol as a novel mean for the effective control of *Botrytis cinerea* in tomato and cucumber plants. *Plants* **2023**, *12*, 1074. [CrossRef]
10. Available online: https://food.ec.europa.eu/plants/pesticides/eu-pesticides-database_en (accessed on 5 September 2023).
11. Chaudhari, A.K.; Singh, V.K.; Das, S.; Dubey, N.K. Nanoencapsulation of essential oils and their bioactive constituents: A novel strategy to control mycotoxin contamination in the food system. *Food Chem. Toxicol.* **2021**, *149*, 112019. [CrossRef]
12. Pavoni, L.; Perinelli, D.R.; Bonacucina, G.; Cespi, M.; Palmieri, G.P. An overview of micro- and nanoemulsions as vehicles for essential oils: Formulation, preparation and stability. *Nanomaterials* **2020**, *10*, 135. [CrossRef]
13. Enayati, S.; Davari, M.; Habibi-Yangjeh, A.; Ebadollahi, A.; Feizpoor, S. Enhancement of the antifungal properties of Zataria multiflora essential oil through nanoencapsulation with ZnO nanomaterial. *Res. Sq.* **2021**, *42*, 503–513. [CrossRef]
14. Akhtari, A.; Davari, M.; Habibi-Yangjeh, A.; Ebadollahi, A.; Feizpour, S. Antifungal activities of pure and ZnO-encapsulated essential oil of Zataria multiflora on *Alternaria solani* as the pathogenic agent of tomato early blight disease. *Front. Plant Sci.* **2022**, *13*, 932475. [CrossRef] [PubMed]
15. Kamou, N.N.; Kalogiouri, N.P.; Tryfon, P.; Papadopoulou, A.; Karamanoli, K.; Dendrinou-Samara, C.; Menkissoglu-Spiroudi, U. Impact of geraniol and geraniol nanoemulsions on *Botrytis cinerea* and effect of geraniol on cucumber plants' metabolic profile analyzed by LC-QTOF-MS. *Plants* **2022**, *11*, 2513. [CrossRef] [PubMed]
16. White, P.J.; Broadley, M.R. Calcium in plants. *Ann. Bot.* **2003**, *92*, 487–511. [CrossRef] [PubMed]
17. McLaughlin, S.; Wimmer, R. Calcium physiology and terrestrial ecosystem processes. *New Phytol.* **1999**, *142*, 373–417. [CrossRef]
18. Parra-Terraza, S.; Villarreal-Romero, M.; Sánchez-Peña, P.; Corrales-Madrid, J.L.; Hernández-Verdugo, S. Effect of calcium and osmotic potential of the nutrient solution on the blossom end rot, mineral composition and yield of tomato. *Interscience* **2008**, *33*, 449–456.
19. Zhang, L.; Du, L.; Poovaiah, W.B. Calcium signaling and biotic defense responses in plants. *Plant Signal Behav.* **2014**, *9*, e973818. [CrossRef] [PubMed]
20. European Food Safety Authority (EFSA). *Outcome of the Consultation with Member States and EFSA on the Basic Substance Application for Approval of Calcium Hydroxide for the Extension of Use in Plant Protection as a Fungicide in Grapevine and Peach, and as Insecticide in Grapevine, Plum, Peach, Apricot, Apple, Pear, Almond and Strawberry*; EFSA Publications: Parma, Italy, 2020.
21. Zhu, J.; Zhang, P.; Ding, J.; Dong, Y.; Cao, Y.; Dong, W.; Zhao, X.; Li, X.; Camaiti, M. Nano Ca(OH)$_2$: A review on synthesis, properties and applications. *J. Cult. Herit.* **2021**, *50*, 25–42. [CrossRef]
22. Nishanthi, R.; Ravindran, V. Role of calcium hydroxide in dentistry: A review. *Int. J. Pharm. Res.* **2020**, *12*, 2822–2827.
23. Harish; Kumari, S.; Parihar, J.; Akash; Kumari, J.; Kumar, L.; Debnath, M.; Kumar, V.; Mishra, R.K.; Gwag, J.S.; et al. Synthesis, characterization, and antibacterial activity of calcium hydroxide nanoparticles against gram-positive and gram-negative bacteria. *ChemistrySelect* **2022**, *7*, e202203094. [CrossRef]

24. Tryfon, P.; Antonoglou, O.; Vourlias, G.; Mourdikoudis, S.; Menkissoglu-Spiroudi, U.; Dendrinou-Samara, C. Tailoring Ca-based nanoparticles by polyol process for use as nematicidals and pH adjusters in Agriculture. *ACS Appl. Nano Mater.* **2019**, *2*, 3870–3881. [CrossRef]

25. Gawande, M.B.; Shelke, S.N.; Zboril, R.; Varma, R.S. Microwave-Assisted Chemistry: Synthetic Applications for Rapid Assembly of Nanomaterials and Organics. *Acc. Chem. Res.* **2014**, *47*, 1338–1348. [CrossRef]

26. Musamali, R.W.; Isa, Y.M. Production of clean hydrogen from methane decomposition over molten NiO-LiOH catalyst systems supported on CaO: Synergistic effect of molten LiOH on catalyst activity. *Asia Pac. J. Chem. Eng.* **2021**, *16*, e2655. [CrossRef]

27. Chen, P.; Wang, Y.; He, S.; Wang, P.; Xu, Y.; Zhang, L. Green synthesis of spherical calcium hydroxide nanoparticles in the presence of tannic acid. *Adv. Mater. Sci. Eng.* **2020**, *2020*, 9501897. [CrossRef]

28. Fuzhi, S.; Wang, Q.; Wang, Y. The effects of crystallinity on the mechanical properties and the limiting PV (pressure × velocity) value of PTFE. *Tribol. Int.* **2016**, *93*, 1–10.

29. Kumar, P.; Kumari, J.; Kumar, L.; Salim, A.; Singhal, R.; Mukhopadhyay, A.K.; Joshi, R.P. Influence of chemical synthesis process on the properties of calcium hydroxide nanoparticles. *Mater. Today Proc.* **2022**, *60*, 153–159.

30. Mourdikoudis, S.; Liz-Marzán, L.M. Oleylamine in nanoparticle synthesis. *Chem. Mater.* **2013**, *25*, 1465–1476. [CrossRef]

31. Guo, H.; Chen, Y.; Cortie, M.B.; Liu, X.; Xie, Q.; Wang, X.; Peng, D.L. Shape-selective formation of monodisperse copper nanospheres and nanocubes via disproportionation reaction route and their optical properties. *J. Phys. Chem. C* **2014**, *118*, 9801–9808. [CrossRef]

32. Kumar, P.; Malhotra, B.; Phalswal, P.; Khanna, P.K.; Salim, A.; Singhal, R.; Mukhopadhyay, A.K. Effect of reaction rate on the properties of chemically synthesized calcium hydroxide nanoparticles. *Mater. Today Proc.* **2020**, *28*, 2305–2310.

33. Khachani, M.; El Hamidi, A.; Halim, M.; Arsalane, S. Non-isothermal kinetic and thermodynamic studies of the dehydroxylation process of synthetic calcium hydroxide Ca(OH)$_2$. *J. Mater. Environ. Sci.* **2014**, *5*, 615–624.

34. Samanta, A.; Podder, S.; Ghosh, C.K.; Bhattacharya, M.; Ghosh, J.; Mallik, A.K.; Dey, A.; Mukhopadhyay, A.K. ROS mediated high anti-bacterial efficacy of strain tolerant layered phase pure nano-calcium hydroxide. *J. Mech. Behav. Biomed. Mater.* **2017**, *72*, 110–128. [CrossRef]

35. Lan, F.; Bai, J.; Wang, H. The preparation of oleylamine modified micro-size sphere silver particles and its application in crystalline silicon solar cells. *RSC Adv.* **2018**, *8*, 16866. [CrossRef] [PubMed]

36. Lenin, R.; Joy, P.A. Role of primary and secondary surfactant layers on the thermal conductivity of lauric acid coated magnetite nanofluids. *J. Phys. Chem. C* **2016**, *120*, 11640–11651. [CrossRef]

37. Vamvakidis, K.; Mourdikoudis, S.; Makridis, A.; Paulidou, E.; Angelakeris, M.; Dendrinou-Samara, C. Magnetic hyperthermia efficiency and MRI contrast sensitivity of colloidal soft/hard ferrite nanoclusters. *J. Colloid Interf. Sci.* **2018**, *511*, 101–109. [CrossRef] [PubMed]

38. Antonoglou, A.; Giannousi, K.; Mourdikoudis, S.; Dendrinou-Samara, C. Magnetic nanoemulsions as candidates for alzheimer's disease dual imaging theranostics. *Nanotechnology* **2020**, *31*, 465702. [CrossRef]

39. Rajabi, H.; Mosleh, M.H.; Mandal, P.; Lea-Langton, A.; Sedighi, M. Emissions of volatile organic compounds from crude oil processing—Global emission inventory and environmental release. *Sci. Total Environ.* **2020**, *727*, 138654. [CrossRef]

40. Singh, D.K. Pesticides and environment. *Pestic. Chem. Toxicol.* **2012**, *1*, 114–122.

41. Zarepour, A.; Zarrabi, A.; Larsen, K.L. Fabricating β-cyclodextrin based pH-responsive nanotheranostics as a programmable polymeric nanocapsule for simultaneous diagnosis and therapy. *Int. J. Nanomed.* **2019**, *14*, 7017–7038. [CrossRef]

42. Ahmed, L.; Atif, R.; Salah Eldeen, T.; Yahya, I.; Omara, A.; Eltayeb, M. Study the using of nanoparticles as drug delivery system based on mathematical models for controlled release. *IJLTEMAS* **2019**, *8*, 52–56.

43. Wu, J.; Zhang, Z.; Gu, J.; Zhou, W.; Liang, X.; Zhou, G.; Han, C.C.; Xu, S.; Liu, Y. Mechanism of a long-term controlled drug release system based on simple blended electrospun fibers. *J. Control. Release* **2020**, *320*, 337–346. [CrossRef]

44. Yegin, Y.; Perez-Lewis, K.; Zhang, M.; Akbulut, M.; Taylor, T.M. Development and characterization of geraniol-loaded polymeric nanoparticles with antimicrobial activity against foodborne bacterial pathogens. *J. Food Eng.* **2016**, *170*, 64–71. [CrossRef]

45. Biondo, F.; Baldassarre, F.; Vergaro, V.; Ciccarella, G. Controlled Biocide Release from Smart Delivery Systems: Materials Engineering to Tune Release Rate, Biointeractions, and Responsiveness. In *Nanotechnology-Based Sustainable Alternatives for the Management of Plant Diseases*; Balestra, G.M., Fortunati, E., Eds.; Elsevier: Amsterdam, The Netherlands, 2022.

46. Cullity, B.D. *Elements of X-ray Diffraction*, 2nd ed.; Addison-Wesley Publishing Co., Ltd.: Reading, MA, USA, 1978.

47. Khan, A.; Toufiq, A.M.; Tariq, F.; Khan, Y.; Hussain, R.; Akhtar, N.; Rahman, S. Influence of Fe doping on the structural, optical, and thermal properties of α-MnO$_2$ Nanowires. *Mat. Res. Express* **2019**, *6*, 065043. [CrossRef]

48. Davis, E.A.; Mott, N.F. Conduction in non-crystalline systems V. Conductivity, optical absorption and photoconductivity in amorphous semiconductors. *Philos. Mag. A* **1970**, *22*, 903–922. [CrossRef]

49. Dash, V. Release kinetic studies of aspirin microcapsules from ethyl cellulose, cellulose acetate phthalate and their mixtures by emulsion solvent evaporation method. *Sci. Pharm.* **2010**, *78*, 93–101. [CrossRef] [PubMed]

50. Lokhandwala, H.; Deshpande, A.; Deshpande, S. Kinetic modeling and dissolution profiles comparison: An overview. *Int. J. Pharm. Bio. Sci.* **2013**, *4*, 728–773.

51. Ramteke, K.H.; Dighe, P.A.; Kharat, A.R.; Patil, S.V. Mathematical models of drug dissolution: A review. *Sch. Acad. J. Pharm.* **2014**, *3*, 388396.

52. Banik, S.; Pérez-de-Luque, A. In vitro effects of copper nanoparticles on plant pathogens, beneficial microbes, and crop plants. *Span. J. Agricult. Res.* **2017**, *15*, e1005. [CrossRef]
53. Tryfon, P.; Kamou, N.N.; Mourdikoudis, S.; Karamanoli, K.; Menkissoglu-Spiroudi, U.; Dendrinou-Samara, C. CuZn and ZnO Nanoflowers as nano-fungicides against *Botrytis cinerea* and *Sclerotinia sclerotiorum*: Phytoprotection, translocation, and impact after foliar application. *Materials* **2021**, *14*, 7600. [CrossRef]

 *inorganics*

MDPI

Article

Alternative Synthesis of MCM-41 Using Inexpensive Precursors for CO_2 Capture

Guillermo D. Aquino [1,2], M. Sergio Moreno [3], Cristian M. Piqueras [4], Germán P. Benedictto [1,2] and Andrea M. Pereyra [1,5,*]

[1] Centro de Investigación y Desarrollo en Ciencias Aplicadas (CINDECA), UNLP-CONICET, La Plata 1900, Argentina; guillermo.aquino.94@gmail.com (G.D.A.); german.benedictto@gmail.com (G.P.B.)
[2] Laboratorio de Investigaciones Químicas, FRA, UTN, Villa Domínico 1874, Argentina
[3] Instituto de Nanociencia y Nanotecnología (INN), CNEA-CONICET-CAB, San Carlos de Bariloche 8402, Argentina; meketo@gmail.com
[4] Planta Piloto de Ingeniería Química (PLAPIQUI), UNS-CONICET, Bahía Blanca 8000, Argentina; cpiqueras@plapiqui.edu.ar
[5] Centro de Investigación y Desarrollo en Tecnología de Materiales (CITEMA), FRLP, UTN-CICPBA, La Plata 1900, Argentina
* Correspondence: andreampereyra@yahoo.com

Abstract: We explore the use of industrial sources of silicon and surfactant for obtaining low-cost MCM-41 materials and evaluate their performances as CO_2 adsorbents. All of them presented a high specific surface area with different structural characteristics and textural properties. Interestingly, the MCM-41 manufactured with the most economical reagents presented a S_{BET} of 1602 $m^2 \cdot g^{-1}$. The template was removed by using thermal treatments in an air atmosphere or a washing process. Preservation of silanol groups proved to be more effective under washing or mild thermal treatment conditions with the advantage of their lower cost and environmental benefit. Surface reactivity against CO_2 was enhanced by anchoring APTS to silanol groups through wet grafting. All amino-functionalized materials showed a performance as CO_2 adsorbents comparable to those reported in the literature, reaching values close to 30 $cm^3 \cdot g^{-1}$ at 25 °C and 760 mmHg. Samples with a higher concentration of silanol groups showed better performance. Our studies indicate that adsorbed CO_2 is retained at least up to 50 °C, and the CO_2 is chemisorbed on the silica modified with amine groups. The chemisorbed gas at very low pressures points to the potential use of these materials for CO_2 storage.

Keywords: MCM-41; industrial reactants; silanol group preservation; CO_2 capture

Citation: Aquino, G.D.; Moreno, M.S.; Piqueras, C.M.; Benedictto, G.P.; Pereyra, A.M. Alternative Synthesis of MCM-41 Using Inexpensive Precursors for CO_2 Capture. *Inorganics* 2023, 11, 480. https://doi.org/10.3390/inorganics11120480

Academic Editors: Roberto Nisticò, Torben R. Jensen, Luciano Carlos, Hicham Idriss and Eleonora Aneggi

Received: 25 October 2023
Revised: 7 December 2023
Accepted: 11 December 2023
Published: 14 December 2023

1. Introduction

The ongoing challenge of global warming is a pressing environmental issue of significant concern. In response to the need to reduce CO_2 emissions, the development of carbon capture and storage (CCS) technologies has emerged as a promising alternative [1]. CCS includes diverse methodologies such as absorption, adsorption, membrane-based systems, and cryogenic separation [2]. In this context, the application of porous materials as adsorbents is gaining substantial attraction due to its advantages, including low energy requirements for adsorbent regeneration, high adsorption capacity, and selectivity for CO_2.

Numerous adsorbents have been proposed and developed, such as activated carbon [3], zeolites [4], mesoporous silicas [5], lamellar double hydroxides [6], metal–organic frameworks [7], graphene oxide [8], and magnesium oxide-based adsorbents [9]. Among these materials, mesoporous silicas have emerged as particularly noteworthy candidates, due to their well-defined pore structures, high surface areas, and chemical surface compositions, suitable for functionalization with reactive species for CO_2 chemisorption. In particular, amine-modified MCM-41 is effective in capturing CO_2, with reported sorption capacities between 0.87 and 2.41 mmol/g under different adsorption conditions [10–18].

Nowadays, a persistent challenge involves improving the efficiency of existing adsorbents while reducing production costs. Achieving this balance is critical for the widespread use of CO_2 adsorption technologies at an industrial scale. In this regard, considerable attention has been given to template agent removal procedures. Thermal treatments are proven to be both fast and effective in removing surfactants, as reported in most of the current literature that analyses MCM-41 silicas as CO_2 adsorbents [10–13]. However, they also produce the condensation of silanol groups into siloxane bonds when applied at temperatures above 270 °C with a detrimental effect on the functionalization capability of the material [19]. Moreover, the combustion of the organic template also leads to CO_2 emission, which is contrary to the potential application of these adsorbent materials [19]. An alternative consists of the use of washing methodologies for template removal. Conversely, employing solvents may offer the advantage of retaining these essential groups and avoiding CO_2 emissions. On the other hand, inefficient extraction might lead to incomplete template removal, resulting in materials with diminished surface areas and restricted accessibility to the silanol groups. In addition, it is necessary to consider the environmental impact of chemical solvents employed in the washing procedure.

Once the porosity of the MCM-41 silicas has been released, surface modification becomes necessary to enhance their efficacy as carbon dioxide adsorbents. Among the different molecules used for this purpose, amines stand out, given their highly selective CO_2 uptake [20]. The 3-aminopropyltrimethoxysilane (APTS) is broadly used, due to its low cost and relatively small size, for its incorporation into the MCM-41 through two methodologies: dry grafting [10,12,14–16] and wet grafting [11,13]. It has been proven that wet conditions promote materials with higher CO_2 adsorption capacities. This was attributed to the role of water in the generation of silanol groups on the silica surface during the grafting process, which serve as new active sites for the anchoring of 3-aminopropyl groups [17].

In this work, the use of low-cost industrial sources of silicon and surfactants for the synthesis of MCM-41 mesoporous silica was explored. An alkaline silicate solution was used as a source of silicon, and the template consisted of a hydroalcoholic solution of CTAC that is currently used as a wetting agent in the plastic and textile industry. Different methodologies for template removal were applied to maximize surface areas while safeguarding the surface-bound silanol groups, minimizing energy consumption, and using environmentally friendly reagents. The materials were surface modified by grafting with APTS to enhance CO_2 adsorption properties, and their retentions were evaluated as functions of temperature by in situ Fourier transform infrared spectra (FTIR).

2. Experimental Procedure

2.1. Materials

The reactants used in this study were tetraethyl orthosilicate (TEOS 98%, Aldrich, St. Louis, MO, USA), industrial sodium silicate solution (ISS, 27.35 wt% SiO_2 and 8.30 wt% Na_2O, Mejorsil, Quilmes, Argentina), hexadecyltrimethylammonium bromide (CTAB 96%, Sigma-Aldrich, St. Louis, MO, USA), industrial N-hexadecyltrimethylammonium chloride (CTAC, 48–50 wt% CTAC, 30–35 wt% EtOH, and less than 1 wt% N,N-dimethylhexadecylamine (DMHA), Meranol, Avellaneda, Argentina), sulfuric acid (H_2SO_4, 98%, Anedra, Los Troncos del Talar, Argentina), ammonia (25%, Merck, Darmstadt, Germany), absolute ethyl alcohol (99.5%, Soria Analytical, Buenos Aires, Argentina), and 3-aminopropyltrimethoxysilane (97%, Sigma-Aldrich).

2.2. Synthesis and Functionalization

Different sources of silicon and structure-directing agents were explored. The samples were identified considering the silicon source, surfactant, and the thermal treatment used to remove the template.

The MCM-41 synthesized from ISS and CTAB or CTAC industrial solution was obtained following the methodology proposed by Edler [21]. In a typical preparation, CTAB or CTAC was dissolved in deionized water under stirring at 30 °C until a clear solution was obtained. Then, ISS was added dropwise to the mixture under vigorous stirring. After 15 min under stirring, the pH was adjusted using a solution of H_2SO_4 (10 wt%). The composition of the obtained gel was 1 SiO_2: 0.29 Na_2O: 0.25–0.26 CTAB or CTAC: <0.006 DMHA: 1.09–1.27 EtOH: 155.47–156.12 H_2O. The mixture was then transferred to a Teflon-lined stainless-steel autoclave and kept in an oven at 100 °C for 4 h. After that, the solid was filtered, washed with deionized water, and dried at 100 °C for 24 h. The samples were called S-B and S-C, respectively.

A reference sample was prepared by using TEOS and CTAB [22]. The molar composition was 1 TEOS: 0.3 CTAB: 11 NH_4OH: 58 EtOH: 144 H_2O. The medium synthesis temperature was 30 °C. The complete dissolution of CTAB was ensure, and, subsequently, the silicon source was added dropwise under stirring at 30 °C for 2 h. The material was recovered by filtration and washed with ethanol and then with distilled water. Finally, excess moisture was removed by drying it at 100 °C for 24 h. The sample was called T-B.

The surfactant was removed using different thermal treatments in air: A long treatment (LT, 540 °C for 10 h, heating rate 1 °C/min) or a short treatment (ST, 510 °C for 2 h, 5 °C/min). Samples were called S-B-LT, S-B-ST, S-C-LT, S-C-ST, T-B-LT, and T-B-ST, respectively.

To avoid the condensation of silanol groups induced by thermal treatments [23], a washing process with ethanol (i.e., solvent extraction) under acidic conditions at 70 °C for 40 min was used to remove the template [24]. This procedure was repeated 4 times, mixing the sample previously extracted with a fresh solvent solution. Finally, the solid was dried at 100 °C for 24 h. The samples were identified by the -W suffix.

Once the surfactant was removed, the materials were functionalized with APTS by the grafting method reported in reference [13]. The MCM-41 was added to anhydrous ethanol under stirring for 10 min. Then, distilled water was added under continuous stirring for 30 min. The APTS was added dropwise at 70 °C. The mixture was stirred and refluxed for 10 h. The APTS-modified material was obtained after washing with anhydrous ethanol and drying at 80 °C for 12 h. Samples were called S-B-ST-g, T-B-ST-g, S-C-ST-g, and S-C-W-g.

Figure 1 illustrates the different synthesis routes followed in this work.

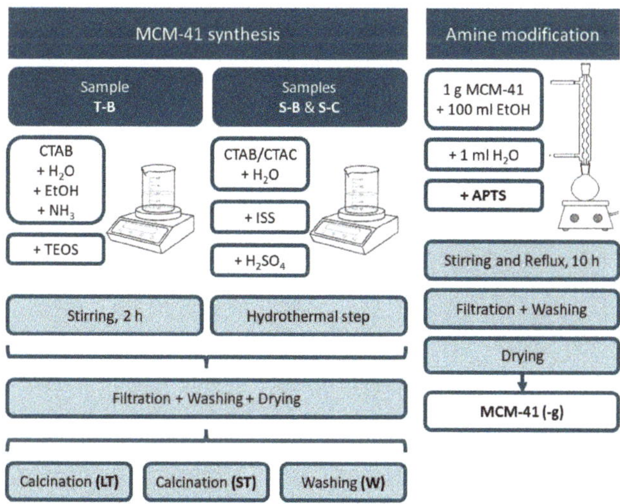

Figure 1. Synthesis and functionalization scheme.

2.3. Characterization

The small angle X-ray diffraction patterns (SAXS) were obtained using XENOCS (Santa Barbara, CA, USA) XEUSS 1.0 equipment with a Pilatus 100K detector with Cu Kα radiation (λ = 1.5405 Å). Scanning electron microscopy (SEM) images were acquired on an FEI (Hillsboro, OH, USA) Inspect S50 microscope operated at 2 kV. The samples were sputtered with a thin film of gold. Transmission electron microscopy (TEM) was performed using a FEI (Hillsboro, OH, USA) Tecnai F20 G2 microscope operated at 200 kV. FTIR spectra were obtained in a Thermo Fisher Scientific (Buenos Aires, Argentina) FTIR Nicolet iS5 spectrophotometer with an iD5 ATR accessory.

The N_2 adsorption–desorption isotherms at 77 K were measured on Micromeritics (Buenos Aires, Argentina) ASAP 2020 apparatus. Prior to the analysis, the samples were degassed at 100 °C for 12 h under a vacuum of 10^{-2} mmHg. The specific surface area was obtained from the gas adsorption isotherms by applying the Brunauer–Emmett–Teller theory (S_{BET}) and non-local density functional theory (S_{DFT}). The total pore volume (V_P) was estimated using the Gurvich rule at a relative pressure of 0.953 [25]. The pore size distribution (PSD) analysis was performed using the NLDFT model applied to cylindrical pores and an oxide surface.

2.4. CO_2 Adsorption Testing

The CO_2 adsorption/desorption isotherms were measured in Quantachrome (Boynton Beach, USA) Autosorb iQ model automatic sorption equipment with its control and data analysis software (ASiQwin 6.0). The samples were introduced into a 12 mm diameter quartz cell provided for the chemisorption mode. The isotherms were fitted using the Langmuir isotherm to determine the moles of CO_2 sorbed in a monolayer covering the surface. CO_2 isotherms were obtained at 25 °C in the pressure range from 10 to 800 Torr.

Diffuse reflectance infrared spectroscopy (DRIFTS) measurements were performed on a Thermo Fisher Scientific (Buenos Aires, Argentina) Nicolet 6700 with a low-temperature high sensitivity Thermo Fischer MCT-A detector. This equipment was assembled with a sealed cell that allowed the treatment of samples with gases at a controlled temperature. Spectra were taken after dosing 5 Torr of CO_2 and then in vacuum at different temperatures up to 100 °C.

3. Results and Discussion

The SAXS diffraction patterns shown in Figure 2 are consistent with the expected 2D hexagonal structure (plane group p6m) and are indexed accordingly. The samples synthesized with ISS exhibited a higher degree of order. Among them, the one obtained using CTAC presented broadened reflections. It can be observed that a longer thermal treatment at higher temperature (LT) produced a slight decrease in the interpore distance from 4.64 to 4.57 nm for the S samples and from 4.02 to 3.83 nm for the T samples. The number of observed reflections indicated high-quality samples in the bulk analysis, confirming a well-ordered hexagonal structure.

The SEM images (Figure 3) showed morphological differences depending on the silicon source. Samples made using TEOS had a spherical shape, whereas those synthesized with sodium silicate presented a more irregular geometry and a smaller particle size.

In Figure 4, representative TEM images of each sample are shown. Different morphologies are observed. Samples obtained using ISS tend to present porous structures organized at different spatial scales. S-C samples presented particles with a bowl shape and shorter mesochannels compared to samples S-B. Mesochannels in S-C materials are oriented more randomly than those in S-B samples due to their lengths. Samples T-B have a spherical shape with a broad particle size distribution and radial mesochannels.

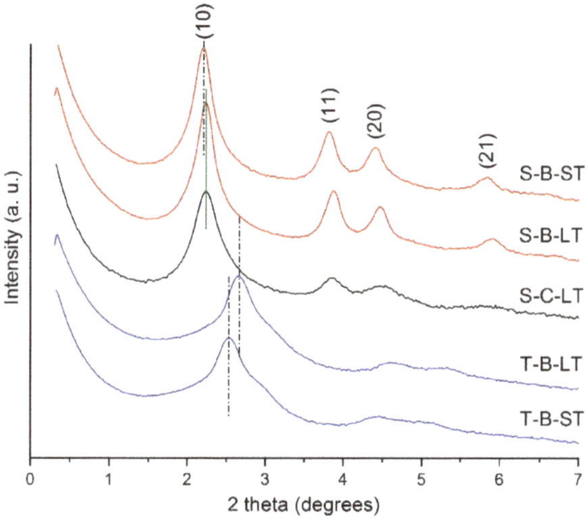

Figure 2. SAXS of MCM-41 samples.

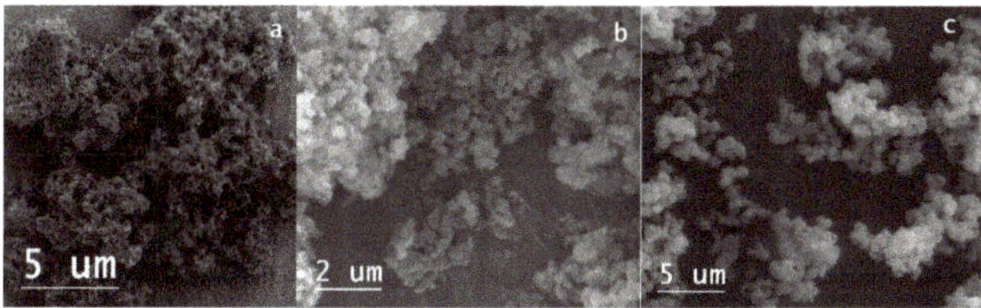

Figure 3. SEM images of samples S-C (**a**), S-B (**b**), and T-B (**c**).

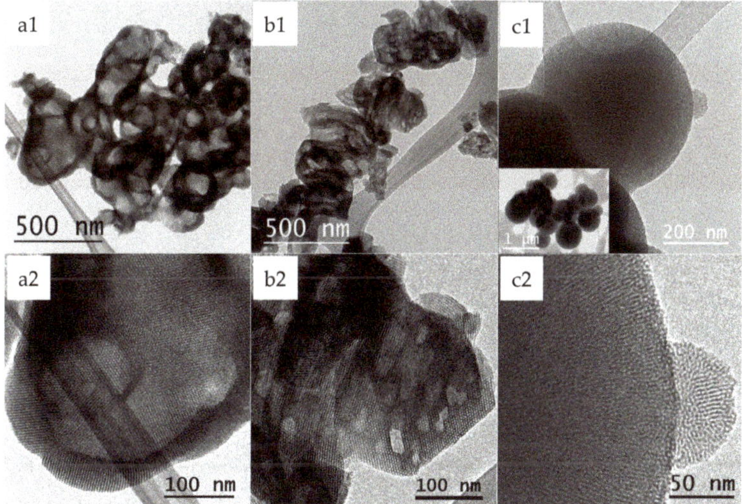

Figure 4. TEM images at different magnifications of samples S-C-ST (**a1,a2**), S-B-ST (**b1,b2**), and T-B-ST (**c1,c2**).

Before textual analysis, the surfactant elimination was verified by FTIR. Spectra of the S-C sample before and after removal treatments are shown in Figure 5 (left). Similar results were obtained for S-B and T-B samples.

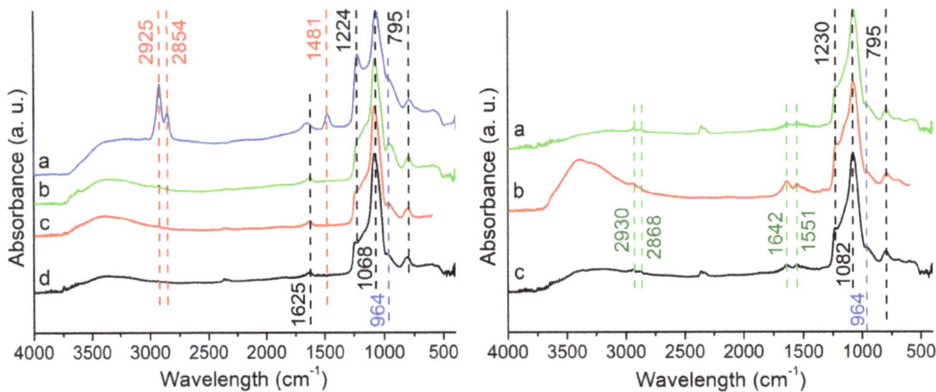

Figure 5. FTIR spectra of S-C samples: (**Left**) (a) as-made, (b) S-C-W, (c) S-C-ST, and (d) S-C-LT; (**Right**) (a) S-C-W-g, (b) S-C-ST-g, and (c) S-C-LT-g. Bands attributed to surfactant (red lines), silanol groups (blue line), aminopropyl groups (green lines), and Si-O modes (black lines) are indicated.

The spectra of as-made samples showed bands attributed to the surfactant at 2854 cm^{-1}, 2925 cm^{-1}, and 1480 cm^{-1} [26,27]. These bands are assigned to C-H asymmetric stretching of alkyl chains of CTA cation and symmetric and asymmetric stretching C-H scissoring vibration of the CH$_3$-N$^+$ group. The absence of these bands after treatments indicated the removal of the template. Other bands observed between 1224 cm^{-1} and 795 cm^{-1} are assigned to different Si-O modes [28,29]. The preservation of silanol groups was confirmed by the band at 964 cm^{-1} (blue line), which corresponds to the stretching vibrations of the surface Si-O- groups [28,30–32]. Bands observed at 1625 cm^{-1} and 3400 cm^{-1} are assigned to OH bending vibrations and -OH units of adsorbed water and silanol groups [28].

From the point of view of preservation of silanol groups, solvent extraction and ST treatment seem to be more efficient than the longer one. The incorporation of APTS molecules was confirmed (Figure 5, right) by the appearance of the bands at 2925 cm^{-1} and 2854 cm^{-1} (C-H bonds) and 1642 cm^{-1} (N-H deformation in RNH$_3^+$) [33].

An indication of the high reactivity of the functionalized material against CO$_2$ adsorption was evidenced by the presence of the band around 1555 cm^{-1} (C=O stretch), which is attributed to carbamate formed by the exposure to air [33–35]. The symmetric and asymmetric stretching modes of CO$_2^-$ could be seen around 2349 cm^{-1}, 1386 cm^{-1}, and 1430 cm^{-1} [20,36].

Nitrogen adsorption and desorption isotherms and the PSD for the synthesized materials subjected to ST treatment and for their functionalized counterparts are shown in Figure 6. The main textural parameters obtained from different methods are compiled in Table 1, with the sample S-C-ST showing the best textural properties.

Nitrogen isotherms from S-C-ST and S-B-ST can be classified as type IV according to the IUPAC classification, which is characteristic of mesoporous materials. When examining the curves, two different hysteresis loops can be seen. The first one corresponds to H1 type and appears at a relative pressure of around 0.35. This can be attributed to a narrow pore size distribution of the open-ended tubular pores from MCM-41 material framework [25]. The second one extends over the relative pressure range of 0.46–0.95. This hysteresis loop can be associated with the existence of the porosity observed at a bigger scale, as revealed in the TEM images. On the other hand, T-B-ST exhibits a type I-like isotherm, as reported elsewhere [37,38]. Samples S-C-ST and S-B-ST presented a narrow PSD with a pore diameter of about 4 nm.

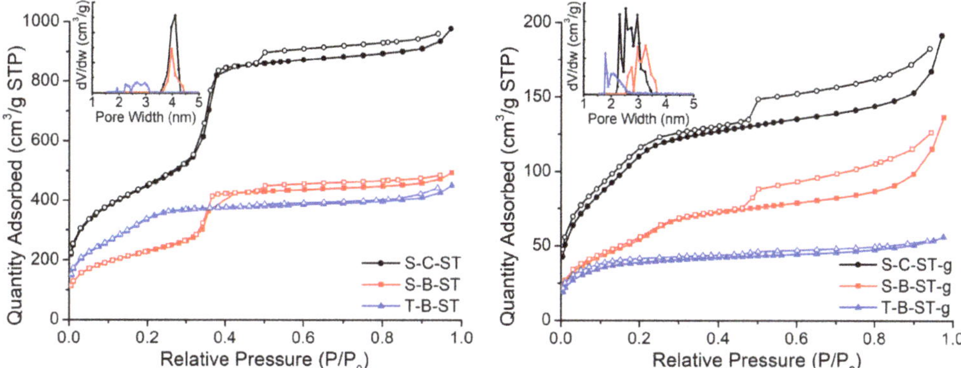

Figure 6. N_2 adsorption (solid symbols) and desorption (open symbols) isotherms of as-made (**left**) and amine-modified (**right**) MCM-41.

Table 1. Textural parameters for MCM-41 materials obtained from N_2 adsorption isotherms.

Sample	S_{BET} (m²/g)	S_{DFT} (m²/g)	V_P (cm³/g)	$D_{P\ NLDFT}$ (nm)
S-C-W	760	608	0.68	3.95
S-C-LT	992	818	0.86	3.95
S-C-ST	1602	1356	1.45	4.10
S-C-ST-g	398	295	0.26	2.24–3.45
S-B-ST	812	673	0.73	3.95
S-B-ST-g	196	152	0.18	2.50–3.60
T-B-ST	1200	886	0.66	2.24–3.10
T-B-ST-g	138	121	0.083	1.77–2.67

After amine modification, the textural parameters decreased significantly. Samples synthesized using ISS (S-C-ST g and S-B-ST-g) presented isotherms similar to type I, preserving the H4 hysteresis loop. This might suggest a reduction in the MCM-41 channel diameter. The second loop seems to be preserved. However, T-B-ST-f exhibited a full type I isotherm commonly attributed to microporous materials, after amine modification, which agrees with the shift of PSD to lower values. In addition, amine-modified samples did not display fully reversible isotherms over the complete relative pressure range. A similar behavior occurs in activated carbons with the only presence of micropores, where the solid wall potential has a great influence on physical adsorption [39].

Estimating the surface area of adsorbents with mesopores in the range of 20 to 40 Å, such as MCM-41 and MCM-48, using the BET approach can pose challenges. In this case, pore filling takes place at pressures very near the range where the formation of monolayer-multilayer structures on the pore walls occurs. This phenomenon is a consequence of the growing potential of pore walls during the condensation process [40]. The presence of these multilayer structures leads to a significant overestimation of the monolayer capacity when conducting BET analysis. As a response to this challenge, we chose to present the surface area reported by the NLDFT model. These models account for the increasing potential of the walls concerning nitrogen adsorption as pore size decreases, thereby yielding more dependable and accurate surface area results.

3.1. CO_2 Adsorption Performance

The CO_2 sorption isotherms are shown in Figure 7. In the upper left panel, the results for different sources of silicon and surfactant for ST samples are compared. The isotherms are similar to those of type I of the IUPAC classification [41], in which the sorption of CO_2 on the surface can be described without the formation of a multilayer for the pressure range up to 800 mmHg. They can be fitted by the Langmuir isotherm, indicating monolayer adsorption. The physisorption isotherms are nearly identical to the sorption ones, suggesting that the adsorption is almost completely of a physical nature. However, all samples exhibited nonreversible isotherms, which might indicate some degree of CO_2 chemisorption. The samples functionalized significantly increased the sorption of CO_2 [42]. The adsorbed volumes are comparable to those found in the literature (see Table 2). All modified samples exhibited similar adsorption curves, with S-C-ST g material demonstrating a slightly better performance at higher pressures. This might be due to its superior textural properties.

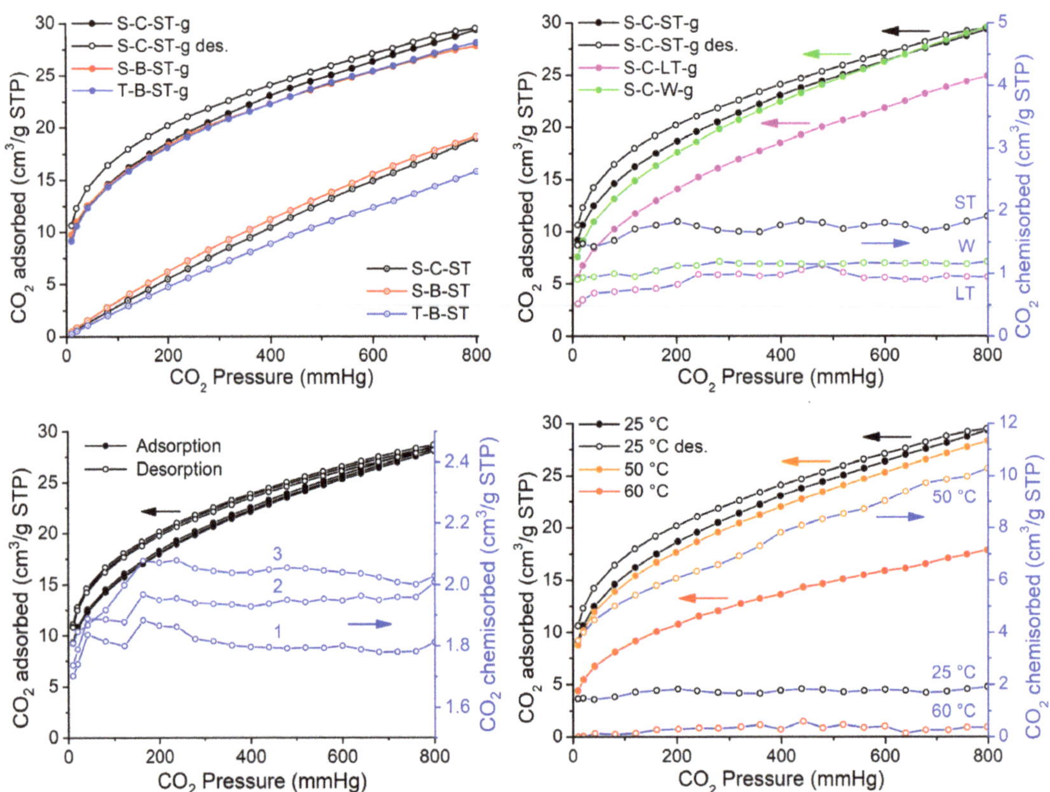

Figure 7. CO_2 total sorption isotherms at 25 °C of as-made and amine-modified MCM-41 (**upper panels**). Sorption isotherms of three consecutive cycles and adsorption/desorption isotherms at different temperatures for S-C-ST g sample (**lower panels**).

Table 2 summarizes the results for our materials and similar ones reported in the literature. When available, it can be seen that there is a great variation in the conditions of the studies and the preparation of reported samples. Our studies presented a slightly better performance than those carried out at similar conditions of pressure and temperature [10,12,14,15]. Other studies performed under dynamic conditions led to higher adsorption values [11,13]. In all cases, textural properties decreased after amine modification.

Table 2. CO_2 adsorption for amine-modified samples and results available in the literature.

Material			Amine-Modified Material		CO_2 Adsorption			Reference
	Si Source	S_{BET} $(m^2/g)/V_P$ (cm^3/g)	Modification	S_{BET} $(m^2/g)/V_P$ (cm^3/g)	Temp. (°C)	CO_2 Pressure (mmHg)	CO_2 Adsorbed (mmol/g)	
MCM-41	ISS	760/0.68	APTS, WG	--/--	25	760	1.29	S-C-W-g
MCM-41	ISS	992/0.86	APTS, WG	--/--	25	760	1.09	S-C-LT-g
MCM-41	ISS	1602/1.45	APTS, WG	398/0.26	25	760	1.28	S-C-ST-g
MCM-41	ISS	812/0.73	APTS, WG	196/0.18	25	760	1.22	S-B-ST-g
MCM-41	TEOS	1200/0.66	APTS, WG	138/0.08	25	760	1.24	T-B-ST-g
MCM-41	--	997/0.90	APTS, WG	958/0.51	70	114	1.88	[13]
MCM-41	TEOS	1059/0.68	APTS, DG	198/0.13	Room temp.	750	1.15	[10]
SBA-15	TEOS	548/0.95	APTS, DG	304/0.55		750	0.93	
MCM-41	TEOS	894/1.28	APTS, DG	544/0.74	45	750	0.87	[14]
MCM-41	TEOS	992/0.69	APTS, WG	736/0.37	25	750	2.41	[11]
MCM-41	--	1031/0.90	APTS, DG	17/0.04	20	750	1.07	[12]
					30	750	0.96	
MCM-41	TEOS	1045/2.59	APTS, DG	--/--	30	750	1.20	[15]
MCM-41	Sodium silicate Merck	864/0.62	APTS, DG	207/0.13	30	3.8	0.39	[16]
SBA-15	TEOS	782/0.73	APTS, DG	280/0.29	60	114	1.06	
SBA-15	TEOS	766/1.32	TRI-s, DG	177/0.49			1.30	[17]
			TRI-s, WG	112/0.31	25	760	1.66	
Bimodal silica	TEOS	612/1.59	TRI-s, DG	203/0.96			1.73	
			TRI-s, WG	71/0.56			1.96	
Hierarchically Ordered Porous Silica	TEOS	1045/0.58	APTS, DG	781/0.45	0	750	1.98	[18]

APTS: aminopropyltrimethoxysilane or aminopropyltriethoxysilane; TRI-s: N1-(3-Trimethoxysilylpropyl)diethylenetriamine; WG: Wet grafting; DG: Dry grafting; '--': No data available.

In the upper right panel, the effects of the method used to remove the surfactant on the CO_2 sorption isotherms are compared. The long treatment has a deleterious effect on the CO_2 adsorption performance. This might be attributed to a diminished concentration of silanol groups on the surface. Interestingly, the S-C-W g sample has a very similar performance to the S-C-ST-g, with considerable economic and environmental benefits. The better performance of -ST and -W samples can be rationalized based on the stronger Si-OH signal observed in FTIR and the employed wet grafting methodology, which induced the formation of new silanol groups helping to create new active sites for aminopropyl anchoring [17].

Figure 7 (left lower panel) shows the sorption isotherms of three consecutive cycles for the S-C-ST g sample. The adsorption and desorption are comparable. A progressive small difference in chemisorption with cycles is observed, which would reflect a greater number of adsorption sites. This result may be attributed to the progressive removal of surface residue from the surface occupied by active sites for adsorption, added to the fact that the binding of the amine responsible for this chemical adsorption is strongly anchored to the surface of the MCM-41, which is consistent with the existing literature [43].

Upon exposing the sample S-C-ST g to adsorption at different temperatures (Figure 7, right lower panel), it is observed that the combined adsorption decreases slightly between 25 and 50 °C, whereas the chemisorption increases fivefold. As the adsorption temperature is raised to 60 °C, the combined adsorption decreases remarkably, and chemical adsorption is reduced almost to zero. The adsorption energy decreases with increasing temperature; however, it appears that there is a competition between this physical phenomenon and the phenomenon of chemical adsorption, the latter being favored in the range of 25–50 °C.

3.2. FTIR with In Situ CO_2 Adsorption

FTIR measurements were carried out to assess CO_2 retention at different temperatures, Figure 8. The amine-modified samples (S-C-ST-g, S-B-ST-g, and T-B-ST-g) were exposed to a controlled atmosphere of CO_2 (Figure 8 left), and then the spectra were acquired under vacuum at different temperatures. The spectra corresponding to 50 °C and 100 °C are shown in Figure 8 (right).

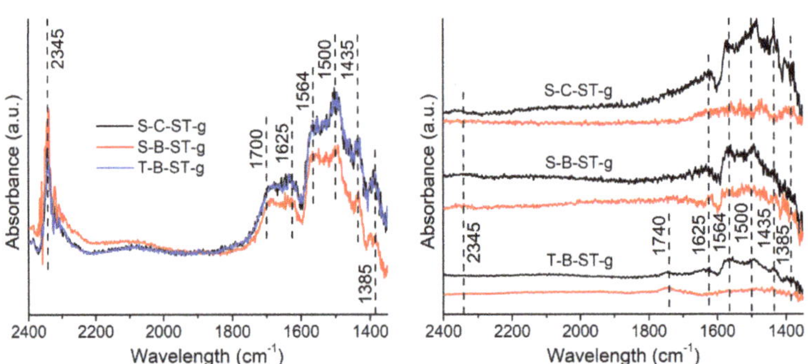

Figure 8. (**Left**) FTIR spectra at 5 Torr of CO_2; (**Right**) Spectra acquired at 50 °C (black) and 100 °C (red).

In all cases, the band attributed to the CO_2 gas phase at 2345 cm^{-1} [36] is observed under the 5 Torr of CO_2 atmosphere. Additional bands are visible in the range 1700–1400 cm^{-1} [20,34,35,44], which could be assigned to the species formed from CO_2 chemisorption on -NH$_2$ groups. Bands at 1500 cm^{-1} and 1625 cm^{-1} are associated with -NH$_3^+$ deformation [35]. In the as-made samples, we did not observe bands in this range. It can be concluded that there is no CO_2 chemisorption on the MCM-41 surface without aminopropyl groups. It is corroborated that CO_2 is retained at least up to 50 °C, in agreement with adsorption measurements.

In summary, the response of the functionalized materials shows sensitivity to CO_2 adsorption at very low pressures. This adsorption is retained after high vacuum evacuation at 25 °C. Furthermore, upon heating the sample to 100 °C, the bands corresponding to chemisorbed CO_2 disappear.

4. Conclusions

In this work, inexpensive MCM-41 materials were synthesized successfully using different sources of silicon and surfactants of industrial origin. Surface reactivity was preserved by using gentle methods of template removal, which are environmentally sustainable and implies reducing costs. All of them presented a high specific surface area, but the structural features depended on the reactant nature. Amine-modified materials presented a CO_2 adsorbent capacity comparable with other systems previously reported. The retention of chemisorbed gas at least up to 50 °C at very low pressures suggests the potential of these materials for CO_2 storage. These results combined whit the simplicity of fabrication of the porous materials for the CO_2 sorption constitute a low-cost alternative.

Author Contributions: Conceptualization, G.D.A., A.M.P., G.P.B. and M.S.M.; methodology, G.D.A., A.M.P., G.P.B. and M.S.M.; investigation, G.D.A., A.M.P., G.P.B., C.M.P. and M.S.M.; writing—original draft preparation, G.D.A., A.M.P., G.P.B., C.M.P. and M.S.M.; writing—review and editing, G.D.A., A.M.P., G.P.B., C.M.P. and M.S.M.; project administration, A.M.P. and M.S.M.; funding acquisition, A.M.P. and M.S.M. All authors have read and agreed to the published version of the manuscript.

Funding: This work was financially supported by the Agencia Nacional de Promoción Científica y Tecnológica (ANPCyT) (PICT 2020-03966), Universidad Tecnológica Nacional (PID 8621TC), and Consejo Nacional de Investigaciones Científicas y Técnicas (CONICET) (PIP 11220200100768CO).

Data Availability Statement: The data presented in this study are available in article.

Acknowledgments: The authors thank Paula Troyón for SEM image acquisition.

Conflicts of Interest: The authors declare no conflict of interest.

References

1. Shu, D.Y.; Deutz, S.; Winter, B.A.; Baumgärtner, N.; Leenders, L.; Bardow, A. The Role of Carbon Capture and Storage to Achieve Net-Zero Energy Systems: Trade-Offs between Economics and the Environment. *Renew. Sustain. Energy Rev.* **2023**, *178*, 113246. [CrossRef]
2. Tan, Y.; Nookuea, W.; Li, H.; Thorin, E.; Yan, J. Property Impacts on Carbon Capture and Storage (CCS) Processes: A Review. *Energy Convers. Manag.* **2016**, *118*, 204–222. [CrossRef]
3. Vorokhta, M.; Nováková, J.; Dopita, M.; Khalakhan, I.; Kopecký, V.; Švábová, M. Activated Three-Dimensionally Ordered Micromesoporous Carbons for CO_2 Capture. *Mater. Today Sustain.* **2023**, *24*, 100509. [CrossRef]
4. Cavallo, M.; Dosa, M.; Porcaro, N.G.; Bonino, F.; Piumetti, M.; Crocella, V. Shaped Natural and Synthetic Zeolites for CO_2 Capture in a Wide Temperature Range. *J. CO2 Util.* **2023**, *67*, 102335. [CrossRef]
5. Medina-Juárez, O.; Rangel-Vázquez, I.; Ojeda-López, R.; García-Sánchez, M.Á.; Rojas-González, F. Importance of the Polarity on Nanostructured Silica Materials to Optimize the Hydrolytic Condensation with Molecules Related to CO_2 Adsorption. *Environ. Sci. Pollut. Res.* **2022**, *29*, 58472–58483. [CrossRef]
6. Rocha, C.; Soria, M.A.; Madeira, L.M. Doping of Hydrotalcite-Based Sorbents with Different Interlayer Anions for CO_2 Capture. *Sep. Purif. Technol.* **2020**, *235*, 116140. [CrossRef]
7. Elsabawy, K.M.; Fallatah, A.M. Synthesis of Newly Wings like Structure Non-Crystalline Ni++-1,3,5-Tribenzyl-1,3,5-Triazine-2,4,6-(1H,3H,5H)-Trione Coordinated MOFs for CO_2-Capture. *J. Mol. Struct.* **2019**, *1177*, 255–259. [CrossRef]
8. Shen, Z.; Song, Y.; Yin, C.; Luo, X.; Wang, Y.; Li, X. Construction of Hierarchically Porous 3D Graphene-like Carbon Material by B, N Co-Doping for Enhanced CO_2 Capture. *Microporous Mesoporous Mater.* **2021**, *322*, 111158. [CrossRef]
9. Alkadhem, A.M.; Elgzoly, M.A.A.; Onaizi, S.A. Novel Amine-Functionalized Magnesium Oxide Adsorbents for CO_2 Capture at Ambient Conditions. *J. Environ. Chem. Eng.* **2020**, *8*, 103968. [CrossRef]
10. Zhao, H.; Hu, J.; Wang, J.; Zhou, L.; Liu, H. CO_2 Capture by the Amine-Modified Mesoporous Materials. *Acta Phys.-Chim. Sin.* **2007**, *23*, 801–806. [CrossRef]
11. Rao, N.; Wang, M.; Shang, Z.; Hou, Y.; Fan, G.; Li, J. CO_2 Adsorption by Amine-Functionalized MCM-41: A Comparison between Impregnation and Grafting Modification Methods. *Energy Fuels* **2018**, *32*, 670–677. [CrossRef]
12. Mello, M.R.; Phanon, D.; Silveira, G.Q.; Llewellyn, P.L.; Ronconi, C.M. Amine-Modified MCM-41 Mesoporous Silica for Carbon Dioxide Capture. *Microporous Mesoporous Mater.* **2011**, *143*, 174–179. [CrossRef]
13. Wang, X.; Chen, L.; Guo, Q. Development of Hybrid Amine-Functionalized MCM-41 Sorbents for CO_2 Capture. *Chem. Eng. J.* **2015**, *260*, 573–581. [CrossRef]
14. Sanz, R.; Calleja, G.; Arencibia, A.; Sanz-Pérez, E.S. CO_2 Capture with Pore-Expanded MCM-41 Silica Modified with Amino Groups by Double Functionalization. *Microporous Mesoporous Mater.* **2015**, *209*, 165–171. [CrossRef]
15. Loganathan, S.; Tikmani, M.; Ghoshal, A.K. Novel Pore-Expanded MCM-41 for CO_2 Capture: Synthesis and Characterization. *Langmuir* **2013**, *29*, 3491–3499. [CrossRef] [PubMed]
16. Chang, F.Y.; Chao, K.J.; Cheng, H.H.; Tan, C.S. Adsorption of CO_2 onto Amine-Grafted Mesoporous Silicas. *Sep. Purif. Technol.* **2009**, *70*, 87–95. [CrossRef]
17. Anyanwu, J.T.; Wang, Y.; Yang, R.T. CO_2 Capture (Including Direct Air Capture) and Natural Gas Desulfurization of Amine-Grafted Hierarchical Bimodal Silica. *Chem. Eng. J.* **2022**, *427*, 131561. [CrossRef]
18. Jin, X.; Ge, J.; Zhang, L.; Wu, Z.; Zhu, L.; Xiong, M. Synthesis of Hierarchically Ordered Porous Silica Materials for CO_2 Capture: The Role of Pore Structure and Functionalized Amine. *Inorganics* **2022**, *10*, 87. [CrossRef]
19. Ghaedi, H.; Zhao, M. Review on Template Removal Techniques for Synthesis of Mesoporous Silica Materials. *Energy Fuels* **2022**, *36*, 2424–2446. [CrossRef]
20. Bacsik, Z.; Atluri, R.; Garcia-Bennett, A.E.; Hedin, N. Temperature-Induced Uptake of CO_2 and Formation of Carbamates in Mesocaged Silica Modified with n-Propylamines. *Langmuir* **2010**, *26*, 10013–10024. [CrossRef]
21. Edler, K.J.; White, J.W. Further Improvements in the Long-Range Order of MCM-41 Materials. *Chem. Mater.* **1997**, *9*, 1226–1233. [CrossRef]
22. Grün, M.; Unger, K.K.; Matsumoto, A.; Tsutsumi, K. Novel Pathways for the Preparation of Mesoporous MCM-41 Materials: Control of Porosity and Morphology. *Microporous Mesoporous Mater.* **1999**, *27*, 207–216. [CrossRef]
23. Araujo, A.S.; Jaroniec, M. Thermogravimetric Monitoring of the MCM-41 Synthesis. *Thermochim. Acta* **2000**, *363*, 175–180. [CrossRef]
24. Marcilla, A.; Beltran, M.; Gómez-Siurana, A.; Martinez, I.; Berenguer, D. Template Removal in MCM-41 Type Materials by Solvent Extraction: Influence of the Treatment on the Textural Properties of the Material and the Effect on Its Behaviour as Catalyst for Reducing Tobacco Smoking Toxicity. *Chem. Eng. Res. Des.* **2011**, *89*, 2330–2343. [CrossRef]
25. Rouquerol, J.; Rouquerol, F.; Llewellyn, P.; Maurin, G.; Sing, K.S.W. *Adsorption by Powders and Porous Solids: Principles, Methodology and Applications*, 2nd ed.; Academic Press: Cambridge, MA, USA, 2014; ISBN 9780080970356.

26. Su, G.; Yang, C.; Zhu, J.J. Fabrication of Gold Nanorods with Tunable Longitudinal Surface Plasmon Resonance Peaks by Reductive Dopamine. *Langmuir* **2015**, *31*, 817–823. [CrossRef] [PubMed]
27. Banjare, R.K.; Banjare, M.K.; Panda, S. Effect of Acetonitrile on the Colloidal Behavior of Conventional Cationic Surfactants: A Combined Conductivity, Surface Tension, Fluorescence and FTIR Study. *J. Solut. Chem.* **2020**, *49*, 34–51. [CrossRef]
28. Huo, C.; Ouyang, J.; Yang, H. CuO Nanoparticles Encapsulated inside Al-MCM-41 Mesoporous Materials via Direct Synthetic Route. *Sci. Rep.* **2014**, *4*, 3682. [CrossRef]
29. La-Salvia, N.; Lovón-Quintana, J.J.; Lovón, A.S.P.; Valença, G.P. Influence of Aluminum Addition in the Framework of MCM-41 Mesoporous Molecular Sieve Synthesized by Non-Hydrothermal Method in an Alkali-Free System. *Mater. Res.* **2017**, *20*, 1461–1469. [CrossRef]
30. Li, L.L.; Sun, H.; Fang, C.J.; Xu, J.; Jin, J.Y.; Yan, C.H. Optical Sensors Based on Functionalized Mesoporous Silica SBA-15 for the Detection of Multianalytes (H^+ and Cu^{2+}) in Water. *J. Mater. Chem.* **2007**, *17*, 4492–4498. [CrossRef]
31. Ganji, S.; Mutyala, S.; Neeli, C.K.P.; Rao, K.S.R.; Burri, D.R. Selective Hydrogenation of the CC Bond of α,β-Unsaturated Carbonyl Compounds over PdNPs–SBA-15 in a Water Medium. *RSC Adv.* **2013**, *3*, 11533–11538. [CrossRef]
32. Medeiros de Paula, G.; do Nascimento Rocha de Paula, L.; Freire Rodrigues, M.G. Production of MCM-41 and SBA-15 Hybrid Silicas from Industrial Waste. *Silicon* **2022**, *14*, 439–447. [CrossRef]
33. Qi, G.; Wang, Y.; Estevez, L.; Duan, X.; Anako, N.; Park, A.H.A.; Li, W.; Jones, C.W.; Giannelis, E.P. High Efficiency Nanocomposite Sorbents for CO_2 Capture Based on Amine-Functionalized Mesoporous Capsules. *Energy Environ. Sci.* **2011**, *4*, 444–452. [CrossRef]
34. Knöfel, C.; Martin, C.; Hornebecq, V.; Llewellyn, P.L. Study of Carbon Dioxide Adsorption on Mesoporous Aminopropylsilane-Functionalized Silica and Titania Combining Microcalorimetry and in Situ Infrared Spectroscopy. *J. Phys. Chem. C* **2009**, *113*, 21726–21734. [CrossRef]
35. Socrates, G. *Infrared and Raman Characteristic Group Frequencies: Tables and Charts*, 3rd ed.; John Wiley & Sons, Ltd: Chichester, UK, 2004; ISBN 978-0-470-09307-8.
36. Dañon, A.; Stair, P.C.; Weitz, E. FTIR Study of CO_2 Adsorption on Amine-Grafted SBA-15: Elucidation of Adsorbed Species. *J. Phys. Chem. C* **2011**, *115*, 11540–11549. [CrossRef]
37. Schmidt, R.; Stöcker, M.; Hansen, E.; Akporiaye, D.; Ellestad, O.H. MCM-41: A Model System for Adsorption Studies on Mesoporous Materials. *Microporous Mater.* **1995**, *3*, 443–448. [CrossRef]
38. Wloch, J.; Rozwadowski, M.; Lezanska, M.; Erdmann, K. Analysis of the Pore Structure of the MCM-41 Materials. *Appl. Surf. Sci.* **2002**, *191*, 368–374. [CrossRef]
39. Lowell, S.; Shields, J.E.; Thomas, M.A.; Thommes, M. *Characterization of Porous Solids and Powders: Surface Area, Pore Size and Density*; Particle Technology Series; Springer: Dordrecht, The Netherlands, 2004; Volume 16, ISBN 978-90-481-6633-6.
40. Kruk, M.; Jaroniec, M.; Sayari, A. Adsorption Study of Surface and Structural Properties of MCM-41 Materials of Different Pore Sizes. *J. Phys. Chem. B* **1997**, *101*, 583–589. [CrossRef]
41. Sing, K.S.W.; Everett, D.H.; Haul, R.A.W.; Moscou, L.; Pierotti, R.A.; Rouquerol, J.; Siemieniewska, T. Reporting Physisorption Data for Gas/Solid Systems with Special Reference to the Determination of Surface Area and Porosity. *Pure Appl. Chem.* **1985**, *57*, 603–619. [CrossRef]
42. Builes, S.; Vega, L.F. Understanding CO_2 Capture in Amine-Functionalized MCM-41 by Molecular Simulation. *J. Phys. Chem. C* **2012**, *116*, 3017–3024. [CrossRef]
43. Chanapattharapol, K.C.; Krachuamram, S.; Youngme, S. Study of CO_2 Adsorption on Iron Oxide Doped MCM-41. *Microporous Mesoporous Mater.* **2017**, *245*, 8–15. [CrossRef]
44. Silverstein, R.M.; Webster, F.X.; Kiemle, D.J.; Bryce, D.L. *Spectrometric Identification of Organic Compounds*, 7th ed.; John Wiley & Sons, Ltd.: Hoboken, NJ, USA, 2005; ISBN 0-471-39362-2.

Article

Continuous and Intermittent Planetary Ball Milling Effects on the Alloying of a Bismuth Antimony Telluride Powder Mixture

Georgios Samourgkanidis * and Theodora Kyratsi

Department of Mechanical and Manufacturing Engineering, University of Cyprus, Nicosia 1678, Cyprus
* Correspondence: samourgkanidis.georgios@ucy.ac.cy

Abstract: This study investigates the effects of continuous and in-steps mechanical alloying of a bismuth antimony telluride powder mixture ($Bi_{0.4}Sb_{1.6}Te_{3.0}$) via the mechanical planetary ball milling (PBM) process as a function of milling time and powder mixture amount. X-ray diffraction (XRD) and scanning electron microscopy (SEM) were used to characterize the phase, composition, and morphology of the alloy. The alloyed powder with the optimum PBM conditions was then hot pressed (HP), and its thermoelectric properties were further investigated. The results on the alloying of the powder mixture showed that due to the high agglomeration tendency of BST during the PBM process, a significant deviation occurs in the development of a single-phase state over time when the powder mixture is milled continuously and in-steps. 'In-steps' refers to the procedure of interrupting the PBM process and detaching the agglomerated powder adhering to the inner walls of the vessel. This task was repeated every hour and a half of the PBM process for a total of 12 h, and the results were compared with those of the 12 h continuous PBM process of the same mixture. In addition, the procedure was repeated with different amounts of mixture (100 g and 150 g) to determine the most efficient method of producing the material as a function of time. As for the thermoelectric profile of the powder, the data showed results in direct agreement with those in the literature.

Keywords: mechanical alloying; planetary ball milling; bismuth antimony telluride; thermoelectric materials

Citation: Samourgkanidis, G.; Kyratsi, T. Continuous and Intermittent Planetary Ball Milling Effects on the Alloying of a Bismuth Antimony Telluride Powder Mixture. *Inorganics* **2023**, *11*, 221. https://doi.org/10.3390/inorganics11050221

Academic Editors: Roberto Nisticò, Torben R. Jensen, Luciano Carlos, Hicham Idriss and Eleonora Aneggi

Received: 20 April 2023
Revised: 15 May 2023
Accepted: 16 May 2023
Published: 20 May 2023

1. Introduction

Over the last two decades, there has been an increasing trend in the study and development of thermoelectric (TE) alloys from submicron- and nano-powders, necessitating the incorporation of mechanical alloying (MA) processes in the field of TE materials. However, the race to create fine and ultra-fine TE material powders using MA started a little earlier [1–4]. In general, the MA process is based on the repetitive process of cold welding and fracturing of powder particles with the rotational kinetic energy of the grinding balls [5]. It is a high-energy process capable of producing alloys that would be impossible to create otherwise, with a typical example being the aluminum–tantalum alloy (Al-Ta), for which the melting method for the alloy's formation becomes restrictive due to the large difference in the melting points of its individual elements (Aluminum at 933 K, Tantalum at 3293 K) [6].

One of the most commonly used MA methods is that of the planetary ball milling (PBM) process [7–11], and it has significant advantages over other conventional methods (mixer mills, stirred mills, etc.). It is effective for wet and dry grinding processes due to the high mechanical stresses reached during the operation, and submicron and nano-sized particles can be achieved with the proper milling balls and milling conditions. The operating principle of this method is based on the rotation of the grinding containers around the common axis of rotation of the main disk (also known as the sun-disk), as well as the vertical axis of rotation of the containers themselves. In this manner, the kinetic energy of the grinding containers, due to rotation and self-rotation, is greatly increased, resulting in high impact energies of the milling balls inside, which in turn crash any powder particles trapped on their surface in between collisions [12–15]. The PBM process is most effective when the milling conditions, such as rotational speed, milling time, ball-to-material ratio, number, size, and material of the grinding balls, are first evaluated [16,17]. In particular, the material of the milling balls should be chosen based on the reactions that occur during the milling process in order to avoid any side effects that could contaminate or change the structure of the grinding mixture. The materials from which grinding balls are usually made of are tungsten carbide (WC), corundum (Al_2O_3), stainless steel (SS), zirconia (ZrO_2), and agate (SiO_2) [18–22].

As previously stated, the introduction of MA in the world of TE materials accelerated the development of new thermoelectric alloys with higher efficiencies and wider temperature ranges [23,24]. The current state of TE materials shows conversion efficiencies ranging from 5% to 20%, a percentage range that can be improved further using suitable methods of alloying, doping, and nanostructuring. The efficiency of a TE material is projected through the dimensionless thermoelectric figure of merit, ZT, which is directly related to the physical properties of the material and is calculated from the relationship $ZT = S^2 * \sigma/\kappa$, where S is the Seebeck coefficient, σ the electrical conductivity, and κ the thermal conductivity. In the case of an ideal TE material (high Seebeck coefficient and electrical conductivity, low thermal conductivity), the ZT value should be greater than one in order for it to have an efficiency greater than 10% [25–29]. One of the the most widely studied and commercialized TE alloys are those of bismuth telluride due to their excellent performance in the temperature range [250 K–400 K] [30–37]. These alloys are easily prepared as fine powders by MA, and their thermoelectric properties are acquired through mechanical and thermal treatments [38,39].

The current study looks into the manufacturing process of a specific bismuth telluride TE alloy, that of $Bi_{0.4}Sb_{1.6}Te_{3.0}$, in two distinct but related ways. This particular p-type alloy possess superior TE performance around room temperature [37], making it a frequently researched and used bismuth telluride alloy. Although the methods used are presented only for the production of the specific material, they can be applied to the general group of TE materials that tend to agglomerate during the alloying process using the MA methods. From an innovation standpoint, this work demonstrates manufacturing techniques that can process materials faster and more efficiently, using less time and energy while producing larger quantities. For example, with the ongoing research in additive manufacturing and 3D printing of TE materials [40–42], the need for cost-effective and efficient methods to produce TE powder materials in large quantities becomes critical.

2. Experimental Procedures

As starting materials, elemental powders of Bi (99.99% metal basis, −200 mesh), Sb (99.5% metal basis, −200 mesh), and Te (99.5% metal basis, −200 mesh) were used and weighed according to the desired stoichiometry. To avoid oxidation phenomena and to protect the user against the toxicity of the materials, the whole procedure (weighing process, powder loading, and unloading) was carried out in a glovebox containing argon. The powder mixture with the correct composition was then loaded into a 250 mL tungsten carbide (WC) vial along with 110 WC milling balls with a diameter of d = 10 mm and milled in an argon atmosphere using the FRITSCH planetary mono mill Pulverisette 6 classic line. All experiments were performed under the same PBM conditions: milling speed at 400 rpm, 10 min of milling time per repetition, and 5 min of pause time per repetition. The reverse mode was always ON during the milling process, as it favors the alloying process of dry materials and improves the homogeneity of the mixture. As the PBM process in this study was performed for two different amounts of material mixture, the ball-to-material ratio varied in each test. In particular, the ratio was 8.6:1 for the 100 g of powder mixture (BST-100) and 5.7:1 for the 150 g of powder mixture (BST-150).

The X-ray diffraction patterns of the powder were acquired using a Rigaku MiniFlex diffractometer, and the structural and phase characteristics were analyzed using the X-Pert highscore plus software program. All X-ray diffraction graphs were taken in the angular range of 10–60 degrees, angular step of 0.02 degree, and scan speed of 1 degree per minute. The acceleration voltage was set to 30 kV and the beam current to 15 mA, resulting in Cu-Ka radiation (k = 1.5406 Å). The chemical composition (EDX) and morphology (SEM) of the powder was examined with the use of TESCAN VEGA-II Model LSH scanning electron microscopy, and the EDX analysis was carried out under identical conditions (magnification, live time, dead time, takeoff angle, etc.) for different regions of the powder sample.

To investigate the thermoelectric properties of the alloy powder under optimal milling time and amount conditions, a small portion of it was HP inside a cylindrical graphite matrix, using a Thermal Technology Graphite Hot Press (Model HP20-4560), to form a 10 mm diameter and 2.5 mm thick disk pellet. The HP conditions were optimal and were based on our previous work [34]. Specifically, the alloy powder was HP for 1 h at the temperature of 420 °C under a mechanical press of 80 MPa (pressure equivalent to 640 kg). The temperature ramp rate was set to 10 °C/min, and the pressing ramp rate was set to 16 kg/min. The average density of the HP sample was ρ_{sample} = 6.022 g/cm^3 when measured using the Archimedes method in ethanol, which is greater than 90% of the bulk material (ρ_{bulk} = 6.564 g/cm^3 as measured experimentally). The thermoelectric evaluation of the HP sample was performed in the temperature range [300 K–500 K] using a Netzsch LFA 457 laser flash apparatus (thermal properties) and a ZEM-3 series ULVAC-Riko system (electrical properties).

3. Results and Discussion

Figure 1 shows the X-ray diffraction patterns versus time of the BST-150 powder mixture PBMed in-steps. The time step here was 1.5 h, for a total of 12 h, and in between steps, the milling container was removed from the milling machine and placed inside an argon-filled glovebox, where the agglomerated powder was extracted, milled with agate mortar, sieved, and then added back to the PBM process (Figure 2). The agate mortar grinding and sieving sub-process was repeated until no powder was left on the sieving basket and all agglomerated powder particles were thoroughly ground. The vertical axis of the intensities is linearly scaled. The alloying process of the elements is clearly visible from the behavior of the BST's base peak (015) (shown in the JCPDS plot). As the milling time increases, the multi-phase material degrades into a single-phase material.

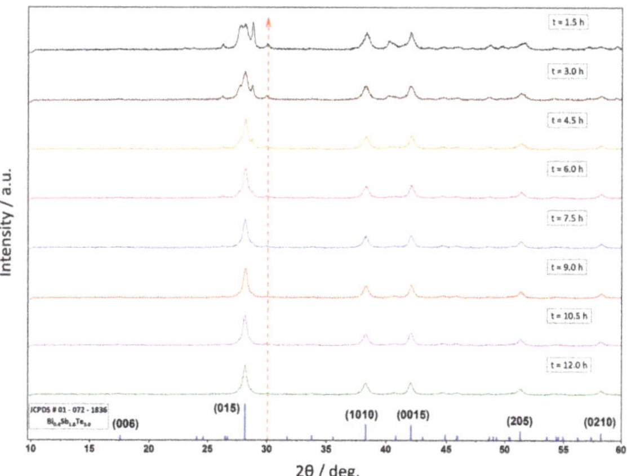

Figure 1. X-ray diffraction patterns of the BST-150 mixture versus time for each step of the PBM process.

Figure 2. Agglomerated powder (**a**) before and (**b**) after agate grinding and sieving; (**c**) milling process diagram for the two distinct methods.

The red dashed line represents the tellurium (Te) powder characteristic peak (Figure 3), which was observed to have the lowest damping rate during the milling steps, indicating tellurium's difficulty in participating in the alloying process. This can be attributed to the higher diffusion rate that tellurium has compared to the other two elements within the agglomerated powder as a result of its entrapment there, without participating in the alloying process at the same rate. At this point, it should be noted that the mechanical ball milling method alloys two or more materials during the collision of the milling balls with each other or with the walls of the milling container. The kinetic energy of the balls is converted into heat at the point of impact, instantly welding together the powder particles present. Thus, in cases where the grinding powder has a tendency to agglomerate on the base of the milling container, it traps a large percentage of the individual chemical elements, which can no longer participate in the alloying process because they never come into contact with the milling balls.

Figure 3 presents the X-ray diffraction pattern of the BST-150 powder mixture PBMed for 24 h in-steps, as well as the X-ray diffraction patterns of its individual elemental powders and the JCPDS diagram. According to the 24 h X-ray diffraction pattern, the characteristic tellurium peak appears to have completely faded, compared to the 12 h plot (Figure 1), and is now at the noise level. This signifies that the concentration of the elemental tellurium in the mixture has decreased significantly, with the majority of it now alloyed with the other elements.

Figure 4 shows the X-ray diffraction pattern of the BST-150 powder mixture for the two different PBM methods, continuous and in-steps, and clearly demonstrates the entrapment mechanism of the individual chemical elements in the agglomerated powder described above. The upper X-ray diffraction pattern shows the PBM process with steps for a total milling time of 3 h; the middle shows the continuous PBM process for a total milling time of 12 h; and the bottom shows the PBM process with steps for a total milling time of 12 h. The main characteristic peak (015) of the BST in the first two plots indicates that the mixture is still elemental and multiphase, and by comparing them, it is possible to conclude that after 3 h of continuous PBM, the mixture inside the vial has already agglomerated, significantly reducing the alloying process. Furthermore, when comparing the 12 h plots, it is clear that the alloying process is much more effective when the milling process is carried out in-steps. However, as previously discussed, the single-phase state for the BST-150 mixture occurs after 24 h of the PBM process in-steps because the characteristic of the tellurium peak (shown with the dashed line) fades completely at that point.

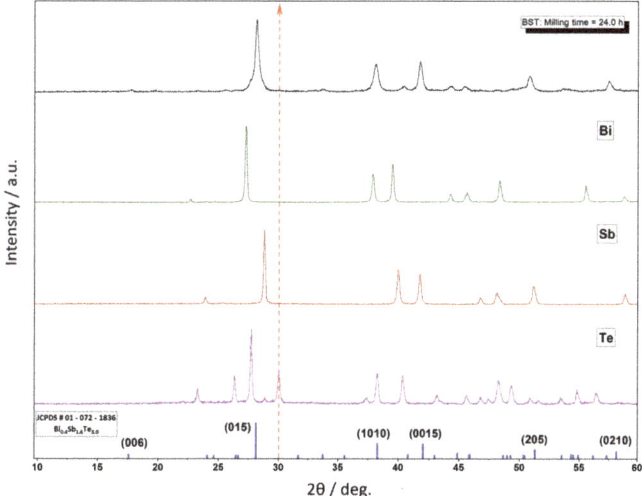

Figure 3. X-ray diffraction pattern of the BST-150 mixture milled in-steps for 24 h, along with the corresponding JCPDS diagram. For comparison, X-ray diffraction patterns of the individual elemental powders are also shown.

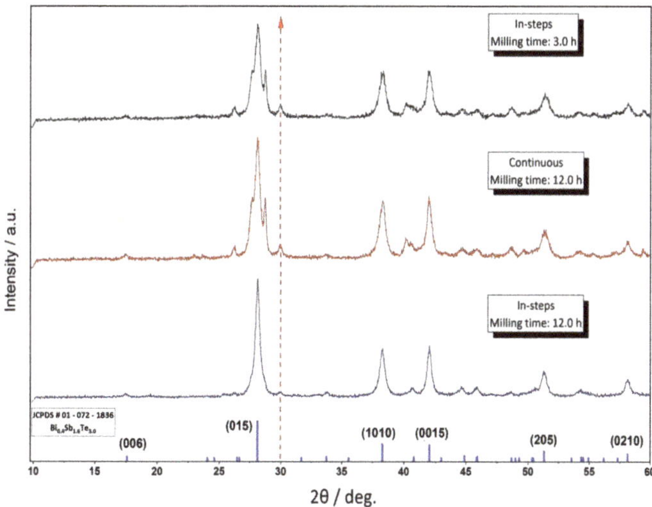

Figure 4. Comparison X-ray diffraction patterns for the BST-150 powder.

The procedures for the BST-100 powder mixture were the same as those for the BST-150. The X-ray diffraction patterns versus time of the BST-100 powder mixture PBMed in-steps and the X-ray diffraction pattern of the same mixture PBMed for 12 h in-steps, along with the X-ray diffraction patterns of its individual elemental powders and the JCPDS diagram, are shown in Figures 5 and 6, respectively. For convenience, the X-ray diffraction patterns in Figure 6 are presented every 3 h. The alloying process is faster in the case of the BST-100 powder mixture because of the larger ball-to-material ratio. In particular, when comparing the 3 h of PBM for the BST-150 and BST-100 powder mixtures, the main characteristic peak of the BST appears to have reconstructed to a smooth peak for the latter. Additionally, in the case of the 12 h PBM, the characteristic peak of the tellurium powder has faded completely for the BST-100 powder mixture, compared to the BST-150 where the same peak fades within 24 h of PBM. Based on these comparison results, it is feasible to conclude that reducing the powder mixture quantity by 50% (and thus increasing the ball-to-material ratio) reduces the optimum milling time for the alloying process by a factor of two.

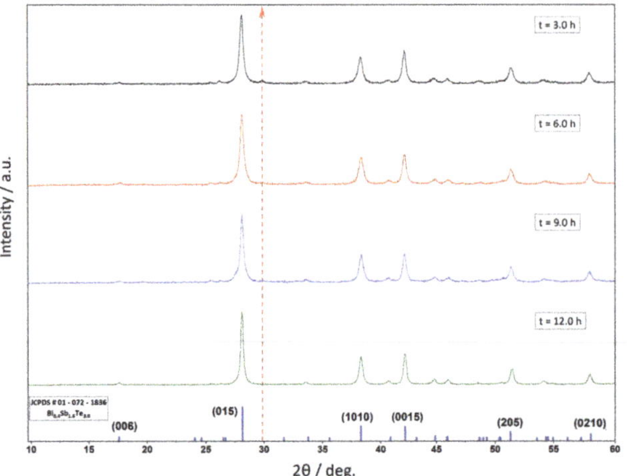

Figure 5. X-ray diffraction patterns of the BST-100 powder mixture versus time for each step of the PBM process.

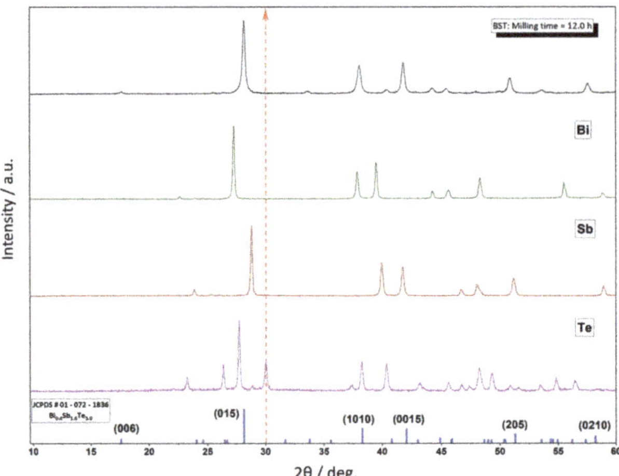

Figure 6. X-ray diffraction pattern of the BST-100 powder milled in-steps for 12 h, along with the corresponding JCPDS diagram. For comparison, X-ray diffraction patterns of the individual elemental powders are also shown.

Shown in Figure 7 is the X-ray diffraction patterns of the BST-100 powder mixture for the two different PBM processes. The upper X-ray diffraction pattern shows the PBM process with steps for a total milling time of 3 h, the middle shows the continuous PBM process for a total milling time of 12 h, and the bottom shows the PBM process with steps for a total milling time of 12 h. When the bottom graph is compared to the middle one, it is clear that the multiphase state is still strongly present in the mixture and comparable to the first graph. This can also be seen in the shape of the BST's base characteristic peak (015) in the middle graph, which is depicted with a red circle in both the graph and the inset, as an extra peak coming up on the side of the main peak.

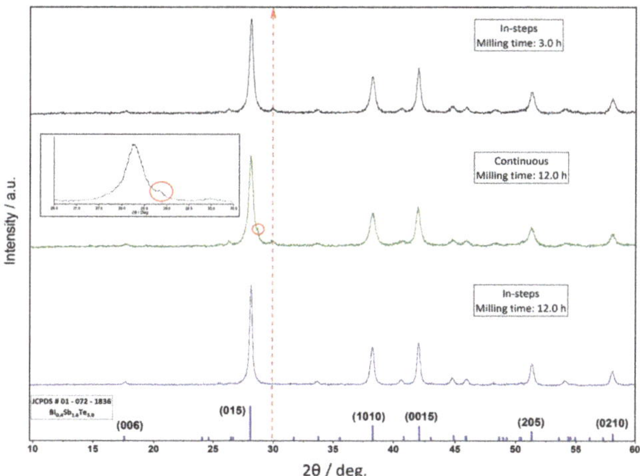

Figure 7. Comparison of X-ray diffraction patterns for the BST-100 powder.

The chemical composition and powder morphology of the produced BST powder with the optimum PBM conditions were determined using the SEM/EDS characterization method. Figure 8 presents the results of the SEM/EDS analysis for the BST-100 powder mixture, which was also the same for the BST-150 powder mixture. In terms of powder morphology, Figure 8a shows the image from the secondary electron detector (SE), where the majority of the particles appear to be smaller than 5 μm in size and have an irregular shape and geometry. Figure 8b shows the image from the back-scattered electron detector (BSE). The contrast of the image is uniformly spread here, implying no secondary phases in the powder mixture and agreeing with the X-ray diffraction results presented above. Finally, the chemical composition of the produced alloy can be seen qualitatively and quantitatively from the energy-dispersive X-ray spectroscopy (EDS) analysis shown in Figure 8c. All three elements are clearly present in the alloy, with their radiated energies and percent atomic values listed in the inset. It should be noted at this point that this analysis was performed on a large agglomerated particle of the alloy in order to avoid radiation from the carbon tape substrate on which the powder was placed.

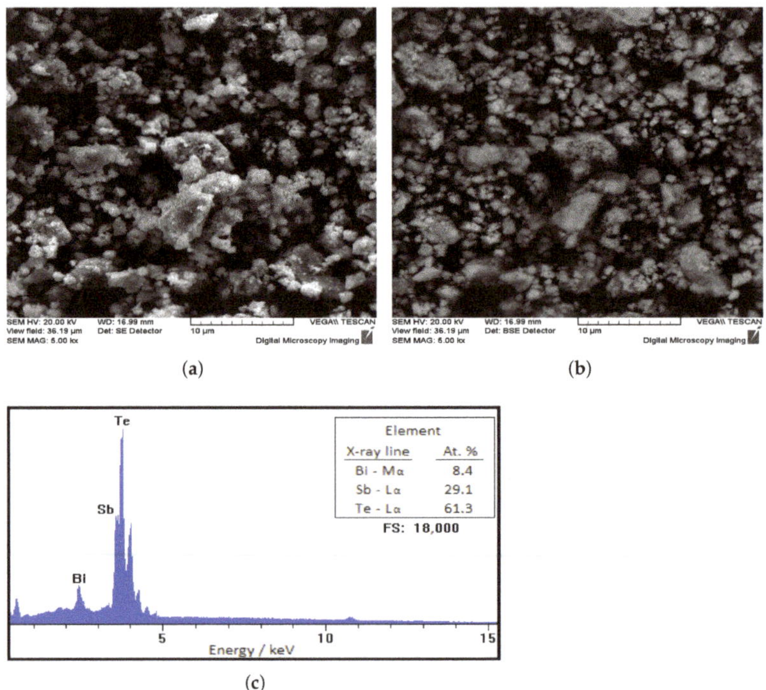

Figure 8. Scanning electron microscopy images: (**a**) SE detector, (**b**) BSE detector, (**c**) EDS spectrum.

As previously stated in Section 2, a small portion of the produced BST powder was hot pressed under optimal pressing and temperature conditions, yielding a cylindrical disk pellet with a thickness of th = 2.5 mm and a diameter of d = 10 mm, the thermoelectric properties of which are shown in Figure 9 along with the results of the studies [36,37]. All measurements and calculations were performed in the temperature range of [300 K–500 K]. Starting with the property of thermal conductivity κ, as shown in Figure 9a, its values vary within the range of [0.7–1.1] $Wm^{-1}K^{-1}$ and have an overall minimum value of κ_{min} = 0.73 $Wm^{-1}K^{-1}$ in the temperature range of [325 K–350 K], which is the lowest when compared to the other two studies. Similarly, the Seebeck coefficient S (Figure 9b) exhibits an overall maximum over the same temperature range, with its maximum value at S_{max} = 260 $\mu V\ K^{-1}$, followed by a steady rapid decline. In comparison to the other two studies, the Seebeck coefficient in this work is higher within the temperature range of [300 K–375 K]. The electrical conductivity σ, on the other hand (Figure 9c), has an overall minimum of σ_{min} = 2.52 × 10^4 S/m in the temperature range of [450 K–475 K], which results from a steep decline and is the lowest value when compared to the other two studies. The power factor of the sample is calculated by combining the values of the Seebeck coefficient and electrical conductivity using the equation PF = $S^2\sigma$, and as shown by the curve in Figure 9d, there is a constant drop as a function of temperature with no maxima or minima, which is consistent with the other two studies. Finally, the figure of merit ZT is shown in Figure 9e and was calculated using the equation ZT = PF/κ. As can be seen from the curve, a total maximum with a value of ZT = 1.02 appears at the temperature of T = 325 K, and represents the thermoelectric material's optimal performance temperature. In comparison to the other two studies, the ZT appears to be very close to the curves as they move across a zone as a whole.

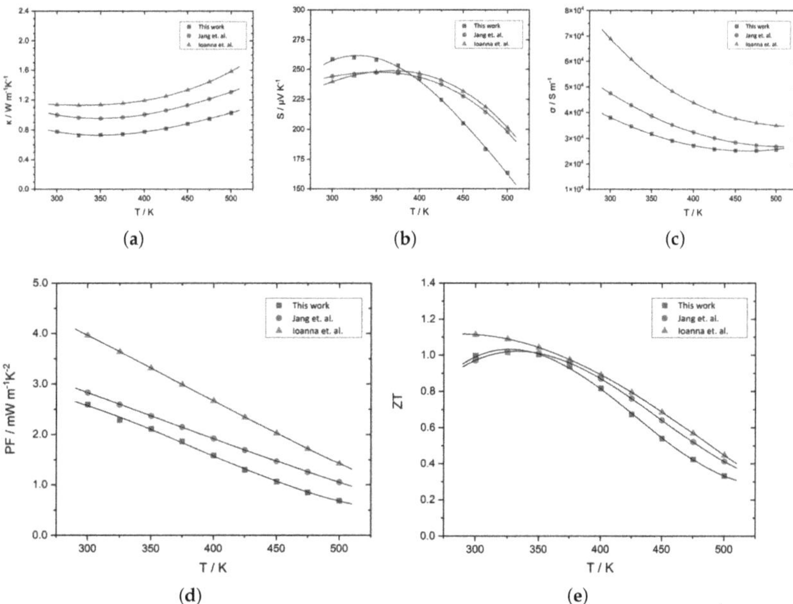

Figure 9. Thermoelectric properties of the produced BST powder compared to the literature (Jang et al. [36] and Ioannou et al. [37]): (**a**) Thermal conductivity, (**b**) Seebeck coefficient, (**c**) Electrical conductivity, (**d**) Power factor, and (**e**) Figure of merit.

4. Conclusions

The impact of both continuous and intermittent planetary ball milling on the alloying of the bismuth antimony telluride ($Bi_{0.4}Sb_{1.6}Te_{3.0}$) thermoelectric material is discussed in the current work. Because of the mixture's high agglomeration tendency, the results revealed a significant difference in the alloying process in terms of milling time and powder mixture quantity, when the powder mixture was PBMed in-steps versus continuously. The term "in-steps" refers to the procedure in which the PBM process is interrupted and the agglomerated powder adhering to the inner walls of the container is detached, ground in an agate mortar, and sieved before the alloying process is resumed. In particular, the results showed that PBMing the mixture continuously for more than 3 h has no effect on the alloying process because most of the elemental compounds are already trapped inside the agglomerated powder and do not participate in the alloying process. Additionally, applying the same procedure for two different amounts of powder mixture (100 g and 150 g) showed that increasing the powder mixture amount by 50% w/w resulted in an increase in the overall time needed to achieve the single-phase condition of the alloy by a factor of 2. Concerning the thermoelectric properties of the produced powder, the results revealed a very good thermoelectric profile, consistent with the literature, with the figure of merit reaching a maximum value of Z = 1.02 at T = 325 K, demonstrating the effectiveness of the presented technique to process materials faster and more efficiently, consuming less time and energy while producing larger quantities.

Author Contributions: Conceptualization, G.S.; methodology, G.S.; software, G.S.; validation G.S.; formal analysis, G.S. and T.K.; investigation, G.S.; resources, T.K.; data curation, G.S.; writing—original draft preparation, G.S.; writing—review and editing, G.S. and T.K.; visualization, G.S.; supervision, T.K.; project administration, T.K.; funding acquisition, T.K. All authors have read and agreed to the published version of the manuscript.

Funding: This work was funded by the University of Cyprus (internal project ADD-THERM).

Institutional Review Board Statement: Not applicable.

Informed Consent Statement: Not applicable.

Data Availability Statement: Data available on request from the authors.

Conflicts of Interest: The authors declare no conflicts of interest.

References

1. Hasezaki, K.; Nishimura, M.; Umata, M.; Tsukuda, H.; Araoka, M. Mechanical alloying of BiTe and BiSbTe thermoelectric materials. *Mater. Trans. JIM* **1994**, *35*, 428–432. [CrossRef]
2. Wunderlich, W.; Pixius, K.; Schilz, J. Microstructure of mechanical alloyed $Si_{76}Ge_{23.95}P_{0.05}$. *Nanostruct. Mater.* **1995**, *6*, 441–444. [CrossRef]
3. Tokiai, T.; Uesugi, T.; Eton, Y.; Tamura, S.; Yoneyama, Y.; Koumoto, K. Thermoelectric properties of p-type bismuth telluride material fabricated by plasma sintering of metal powder mixture. *J. Ceram. Soc. Jpn.* **1996**, *104*, 837–843. [CrossRef]
4. Sugiyama, A.; Kobayashi, K.; Ozaki, K.; Nishio, T.; Matsumoto, A. Preparation of functionally graded Mg2Si-FeSi2 thermoelectric material by mechanical alloying-pulsed current sintering process. *Nippon Kinzoku Gakkaishi (1952)* **1998**, *62*, 1082–1087.
5. Suryanarayana, C. Mechanical alloying and milling. *Prog. Mater. Sci.* **2001**, *46*, 1–184. [CrossRef]
6. El-Eskandarany, M.S. *Mechanical Alloying: For Fabrication of Advanced Engineering Materials*; William Andrew: Norwich, NY, USA, 2001.
7. Burmeister, C.F.; Kwade, A. Process engineering with planetary ball mills. *Chem. Soc. Rev.* **2013**, *42*, 7660–7667. [CrossRef]
8. Zheng, Y.; Liu, C.; Miao, L.; Lin, H.; Gao, J.; Wang, X.; Chen, J.; Wu, S.; Li, X.; Cai, H. Cost effective synthesis of p-type Zn-doped MgAgSb by planetary ball-milling with enhanced thermoelectric properties. *RSC Adv.* **2018**, *8*, 35353–35359. [CrossRef]
9. Bumrungpon, M.; Hirota, K.; Takagi, K.; Hanasaku, K.; Hirai, T.; Morioka, I.; Yasufuku, R.; Kitamura, M.; Hasezaki, K. Synthesis and thermoelectric properties of bismuth antimony telluride thermoelectric materials fabricated at various ball-milling speeds with yttria-stabilized zirconia ceramic vessel and balls. *Ceram. Int.* **2020**, *46*, 13869–13876. [CrossRef]
10. Balasubramanian, P.; Battabyal, M.; Bose, A.C.; Gopalan, R. Effect of ball-milling on the phase formation and enhanced thermoelectric properties in zinc antimonides. *Mater. Sci. Eng. B* **2021**, *271*, 115274. [CrossRef]
11. Im, H.J.; Koo, B.; Kim, M.S.; Lee, J.E. Optimization of high-energy ball milling process for uniform p-type Bi-Sb-Te thermoelectric material powder. *Korean J. Chem. Eng.* **2022**, *39*, 1227–1231. [CrossRef]
12. Fokina, E.; Budim, N.; Kochnev, V.; Chernik, G. Planetary mills of periodic and continuous action. *J. Mater. Sci.* **2004**, *39*, 5217–5221. [CrossRef]
13. Mio, H.; Kano, J.; Saito, F.; Kaneko, K. Optimum revolution and rotational directions and their speeds in planetary ball milling. *Int. J. Miner. Process.* **2004**, *74*, S85–S92. [CrossRef]
14. Bruckmann, A.; Krebs, A.; Bolm, C. Organocatalytic reactions: Effects of ball milling, microwave and ultrasound irradiation. *Green Chem.* **2008**, *10*, 1131–1141. [CrossRef]
15. Stolle, A.; Szuppa, T.; Leonhardt, S.E.; Ondruschka, B. Ball milling in organic synthesis: Solutions and challenges. *Chem. Soc. Rev.* **2011**, *40*, 2317–2329. [CrossRef]
16. Schneider, F.; Stolle, A.; Ondruschka, B.; Hopf, H. The Suzuki- Miyaura reaction under mechanochemical conditions. *Org. Process. Res. Dev.* **2009**, *13*, 44–48. [CrossRef]
17. Szuppa, T.; Stolle, A.; Ondruschka, B.; Hopfe, W. Solvent-free dehydrogenation of γ-terpinene in a ball mill: Investigation of reaction parameters. *Green Chem.* **2010**, *12*, 1288–1294. [CrossRef]
18. Sasaki, K.; Masuda, T.; Ishida, H.; Mitsuda, T. Structural degradation of tobermorite during vibratory milling. *J. Am. Ceram. Soc.* **1996**, *79*, 1569–1574. [CrossRef]
19. Balema, V.P.; Wiench, J.W.; Pruski, M.; Pecharsky, V.K. Solvent-free mechanochemical synthesis of phosphonium salts. *Chem. Commun.* **2002**, *7*, 724–725. [CrossRef]
20. Rodríguez, B.; Bruckmann, A.; Bolm, C. A highly efficient asymmetric organocatalytic aldol reaction in a ball mill. *Chem.–Eur. J.* **2007**, *13*, 4710–4722. [CrossRef]
21. Patil, P.R.; Kartha, K.R. Solvent-free synthesis of thioglycosides by ball milling. *Green Chem.* **2009**, *11*, 953–956. [CrossRef]
22. Tadier, S.; Le Bolay, N.; Rey, C.; Combes, C. Co-grinding significance for calcium carbonate–calcium phosphate mixed cement. Part I: Effect of particle size and mixing on solid phase reactivity. *Acta Biomater.* **2011**, *7*, 1817–1826. [CrossRef] [PubMed]
23. Gayner, C.; Kar, K.K. Recent advances in thermoelectric materials. *Prog. Mater. Sci.* **2016**, *83*, 330–382. [CrossRef]
24. Liu, W.; Hu, J.; Zhang, S.; Deng, M.; Han, C.G.; Liu, Y. New trends, strategies and opportunities in thermoelectric materials: A perspective. *Mater. Today Phys.* **2017**, *1*, 50–60. [CrossRef]
25. Chasmar, R.; Stratton, R. The thermoelectric figure of merit and its relation to thermoelectric generators. *Int. J. Electron.* **1959**, *7*, 52–72. [CrossRef]
26. Rowe, D.; Min, G. Evaluation of thermoelectric modules for power generation. *J. Power Sources* **1998**, *73*, 193–198. [CrossRef]
27. Rowe, D.M. *CRC Handbook of Thermoelectrics: Macro to Nano*; CRC Taylor & Francis: Boca Ratcon, FL, USA, 2006.
28. Ismail, B.I.; Ahmed, W.H. Thermoelectric power generation using waste-heat energy as an alternative green technology. *Recent Patents Electr. Electron. Eng. (Former. Recent Patents Electr. Eng.)* **2009**, *2*, 27–39. [CrossRef]

29. Fleurial, J.P. Thermoelectric power generation materials: Technology and application opportunities. *JOM* **2009**, *61*, 79–85. [CrossRef]

30. Yamashita, O.; Tomiyoshi, S.; Makita, K. Bismuth telluride compounds with high thermoelectric figures of merit. *J. Appl. Phys.* **2003**, *93*, 368–374. [CrossRef]

31. Hong, S.J.; Chun, B.S. Microstructure and thermoelectric properties of extruded n-type 95% Bi_2Te_2–5% Bi_2Se_3 alloy along bar length. *Mater. Sci. Eng. A* **2003**, *356*, 345–351. [CrossRef]

32. Chung, D.Y.; Hogan, T.P.; Rocci-Lane, M.; Brazis, P.; Ireland, J.R.; Kannewurf, C.R.; Bastea, M.; Uher, C.; Kanatzidis, M.G. A new thermoelectric material: CsBi4Te6. *J. Am. Chem. Soc.* **2004**, *126*, 6414–6428. [CrossRef]

33. Kanatzia, A.; Papageorgiou, C.; Lioutas, C.; Kyratsi, T. Design of ball-milling experiments on Bi_2Te_3 thermoelectric material. *J. Electron. Mater.* **2013**, *42*, 1652–1660. [CrossRef]

34. Symeou, E.; Nicolaou, C.; Delimitis, A.; Androulakis, J.; Kyratsi, T.; Giapintzakis, J. High thermoelectric performance of $Bi_{2-x}Sb_xTe_3$ bulk alloys prepared from non-nanostructured starting powders. *J. Solid State Chem.* **2019**, *270*, 388–397. [CrossRef]

35. Symeou, E.; Nicolaou, C.; Kyratsi, T.; Giapintzakis, J. Enhanced thermoelectric properties in vacuum-annealed $Bi_{0.5}Sb_1Te_3$ thin films fabricated using pulsed laser deposition. *J. Appl. Phys.* **2019**, *125*, 215308. [CrossRef]

36. Jang, K.W.; Kim, H.J.; Jung, W.J.; Kim, I.H. Charge transport and thermoelectric properties of p-type $Bi_{2-x}Sb_xTe_3$ prepared by mechanical alloying and hot pressing. *Korean J. Met. Mater.* **2018**, *56*, 66–71.

37. Ioannou, I.; Ioannou, P.S.; Kyratsi, T.; Giapintzakis, J. Low-cost preparation of highly-efficient thermoelectric $Bi_xSb_2Te_3$ nanostructured powders via mechanical alloying. *J. Solid State Chem.* **2023**, *319*, 123823. [CrossRef]

38. Zakeri, M.; Allahkarami, M.; Kavei, G.; Khanmohammadian, A.; Rahimipour, M. Synthesis of nanocrystalline Bi_2Te_3 via mechanical alloying. *J. Mater. Process. Technol.* **2009**, *209*, 96–101. [CrossRef]

39. Jimenez, S.; Perez, J.G.; Tritt, T.M.; Zhu, S.; Sosa-Sanchez, J.L.; Martinez-Juarez, J.; López, O. Synthesis and thermoelectric performance of a p-type $Bi_{0.4}Sb_{1.6}Te_3$ material developed via mechanical alloying. *Energy Convers. Manag.* **2014**, *87*, 868–873. [CrossRef]

40. Shi, J.; Chen, H.; Jia, S.; Wang, W. 3D printing fabrication of porous bismuth antimony telluride and study of the thermoelectric properties. *J. Manuf. Process.* **2019**, *37*, 370–375. [CrossRef]

41. Yan, Y.; Ke, H.; Yang, J.; Uher, C.; Tang, X. Fabrication and thermoelectric properties of n-type $CoSb_{2.85}Te_{0.15}$ using selective laser melting. *ACS Appl. Mater. Interfaces* **2018**, *10*, 13669–13674. [CrossRef]

42. Qiu, J.; Yan, Y.; Luo, T.; Tang, K.; Yao, L.; Zhang, J.; Zhang, M.; Su, X.; Tan, G.; Xie, H.; et al. 3D Printing of highly textured bulk thermoelectric materials: Mechanically robust BiSbTe alloys with superior performance. *Energy Environ. Sci.* **2019**, *12*, 3106–3117. [CrossRef]

inorganics

Article

Tetragonal Nanosized Zirconia: Hydrothermal Synthesis and Its Performance as a Promising Ceramic Reinforcement

Shikai Liu *, Jialin Wang, Yingxin Chen, Zhijian Song, Bibo Han, Haocheng Wu, Taihang Zhang and Meng Liu

School of Materials Science and Engineering, Henan University of Technology, Zhengzhou 450000, China; 2021920438@haut.edu.cn (J.W.); c1228892791@163.com (Y.C.); s18860366393@163.com (Z.S.); hanbibo2022@163.com (B.H.); 15993592113@163.com (H.W.); taihang_zhang@163.com (T.Z.); meng_liu@haut.edu.cn (M.L.)
* Correspondence: shikai_liu@haut.edu.cn; Tel.: +86-371-67758737

Abstract: In this study, we produced zirconia nanoparticles with a pure tetragonal phase, good dispersion, and an average particle size of approximately 7.3 nm using the modified hydrothermal method. Zirconium oxychloride ($ZrOCl_2 \cdot 8H_2O$) was used as zirconium source, while propanetriol was used as an additive. The influence of propanetriol content, sonication time, hydrothermal temperature, and type of dispersant on the physical phase and dispersibility of zirconia nanoparticles was investigated. Monoclinic zirconia was found to completely transform into a tetragonal structure when the mass fraction of glycerol was increased to 5 wt%. With the increase in the mechanical stirring time under ultrasonic conditions, the size distribution range of the prepared particles became narrower and then wider, and the particle size became first smaller and then larger. Ultrasonic and mechanical stirring for 5 min had the best effect. When comparing the effects of different dispersants (PEG8000, PVP, and CTAB), it was found that the average particle size of zirconia nanoparticles prepared with 0.5 wt% PVP was the smallest. Furthermore, by adding different concentrations of pure tetragonal phase nanozirconia to 3Y-ZrO$_2$ as reinforcement additives, the bending strength of the prepared ceramics increased first and then decreased with increasing addition amounts. When the amount of addition was 1 wt% and the ceramic was calcined at 1600 °C, the flexural strength of the ceramic increased significantly, which was about 1.6 times that of the unadded ceramic. The results are expected to provide a reference for the reinforcement of high-purity zirconia ceramics.

Keywords: nanosized zirconia; tetragonal phase; hydrothermal method; propanetriol; ceramic reinforcement

Citation: Liu, S.; Wang, J.; Chen, Y.; Song, Z.; Han, B.; Wu, H.; Zhang, T.; Liu, M. Tetragonal Nanosized Zirconia: Hydrothermal Synthesis and Its Performance as a Promising Ceramic Reinforcement. *Inorganics* **2023**, *11*, 217. https://doi.org/10.3390/inorganics11050217

Academic Editors: Roberto Nisticò, Torben R. Jensen, Luciano Carlos, Hicham Idriss and Eleonora Aneggi

Received: 11 April 2023
Revised: 12 May 2023
Accepted: 15 May 2023
Published: 17 May 2023

1. Introduction

Zirconia nanoparticles (ZrO_2) possess exceptional physical properties, including remarkable strength, hardness, and resistance to wear. It also exhibits low thermal conductivity, excellent thermal insulation, and superior thermal shock resistance. Additionally, ZrO_2 nanoparticles demonstrate outstanding biocompatibility and chemical stability, displaying favorable acid and alkali corrosion resistance in both oxidizing and reducing atmospheres [1–3]. These unique characteristics make it highly versatile, finding applications in diverse fields ranging from electronic products to oxygen sensors, catalysis, special ceramics, and refractory materials [4–9]. Consequently, it represents a focal point of current research within the realm of inorganic materials.

Superior nano ZrO_2 powders are required to meet specific criteria, including small grain size, narrow particle size distribution, controllable physical phase and morphology, good monodispersity, and minimal agglomeration [10]. ZrO_2 exists in three crystalline structures, namely monoclinic (m-ZrO_2), tetragonal (t-ZrO_2), and cubic (c-ZrO_2) phases [11]. Nanozirconia with different physical phases possesses unique properties and applications. Particularly, the tetragonal-phase nanozirconia can serve as a reinforcement for compounding with other materials to produce martensitic phase transformation for ceramic toughening. This comes from its phase transformation toughening effect [12]. For example, alumina ceramics can be toughened using tetragonal zirconia, and ZrO_2 plays a critical role in the polycrystalline transformation of the composite [13]. Additionally, t-ZrO_2 exhibits exceptional biocompatibility, high shear strength, and great potential in the field of biomaterials [14,15]. Owing to these outstanding physicochemical properties, tetragonal-phase zirconia nanoparticles present promising applications in advanced ceramics and new composite materials.

The preparation of t-ZrO_2 nanopowder particles in a convenient and efficient manner has significant academic and practical value. Several methods are commonly used to prepare such powders, including the sol-gel, microwave, and hydrothermal routes [16–19]. For example, Chao Yang et al. [20] successfully synthesized pure tetragonal zirconia (t-ZrO_2) with an average particle size of approximately 7 nm via a modified hydrothermal method using zirconium nitrate ($Zr(NO_3)_4$) as the zirconium precursor and glycerol as an additive. Similarly, Hongju Qiu et al. [21] used a novel sol-gel flux method to synthesize tetragonal zirconia nanopowders stabilized by 3 mol% Y_2O_3, producing single tetragonal-phase 3Y-TZP nanoparticles in the binary system NaCl + KCl when the Y_2O_3 content in the raw material was 5 mol%. In comparison to the other methods, hydrothermal synthesis demonstrates superior economic efficiency in the fabrication of ZrO_2 nanoparticles that possess advantageous characteristics such as fine particle sizes, small agglomeration strengths, and a narrow distribution of particle sizes [22]. Currently, doping with cations during preparation is the primary method to improve the stability of tetragonal-phase zirconia in zirconia systems. Examples of such dopants include yttrium oxide-stabilized tetragonal-phase zirconia (Y-TZP) and cerium oxide-stabilized tetragonal-phase zirconia (Ce-TZP) [23–25]. However, the major drawback of Y-TZP is that these ceramics are prone to aging, which means that metastable tetragonal grains are very susceptible to slowly transforming to a monoclinic phase triggered by water-derived species [26]. In addition, existing issues such as the dearth of tetragonal phase content, exorbitant reaction costs, stringent equipment demands, and paltry yields are also present. Moreover, nanoparticles exhibit a proclivity for agglomeration because of their elevated surface energy, and the existence of agglomerates hinders interparticle flow and diminishes sintering activity, ultimately leading to an immense depletion in powder quality. According to available studies, changing the surface charge of the powder by adding dispersants is an effective method to improve the agglomeration of zirconia nanopowders [27]. However, different dispersants also have different effects on zirconia [16,28]. Therefore, it is especially important to prepare zirconia nanopowders with good monodispersity and high tetragonal phase content by adjusting the experimental parameters.

In this paper, we present a systematic investigation of the impact of various preparation conditions on the synthesis of tetragonal-phase zirconia nanoparticles via the hydrothermal method using a propanetriol system. We aim to obtain pure tetragonal-phase zirconia nanoparticles with small particle sizes and good monodispersity and to perform relevant tests on their properties.

2. Results and Discussion

2.1. Glycerol Content

The results of the laser particle size analysis of samples with different mass fractions of the propanetriol system are presented in Figure 1. As shown in the figure, the particle size of the samples showed a trend of an increase and then a decrease with the increase in the mass fraction of propanetriol, and the median diameter (D50) was 518, 692, 449, and 302 nm, respectively. The particle size distribution curves of samples a (1 wt%), b (2.5 wt%), and c (5 wt%) are similar. However, when the addition of propanetriol reached 10 wt%, the particle size and distribution differed significantly, with a concentration of particle size. This observation suggests that the excessive addition of propanetriol might hinder the zirconia crystallization process, leading to incomplete crystallization and the production of other substances. However, it should be noted that the particle size distribution was measured using a laser particle size analyzer, and the refractive index parameter was set on the basis of the refractive index of zirconia. Therefore, the significance of the reference may not be significant. Based on the results, the sample prepared with 5 wt% propanetriol had the smallest median diameter and the best particle size distribution.

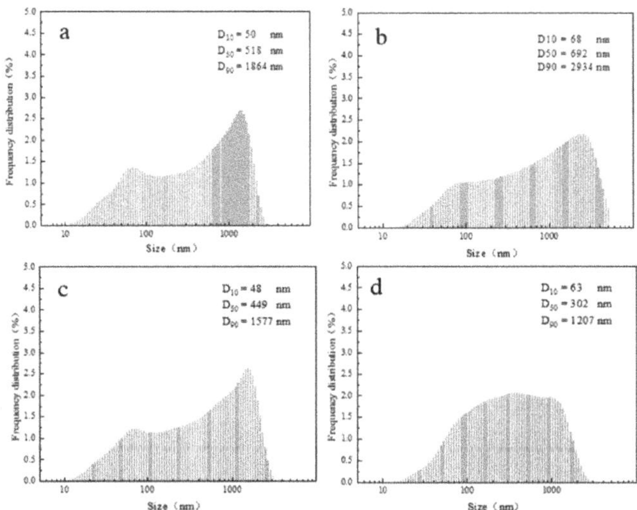

Figure 1. Particle size distribution of ZrO_2 with different mass fractions of propanetriol. (**a**) 1 wt%; (**b**) 2.5 wt%; (**c**) 5 wt%; (**d**) 10 wt%.

The zirconia synthesized through the hydrothermal method should have a particle size in the nanometer range. However, the measured particle size of the samples was found to be at the submicrometer level. This discrepancy may be attributed to the fact that the laser particle size analyzer primarily measures particle sizes in the micrometer range and may not accurately measure particle sizes in the nanometer range. To verify the particle size of the hydrothermal product and the accuracy of the laser particle size analyzer effect pattern on the measured samples, we conducted TEM testing on four groups of samples. The results, as shown in Figure 2, indicate that the prepared samples have good monodispersity with clear boundaries between the particles. However, multiple particle stacking was observed in all of the samples, which may be due to problems arising from the preparation of the samples during TEM testing. After statistical analysis, the average particle sizes of the four groups of samples were found to be approximately 9.3 nm, 10.1 nm, 8.2 nm, and 8.0 nm, respectively. The addition of 5 wt% propanetriol resulted in a relatively good dispersion. The changing pattern of the particle size also followed an increasing and then decreasing trend, consistent with the results obtained from the laser particle size distribution meter. Therefore, while the laser particle size analyzer may not provide exact particle size measurements for nanometer-sized particles, it can still be used for qualitative analysis and to study the impact of different conditions on particle size distribution.

Figure 3 shows the XRD patterns of propanetriol systems with varying mass fractions. The results demonstrate that the physical phase is significantly influenced by the content of propanetriol. Comparison with the standard card of the tetragonal phase (JCPDS NO.50-1089) reveals that the first three sets of experiments at $2\theta = 30.12°$, $34.62°$, $35.11°$, $50.22°$, $50.57°$, $59.44°$, $60.06°$, $62.84°$, and $74.41°$ exhibit tetragonal phase diffraction peaks corresponding to (011), (002), (110), (112), (020), (013), (121), (202), and (220) crystallographic planes, respectively. The crystalline intensity varies from weak to strong, and then from weak to weak. The diffraction peaks at $2\theta = 35.3°$ and $60.2°$ are slightly asymmetrical due to the broadening superposition of double peaks with different intensities. This feature distinguishes t-ZrO_2 from c-ZrO_2, which only exhibits symmetric diffraction peaks.

Therefore, the sample prepared under this condition is confirmed to be t-ZrO_2 [29–31]. When the addition of propanetriol is 1 wt%, a diffraction peak appears at $2\theta = 28.16°$, and the standard control monoclinic phase card (JCPDS NO.37-1484) is the diffraction peak of zirconium oxide monoclinic phase in the crystalline plane. In contrast, when zirconium oxychloride is used as the zirconium source and ammonia as the precipitant, the structure is usually mixed-phase without adding propanetriol. The diffraction peak of the monoclinic phase is relatively strong, indicating that the addition of propanetriol facilitates the conversion of the monoclinic phase to the tetragonal phase. When the addition of propanetriol was increased to 2.5 wt%, the diffraction peaks of tetragonal phase zirconia were sharpened by XRD full spectrum fitting, indicating an increase in the crystallinity and grain size of tetragonal phase zirconia with the addition of propanetriol. With a further increase in the mass fraction of propanetriol to 5 wt%, the crystallinity decreased slightly, but the monoclinic peak at $2\theta = 28.16°$ disappeared completely. All diffraction peaks corresponded to those of tetragonal zirconia, indicating the formation of a pure tetragonal structure. Furthermore, the diffraction peaks broadened, the width of the half-height increased, and the grain size decreased. With a further increase in the amount of propanetriol to 10 wt%, the diffraction peak intensity of zirconia was very low, and no zirconia with good crystallinity was formed. Therefore, increasing the mass fraction of propanetriol favors the generation of the tetragonal phase. Under the condition of 5 wt%, the monoclinic phase has completely transformed into the tetragonal phase. However, as the amount of propanetriol continues to increase, the crystallinity of the hydrothermal products increases and then decreases. To a certain extent, continuing to increase is not conducive to crystallization. Calculated by the Scherrer equation, the current system exhibited average grain sizes of 6.89 nm, 7.82 nm, 6.25 nm, and 6.56 nm with increasing mass fractions of propanetriol. These values were consistent with the pattern of intensity change observed for diffraction peaks.

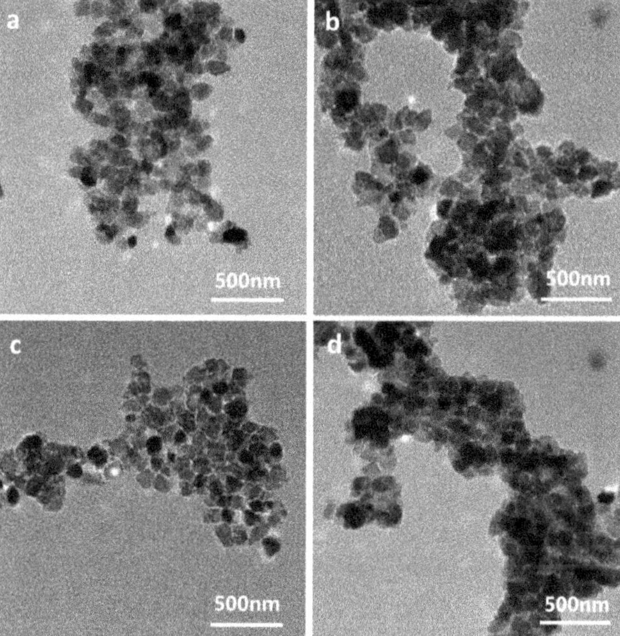

Figure 2. TEM images of ZrO_2 material with different mass fractions of glycerol: (**a**) 1 wt%; (**b**) 2.5 wt%; (**c**) 5 wt%; (**d**) 10 wt%.

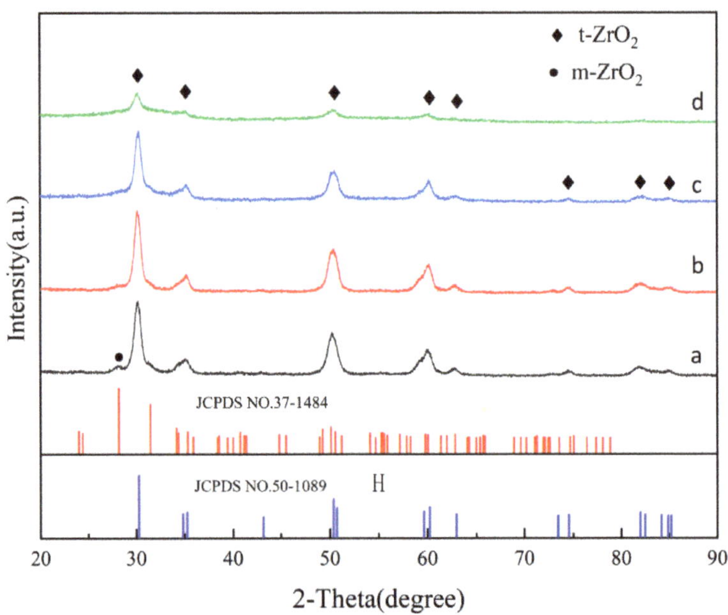

Figure 3. Effect of different mass fractions of propanetriol on ZrO$_2$ crystals: (**a**) 1 wt%; (**b**) 2.5 wt%; (**c**) 5 wt%; (**d**) 10 wt%.

To further verify the particle size and microscopic morphology of the samples, the samples prepared under 2.5 and 5 wt% propanetriol conditions were characterized by SEM, as shown in Figure 4. SEM images of the two groups of samples were observed, and it was found that the particle sizes belonged to the nanometer level and the average particle sizes ranged from about 15 to 25 nm. However, the particle agglomeration was more severe in Figure 4a, and the particle boundaries were blurred and the grains were fine. With the increase in propanetriol's mass fraction, the edges of the particles in Figure 4b are clearer, and the grains are well developed. Because of their smaller grains, it leads to the presence of a larger surface energy and the appearance of agglomeration. Therefore, the above analysis shows that the crystallinity of the product improves and the particle size decreases with the increasing addition of propanetriol. Therefore, the optimal amount of addition was 5 wt% propanetriol, the prepared sample had relatively good dispersion and uniform particle size, and the product had a pure tetragonal phase structure.

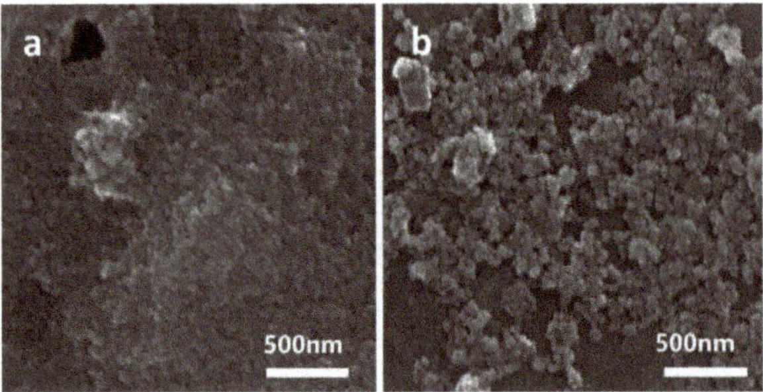

Figure 4. SEM images of ZrO$_2$ material with different mass fractions of propanetriol: (**a**) 2.5 wt%; (**b**) 5 wt%.

2.2. Effect of the Ultrasonic Mode on Tetragonal Nano-Zirconia

The agglomeration of zirconia nanoparticles prepared by the hydrothermal method is a common problem due to the high surface activity of small particles. However, reasonable sonication can improve the dispersion of ZrO_2 nanocrystals [32]. The particle size distributions were measured at different sonication times and are shown in Figure 5. The results showed that when only stirring was performed without sonication, the particle size increased significantly, and the D50 value was too large with a wide particle size distribution, leading to irrational test results. However, sonication for 2 min significantly refined the particles, resulting in a decrease in D50 and a narrower particle size distribution. This indicates that sonication is a crucial step in the preparation process. When sonication was combined with mechanical stirring for 5 min, the effect was even more significant, resulting in a smaller median diameter and a narrower size distribution. However, with a further extension of the sonication time, the median diameter increased, indicating that the ultrasonic time should not be too long. The vibration caused by the ultrasonic frequency may regroup the originally dispersed small particles because of their high surface energy, which is not conducive to powder dispersion. Therefore, the ultrasonic and mechanical stirring conditions for 5 min were selected as the best conditions for future studies in this group of experiments.

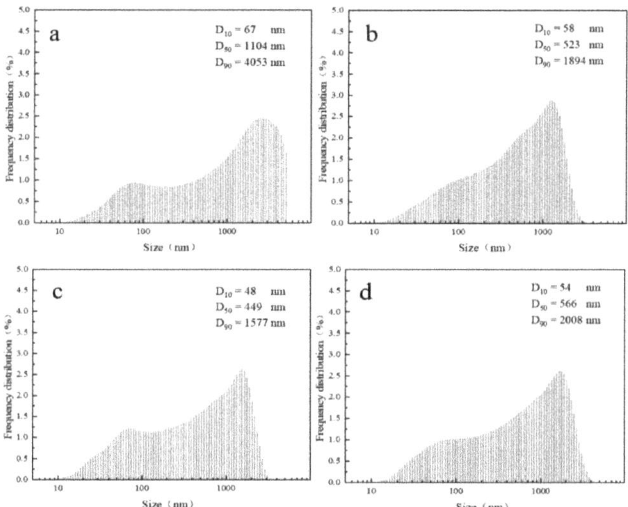

Figure 5. Particle size distribution of ZrO_2 with different ultrasound + mechanical stirring times: (**a**) mechanical stirring 2 min; (**b**) 2 min; (**c**) 5 min; (**d**) 10 min.

2.3. Hydrothermal Reaction Temperature

Figure 6 displays the particle size distribution obtained at various reaction temperatures using zirconium oxychloride and propanetriol at concentrations of 10 and 5 wt%, respectively. As illustrated in the figure, an overall increasing trend in particle size is observed with increasing hydrothermal temperature, with D50 increasing from 121 to 449 nm. The observed trend can be attributed to the fact that hydrothermal temperature is a critical factor affecting crystal growth. Specifically, the increase in hydrothermal temperature leads to greater solute adsorption onto the crystal, resulting in decreased crystal surface energy and facilitating grain growth.

Figure 7 presents the XRD analysis of the prepared samples at various temperatures within the propanetriol system. As shown in the figure, when the reaction temperature is set at 160 °C, weak diffraction peaks are observed for the products, indicating that zirconia is in a newly crystallized state and still contains many amorphous regions. As the temperature increases, the intensity of the diffraction peaks increases, signifying an increase in crystallinity. However, in comparison with the monoclinic phase standard card (JCPDS NO.37-1484), it is found that monoclinic phase diffraction peaks appear at $2\theta = 28.16°$ when the reaction temperature is set to 180 °C. At 200 °C, these monoclinic peaks disappear completely, and all are tetragonal phase reaction peaks with good crystallinity. It is apparent that an increase in reaction temperature not only facilitates grain growth and enhances crystallinity but also encourages the transformation from the monoclinic phase to the tetragonal phase. The average grain sizes of the hydrothermal products prepared at 180 °C and 200 °C were calculated to be 5.56 nm and 6.25 nm, respectively, according to the Scherrer equation.

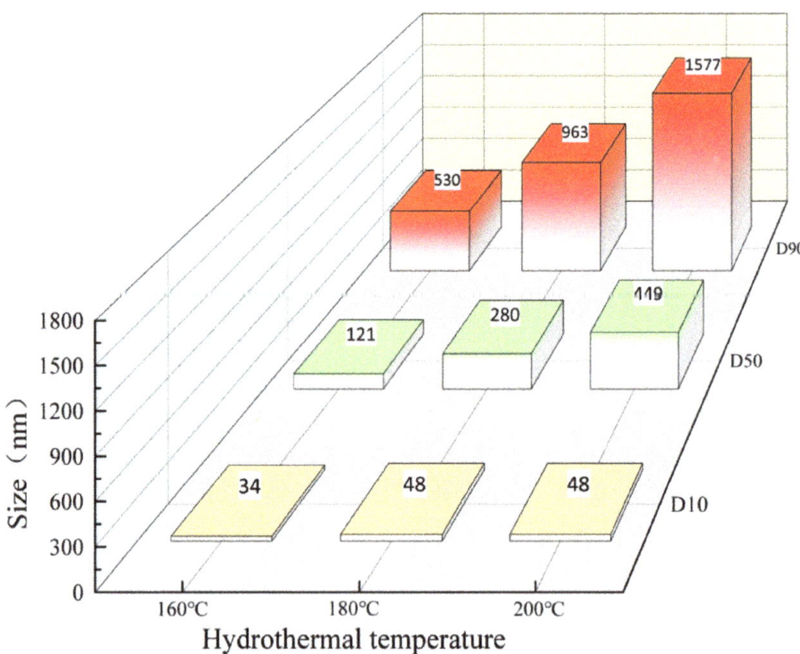

Figure 6. Distribution of the particle size of ZrO_2 at different hydrothermal temperatures.

Figure 8 shows SEM images of the samples prepared at hydrothermal temperatures of 160 °C, 180 °C, and 200 °C within the glycerol system. As can be observed in the figure, the driving force provided at a hydrothermal temperature of 160 °C is insufficient to induce crystallization in most crystals, which remain in the amorphous state. As the temperature increases, the crystallinity of the samples gradually increases, along with the driving force for crystallization, and when the temperature reaches 200 °C, good crystallinity is obtained, with relatively uniform grain sizes and a sphere-like granular morphology. Based on these observations, 200 °C is determined to be the optimal hydrothermal temperature. However, samples prepared at 200 °C still exhibit particle agglomeration, which was addressed by adding dispersant to mitigate the agglomeration effect.

2.4. Dispersant Type

Figure 9 displays the particle size distribution for the optimal addition of the three dispersants. In the supplementary materials (Figures S1–S3), detailed control experiments were conducted on the optimal dosage of three dispersants. Here, the optimal dosage of each dispersant was directly selected for research. As shown in the figure, all three dispersants effectively reduce the size of the sample particle. Figure 10 presents XRD test plots of samples prepared with the optimal content of each of the three dispersants based on the particle size distribution. Specifically, Figure 10a–c correspond to the addition of 0.5 wt% PEG8000, 0.5 wt% PVP, and 0.3 wt% CTAB, respectively. It is observable from these plots that dispersants have almost no impact on the physical phase, with patterns similar to those without dispersants. In Figure 10c, the diffraction peaks of the tetragonal-phase zirconium oxide are observed to have the highest intensity and the narrowest half-height width, indicating a relatively larger grain size. Meanwhile, the other two groups of samples exhibit little difference from each other. Average grain sizes were calculated using the Scherrer equation as 3.75 nm, 3.69 nm, and 3.91 nm, respectively. The general morphological characteristics of the three sets of samples were then observed by scanning electron microscopy.

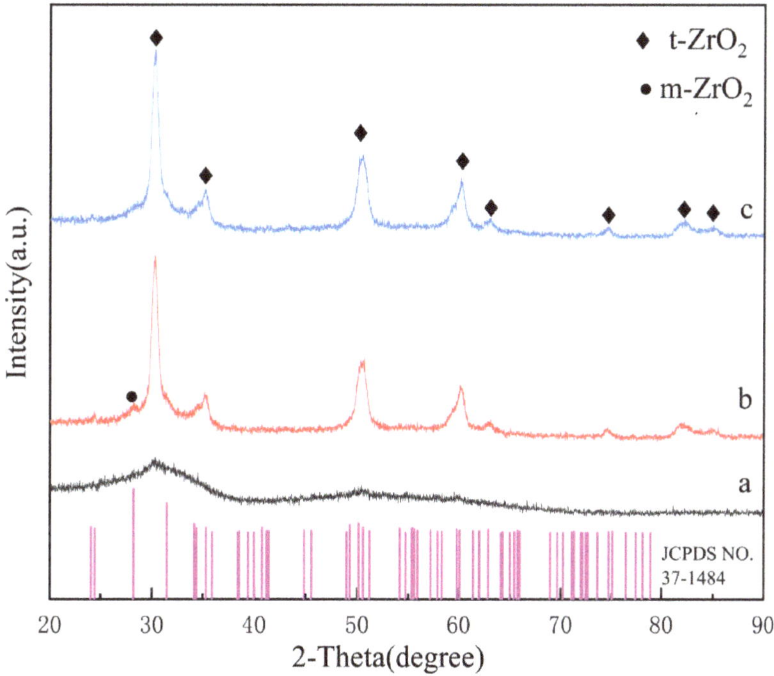

Figure 7. Influence of different hydrothermal temperatures on ZrO$_2$ crystals: (**a**) 160 °C; (**b**) 180 °C; (**c**) 200 °C.

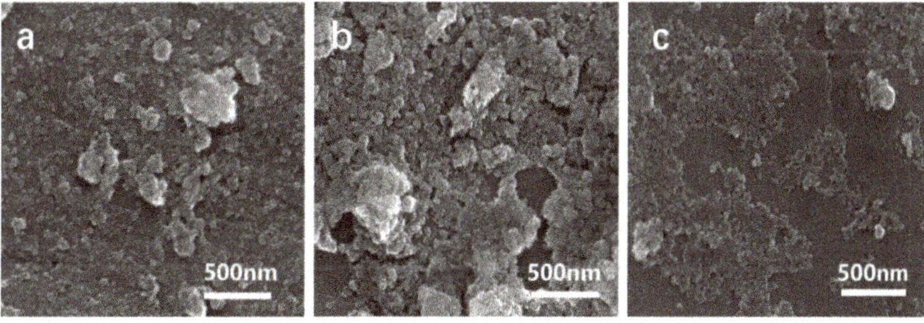

Figure 8. SEM images of the ZrO$_2$ material at different hydrothermal temperatures for the glycerol system: (**a**) 160 °C; (**b**) 180 °C; (**c**) 200 °C.

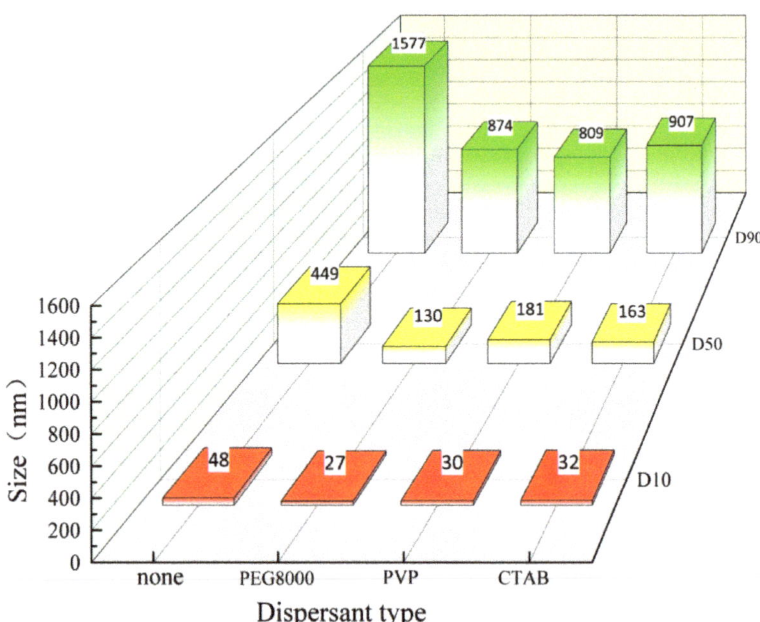

Figure 9. Particle size distribution of ZrO_2 prepared under the conditions of optimal addition of different types of dispersants: 0.5 wt% PEG8000; 0.5 wt% PVP; 0.3 wt% CTAB.

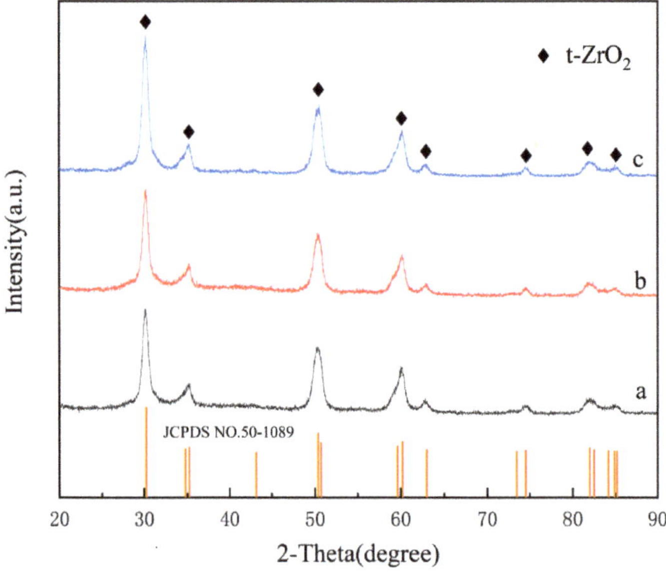

Figure 10. Effect of different types of dispersants on ZrO_2 crystals under optimal addition conditions: (**a**) 0.5 wt% PEG8000; (**b**) 0.5 wt% PVP; (**c**) 0.3 wt% CTAB.

As indicated in Figure 11, the morphology of the three sample groups with the addition of dispersant is significantly different from that of samples without dispersant, converting from an irregular to a sphere-like shape, accompanied by a notable improvement in the phenomenon of agglomeration and a refinement of the particle size to about 10~20 nm.

However, the effects of different types of dispersants are not identical. Samples prepared with PEG8000 dispersant exhibit higher sphericity but relatively larger particle sizes; samples prepared with PVP exhibit the smallest particle size but relatively poor dispersibility; in contrast, CTAB exhibits the best dispersibility and moderate particle size.

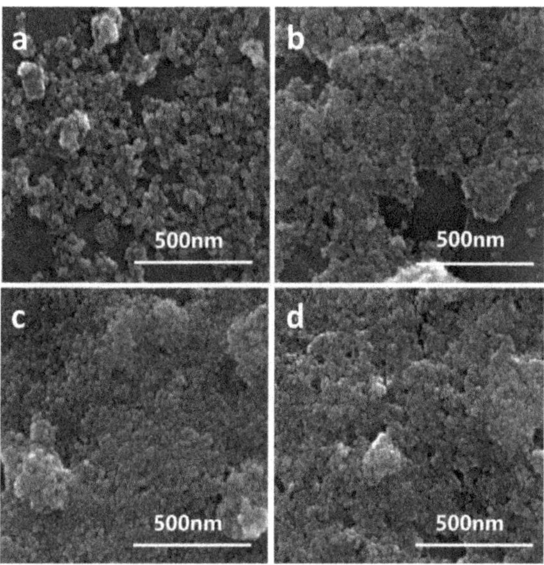

Figure 11. SEM images of ZrO_2 prepared by different kinds of dispersants of the propanetriol system under optimal addition conditions: (**a**) none; (**b**) 0.5 wt% PEG8000; (**c**) 0.5 wt% PVP; (**d**) 0.3 wt% CTAB.

To further investigate the effect of different types of dispersant on particle size, samples prepared with optimal addition of the three groups of dispersants were subjected to TEM tests, and the results are presented in Figure 12. When no dispersant was added, the sample particles exhibited poor dispersion, with large agglomerates of fine particles having serious agglomeration. With the addition of a dispersant, the sample was well dispersed, characterized by relatively uniform particle isometric axes and the disappearance of large agglomerates. By measurement and calculation, the average particle size of zirconia nanoparticles prepared with 0.5 wt% PEG8000 was about 9.2 nm, although some particles still agglomerated; the average particle size of zirconia nanoparticles prepared with 0.5 wt% PVP was about 7.3 nm, exhibiting the smallest particle size; and the average particle size of zirconia nanoparticles prepared with 0.3 wt% CTAB was 8.7 nm, featuring uniform isometric morphology and good dispersion.

Glycerol is a polyol composed of three hydrophilic hydroxyl groups. When used as an additive, it exhibits good mutual solubility with water, uniformly disperses and adsorbs around fully hydrolyzed zirconium oxychloride, and plays a crucial role in the regulation of the physical phase, which contributes to the stable existence of the tetragonal phase [33]. By regulating the mass fraction of propanetriol and hydrothermal temperature, pure tetragonal phase nanozirconia can be prepared by adding 5 wt% propanetriol and 10 wt% zirconium chloride, appropriate amounts of dispersant, and reacting under ultrasonic and mechanical stirring for 5 min at a hydrothermal temperature of 200 °C. However, it should be noted that when zirconia phase analysis is performed by XRD, the diffraction peaks corresponding to the tetragonal and cubic phases may overlap. Although the diffraction peaks at $2\theta = 35.3°$ and $60.2°$ possess slight asymmetry due to the broadening superposition of double peaks with different intensities, which can be used to identify whether the phase is pure tetragonal or not, more reliable phase information can be obtained by adding 0.5 wt% PVP to the sample and subjecting it to the Raman tests. The results are presented in Figure 13.

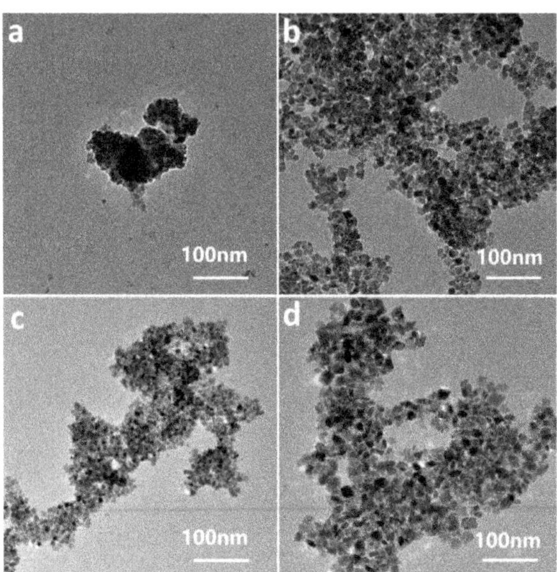

Figure 12. TEM images of ZrO_2 prepared by different types of dispersants of the glycerol system under optimal addition conditions: (**a**) none; (**b**) 0.5 wt% PEG8000; (**c**) 0.5 wt% PVP; (**d**) 0.3 wt% CTAB.

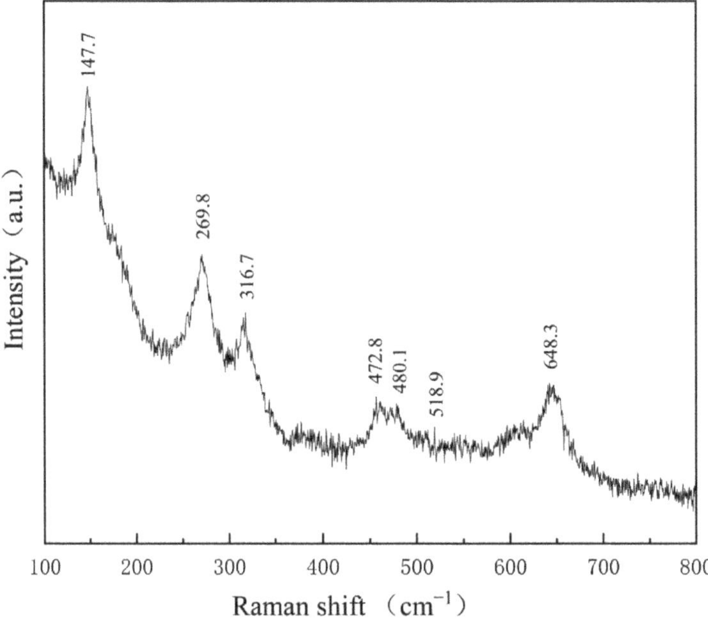

Figure 13. Raman spectra of samples under optimal conditions of the glycerol system.

Figure 13 exhibits the Raman scattering spectra of tetragonal-phase zirconia prepared under optimized conditions in the glycerol system. The figure mainly demonstrates seven distinct peaks in the Raman spectrum, wherein six are located at 147.7 cm^{-1}, 269.8 cm^{-1}, 316.7 cm^{-1}, 472.8 cm^{-1}, 518.9 cm^{-1}, and 648.3 cm^{-1} in order, respectively, and match with the six primary Raman activation modes of the tetragonal-phase zirconia analyzed by the entropy group theory, belonging to $B1_g$, E_g, B_{1g}, E_g, A_{1g}, and E_g symmetry, respectively. The other scattering peak at 480.1 cm^{-1} may overlap with the shoulder peak at 472.8 cm^{-1} and can be attributed to the E_g of the tetragonal phase. Compared to uniform standards reported in the article [34], this result is very close to the theoretical calculation of the laser Raman scattering peak position of the tetragonal-phase zirconia. However, some differences still exist that could be attributed to the imperfect crystallinity of the sample prepared by hydrothermal reaction, the presence of numerous crystal defects, etc. In addition, it can be seen from the observed spectrum that the Raman peak has a more obvious broadening phenomenon, the baseline of the spectrum has a more obvious fluorescence background, and there is no obvious polymer residue in the sample after full washing. These should be due to the introduction of defects in the sample. Moreover, since the tetragonal-phase zirconia and the cubic phase severely overlap in XRD patterns, making them difficult to distinguish, the unique peak position corresponding to the cubic phase zirconia and where the Raman activation mode F_{2g} is located was not observed in the Raman peak, which is typically located close to 490 cm^{-1} [35]. On the basis of the uniform standards in the articles and the above analysis, it can be concluded that zirconia prepared under optimal conditions in this system exhibits a pure tetragonal phase structure.

2.5. Effect of Adding Nano-Zirconia on Flexural Strength of 3Y-ZrO$_2$

Figure 14 shows the flexural strength of the prepared pure tetragonal-phase zirconia nanoparticles added to 3Y-ZrO$_2$ (3 mol yttrium oxide-stabilized zirconia) at different contents in the glycerol system after calcination at different temperatures. In general, the calcination temperature required to achieve maximum flexural strength in zirconia ceramics is greater than or equal to the sintering temperature of the ceramics. On the basis of the lateral factor analysis of the graph, it can be observed that as the temperature increases, the overall flexural strength gradually increases when nano-zirconia is added at 0 wt%, 0.5 wt%, 1 wt%, 3 wt%, and 5 wt%. When the calcination temperature reaches 1500 °C and continues to increase, the flexural strength of the samples with nanoaddition increases significantly faster, indicating that the sintering temperature of 3Y-ZrO$_2$ with the addition of nanozirconia is greater than 1500 °C. Analysis of different addition factors in the longitudinal direction reveals that in 3Y-ZrO$_2$ samples with the addition of nano-zirconia, the flexural strength exhibits a trend of increase and then decrease with the increase in nanozirconia content, with the highest flexural strength achieved at 1600 °C with the addition of 1 wt% (286.88 Mpa). However, when the calcination temperature is less than or equal to 1500 °C, the flexural strength of the added group is generally not as high as the control group, as the sintering temperature has not yet been reached and the flexural strength continues to increase. At 1600 °C, the flexural strengths of the samples with 0.5 wt%, 1 wt%, and 3 wt% additions were greater than those of the control group, indicating that the addition of zirconia nanoparticles was beneficial to strength improvement, as zirconia nanoparticles themselves act as glass network intermediates and complement the mechanical properties of 3Y-ZrO$_2$. The cross-sectional SEM image of the sample with 1 wt% addition is shown in Figure S4 in the supplementary material. It was shown in that a high flexural strength can be reached by the formation of a comparatively high percentage of fine tetragonal zirconia grains in 3Y-ZrO$_2$ ceramics. Such microstructure provides the implementation of a high energy-consuming fracture micromechanism in flexure due to crack growth along the boundaries of the fine grains or their agglomerates, with rarely occurring cleavage fracture areas. In this case, the crack path undergoes multiple branching occurrences, which slow down the speed of fracture propagation and improve the fracture toughness and strength of ceramics [36,37]. The flexural strength of the samples continued to increase without showing any decreasing trend as the calcination temperature increased. However, according to the report, as the temperature increases in the range of 1200~1600 °C, although the fracture strength of ceramics continues to increase, the fracture toughness of ceramics reaches a maximum at 1500 °C and starts decreasing as the temperature increases further, exhibiting a trend of an increase first and then a decrease. Unfortunately, because of time constraints, no further studies have been conducted at higher sintering temperatures.

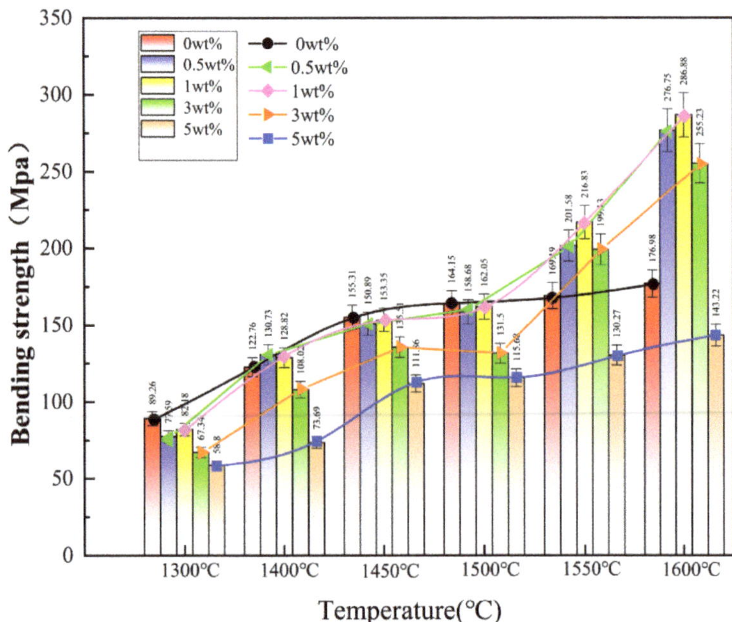

Figure 14. Flexural strength of the sample strips after calcination by adding the prepared zirconia to 3Y-ZrO_2.

3. Experimental

3.1. Materials and Methods

Using the hydrothermal method, the influencing factors regarding the preparation of t-ZrO_2 nanopowders were systematically investigated by adding different types and contents of dispersants, with zirconium oxychloride as the zirconium source, ammonia with a mass fraction of 28% as the precipitant, and propanetriol as an additive. Polyethylene glycol 8000 (PEG8000), polyvinylpyrrolidone (PVP), and cetyl trimethyl ammonium bromide (CTAB) were used as dispersants. The contents of the dispersants are shown in Table 1. The zirconium hydroxide precursors were formed by adding propanetriol to the zirconium oxychloride solution under ultrasonic and mechanical stirring conditions, completely dissolving it and then adding dispersant, followed by twice the molar amount of ammonia of zirconium oxychloride, and fully ultrasonic and stirring. To disperse the raw material as homogeneously as possible, we performed a systematic control test on the sonication time and method, setting the power at 80 W and the rotational speed at 1000 RPM, and finally obtained 5 min of sonication and mechanical stirring as the optimal choice. The precursor was transferred to a reactor with a capacity of 50 mL, filled with 75%, and reacted at 200 °C for 12 h. The synthesized sample was allowed to stand in deionized water, and then 0.1 mol/L of silver nitrate solution was added dropwise to the supernatant until no precipitation was produced. Then, the impurity ions were removed by centrifugation three times, dried, and ground to obtain the final sample. The experimental procedure is shown in Figure 15.

Table 1. Factor levels at different dispersant mass fractions in the preparation of ZrO_2 nanoparticles.

Sample Number	PEG8000 (wt%)	PVP (wt%)	CTAB (wt%)	Zirconium Oxychloride:Glycerol	Hydrothermal Reaction Temperature (°C)
1	/	/	/	2:1	200
2	0.1	0.1	0.1	2:1	200
3	0.3	0.3	0.3	2:1	200
4	0.5	0.5	0.5	2:1	200
5	1	1	1	2:1	200
6	2	2	2	2:1	200

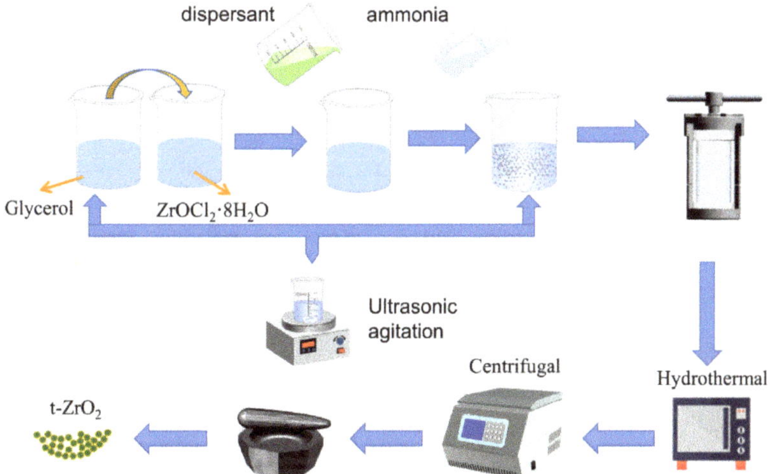

Figure 15. Flow chart of the preparation of tetragonal phase nano-zirconia powder.

3.2. Characterization

The crystal structure and phase composition of the materials were measured by XRD (D8 ADVANCE CEO, BRUKER) under the conditions of Cu Kα radiation and a 2θ range of 20–90°. The crystallite size was estimated from the peak broadening of the (011) reflection in the XRD pattern using the Scherrer equation as follows:

$$D = \frac{K\lambda}{\beta \cos \theta} \tag{1}$$

The particle size and microscopic morphology of ZrO_2 nanopowders were analyzed using field emission scanning electron microscopy (FESEM, INSPECT F50, FEI, USA) to observe their morphological characteristics. The particle size of the particles is measured using a laser particle size analyzer (LT3600, Light Truth) with a controlled refractive index range of 5–8% of the sample at the time of measurement. The microstructure and dispersion of the nanoparticles were observed by transmission electron microscopy (TEM) (JEOL 1400 Flash, JEOL, Japan). The molecular structure of the powder can be characterized by Raman spectroscopy (Lab RAM Armis, HORIBA Jobin Yvon). The three-point bending method (as shown in Figure 16) was used to measure the flexural strength of the ceramics (International Standard ISO 6872:2015 Dentistry—Ceramic materials; ISO, 2015). After the corresponding parameters were set in the equipment, the span distance was adjusted to

20 mm, and the indenter loading speed was 0.5 mm/min. The flexural strength (MPa) was calculated using the following equation:

$$\sigma = 3FL/2bh^2 \tag{2}$$

where F is the fracture load (N), L is the distance between the supporting rollers (mm), b is the width of the sample (mm) and h is the thickness of the sample (mm).

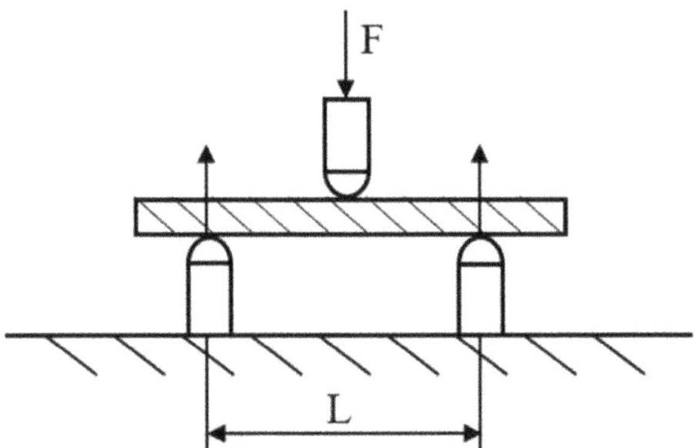

Figure 16. Schematic diagram of the three-point bending method.

4. Conclusions

Pure tetragonal-phase zirconia nanoparticles with good dispersion and an average particle size of 7.3 nm were prepared by the hydrothermal method with the addition of 5 wt% propanetriol, 10 wt% zirconium chloride, 0.5 wt% PVP, ultrasonication and mechanical stirring for 5 min, and a hydrothermal temperature of 200 °C. In the preparation of tetragonal-phase zirconia nanoparticles by propanetriol modulation, it was found that an appropriate amount of propanetriol was favorable for the generation of tetragonal-phase zirconia nanoparticles, and 5 wt% propanetriol induced the complete transformation of the monoclinic phase to the tetragonal phase. The increasing hydrothermal temperature gradually transformed the monoclinic phase into the tetragonal phase. Ultrasonication and mechanical stirring had a certain effect on particle size and distribution, with the optimal effect achieved at 5 min. The addition of dispersant effectively alleviated the agglomeration problem, with 0.5 wt% PVP being particularly favorable for reducing the particle size.

Pure zirconia nanoparticles in the tetragonal phase prepared using propanetriol were added to 3Y-ZrO$_2$ at different contents, and their flexural strengths were tested after calcination at varying temperatures. As the calcination temperature increased to above 1500 °C, the flexural strength of the 3Y-ZrO$_2$ samples with the addition of zirconia nanoparticles showed a trend of initial increase and then decrease with increasing nanoparticle content, and the flexural strength reached its maximum value of 286.88 Mpa at 1600 °C with 1 wt% zirconia nanoparticles added. The performance of 3Y-ZrO$_2$ ceramics after the addition of t-ZrO$_2$ was significantly increased, which provides important academic value and practical significance for the development and preparation of high-performance zirconia nanopowders.

Supplementary Materials: The following supporting information can be downloaded at: https://www.mdpi.com/article/10.3390/inorganics11050217/s1, Figure S1: Particle size distribution of ZrO$_2$ prepared with different mass fractions of PEG8000: (a) 0; (b) 0.1 wt%; (c) 0.3 wt%; (d) 0.5 wt%; (e) 1 wt%; (f) 2 wt%; Figure S2: Particle size distribution of ZrO$_2$ prepared with different mass fractions of PVP: (a) 0; (b) 0.1 wt%; (c) 0.3 wt%; (d) 0.5 wt%; (e) 1 wt%; (f) 2 wt%; Figure S3: Particle size distribution of ZrO$_2$ prepared with different mass fractions of CTAB: (a) 0; (b) 0.1 wt%; (c) 0.3 wt%;

(d) 0.5 wt%; (e) 1 wt%; (f) 2 wt%; Figure S4: SEM image of the cross-section of 3Y-ZrO$_2$ calcined sample: (a) 0; (b) 1 wt%.

Author Contributions: Conceptualization and methodology, S.L.; formal analysis and writing—original draft preparation, J.W.; methodology and investigation, Y.C.; formal analysis and data curation, Z.S.; data curation and validation, B.H. and H.W.; validation, T.Z.; methodology, M.L. All authors have read and agreed to the published version of the manuscript.

Funding: This research was funded by the Science and Technology Program of Henan Province, China (222102230034).

Acknowledgments: We thank L. P. Guo from the School of Foreign Languages at Henan University of Technology for help.

Conflicts of Interest: The authors declare no conflict of interest.

References

1. Basu, B. Toughening of Yttria-Stabilised Tetragonal Zirconia Ceramics. *Int. Metall. Rev.* **2005**, *50*, 239–256. [CrossRef]
2. You, X.; Chen, F.; Zhang, J.; Anpo, M. A Novel Deposition Precipitation Method for Preparation of Ag-Loaded Titanium Dioxide. *Catal. Lett.* **2005**, *102*, 247–250. [CrossRef]
3. Huang, W.; Qiu, H.; Zhang, Y.; Nan, L.; Gao, L.; Chen, J.; Omran, M.; Chen, G. Preparation of Nano Zirconia by Binary Doping: Effect of Controlled Sintering on Structure and Phase Transformation. *Ceram. Int.* **2022**, *48*, 25374–25381. [CrossRef]
4. Albayrak, S.; Becker-Willinger, C.; Aslan, M.; Veith, M. Influence of Nano-Scaled Zirconia Particles on the Electrical Properties of Polymer Insulating Materials. *IEEE Trans. Dielectr. Electr. Insul.* **2012**, *19*, 76–82. [CrossRef]
5. Rehman, M.; Noor, T.; Iqbal, N. Effect of Zirconia on Hydrothermally Synthesized Co$_3$O$_4$/TiO$_2$ Catalyst for NOx Reduction from Engine Emissions. *Catalysts* **2020**, *10*, 209. [CrossRef]
6. Hori, C. Thermal Stability of Oxygen Storage Properties in a Mixed CeO$_2$-ZrO$_2$ System. *Appl. Catal. B Environ.* **1998**, *16*, 105–117. [CrossRef]
7. Mandal, S.; Kumar, C.; Kumar, D.; Syed, K.; Ende, M.; Jung, I.; Finkeldei, S.C.; Bowman, W.J. Designing Environment-friendly Chromium-free Spinel-Periclase-Zirconia Refractories for Ruhrstahl Heraeus Degasser. *J. Am. Ceram. Soc.* **2020**, *103*, 7095–7114. [CrossRef]
8. Jiang, L.; Guo, S.; Bian, Y.; Zhang, M.; Ding, W. Effect of Sintering Temperature on Mechanical Properties of Magnesia Partially Stabilized Zirconia Refractory. *Ceram. Int.* **2016**, *42*, 10593–10598. [CrossRef]
9. Liu, C.X.; Gao, N. An Enzymesensor Based on Strongly Negative Nano-Zirconia-Embedded Carben Paste Electrode for H_2O_2 Assay. *Chem. Sens.* **2005**, *25*, 17–20. [CrossRef]
10. Li, N.; Yu, N.; Yi, Z.; An, D.; Xie, Z. CeO$_2$-Stabilised ZrO$_2$ Nanoparticles with Excellent Sintering Performances Synthesized by Sol-Gel-Flux Method. *J. Eur. Ceram. Soc.* **2022**, *42*, 1645–1655. [CrossRef]
11. Cousland, G.P.; Cui, X.Y.; Smith, A.E.; Stampfl, A.P.J.; Stampf, C.M. Mechanical Properties of Zirconia, Doped and Undoped Yttria-Stabilized Cubic Zirconia from First-Principles. *J. Phys. Chem. Solids* **2018**, *122*, 51–71. [CrossRef]
12. Kelly, J.R.; Denry, I. Stabilized Zirconia as a Structural Ceramic: An Overview. *Dent. Mater.* **2008**, *24*, 289–298. [CrossRef] [PubMed]
13. Pezzotti, G.; Zhu, W.; Zanocco, M.; Marin, E.; Sugano, N.; Mcentire, B.J.; Bal, B.S. Reconciling in Vivo and in Vitro Kinetics of the Polymorphic Transformation in Zirconia-Toughened Alumina for Hip Joints: II. Theory. *Mater. Sci. Eng. C* **2017**, *72*, 446–451. [CrossRef] [PubMed]
14. Guo, H.; Khor, K.A.; Yin, C.B.; Miao, X. Laminated and Functionally Graded Hydroxyapatite/Yttria Stabilized Tetragonal Zirconia Composites Fabricated by Spark Plasma Sintering. *Biomaterials* **2003**, *24*, 667–675. [CrossRef]
15. Torres-Lagares, D. A Review on CAD/CAM Yttria-Stabilized Tetragonal Zirconia Polycrystal (Y-TZP) and Polymethyl Methacrylate (PMMA) and Their Biological Behavior. *Polymers* **2022**, *14*, 906. [CrossRef]
16. Guo, M.; Wang, G.; Zhao, Y.; Li, H.; Burgess, K. Preparation of Nano-ZrO$_2$ Powder via a Microwave-Assisted Hydrothermal Method. *Ceram. Int.* **2021**, *47*, 12425–12432. [CrossRef]
17. Salas, P.; Rosa-Cruz, E.; Mendoza-Anaya, D.; González, P.; Rodrguez, R.; Castao, V.M. High Temperature Thermoluminescence Induced on UV-Irradiated Tetragonal ZrO$_2$ Prepared by Sol–Gel. *Mater. Lett.* **2000**, *45*, 241–245. [CrossRef]
18. Song, J.; Zhang, H.; Feng, Z.; Zhang, J.; Huang, X. Controllable Preparation of 5mol% Y$_2$O$_3$-Stabilized Tetragonal ZrO$_2$ by Hydrothermal Method. *J. Alloys Compd.* **2020**, *856*, 156766. [CrossRef]
19. Yurdakul, A.; Gocmez, H. One-Step Hydrothermal Synthesis of Yttria-Stabilized Tetragonal Zirconia Polycrystalline Nanopowders for Blue-Colored Zirconia-Cobalt Aluminate Spinel Composite Ceramics. *Ceram. Int.* **2019**, *45*, 5398–5406. [CrossRef]
20. Yang, C.; Jin, Z.; Chen, X.; Fan, J.; Fan, Y. Modified Wet Chemical Method Synthesis of Nano-ZrO$_2$ and Its Application in Preparing Membranes. *Ceram. Int.* **2021**, *47*, 13432–13439. [CrossRef]
21. Qiu, H.; Zhang, Y.; Huang, W.; Peng, J.; Chen, J.; Gao, L.; Omran, M.; Li, N.; Chen, G. Sintering Properties of Tetragonal Zirconia Nanopowder Preparation of the NaCl + KCl Binary System by the Sol–Gel–Flux Method. *ACS Sustain. Chem. Eng.* **2023**, *11*, 1067–1077. [CrossRef]

22. Duran, C.; Jia, Y.; Sato, K.; Hotta, Y.; Watari, K. Hydrothermal Synthesis of Nano ZrO_2 Powders. *Key Eng. Mater.* **2006**, *317–318*, 195–198. [CrossRef]
23. Hsu, Y.-W.; Yang, K.-H.; Chang, K.-M.; Yeh, S.-W.; Wang, M.-C. Synthesis and Crystallization Behavior of 3 Mol% Yttria Stabilized Tetragonal Zirconia Polycrystals (3Y-TZP) Nanosized Powders Prepared Using a Simple Co-Precipitation Process. *J. Alloys Compd.* **2011**, *509*, 6864–6870. [CrossRef]
24. Chevalier, J.; Gremillard, L.; Virkar, A.V.; Clarke, D.R. The Tetragonal-Monoclinic Transformation in Zirconia: Lessons Learned and Future Trends. *J. Am. Ceram. Soc.* **2010**, *92*, 1901–1920. [CrossRef]
25. Gionco, C.; Hernández, S.; Castellino, M.; Gadhi, T.A.; Paganini, M.C. Synthesis and Characterization of Ce and Er Doped ZrO_2 Nanoparticles as Solar Light Driven Photocatalysts. *J. Alloys Compd.* **2018**, *775*, 896–904. [CrossRef]
26. Morales–Rodríguez, A.; Poyato, R.; Gutiérrez–Mora, F.; Muñoz, A.; Gallardo–López, A. The Role of Carbon Nanotubes on the Stability of Tetragonal Zirconia Polycrystals. *Ceram. Int.* **2018**, *44*, 17716–17723. [CrossRef]
27. Sun, J.; Gao, L.A.; Guo, J.K. Effect of Dispersant on the Measurement of Particle Size Distribution of Nano Size Y-TZP. *J. Inorg. Mater.* **1999**, *14*, 465–469. [CrossRef]
28. Yu, J.; Yu, F.-X.; Wang, S.; Zhang, J.-F.; Fan, F.-Q.; Long, Q. Effect of Dispersant Content and Drying Method on $ZrO_2@Al_2O_3$ Multiphase Ceramic Powders. *Ceram. Int.* **2018**, *44*, 17630–17634. [CrossRef]
29. Lin, C.; Zhang, C.; Lin, J. Phase Transformation and Photoluminescence Properties of Nanocrystalline ZrO_2 PowdersPrepared via the Pechini-Type SolGel Process. *J. Phys. Chem. C* **2007**, *111*, 3300–3307. [CrossRef]
30. Noh, H.J.; Seo, D.S.; Kim, H.; Lee, J.K. Synthesis and Crystallization of Anisotropic Shaped ZrO_2 Nanocrystalline Powders by Hydrothermal Process. *Mater. Lett.* **2003**, *57*, 2425–2431. [CrossRef]
31. Chang, Q.; Zhou, J.; Wang, Y.; Meng, G. Formation Mechanism of Zirconia Nano-Particles Containing Pores Prepared via Sol-Gel-Hydrothermal Method. *Adv. Powder Technol.* **2010**, *21*, 425–430. [CrossRef]
32. Chao, Y.A.; Jw, A.; Xc, A.; Xd, A.; Mq, A.; Hv, B.; Yf, A. Modified Hydrothermal Treatment Route for High-Yield Preparation of Nanosized ZrO_2. *Ceram. Int.* **2020**, *46*, 19807–19814. [CrossRef]
33. Zang, S.; Yang, Q.; He, N.; Ji, Y.; Yang, H. Influence of Additives on the Tetragonal Phase Purity and Grain Size of Zirconia. *J. Clin. Rehabil. Tissue Eng. Res.* **2018**, *22*, 5445–5451. [CrossRef]
34. Mondal, A.; Zachariah, A.; Nayak, P.; Nayak, B.B. Synthesis and Room Temperature Photoluminescence of Mesoporous Zirconia with a Tetragonal Nanocrystalline Framework. *J. Am. Ceram. Soc.* **2010**, *93*, 387–392. [CrossRef]
35. Thackeray, D. The Raman Spectrum of Zirconium Dioxide. *Spectrochim. Acta Part A Mol. Spectrosc.* **1974**, *30*, 549–550. [CrossRef]
36. Kulyk, V.; Duriagina, Z.; Vasyliv, B.; Vavrukh, V.; Kovbasiuk, T.; Lyutyy, P.; Vira, V. The Effect of Sintering Temperature on the Phase Composition, Microstructure, and Mechanical Properties of Yttria-Stabilized Zirconia. *Materials* **2022**, *15*, 2707. [CrossRef]
37. Kulyk, V.; Duriagina, Z.; Kostryzhev, A.; Vasyliv, B.; Vavrukh, V.; Marenych, O. The Effect of Yttria Content on Microstructure, Strength, and Fracture Behavior of Yttria-Stabilized Zirconia. *Materials* **2022**, *15*, 5212. [CrossRef]

inorganics

Article

Modification of Graphite/SiO₂ Film Electrodes with Hybrid Organic–Inorganic Perovskites for the Detection of Vasoconstrictor Bisartan 4-Butyl-N,N-bis{[2-(2H-tetrazol-5-yl)biphenyl-4-yl]methyl}imidazolium Bromide

Georgios Papathanidis [1], Anna Ioannou [1], Alexandros Spyrou [1], Aggeliki Mandrapylia [2], Konstantinos Kelaidonis [2], John Matsoukas [2], Ioannis Koutselas [1] and Emmanuel Topoglidis [1,*]

[1] Materials Science Department, University of Patras, 26504 Patras, Greece; up1022671@ac.upatras.gr (G.P.); ioannou.p.anna@gmail.com (A.I.); up1061312@upnet.gr (A.S.); ikouts@upatras.gr (I.K.)

[2] NewDrug, P.C., Patras Science Park, 26504 Patras, Greece; aggelikimandrapulia@gmail.com (A.M.); k.kelaidonis@gmail.com (K.K.); imats@upatras.gr (J.M.)

* Correspondence: etop@upatras.gr

Abstract: In the present work, a hybrid organic–inorganic semiconductor (HOIS) has been used to modify the surface of a graphite paste/silica (G–SiO₂) film electrode on a conducting glass substrate to fabricate a promising, sensitive voltammetric sensor for the vasoconstrictor bisartan BV6, which could possibly treat hypertension and COVID-19. The HOIS exhibits exceptional optoelectronic properties with promising applications not only in light-emitting diodes, lasers, or photovoltaics but also for the development of voltammetric sensors due to the ability of the immobilized HOIS lattice to interact with ions. This study involves the synthesis and characterization of an HOIS and its attachment on the surface of a G–SiO₂ film electrode in order to develop a nanocomposite, simple, sensitive with a fast-response, low-cost voltammetric sensor for BV6. The modified HOIS electrode was characterized using X-ray diffraction, scanning electron microscopy, and optical and photoluminescence spectroscopy, and its electrochemical behavior was examined using cyclic voltammetry. Under optimal conditions, the modified G–SiO₂ film electrode exhibited a higher electrocatalytic activity towards the oxidation of BV6 compared to a bare graphite paste electrode. The results showed that the peak current was proportional to BV6 concentration with a linear response range from 0 to 65×10^{-6} (coefficient of determination, 0.9767) and with a low detection limit of 1.5×10^{-6} M (S/N = 3), estimated based on the area under a voltammogram, while it was 3.5×10^{-6} for peak-based analysis. The sensor demonstrated good stability and reproducibility and was found to be appropriate for the determination of drug compounds such as BV6.

Keywords: graphite/SiO₂ film; hybrid organic–inorganic semiconductor; bisartan BV6; cyclic voltammetry; sensor

Citation: Papathanidis, G.; Ioannou, A.; Spyrou, A.; Mandrapylia, A.; Kelaidonis, K.; Matsoukas, J.; Koutselas, I.; Topoglidis, E. Modification of Graphite/SiO₂ Film Electrodes with Hybrid Organic–Inorganic Perovskites for the Detection of Vasoconstrictor Bisartan 4-Butyl-N,N-bis{[2-(2H-tetrazol-5-yl)biphenyl-4-yl]methyl}imidazolium Bromide. *Inorganics* 2023, 11, 485. https://doi.org/10.3390/inorganics11120485

Academic Editors: Roberto Nisticò, Torben R. Jensen, Shuang Xiao, Luciano Carlos, Hicham Idriss and Eleonora Aneggi

Received: 8 July 2023
Revised: 1 December 2023
Accepted: 8 December 2023
Published: 18 December 2023

1. Introduction

Herein, we report the electrochemical properties and sensing of a drug molecule, 4-butyl-N,N-bis{[2-(2H-tetrazol-5-yl)biphenyl-4-yl]methyl}imidazolium bromide (BV6), an angiotensin II receptor blocker, usually referred to as bisartans. This work is in the realm of exploring electrochemical methods for drug detection [1]. Furthermore, similar efforts were used for the detection of a similar compound, losartan, the first oral non-peptide mimetic of angiotensin II hypertension medicine [2]. It is expected that the sensing electrodes used for this molecule could be applied for other structurally similar molecular drugs, whilst the choice of BV6 is based on its proposed applicability in disease prevention. Electrochemical techniques such as voltammetry and potentiometry enable quick drug detection in biological fluids, pills, or suspensions and ensure drug quality and dose. These methods offer specificity, high sensitivity, low detection limits in the μM/nM range, the possibility of real-time results, ease of preparation and operation, and cost-effectiveness when compared with other analytical methods used for drug detection [3,4].

As illustrated in Scheme 1, the structure of BV6 consists of two biphenyl moieties (either ortho-substituted tetrazole or carboxylate) at the N-1 and N-3 nitrogens of imidazole. Bisartans regulate the components of the renin angiotensin system, a prime target for cardiovascular disease therapy. According to the literature, in in-silico studies, these appear to strongly bind to the SARS-CoV-2 RBD/ACE2 complex spike [5–11].

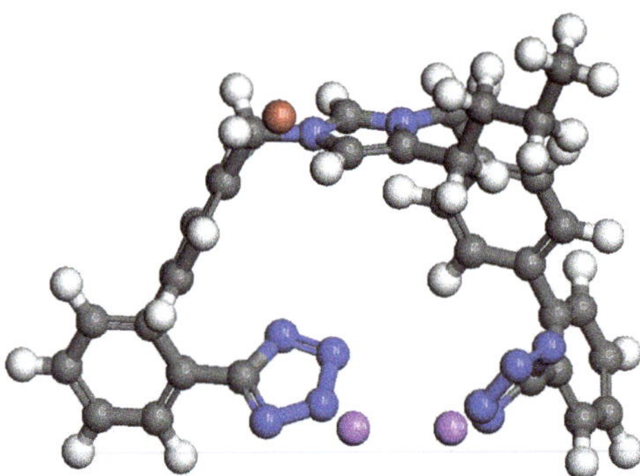

Scheme 1. Structure of BV6. White spheres are H, gray C, blue N, purple Na, and red Br.

The materials that have been used for the fabrication of electrodes for drug determination exhibit a large variety of electrical and electronic properties. In this work, naturally formed hybrid organic–inorganic semiconductors (HOISs), also named as short perovskite materials [12], have been selected as they exhibit enhanced quantum and dielectric confinement of their direct band gap excitons, as well as high mobility and diffusion lengths for their electrons and holes.

In general, HOISs form low-dimensional semiconducting structures, whose dimensionality strongly depends on the starting stoichiometry and precursor selection. The dimensionality refers to that of the active semiconducting inorganic network—usually a metal halide; this last network can take a range of dimensionalities from 0D up to 3D. The dimensionality also depends on the choice of the organic part content. Such networks are shown in Scheme 2, spanning from two-dimensional up to three-dimensional cases; 2D, left (n = 1), to 3D, right (n = ∞) [13]. The enhanced mobility of the charged carriers, i.e., electrons and holes, as well as their increased diffusion length, allows HOISs to serve as sensors [14], photovoltaics [15], light-emitting diodes [16–19], and chemielectroluminescent materials. Additionally, their cost and fabrication simplicity places them as suitable for use as electrode materials. Also, their inherent optical absorption (OA) and photoluminescence (PL) excitonic peaks are strong and tunable by chemistry, as has already been evidenced by numerous publications [20,21]. Also, the excitons in HOISs are distinct electronic states at room temperature, including the perovskite used in this work. The aforementioned improved optical and electronic properties of HOISs, as well as their flexible nature, allows them to be used in novel devices. However, a unique disadvantage these materials exhibit is that they are usually degraded by water; thus, organic solvents have been used in the present study.

Graphite paste/silica (G–SiO$_2$) film electrodes, present great advantages as opposed to standard carbon electrodes (glassy or metallic) and their use in electrochemical studies is constantly increasing [3,22]. Previous studies have shown that they exhibit high conductivity, high electrocatalytic activity for many redox reactions, and resistance to surface fouling. Additionally, they can remain stable in organic solvents for a relatively long time (several hours), perform at lower overpotentials, and are reusable [23]. Adding to the aforementioned advantages, the ease of manufacturing, the lower cost, and the possibility of chemical modification make the G–SiO$_2$ film electrodes even more appealing [24].

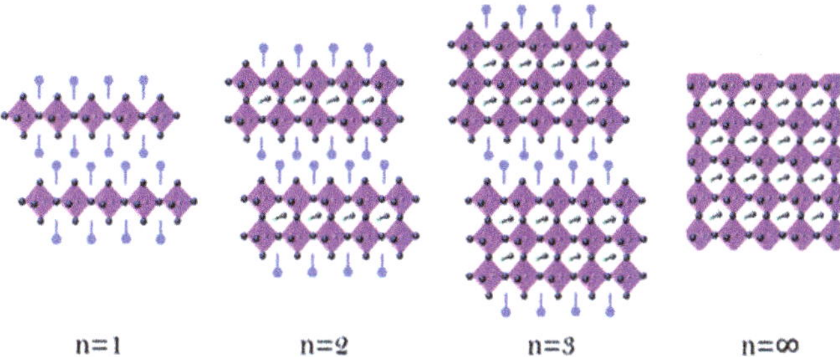

Scheme 2. Visual representation of organic–inorganic hybrid perovskite networks as n = 1 (2D), n = 2, and n = 3 (quasi–2D), n = ∞ (3D); a specific yet general example has chemical formula of A$_2$M$_{n-1}$Pb$_n$X$_{3n+1}$, where usually X = Cl, Br, I, M a small amine or Cs, and A a long amine.

Over the last 30 years, material scientists have produced a variety of electrodes with unique electronic and structural characteristics, with the ability to mediate fast electron transfer for a wide range of electroactive compounds. These electrodes also exhibit electrocatalytic activity towards agricultural or food-related hazardous substances and pharmaceutical organic compounds and guide the development of new voltammetric and amperometric quantitative sensors [1,2,4,14,25]. Most electrochemical methods rely on the modification of electrodes, such as the graphite paste or the glassy carbon electrodes, by nanostructure or porous materials to enhance and increase their electroactive surface area and to improve their mechanical strength and chemical stability. By using a modifier (such as conducting polymers, nanoparticles, etc.), it is possible to dramatically improve the sensitivity and selectivity of an electrode [2,24,26]. Therefore, in this work, an HOIS has been synthesized and used as a modifier on the surface of a G–SiO$_2$ film electrode on a fluorine-doped tin oxide (FTO) glass substrate. More specifically, the perovskite has been adsorbed steadily on the surface of a G–SiO$_2$ film electrode for the electrochemical characterization and quantitative determination of BV6 for the first time.

2. Results and Discussion

2.1. X-ray Diffraction

In Figure 1, the XRD patterns of all the materials are presented. For the perovskite (Figure 1b) employed in this work, no precursor PbBr$_2$-associated peaks were found in its XRD pattern, showing the complete reaction towards the final perovskite.

For the composite films (Figure 1d), before being used for the electrochemical measurements, the characteristic peaks of the perovskite at 14.85°, 21.35°, and 30.17° on the G–SiO$_2$/FTO/perovskite film are observed, confirming its successful immobilization on the G–SiO$_2$ surface without its structure being altered in a significant manner. In this point, it is important to note that the pattern (1d) shows several small peaks at 13.69°, 27.34°, 35.96°, and 44.59° which may have arisen from recrystallization of the perovskite, involving Na$^+$ ions and possibly SiO$_2$ units on graphite. The pattern of the G–SiO$_2$ on the FTO glass substrate presents three slightly intense peaks at 26.54°, 38°, and 51.43° which correspond to the plane indices (110), (200), and (211) of the tetragonal SnO$_2$ structure due to the FTO glass slide. Smaller peaks which correspond to the plane indices (101), (220), (310), and (301) also confirm the previous pattern. The amorphous character of the G–SiO$_2$ on FTO is only presented as a broad peak (Figure 1a) centered at about 26°. After the cyclic voltammetric measurements, the XRD pattern of the used composite material exhibited the peaks of the perovskite, as well as a double peak at 11.63° and a single peak at 35.33°. To determine the origin of these peaks, which could be due to some reaction or degradation of the perovskite, another pattern (Figure 1c) has been added. It corresponds to a reaction of the perovskite and BV6 after being dissolved in DMF, while this time BV6 has a much greater concentration than the one used in the sensing experiments. Thus, it is concluded that the peaks at low angles may be due to the formation of low-dimensional perovskites. This will become more evident in the optical measurements as depicted later on.

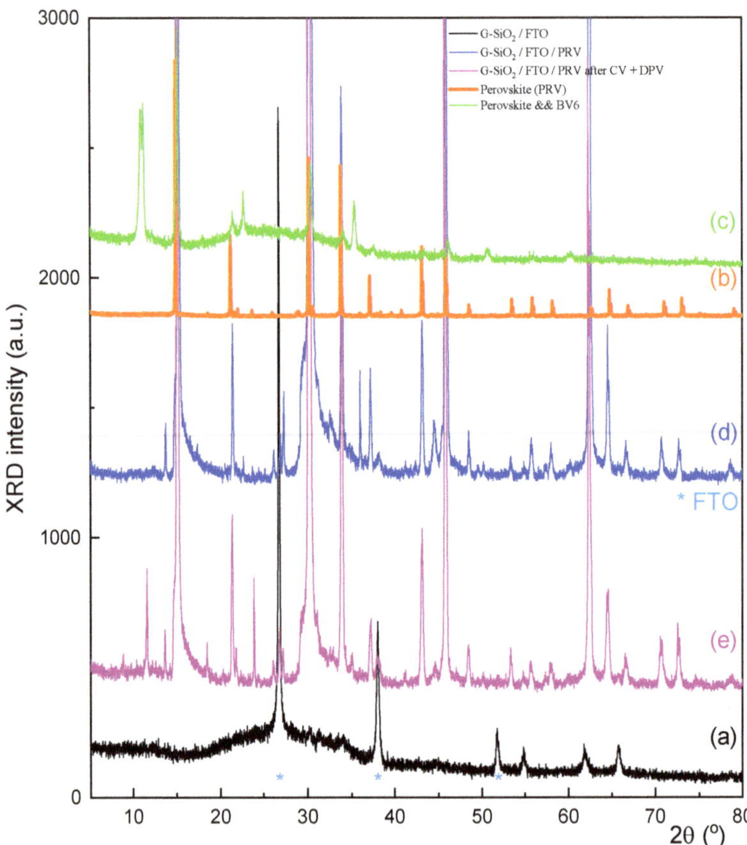

Figure 1. XRD patterns of (**a**) G-SiO$_2$/FTO film; (**b**) Perovskite; (**c**) Perovskite with BV6; (**d**) Perovskite/G–SiO$_2$/FTO film before the sensing of BV6; and (**e**) Perovskite/G–SiO$_2$/FTO film after the sensing of BV6. (* is used for the peaks of FTO glass).

In effect, it is safe to assume that the perovskite remains mostly intact during the electrochemical measurements, while only a small part of it is consumed by its embedding in the amorphous matrix, as well as due to its electrochemical interaction with the BV6. As will be discussed later, the optical properties, especially the UV absorption spectra, related to the band gap and exciton remain intact after usage. It is interesting that BV6 can alter the perovskite's structure, most probably due to the presence of the bromine in its structure, or in general its halogen content, which probably allows BV6 to intercalate within PbBr$_x$ layers, giving rise to the low angle peak (Figure 1c), which differs though by 0.6° from that exhibited from the low angle peak of the used sample (Figure 1e). Due to the complexity of the XRD patterns' peaks among the samples that have interacted with BV6, it is conjectured that any electrochemical redox reaction leads to new BV6/perovskite complexes which under the applied oxidation potential give rise to electrochemical peaks; this is conjectured to take place by donating redox-generated electrons while any by-products revert to the original perovskite structure.

2.2. *SEM Imaging of Surface Topography*

To evaluate the perovskite/G–SiO$_2$ films on FTO glass slides, before and after the immobilization of the perovskite on their surface, scanning electron microscopy (SEM)/energy-dispersive X-ray (EDX) was also used. The same technique was used to examine any modification on the surface of the films after the cyclic voltammetry (CV) measurements and after the additions of BV6 concentrations in the organic electrolyte solution. This allows us to monitor if the applied biases or the BV6 concentration could alter the morphology, stoichiometry, and structure of the immobilized perovskite.

Figure 2a,b show the SEM images of the perovskite before its immobilization on the G–SiO$_2$/FTO films. After the reaction of BV6 with perovskite, as depicted in Figure 2c,d, the formation of a cube-like structure was observed at the center of the grains of the perovskite, while away from the fractal-like and cuboid structure no Pb was detected, implying the existence of BV6. The fractal-like growth is possibly linked to the planar form of the BV6 molecule and its capability to carry charge of different polarity in different regions, leading to the formation of some new perovskite phases or shapes.

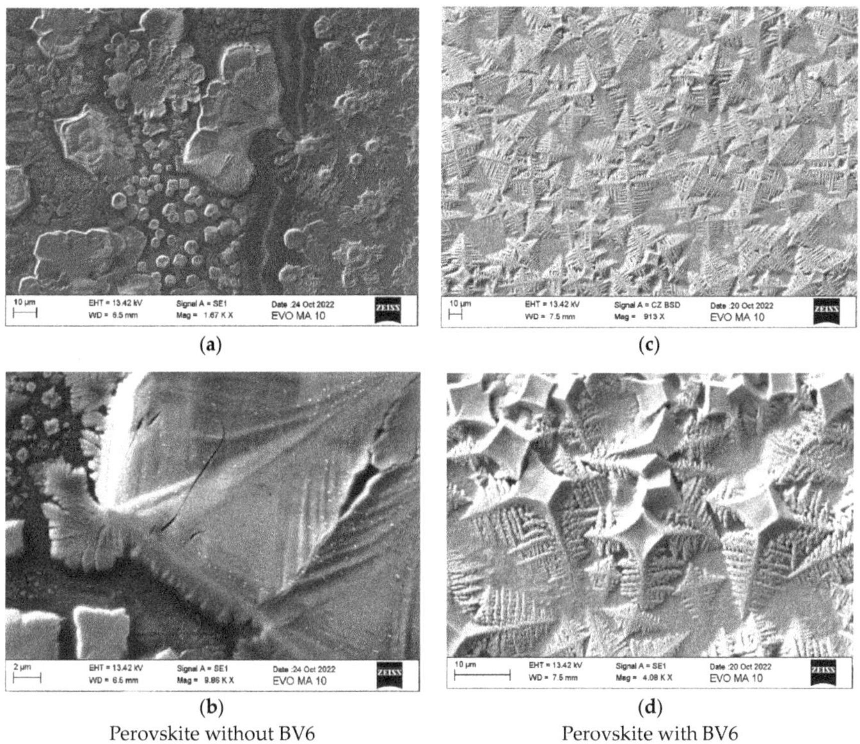

(a) (c)

(b) (d)

Perovskite without BV6 Perovskite with BV6

Figure 2. SEM images of (**a**,**b**) the sensor before reaction with BV6; (**c**,**d**) the sensor surface after reaction with BV6. Images b and d show increased detail of the film characteristics.

The addition of perovskite in the G–SiO$_2$ paste exhibits a homogenous nanostructure with a high effective surface consisting of visually numerous wrinkles and a highly rough surface, in favor of high perovskite loading and the introduction of multiple potential electroactive sites that provide a clear advantage for the sensing of BV6. Such an image acquired before the electrochemical measurements is shown in Figure 3 (cross section), while Figure 4 shows the images of the film before and after the electrochemical measurements.

Figure 3 shows that the top active film has a thickness of about 1–2 μm, which is of the same scale as the perovskite crystallites retained on top of the film.

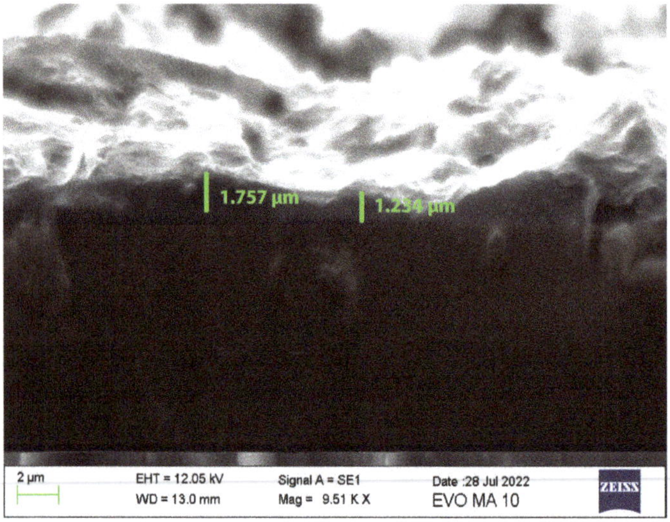

Figure 3. SEM image of a cross section of the perovskite/G–SiO$_2$/FTO film before use for BV6 sensing.

Finally, the SEM images for the electrodes' surface, after their usage for the electrochemical measurements, are shown in Figure 4. EDX elemental analysis for the atomic stoichiometry of Pb:Br shows that the observed cuboidal form of the 3D perovskite is at a ratio in excess of 3, even up to 3.5, depending on the region of the surface, verifying the existence of the 3D perovskite form instead of any PbBr$_2$ by-products, which would have a ratio of 2.

2.3. Optical Measurements

In Figure 5, the UV–Vis optical absorption spectra of all materials used in this work are presented. The pristine perovskite (Figure 5a) shows an energy band gap along with its accompanying exciton absorption peak at 522 nm, where both are convoluted, in accordance with previous works [20]. BV6 shows no absorption peaks in the region of interest (Figure 5b). Before the use of the composite electrode for sensing BV6, the combined G–SiO$_2$/perovskite optical absorption peak appears to shift to lower wavelengths (Figure 5d), ca. 509 nm, probably due to some low dimensional species that have been afforded by the perovskite growth within the spaces afforded by the amorphous paste. After the electrochemical measurements (Figure 5e), the peak further blue shifts towards 503 nm. Similarly, in order to further investigate the experiments performed by reacting BV6 and perovskite on a composite electrode, another similar reaction product was formed by reacting substantial quantities of BV6 with perovskites, in DMF, as a standalone material. In this case, the perovskite-BV6 product (Figure 5c) appears to exhibit a peak at 514 nm, which again shows that BV6 can intercalate at sufficiently large quantities within the perovskite structure and yield other intermediate structures which will be reported elsewhere.

In Figure 6, the corresponding PL spectra of the compounds are provided. The pristine perovskite shows a peak, slightly red-shifted with regard to the absorption peak, at 545 nm (Figure 6a). BV6 shows only high-energy luminescence (Figure 6b) when excited with 350 nm with its peaks appearing at 408, 433, and 461 nm again away from the perovskite peak and in summary has no low energy peaks. Figure 6c shows the PL of the reaction among BV6 and perovskites when using large amounts of both perovskite and BV6, showing in accordance with the optical absorption spectra the possibility of forming low-dimensional structures, since the PL has blue-shifted with regard to the pristine perovskite compound.

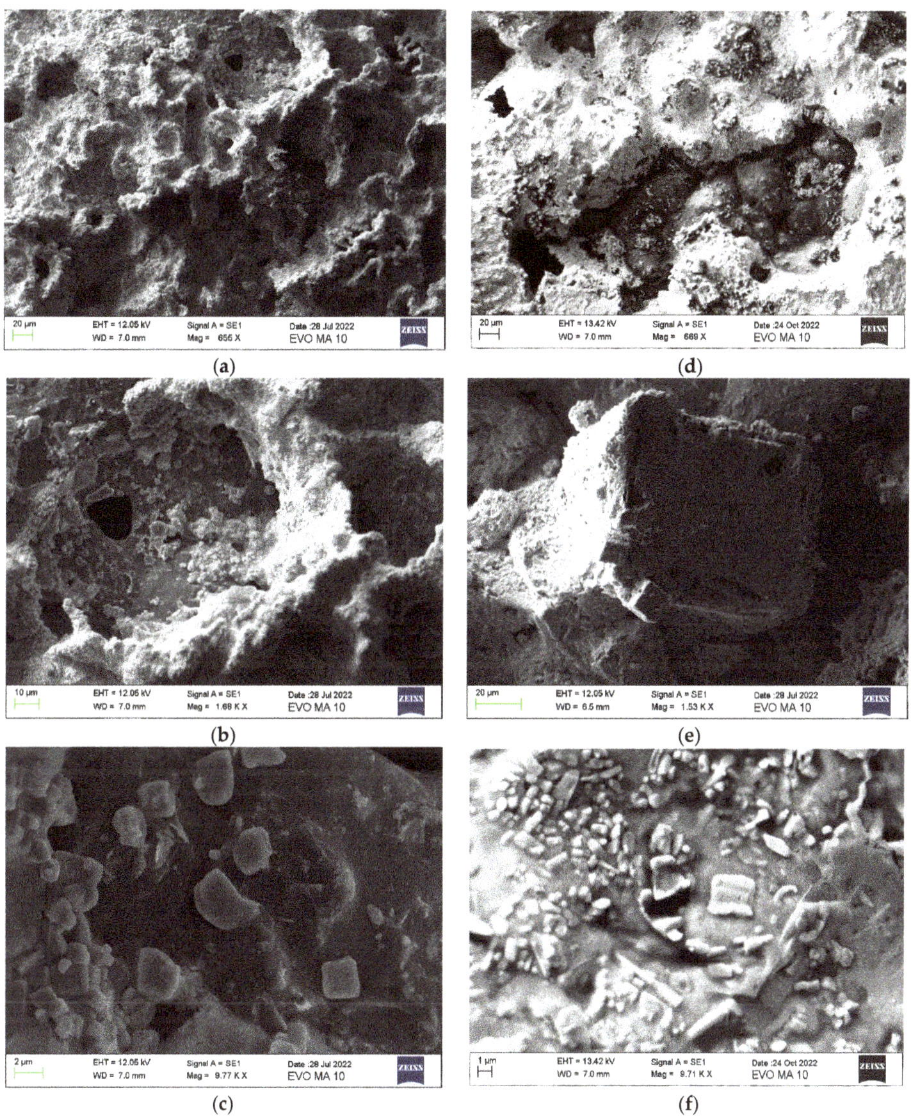

Figure 4. SEM images of (**a–c**) Perovskite/G–SiO$_2$/FTO before BV6 sensing and (**d–f**) Perovskite/G–SiO$_2$/FTO film after BV6 sensing.

Before sensing, the perovskite OA peak at 545 nm is shifted to 530 nm, see Figure 6d, once being mixed with the G–SiO$_2$ paste, which corresponds to the lower peak as seen in the UV–Vis spectra. This is in accordance with the blue-shift evident in the optical absorption spectra of the perovskite/composite material, which shows some strong quantum confinement of the perovskite structures. After sensing, Figure 6e, the peak at 415 nm corresponds to the BV6 molecule that has been absorbed by the perovskite layer, while the used electrode also exhibits an extra PL peak at 554 nm, red-shifted with regard to the pristine perovskite, possibly due to perovskite with trapped excitons and/or defects.

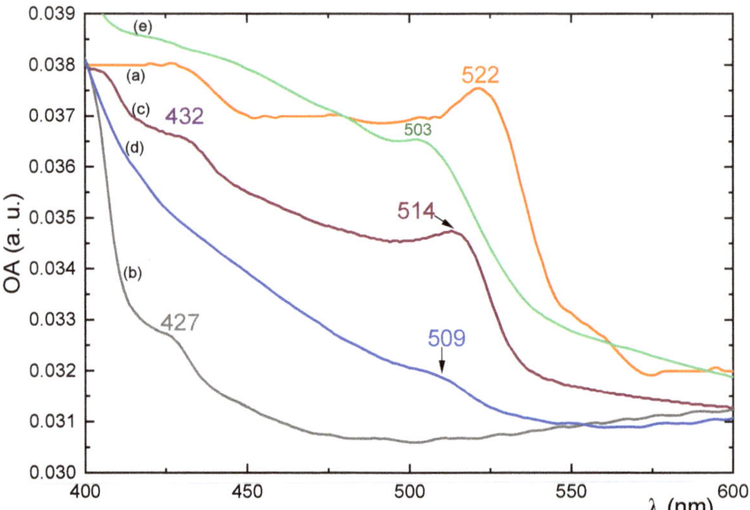

Figure 5. Optical absorption spectra of (**a**) Perovskite; (**b**) BV6; (**c**) Perovskite with BV6; (**d**) Perovskite/G–SiO$_2$/FTO film before the sensing of BV6; and (**e**) Perovskite/G–SiO$_2$/FTO film after the sensing of BV6.

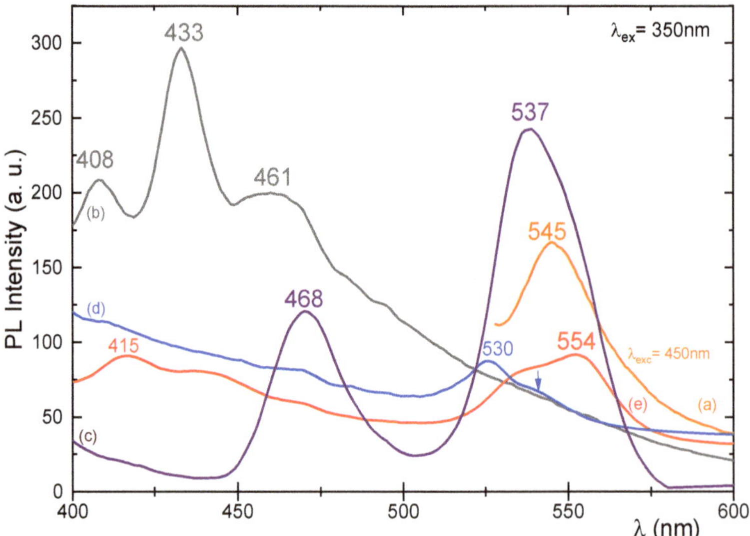

Figure 6. PL spectra of (**a**) Perovskite; (**b**) BV6; (**c**) Perovskite with BV6; (**d**) Perovskite/G–SiO$_2$/FTO film before the sensing of BV6; and (**e**) Perovskite/G–SiO$_2$/FTO film after the sensing of BV6. Arrow shows the position of the small excitonic at 540 nm, before sensing.

The used composite electrode was stored at room temperature with a 50% relative humidity for four months and underwent electrochemical sensing measurements again. Its optical properties, as determined by spectroscopic methods, are depicted in Figure 7, which verify that after this time, the UV–Vis absorption spectrum remains intact, yet the PL signal shifts the excitonic degradation peak, at 554 nm, back to 543 and 507 nm, even after this long period. It appears as if another low-dimensional adduct has formed after this long time and a pristine 3D perovskite has recovered from the sensing experiments, with any chemical/redox reaction with BV6.

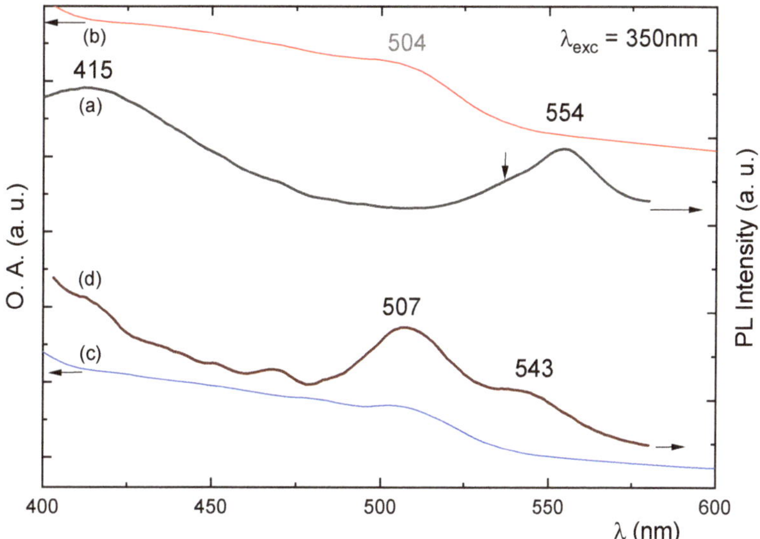

Figure 7. PL (**a,c**) and UV–Vis optical absorption (**b,d**) spectra, respectively, of the composite film after the 1st use before storage (**a,c**) and, after the 2nd use and 4 months of storage (**b,d**). Arrow shows the position of the excitonic peak before sensing; here, only a shoulder is observed.

2.4. Electrochemical Behavior of Perovskite/G–SiO$_2$/FTO Film Electrode

The electrochemical behavior of the composite perovskite/G–SiO$_2$/FTO film electrode was investigated by CV in DCM/TBAH electrolyte, where the perovskite appears to be stable and does not desorb from the surface of the film, as evidenced by the color of the electrolyte solution and the observation of the excitonic OA and PL peaks for the used films. The CVs of the FTO substrate and of the G–SiO$_2$/FTO film electrode, before and after the immobilization of the perovskite on its surface, are presented in Figure 8. The bare film electrode exhibited a large background current and an almost rectangular CV plot without the presence of any oxidation or reduction peaks (Figure 8b). This shape of the CV plot is expected for double-layer and pseudo-capacitive materials. Also, in Figure 8a, the CV plot of the FTO glass substrate does not exhibit any noticeable redox response and the respective capacitive current is negligible. After the attachment of the perovskite on the surface of the G–SiO$_2$/FTO film electrode, the background current is higher due to the larger surface area of the modified film electrode and, in addition, a small irreversible anodic peak at +0.31 V is observed (Figure 8c) due to the presence of the adsorbed perovskite. As shown in Figure 9a, this peak disappeared once BV6 was added. However, a similar peak appeared at +0.21V that could be related to the peak at +0.31 V. The shape and maximum value of this second peak will be affected by the progressive addition of BV6 to the electrolyte solution, allowing in this way the detection of BV6. Preliminarily, this suggests that the perovskite actively interferes with the BV6 towards its detection, or else this peak would have remained invariant to the added drug concentration.

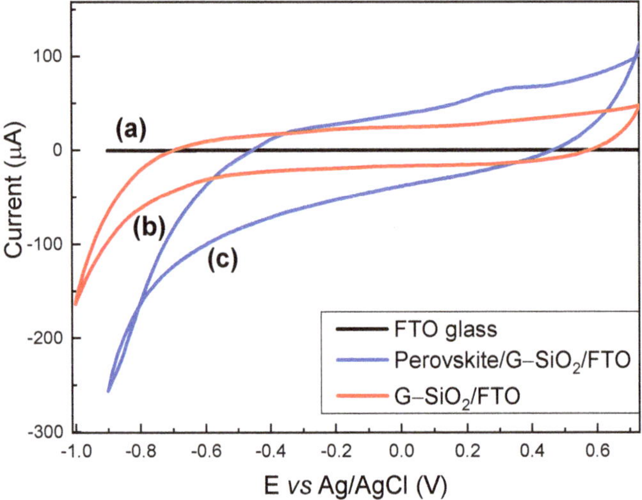

Figure 8. CVs at 0.1 V/s of (**a**) FTO glass; (**b**) G–SiO$_2$/FTO film electrode; (**c**) Perovskite/G–SiO$_2$/FTO hybrid electrode 0.1 M TBAH electrolyte in DCM.

A high perovskite loading was possible to attain on the surface of the G–SiO$_2$/FTO film electrode due to the high surface area of the film electrode and the molecular size of the perovskite used in this study. Adsorption of 40 µL of the resultant perovskite solution onto the film's surface caused its dark yellow-orange coloration, which suggests a high surface coverage. It is plausible that due to the organic–inorganic nature of the perovskite, its active materials could exhibit a similar immobilization behavior as is typically the case for proteins on the surface of the host G–SiO$_2$/FTO film electrode.

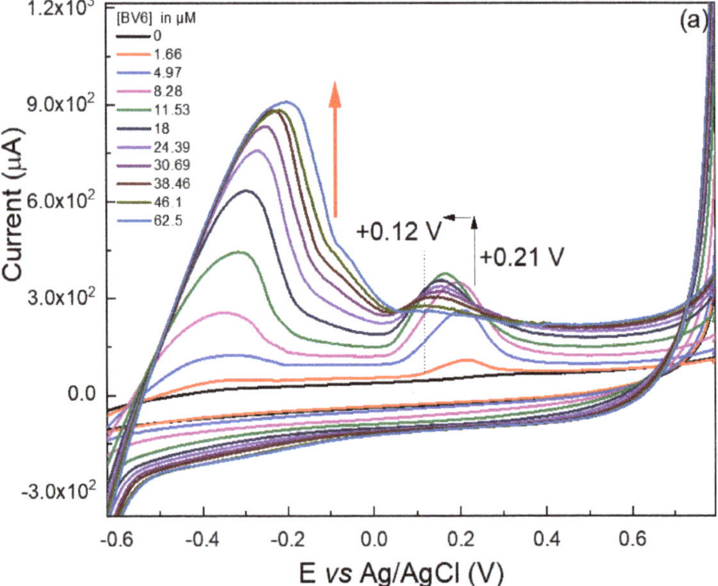

Figure 9. *Cont.*

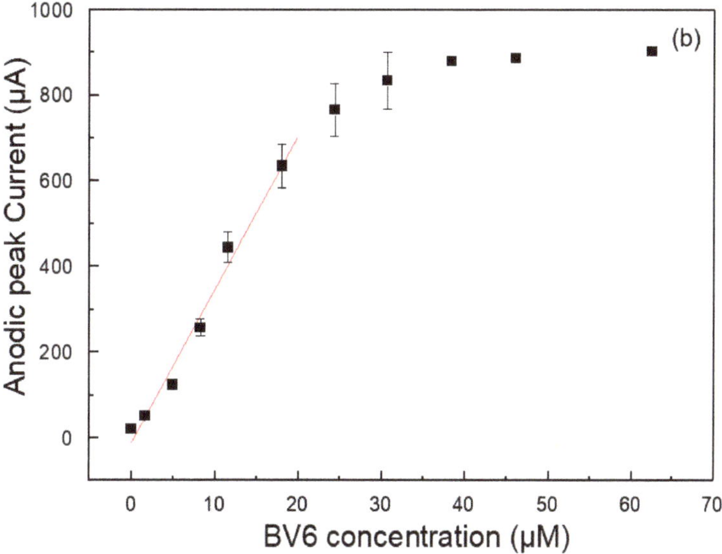

Figure 9. (**a**) CV scans at 0.1 V/s of a perovskite/G–SiO$_2$/FTO film electrode in the absence and presence of increasing concentrations of BV6 in 0.1 M TBAH in DCM. Red arrow shows the progressive increase in the anodic peak. (**b**) Plot of background corrected peak current at −0.36 V versus BV6 concentration and in particular the linear segment of the calibration plot. Error bars are standard deviations of the mean with n = 3.

In a comparable study [2] wherein a different type of perovskite was adsorbed on the surface of the same film electrode material, it was observed that the shape of the CV of the composite film was different on the return scan in contrast to the one presented above. This is likely due to the different chemical composition of the adsorbed perovskite and the possible partial oxidation of the previously reported immobilized perovskite.

The absence of a diffusion-based reversible wave in both studies suggests that these particular 3D-like perovskites are incapable of shifting by diffusion through the G–SiO$_2$/FTO film electrode structure, if perovskites are to undergo a redox reaction. This suggests that no electron hopping occurs among neighboring perovskite nanoparticles immobilized on the surface of the film electrode.

2.5. Voltammetric BV6 Sensor

The quantitative voltammetric detection of BV6 on the perovskite/G–SiO$_2$/FTO film electrode was conducted via CV and the results obtained are shown in Figure 9a. Upon consecutive additions of increasing amounts of BV6, two large anodic peaks at −0.36 V and at +0.21 V can be observed. A peak at −0.36 V starts to appear and progressively increases and shifts to less negative biases, to −0.19 V. Similarly, the small anodic peak of the perovskite-modified film electrode at +0.21 V, see Figure 8c, starts to increase as well and shifts to less positive biases, to +0.15 V. Both observations are due to the high specific area and increased conductivity of the film with the adsorbed perovskite.

The current response after each BV6 addition is relatively fast, reaching the steady-state value within a minute. Both anodic peaks are probably due to the oxidation of BV6 on the surface of the modified film electrode and are irreversible. It seems that these anodic peaks depend on the immobilized perovskite and the amount of BV6 added. It is also essential to note that during the CV measurements the perovskite remains fixed and stable on the surface of the film electrode that is immersed in the organic solvent. Perovskite desorption does not occur, as this would have been evident from the coloration of the electrolyte solution that does not take place.

Like in a recent study related to the oxidation of losartan [2] on the surface of a perovskite/G–SiO$_2$/ITO film, it is suggested that also in this case the immobilized perovskite serves as an oxidation center for BV6. It is possible that this was facilitated by the transformation of the perovskite into another type from its pristine form. It is probable that BV6 redox reactions could be catalyzed by the perovskite itself. Also, supplementary experiments have been performed to elucidate possible electrochemical reaction steps, such as high-performance liquid chromatographic differentiation of the used solutions under acetonitrile/water, in order to detect substantial degradation species of BV6 or any other perovskite/BV6 species, which were not found. Furthermore, based on the facts that electrodes without perovskite did not show any BV6 detection, neither did perovskite-containing electrodes under blank solutions, it is therefore suggested that the BV6 was likely being detected through a more complex procedure in the organic solvent. As the negative electrode attracts BV6 molecules, containing Br counter ions, these are mixed with the perovskite's bromine. The net result is that upon the voltage-forced oxidation, the complex BV6/perovskite species undergo a redox reaction, providing the detected peaks. The proposed catalytic function of BV6 is definitely related to the increase, as well as to the shift of both anodic peaks and particularly to the one at +0.21 V that shifts to +0.12 V upon the addition of increasing BV6 concentrations. Before the addition of BV6, the perovskite-modified G–SiO$_2$/FTO film electrode exhibits a small ill-defined anodic peak at +0.21 V for at least five consecutive CV scans, suggesting that maybe part of the perovskite is consumed or oxidized on the surface of the film electrode, as was discussed by us and others in the literature [2,27].

Figure 9b displays a background corrected peak current at −0.36 V versus the BV6 concentration plot, the gradual increase in the electrocatalytic anodic current maximum of the first anodic peak at −0.36 V upon increasing additions of BV6 in the electrolyte DCM + TBAH in the range of 0–65 × 10^{-6} M fitted the apparent linear with a computed correlation coefficient of 0.9767; the CV experimental linear range was set to be from 1 to 18 × 10^{-6} M and is displayed clearly in Figure 9b, after background current subtraction. The detection limit LOD was estimated as 3.5 × 10^{-6} M. Using the respective areas under the curves from −0.52 V up to 0.04 V, instead of the peak values, provides a respective LOD of 1.5 × 10^{-6} M, where the endpoint of each curve has been subtracted as baseline.

Figure 10 shows the CVs of a bare G–SiO$_2$/FTO film electrode before and after the progressive addition of increasing concentrations of BV6. A slight progressive increase in the anodic current occurs at +0.3 V due to the addition of 25–75 × 10^{-6} M (there is no increase for lower concentrations) of BV6 in DCM + TBAH electrolyte solution. The quite weak anodic peak is probably also due to the oxidation of BV6 on the surface of the bare film electrode which could also slightly contribute to this. This suggests that the supporting material (the conducting glass and the film of graphite paste and SiO$_2$) is also slightly catalytic toward the oxidation of BV6. This is quite advantageous as there is a synergy between the bare film electrode and the specificity of the perovskite catalyst, producing a hybrid catalyst which is more active than the individual materials.

The reproducibility of the perovskite-modified film electrodes was also tested. Five separate films were prepared under the same conditions and were tested for the voltammetric determination of BV6 and gave very similar results with a relative standard deviation of 4%.

Finally, the stability of the perovskite-modified film electrodes was tested. After their use as a sensor for the detection of BV6, they were stored at room temperature for up to four months without losing their voltammetric response towards BV6 detection. They maintained their OA absorption spectra, while the PL showed some defect states, which were reverted after a long storage period. In Figure S1, the CVs of a four-month-old perovskite-modified film electrode after the addition of increasing BV6 concentrations are exhibited. Both anodic peaks due to the addition of BV6 could be observed. However, it is not recommended to reuse these composite films as their prolonged use could trigger issues like desorption of the perovskite from the surface of the film, due to its Pb content. Also, there is the possibility that there are minute traces of BV6 adsorbed at the composite electrode, rendering it less sensitive after multiple usage. Through observations, the active perovskite compound allows BV6 electrochemical measurements even after a period of four months, as evidenced by the OA and PL patterns, while even the four-month-old perovskite can still perform BV6 detection measurements as evidenced in the Supplementary Information Figure S1. This is the first study of the use of a modified electrode for the electrochemical determination of BV6.

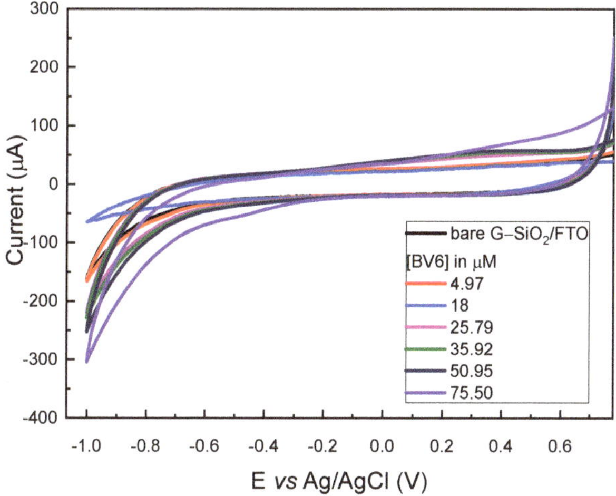

Figure 10. CVs at 0.1 V/s of a bare GSiO$_2$/FTO film electrode in DCM 0.1 M TBAH in the absence and presence of increasing concentrations of BV6.

3. Materials and Methods

3.1. Chemicals

Lead bromide (PbBr$_2$, 99.999% trace metals basis), methyl amine (CH$_3$NH$_2$ abbr. MA, 40% solution in water), N,N'-dimethylformamide (DMF, 99.8%), dimethyl sulfoxide (abbr. DMSO, anhydrous, >99.9%), hydrobromic acid (HBr, ACS reagent, 48%), and tetra-n-butylammonium hexafluorophosphate (TBAH, 98%) were purchased from Acros Organics (Geel, Belgium); dichloromethane (DCM), and sodium silicate (Na$_2$SiO$_3$) were supplied by Fisher Chemicals (Hampton, NH, USA); FTO TEC15 glass slides (2.2 mm thick and 15 Ohm/sqr sheet resistance) were purchased from XOP Glass, Castellón, Spain. Graphite powder (synthetic, APS 7–11 micron, 99%) was obtained from Alfa Aesar (Kandel, Germany). BV6 (4-butyl-N,N -bis{[2-(2H-tetrazol-5-yl)biphenyl-4-yl]methyl}imidazolium bromide) was supplied by NewDrug P.C., Patras Science Park (Patras, Greece), which also provided any needed nuclear magnetic resonance data or high-performance liquid chromatographic results.

3.2. Preparation of Graphite/SiO$_2$ Film Electrodes

The G–SiO$_2$ film electrodes were prepared as reported previously [2]. In brief, 300 mg of silicate liquid polymer (50% Na$_2$SiO$_3$, pH 12–13) was mixed with 20% of its weight in graphite powder and placed under stirring for homogenization at 40 °C, acquiring a paste-like texture. The paste was then sonicated for 2 min. After that, 100 µL of the silicate/graphite paste was applied on the surface of a clean conducting FTO glass slide by the "doctor blade" technique (coating process). Before the application of the paste, the FTO glass slide was masked with 3M Magic Scotch tape (type 810, thickness 62.5 µm) for the control of the thickness and width of the area spread. The resulting G–SiO$_2$ films were then allowed to dry at room temperature before being heated at 330 °C for 100 min. The final G–SiO$_2$/FTO films were divided in 2.5 × 1 cm^2 samples.

3.3. Perovskite Synthesis and Its Immobilization on the G–SiO$_2$/FTO Electrode

Synthesis of the 3D CH$_3$NH$_3$PbBr$_3$ was conducted by dissolving 367 mg of PbBr$_2$ in 3 mL DMF with the continuous addition of 0.5 mL HBr, while stirring at 80 °C until an optically clear solution was obtained. In a different vial, 31 mg of methylamine was mixed in a solution containing 1 mL HBr and 2 mL DMF and left under stirring, without heating, for 1 min. The two solutions were then mixed, yielding a bright orange precipitate. Afterwards, the mixture was filtered to obtain the orange crystals. The rest of the solution was placed in a Petri dish at 80 °C until more crystals were formed and the solvents were fully evaporated. For the preparation of the film, 60 mg of 3D Br perovskite was dissolved in 200 µL DMF, yielding a transparent solution. Afterwards, 40 µL of the perovskite dissolved solution was drop-casted on the surface of the G–SiO$_2$/FTO electrode (size: 2.5 × 1 cm^2). Finally, these film electrodes were left to dry at 60 °C for 24 h.

3.4. Preparation of the BV6 Solution

BV6 (4-butyl-N,N-bis{[2-(2H-tetrazol-5-yl)biphenyl-4-yl]methyl}imidazolium bromide) is a synthetic designed molecule synthesized by NewDrug P.C. [28].

3.5. Characterization of Perovskite/G–SiO$_2$/FTO Hybrid Electrode Films

The G–SiO$_2$ films on FTO glass substrates, with or without perovskite attached on their surface, were characterized by power X-ray diffraction (XRD), within the 2θ value range of 2–80°, using a Bruker D8 Advanced X-ray diffractometer (Bruker AXS GmbH, Karlsruhe, Germany) equipped with a Lynx Eye detector and Ni filtered Cu Kα radiation. The surface morphology and thickness of the perovskite/G–SiO$_2$ films on FTO glass, as well as their chemical composition, were analyzed by a ZEISS EVO MA 10 scanning electron microscope equipped with an energy dispersive spectroscopy (EDS) analyzer (Oxford Instrument, Abingdon, UK) with 127 eV resolution.

3.6. Optical Measurements

All measurements were performed under air at room temperature. Specifically, the UV–Vis optical absorption spectra were recorded on a UV-1800 Shimadzu spectrophotometer (Shimadzu Europa, Duisburg, Germany) in the range of 200–800 nm, at a sampling step of 0.5 nm, using 1.5 nm slits. As a light source, a combination of halogen and deuterium (D$_2$) lamps were used. The samples were measured as thin or thick spin-coated films on quartz substrates. The same samples were mounted in a Hitachi F-2500 fluorescence spectrophotometer (Hitachi Ltd., Tokyo, Japan) employing a xenon 150 W lamp and an R928 photomultiplier, and their photoluminescence spectra were recorded.

3.7. Electrochemical Measurements

The electrochemical behavior of the film electrodes was investigated by cyclic voltammetry using an Autolab PGSTAT101 potentiostat (Metrohm Autolab, Utrecht, The Netherlands) and a 5 mL conventional three-electrode electrochemical cell, with a platinum wire counter electrode, a Ag/AgCl/KCl$_{sat}$ reference electrode, and a perovskite/G–SiO$_2$ film on FTO glass as the working electrode. The electrolyte solution used was 5 mL DCM with 0.1 M of TBAH. The same experimental setup and conditions were used in the determination of BV6 in a solution. An aqueous BV6 stock solution (dissolving 15 mg of BV6 in 1 mL distilled water) was spiked into the cell using a micro-syringe. Cyclic voltammograms (CVs) were recorded 1 min after each BV6 addition. All electrochemical measurements were carried out at room temperature.

4. Conclusions

In the present work, an HOIS was used successfully to modify the surface of a G–SiO$_2$ film electrode on a conducting glass FTO substrate to fabricate a promising, sensitive voltammetric sensor for bisartan BV6, a possible agent for treating hypertension and COVID-19. The perovskite-modified film electrode exhibits an enhanced electrocatalytic activity towards the oxidation of BV6 with a relatively low LOD (1.5 × 10^{-6} M), high reproducibility, and stability towards BV6 determination. This approach could easily be extendable to many other drug analytes by selecting the proper perovskite material as the modifier, and this material could be used to develop low-cost modified electrodes for disposable sensing.

Supplementary Materials: The following supporting information can be downloaded at: https://www.mdpi.com/article/10.3390/inorganics11120485/s1, Figure S1: CV scans at 0.1 V/s of a four-month-old perovskite-modified G–SiO$_2$/FTO hybrid film electrode in the absence and presence of increasing concentrations of BV6 in DCM and 0.1 M TBAH electrolyte.

Author Contributions: Conceptualization, E.T., I.K. and J.M.; methodology, E.T. and I.K.; investigation, A.S., G.P., A.I., A.M., E.T. and I.K.; formal analysis, E.T., I.K., G.P., A.S. and A.I.; resources, E.T., I.K. and J.M.; writing—original draft preparation, G.P., E.T., A.I., K.K., J.M. and I.K.; supervision I.K., E.T. and J.M. All authors have read and agreed to the published version of the manuscript.

Funding: This research received no external funding.

Data Availability Statement: Data are contained within the article and the Supplementary Materials.

Acknowledgments: John Matsoukas would like to thank the Patras Science Park and the Region of Western Greece (Research and Technology) for their invaluable support throughout all the necessary steps in the implementation of the current research.

Conflicts of Interest: John Matsoukas, Konstantinos Kelaidonis and Aggeliki Mandrapylia were employed by the NewDrug company. The NewDrug company was not involved in the study design, collection, analysis, interpretation of data, the writing of this article or the decision to submit it for publication. The remaining authors of the paper declare that the research was conducted in the absence of any commercial or financial relationships that could be construed as a potential conflict of interest.

References

1. Ferreira, L.M.; Martins, P.R.; Araki, K.; Angnes, L. Tuning Selectivity and Sensitivity of Mixed-polymeric Tetraruthenated Metalloporphyrins Modified Electrodes as Voltammetric Sensors of Chloramphenicol. *Electroanalysis* **2019**, *31*, 688–694. [CrossRef]
2. Nikolaou, P.; Vareli, I.; Deskoulidis, E.; Matsoukas, J.; Vassilakopoulou, A.; Koutselas, I.; Topoglidis, E. Graphite/SiO$_2$ film electrode modified with hybrid organic-inorganic perovskites: Synthesis, optical, electrochemical properties and application in electrochemical sensing of losartan. *J. Solid State Chem.* **2019**, *273*, 17–24. [CrossRef]
3. Deskoulidis, E.; Petrouli, S.; Apostolopoulos, V.; Matsoukas, J.; Topoglidis, E. The use of electrochemical voltammetric techniques and high-pressure liquid chromatography to evaluate conjugation efficiency of multiple sclerosis peptide-carrier conjugates. *Brain Sci.* **2020**, *10*, 577. [CrossRef] [PubMed]
4. Guo, W.; Umar, A.; Alsaiari, M.A.; Wang, L.; Pei, M. Ultrasensitive and selective label-free aptasensor for the detection of penicillin based on nanoporous PtTi/graphene oxide-Fe$_3$O$_4$/MWCNT-Fe$_3$O$_4$ nanocomposite. *Microchem. J.* **2020**, *158*, 105270. [CrossRef]
5. Reyes, C.; Pistillo, A.; Fernández-Bertolín, S.; Recalde, M.; Roel, E.; Puente, D.; Sena, A.G.; Blacketer, C.; Lai, L.; Alshammari, T.M.; et al. Characteristics and outcomes of patients with COVID-19 with and without prevalent hypertension: A multinational cohort study. *BMJ Open* **2021**, *11*, e057632. [CrossRef]
6. Zhang, P.; Zhu, L.; Cai, J.; Lei, F.; Qin, J.J.; Xie, J.; Liu, Y.M.; Zhao, Y.C.; Huang, X.; Lin, L.; et al. Association of inpatient use of angiotensin-converting enzyme inhibitors and angiotensin II receptor blockers with mortality among patients with hypertension hospitalized with COVID-19. *Circ. Res.* **2020**, *126*, 1671–1681. [CrossRef] [PubMed]
7. Ridgway, H.; Moore, G.J.; Mavromoustakos, T.; Tsiodras, S.; Ligielli, I.; Kelaidonis, K.; Chasapis, C.T.; Gadanec, L.K.; Zulli, A.; Apostolopoulos, V.; et al. Discovery of a new generation of angiotensin receptor blocking drugs: Receptor mechanisms and in silico binding to enzymes relevant to SARS-CoV-2. *Comput. Struct. Biotechnol. J.* **2022**, *20*, 2091–2111. [CrossRef]
8. Ridgway, H.; Chasapis, C.T.; Kelaidonis, K.; Ligielli, I.; Moore, G.J.; Gadanec, L.K.; Zulli, A.; Apostolopoulos, V.; Mavromoustakos, T.; Matsoukas, J.M. Understanding the Driving Forces that Trigger Mutations in SARS-CoV-2: Mutational Energetics and the Role of Arginine Blockers in COVID-19 Therapy. *Viruses* **2022**, *14*, 1029. [CrossRef]

9. Moore, G.J.; Pires, J.M.; Kelaidonis, K.; Gadanec, L.K.; Zulli, A.; Apostolopoulos, V.; Matsoukas, J.M. Receptor interactions of angiotensin II and angiotensin receptor blockers—Relevance to COVID-19. *Biomolecules* **2021**, *11*, 979. [CrossRef]

10. Gao, C.; Cai, Y.; Zhang, K.; Zhou, L.; Zhang, Y.; Zhang, X.; Li, Q.; Li, W.; Yang, S.; Zhao, X.; et al. Association of hypertension and antihypertensive treatment with COVID-19 mortality: A retrospective observational study. *Eur. Heart J.* **2020**, *41*, 2058–2066. [CrossRef]

11. Almutlaq, M.; Mansour, F.A.; Alghamdi, J.; Alhendi, Y.; Alamro, A.A.; Alghamdi, A.A.; Alamri, H.S.; Alroqi, F.; Barhoumi, T. Angiotensin II Exaggerates SARS-CoV-2 Specific T-Cell Response in Convalescent Individuals following COVID-19. *Int. J. Mol. Sci.* **2022**, *23*, 8669. [CrossRef] [PubMed]

12. Anyfantisa, G.C.; Papavassiliou, G.C.; Aloukos, P.; Couris, S.; Weng, Y.F.; Yoshino, H.; Murata, K. Unsymmetrical single-component nickel 1, 2-dithiolene complexes with extended tetrachalcogenafulvalenedithiolato ligands. *Z. Naturforsch B* **2007**, *62*, 200–204. [CrossRef]

13. Cohen, B.E.; Wierzbowska, M.; Etgar, L. High efficiency quasi 2D lead bromide perovskite solar cells using various barrier molecules. *Sustain. Energy Fuels* **2017**, *1*, 1935–1943. [CrossRef]

14. Nikolaou, P.; Vassilakopoulou, A.; Papadatos, D.; Topoglidis, E.; Koutselas, I. A chemical sensor for CBr_4 based on quasi-2D and 3D hybrid organic–inorganic perovskites immobilized on TiO_2 films. *Mater. Chem. Front.* **2018**, *2*, 730–740. [CrossRef]

15. Kojima, A.; Teshima, K.; Shirai, Y.; Miyasaka, T. Organometal halide perovskites as visible-light sensitizers for photovoltaic cells. *J. Am. Chem. Soc.* **2009**, *131*, 6050–6051. [CrossRef] [PubMed]

16. Lin, K.; Xing, J.; Quan, L.N.; de Arquer, F.P.; Gong, X.; Lu, J.; Xie, L.; Zhao, W.; Zhang, D.; Yan, C.; et al. Perovskite light-emitting diodes with external quantum efficiency exceeding 20 per cent. *Nature* **2018**, *562*, 245–248. [CrossRef] [PubMed]

17. Cho, H.; Jeong, S.H.; Park, M.H.; Kim, Y.H.; Wolf, C.; Lee, C.L.; Heo, J.H.; Sadhanala, A.; Myoung, N.; Yoo, S.; et al. Overcoming the electroluminescence efficiency limitations of perovskite light-emitting diodes. *Science* **2015**, *350*, 1222–1225. [CrossRef] [PubMed]

18. Shan, Q.; Song, J.; Zou, Y.; Li, J.; Xu, L.; Xue, J.; Dong, Y.; Han, B.; Chen, J.; Zeng, H. High performance metal halide perovskite light-emitting diode: From material design to device optimization. *Small* **2017**, *13*, 1701770. [CrossRef]

19. Fakharuddin, A.; Gangishetty, M.K.; Abdi-Jalebi, M.; Chin, S.H.; bin Mohd Yusoff, A.R.; Congreve, D.N.; Tress, W.; Deschler, F.; Vasilopoulou, M.; Bolink, H.J. Perovskite light-emitting diodes. *Nat. Electron.* **2022**, *5*, 203–216. [CrossRef]

20. Koutselas, I.B.; Ducasse, L.; Papavassiliou, G.C. Electronic properties of three-and low-dimensional semiconducting materials with Pb halide and Sn halide units. *J. Condens. Matter Phys.* **1996**, *8*, 1217. [CrossRef]

21. Ali, S.M. Smart perovskite sensors: The electrocatalytic activity of $SrPdO_3$ for hydrazine oxidation. *J. Electrochem. Soc.* **2018**, *165*, B345. [CrossRef]

22. Nikolaou, P.; Deskoulidis, E.; Topoglidis, E.; Kakoulidou, A.T.; Tsopelas, F. Application of chemometrics for detection and modeling of adulteration of fresh cow milk with reconstituted skim milk powder using voltammetric fingerpriting on a graphite/SiO_2 hybrid electrode. *Talanta* **2020**, *206*, 120223. [CrossRef] [PubMed]

23. Reyes-Gomez, J.; Medina, J.A.; Jeerage, K.M.; Steen, W.A.; Schwartz, D.T. High capacity SiO_2-graphite composite electrodes with chemically incorporated metal MHCFs for electrochemically switched alkaline cation exchange. *J. Electrochem. Soc.* **2004**, *151*, D87. [CrossRef]

24. Liu, A.; Wei, M.D.; Honma, I.; Zhou, H. Biosensing properties of titanatenanotube films: Selective detection of dopamine in the presence of ascorbate and uric acid. *Adv. Funct. Mater.* **2006**, *16*, 371–376. [CrossRef]

25. Zhao, W.; Zhang, J.; Kong, F.; Ye, T. Application of Perovskite Nanocrystals as Fluorescent Probes in the Detection of Agriculture- and Food-Related Hazardous Substances. *Polymers* **2023**, *15*, 2873. [CrossRef] [PubMed]

26. Topoglidis, E.; Campbell, C.J.; Cass, A.E.; Durrant, J.R. Factors that affect protein adsorption on nanostructured titania films. A novel spectroelectrochemical application to sensing. *Langmuir* **2001**, *17*, 7899–7906. [CrossRef]

27. Agelis, G.; Resvani, A.; Koukoulitsa, C.; Tůmová, T.; Slaninová, J.; Kalavrizioti, D.; Spyridaki, K.; Afantitis, A.; Melagraki, G.; Siafaka, A.; et al. Rational design, efficient syntheses and biological evaluation of N, N′-symmetrically bis-substituted butylimidazole analogs as a new class of potent Angiotensin II receptor blockers. *Eur. J. Med. Chem.* **2013**, *62*, 352–370. [CrossRef]

28. Samu, G.F.; Scheidt, R.A.; Kamat, P.V.; Janáky, C. Electrochemistry and spectroelectrochemistry of lead halide perovskite films: Materials science aspects and boundary conditions. *Chem. Mater.* **2018**, *30*, 561–569. [CrossRef]

Article

Enhanced Thermal Stability of Sputtered TiN Thin Films for Their Applications as Diffusion Barriers against Copper Interconnect

Abdullah Aljaafari [1,*], Faheem Ahmed [1], Nagih M. Shaalan [1,2], Shalendra Kumar [1,3] and Abdullah Alsulami [4]

[1] Department of Physics, College of Science, King Faisal University, P.O. Box 400, Al-Ahsa 31982, Saudi Arabia
[2] Physics Department, Faculty of Science, Assiut University, Assiut 71516, Egypt
[3] Department of Physics, School of Engineering, University of Petroleum & Energy Studies, Dehradun 248007, India
[4] Physics Department, College of Sciences and Art at ArRass, Qassim University, ArRass 51921, Saudi Arabia
* Correspondence: aaljaafari@kfu.edu.sa

Abstract: In this work, the deposition of titanium nitride (TiN) thin film using direct current (DC) sputtering technique and its application as diffusion barriers against copper interconnect was presented. The deposited film was analyzed by using X-ray diffraction (XRD), field-emission scanning electron microscopy (FESEM), and X-ray photoelectron spectroscopy (XPS) techniques. XRD patterns showed the face-centered cubic (FCC) structure for the $TiN/SiO_2/Si$ film, having (111) and (200) peaks and TiN (111), Cu(111), and Cu(200) peaks for $Cu/TiN/SiO_2/Si$ film. FESEM images revealed that the grains were homogeneously dispersed on the surface of the TiN film, having a finite size. XPS study showed that Ti2p doublet with peaks centered at 455.1 eV and 461.0 eV for TiN film was observed. Furthermore, the stoichiometry of the deposited TiN film was found to be 0.98. The sheet resistance of the TiN film was analyzed by using a four-point probe method, and the resistivity was calculated to be 11 $\mu\Omega$ cm. For the utilization, TiN film were tested for diffusion barrier performance against Cu interconnect. The results exhibited that TiN film has excellent performance in diffusion barrier for copper metallization up to a temperature of 700 °C. However, at a higher annealing temperature of 800 °C, the formation of Cu_3Si and $TiSi_2$ compounds were evident. Thus, stoichiometric TiN film with high thermal stability and low resistivity produced in this study could be applied for the fabrication of microelectronic devices.

Keywords: thin films; diffusion barriers; TiN; Sputtering; XRD

Citation: Aljaafari, A.; Ahmed, F.; Shaalan, N.M.; Kumar, S.; Alsulami, A. Enhanced Thermal Stability of Sputtered TiN Thin Films for Their Applications as Diffusion Barriers against Copper Interconnect. *Inorganics* **2023**, *11*, 204. https://doi.org/10.3390/inorganics11050204

Academic Editor: Christian Julien

Received: 29 March 2023
Revised: 20 April 2023
Accepted: 4 May 2023
Published: 9 May 2023

1. Introduction

While metallization is a strong technique for manufacturing integrated circuits built on silicon substrates, interconnect materials, such as copper, are favorable for ultra-large scale integration (ULSI) circuits, owing to higher resistance to electro-migration than aluminum and its alloys and low electrical resistivity [1]. However, the electrical performance of ULSIs is severely damaged or degraded by the copper and silicon contact because copper diffuses into the silicon or SiO_2, even at low temperatures [2]. Therefore, a barrier between the copper and the silicon is of high importance to separate the two layers. Besides, the required barriers should not only have the property of resisting copper penetration, but they should also be highly adhesive [3]. So far, refractory metal nitrides are being used as promising diffusion inhibitors, owing to their high thermal stability, conductivity, and melting point.

Different refractory metal nitrides, such as Ti-N [4–6], W-N [7], Zr-N [8], Ta-N [9–11], and TiZr-N [12–15] have been used for their applications as diffusion barriers in copper metallization. Nanocrystalline TiN, however, has gained substantial attention due to its hard metallurgical coatings [16]. Additionally, it has the property of showing different colors on different types of surfaces. Furthermore, the deposition of TiN on any type of surface is relatively easy, as compared to other nitrides, providing a chemically stable coating. Stoichiometric crystalline TiN films with a NaCl-type face-centered cubic structure are found to be good conductors of electricity, as they show golden color and are found to be ideal candidates as diffusion barriers [17]. Furthermore, the stoichiometric TiN films resulted in low resistivity. Therefore, TiN films deposited with different deposition processes and at different deposition conditions possess diverse microstructural properties [1,2]. Therefore, in order to achieve better performance of TiN thin films, deposition process and deposition parameters need to be optimized. Much research has been performed so far to understand the effect of deposition conditions on the properties and structure of TiN films [2,3]. Additionally, several deposition methods have been employed so far for the deposition of nanocrystalline TiN films [6–10]. Direct-current (DC) sputtering, however, is considered to be most promising deposition technique for obtaining stoichiometric TiN films among the various physical vapor deposition (PVD) techniques. The sputter method has the advantage of low gaseous contamination, compositional homogeneity, and good adhesion between the substrate and deposited films. Additionally, films of different morphological and crystal structures can be produced [11]. A study [11] shows that thinner diffusion barriers employed in metallization, showing stability at high temperatures, are the future research direction. The effect of thickness and resistivity of diffusion barriers on copper interconnect resistivity is given by the relation [18].

$$\rho(Cu) = \frac{1.7 \; \mu\Omega \; cm}{\left[\left(1 - \frac{B}{M2}\right)\left(1 - 2\frac{B}{M1}\right) + \left(\frac{1.7}{\rho b}\right)\left(\frac{B}{M1}\right)\left(2 + \frac{M1}{M2} - 2\frac{B}{M2}\right)\right]} \tag{1}$$

where, height and width of metal interconnects are given by $M1$ and $M2$, and B and ρb are the thickness and resistivity of diffusion barriers, respectively. Apart from thickness effect, copper interconnect resistivity is also affected by the resistivity of diffusion barriers based on Equation (1). Therefore, low resistive diffusion barrier yields a low resistive copper interconnect. Therefore, in order to have an effective application of TiN for diffusion barrier in microelectronic devices, we need to focus on the thermal stability of the low resistive TiN films. Thus, TiN films with low resistivity, good stoichiometry (golden color film), and high thermal stability are highly desired.

The aim of this work was to produce a high temperature stable stoichiometric TiN film with low resistivity using DC sputtering technique. The diffusion barrier performance of the TiN film for copper interconnect has been investigated at high temperature of ~700 °C for its successful utilization in microelectronic devices. This study will provide a clear direction towards the fabrication of microelectronic devices with high thermal stability and high performance.

2. Results and Discussion

The XRD pattern of as-deposited TiN film is shown in Figure 1a, where single phase and face centered cubic (FCC) symmetry, with (111) and (200) crystal planes, having a (111) plane, as preferred its orientation, can be observed. The calculated d-spacing values are in good agreement with the standard JCPDS (65-5759) for TiN structure. Scherrer's formula [2] was used to calculate the grain size (D) of the film, and the grain size was found to be ~42 nm.

Figure 1b shows the XRD patterns acquired from the $Cu/TiN/SiO_2/Si$ film. In Figure 1b, the diffraction peaks of as-deposited film of $Cu/TiN/SiO_2/Si$ and annealed at 700 °C and 800 °C can be clearly seen. It could be observed, from Figure 1b, that, for as-deposited $Cu/TiN/SiO_2/Si$ film, the peaks correspond to TiN (111), Cu (111), and Cu (200) are detected. On annealing at a temperature of 700 °C, the intensity of Cu (111) and Cu (200) peak increases. However, on further annealing the film at 800 °C, there is a sharp decrease in intensity of Cu (111) peak, and subsequently diffraction peaks of $TiSi_2$

and Cu_3Si peak appear, which indicate intermixing of Cu and Si through the TiN film. Therefore, it can be concluded that, above 700 °C, the barrier fails. Additionally, the Cu peak in the XRD pattern (Figure 1b) disappears, which is an indication of Cu diffusion through the barrier film. The Cu becomes diffused into the Si to form Cu_3Si.

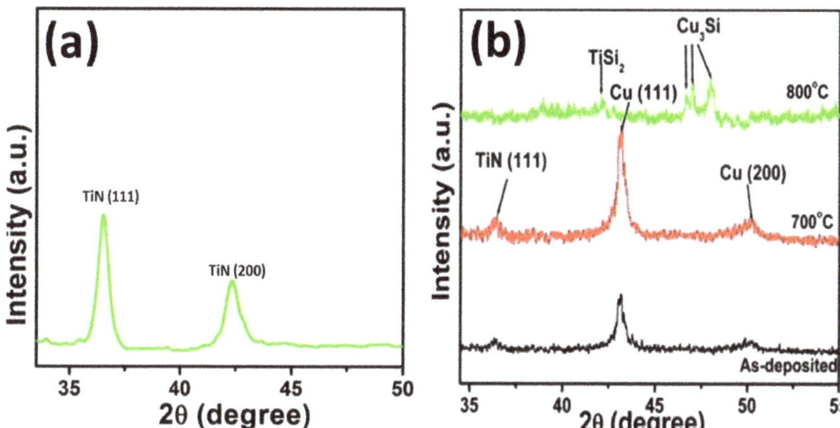

Figure 1. XRD patterns of (**a**) TiN film deposited on Si/SiO_2 substrates at 3 sccm nitrogen flow, (**b**) Cu/TiN films deposited on Si/SiO_2 substrates for as-deposited, annealed at 700 °C, and 800 °C.

XPS survey spectra for TiN film (not shown here) indicate that the film is not only composed of Ti and N, but also a small amount of Oxygen (O) and Carbon (C). However, the presence of O and C in the measurement could be attributed to the existence of ambient atmosphere in the analysis chamber during the XPS analysis [19]. XPS core level spectra in the Ti2p and N1s regions for the sputtered TiN film are shown in Figure 2. The presence of the binding energy doublet peaks for TiN at 455.1 eV and 461.0 eV [20] can be observed from Figure 2a. Deconvoluted Ti2p and N1s spectra for deposited TiN film are shown in Figure 3.

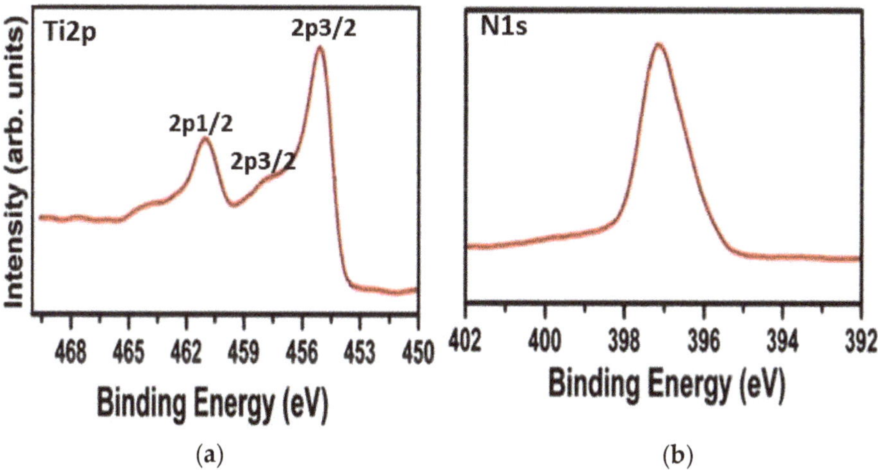

Figure 2. XPS spectra of (**a**) Ti2p, and (**b**) N1s for the deposited TiN film at 3sccm nitrogen gas flow.

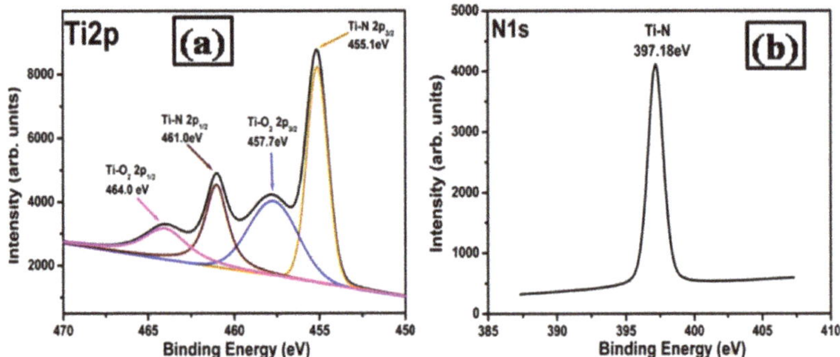

Figure 3. Deconvoluted XPS spectra of (**a**) Ti2p and (**b**) N1s for TiN film deposited at 3 sccm nitrogen tflow.

It can be seen, in Figure 3a, that the Ti2p$_{3/2}$ peaks can be fitted with two mixed Gaussian–Lorentzian curves, with binding energy of the first peak centered at 455.1 eV for TiN [20] and the second peak corresponds to TiO$_2$ and is observed at 457.8 eV [20]. However, it was noted that Ti2p$_{1/2}$ has only one peak at 461.0 eV for TiN [20]. The N1s fitting shows only one peak, as shown in Figure 3b, with a peak centered at 397.1 eV [20]. The N1s photoelectron peak position of TiN film is found to vary from 396.3 eV to 397.7 eV, as mentioned in the earlier reports [21]. However, reportedly, the N1s spectrum of pure nitride has full width at half maxima (FWHM) of about 1.6–1.9 eV and is of simple Gaussian shape [22]. The presence of oxygen in the film may be due to the use of commercial nitrogen gas, which can have some amount of oxygen impurities [23]. The stoichiometry of the TiN film was found to be ~0.98, as obtained from XPS analysis. It has been reported earlier that [17] the nitrogen vacancy is the most significant defect in sub-stoichiometric compositions, and for over-stoichiometric case, excess nitrogen acted as an interstitial defect. In the present work, only nearly stoichiometric nitride film was grown and investigated to avoid any interference from the composition variations of nitride film.

Figure 4 shows the FESEM image, EDX spectrum, and cross-sectional FESEM image for the as-deposited TiN film. It can be seen from the FESEM image (Figure 4a) that the grains of finite size are uniformly distributed on the surface of the film. The grain size is found to be ~40 nm, as shown in Figure 4a. EDX analysis gives the percentage composition so as to obtain the stoichiometric of TiN film. The stoichiometry of the TiN film was found to be 0.98. The stoichiometry obtained by EDX (Figure 4b) is found to be comparable with the XPS result.

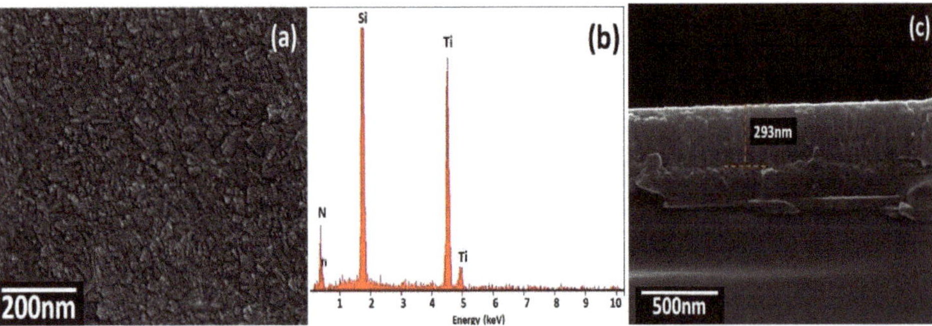

Figure 4. (**a**) FESEM and the corresponding (**b**) EDX and (**c**) cross-sectional FESEM images of the TiN film at 3 sccm nitrogen gas flow rate.

The cross-sectional image of the TiN film can be seen in Figure 4c. As can be seen from Figure 4c, that film is dense, with columnar structures, which is often the uniqueness of the DC sputtered films. The thickness of the film was found to be ~290 nm.

Figure 5 shows the FESEM images of the Cu surface for as-deposited and annealed $Cu/TiN/SiO_2/Si$ films. As can be seen from FESEM image in Figure 5a, the surface of as-deposited Cu film remained smooth, and there were no visible defects, which can also be confirmed from EDX spectrum (Figure 5b). However, at 800 °C (see Figure 5c), Cu almost disappeared from the surface, and large conglomerations are formed, having many defects and voids. From the EDX spectrum (see Figure 5d), it can be seen that these conglomerations mostly comprise Cu and Si, which confirms that the conglomerations are mainly Cu_3Si. The results obtained by EDX analysis are in good agreement with the results attained by XRD.

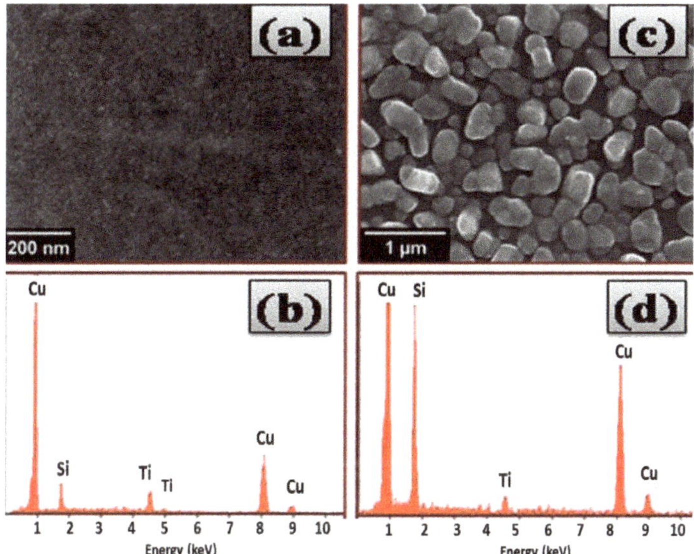

Figure 5. (a) FESEM and corresponding (b) EDX spectrum of as deposited $Cu/TiN/SiO_2/Si$ system. (c) FESEM and the corresponding (d) EDX spectrum of $Cu/TiN/SiO_2/Si$ system annealed at 800 °C.

The electrical resistivity of the obtained TiN film was found to be 11 $\mu\Omega$ cm. Variation in the resistivity of the film can be observed due to many parameters, such as thickness, structure, texture, stoichiometry (N/Ti), and impurity concentrations in various layers [24]. The resistivity of CVD-deposited TiN film has been reported to be of the order of 100 $\mu\Omega$ cm or higher [25], while a resistivity of 36 $\mu\Omega$ cm has been reported for the reactive sputtered TiN films with a substrate bias of -100 V [3]. A 220 $\mu\Omega$ cm resistivity for reactively sputtered films at 400 °C and a substrate bias of -40 V has also been reported earlier [17,26,27]. A resistivity of 11 $\mu\Omega$ cm with a nitrogen flow of 3 sccm is obtained in this study, which was found to be lower than the resistivities obtained in other studies [17,26,27]. As reported earlier, a resistivity lower than 300 $\mu\Omega$ cm for TiN film is required in order to have its diffusion barrier applications for ULSI generation, which clearly is an indication that TiN film deposited here can be an ideal candidate for its successful usage as a diffusion barrier in future applications [4].

Cu sheet resistance as a function of the annealing temperature was used to observe the ability of diffusion barrier against copper diffusion. Figure 6 presents the variation percentage of sheet resistance as a function of annealing temperature for the $Cu/TiN/SiO_2/Si$ film and is compared with the $Cu/SiO_2/Si$, as calculated using Equation (2). The resistivity of the as-deposited Cu film was found to be ~2.3 $\mu\Omega$ cm, which is slightly higher, as compared to the bulk value (bulk value: 1.7 $\mu\Omega$ cm). For both the films, the average value of sheet resistance initially decreases after annealing with respect to that of as-deposited films. This could be due to the grain growth and defect annihilation of the Cu films during annealing. The sheet resistance of Cu/SiO$_2$/Si film is found to increase drastically at a temperature above 200 °C, concluding that the film without a TiN barrier cannot withstand at high annealing temperature. However, the sheet resistance for Cu/TiN/SiO$_2$/Si barrier film starts to increase abruptly at an annealing temperature above 700 °C. Additionally, the color of Cu changes from reddish–yellow (the color of Cu) to gray. This increase in the resistivity is associated with the fact that thermally activated Cu atoms begin to diffuse into the Si substrate through the TiN barrier. Therefore, with increasing annealing temperature, Cu atoms create larger amounts of defects, i.e., the formation of Cu$_3$Si and the sheet resistance of Cu films tends to increase drastically, resulting in the increase in electrical resistance. The outcome reveals that TiN barrier film is stable up to an annealing temperature of 700 °C for 60 min. However, drastic increases in the sheet resistance were observed after annealing above 700 °C for the Cu/TiN/SiO$_2$/Si film. Thus, TiN can be used as a diffusion barrier for Cu metallization and can meet the future research direction of diffusion barriers. These results of sheet resistance are also in good agreement with the results obtained by XRD analysis. The disappearance of Cu in the XRD pattern (Figure 1b) is a clear indication of a large amount of Cu diffusing through the barrier film into Si substrate where it reacts with Si to form Cu$_3$Si on annealing at high temperature, which is also further confirmed by FESEM images. Therefore, the development of high resistivity Cu$_3$Si corresponds to the drastic increase in sheet resistance of Cu film. The failure mechanism of the Cu/TiN/SiO$_2$/Si system has been studied previously by I. Chen and J.L. Wang [28]. The suggested mechanism is that Cu breaks Si-Si bonds as it reaches the TiN/SiO$_2$/Si interface and Si point defects generated there, which resulted in the formation of Cu$_3$Si and TiSi$_2$. Therefore, it is highly anticipated that, as copper promotes the formation of silicides, Cu may act as a catalyst for the silicidation of TiN in this study. In addition, the concentration of these point defects is greatly increased during thermal annealing and supports the reaction of Si with TiN, and then the formation of Cu$_3$Si and TiSi$_2$. Therefore, the failure of a multilayered system can be determined when the high resistivity of the film is obtained.

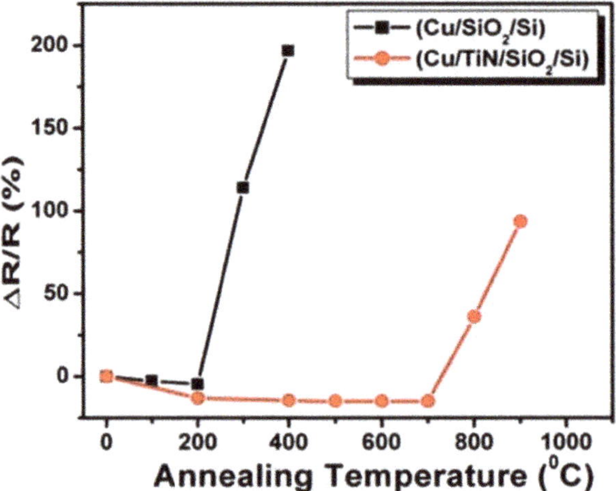

Figure 6. Plot of sheet resistance (%) vs. annealing temperature for Cu and Cu/TiN films deposited on Si/SiO$_2$ substrates.

A compilation of diffusion studies reported earlier has been shown in Figure 7. The results of this work have been compared with these previous reports. L. Slot et al. [29] performed in situ measurement of sheet resistance, backscattering, and XRD analysis to monitor the reaction between Cu and Si and to obtain the results of overall compositions. It was observed that Cu$_3$Si phase with various concentration of Si ranging from 17–20 at.% was formed at temperature of about 200 °C on crystalline Si. Shin et al. [30] reported the deposition of TiN barrier film using flow modulation chemical vapor deposition (FMCVD) technique with titanium tetrachloride and ammonia. Diffusion barrier results showed that, at 400 °C, Cu became diffused through the TiN layer, resulting in copper silicides. In order to improve the diffusion barrier property, Shin et al. introduced a monolayer of Al atoms between the two TiN films. Their results showed that, for TiN films with a Al interlayer, there is enhancement in the diffusion barrier, and the diffusion of Cu through the barrier occurred at a temperature of 500 °C, which is higher than that of their TiN film without a Al interlayer. In general, at a temperature as low as 200 °C, the formation of Cu-Si compounds occurs, which resulted in the increase in sheet resistance of Cu film in the Cu/barrier/Si samples. In another work, Chaipyang et al. and Kim et al. [31,32] showed that the sheet resistance of Cu/TiN/Si film remains stable after annealing up to a temperature of 600 °C. However, the sheet resistance was found to increase drastically after annealing above 600 °C due to the formation of Cu$_3$Si. Therefore, it can be seen clearly that the thermal stability of TiN film obtained in this work is higher (~700 °C) than the other reported works, and the TiN film produced in this work is stoichiometric and has lower resistivity.

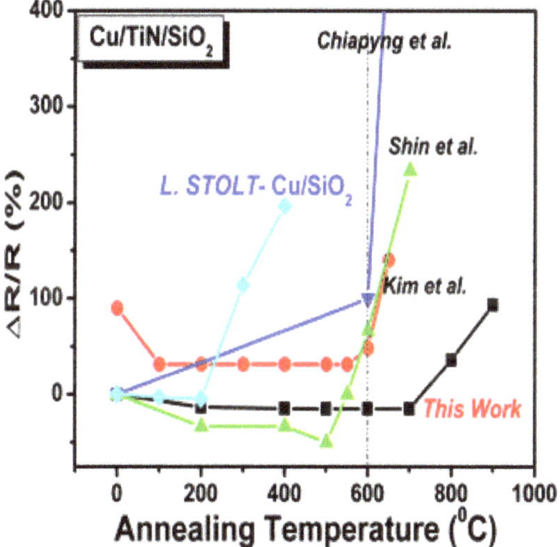

Figure 7. Plot of sheet resistance (%) vs. annealing temperature for TiN films deposited on Si/SiO$_2$ substrates and its comparison with earlier reports [29–31].

3. Experimental Details

In order to have a successful deposition of TiN film, a DC magnetron sputtering technique with a commercial titanium (Ti) target (purity 99.99%, 100 mm and 5 mm diameter and thickness, respectively) was used. Prior to the deposition, Si/SiO$_2$ (100) substrate was oxidized for 1 h in a furnace at 1000 °C, and it was further cleaned with acetone and methanol in ultrasonic baths. The substrate was clamped on the substrate holder at a distance of 12 cm from the target. The chamber was maintained at 1×10^{-6} torr vacuum condition. Once the desired vacuum was achieved, the chamber was then filled with argon and nitrogen gas, as per the required working gas flow. The target was pre-sputtered for 5 min in argon gas before introducing nitrogen gas into the chamber. The argon and nitrogen gas flow rates were fixed at 10 sccm and 3 sccm, respectively, during the deposition process. The substrate temperature of 200 °C and DC power of ~350 W were maintained throughout the deposition process.

The thermal stability of the TiN film for diffusion barrier was studied against Cu/Si metallization. DC sputtering technique described above with a commercial copper (Cu) (purity 99.99%, 100 mm and 5 mm diameter and thickness, respectively) was used for the deposition of Cu film. The Cu film was deposited on a TiN/SiO$_2$/Si substrate, and the substrate was fixed to a substrate holder at a distance of ~12 cm from the target. The chamber was maintained at 1×10^{-6} torr vacuum condition. The DC power was fixed at ~350 W, and argon gas flow rate was maintained at 10 sccm throughout the deposition process. The target was pre-sputtered for 5 min in argon gas before the deposition. The deposition of the film was performed for 8 min, and, to avoid any type of diffusion within the layers, the substrate was kept at room temperature. After deposition of Cu film of ~350 nm, the heat treatment of Cu/TiN/SiO$_2$/Si sample at 200–1000 °C for 60 min was performed in vacuum with pressure maintained at 6.7×10^5 torr during each annealing condition.

Cu sheet resistance as a function of the annealing temperature was used to observe the ability of the diffusion barrier against Cu diffusion. The variation percentage of sheet resistance ($\Delta R_s/R_s\%$) is defined in Equation (2):

$$\frac{\Delta R_s}{R_s}\% = \frac{R_{s,after-anneal} - R_{s,as-deposited}}{R_{s,as-deposited}} \times 100\% \tag{2}$$

The crystallographic structure and texture of the TiN/SiO$_2$/Si and Cu/TiN/SiO$_2$/Si films were studied within the scan range of 35 to 50° using X-ray powder diffractometer (XRD) (X'pert MPD 3040) with 40 kV and 30 mA operating parameters, using CuKα (1.541 Å) radiation. The chemical and phase composition of TiN film was studied using X-ray photoelectron spectroscopy (XPS) (ESCALAB 250 XPS) with a VG-Scientific Sigma Probe spectrometer, having a monochromated Al-Kα source with hemispherical analyzer. The sample was etched using a 3 keV argon ion beam prior to the analysis until the 1 s peak of oxygen reached a minimum stable value. The morphological study and the thickness of the deposited TiN/SiO$_2$/Si and Cu/TiN/SiO$_2$/Si films were performed using field-emission scanning electron microscopy (FESEM) (TESCAN; MIRA II LMH microscope) attached with Energy Dispersive X-ray (EDX) analysis. The resistivity study for the deposited films were performed using four-point probe (Keithley-2002) method operated at room temperature.

4. Conclusions

In summary, TiN film has been successfully deposited on SiO$_2$/Si substrate using DC sputtering technique at 3 sccm nitrogen gas flow rate, and its barrier performances and failure mechanisms were studied. The structural, morphological, and electrical properties of the as-deposited TiN/SiO$_2$/Si and Cu/TiN/SiO$_2$/Si films were analyzed. The XRD patterns revealed the presence of face centered cubic (FCC) phase for TiN/SiO$_2$/Si, having (111) preferred orientation and a grain size of ~42 nm, was obtained. The existence of different phases on the film surface was observed from the XPS analysis. Doublet peaks of Ti2p, corresponding to TiN and TiO$_2$, were observed in the spectra of TiN film with stoichiometry of 0.98 for TiN. The FESEM images show a uniform distribution of finite-sized grain on the surface of film with a columnar structure. The electrical resistivity of ~11 $\mu\Omega$ cm was achieved. In particular, high crystallinity, high stoichiometry, and lower resistivity TiN film have been obtained. The diffusion barrier property shows that TiN film can be used successfully as diffusion barrier for metallization in Cu up to a temperature of 700 °C, and, upon further increase in temperature to 800 °C, the diffusion of Cu through the barrier occurred. The failure of diffusion barrier at higher annealing temperature is due to the formation of various copper silicide phases. The present study shows that the structural stability of TiN film at high temperatures possesses key parameters that improve the barrier properties.

Author Contributions: Conceptualization, A.A. (Abdullah Aljaafari) and F.A.; methodology, F.A. and A.A. (Abdullah Aljaafari); validation, S.K. and N.M.S.; formal analysis, S.K. and A.A. (Abdullah Alsulami); investigation, F.A., S.K. and N.M.S.; resources, A.A. (Abdullah Aljaafari); data curation, A.A. (Abdullah Alsulami); N.M.S.; writing—original draft preparation, A.A. (Abdullah Aljaafari) and F.A.; writing—review and editing, S.K., N.M.S. and A.A. (Abdullah Alsulami); visualization, S.K.; supervision, A.A. (Abdullah Aljaafari).; project administration, A.A. (Abdullah Aljaafari); funding acquisition, A.A. (Abdullah Aljaafari). All authors have read and agreed to the published version of the manuscript.

Funding: This work was supported by the Deputyship for Research and Innovation, Ministry of Education in Saudi Arabia, Project number INST020. And The APC was funded by [INST020].

Institutional Review Board Statement: Not applicable.

Informed Consent Statement: Not applicable.

Data Availability Statement: Available upon reasonable request.

Acknowledgments: The authors extend their appreciation to the Deputyship for Research and Innovation, Ministry of Education in Saudi Arabia for funding this research work through the project number INST020.

Conflicts of Interest: The authors declare no conflict of interest.

References

1. Petrov, I.; Hultman, L.; Helmersson, U.; Sundgren, J.E.; Greene, J.E. Microstructure modification of TiN by ion bombardment during reactive sputter deposition. *Thin Solid Film.* **1989**, *169*, 299–314. [CrossRef]
2. Arshi, N.; Lu, J.; Joo, Y.K.; Lee, C.G.; Yoon, J.H.; Ahmed, F. Study on structural, morphological and electrical properties of sputtered titanium nitride films under different argon gas flow. *Mater. Chem. Phys.* **2012**, *134*, 839–844. [CrossRef]
3. Gagnon, G.; Currie, J.F.; Beique, G.; Brebner, J.L.; Gujrathi, S.C.; Ouellet, L. Characterization of reactively evaporated TiN layers for diffusion barrier applications. *J. Appl. Phys.* **1994**, *75*, 1565–1570. [CrossRef]
4. Chung, S.H.; Lachab, M.; Wang, T.; Lacroix, Y.; Basak, D.; Fareed, Q.; Kawakami, Y.; Nishino, K.; Sakai, S. Effect of oxygen on the activation of Mg acceptor in GaN epilayers grown by metalorganic chemical vapor deposition. *Jpn. J. Appl. Phys.* **2000**, *39*, 4749. [CrossRef]
5. Patsalas, P.; Charitidis, C.; Logothetidis, S. The effect of substrate temperature and biasing on the mechanical properties and structure of sputtered titanium nitride thin films. *Surf. Coat. Technol.* **2000**, *125*, 335–340. [CrossRef]
6. Bae, Y.W.; Lee, Y.W.; Bessman, T.S.; Blau, T.J. Nanoscale hardness and microfriction of titanium nitride films deposited from the reaction of tetrakis (dimethylamino) titanium with ammonia. *Appl. Phys. Lett.* **1995**, *65*, 1895–1896. [CrossRef]
7. Arshi, N.; Lu, J.; Koo, B.H.; Lee, C.G.; Ahmed, F. Effect of nitrogen flow rate on the properties of TiN film deposited by e beam evaporation technique. *Appl. Surf. Sci.* **2012**, *258*, 8498–8505. [CrossRef]
8. Yokota, K.; Nakamura, K.; Kasuya, T.; Tamura, S.; Sugimoto, T.; Akamatsu, K.; Nakao, K.; Miyashita, F. Relationship between hardness and lattice parameter for TiN films deposited on SUS 304 by an IBAD technique. *Surf. Coat. Technol.* **2002**, *158–159*, 690–693. [CrossRef]
9. Cheng, H.E.; Hon, M.H. Texture formation in titanium nitride films prepared by chemical vapor deposition. *J. Appl. Phys.* **1996**, *79*, 8047–8053. [CrossRef]
10. Chatterjee, S.; Chandrashekhar, S.; Sudarshan, T.S. Deposition processes and metal cutting applications of TiN coatings. *J. Mater. Sci.* **1992**, *27*, 3409–3423. [CrossRef]
11. Banerjee, R.; Chandra, R.; Ayyub, P. Influence of the sputtering gas on the preferred orientation of nanocrystalline titanium nitride thin films. *Thin Solid Films* **2002**, *405*, 64–72. [CrossRef]
12. McKenzie, D.R.; Yin, Y.; McFall, W.D.; Hoang, N.H. The orientation dependence of elastic strain energy in cubic crystals and its application to the preferred orientation in titanium nitride thin films. *J. Phys. Condens. Matter* **1996**, *8*, 5883. [CrossRef]
13. Pelleg, J.; Zevin, L.Z.; Lungo, S.; Croitoru, N. Reactive-sputter-deposited TiN films on glass substrates. *Thin Solid Film.* **1991**, *197*, 117–128. [CrossRef]
14. Chou, T.C. Comment on Formation of polyhedral N2 bubbles during reactive sputter deposition of epitaxial TiN (100) films. *J. Appl. Phys.* **1990**, *67*, 2670–2672. [CrossRef]
15. Ensinger, W. Low energy ion assist during deposition—An effective tool for controlling thin film microstructure. *Nucl. Instrum. Methods Phys. Res. Sect. B Beam Interact. Mater. At.* **1997**, *127*, 796–808. [CrossRef]
16. Veprek, S.; Veprek-Heijman, M.G.; Karvankova, P.; Prochazka, J. Different approaches to superhard coatings and nanocomposites. *Thin Solid Film.* **2005**, *476*, 1–29. [CrossRef]
17. Patsalas, P.; Charitidis, C.; Logothetidis, S.; Dimitriadis, C.A.; Valassiades, O. Combined electrical and mechanical properties of titanium nitride thin films as metallization materials. *J. Appl. Physic* **1999**, *86*, 5296–5298. [CrossRef]
18. Sundgren, J.E. Structure and properties of TiN coatings. *Thin Solid Film.* **1985**, *128*, 21–44. [CrossRef]

19. Sung, Y.M.; Kim, H.J. Optimum substrate bias condition for TiN thin film deposition using an ECR sputter system. *Surf. Coat. Technol.* **2003**, *171*, 75–82. [CrossRef]
20. Zhao, J.; Garza, E.G.; Lam, K.; Jones, C.M. Comparison study of physical vapor-deposited and chemical vapor-deposited titanium nitride thin films using X-ray photoelectron spectroscopy. *Appl. Surf. Sci.* **2000**, *158*, 246–251. [CrossRef]
21. Guemmaz, M.; Moraitis, G.; Mosser, A.; Khan, M.A.; Parlebas, J.C. Band structure of substoichiometric titanium nitrides and carbonitrides: Spectroscopical and theoretical investigations. *J. Phys. Condens. Matter* **1997**, *40*, 8453. [CrossRef]
22. Bertoti, I. Characterization of nitride coatings by XPS. *Surf. Coat. Technol.* **2002**, *151*, 194–203. [CrossRef]
23. Wu, H.Z.; Chou, T.C.; Mishra, A.; Anderson, D.R.; Lampert, J.K.; Gujrathi, S.C. Characterization of titanium nitride thin films. *Thin Solid Film.* **1990**, *191*, 55–67. [CrossRef]
24. Huang, H.H.; Hon, M.H. Effect of N_2 addition on growth and properties of titanium nitride films obtained by atmospheric pressure chemical vapor deposition. *Thin Solid Film.* **2002**, *416*, 54–61. [CrossRef]
25. Charatan, R.M.; Gross, M.E.; Eaglesham, D.J. Plasma enhanced chemical vapor deposition of titanium nitride thin films using cyclopentadienyl cycloheptatrienyl titanium. *J. Appl. Phys.* **1994**, *76*, 4377–4382. [CrossRef]
26. Saoula, N.; Djerourou, S.; Yahiaoui, K.; Henda, K.; Kesri, R.; Erasmus, R.M.; Comins, J.D. Study of the deposition of Ti/TiN multilayers by magnetron sputtering. *Surf. Interface Anal.* **2010**, *42*, 1176–1179. [CrossRef]
27. Subramanian, B.; Ananthakumar, R.; Jayachandran, M. Structural and tribological properties of DC reactive magnetron sputtered titanium/titanium nitride (Ti/TiN) multilayered coatings. *Surf. Coat. Technol.* **2011**, *205*, 3485–3492. [CrossRef]
28. Chen, J.S.; Wang, J.L. Diffusion barrier properties of sputtered TiB_2 between Cu and Si. *J. Electrochem. Soc.* **2000**, *147*, 1940. [CrossRef]
29. Stolt, L.; d'Heurle, F.M.; Harper, J.M.E. On the formation of copper-rich copper silicides. *Thin Solid Film.* **1991**, *200*, 147–156. [CrossRef]
30. Shin, Y.H.; Shimogaki, Y. Diffusion barrier property of TiN and TiN/Al/TiN films deposited with FMCVD for Cu interconnection in ULSI. *Sci. Technol. Adv. Mater.* **2004**, *5*, 399. [CrossRef]
31. Lee, C.; Kuo, Y.L. The evolution of diffusion barriers in copper metallization. *JOM* **2007**, *59*, 44–49. [CrossRef]
32. Kim, T.H.; Howlader, M.M.R.; Itoh, T.; Suga, T. Room temperature Cu–Cu direct bonding using surface activated bonding method. *J. Vac. Sci. Technol. A Vac. Surf. Film.* **2003**, *21*, 449–453. [CrossRef]

Article

Combination of Multiple Operando and In-Situ Characterization Techniques in a Single Cluster System for Atomic Layer Deposition: Unraveling the Early Stages of Growth of Ultrathin Al$_2$O$_3$ Films on Metallic Ti Substrates

Carlos Morales [1], Ali Mahmoodinezhad [1], Rudi Tschammer [1], Julia Kosto [1], Carlos Alvarado Chavarin [2], Markus Andreas Schubert [2], Christian Wenger [2], Karsten Henkel [1] and Jan Ingo Flege [1,*]

[1] Applied Physics and Semiconductor Spectroscopy, Brandenburg University of Technology Cottbus-Senftenberg, 03046 Cottbus, Germany
[2] IHP—Leibniz Institut fuer Innovative Mikroelektronik, 15236 Frankfurt, Germany
* Correspondence: janingo.flege@b-tu.de

Abstract: This work presents a new ultra-high vacuum cluster tool to perform systematic studies of the early growth stages of atomic layer deposited (ALD) ultrathin films following a surface science approach. By combining operando (spectroscopic ellipsometry and quadrupole mass spectrometry) and in situ (X-ray photoelectron spectroscopy) characterization techniques, the cluster allows us to follow the evolution of substrate, film, and reaction intermediates as a function of the total number of ALD cycles, as well as perform a constant diagnosis and evaluation of the ALD process, detecting possible malfunctions that could affect the growth, reproducibility, and conclusions derived from data analysis. The homemade ALD reactor allows the use of multiple precursors and oxidants and its operation under pump and flow-type modes. To illustrate our experimental approach, we revisit the well-known thermal ALD growth of Al$_2$O$_3$ using trimethylaluminum and water. We deeply discuss the role of the metallic Ti thin film substrate at room temperature and 200 °C, highlighting the differences between the heterodeposition (<10 cycles) and the homodeposition (>10 cycles) growth regimes at both conditions. This surface science approach will benefit our understanding of the ALD process, paving the way toward more efficient and controllable manufacturing processes.

Keywords: ALD; in-situ; operando; XPS; ellipsometry; QMS

Citation: Morales, C.; Mahmoodinezhad, A.; Tschammer, R.; Kosto, J.; Alvarado Chavarin, C.; Schubert, M.A.; Wenger, C.; Henkel, K.; Flege, J.I. Combination of Multiple Operando and In-Situ Characterization Techniques in a Single Cluster System for Atomic Layer Deposition: Unraveling the Early Stages of Growth of Ultrathin Al$_2$O$_3$ Films on Metallic Ti Substrates. *Inorganics* **2023**, *11*, 477. https://doi.org/10.3390/inorganics11120477

Academic Editors: Roberto Nisticò, Torben R. Jensen, Luciano Carlos, Hicham Idriss and Eleonora Aneggi

Received: 2 November 2023
Revised: 1 December 2023
Accepted: 8 December 2023
Published: 14 December 2023

1. Introduction

Appeals from national and supranational institutions have become more frequent and imperative regarding the scarcity of critical materials and the effects of anthropogenic climate change. To minimize the global hazardous impact of these threats, the path towards the so-called green energy transition requires not only novel materials and technologies but also higher efficiency and sustainability of current processing techniques and devices.

In this context, atomic layer deposition (ALD) has gained increased attention in the last decades due to its potential use in microelectronics for device miniaturization, thanks to the excellent control of thickness and conformality of structures with high aspect ratio, the use of relatively low temperatures—from room temperature (25 °C, RT) to 400 °C—and the possibility of mixing different elements to grow mixed compounds, layers with different doping levels, and complex heterostructures [1–4]. Moreover, the flexibility offered by ALD has become very attractive for non-ALD specialists [5], finding multiple applications in a great variety of fields away from the more traditional gate oxides in microelectronics, such as photovoltaics [6,7], sensing [8,9], catalysis [10,11], or energy storage [12,13]. The fast-spreading of ALD can be understood considering the basics of its fundamental principles as well as its easy scalability and implementation in industrial processes; the ALD mentioned strengths come from its self-limiting nature, achieved by subsequent pulses of reactants (organometallic precursors and corresponding co-reactants) that can only undergo a self-limiting reaction with the available active surface sites within each sub-cycle [2,14]. Furthermore, the so-called area-selective ALD takes advantage of this self-limiting nature of the ALD reaction mechanism by controlling the area where the growth occurs using passivation strategies that inhibit the precursor-surface reaction [15–17].

Considering the capital importance of ALD chemistry, the rational design of ALD organometallic precursors has become a hot research topic [18–21]. The requirements include (1) a sufficiently high vapor pressure to minimize the pulse lengths together with a high decomposition temperature, (2) a high reactivity towards the substrate surface to minimize any delays in the nucleation process and significant deviation from the expected layer-by-layer growth, and (3) no possibility of self-reaction when adsorbed on the surface or with the byproducts. The precursors should also present low toxicity, low cost, and the possibility for scale-up.

However, searching for novel ALD precursors is not the only challenge; the reaction mechanism for many ALD processes is not deeply understood yet, which limits the use of optimized and controllable ALD processes in industrial applications. As stated by H. Sønsteby and coworkers [22], even for well-known processes such as the growths of Al_2O_3 or Fe_2O_3 on hydroxylated Si surfaces using trimethylaluminum (TMA) and water or bis(cyclopentadienyl) iron [Fe(cp)$_2$] and O_3, the reported growth-per-cycle (GPC) values vary significantly between 1 to 3 and 0.2 to 1.4 Å/cycle, respectively. This experimentally observed lack of reproducibility within the same process and under seemingly equivalent experimental conditions is not consistent with a self-limited process. However, it must rather be related to (1) a lack of control over the experimental parameters, i.e., uncontrolled dosing of precursors, inadequate purging, real or virtual leaks in the reactor, reactor design, etc., and (2) the role of the substrate. To diagnose the ALD process, several operando devices and techniques can be used, such as quartz crystal microbalance (QCM) [23–25], ellipsometry [22,25–27], or quadrupole mass spectrometry (QMS) [22]. In the first two cases, the GPC can be monitored by indirect measurements of the thickness through changes in mass or light polarization, respectively, whereas QMS allows good control of the precursor dose while enabling cross-checking of possible reactor malfunction, e.g., the presence of vacuum leaks. Furthermore, these characterization techniques can also be employed to study the reaction mechanism by precisely following the evolution of the GPC under different conditions (substrate, temperature, pressure, dose and purging time, etc.) [27,28], the presence and nature of byproducts [29–31], and the evolution of film properties [27]. Similarly, the adsorbed intermediate states on the surface can be followed by infrared spectroscopy under operando conditions [31,32].

Most of these techniques, however, are frequently used to study what S. Elliott calls the homo-deposition regime, i.e., when the organometallic molecule interacts with the relatively thick grown film and not during the hetero-deposition regime, i.e., when the reaction takes place between the precursor and the substrate surface [16]. The differences between these two situations, closely related to the steady-state and nucleation regimes, respectively, are more significant when dealing with ultrathin deposits (below 10 nm thickness), where the substrate plays a crucial role in terms of interfacial effects affecting the ALD reaction mechanism and the ALD-layer properties.

Thanks to its probe depth (~10 nm maximum in conventional laboratory equipment) and sensitivity to elemental oxidation state and without requiring a change in the dipole moment of the probed species (as for infrared spectroscopy), X-ray photoemission spectroscopy (XPS) constitutes a potentially powerful tool to characterize the early stages of growth, especially considering the evolution of the interface between film and substrate and the intermediate species after each sub-cycle. Although the relatively high pressures present in the ALD process (10^{-4} to 10 mbar), the comparably high amount of impurities, and the typical lack of a high crystalline order have prevented a classic surface science approach like that applied to deposits grown using physical vapor deposition techniques (e.g., molecular beam epitaxy), an increasing number of operando and in situ (also referred as in vacuo) studies, have been reported in the last years. For instance, operando studies have been recently performed using flow-type reaction cells in differentially pumped X-ray photoelectron spectroscopy (NAP-XPS) devices [33–35]. Due to the characteristic high pressures of ALD processes, especially when a carrier gas is used, these operando and time-resolved XPS experiments are typically limited to synchrotron facilities, although some experiments have been carried out at standard NAP-XPS setups [36,37]. A more typical scenario is the in situ approach, where the film is transferred under controlled conditions, i.e., high or ultra-high vacuum conditions, from the ALD reactor to the analysis chamber, thus preventing film/surface modification or the deposit of contaminants, e.g., adventitious carbon, after exposure to atmosphere. In this regard, multiple examples in the literature can be found using lab-based tools [25,26,38,39] or synchrotron radiation [40–42], where other X-ray-based techniques, apart from XPS, are used [43], such as X-ray reflectivity (XRR), grazing incidence small angle X-ray scattering (GISAXS), X-ray fluorescence (XRF), and X-ray absorption (XAS). As a drawback, cross-contamination between the ALD reactor and the rest of the vacuum system due to the use of organic reactants at relatively high pressures can be expected, thus making it necessary to either dedicate the system entirely to ALD-based experiments or thoroughly clean the whole system after them. This brief overview intends to show how the gap between the ALD and the surface science communities has gradually dissipated in the last few years.

In the present work, we introduce our new cluster tool capable of combining operando (spectroscopic ellipsometry and QMS) and in situ (XPS) characterization techniques to study the complete ALD process, with particular emphasis on the very early stages of growth. The homemade ALD reactor allows the use of organometallic precursors with different vapor pressures and, particularly, the operation in the so-called flow (1–10 mbar) and pump-type (10^{-4}–10^{-3} mbar) modes, i.e., with and without carrier gas, respectively. To prove the capabilities of the system and exemplify the surface science strategy applied in this kind of experiment, we have revisited the well-known thermal ALD (T-ALD) deposition of Al_2O_3 using TMA and water. By performing a quantitative XPS analysis and comparing it with the results of ellipsometric measurements, we discuss the initial variations in the GPC, emphasizing the influence of experimental factors on estimating the GPC values. We compare the early stages of growth of Al_2O_3 on polycrystalline Ti at two different temperatures (RT and 200 °C) and discuss the role of TMA in the early passivation of the easily oxidizing metallic Ti surface. To the authors' knowledge, there are no systematic XPS studies targeting the TMA/H_2O interaction with metallic Ti surfaces and the influence of T-ALD processes on modifying the film/substrate interface during the very early stages of growth, which could have important implications for the growth of passivating coatings of highly reactive surfaces [44–47]. Furthermore, ex situ characterization by transmission electron microscopy (TEM) performed on nanostructured substrates will help to elucidate the influence of other factors on ALD growth and demonstrate that special care must be taken when comparing the same processes applied to substrates with different aspect ratios. We believe that this surface science approach will improve our understanding of the fundamentals of the ALD reaction mechanisms, thus paving the way to more efficient and controllable manufacturing processes.

The manuscript is divided into three main sections, followed by final conclusions. Section 2 is divided into two subsections, referred to as the presentation and description of the operando (ellipsometry and QMS) and in situ (XPS) characterization of the thermal ALD (T-ALD) growth of Al_2O_3 ultrathin films ($\leq$10 nm) in both cases complemented with ex situ TEM measurements. Subsequently, Section 3 will present a comprehensive discussion of the previous measurements, highlighting the synergies from the combination of multiple characterization techniques and how they can be used to cross-validate the scientific results following a more classic surface science approach. The details of the experimental setup and the ALD procedure are described in depth in Section 4.

2. Results

2.1. Operando Characterization

The T-ALD growth of Al_2O_3 was first characterized through ellipsometry as a diagnostic technique to confirm the self-limiting nature of the reaction mechanism. Figure 1a shows the evolution of thickness as a function of time for a growth performed at room temperature (RT) on Si substrates with about 2 nm thick natural SiO_2 adlayer, showing an almost ideal linear behavior indicating a constant GPC. The initial 3 Å step after the first TMA dose is probably related to the first adsorption of a complete layer of Al-CH$_3$ on the substrate surface (the Al-C bond distance is ~2 Å). Furthermore, the top left inset showcases the typical modified step-like behavior associated with the ALD cycles. Here, the thickness increases with the TMA dose due to the adsorption of the metallic cation together with the remaining ligands and decreases after the H_2O pulse due to Al-O bond formation and the release of the methyl groups. The GPC estimated from these individual steps (taken at the steady growth stage after more than 60 cycles) is about 1.0 Å/cycle, corresponding to the minimum values reported in the literature for equivalent ALD processes. Unlike laser-based ellipsometers [27], the time between consecutive measurements and signal-to-noise ratio does not allow for characterizing the steady states of the purging steps, i.e., when the thickness should be constant. This limitation prevents further analysis of possible leaks or insufficient purging of the gas lines that would induce uncontrolled adsorption or oxidation of the precursor, as shown elsewhere, e.g., for the growth of Ga_2O_3 using TMGa and O_2 plasma [48]. Moreover, the bottom right inset displays a cross-sectional TEM image of a nanostructured silicon substrate onto which a 5 nm thick, homogeneous Al_2O_3 layer has been deposited under the same conditions, demonstrating a high conformality that is especially remarkable at the edges.

The linear fitting of the ellipsometric measurements (Figure 1a, red) shows an average GPC value of 0.65 Å/cycle for the first 60 cycles, whereas if the fit is limited to the 10 first cycles (blue), the GPC is only 0.56 Å/cycle, a value much lower than those typically reported in the literature, or even the one estimated above for a single step after more than 60 cycles. This finding points toward a distinct difference in chemistry between the homodeposition and heterodeposition cases that notably influence the growth rate. Therefore, we have complemented this initial characterization with the XPS measurements performed after subsequent complete ALD cycles.

Figure 1b shows the evolution of the substrate (Si 2p) and film (Al 2p) intensities as a function of the total number of ALD cycles. By fitting the experimental data with a layer-by-layer growth model of the form $exp(-d/\lambda)$ and $(1 - exp(-d/\lambda))$, respectively [49] (where d is the film thickness and λ is the inelastic mean free path of the photoelectrons), we obtain average GPC values of around 0.9 Å/cycle for both substrate and film signals, as expected. The previous growth model expressions allow calculating the instantaneous (i.e., referred to the individual deposition performed between subsequent XPS measurements) and accumulated (i.e., considering the total number of ALD cycles until the specific XPS measurement) GPC at individual points, as shown in Figure 1c (where the dashed lines correspond to the average GPC values estimated in Figure 1b). Although the GPC converges to the expected value above 60 cycles, there are some important deviations before 30 cycles. During the first 20 cycles, the GPC values estimated from both the substrate and the film signals are lower, in the range between 0.5 and 0.7 Å/cycle, in line with those obtained from the ellipsometry data fit (Figure 1a, blue line), also validating the optical modeling incorporating values of the indexes of refraction known for the bulk materials. Moreover, there seems to be an abrupt increase after 30 cycles, probably related to a complete coalescence of the alumina film (the XPS fitting model assumes a complete surface coverage, which can result in slight deviations in the presence of initial nucleation delays or film pinholes). It is worth mentioning that these sorts of singular deviations, especially when considering a low number of cycles, could be critically influenced by insufficient conditioning of the ALD reactor and the gas lines before deposition. This technical issue can artificially modify the GPC during the first or second cycles and can easily be identified by in situ ellipsometry measurements. In this case, however, we have not detected any unusual behavior, and thus, the deviation of the instantaneous GPC estimated by XPS is probably related to the applied model, as mentioned above.

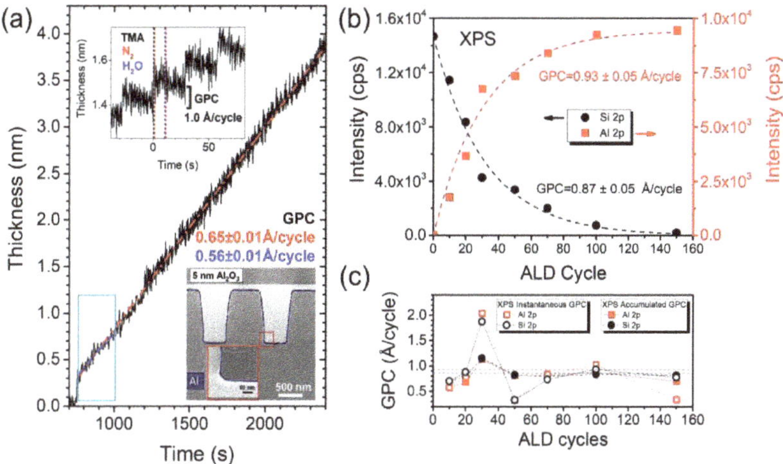

Figure 1. (**a**) Spectroscopic ellipsometry measurements showing the thickness evolution of T-ALD Al_2O_3 on Si wafers using TMA and H_2O at room temperature. The dashed red and blue lines (with the corresponding values) show the fit performed for the first 60 and 10 cycles, respectively. The semi-transparent blue box indicates the first 10 cycles where the corresponding fit is performed. The top left inset depicts a magnified view of the ellipsometric data, showing the step-like behavior of the ALD process and indicating the TMA, N_2, and H_2O doses (time has been set to 0 at the beginning of the cycle). The bottom right inset shows a cross-sectional TEM image of a Si-based nanostructured substrate with a 5 nm thick Al_2O_3 layer deposited under the same conditions. (**b**) X-ray photoemission spectroscopy (XPS) intensities of the Al 2p (squares) and Si 2p (circles) peaks as a function of the total amount of ALD cycles and corresponding fittings (dashed lines) in red and black, respectively. (**c**) Instantaneous (open symbols) and accumulated (filled symbols) GPC values as a function of the total amount of ALD cycles estimated from XPS measurements.

Therefore, the GPC estimated from two different techniques (operando and in situ) shows similar values and trends, highlighting the reproducibility of the process (measurements taken on different samples) and the possibility of obtaining complementary information, especially during the early stages of growth where more significant deviations from the ideal linear growth are expected owing to the special chemistry at the interface. It is worth noting that the GPC values obtained from both techniques depend on the applied model, in both cases assuming a continuous layer from the beginning. In particular, in the applied ellipsometric model, the refractive index and extinction coefficient (see Section 4) were derived from thicker films and assumed to remain constant while analyzing thinner films. This approach could slightly modify the estimated GPC for low coverages.

The ALD process at RT and 200 °C has also been followed by QMS in multiple ion detection modes, as shown in Figure 2 for the $m/z = 18$ (H_2O) and $m/z = 16$ (CH_4) signals as a function of time. As a result of the reaction between the TMA molecule with the substrate (and reactor walls) and subsequent oxidation with water, two pulses of residual CH_4 are measured just after the TMA and H_2O dose, as extensively reported previously in the literature [29,31]. In particular, on SiO_2 surfaces, it has been shown that increasing the temperature to 200 °C increases the release of TMA ligands, i.e., methyl groups, from approximately one to two [50–52]. This effect is well reproduced by the ratio of the CH_4 integrated areas of the TMA and H_2O sub-cycles, giving 1.4 and 2 for the RT and 200 °C growth, respectively. Moreover, in the case of treating with a well-known system where the reaction mechanism is well established [53], QMS can also be used to diagnose the ALD process. By comparing the processes at (a) RT and (b) 200 °C, we notice an H_2O signal when the TMA is dosed at a substrate temperature of 200 °C compared to the RT growth.

As the growth at high temperature (shown in Figure 2b) was performed immediately after the RT process on a second substrate, we believe this water comes from a virtual leak due to cold spots on the reactor walls (which were nominally heated up to 120 °C), leading to H_2O signals in the QMS measurements only if there is some other gas, TMA or N_2, acting as carrier gas (thus explaining also the increase in water after the second N_2 purge pulse). The presence of this extra water could explain why the estimated GPC at 200 °C by ellipsometry is slightly higher than expected, around 1.6 Å/cycle compared to the theoretical 1.3 Å/cycle [52]. Nevertheless, this technical issue can easily be minimized by baking out the ALD reactor before further depositions, mostly eliminating the water excess (as verified by QMS). Despite this virtual leak, however, the ALD process preserves its self-limiting behavior without becoming a CVD process, as indicated by the XPS and ellipsometric measurements and the GPC saturation when increasing the TMA pulse duration.

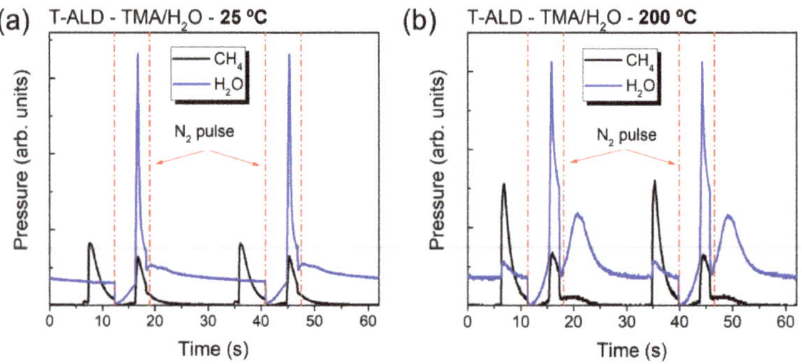

Figure 2. Quadrupole mass spectrometry (QMS) measurements for T-ALD of Al_2O_3 on Si wafers at (**a**) RT and (**b**) 200 °C following the $m/z = 18$ (H_2O, blue) and $m/z = 16$ (CH_4, black) signals as a function of time.

We note that no signal from Al-containing species was detected, possibly indicating that all the TMA is completely consumed at the sample surface and reactor walls. As the increase in TMA dose does not lead to a higher GPC as measured by ellipsometry, i.e., saturation of the process at the sample surface, the saturation on the reactor walls between the sample and the QMS is not fully achieved (and thus, the ALD chemistry is not exactly the same in all exposed surfaces). This fact implies that the current QMS setup (see Section 4 Materials and Methods) properly works for ALD-process monitoring but presents limitations for detailed mechanistic studies. Due to space limitations, the current QMS configuration without the use of a capillary [29] or an orifice [30] directly positioned above the sample surface is not ideal for these thermal-ALD experiments using TMA/H_2O. The wide temperature window of the process implies that the entire inner reactor surface will contribute to the byproduct signal in the QMS measurements as it is coated with alumina, limiting the sample specificity in this test case. This situation is critical if different reactions take place on different surfaces. Therefore, the shown QMS measurements prove the possibility of combining multiple characterization techniques simultaneously, not only for ALD-process characterization but also for diagnosing the experimental setup.

2.2. In Situ Characterization

To prove the relevance of the XPS technique for understanding the initial stages of growth, we shift to metallic Ti substrates in this subsection (thermally evaporated Ti films on Si wafers at RT, see Section 4), as we expect a more complex interaction between the substrate, film, and ALD precursors due to the high reactivity and facile oxidation of metallic Ti compared to SiO_x/Si surfaces.

Although reliable ellipsometry measurements cannot be performed when using these substrates due to their relatively high roughness, quantitative XPS analysis using the same treatment as described before shows GPC values in the order of 1 Å/cycle, confirming that the T-ALD process remains basically the same. Moreover, Figure 3 shows the atomic concentration of Ti, O, Al, and C estimated from XPS measurements as a function of the total number of ALD cycles. As expected, we observe an exponential decrease in the Ti concentration and a simultaneous increase in the Al content. More interesting are the behaviors of oxygen and carbon. On the one hand, the RT growth shows an almost parallel increase in the O and Al signals, whereas at 200 °C the oxygen increases more quickly than the Al concentration until an O/Al ratio of 2.3 for both temperatures is reached, indicating an excess of oxygen in the alumina film probably due to a hydroxyl-terminated surface and potential diffusion of water molecules on the top layer [54–56]. The differences between the interface regions are related to the partial oxidation of the Ti substrate at higher temperatures, as described in the following paragraphs. On the other hand, the carbon concentration starts increasing in both cases (in line with the initial adsorption of TMA at the surface, as seen in the ellipsometry measurements in Figure 1a), almost disappearing afterward at high temperatures whereas it stabilizes at about 10% at RT, pointing to a lower efficiency of the oxidation step owing to the incomplete removal of the methyl ligands. Furthermore, the initial increase in carbon in both cases indicates that the reaction mechanism slightly differs between the heterodeposition, i.e., the reaction of the organometallic precursor with the Ti substrate, and the homodeposition (i.e., TMA on Al_2O_3) regimes. We will analyze these differences in more depth in the following paragraphs.

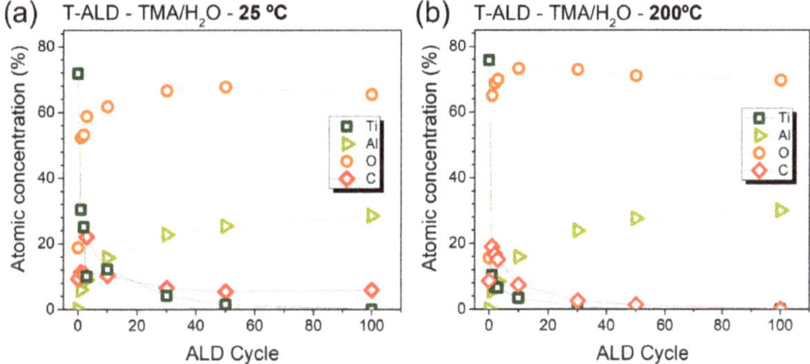

Figure 3. Evolution of Ti, Al, O, and C atomic concentrations as a function of the total amount of ALD cycles for the T-ALD of TMA and water on a Ti substrate process performed at (**a**) RT and (**b**) 200 °C.

The evolution of the Ti oxidation state during the early stages of growth is shown in Figure 4. For comparison purposes, we first expose a bare $Ti/SiO_x/Si$ substrate (Ti film thickness is about 10 nm) at RT to H_2O doses of the same duration as the H_2O pulses during the T-ALD process and follow the changes in the Ti 2p XPS spectra (see Figure 4a). The initial surface is almost metallic, as clearly documented by an asymmetric Ti^0 component at ~453.7 eV for the Ti $2p_{3/2}$ core level accompanied by a spin-orbit splitting (ΔE) of 6.05 eV along with tiny, symmetrical Ti^{2+} and Ti^{3+} $2p_{3/2}$ components at 455.3 and 457.1 eV, respectively (with corresponding ΔE of 5.6 and 5.2 eV) [57]. From the first H_2O dose, we observe a gradual oxidation of the Ti surface, with increasing Ti^{2+} and Ti^{3+} components and two extra, symmetric Ti^{4+} components appearing (Ti $2p_{3/2}$ component at 458.8 eV with ΔE of 5.7 eV). The changes are clearly evident after 4 pulses, when the Ti^{4+} components, especially for the Ti $2p_{1/2}$ level, start to become visible. In contrast, during the T-ALD growth at RT (see Figure 4b), the Ti 2p region does not show any visible change regardless of the number of ALD cycles, particularly in the region where the Ti^{4+} components are expected. Moreover, the first pulse of TMA and subsequent chemisorption of the molecule on the Ti surface seems to passivate the substrate, limiting its oxidation. Figure 4c summarizes the Ti oxidation trend extracted from the XPS fitting of the Ti 2p spectra as a function of H_2O dose (each cycle refers to an H_2O pulse of 0.5 s; see Section 4).

While the bare Ti surface is easily oxidized with H_2O, drastically decreasing the metallic Ti^0 component during the first 10 pulses, the Ti surfaces show more moderate oxidation during the T-ALD process, especially in the case of the RT growth. The differences between RT and 200 °C T-ALD depositions are likely related to the residual water in the ALD chamber prior to the growth (see also Figure 2), which could promote faster and deeper Ti oxidation during the heating ramp of the 200 °C growth (Figure 4c). Therefore, it can be inferred that the initial chemisorption of TMA creates a shielding layer against Ti oxidation, thus explaining why even ultrathin Al_2O_3 films can passivate surfaces, as shown in the case of Si substrates [58,59] or perovskites solar cells [60–62].

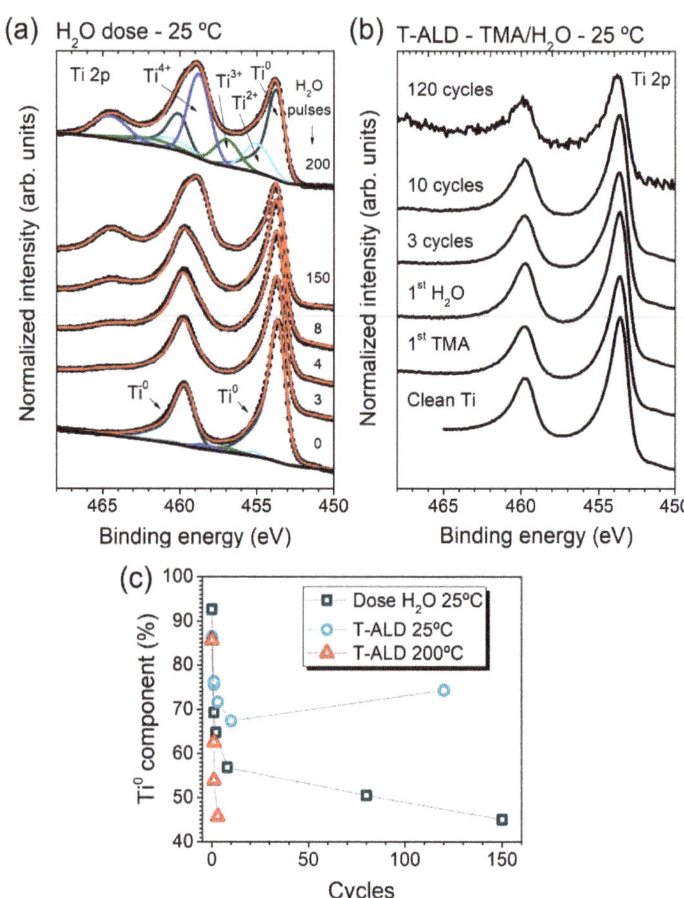

Figure 4. (a) Ti 2p X-ray photoemission spectroscopy (XPS) data of a bare metallic Ti surface as a function of H_2O pulses of 0.5 s. Black circle symbols and continuous red lines represent the raw data and the fitted curve, respectively. The dark and light green and blue lines correspond to the different Ti $2p_{3/2}$ and $2p_{1/2}$ components, as indicated. (b) Ti 2p XPS spectra of the T-ALD growth performed at RT, as labeled. (c) Evolution of the metallic Ti^0 component (%) as a function of H_2O pulses for the H2O dose at RT (squares), T-ALD growth at RT (circles), and T-ALD at 200 °C (triangles).

As discussed in Figure 3, the amount of carbon increases during the first ALD cycle and finally decreases and stabilizes from about 10 cycles. This evolution and the slower GPC identified above for the operando characterization constitute a fingerprint of the two regimes previously introduced, the hetero and homo-deposition stages. Figure 5 shows the spectral evolution of the XPS C 1s data for the two T-ALD processes at (a) RT and (b) 200 °C. The initial as-grown Ti surfaces show some carbon (<5 at%) from the residual gas within the preparation chamber (see Section 4), which is probably due to cross-contamination with the ALD reactor, as both chambers are (if only for a short time) connected during sample transfer to or from the analysis chamber. The XPS survey spectra exclude the presence of any other element resulting from the cross-contamination. Three components can be distinguished in the C 1s, the first at ~284.7 eV related to sp^3 hybridization of C-C and C-H bonds [63–65], and the other two, labeled as TiC (~281.7) and TiC* (~282.5 eV), referred to as titanium carbide [66–68]. In particular, the TiC* component has been reported as an interfacial effect between the metallic carbide and the carbon deposits or attributed to a disordered structure [68]. As there seems to be a strong relationship between both TiC and TiC* components, we have fixed their relative intensity ratio, energy shift, and corresponding full width half maximum (FWHM) in the fitting process, thus understanding the changes in the C 1s spectra not from variations of the substrate carbide, but from the chemical interaction with the TMA and H$_2$O. The first dose of TMA shows the appearance of two new components at ~286 eV and ~283.5 eV, associated with C-O [65,69] bonds and the TMA molecule adsorbed at the surface [70,71], respectively. We note that the TMA component, compared to the C-C peak, is more pronounced at RT than at 200 °C, which is likely related to the release of more methyl ligands at higher temperatures and probably also to the non-negligible amount of water due to the virtual leak mentioned previously. The second sub-cycle, i.e., the H$_2$O dose, showcases how the oxidation step is more efficient at higher temperatures, featuring a completely disappearing TMA component, while a higher amount of carbon bonded to oxygen species remains at the surface. A similar trend is observed for the third TMA and H$_2$O sub-cycles. Interestingly, after ten cycles, i.e., at a thickness of about 1 nm, the signal of TMA after a complete cycle is higher than after the first and third cycles, especially at RT, while the C-C component associated with residual carbon decreases. This intensity evolution indicates a potential interface effect on the reaction mechanism, related to a different amount of hydroxyl groups between the initial Ti surface and the grown alumina and, more complex, to the presence of side-reactions initially promoted by the Ti with the TMA [16]. A full understanding of the heterodeposition regime would require complementary operando measurements, such as infrared spectroscopy, as well as theoretical simulations. Finally, thick deposits (>10 nm) show the presence of only C-C component and tiny traces of C-O and aluminium carbonate (~290 eV) in both cases [72].

Commonly in the literature, particularly regarding ex situ XPS studies, the surface carbon residue is considered an indicator of the whole film quality. Nevertheless, previous data from Figures 3 and 5 point to a more complex evolution of the carbon amount as a function of the total number of ALD cycles, the role of the substrate, and the ALD process conditions. Figure 6 shows the XPS (a) C 1s and (b) O 1s spectra of the as-grown ~10 nm Al$_2$O$_3$ ALD film on Ti at RT before and after 10 min of Ar$^+$ sputtering. The carbon residue is located in the top surface region in the form of sp^3-hybridized C-C and C-H bonds, whereas the bulk contains residual carbon bonded to oxygen (C-O and C=O species) and aluminum (carbonates, Al-O-C). The C1 s spectrum of the sputtered film prepared at RT resembles the spectra of the as-grown layer deposited at 200 °C with 120 ALD cycles, where the aluminate species were visible, and the residual C-C component was much lower. This similarity indicates that the intensity of the C-C component is inversely related to the efficiency of the TMA oxidation process, which is enhanced at higher temperatures. Similar results in terms of carbon species and their film distribution have previously been reported for plasma-enhanced ALD Al$_2$O$_3$ films [73]. The O 1s spectrum (Figure 6b) also reflects some changes after sputtering, decreasing the amount of OH groups and thus explaining the excess of oxygen estimated by XPS for the as-grown sample (see Figure 3) [73–75].

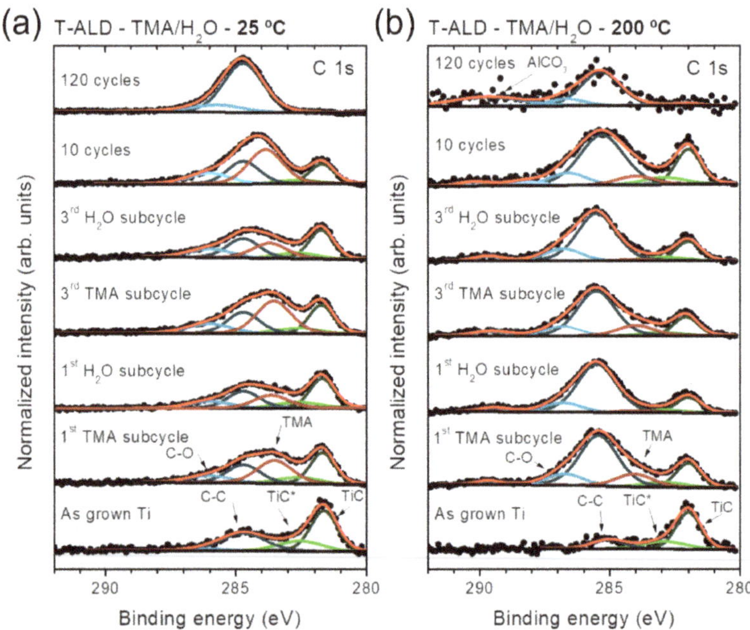

Figure 5. C 1s X-ray photoemission spectroscopy (XPS) spectra taken at different stages of the Al$_2$O$_3$ T-ALD process performed at (**a**) RT and (**b**) 200 °C, as labeled. Black circle symbols and continuous red lines represent the raw data and the fitted curve, respectively. Dark and light green lines correspond to TiC-related components, red to TMA, and blue lines to C-C and C-O species, as indicated.

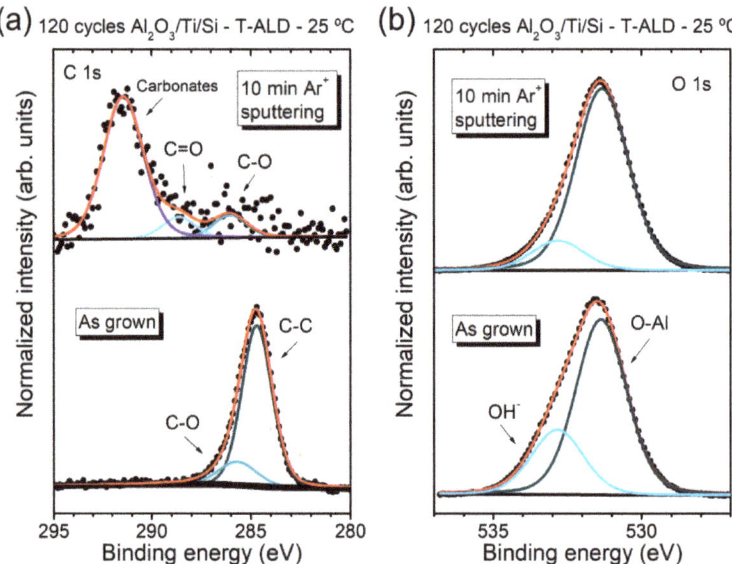

Figure 6. In-situ X-ray photoemission spectroscopy (XPS) spectra of the (**a**) C 1s and (**b**) O 1s regions of the Al$_2$O$_3$ T-ALD process performed at RT before and after Ar$^+$ sputtering. Black circle symbols and continuous red lines represent the raw data and the fitted curve, respectively.

Up to this point, all the growths discussed in detail have been performed on flat substrates. Figure 7 depicts a cross-sectional TEM image of a 15 nm thick T-ALD Al_2O_3 film deposited at RT on a Si-based nanostructured substrate with varying distances between individual stripes, i.e., different aspect ratios. Particularly in the energy-dispersive X-ray compositional mapping of carbon (Figure 7c), we observe a significant influence of the nanostructures' aspect ratio. With an increasing aspect ratio, i.e., with decreasing width while the height remains fixed, carbon residue accumulates in the pits, which becomes significantly higher than at the top of the structures. Similarly, the Al_2O_3 deposit is comparatively thicker in those high and narrow trenches. As indicated by V. Cremers and coworkers [2], these deviations from the ALD growth on flat surfaces are related to the mean free path of the reactant molecules and the design of the ALD reactor, which ultimately will determine the flow regime (molecular or viscous), with important consequences on the conformality and homogeneity of the ALD deposit.

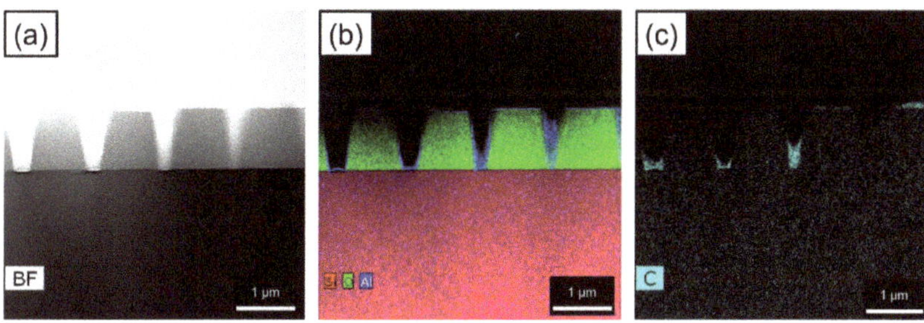

Figure 7. Transmission electron microscopy (TEM) cross-section image of a 15 nm thick T-ALD Al_2O_3 film deposited at RT on Si-based nanostructured substrates. (**a**) Bright-field (BF) image; (**b**) BF image superimposed to energy dispersive X-ray spectroscopy (EDX) compositional mapping of silicon, oxygen, and aluminum; and (**c**) EDX mapping of carbon in the same region.

3. Discussion

In the previous section, we analyzed the Al_2O_3 T-ALD process using TMA and H_2O by combining operando and in situ characterization techniques in the same UHV cluster tool, following a classic surface science approach not so explored before by the majority of the ALD community.

In particular, both the results from spectroscopic ellipsometry and XPS point to the existence of different growth regimes depending on the total amount of ALD cycles and, therefore, on the interaction of the organometallic precursor with the substrate surface (if the ALD-deposited film is thin enough so there are significant interface effects), or with the surface of a film that may not be considered bulk-like. These two hetero and homo-deposition regimes are seen to affect the GPC and the efficiency of the ALD reaction mechanism, translating into higher carbon deposits and lower growth ratios near the interface. Precisely, these chemical differences can also affect the early stages of growth in terms of nucleation and growth delay, as has been extensively reported in the literature [14,22]. These facts are of particular relevance for ultrathin deposits (<10 nm), where the cross-interactions between the substrate, film, and ALD precursors play a key role in the final chemical and physical properties of ALD films. Given the ongoing drive towards device miniaturization, the combined expertise of the ALD and surface science communities should provide new insights that will promote our understanding of the substrate/film interaction, help us identify the role of the surface on the ALD reaction mechanism, and serve to answer the question how the ALD process could be tailored by controlling the surface properties of the substrate. The last issue requires sophisticated substrate surface preparation, which constitutes a highly demanding step (particularly regarding the conservation of clean surfaces until the organometallic dose). Hence, for this purpose, different approaches have been followed, employing single crystals [76,77], crystalline nanoparticles [78,79], or free-standing

2D materials [73,80], which may be combined with more complex and realistic substrates in view of potential applications. In this framework, in situ photoelectron spectroscopies, using both commercial X-ray and synchrotron radiation sources, play a fundamental role in accessing the elemental composition and oxidation states of the substrate, film, interfaces, and reaction intermediates on the surface. Furthermore, by comparing the defective, not well-ordered ALD deposits with their crystalline counterparts (e.g., epitaxial ultrathin films grown by molecular beam epitaxy), we will gain a better understanding of the role of defects in modifying the properties of thin ALD films.

Moreover, this systematic combination of the operando and in-situ approach also improves the reproducibility of the ALD process, making it possible to understand the influence of the ALD reactor design (or its malfunction) on the properties of the ALD deposit [22]. In this sense, the water excess identified based on the QMS measurements of the T-ALD process at 200 °C had important consequences on interpreting the XPS data. Instead of (erroneously) deducing that the TMA is decomposing at the surface, raising the amount of residual C-C and C-H due to a catalytic effect of the metallic Ti, we could establish that the excess of water is likely related to a virtual leak from the cold spots in the reactor walls that partially oxidize the TMA during the first ALD sub-cycle. Therefore, special attention must be paid to the characterization of ultrathin ALD deposits, relying on the combination of multiple characterization techniques to enable reliable cross-checking between them. Furthermore, the use of other equipment and techniques, such as a quartz crystal microbalance or infrared spectroscopy, would allow for an even more profound discussion in terms of film nucleation (e.g., up to what extent the lower GPC is due to a nucleation delay and thus incomplete coverage) or intermediates states under operando conditions without interrupting the growth process because of the need to transfer the sample, which may affect the exact deposition conditions (e.g., via decreasing the substrate temperature before transferring the sample to the XPS analyzer chamber and ramping it up afterward for continued deposition).

Finally, and related to the use of different types of substrates, we have shown how the properties of the ALD deposits, in terms of thickness and amount of residual carbon, depend on the substrate, not only in terms of chemical properties but on its morphology and, particularly, on the aspect ratio of nanostructures. Furthermore, the specific ALD reactor design also has a crucial influence on these processes, as extensively discussed by V. Cremers and coworkers [2]. Thus, the surface science approach applied to well-known, simple, and (usually) flat surfaces must be adapted to more realistic scenarios and applications where a complementary engineering approach is required to finally optimize the ALD processes by focusing on the flow regime, local pressure distribution, and reactor design.

4. Materials and Methods

4.1. ALD-XPS Cluster

The in situ ALD-XPS cluster comprises a homemade ALD reactor compatible with ultra-high vacuum conditions (UHV) attached to a state-of-the-art UHV-XPS system, see Figure 8.

The homemade ALD reactor consists of a stainless steel UHV-compatible chamber (Pfeiffer, Assla, Germany) with a base pressure of 10^{-8} mbar, which is directly attached to the preparation chamber of the XPS system. After the ALD growth, and once the pressure is below 10^{-7} mbar at the ALD reactor, the sample is transferred to the analysis chamber, which takes around 15 min. Continuing with the ALD reactor, the sample, 10×10 mm maximum in a conventional flag-type sample holder, is placed at the center (focal point) of the chamber, below the entrance of the reactive gases, held in a sample stage capable of heating the sample to a maximum temperature of 1000 °C by the use of a silicon nitride ceramic heater compatible with UHV conditions as well as atmospheric pressures and highly oxidative environments (343-HEATER-SIN-8X10, Allectra, Berlin, Germany). The temperature is monitored through a K-type thermocouple in close proximity to the sample. The temperature measured by the thermocouple is around 10% lower than that estimated on the surface of Si(100) by applying a temperature-dependent optical model on the ellipsometry measurements between 100 and 200 °C. The position of the sample can be precisely varied and controlled through a 3-axis linear manipulator with a θ rotator, including correction of the sample tilt with respect to the incident light beam from the ellipsometer. The spectroscopic ellipsometer (SER 801 UV-VIS, SENTECH, Berlin, Germany) is installed at $70°$ to the sample surface normal, as shown in Figure 8b. The stress-free viewports, where the ellipsometer arms are mounted, are protected by pneumatic shutters to minimize undesired coatings. The QMS (HAL/3F 301 RC, Hidden Analytical, Warrington, UK) is separated from the ALD reactor by an elbow gate valve and a blind DN40 CF flange with a 500 μm ($\varnothing$) aperture to constrain the pressure in the QMS area to $<10^{-4}$ mbar via differential pumping (see Figure 8c). Depending on the total pressure applied during the ALD process (i.e., pump-type or flow-type mode), membranes of different porous sizes can be installed on the aperture to regulate the final pressure. The aperture is maintained at room temperature, stopping the chamber heating at the elbow gate valve. No condensation issues have been detected during these experiments. The QMS is differentially pumped through a secondary turbopump connected to the bypassed turbopump at the load-lock chamber (pump-type operation mode) or directly through the turbopump of the ALD reactor (flow-type operation mode). A scheme of both configurations is shown in Figure 9a. The ALD reactor chamber can be pumped in two ways: through a turbopump (67 L/s) or a scroll pump bypassing the turbopump (3.3 L/s), for the pump-type (10^{-4}–10^{-3} mbar) and flow-type (1–10 mbar) operation modes, respectively (see Figure 9a). During growths performed at high temperatures, the reactor walls can be heated up to a maximum of 200 °C. The configuration of the ALD reactor chamber is highly flexible, allowing the installation of more instruments or the rearrangement of the existing ones.

The ALD gas lines are schematically represented in Figure 9b. The lines are based on VCR® components, allowing a quick modification if required and easy purging down to a base pressure of 10^{-7} mbar before the growth. The system comprises three lines for precursors (left line in Figure 9b), oxidants/reactants (right line), and purging gas (middle line), each regulated by a mass flow controller (F-111B 200, Bronkhorst, Ruurlo, The Netherlands) and connected to a shared N_2 supply. The precursor line allows the installation of three different kinds of precursors depending on their vapor pressure (v_p): high (the quantity of generated vapor is high enough to be pumped directly into the chamber without the use of a carrier gas, use of cylinder container), medium (the cylinder container output is connected to a 3-way pneumatic ALD valve to allow the use of carrier gas on its transportation to the reactor), and low (use of electropolished stainless steel bubbler). This configuration allows complex ALD super-cycles to deposit mixed compounds. Moreover, the oxidant/reactant line configuration follows a similar design, with three different sublines for H_2O, O_3/O_2 (directly connected to an ozone generator fed with pure O_2, OXP-30 Ozone Generator from Oxidation Technologies, Inwood, IA, USA), and a third subline allowing for the connection to other gas sources. Finally, the third line is used for purging, particularly during pump-type operation, when no carrier gas is used in the precursor and oxidant lines. The lines can be heated up to 90 °C using heating wires. The ALD process is controlled by ALD pneumatic valves (Swagelok, Berlin, Germany) using homemade LabVIEW-based (2020 SP1) software.

The X-ray photoelectron spectroscopy (XPS) system consists of a load-lock chamber (base pressure of high 10^{-8} mbar) with sample storage capability. The preparation chamber (10^{-9} mbar range) allows pre and post-growth treatments and/or experiments via sample heating (up to 300 °C), exposure to different gases through corresponding leak valves, and monitoring of atmosphere composition using a QMS (e-Vision 2 EV2-110-000FT, MKS, Munich, Germany). Several e-beam or Knudsen cell evaporators can also be installed (e.g., as the one used for the Ti evaporation). The XPS analysis chamber (10^{-10} mbar range) consists of an Omicron EA 125 hemispherical electron analyzer with a non-monochromatized twin X-ray anode (Al/Mg) and a five-channeltron detector for efficient counting. At a pass energy of 20 eV, the overall spectral resolution is about 1.1 and 1.0 eV, respectively. Finally, it is possible to transfer the ALD samples through the load-lock chamber in a vacuum suitcase to a glovebox or other UHV systems at our home lab facilities for additional treatments and/or characterization (low energy electron diffraction (LEED), ultraviolet photoelectron spectroscopy (UPS), hard X-ray photoelectron spectroscopy (HAXPES), and scanning probe microscopy (SPM)).

Our lab facilities count on a second ex situ ALD reactor with the same kind of gas line configuration and operation modes. This system allows simultaneous growth on several substrates or 2-inch wafers when the deposition recipe is well established using the more complex in situ ALD-XPS cluster.

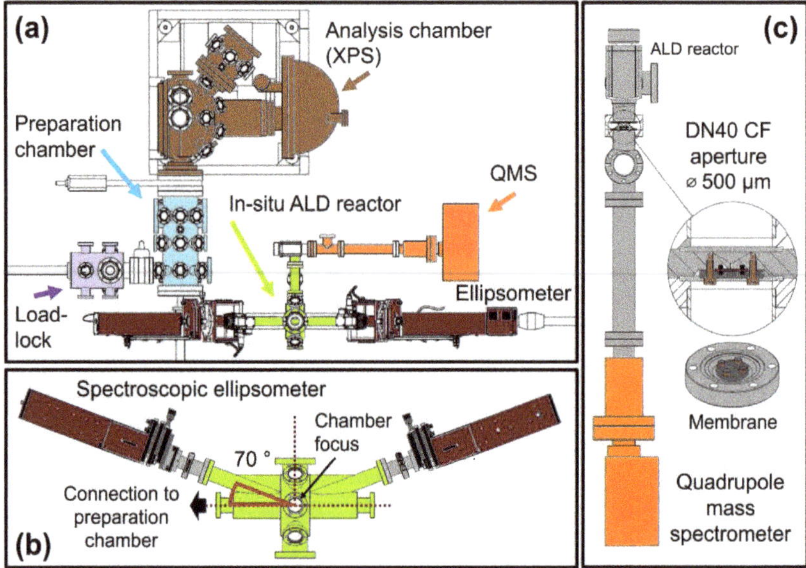

Figure 8. (**a**) Overall top-view scheme of the complete ALD-XPS system, consisting of a load-lock chamber (purple), a preparation chamber equipped with a quadrupole mass spectrometer, a radiative sample heating stage (up to 300 °C), gas input lines, and the possibility to install several evaporators (light blue); an analysis chamber equipped with an XPS analyzer and a twin non-monochromatized X-ray source with Al and Mg anodes (brown); and the homemade ALD reactor (green) where the spectroscopic ellipsometer (dark red) and the differentially pumped mass spectrometer (orange) are installed. For clarity, the manipulator of the heating stage at the ALD reactor is not included. (**b**) Lateral view of the ALD reactor, including the ellipsometer at an angle of 70° with respect to the normal of the sample surface. The manipulator of the heating stage of the ALD reactor, where the sample is held, is installed on the front window in this view, allowing fine correction of the sample position and tilt with respect to the incident polarized light. The connection to the preparation chamber, hidden in (**a**), is indicated by the black arrow on the left side. (**c**) Detailed view of the quadrupole mass spectrometer. The entrance of the QMS is separated from the ALD reactor by an angle valve and a DN40 CF dummy flange with a 500 µm aperture where a membrane of different pore sizes can (depending on the usage of pump or flow type mode) be installed to constrain the pressure in the QMS area to $<10^{-4}$ mbar.

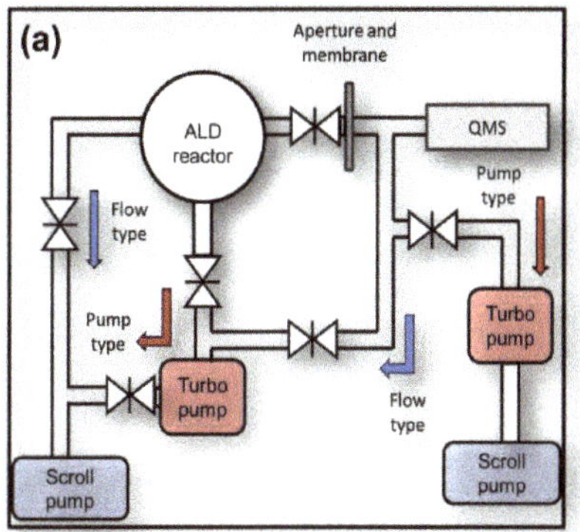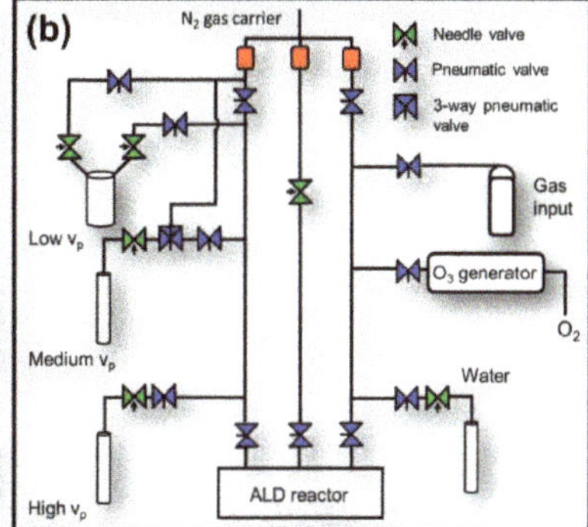

Figure 9. (**a**) Pump-lines scheme of the ALD reactor and associated differentially pumped QMS. The blue and red arrows indicate the open valve configuration under the flow and pump-type operation modes, respectively. (**b**) ALD-gas lines scheme, where the blue and green valves correspond to the ALD pneumatic and manual valves, respectively, and the top red boxes represent the mass flow controllers connected to a shared N_2 supply.

4.2. Thermal ALD Process

The alumina (Al_2O_3) films were deposited using thermal ALD (T-ALD) using the commercial ALD precursor trimethylaluminum (TMA) from Sigma Aldrich (Taufkirchen, Germany), in combination with ultrapure water from Alfa Aesar, spectrophotometric degree. The ALD reactor operates in pump-type mode using N_2 (99.9999%, Air Liquide) as purging gas (at a pressure of 10^{-2} mbar) between precursor (10^{-4} mbar) and oxidant (10^{-3} mbar) pulses and controlled by an independent mass-flow controller (F-111B 200, Bronkhorst, Ruurlo, The Netherlands). The TMA and water containers were kept at RT during the growths, providing sufficient vapor pressure to pump the precursor and the reactant directly into the reactor using the reactor turbopump (see Figure 9). During the growth at 200 °C, the temperature of the precursor and H_2O lines were kept at 90 °C, whereas the walls of the ALD reactor were ramped up to 120 °C. The substrate temperature was set to ~25 °C (RT) and 200 °C, respectively. The ALD recipe consisted of a 0.5 s TMA pulse followed by 0.5 s of N_2 purging flow, continued by an H_2O dose of 0.5 s and subsequent 0.5 s of N_2 for purging. The reactor was purged for 15 s between cycles by pumping with the turbopump ($<3 \cdot 10^{-5}$ mbar). The N_2 flux was set to 60 sccm. The gas input was regulated by ALD pneumatic valves (Swagelok, Berlin, Germany) controlled by LabVIEW-based (2020 SP1) software. As discussed in Section 2, once the growth is stabilized, the estimated average growth per cycle (GPC) is 0.9 ± 0.1 Å/cycle. Two types of substrates were used. The first was p-type Si (100) single crystals cut from 3″ wafers covered by native oxide (SiO_x/Si) from CrysTec. Prior to the ALD deposition, they were annealed at 250 °C in UHV to remove adventitious carbon and subsequently characterized using XPS. The second substrate consisted of Ti thin films deposited by thermal evaporation at RT under UHV conditions in the preparation chamber (see Figure 8) on similar SiO_x/Si wafer pieces. Before the ALD deposition, the $Ti/SiO_x/Si$ substrates were also routinely characterized using XPS.

4.3. ALD Film Characterization

The operando characterization of the T-ALD process was performed using spectroscopic ellipsometry, SER 801 model from SENTECH, and quadrupole mass spectrometry, HAL/3F 301 RC from Hiden Analytical. The bare SiO_x/Si substrate was used for the ellipsometry measurements, performed with a maximum spectral range from 240 to 1000 nm (UV-VIS). The ellipsometric modeling and parameter fitting were performed with the SpectraRay/4 (6.0.8.2) software, considering an air/Al_2O_3/SiO_x/Si multilayer system where the atomic layer deposited alumina was fitted using a Cauchy model, and the initial thickness of the native oxide was set to 2.3 nm, as extracted from an as-introduced SiO_x/Si reference sample. The QMS was operated in multiple ion detection mode (MID), following the CH_4 ($m/z = 16$), H_2O ($m/z = 18$), O_2 ($m/z = 32$), and TMA ($m/z = 57$) signals as a function of time.

In situ X-ray photoelectron spectroscopy measurements (XPS) were performed with an Omicron EA 125 hemispherical electron analyzer using non-monochromatized Mg K_α radiation. The pass energy was set to 20 eV, yielding an overall spectral resolution of about 1.0 eV. The sample charging was corrected considering the Si 2p (Si^0) and Ti 2p (Ti^0) contributions from the SiO_x/Si and $Ti/SiO_x/Si$ substrates as an internal reference, respectively. The spectra have been fitted using the XPSPeak software, version 4.1, whereas the electron inelastic mean free path (IMFP) through the alumina matrix was calculated using the Tanuma, Powell, and Penn formula IMFP-TPP2M [81]. As-grown ~10 nm thick Al_2O_3 films were gently sputtered by Ar^+ cations (cold cathode ion source ISE 5, Scienta Omicron, Taunusstein, Germany) with an accelerating voltage of 500 eV at a pressure of 4×10^{-6} mbar, yielding an estimated sputter rate of around 0.3 nm/min.

The Al_2O_3 films were finally characterized ex situ using transmission electron microscopy (TEM) and energy dispersive X-ray spectroscopy (EDX), performed with an FEI Tecnai Osiris instrument operated at 200 kV, using nanostructured Si substrates fabricated at the Leibniz-Institut für innovative Mikroelektronik (IHP).

5. Conclusions

We have presented a new in situ ultra-high vacuum cluster tool where multiple operando (spectroscopic ellipsometry and quadrupole mass spectrometry) and in situ (X-ray photoelectron spectroscopy) techniques are combined to characterize the early stages of growth of atomic layer deposited films.

To show the capabilities of the new system, we have revisited the well-known thermal ALD growth of Al_2O_3 using TMA and H_2O, with particular emphasis on distinguishing the hetero and homo-deposition regimes. Indeed, we find that the ALD reaction mechanism is slightly modified in terms of TMA oxidation efficiency and growth per cycle for deposits up to 10 nm. In contrast to what is generally believed, the ALD reaction mechanism critically depends on the nature of the surface, particularly on the reactivity between the selected precursor and the bare substrate and, subsequently, with the ALD deposit. The growth might be influenced, especially for ultra-thin films, by the cross-interaction between the film and substrate. Moreover, the combination of multiple characterization techniques allows a reliable cross-check between them, enabling us to characterize the ALD process and diagnose possible malfunctions of the ALD reactor at the same time.

The operando and in situ characterization of ALD materials using a more traditional surface science approach allows for studying the complex interactions between the substrate, film, and reactants as well as the inter-relation with ALD process parameters in a systematic manner, thereby opening the door to a deep understanding of the relationship between the substrates and ALD reaction mechanism during the early stages of growth. The precise control of the substrate surface properties and its interface with the ALD material will allow tailoring the ALD film properties and optimizing the ALD process by rational design, ultimately decreasing the amount of wasted precursor and associated costs, thus paving the way for using ALD more effectively in existing and new application areas.

Author Contributions: Conceptualization, C.M. and J.I.F.; methodology, C.M. and J.I.F.; validation, C.M., A.M., R.T. and J.K.; formal analysis, C.M., A.M., R.T. and J.K.; investigation, C.M., A.M., R.T. and J.K.; resources, C.W., K.H. and J.I.F.; data curation, C.M., A.M., R.T., J.K., C.A.C., M.A.S. and K.H.; writing—original draft preparation, C.M.; writing—review and editing, K.H. and J.I.F.; supervision, J.I.F.; project administration, J.I.F.; funding acquisition, K.H., C.W. and J.I.F. All authors have read and agreed to the published version of the manuscript.

Funding: This work has been funded by the Federal Ministry of Education and Research of Germany (BMBF) within the iCampus2 project, grant number 16ME0420K, and the European Regional Development Fund (ERDF 2014-2020), contract number 85053620. C. Morales thanks the Postdoc Network Brandenburg for a PNB individual grant. R. Tschammer acknowledges the support by BTU/BAM in the framework of the BTU-BAM Graduate School »Trustworthy Hydrogen«.

Data Availability Statement: Raw data are available upon request.

Acknowledgments: We thank Guido Beuckert for the technical support.

Conflicts of Interest: The authors declare no conflict of interest.

References

1. Coll, M.; Napari, M. Atomic Layer Deposition of Functional Multicomponent Oxides. *APL Mater.* **2019**, *7*, 110901. [CrossRef]
2. Cremers, V.; Puurunen, R.L.; Dendooven, J. Conformality in Atomic Layer Deposition: Current Status Overview of Analysis and Modelling. *Appl. Phys. Rev.* **2019**, *6*, 021302. [CrossRef]
3. Oviroh, P.O.; Akbarzadeh, R.; Pan, D.; Coetzee, R.A.M.; Jen, T.-C. New Development of Atomic Layer Deposition: Processes, Methods and Applications. *Sci. Technol. Adv. Mater.* **2019**, *20*, 465–496. [CrossRef] [PubMed]
4. Vasiliev, V.Y. Composition, Structure, and Functional Properties of Thin Silicon Nitride Films Grown by Atomic Layer Deposition for Microelectronic Applications (Review of 25 Years of Research). *J. Struct. Chem.* **2022**, *63*, 1019–1050. [CrossRef]
5. Alvaro, E.; Yanguas-Gil, A. Characterizing the Field of Atomic Layer Deposition: Authors, Topics, and Collaborations. *PLoS ONE* **2018**, *13*, e0189137. [CrossRef] [PubMed]
6. Xing, Z.; Xiao, J.; Hu, T.; Meng, X.; Li, D.; Hu, X.; Chen, Y. Atomic Layer Deposition of Metal Oxides in Perovskite Solar Cells: Present and Future. *Small Methods* **2020**, *4*, 2000588. [CrossRef]
7. Ghosh, S.; Yadav, R. Future of Photovoltaic Technologies: A Comprehensive Review. *Sustain. Energy Technol. Assess.* **2021**, *47*, 101410. [CrossRef]
8. Marichy, C.; Pinna, N. Atomic Layer Deposition to Materials for Gas Sensing Applications. *Adv. Mater. Interfaces* **2016**, *3*, 1600335. [CrossRef]
9. Xu, H.; Akbari, M.K.; Kumar, S.; Verpoort, F.; Zhuiykov, S. Atomic Layer Deposition—State-of-the-Art Approach to Nanoscale Hetero-Interfacial Engineering of Chemical Sensors Electrodes: A Review. *Sens. Actuators B Chem.* **2021**, *331*, 129403. [CrossRef]
10. Cao, K.; Cai, J.; Liu, X.; Chen, R. Review Article: Catalysts Design and Synthesis via Selective Atomic Layer Deposition. *J. Vac. Sci. Technol. A Vac. Surf. Films* **2018**, *36*, 010801. [CrossRef]
11. Xu, D.; Yin, J.; Gao, Y.; Zhu, D.; Wang, S. Atomic-Scale Designing of Zeolite Based Catalysts by Atomic Layer Deposition. *ChemPhysChem* **2021**, *22*, 1287–1301. [CrossRef]
12. Zhao, Z.; Kong, Y.; Zhang, Z.; Huang, G.; Mei, Y. Atomic Layer–Deposited Nanostructures and Their Applications in Energy Storage and Sensing. *J. Mater. Res.* **2020**, *35*, 701–719. [CrossRef]
13. Zhao, Y.; Zhang, L.; Liu, J.; Adair, K.; Zhao, F.; Sun, Y.; Wu, T.; Bi, X.; Amine, K.; Lu, J.; et al. Atomic/Molecular Layer Deposition for Energy Storage and Conversion. *Chem. Soc. Rev.* **2021**, *50*, 3889–3956. [CrossRef]
14. Richey, N.E.; de Paula, C.; Bent, S.F. Understanding Chemical and Physical Mechanisms in Atomic Layer Deposition. *J. Chem. Phys.* **2020**, *152*, 040902. [CrossRef]
15. Chen, R.; Kim, H.; McIntyre, P.C.; Porter, D.W.; Bent, S.F. Achieving Area-Selective Atomic Layer Deposition on Patterned Substrates by Selective Surface Modification. *Appl. Phys. Lett.* **2005**, *86*, 191910. [CrossRef]
16. Elliott, S.D. Atomic-Scale Simulation of ALD Chemistry. *Semicond. Sci. Technol.* **2012**, *27*, 074008. [CrossRef]
17. Mackus, A.J.; Merkx, M.J.; Kessels, W.M. From the Bottom-up: Toward Area-Selective Atomic Layer Deposition with High Selectivity. *Chem. Mater.* **2018**, *31*, 2–12. [CrossRef]
18. Devi, A. 'Old Chemistries' for New Applications: Perspectives for Development of Precursors for MOCVD and ALD Applications. *Coord. Chem. Rev.* **2013**, *257*, 3332–3384. [CrossRef]
19. Hatanpää, T.; Ritala, M.; Leskelä, M. Precursors as Enablers of ALD Technology: Contributions from University of Helsinki. *Coord. Chem. Rev.* **2013**, *257*, 3297–3322. [CrossRef]
20. Shahmohammadi, M.; Mukherjee, R.; Takoudis, C.G.; Diwekar, U.M. Optimal Design of Novel Precursor Materials for the Atomic Layer Deposition Using Computer-Aided Molecular Design. *Chem. Eng. Sci.* **2021**, *234*, 116416. [CrossRef]
21. Oh, I.-K.; Sandoval, T.E.; Liu, T.-L.; Richey, N.E.; Bent, S.F. Role of Precursor Choice on Area-Selective Atomic Layer Deposition. *Chem. Mater.* **2021**, *33*, 3926–3935. [CrossRef]
22. Sønsteby, H.H.; Yanguas-Gil, A.; Elam, J.W. Consistency and Reproducibility in Atomic Layer Deposition. *J. Vac. Sci. Technol. A Vac. Surf. Films* **2020**, *38*, 020804. [CrossRef]
23. Elam, J.W.; Schuisky, M.; Ferguson, J.D.; George, S.M. Surface Chemistry and Film Growth during TiN Atomic Layer Deposition Using TDMAT and NH3. *Thin Solid. Films* **2003**, *436*, 145–156. [CrossRef]
24. Meng, X.; Cao, Y.; Libera, J.A.; Elam, J.W. Atomic Layer Deposition of Aluminum Sulfide: Growth Mechanism and Electrochemical Evaluation in Lithium-Ion Batteries. *Chem. Mater.* **2017**, *29*, 9043–9052. [CrossRef]

25. Nieminen, H.-E.; Chundak, M.; Heikkilä, M.J.; Kärkkäinen, P.R.; Vehkamäki, M.; Putkonen, M.; Ritala, M. *In Vacuo* Cluster Tool for Studying Reaction Mechanisms in Atomic Layer Deposition and Atomic Layer Etching Processes. *J. Vac. Sci. Technol. A* **2023**, *41*, 022401. [CrossRef]

26. Schmidt, D.; Strehle, S.; Albert, M.; Hentsch, W.; Bartha, J.W. Top Injection Reactor Tool with in Situ Spectroscopic Ellipsometry for Growth and Characterization of ALD Thin Films. *Microelectron. Eng.* **2008**, *85*, 527–533. [CrossRef]

27. Naumann, F.; Reck, J.; Gargouri, H.; Gruska, B.; Blümich, A.; Mahmoodinezhad, A.; Janowitz, C.; Henkel, K.; Flege, J.I. In Situ Real-Time and Ex Situ Spectroscopic Analysis of Al_2O_3 Films Prepared by Plasma Enhanced Atomic Layer Deposition. *J. Vac. Sci. Technol. B* **2020**, *38*, 014014. [CrossRef]

28. Sønsteby, H.H.; Bratvold, J.E.; Weibye, K.; Fjellvåg, H.; Nilsen, O. Phase Control in Thin Films of Layered Cuprates. *Chem. Mater.* **2018**, *30*, 1095–1101. [CrossRef]

29. Juppo, M.; Rahtu, A.; Ritala, M.; Leskelä, M. In Situ Mass Spectrometry Study on Surface Reactions in Atomic Layer Deposition of Al_2O_3 Thin Films from Trimethylaluminum and Water. *Langmuir* **2000**, *16*, 4034–4039. [CrossRef]

30. Matero, R.; Rahtu, A.; Ritala, M. In Situ Quadrupole Mass Spectrometry and Quartz Crystal Microbalance Studies on the Atomic Layer Deposition of Titanium Dioxide from Titanium Tetrachloride and Water. *Chem. Mater.* **2001**, *13*, 4506–4511. [CrossRef]

31. Goldstein, D.N.; McCormick, J.A.; George, S.M. Al_2O_3 Atomic Layer Deposition with Trimethylaluminum and Ozone Studied by in Situ Transmission FTIR Spectroscopy and Quadrupole Mass Spectrometry. *J. Phys. Chem. C* **2008**, *112*, 19530–19539. [CrossRef]

32. Cabrera, W.; Halls, M.D.; Povey, I.M.; Chabal, Y.J. Surface Oxide Characterization and Interface Evolution in Atomic Layer Deposition of Al_2O_3 on InP(100) Studied by in Situ Infrared Spectroscopy. *J. Phys. Chem. C* **2014**, *118*, 5862–5871. [CrossRef]

33. Kokkonen, E.; Kaipio, M.; Nieminen, H.-E.; Rehman, F.; Miikkulainen, V.; Putkonen, M.; Ritala, M.; Huotari, S.; Schnadt, J.; Urpelainen, S. Ambient Pressure X-ray Photoelectron Spectroscopy Setup for Synchrotron-Based *In Situ* and *Operando* Atomic Layer Deposition Research. *Rev. Sci. Instrum.* **2022**, *93*, 013905. [CrossRef] [PubMed]

34. Shavorskiy, A.; Kokkonen, E.; Redekop, E.; D'Acunto, G.; Schnadt, J.; Knudsen, J. Time-Resolved APXPS with Chemical Potential Perturbations: Recent Developments at the MAX IV Laboratory. *Synchrotron Radiat. News* **2022**, *35*, 4–10. [CrossRef]

35. D'Acunto, G.; Shayesteh, P.; Kokkonen, E.; Boix De La Cruz, V.; Rehman, F.; Mosahebfard, Z.; Lind, E.; Schnadt, J.; Timm, R. Time Evolution of Surface Species during the ALD of High-k Oxide on InAs. *Surf. Interfaces* **2023**, *39*, 102927. [CrossRef]

36. Head, A.R.; Chaudhary, S.; Olivieri, G.; Bournel, F.; Andersen, J.N.; Rochet, F.; Gallet, J.-J.; Schnadt, J. Near Ambient Pressure X-ray Photoelectron Spectroscopy Study of the Atomic Layer Deposition of TiO_2 on RuO_2 (110). *J. Phys. Chem. C* **2016**, *120*, 243–251. [CrossRef]

37. Temperton, R.H.; Gibson, A.; O'Shea, J.N. *In Situ* XPS Analysis of the Atomic Layer Deposition of Aluminium Oxide on Titanium Dioxide. *Phys. Chem. Chem. Phys.* **2019**, *21*, 1393–1398. [CrossRef]

38. Strehle, S.; Schumacher, H.; Schmidt, D.; Knaut, M.; Albert, M.; Bartha, J.W. Effect of Wet Chemical Substrate Pretreatment on the Growth Behavior of Ta(N) Films Deposited by Thermal ALD. *Microelectron. Eng.* **2008**, *85*, 2064–2067. [CrossRef]

39. Fukumizu, H.; Sekine, M.; Hori, M.; McIntyre, P.C. Initial Growth Analysis of ALD Al_2O_3 Film on Hydrogen-Terminated Si Substrate via in Situ XPS. *Jpn. J. Appl. Phys.* **2020**, *59*, 016504. [CrossRef]

40. Tallarida, M.; Karavaev, K.; Schmeisser, D. The Initial Atomic Layer Deposition of HfO_2/Si(001) as Followed *In Situ* by Synchrotron Radiation Photoelectron Spectroscopy. *J. Appl. Phys.* **2008**, *104*, 064116. [CrossRef]

41. Kolanek, K.; Tallarida, M.; Michling, M.; Schmeisser, D. In Situ Study of the Atomic Layer Deposition of HfO_2 on Si. *J. Vac. Sci. Technol. A Vac. Surf. Films* **2012**, *30*, 01A143. [CrossRef]

42. Tallarida, M.; Schmeisser, D. In Situ ALD Experiments with Synchrotron Radiation Photoelectron Spectroscopy. *Semicond. Sci. Technol.* **2012**, *27*, 074010. [CrossRef]

43. Devloo-Casier, K.; Ludwig, K.F.; Detavernier, C.; Dendooven, J. In Situ Synchrotron Based X-ray Techniques as Monitoring Tools for Atomic Layer Deposition. *J. Vac. Sci. Technol. A Vac. Surf. Films* **2014**, *32*, 010801. [CrossRef]

44. Dingemans, G.; Kessels, W.M.M. Status and Prospects of Al_2O_3-Based Surface Passivation Schemes for Silicon Solar Cells. *J. Vac. Sci. Technol. A Vac. Surf. Films* **2012**, *30*, 040802. [CrossRef]

45. Zardetto, V.; Williams, B.L.; Perrotta, A.; Di Giacomo, F.; Verheijen, M.A.; Andriessen, R.; Kessels, W.M.M.; Creatore, M. Atomic Layer Deposition for Perovskite Solar Cells: Research Status, Opportunities and Challenges. *Sustain. Energy Fuels* **2017**, *1*, 30–55. [CrossRef]

46. Banerjee, S.; Das, M.K. A Review of Al_2O_3 as Surface Passivation Material with Relevant Process Technologies on C-Si Solar Cell. *Opt. Quant. Electron.* **2021**, *53*, 60. [CrossRef]

47. Ghosh, S.; Pariari, D.; Behera, T.; Boix, P.P.; Ganesh, N.; Basak, S.; Vidhan, A.; Sarda, N.; Mora-Seró, I.; Chowdhury, A.; et al. Buried Interface Passivation of Perovskite Solar Cells by Atomic Layer Deposition of Al_2O_3. *ACS Energy Lett.* **2023**, *8*, 2058–2065. [CrossRef]

48. Mahmoodinezhad, A.; Janowitz, C.; Naumann, F.; Plate, P.; Gargouri, H.; Henkel, K.; Schmeißer, D.; Flege, J.I. Low-Temperature Growth of Gallium Oxide Thin Films by Plasma-Enhanced Atomic Layer Deposition. *J. Vac. Sci. Technol. A Vac. Surf. Films* **2020**, *38*, 022404. [CrossRef]

49. Briggs, D.; Seah, M.P. (Eds.) *Practical Surface Analysis*, 2nd ed.; Chapter 5; Wiley: New York, NY, USA, 1990; Volume 1, ISBN 978-0-471-92081-6.

50. Vandalon, V.; Kessels, W.M.M. What Is Limiting Low-Temperature Atomic Layer Deposition of Al_2O_3? A Vibrational Sum-Frequency Generation Study. *Appl. Phys. Lett.* **2016**, *108*, 011607. [CrossRef]

51. Sperling, B.A.; Kalanyan, B.; Maslar, J.E. Atomic Layer Deposition of Al_2O_3 Using Trimethylaluminum and H_2O: The Kinetics of the H_2O Half-Cycle. *J. Phys. Chem. C* **2020**, *124*, 3410–3420. [CrossRef]

52. Gu, B.; Le Trinh, N.; Nguyen, C.T.; Yasmeen, S.; Gaiji, H.; Kang, Y.; Lee, H.-B.-R. Computational Modeling of Physical Surface Reactions of Precursors in Atomic Layer Deposition by Monte Carlo Simulations on a Home Desktop Computer. *Chem. Mater.* **2022**, *34*, 7635–7649. [CrossRef]

53. Puurunen, R.L. Surface Chemistry of Atomic Layer Deposition: A Case Study for the Trimethylaluminum/Water Process. *J. Appl. Phys.* **2005**, *97*, 121301. [CrossRef]

54. Renault, O.; Gosset, L.G.; Rouchon, D.; Ermolieff, A. Angle-Resolved X-ray Photoelectron Spectroscopy of Ultrathin Al_2O_3 Films Grown by Atomic Layer Deposition. *J. Vac. Sci. Technol. A Vac. Surf. Films* **2002**, *20*, 1867–1876. [CrossRef]

55. Zhang, L.; Jiang, H.C.; Liu, C.; Dong, J.W.; Chow, P. Annealing of Al_2O_3 Thin Films Prepared by Atomic Layer Deposition. *J. Phys. D Appl. Phys.* **2007**, *40*, 3707–3713. [CrossRef]

56. Kääriäinen, T.O.; Cameron, D.C. Plasma-Assisted Atomic Layer Deposition of Al_2O_3 at Room Temperature. *Plasma Process. Polym.* **2009**, *6*, S237–S241. [CrossRef]

57. Biesinger, M.C.; Lau, L.W.M.; Gerson, A.R.; Smart, R.S.C. Resolving Surface Chemical States in XPS Analysis of First Row Transition Metals, Oxides and Hydroxides: Sc, Ti, V, Cu and Zn. *Appl. Surf. Sci.* **2010**, *257*, 887–898. [CrossRef]

58. Richter, A.; Benick, J.; Hermle, M.; Glunz, S.W. Excellent Silicon Surface Passivation with 5 Å Thin ALD Al_2O_3 Layers: Influence of Different Thermal Post-Deposition Treatments: Excellent Silicon Surface Passivation with 5 Å Thin ALD Al_2O_3 Layers: Influence of Different Thermal Post-Deposition Treatments. *Phys. Status Solidi RRL* **2011**, *5*, 202–204. [CrossRef]

59. Pain, S.L.; Khorani, E.; Niewelt, T.; Wratten, A.; Paez Fajardo, G.J.; Winfield, B.P.; Bonilla, R.S.; Walker, M.; Piper, L.F.J.; Grant, N.E.; et al. Electronic Characteristics of Ultra-Thin Passivation Layers for Silicon Photovoltaics. *Adv. Mater. Inter.* **2022**, *9*, 2201339. [CrossRef]

60. Kot, M.; Das, C.; Wang, Z.; Henkel, K.; Rouissi, Z.; Wojciechowski, K.; Snaith, H.J.; Schmeisser, D. Room-Temperature Atomic Layer Deposition of Al_2O_3: Impact on Efficiency, Stability and Surface Properties in Perovskite Solar Cells. *ChemSusChem* **2016**, *9*, 3401–3406. [CrossRef]

61. Ramos, F.J.; Maindron, T.; Béchu, S.; Rebai, A.; Frégnaux, M.; Bouttemy, M.; Rousset, J.; Schulz, P.; Schneider, N. Versatile Perovskite Solar Cell Encapsulation by Low-Temperature ALD-Al_2O_3 with Long-Term Stability Improvement. *Sustain. Energy Fuels* **2018**, *2*, 2468–2479. [CrossRef]

62. Kruszyńska, J.; Ostapko, J.; Ozkaya, V.; Surucu, B.; Szawcow, O.; Nikiforow, K.; Hołdyński, M.; Tavakoli, M.M.; Yadav, P.; Kot, M.; et al. Atomic Layer Engineering of Aluminum-Doped Zinc Oxide Films for Efficient and Stable Perovskite Solar Cells. *Adv. Mater. Inter.* **2022**, *9*, 2200575. [CrossRef]

63. Schier, V.; Michel, H.-J.; Halbritter, J. ARXPS-Analysis of Sputtered TiC, SiC and Ti0.5Si0.5C Layers. *Fresenius J. Anal. Chem.* **1993**, *346*, 227–232. [CrossRef]

64. Rousseau, B.; Estrade-Szwarckopf, H.; Thomann, A.-L.; Brault, P. Stable C-Atom Displacements on HOPG Surface under Plasma Low-Energy Argon-Ion Bombardment. *Appl. Phys. A* **2003**, *77*, 591–597. [CrossRef]

65. Morales, C.; Díaz-Fernández, D.; Mossanek, R.J.O.; Abbate, M.; Méndez, J.; Pérez-Dieste, V.; Escudero, C.; Rubio-Zuazo, J.; Prieto, P.; Soriano, L. Controlled Ultra-Thin Oxidation of Graphite Promoted by Cobalt Oxides: Influence of the Initial 2D CoO Wetting Layer. *Appl. Surf. Sci.* **2020**, *509*, 145118. [CrossRef]

66. Luthin, J.; Plank, H.; Roth, J.; Linsmeier, C. Ion Beam-Induced Carbide Formation at the the Titanium–Carbon Interface. *Nucl. Instrum. Methods Phys. Res. Sect. B Beam Interact. Mater. At.* **2001**, *182*, 218–226. [CrossRef]

67. Luthin, J.; Linsmeier, C. Characterization of Electron Beam Evaporated Carbon Films and Compound Formation on Titanium and Silicon. *Phys. Scr.* **2001**, *T91*, 134. [CrossRef]

68. Lewin, E.; Persson, P.O.Å.; Lattemann, M.; Stüber, M.; Gorgoi, M.; Sandell, A.; Ziebert, C.; Schäfers, F.; Braun, W.; Halbritter, J.; et al. On the Origin of a Third Spectral Component of C1s XPS-Spectra for Nc-TiC/a-C Nanocomposite Thin Films. *Surf. Coat. Technol.* **2008**, *202*, 3563–3570. [CrossRef]

69. Yang, D.; Velamakanni, A.; Bozoklu, G.; Park, S.; Stoller, M.; Piner, R.D.; Stankovich, S.; Jung, I.; Field, D.A.; Ventrice, C.A.; et al. Chemical Analysis of Graphene Oxide Films after Heat and Chemical Treatments by X-ray Photoelectron and Micro-Raman Spectroscopy. *Carbon* **2009**, *47*, 145–152. [CrossRef]

70. Lubben, D.; Motooka, T.; Greene, J.E.; Wendelken, J.F.; Sundgren, J.-E.; Salaneck, W.R. Xps, Ups, and Hreels Studies of Excimer-Laser-Induced Dissociation of Al_2 (Ch_3) Adsorbed on Si(100) Surfaces. *MRS Proc.* **1987**, *101*, 151. [CrossRef]

71. Strongin, D.R.; Moore, J.F.; Ruckman, M.W. Synchrotron Radiation Assisted Deposition of Aluminum Oxide from Condensed Layers of Trimethylaluminum and Water at 78 K. *Appl. Phys. Lett.* **1992**, *61*, 729–731. [CrossRef]

72. Gougousi, T.; Barua, D.; Young, E.D.; Parsons, G.N. Metal Oxide Thin Films Deposited from Metal Organic Precursors in Supercritical CO_2 Solutions. *Chem. Mater.* **2005**, *17*, 5093–5100. [CrossRef]

73. Lu, Y.-H.; Morales, C.; Zhao, X.; Van Spronsen, M.A.; Baskin, A.; Prendergast, D.; Yang, P.; Bechtel, H.A.; Barnard, E.S.; Ogletree, D.F.; et al. Ultrathin Free-Standing Oxide Membranes for Electron and Photon Spectroscopy Studies of Solid–Gas and Solid–Liquid Interfaces. *Nano Lett.* **2020**, *20*, 6364–6371. [CrossRef] [PubMed]

74. Gosset, L.G.; Damlencourt, J.-F.; Renault, O.; Rouchon, D.; Holliger, P.; Ermolieff, A.; Trimaille, I.; Ganem, J.-J.; Martin, F.; Séméria, M.-N. Interface and Material Characterization of Thin Al_2O_3 Layers Deposited by ALD Using TMA/H2O. *J. Non-Cryst. Solids* **2002**, *303*, 17–23. [CrossRef]

75. Haeberle, J.; Henkel, K.; Gargouri, H.; Naumann, F.; Gruska, B.; Arens, M.; Tallarida, M.; Schmeißer, D. Ellipsometry and XPS Comparative Studies of Thermal and Plasma Enhanced Atomic Layer Deposited Al_2O_3-Films. *Beilstein J. Nanotechnol.* **2013**, *4*, 732–742. [CrossRef] [PubMed]

76. Gharachorlou, A.; Detwiler, M.D.; Gu, X.-K.; Mayr, L.; Klötzer, B.; Greeley, J.; Reifenberger, R.G.; Delgass, W.N.; Ribeiro, F.H.; Zemlyanov, D.Y. Trimethylaluminum and Oxygen Atomic Layer Deposition on Hydroxyl-Free Cu(111). *ACS Appl. Mater. Interfaces* **2015**, *7*, 16428–16439. [CrossRef]

77. Paul, R.; Reifenberger, R.G.; Fisher, T.S.; Zemlyanov, D.Y. Atomic Layer Deposition of FeO on Pt(111) by Ferrocene Adsorption and Oxidation. *Chem. Mater.* **2015**, *27*, 5915–5924. [CrossRef]

78. Cao, K.; Shi, L.; Gong, M.; Cai, J.; Liu, X.; Chu, S.; Lang, Y.; Shan, B.; Chen, R. Nanofence Stabilized Platinum Nanoparticles Catalyst via Facet-Selective Atomic Layer Deposition. *Small* **2017**, *13*, 1700648. [CrossRef]

79. Li, H.; Yu, P.; Lei, R.; Yang, F.; Wen, P.; Ma, X.; Zeng, G.; Guo, J.; Toma, F.M.; Qiu, Y.; et al. Facet-Selective Deposition of Ultrathin Al_2O_3 on Copper Nanocrystals for Highly Stable CO_2 Electroreduction to Ethylene. *Angew. Chem. Int. Ed.* **2021**, *60*, 24838–24843. [CrossRef]

80. Hong, H.-K.; Jo, J.; Hwang, D.; Lee, J.; Kim, N.Y.; Son, S.; Kim, J.H.; Jin, M.-J.; Jun, Y.C.; Erni, R.; et al. Atomic Scale Study on Growth and Heteroepitaxy of ZnO Monolayer on Graphene. *Nano Lett.* **2017**, *17*, 120–127. [CrossRef]

81. Tanuma, S.; Powell, C.J.; Penn, D.R. Calculations of Electron Inelastic Mean Free Paths. V. Data for 14 Organic Compounds over the 50–2000 eV Range. *Surf. Interface Anal.* **1994**, *21*, 165–176. [CrossRef]

inorganics

Article

Gd₂O₃ Doped UO₂(s) Corrosion in the Presence of Silicate and Calcium under Alkaline Conditions

Sonia García-Gómez [1,*], Javier Giménez [1], Ignasi Casas [1], Jordi Llorca [1] and Joan De Pablo [1,2]

[1] Department of Chemical Engineering, EEBE and Barcelona Research Center in Multiscale Science and Engineering, Universitat Politècnica de Catalunya (UPC), Eduard Maristany, 10-14, 08019 Barcelona, Spain; francisco.javier.gimenez@upc.edu (J.G.); ignasi.casas@upc.edu (I.C.); jordi.llorca@upc.edu (J.L.); joan.de.pablo@upc.edu (J.D.P.)

[2] EURECAT, Centre Tecnològic de Catalunya, Plaça de la Ciència 2, 08243 Manresa, Spain

* Correspondence: sonia.garcia.gomez@upc.edu

Abstract: The anodic reactivity of UO_2 and UO_2 doped with Gd_2O_3 was investigated by electrochemical methods in slightly alkaline conditions in the presence of silicate and calcium. At the end of the experiments, the electrodes were analysed by X-ray photoelectron spectroscopy to determine the oxidation state of the uranium on the surface. The experiments showed that the increase in gadolinia doping level led to a reduction in the reactivity of UO_2, this effect being more marked at the highest doping level studied (10 wt.% Gd_2O_3). This behaviour could be attributed to the formation of dopant-vacancy clusters (Gd^{III}-Ov), which could limit the accommodation of excess O^{2-} into the UO_2 lattice. In addition, the presence of Ca^{2+} and SiO_3^{2-} decreased the anodic dissolution of UO_2. In summary, the Gd_2O_3 doping in presence of silicate and calcium was found to strongly decrease the oxidative dissolution of UO_2, which is a beneficial situation regarding the long-term management of spent nuclear fuel in a repository.

Keywords: UO_2(s); gadolinia doping; anodic oxidation; X-ray photoelectron spectroscopy; silicate and calcium ions

Citation: García-Gómez, S.; Giménez, J.; Casas, I.; Llorca, J.; De Pablo, J. Gd₂O₃ Doped UO₂(s) Corrosion in the Presence of Silicate and Calcium under Alkaline Conditions. *Inorganics* **2023**, *11*, 469. https://doi.org/10.3390/inorganics11120469

Academic Editors: Roberto Nisticò, Torben R. Jensen and Luciano Carlos

Received: 7 October 2023
Revised: 25 November 2023
Accepted: 29 November 2023
Published: 1 December 2023

1. Introduction

The repository concept is an internationally accepted approach for the long-term management of spent nuclear fuel, where the fuel would be sealed in containers. However, in the worst-case-scenario where the containers fail, the contact of the fuel with groundwater could be possible. Since the majority of the radionuclides are located within the UO_2 matrix and taking into account that U^{VI} is more soluble than U^{IV}, the release of the radionuclides to the groundwater would be controlled by the fuel corrosion/dissolution rate, which would be strongly affected by the redox conditions. Under oxidizing aqueous conditions, the corrosion of UO_2 occurs via two steps process. First $U^{IV}O_2$ is oxidized to $U^{IV}_{1-2x}U^{V}_{2x}O_{2+x}$ forming a thin oxidized surface layer, while in the second step soluble U^{VI} is released to the solution [1]. In alkaline conditions, dissolved uranium could reprecipitate, forming corrosion product deposits ($UO_3 \cdot yH_2O$).

Some of the main parameters affecting the corrosion of the UO_2 matrix in a repository are the rare earth (RE) doping of the fuel [2,3] and the composition of the groundwater [4]. Nowadays, UO_2 is commonly doped with rare earth elements such as Gd (in the form of Gd_2O_3) [5] for its critical role as a burnable neutron absorber. The high neutron absorption cross section of ^{155}Gd and ^{157}Gd helps to counterbalance the excess of reactivity in the reactor core during the initial stages of operation, improving overall reactor performance.

The influence of the incorporation of Gd to UO_2 has been extensively studied to determine lattice parameters [6,7], UO_2 oxidation in dry experiments [8,9], UO_2 electrochemical reactivity [1,7,10] and UO_2 dissolution [11,12].

Air oxidation kinetics of UO_2 at high temperature (325 °C) indicated that doping with Gd stabilizes the first oxidation product and inhibits the oxidation to U_3O_8 being the kinetic mechanism dependent on the Gd content [8]. He et al. studied the effect of fission product doping on the anodic reactivity of UO_2 (they worked with SIMFUEL) and concluded that doping with such elements distorts the lattice structure of UO_2. This fact affects its corrosion process, since the oxidation was suppressed as the extent of doping increased [1]. Liu et al. showed that Gd doping lead to a contraction of the fluorite lattice, reducing the electrochemical reactivity of $(U_{1-y}Gd_y)O_2$ [7]. Razdan and Shoesmith studied the influence of trivalent-dopants (such as Gd^{III}) on the electrochemical behaviour of UO_2 [2]. They found that while the mechanism of oxidation/dissolution of Gd-doped UO_2 is comparable to that of SIMFUEL, the reactivity is lower. The effect of Gd-doping was also observed on dissolution experiments, showing a reduction of the oxidative dissolution rate [11–13]. This behaviour was commonly attributed to the formation of Gd^{III}-Ov clusters [14] in order to maintain electroneutrality, limiting the accommodation of O^{2-} ions during oxidation [15]. Therefore, the Gd^{III} influence on fuel reactivity under repository conditions is of great importance.

Concrete has been proposed as a construction material in some deep geological repositories [16]. Water in contact with concrete presents high pH values and contains silicate and calcium ions [17]. Flow-through dissolution experiments with UO_2 performed by Wilson and Gray [4] revealed a considerable reduction in the concentration of U measured when Ca and Si were present in the solution. Santos et al. investigated the effect of SiO_3^{2-} [18] and Ca^{2+} on the corrosion of SIMFUEL [19]. The authors concluded that these ions had little influence on the first oxidation step, but they inhibited/retarded the formation of U^{VI}, probably due to the adsorption of silicate and calcium on the fuel surface. Espriu et al. studied the corrosion of SIMFUEL at high pH and in the presence of calcium and silicate [20]. Their X-ray photoelectron spectroscopy (XPS) analysis evidenced that the presence of both ions in a high alkaline solution have a significant impact on the oxidation of the surface of SIMFUEL, since the surface was less oxidized than that of the experiments conducted in a solution without such ions. However, the corrosion of Gd_2O_3 doped UO_2 under cementitious water is not well known.

In this work, both undoped UO_2 and UO_2 doped with 5 and 10 wt.% Gd_2O_3 pellets were examined electrochemically in order to study the influence of Gd_2O_3 doping on the oxidative reactivity of UO_2 at slightly alkaline groundwater in the presence of silicate and calcium.

2. Results and Discussion

2.1. Characterisation of the Gd_2O_3 Doped UO_2 Pellets

The surface morphology of the sintered UO_2 pellets doped with 5 wt.% and 10 wt.% Gd_2O_3 (Figure 1), exhibited clearly defined grain boundaries and low-porosity microstructure. Furthermore, no unreacted Gd_2O_3 was found and EDX analysis revealed a homogeneous distribution of gadolinium within the UO_2 matrix.

The examination of the samples using X-ray diffraction (XRD) revealed the fluorite crystal structure of UO_2 (Figure 2). The peaks observed in the Gd_2O_3-doped samples were found to be slightly shifted toward greater angles compared to the UO_2 sample. This shift was caused by the lattice contraction as gadolinium concentration increased, as previously observed in previous studies [7].

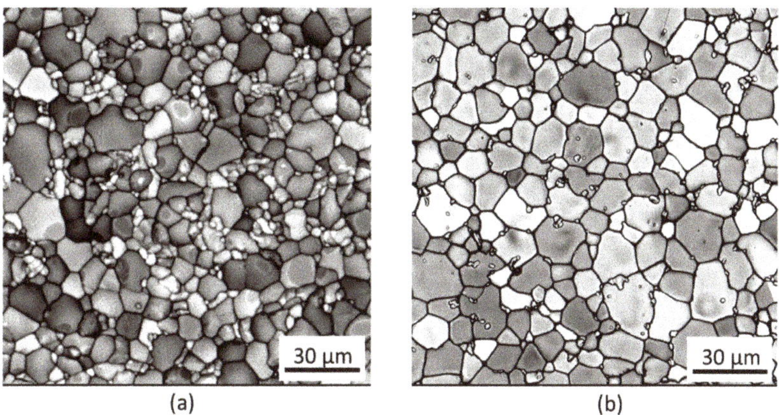

Figure 1. Scanning electron micrograph of UO_2 doped with 5 wt.% Gd_2O_3 (**a**) and 10 wt.% (**b**) Gd_2O_3 after sintering.

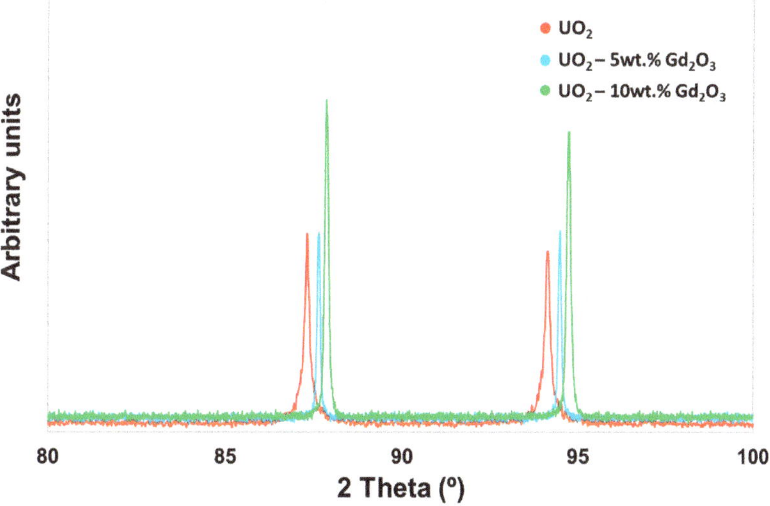

Figure 2. Close-up of the XRD spectra of non-doped UO_2 and UO_2 doped with 5 wt.% and 10 wt.% Gd_2O_3.

2.2. Cyclic Voltammetry

Figure 3 shows a series of cyclic voltammograms (CVs) recorded in silicate and calcium solution on the undoped and Gd_2O_3 doped UO_2 electrodes. Similar stages of both oxidation and reduction regions were previously observed for SIMFUEL [21] and undoped UO_2 [22]. In the forward scan, two different regions can be observed. In region 1 (at the potential of around -0.2 V), the increase in the anodic oxidation current observed is attributed to the formation of a mixed U^{IV}/U^{V} thin surface oxide layer with a thickness limited by the O^{2-} injection into the UO_2 fluorite lattice. It can be observed that the onset of the oxidation in region 1 is independent of the Gd_2O_3 doping level. However, there is a significant difference in the anodic current observed, registering lower current density values as the doping level of Gd_2O_3 increases. This behaviour was observed in [10] for hyperstoichiometric $U_{1-y}Gd_yO_{2+x}$ and can be attributed to the ability of Gd to retard the oxidation of UO_2 matrix surface by the formation of Gd^{III}-Ov clusters. In region 2 (>0.2 V), the thin surface layer is further oxidized to soluble U^{VI} and then the dissolution as UO_2^{2+} could take place. Under alkaline conditions, the solubility of UO_2^{2+} is limited [1], therefore it could be deposited as insoluble $U^{VI}O_3 \cdot yH_2O$ on the electrode surface. The formation of this deposit could explain the suppression of the current observed in Figure 3 at higher voltages (>0.3 V). Another possibility is that the silicate and calcium in solution could be adsorbed on the electrode surface, suppressing the further oxidation of the thin surface layer [19]. The stabilization of the UO_2 matrix in presence of silicate and calcium was previously observed on SIMFUEL [18–20].

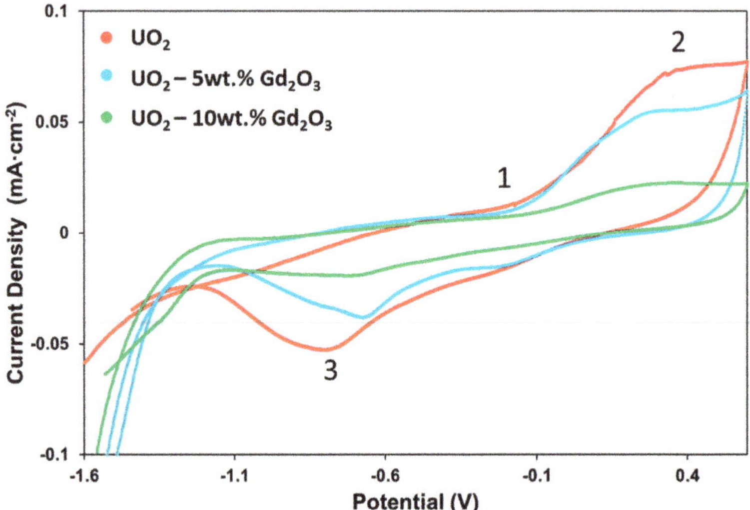

Figure 3. CVs recorded on the three different electrodes (rotation rate of 1000 rpm) in a solution that contains 0.1 M NaCl, 10^{-2} M SiO_3^{2-} and 10^{-3} M Ca^{2+}, pH = 10, at a scan rate of 10 mV·s^{-1}. The potential was scanned from -1.6 V to 0.6 V and back.

On the reverse scan, a cathodic reduction peak (region 3) was observed with a maximum at around -0.7 V. This reduction peak is commonly attributed to the reduction of $U^{IV}_{1-2x}/U^{V}_{2x}O_{2+x}/U^{VI}O_3 \cdot yH_2O$ surface layer [2]. The presence of this reduction peak for all electrodes indicate that an oxidized surface layer is formed despite the presence of Gd, however, the current associated with the reduction peak vary significantly with doping level. This confirms that the thickness of the thin $U^{IV}_{1-2x}/U^{V}_{2x}O_{2+x}$ layer differs among them, which is in agreement with the current densities registered in regions 1 and 2.

In contrast, CV experiments performed on 5 wt.% Gd_2O_3-doped UO_2 electrode in presence of bicarbonate (Figure 4) registered higher anodic oxidation currents in regions 1 and 2 than in solutions that contained silicate and calcium. This behaviour can be attributed to the ability of bicarbonate to form soluble uranyl carbonate complex ions, which avoided the formation of secondary phases on the electrode surface and enhanced the oxidation and dissolution process of UO_2 [23]. This is consistent with the smaller reduction peak observed in region 3 in bicarbonate solution.

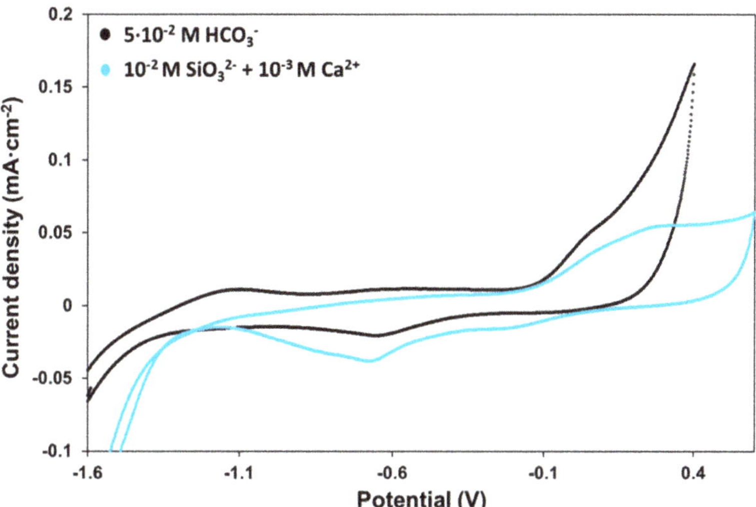

Figure 4. CVs recorded at a scan rate of 10 mV·s^{-1} (rotation rate of 1000 rpm) on 5 wt.% Gd$_2$O$_3$-doped UO$_2$ electrode in a solution that contains 0.1 M NaCl, 10^{-2} M SiO$_3{}^{2-}$ and 10^{-3} M Ca^{2+} (blue) and in a solution that contains 0.1 M NaCl and 5×10^{-2} M HCO$_3{}^-$ (black). The potential was scanned from -1.6 V to 0.6 V and back. Both dissolutions were adjusted at pH 10.

The suppression of the anodic dissolution could be probably attributed to the adsorption of silicate and calcium to the electrode surface. These results are in good agreement with the findings of Wilson and Gray [4], as they detected in leaching experiments a thin layer of Ca and Si, which inhibited the dissolution of UO$_2$. Further investigation in solutions containing Ca, Si and carbonates, which better mimic cementitious water, may be needed to determine whether the predominant factor is the formation of soluble ternary complexes Ca-U(VI)-CO$_3$ [24] or the inhibitory effects of Ca and Si.

2.3. Potentiostatic Oxidation

Since CVs are rapid experiments, to better observe differences in reactivity due to doping level, potentiostatic oxidation experiments were performed. Figure 5 shows a series of current density vs. time plots recorded at an oxidative potential of 0.4 V for 1 h for the different electrodes in a solution with silicate and calcium. In general, for all electrodes, the anodic current decreases linearly on the logarithm scale with time, which is consistent with a loss of surface reactivity due to the growth of a U$^{IV/V}$O$_{2+x}$ surface layer [19].

For the non-doped electrode, the current density registered was considerably higher than in doped electrodes. After 1000 s the current tends toward a steady-state value (reaches a plateau). Regarding doped electrodes, 5 wt.% Gd$_2$O$_3$-doped UO$_2$ electrode registered at short times higher current densities compared to 10 wt.% Gd$_2$O$_3$-doped UO$_2$ electrode, while at long times, the currents continued decreasing, reaching very low values for both electrodes.

Figure 6 shows the cumulative integration of the potentiostatic current densities (from Figure 5) as a function of time for the three electrodes. The general trend observed is that the charge accumulated increases as the doping level decreases. Whereas the total charge for the non-doped electrode continues increasing with time in the period of time registered (3600 s), for doped electrodes their respective total charges tend to reach a stable state.

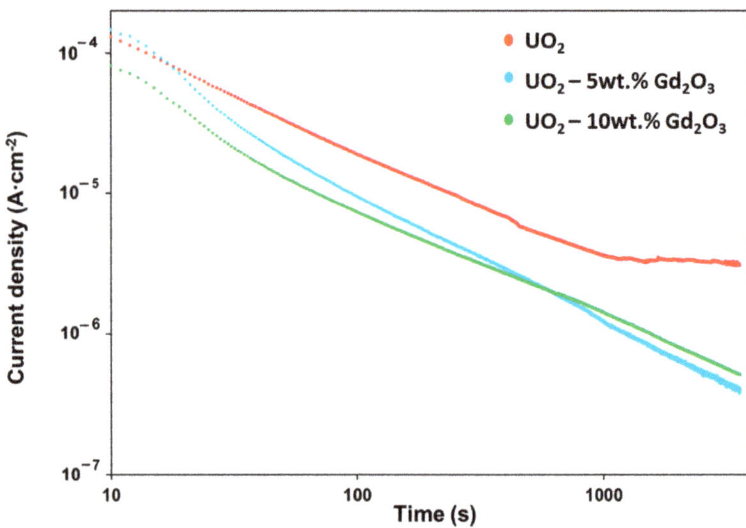

Figure 5. Potentiostatic current-time curves recorded at 0.4 V for 1 h on the three different electrodes in a solution that contains 0.1 M NaCl, 10^{-2} M SiO_3^{2-} and 10^{-3} M Ca^{2+}.

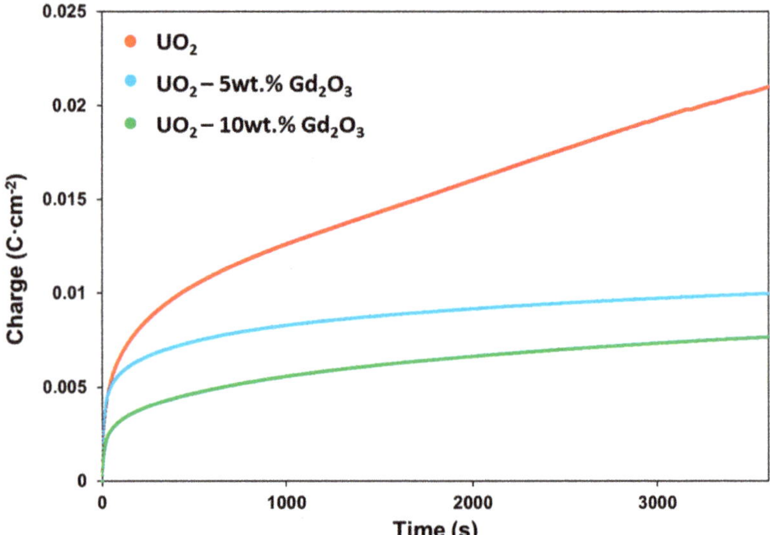

Figure 6. Cumulative anodic charges obtained by integration of the current densities registered from potentiostatic experiments in a solution that contains 0.1 M NaCl, 10^{-2} M SiO_3^{2-} and 10^{-3} M Ca^{2+} at 0.4 V for 1 h on the three electrodes as a function of time.

XPS was performed after potentiostatic experiments to determine the composition of the electrode surface as a function of the extent of doping. Figure 7 shows the deconvoluted spectra for the U $4f_{5/2}$ and U $4f_{7/2}$ regions and their corresponding satellite structures for the three different electrodes while Table 1 collects the percentages of U^{IV}, U^{V} and U^{VI} on the electrodes surface. In all cases, as expected, traces of U^{IV} were detected, whereas U^{V} and U^{VI} states dominate the surface composition of the electrodes, since an oxidizing potential was applied. This is consistent with the formation of an underlying $U^{IV}_{1-2x}U^{V}_{2x} O_{2+x}$

layer and the following formation of UO_2^{2+}, which can be accumulated as a $U^{VI}O_3 \cdot yH_2O$ deposit. It can be observed that as the doping level increases, the U^V content increases, at the expense of U^{VI}. Although the effect of doping with 5 wt.% Gd_2O_3 is small, in the 10 wt.% Gd_2O_3-UO_2 electrode, a strong U^V signal was obtained, which is consistent with a marked decrease in reactivity at high doping levels (10%), as already reported in the literature [7]. The differences observed in U^{VI} concentration provide evidence that doping with Gd_2O_3 slows down the second oxidation step ($U^{IV}_{1-2x}U^V_{2x}O_{2+x}$ to $U^{VI}O_3 \cdot yH_2O$). These results are in agreement with the potentiostatic data, since lower anodic charges lead to lower amount of U^{VI} and hence, lower oxidation.

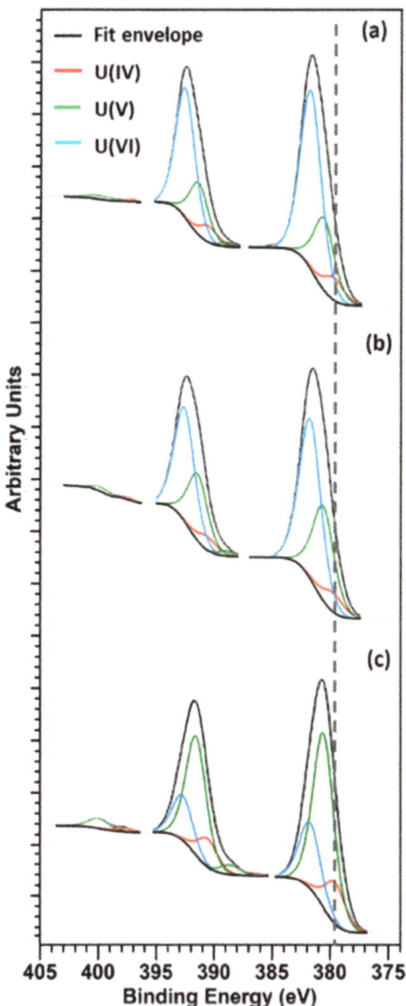

Figure 7. U 4f XPS spectra resolved into U^{IV}, U^V and U^{VI} contributions recorded after potentiostatic oxidation at 0.4 V on (**a**) undoped UO_2 electrode, (**b**) 5 wt.% Gd_2O_3-doped UO_2 electrode and (**c**) 10 wt.% Gd_2O_3-doped UO_2 electrode. The vertical dashed line represents the position of U^{IV} at 379.7 eV.

Table 1. Relative fractions of U^{IV}, U^{V} and U^{VI} on the surface of the different samples from XPS results after potentiostatic oxidation at 0.4 V.

Sample	U^{IV}(%)	U^{V}(%)	U^{VI}(%)
UO_2	8 ± 5	26 ± 5	66 ± 5
UO_2—5 wt.% Gd_2O_3	7 ± 5	35 ± 5	58 ± 5
UO_2—10 wt.% Gd_2O_3	15 ± 5	62 ± 5	23 ± 5

This observation is consistent with the results reported in air oxidation by Scheele et al. [25], who observed a delay on the onset of the second oxidation to U_3O_8 on Gd_2O_3-UO_2 doped samples. XPS analysis performed after dissolution experiments in presence of carbonate/bicarbonate [12] revealed a big contribution of U(V) on the surface of the Gd_2O_3-UO_2 samples, and low U concentration on the leaching, indicating that Gd-doping enhanced the stability of U(V) and inhibited the subsequent oxidation to U(VI). A decrease in the uranium concentration on Gd-doped UO_2 pellets was also reported in leaching experiments in H_2O_2 and γ-irradiation [13]. Similar behaviour was found in the literature for Nd-doped UO_2 in presence of H_2O_2 [26]. This found could indicate that doping UO_2 with trivalent dopants (like Gd^{3+} and Nd^{3+}) would lead to a comparable stabilization effect on the UO_2 structure.

2.4. Cathodic Stripping Voltammetry

Additional potentiostatic experiments were performed at the same conditions (0.4 V for 1 h), but this time followed by cathodic stripping voltammetry (CSV) in order to reduce the surface layer and the deposit. Figure 8 shows the CSV recorded for the three electrodes. Two reduction peaks were observed in this region, which are commonly assigned to the reduction of the dual phase nature of $U^{IV}_{1-2x}U^{V}_{2x}O_{2+x}/U^{VI}O_3 \cdot yH_2O$. The reduction peak observed at -0.7 V (peak 1 in Figure 8) was attributed previously to the reduction of the underlying $U^{IV}_{1-2x}U^{V}_{2x}O_{2+x}$ layer [21,22], whereas the second peak located at -0.8 V (peak 2 in Figure 8) was assigned to the reduction of $UO_3 \cdot yH_2O$ deposit [15].

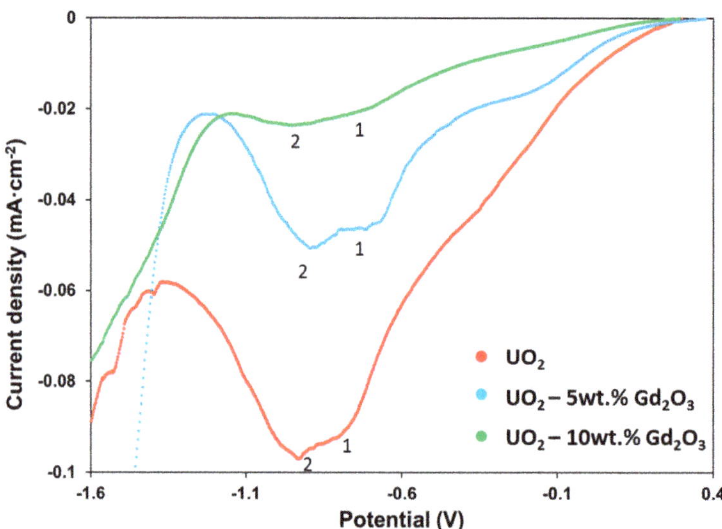

Figure 8. Cathodic stripping voltammograms (CSV) recorded on the different electrodes after potentiostatic oxidation in a solution that contains 0.1 M NaCl, 10^{-2} M SiO_3^{2-} and 10^{-3} M Ca^{2+}. The potential was scanned from 0.4 V to -1.6 V at a scan rate of 10 mV·s^{-1}.

The current peaks can be considered as a measure of the thickness of the oxidized surface layer. Hence, the thickness of the $U^{IV}_{1-2x}U^{V}_{2x}O_{2+x}/U^{VI}O_3 \cdot yH_2O$ layer was lower as the Gd_2O_3 doping level increases. These results are in agreement with the current densities registered in the potentiostatic oxidation experiments (Figure 5).

2.5. Corrosion Potential

Figure 9 shows the corrosion potential (E_{CORR}) registered on each electrode. The horizontal line at -0.4 V represents the threshold for the onset of fuel corrosion. For all electrodes, the values of E_{CORR} increased quickly from the cathodic cleaning potential (-1.6 V) until achieving a steady state value. The values of E_{CORR} recorded after 24 h were -0.10 V, -0.14 V and -0.27 V for undoped UO_2 electrode, 5 wt.% Gd_2O_3 and 10 wt.% Gd_2O_3 doped UO_2 electrodes, respectively, which are substantially more negative as the doping level increases, indicating an influence of the electrode composition on the E_{CORR}.

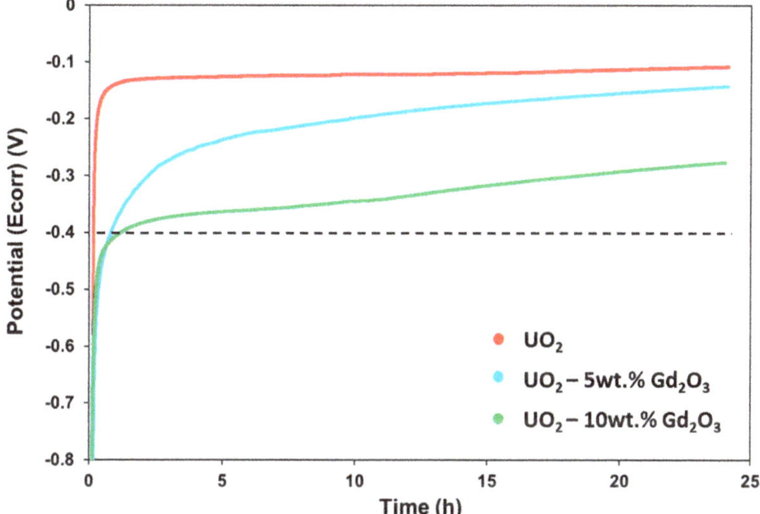

Figure 9. E_{CORR} measurements recorded on the three different electrodes in a solution that contains 0.1 M NaCl, 10^{-2} M SiO_3^{2-} and 10^{-3} M Ca^{2+}. The horizontal dashed line at -0.4 V indicates the threshold potential for the beginning of oxidation of the UO_2.

Figure 10 shows the XPS analysis performed after E_{CORR} measurements and Table 2 displays the fraction of the individual oxidation states of uranium (expressed as percentages) on the electrode surfaces. U^{IV} is the dominant oxidation state present in the 10 wt.% Gd_2O_3 doped UO_2 electrode surface (around 87%), whereas both undoped and 5 wt.% Gd_2O_3 doped electrodes surfaces are predominantly composed of the oxidation states U^{IV} and U^{V}, being the U^{IV} content slightly higher in the doped electrode than in the undoped one. A low amount of U^{VI} was detected in UO_2 electrode (14%) while in doped electrodes the amount was almost negligible (below 4%). As mentioned before, silicate and calcium could be adsorbed on the electrode surface, which could partially inhibit the further oxidation of the thin surface layer [19] at such conditions.

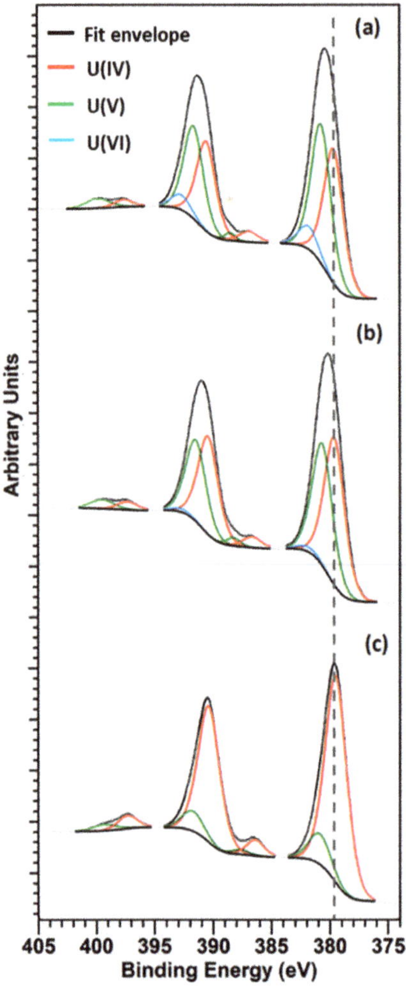

Figure 10. U 4f XPS spectra resolved into U^{IV}, U^{V} and U^{VI} contributions recorded after E_{CORR} experiments on (**a**) undoped UO_2 electrode, (**b**) 5 wt.% Gd_2O_3-doped UO_2 electrode and (**c**) 10 wt.% Gd_2O_3-doped UO_2 electrode. The vertical dashed line represents the position of U^{IV} at 379.7 eV.

Table 2. Relative fractions of U^{IV}, U^{V} and U^{VI} on the surface of the different samples from XPS results after corrosion potential experiments.

Sample	U^{IV}(%)	U^{V}(%)	U^{VI}(%)
UO_2	38 ± 5	48 ± 5	14 ± 5
UO_2—5 wt.% Gd_2O_3	49 ± 5	47 ± 5	4 ± 5
UO_2—10 wt.% Gd_2O_3	87 ± 5	13 ± 5	0 ± 5

Shoesmith et al. reported a reaction sequence on unirradiated UO_2 electrodes in basic solutions in terms of E_{CORR} [27]. For $E_{CORR} < -0.1$ V, the oxidation of UO_2 surface occurs, as a thin surface layer U^{IV}/U^{V} is being formed. Therefore, it is expected that the oxidized states U^{IV} and U^{V} dominates the surface composition. The surface becomes more oxidized as E_{CORR} increases. For $E_{CORR} \geq -0.1$ V, the oxidative dissolution of the oxidized layer (as UO_2^{2+}) occurs, hence the presence of U^{VI} is expected as no complexing agents like bicarbonate are present in the solution. For $E_{CORR} > 0$ V, the formation of secondary phases (in the form of $UO_3 \cdot H_2O$) is expected to take place on the electrode surface.

According to our XPS measurements, the composition of the surface obtained after the corrosion experiments are consistent with E_{CORR} values registered at steady state. For undoped UO_2 electrode, the presence of U^{VI} can be explained due to oxidative dissolution explained for $E_{CORR} \geq -0.1$ V. For doped electrodes, only the oxidation of UO_2 surface occurs, as no U^{VI} was significantly measured ($E_{CORR} < -0.1$ V). However, in the 10 wt.% Gd_2O_3 electrodes, only a slight oxidation of the UO_2 surface (U^V composition of 13%) was observed, which is in agreement with the low E_{CORR} value obtained (-0.27 V, close to the oxidation threshold). On the contrary, the 5 wt.% Gd_2O_3 doped UO_2 electrode registered a higher E_{CORR} (-0.14 V) and a higher degree of oxidation (U^V composition of 47%).

These results would confirm that the formation of the thin surface layer U^{IV}/U^V was influenced by Gd^{III}-doping in the presence of silicate and calcium. Such retardation of the oxidation process can be attributed to the ability of Gd to delay the incorporation of O^{2-} into interstitial sites within the fluorite lattice of the UO_2 surface [7,15]. This effect was particularly marked at high doping levels (10 wt.% Gd_2O_3). However, despite the similar E_{CORR} values achieved for undoped UO_2 and UO_2 doped with 5 wt.% Gd_2O_3, a more rapid oxidation on the undoped UO_2 electrode was observed (the rate of rise of E_{CORR} was higher for the undoped electrode), evidencing a clear decrease in reactivity due to Gd_2O_3.

3. Materials and Methods

3.1. Electrode Preparation

Undoped UO_2 and UO_2 doped with 5 wt.% and 10 wt.% Gd_2O_3 pellets were fabricated by mixing fine-grained powders of UO_2 (supplied by ENUSA, Madrid, Spain) and Gd_2O_3 powders (99.9%, supplied by Sigma Aldrich, St. Louis, MO, USA). The powders were compacted into pellets by means of a hydraulic press and sintered in a horizontal tube furnace (ST196030 HG from Hobersal, Barcelona, Spain) at 1740 °C for 9 h in a reducing atmosphere (5% hydrogen in 95% argon), based on the methodology explained by Baena et al. [28]. This methodology was confirmed to be adequate for obtaining successfully sintered Gd_2O_3-doped UO_2 pellets [12]. The pellets, with a thickness of 3 mm and a diameter of 12 mm, were characterised by scanning electron microscopy (SEM) and X-ray diffraction (XRD). Then, to study the electrochemical behaviour of the pellets, electrodes were prepared from the pellets by following the procedure described in the literature [29]. First, in order to improve the electrical conductivity, one face of the pellet was electroplated with copper. For that purpose, the pellet was fitted at the end of a rubber tube, with the face to be plated exposed to a solution of 0.1 mol$\cdot$L^{-1} $CuSO_4$ and a Cu metal wire immersed as a cathode. Mercury (Thermo Scientific, Waltham, MA, USA) was poured into the rubber tube to make an electrical contact between the pellet and a stainless-steel wire, acting as an anode. A 10 mA current was applied for 5 min by using a laboratory power supply Manson EP-613 (Kwai Chung, Hong Kong). The Cu-plated side of the pellet was connected to a stainless-steel threaded post by using a conductive silver epoxy (8331-14G). The pellet was embedded with a non-conductive epoxy EE 9466 resin so that only one side of the electrode is in contact with the solution. The electrode was left to cure overnight in a vacuum desiccator. The threaded stainless-steel post was screwed onto the shaft of a rotating disk electrode (Metrohm, Herisau, Switzerland). Both the electrode and stainless-steel shaft were tightly wrapped with poly(tetrafluoroethylene) (PTFE) to avoid entering in contact with the solution. Prior to any experiment, the electrode was abraded using wet SiC paper (1200 grit) and rinsed with distilled deionized water to remove any oxidized surface phase.

3.2. Electrochemical Cell and Equipment

Electrochemical experiments were conducted in a single compartment corrosion cell Autolab 1 L model from Metrohm (Herisau, Switzerland). A standard three-electrode set-up was used with some electrodes of UO_2 (doped with 0, 5 and 10 wt.% Gd_2O_3) as working electrode, the counter electrode was a platinum sheet (surface area ~ 1 cm^2) and the reference electrode was an Ag/AgCl (3 M KCl) electrode. All potential values are recorded versus the Ag/AgCl standard reference electrode (+0.208 V, 25 °C vs. standard hydrogen electrode (SHE)). Experiments were conducted at room temperature. All electrochemical experiments were performed using Multi Autolab Potentiostat-Galvanostat M204 from Metrohm, (Herisau, Switzerland). NOVA software 2.1 (Metrohm, Herisau, Switzerland) was used to record and to analyse the data.

3.3. Electrochemical Techniques

Four general types of electrochemical techniques were carried out: cyclic voltammetry (CV), potentiostatic oxidation, cathodic stripping voltammetry (CSV) and corrosion potential (E_{CORR}). Prior to any experiment, the electrode was cathodically cleaned for 120 s at -1.6 V to remove U^{VI} oxides present on the surface. CV is a potential tool to investigate the potential at which UO_2 undergoes oxidation or reduction in a solution. In that regard, the potential (V) was scanned from -1.6 V to 0.6 V and back at a rate of 10 mV·s^{-1} while recording the current. The electrode rotation rate was held constant at 1000 rpm. In potentiostatic oxidation experiments, a constant anodic potential of 0.4 V was applied to the electrode for an hour and current-time profiles were recorded. This technique is capable of identifying variations in oxidation rates and levels of the electrode at a specific oxidizing potential due to Gd doping. After the anodic oxidation, the surface of the electrode was analysed by XPS. Immediately after a second set of identical potentiostatic oxidation experiments, cathodic stripping voltammetry was conducted in order to reduce the oxidized surface layer formed on the electrode. The cathodic charge measured can serve as an indicator of the layer thickness. This was performed by scanning the potential from the potentiostatic oxidation potential (0.4 V) to -1.6 V at a scan rate of 10 mV·s^{-1}, while the current was registered. Finally, in the corrosion potential experiments, the working electrode was immersed in the solution for 24 h, while the potential that exists across the electrodes and electrolyte was measured in the absence of any current flow. This measurement serves as an indicator of the tendency of an electrode to corrode in a specific environment and XPS analysis was performed after the E_{CORR} measurement.

3.4. Solution

The solution used for all experiments was 0.1 mol·dm^{-3} NaCl (99.0%, Sigma Aldrich, St. Louis, MO, USA) as a supporting electrolyte, containing 10^{-2} mol·dm^{-3} SiO$_3^{2-}$ (Na$_2$SiO$_3$·5H$_2$O (99.0%, Sigma Aldrich, St. Louis, MO, USA)) and 10^{-3} mol·dm^{-3} Ca^{2+} (CaCl$_2$ (99.0%, Sigma Aldrich, St. Louis, MO, USA)). In addition, for testing the performance of the electrodes in cyclic voltammetry experiments, a solution that contained 0.1 mol·dm^{-3} NaCl (99.0%, Sigma Aldrich, St. Louis, MO, USA) and $5 \cdot 10^{-2}$ mol·dm^{-3} of NaHCO$_3$ (99.0%, Sigma Aldrich, St. Louis, MO, USA) was prepared. All solutions were prepared using ultrapure water obtained from a Milli-Q water purification system (Merck, Darmstadt, Germany). The pH of the solutions was adjusted to 10 using HCl or NaOH and measured with an Orion model 720A pH meter (Waltham, MA, USA). The pH was measured periodically and did not vary more than 1% throughout the experiments. The solutions were bubbled with N$_2$ (99.99%, Nippon Gases, Tokyo, Japan) before any experiment for 1 h to purge and during the experiments.

3.5. XPS Analysis

XPS analyses were performed to observe changes in the oxidation of the electrode surface after potentiostatic oxidation and corrosion potential experiments. A SPECS system equipment (Berlin, Germany) with an Al anode XR50 source operating at 150 W and a Phoibos MCD-9 detector was used. The sensitivity of the equipment was ±0.1 eV of binding energy (BE). The spectra were recorded at a pressure below 10^{-8} mbar. The spectra were evaluated using CasaXPS software (Casa Software Ltd., Teignmouth, UK) version 2.3.25. C 1s peak was set at 285.0 eV to correct surface charging. The spectra were fitted using a 30% of Lorentzian curve and 70% of Gaussian curve with a Shirley background correction. The $U4f_{7/2}$ and $U4f_{5/2}$ regions were used to determine the total amount of each uranium oxidation state (U^{IV}, U^{V} and U^{VI}) by the deconvolution of the peaks [30–34]. The position of the satellite peaks was used to validate the curve-fitting procedure.

Before transferring to the XPS spectrometer for XPS measurements, electrodes were removed from the electrochemical cell, gently rinsed with MilliQ water to remove electrolyte residue and dried in a desiccator.

3.6. SEM Measurements

A Phenom XL scanning electron microscope coupled to an energy dispersive X-ray detector (EDX) from PhenomWorld (Waltham, MA, USA) was used to analyse the microstructure of the samples.

3.7. X-ray Diffraction (XRD)

A Bruker D8 diffractometer (Karlsruhe, Germany) with an X-ray source of Cu Kα radiation (λ = 1.54056 Å) at 45 kV and 35 mA was used to analyse the crystal structure of the samples.

4. Conclusions

The effect of Gd (5 wt.% and 10 wt.% Gd_2O_3) on the electrochemical behaviour of UO_2 was investigated in a slightly alkaline solution in the presence of silicate and calcium. Cyclic voltammetry experiments showed that the anodic oxidation/dissolution mechanism on Gd_2O_3-UO_2 was similar to that on undoped UO_2 electrode. However, the corrosion process of doped UO_2 decreases, as lower current densities were registered. Both potentiostatic and E_{CORR} experiments showed that such retardation effect was small when doping with 5 wt.% Gd_2O_3, however, it was strongly enhanced when Gd_2O_3 content is around 10 wt.% Gd_2O_3. This behaviour is commonly attributed to the reduction in the availability of oxygen vacancies to accommodate O^{2-} ions during oxidation in the UO_2 lattice, as Gd^{III}-O vacancy clusters might be formed.

In addition, the presence of silicate and calcium in solution was found to have influence on the oxidation of UO_2, leading to a suppression of anodic dissolution probably due to the adsorption of these ions on the electrode surface. This is in contrast to solutions that contain bicarbonate ions, which are known to form soluble complexes with UO_2^{2+} and to accelerate UO_2 corrosion. Additional experiments with solution containing Ca, Si and carbonate would be required to investigate which effect would predominate.

In conclusion, both Gd_2O_3 doping and the presence of silicate and calcium in solution are factors that decrease the reactivity of UO_2, which is beneficial for the assessment of repository safety, as an important fraction of the radionuclide release would be inhibited.

Author Contributions: Conceptualization, I.C. and J.D.P.; methodology, I.C.; writing—original draft, S.G.-G.; writing—review and editing, I.C., J.D.P., J.G. and J.L.; validation, S.G.-G.; funding acquisition, J.D.P. All authors have read and agreed to the published version of the manuscript.

Funding: This research was funded by Ministerio de Economía y Competitividad (Spain) with the projects ENE2017-83048-R and PID2020-116839RB-I00. ENRESA is also acknowledged for its interest and financial support (Project 079-CO-IA-2020-0001 and CO-IA-22-010). S. García-Gómez wants to acknowledge the fellowship with reference code PRE2018-085618. Financial support was also received from the Catalan AGAUR Agency through the Research Groups Support program (grant number: 2021-SGR-GRC-00596).

Data Availability Statement: Data is available on request.

Acknowledgments: J. Llorca is a Serra Húnter Fellow and is grateful to the ICREA Academia program.

Conflicts of Interest: The authors declare no conflict of interest.

References

1. He, H.; Keech, P.G.; Broczkowski, M.E.; Noël, J.J.; Shoesmith, D.W. Characterization of the Influence of Fission Product Doping on the Anodic Reactivity of Uranium Dioxide. *Can. J. Chem.* **2007**, *85*, 702–713. [CrossRef]
2. Razdan, M.; Shoesmith, D.W. Influence of Trivalent-Dopants on the Structural and Electrochemical Properties of Uranium Dioxide (UO$_2$). *J. Electrochem. Soc.* **2014**, *161*, H105–H113. [CrossRef]
3. Kim, J.G.; Ha, Y.K.; Park, S.D.; Jee, K.Y.; Kim, W.H. Effect of a Trivalent Dopant, Gd^{3+}, on the Oxidation of Uranium Dioxide. *J. Nucl. Mater.* **2001**, *297*, 327–331. [CrossRef]
4. Wilson, C.N.; Gray, W.J. Measurement of Soluble Nuclide Dissolution Rates from Spent Fuel. *MRS Proc.* **1989**, *176*, 489–498. [CrossRef]
5. Leinweber, G.; Barry, D.P.; Trbovich, M.J.; Burke, J.A.; Drindak, N.J.; Knox, H.D.; Ballad, R.V.; Block, R.C.; Danon, Y.; Severnyak, L.I. Neutron Capture and Total Cross-Section Measurements and Resonance Parameters of Gadolinium. *Nucl. Sci. Eng.* **2006**, *154*, 261–279. [CrossRef]
6. Lee, J.; Kim, J.Y.G.J.; Youn, Y.S.; Liu, N.; Kim, J.Y.G.J.; Ha, Y.K.; Shoesmith, D.W.; Kim, J.Y.G.J. Raman Study on Structure of U$_{1-y}$Gd$_y$O$_{2-x}$ (Y = 0.005, 0.01, 0.03, 0.05 and 0.1) Solid Solutions. *J. Nucl. Mater.* **2017**, *486*, 216–221. [CrossRef]
7. Liu, N.; Kim, J.; Lee, J.; Youn, Y.S.; Kim, J.G.; Kim, J.Y.; Noël, J.J.; Shoesmith, D.W. Influence of Gd Doping on the Structure and Electrochemical Behavior of UO$_2$. *Electrochim. Acta* **2017**, *247*, 496–504. [CrossRef]
8. Scheele, R.D.; Hanson, B.D.; Casella, A.M. Effect of Added Gadolinium Oxide on the Thermal Air Oxidation of Uranium Dioxide. *J. Nucl. Mater.* **2021**, *552*, 153008. [CrossRef]
9. García-Gómez, S.; Giménez, J.; Casas, I.; Llorca, J.; De Pablo, J. X-ray Photoelectron Spectroscopy (XPS) Study of Surface Oxidation of UO$_2$ Doped with Gd$_2$O$_3$ at Different Temperatures and Atmospheres. *Appl. Surf. Sci.* **2023**, *629*, 157429. [CrossRef]
10. Kim, J.; Lee, J.; Youn, Y.S.; Liu, N.; Kim, J.G.; Ha, Y.K.; Bae, S.E.; Shoesmith, D.W.; Kim, J.Y. The Combined Influence of Gadolinium Doping and Non-Stoichiometry on the Structural and Electrochemical Properties of Uranium Dioxide. *Electrochim. Acta* **2017**, *247*, 942–948. [CrossRef]
11. Casella, A.; Hanson, B.; Miller, W. The Effect of Fuel Chemistry on UO$_2$ Dissolution. *J. Nucl. Mater.* **2016**, *476*, 45–55. [CrossRef]
12. García-Gómez, S.; Giménez, J.; Casas, I.; Llorca, J.; De Pablo, J. Oxidative Dissolution Mechanism of Both Undoped and Gd$_2$O$_3$-Doped UO$_2$(s) at Alkaline to Hyperalkaline pH. *Dalton Trans.* **2023**, *52*, 9823–9830. [CrossRef] [PubMed]
13. Barreiro Fidalgo, A.; Jonsson, M. Radiation Induced Dissolution of (U, Gd)O$_2$ Pellets in Aqueous Solution—A Comparison to Standard UO$_2$ Pellets. *J. Nucl. Mater.* **2019**, *514*, 216–223. [CrossRef]
14. Park, K.; Olander, D.R. Defect Models for the Oxygen Potentials of Gadolinium-and Europium-Doped Urania. *J. Nucl. Mater.* **1992**, *187*, 89–96. [CrossRef]
15. Razdan, M.; Shoesmith, D.W. The Electrochemical Reactivity of 6.0 Wt% Gd-Doped UO$_2$ in Aqueous Carbonate/Bicarbonate Solutions. *J. Electrochem. Soc.* **2014**, *161*, H225–H234. [CrossRef]
16. Heath, T.; Schofield, J.; Shelton, A. Understanding Cementitious Backfill Interactions with Groundwater Components. *Appl. Geochem.* **2020**, *113*, 104495. [CrossRef]
17. Huertas, F.J.; Hidalgo, A.; Rozalén, M.L.; Pellicione, S.; Domingo, C.; García-González, C.A.; Andrade, C.; Alonso, C. Interaction of Bentonite with Supercritically Carbonated Concrete. *Appl. Clay Sci.* **2009**, *42*, 488–496. [CrossRef]
18. Santos, B.G.; Noël, J.J.; Shoesmith, D.W. The Influence of Calcium Ions on the Development of Acidity in Corrosion Product Deposits on SIMFUEL, UO$_2$. *J. Nucl. Mater.* **2006**, *350*, 320–331. [CrossRef]
19. Santos, B.G.; Noël, J.J.; Shoesmith, D.W. The Influence of Silicate on the Development of Acidity in Corrosion Product Deposits on SIMFUEL (UO$_2$). *Corros. Sci.* **2006**, *48*, 3852–3868. [CrossRef]
20. Espriu-Gascon, A.; Shoesmith, D.W.; Giménez, J.; Casas, I.; de Pablo, J. Study of SIMFUEL Corrosion under Hyper-Alkaline Conditions in the Presence of Silicate and Calcium. *MRS Adv.* **2017**, *2*, 543–548. [CrossRef]
21. Shoesmith, D.W. Used Fuel and Uranium Dioxide Dissolution Studies—A Review. In Proceedings of the Corrosion Conference and Expo (Corrosion 2008), New Orleans, LA, USA, 16–20 March 2008; p. 56.
22. Shoesmith, D.W. Fuel Corrosion Processes under Waste Disposal Conditions. *J. Nucl. Mater.* **2000**, *282*, 1–31. [CrossRef]
23. Keech, P.G.; Goldik, J.S.; Qin, Z.; Shoesmith, D.W. The Anodic Dissolution of SIMFUEL (UO$_2$) in Slightly Alkaline Sodium Carbonate/Bicarbonate Solutions. *Electrochim. Acta* **2011**, *56*, 7923–7930. [CrossRef]
24. Maia, F.M.S.; Ribet, S.; Bailly, C.; Grivé, M.; Madé, B.; Montavon, G. Evaluation of Thermodynamic Data for Aqueous Ca-U(VI)-CO$_3$ Species under Conditions Characteristic of Geological Clay Formation. *Appl. Geochem.* **2021**, *124*, 104844. [CrossRef]
25. Scheele, R.D.; Hanson, B.D.; Cumblidge, S.E.; Jenson, E.D.; Kozelisky, A.E.; Sell, R.L.; Macfarlan, P.J.; Snow, L.A. Effect of Gadolinium Doping on the Air Oxidation of Uranium Dioxide. *Mat. Res. Soc. Symp. Proc.* **2004**, *824*, CC8.8.1–CC8.8.6. [CrossRef]
26. Kegler, P.; Neumeier, S.; Klinkenberg, M.; Bukaemskiy, A.; Deissmann, G.; Brandt, F.; Bosbach, D. Accelerated Dissolution of Doped UO$_2$-Based Model Systems as Analogues for Modern Spent Nuclear Fuel under Repository Conditions. *MRS Adv.* **2023**, *8*, 255–260. [CrossRef]
27. Shoesmith, D.W.; Sunder, S.; Bailey, M.G.; Miller, N.H. Corrosion of Used Nuclear Fuel in Aqueous Perchlorate and Carbonate Solutions. *J. Nucl. Mater.* **1996**, *227*, 287–299. [CrossRef]
28. Baena, A.; Cardinaels, T.; Vos, B.; Binnemans, K.; Verwerft, M. Synthesis of UO$_2$ and ThO$_2$ Doped with Gd$_2$O$_3$. *J. Nucl. Mater.* **2015**, *461*, 271–281. [CrossRef]

29. Ofori, D.; Keech, P.G.; Noël, J.J.; Shoesmith, D.W. The Influence of Deposited Films on the Anodic Dissolution of Uranium Dioxide. *J. Nucl. Mater.* **2010**, *400*, 84–93. [CrossRef]
30. Ilton, E.S.; Bagus, P.S. XPS Determination of Uranium Oxidation States. *Surf. Interface Anal.* **2011**, *43*, 1549–1560. [CrossRef]
31. Ilton, E.S.; Boily, J.F.; Bagus, P.S. Beam Induced Reduction of U(VI) during X-Ray Photoelectron Spectroscopy: The Utility of the U4f Satellite Structure for Identifying Uranium Oxidation States in Mixed Valence Uranium Oxides. *Surf. Sci.* **2007**, *601*, 908–916. [CrossRef]
32. Teterin, Y.A.; Popel, A.J.; Maslakov, K.I.; Teterin, A.Y.; Ivanov, K.E.; Kalmykov, S.N.; Springell, R.; Scott, T.B.; Farnan, I. XPS Study of Ion Irradiated and Unirradiated UO$_2$ Thin Films. *Inorg. Chem.* **2016**, *55*, 8059–8070. [CrossRef] [PubMed]
33. Schindler, M.; Hawthorne, F.C.; Freund, M.S.; Burns, P.C. XPS Spectra of Uranyl Minerals and Synthetic Uranyl Compounds. I: The U 4f Spectrum. *Geochim. Cosmochim. Acta* **2009**, *73*, 2471–2487. [CrossRef]
34. Boily, J.F.; Ilton, E.S. An Independent Confirmation of the Correlation of Uf4 Primary Peaks and Satellite Structures of U^{VI}, U^{V} and U^{IV} in Mixed Valence Uranium Oxides by Two-Dimensional Correlation Spectroscopy. *Surf. Sci.* **2008**, *602*, 3637–3646. [CrossRef]

Article

Incorporation of Antimony Ions in Heptaisobutyl Polyhedral Oligomeric Silsesquioxanes

Stefano Marchesi [1], Chiara Bisio [1,2], Fabio Carniato [1,*] and Enrico Boccaleri [3,*]

[1] Dipartimento di Scienze e Innovazione Tecnologica, Università del Piemonte Orientale, Viale Teresa Michel, 11, 15121 Alessandria, Italy; stefano.marchesi@uniupo.it (S.M.); chiara.bisio@uniupo.it (C.B.)
[2] CNR-SCITEC Istituto di Scienze e Tecnologie Chimiche "G. Natta", Via C. Golgi 19, 20133 Milano, Italy
[3] Dipartimento per lo Sviluppo Sostenibile e la Transizione Ecologica, Università del Piemonte Orientale, Piazza Sant'Eusebio, 5, 13100 Vercelli, Italy
* Correspondence: fabio.carniato@uniupo.it (F.C.); enrico.boccaleri@uniupo.it (E.B.); Tel.: +39-0321360217 (F.C.); +39-0131360264 (E.B.)

Abstract: The direct incorporation of Sb(V) ions into a polycondensed silsesquioxane network based on heptaisobutyl POSS units (Sb(V)-POSSs) through a corner-capping reaction is reported for the first time in this work. As a reference sample, a completely condensed monomeric Sb(III)-POSS was prepared using a similar synthetic protocol. The chemical properties of both Sb-containing POSSs were investigated with different analytical and spectroscopic techniques. The analyses confirm the success of the corner-capping reaction for both samples and indicate that an Sb(V)-POSS sample is characterized by a heterogenous multimeric arrangement with an irregular organization of POSS cages linked to Sb(V) centers, and has a more complex structure with respect to the well-defined monomeric Sb(III)-POSS.

Keywords: antimony; silsesquioxane; polyhedral oligomeric silsesquioxane; POSS; corner-capping reaction

Citation: Marchesi, S.; Bisio, C.; Carniato, F.; Boccaleri, E. Incorporation of Antimony Ions in Heptaisobutyl Polyhedral Oligomeric Silsesquioxanes. *Inorganics* **2023**, *11*, 426. https://doi.org/10.3390/inorganics11110426

Academic Editors: Torben R. Jensen, Eleonora Aneggi, Hicham Idriss, Roberto Nisticò and Luciano Carlos

Received: 29 September 2023
Revised: 24 October 2023
Accepted: 25 October 2023
Published: 30 October 2023

1. Introduction

Polyhedral oligomeric silsesquioxanes (POSSs) are a unique class of condensed oligomeric organosilicon compounds consisting of an inorganic cage and pendant arms (e.g., H or organic moieties) bound to its apexes. The silicon atoms of the cage are covalently bound to $1/2$ oxygen (*sesqui-*) and hydrogen or hydrocarbon units (*-ane*), globally forming a 3D siloxane (Si-O-Si) skeleton consisting of tetrahedral base units with a generic chemical formula $(RSiO_{3/2})_n$, where $n = 4\text{–}18$ and R = H or organic substituents [1–4]. Because of their unique nature, they are considered to be excellent models for silica surface sites [5,6] and are very interesting building blocks for the development of hybrid organic–inorganic materials with well-defined physicochemical properties exploitable in many scientific and technological fields [3,7–11].

In addition to completely condensed POSSs with a cubic $R_8Si_8O_{12}$ structure and an inorganic cage diameter of 0.5 ± 0.7 nm [1,2], incompletely condensed POSSs containing reactive silanol groups (Si-OH) (e.g., trisilanol $R_7Si_7O_9(OH)_3$) in their core framework [12,13] represent a more interesting class of silsesquioxanes, owing to their ability to bind different functionalities to the inorganic core through reactions with a wide range of organosilanes or heteroelement precursors, the latter in the form of metal halides or alkoxides [13–16]. In the last few years, POSSs have been combined with several elements, from alkali to alkaline earth metals [17–21], metalloids [22–25], transition metals [11,16,25,26], and some lanthanides and actinides [19,27–34]. The construction of metal-containing POSSs [35–37] is achievable through different strategies: (i) a corner-capping reaction, one of the most studied in the literature, in which the Si-OH groups of partially condensed POSSs react with metal ions to generate fully condensed silsesquioxanes [2,38,39]; (ii) a complexation reaction, where the hydrocarbon groups bonded to the silicon atoms of the cage coordinate the metals [40–42]; and (iii) an interaction between the Si-O$^-$ units of POSS molecules and the metal ions [32,34].

Over the years, several elements in group 15 (pnictogens) of the periodic table have been studied in combination with polysilsesquioxanes and molecular POSSs [43]. Among them, antimony-containing silsesquioxanes are the most investigated, in particular by Feher et al. during the decade 1990–2000 [43–45]. In their work, a series of open-corner cyclohexyl-POSSs were reacted with either (i) a trivalent antimony salt (SbCl$_3$) through a base-catalyzed corner-capping reaction in benzene at room temperature, giving the corre-spondent a completely condensed cubic pnictite ester (c-C$_6$H$_{11}$)$_7$Si$_7$O$_{12}$Sb^{III}) [44], or (ii) 1–3 equivalents of the organometallic pentamethylantimony (Me$_5$Sb), leading to the formation of corresponding mono-, di-, and tri- SbMe$_4$-subistited, open-cage silsesquioxides, thus providing an alternative pathway for the preparation of these types of M-POSSs [45]. For Sb(V)-POSS, a monomeric structure with three methylated Sb(V) ions directly attached to three oxygen atoms of the partially condensed silica cage was obtained [45]. Attempted oxidation of Sb(III) to Sb(V) via ozonolysis in cyclohexyl-POSS samples did not lead to any experimental evidence regarding its formation [43,44]. To the best of our knowledge, no attempts have been made to incorporate in a simple synthetic way, without using expensive organic precursor Sb(V) ions directly in the POSS molecular structure.

The inclusion of Sb(V) and Sb(III) ions in POSS structures could be beneficial for the preparation of novel flame-retardant and/or fire-resistant micro- and nano-composite materials. Indeed, silsesquioxanes represent a fascinating 3D hybrid platform for the preparation of polymer/POSS composite materials with tailored fire-related properties (i.e., thermal inertia, ignition temperature, etc.) [17,46–49]. Antimony tri- and pentox-ides, along with trihalides and oxyhalides, are well-known for their fire retardancy or resistance properties either in combination with halogenated compounds or blended into suitable composite matrices [50–52] (as halogen-functionalized clays [53] or organoclay combined with plasticizers in thermoplastic polymers [54]), thanks to the establishment of synergistic effects.

Based on these considerations, in this work, we propose the direct incorporation of Sb(V) ions into a polycondensed silsesquioxane matrix through a reaction between trisilanol heptaisobutyl POSS and antimony pentachloride salt in equimolar amounts (Sb(V)-POSS, Scheme 1). The preparation was carried out following a corner-capping reaction procedure [27,43–45,55,56] and realized in anhydrous tetrahydrofuran in the presence of triethylamine. A completely condensed monomeric Sb(III)-POSS was also synthetized with the same procedure as a reference sample.

Scheme 1. *Cont.*

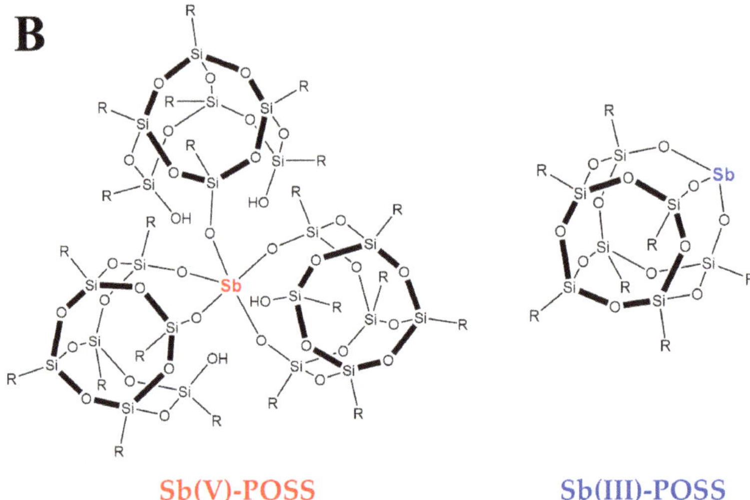

Scheme 1. (**A**) Schematic representation of the synthesis procedure of Sb(V)−POSS and Sb(III)−POSS samples. (**B**) Graphical representation of the structures of Sb(V)−POSS and Sb(III)−POSS.

2. Results and Discussion

A pentavalent antimony-POSS sample (Sb(V)-POSS) was prepared by adapting a corner-capping procedure already used for the preparation of Sb(III)-containing silsesquioxanes [27,43–45,55,56]. The experimental parameters, such as temperature and reaction time, were optimized to achieve a high yield of the materials prepared in this study, while maintaining a stoichiometric ratio of 1:1 between the starting reactants. In detail, the preparation of Sb(V)-POSS was performed in a single step, exploiting a base-assisted corner-capping reaction between the partially condensed trisilanol heptaisobutyl POSS (hib-POSS) and the Sb(V) pentachloride salt (SbCl$_5$) in an equimolar ratio. The reaction was carried out at 50 °C for 4 h in an inert atmosphere (under N$_2$ flow). The reaction mixture was carefully purified and a final oily orange product was obtained. The same experimental protocol was used in the preparation of the trivalent Sb(III)-POSS reference.

The chemical composition of the two samples was evaluated by combining elemental CHN and ICP-AES data. For Sb(III)-POSS, the amounts of Sb(III) and carbon corresponded to 1.1 and 31.3 mmol/g, respectively. Considering the organic composition of the POSS unit, it was possible to conclude that each POSS cage was bound to a single Sb(III) ion, with the formation of a fully condensed monomeric structure, in agreement with the results obtained by Feher et al. and Alphazan et al. for the preparation of cyclohexyl-POSS and heptaisobutyl-POSS containing Sb(III) ions through a corner-capping reaction [43–45,55,56]. By contrast, the stoichiometry of Sb(V)-POSS is completely different; the elemental analysis indicated a carbon/Sb(V) molar ratio value of 88.9, thus suggesting a multimeric structure of the sample with three POSS cages coordinated with one Sb(V) center.

The structural features of Sb(V)-POSS were analyzed via X-ray powder diffraction (XRPD) and the diffractogram was compared to that of both Sb(III)-POSS and the open-corner hib-POSS reactant. The X-ray pattern of Sb(V)-POSS (Figure 1c) was characterized by two main broad signals centered at 7 and 19° 2θ. Compared to the X-ray profile of hib-POSS (Figure 1a), most of the intense and well-resolved crystallographic signals were lost after the reaction. This indicated a loss of order in the arrangement of the silsesquioxane molecules, thus suggesting a mostly irregular rearrangement in the structural network of the POSS cage linked to Sb(V). A similar behavior was also observed for POSS-based polysilsesquioxanes functionalized with Eu^{3+} and/or Tb^{3+} ions and, more generally, in silica-based nanostructures possessing some degree of amorphous components [32,34,57].

Instead, Sb(III)-POSS (Figure 1b) showed a different profile, with the presence of a well-defined reflection at 8.3° 2θ along with several less intense bands at high 2θ values. The presence of these signals suggested a more orderly assembly of the monomeric Sb(III)-POSS with respect to the Sb(V)-POSS sample.

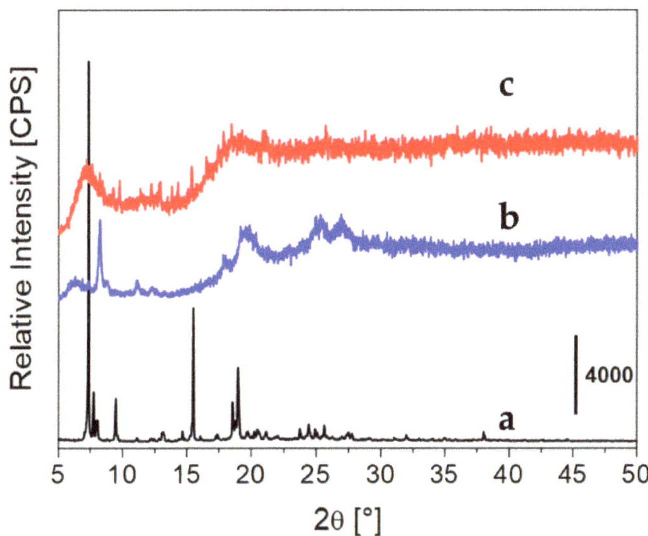

Figure 1. X-ray profiles of hib−POSS (**a**), Sb(III)−POSS (**b**), and Sb(V)−POSS (**c**).

The occurrence of the corner-capping reaction was confirmed through FT-IR spectroscopy by dispersing the solids in a KBr matrix (0.5 wt.%). The vibrational spectrum of Sb(V)-POSS (Figure 2b) was compared to that of both Sb(III)-POSS (Figure 2c) and the reactant hib-POSS (Figure 2a). The IR spectrum of hib-POSS showed two bands at 3250 cm^{-1} and at 890 cm^{-1}, assigned to the stretching of OH and Si-OH groups, respectively. Absorption in the 3000–2800 and 1500–1200 cm^{-1} ranges were attributed to the stretching and bending vibrations of the isobutyl fractions bound to the POSS cage. An intense band centered at 1100 cm^{-1}, due to the vibration modes of the Si-O-Si units, was also detected [58]. After reacting with the metal precursors, the vibrational modes of isobutyl moieties remained unchanged, denoting a preservation of the organic fraction of the POSS unit [27,33,56]. Furthermore, the band attributed to the stretching of Si-O-Si groups appeared to be less intense. This was mainly evident in the spectrum of the Sb(V)-POSS sample. This latter feature is indicative of a local modification of silsesquioxane cage symmetry due to coordination with Sb(III) or Sb(V) ions, consistent with existing literature data on the combination of POSS with several metal centers [29,32,34,59,60]. Finally, the absorptions at 3250 and 890 cm^{-1} were eroded and an intense band at 920 cm^{-1} and 985 cm^{-1}, attributed to the stretching of Si-O-Sb bonds for Sb(III)-POSS and Sb(V)-POSS, respectively, appeared evident in the spectra [56]. In general, this information indicates the success of the corner-capping reaction, with the production of a condensed structure [37,43,61,62]. In the case of the Sb(III)-POSS sample (Figure 2b), the band at approx. 3400 cm^{-1} was assigned to the stretching modes of physisorbed water in the matrix. Signals typical of triethylamine hydrochloride were not present in the IR profiles of either Sb-POSS compound, thus indicating a complete purification of both reaction products.

The XRPD and FT-IR results are well supported by NMR spectroscopy applied to ^{1}H and ^{29}Si nuclei. Samples were thoroughly dissolved in CDCl$_3$ for all analyses (see Section 3). The high-resolution ^{1}H NMR spectrum of Sb(V)-POSS (Figure 3c) was compared to that of both Sb(III)-POSS (Figure 3b) and hib-POSS (Figure 3a). The ^{1}H NMR spectrum of hib-POSS showed three main peaks ascribed to the -CH (1.87 ppm), -CH_2 (0.61 ppm), and -CH_3 (0.98 ppm) protons of the isobutyl groups bound to the POSS cage (Figure 3a) [13,63]. The same profile was also detected for Sb(III)-POSS (Figure 3b), in agreement with previous observations by Feher et al. [44,45] and the ^{1}H NMR spectrum of Sb(V)-POSS (Figure 3c) showed a significant line broadening of the peaks. In addition, a second group of less intense peaks appeared at low ppm (0.68 ppm), which could suggest the presence of two different sets of chemically equivalent methylene groups in different chemical surroundings. The complexity of the structure of Sb(V)-POSS was also confirmed by ^{29}Si NMR data.

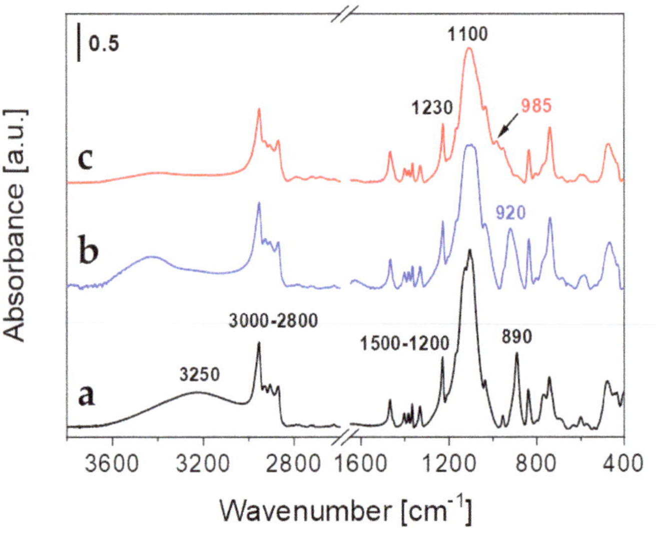

Figure 2. FT–IR spectra of hib−POSS (**a**), Sb(III)−POSS (**b**), and Sb(V)−POSS (**c**) dispersed in a KBr matrix (0.5 wt.%) measured at room temperature (r.t.). Spectra were normalized to the intensity of d isobutyl groups between 1500 and 1200 cm^{-1}.

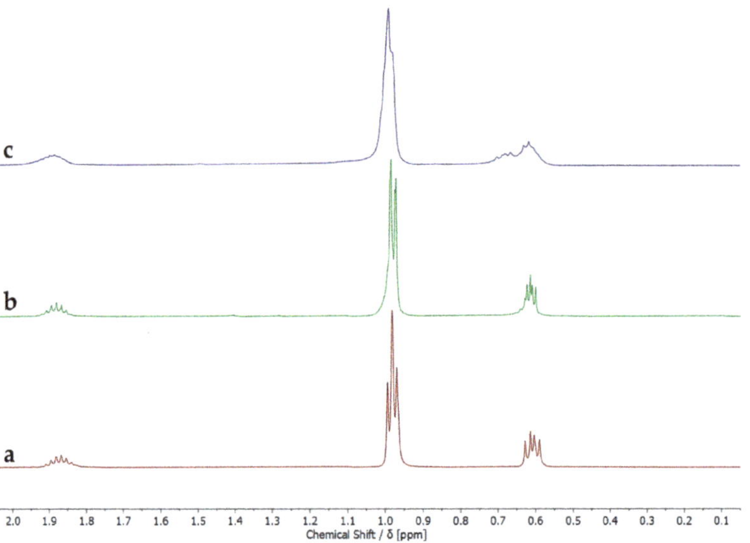

Figure 3. ^{1}H NMR spectra in CDCl$_3$ of hib−POSS (**a**), Sb(III)−POSS (**b**), and Sb(V)−POSS (**c**).

High-resolution ^{29}Si NMR spectra of all compounds provided additional information on the distribution of silicon sites before and after the corner-capping reaction with antimony ions. As a general note, the ^{29}Si-NMR spectrum of hib-POSS (Figure 4a) showed three well-defined signals at -58.8, -67.4, and -68.6 ppm (3:1:3 ratio). The peak at -58.8 ppm was assigned to the three silicon atoms bound to the hydroxyl groups (Si-OH), whereas the other peaks were ascribed to the remaining silicon sites of the cage [13,27,59,64].

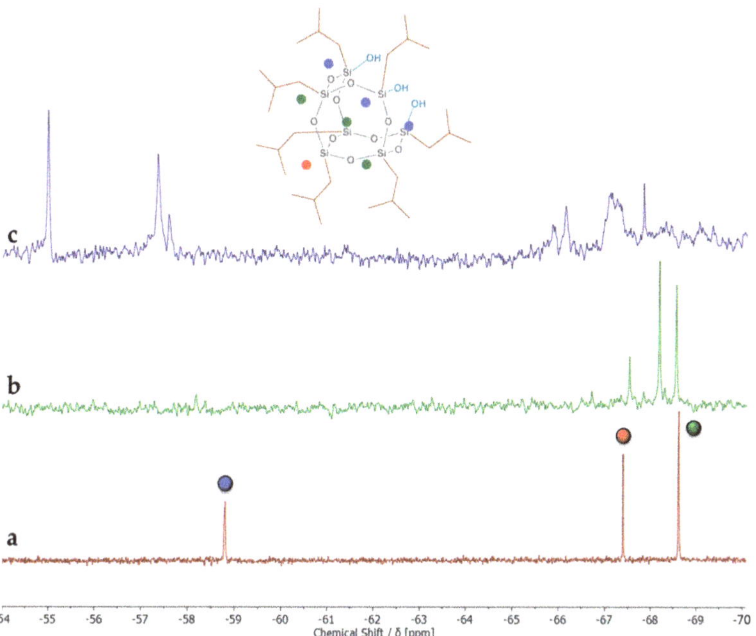

Figure 4. ^{29}Si NMR spectra in CDCl$_3$ of hib$-$POSS (**a**), Sb(III)$-$POSS (**b**), and Sb(V)$-$POSS (**c**).

The absence of the signal at -58.8 ppm in the ^{29}Si NMR spectrum of Sb(III)-POSS (Figure 4b), along with the presence of a new peak at -68.2 ppm attributed to the Si-O-Sb(III) sites, verified the formation of a fully closed monomeric structure (Figure 5A), in agreement with previous literature studies on the preparation of Sb(III)-doped POSS compounds [44,45,55,56]. Instead, the ^{29}Si NMR spectrum of Sb(V)-POSS (Figure 4c) appeared to be more complicated, with the presence of several peaks. The signals at low ppm (from -65 to -70 ppm) were associated with polycondensed silicon sites in different chemical surroundings, while those between -54 and -60 ppm could be assigned to a fraction of residual silanols (not clearly detectable in the IR spectrum), which were partially involved in coordination with Sb(V) [59]. These results supported the elemental, diffraction, and IR data and confirmed that Sb(V)-POSS is characterized by a more heterogeneous multimeric structure with different stoichiometry, compared to that of monomeric Sb(III)-POSS and to the structures reported in the literature [45]. A hypothetical arrangement of the POSS cages around the Sb(V) site, in analogy to other metal-containing POSS structures previously reported [64,65], is shown in Figure 5B. This representation is based on the Sb/C elemental ratio calculated by CHN analysis and corroborated by the NMR DOSY results. The structures were generated using the Avogadro software (ver. 1.2), optimizing their geometries with the Universal Force Field (UFF).

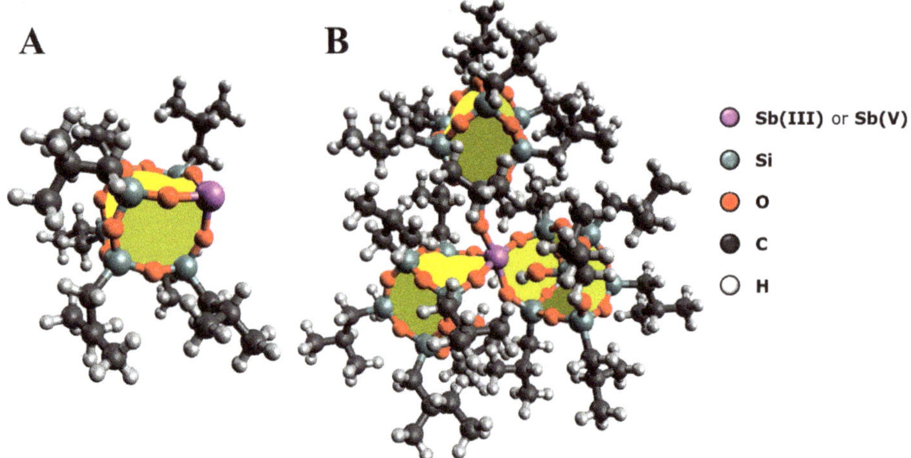

Figure 5. Schematic representation of the structures of Sb(III)−POSS (**A**) and Sb(V)−POSS (**B**).

These considerations were also proven with 2D diffusion-ordered spectroscopy (DOSY) ^{1}H NMR spectra collected for the reference hib-POSS and Sb(V)-POSS samples (Figure 6a and Figure 6b, respectively). The DOSY spectra were recorded in order to determine the molecular translation diffusion coefficients and to estimate the particle size of both samples by applying the Stokes–Einstein Equation (1) defined below:

$$D_t\left[\mathrm{m^2/s}\right] = \frac{k_b \cdot T}{6\pi \cdot \eta \cdot r} \tag{1}$$

where D_t is the translational diffusion coefficient of the sample analyzed (m^2/s), h is the viscosity of the solvent used in the NMR experiments (CDCl$_3$, 0.510 mPa·s), k_b is the Boltzmann constant (1.380649·10^{23} J/K), T is the temperature (300 K), and r is the radius of the spheroidal molecule (Å) [66–68].

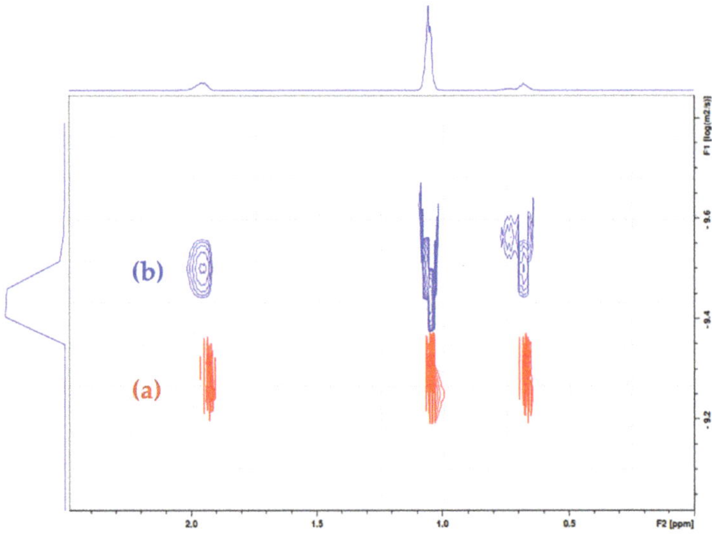

Figure 6. Combined 2D DOSY ^{1}H NMR spectra in CDCl$_3$ of hib−POSS (**a**) and Sb(V)−POSS (**b**).

The ^{1}H spectrum of Sb(V)-POSS is reported on the F2 axis and the translational diffusion constants on the F1 axis (in logarithmic scale, m^2/s). For the sake of clarity, we decided to plot only the ^{1}H NMR spectrum of Sb(V)-POSS on the F2 axis; however, the data in the 2D image are related to both Sb(V)-POSS and hib-POSS samples. The spectra in Figure 6 show a set of three signals for both hib-POSS (Figure 6a) and Sb(V)-POSS (Figure 6b), with a double component visible in the case of the latter. The D_t values extrapolated by the DOSY NMR data were found to be equal to $5.13 \cdot 10^{-10}$ and $3.16 \cdot 10^{-10}$ m^2/s for hib-POSS and Sb(V)-POSS, respectively, corresponding to a particle radius of 8.4 Å for hib-POSS and of 13.6 Å for Sb(V)-POSS. The mean particle size of the Sb(V)-POSS compound was around twice that of the POSS reactant. This result confirmed the existence of a high-molecular-weight multimeric structure for Sb(V)-POSS, compatible with the presence of three POSS units attached to the metal center (Figure 5B). Moreover, a second minor component with a lower diffusion coefficient at 0.68 ppm, already observed in the ^{1}H NMR spectrum, was also observed (Figure 6b).

3. Materials and Methods

3.1. Reactants

Trisilanol heptaisobutyl silsesquioxane (hib-POSS) was purchased from Hybrid Plastics Inc. (Hattiesburg, MI, USA) and stored at 277 K in the refrigerator. Other chemicals were purchased from Sigma-Aldrich/Merck KGaA (Darmstadt, Germany) and stored at room temperature (r.t.), apart from deuterated chloroform ($CDCl_3$, 99.8 atom %D), which was stored at 277 K in the refrigerator. All compounds were used without further purification, unless stated otherwise.

3.2. Materials

3.2.1. Synthesis of Sb(III)-POSS

The completely condensed Sb(III)-doped heptaisobutyl POSS was prepared through a corner-capping reaction, inspired by several synthetic procedures adopted in the literature for the preparation of trivalent antimony-substituted silsesquioxanes [27,43–45,55,56].

In detail, 0.151 g of $SbCl_3$ (0.66 mmol; ≥99.95%) was added with vigorous stirring to a solution of 0.500 g of hib-POSS (0.63 mmol) in 30 mL of anhydrous distilled tetrahydrofuran (THF; 0.37 mol; ≥99.9%) in the presence of 544 mL of triethylamine (Et_3N; 5.22 mmol; ≥99.5%). The reaction was carefully purged with nitrogen for 10 min. The temperature was increased to 50 °C and the mixture was stirred for 4 h. Afterwards, the reaction mixture was filtered to remove unreacted reagents and by-products. The filtered solution was evaporated in vacuo and the resulting solid sample was extracted in chloroform ($CHCl_3$; ≥99.8%) (10 mL + 10 mL of ultrapure water, 3 times); an appropriate amount of sodium sulphate (Na_2SO_4; ≥99.0%) was then used to remove any traces of residual water. The extracted sample was filtered again, evaporated in vacuo, and finally dried overnight in an oven at 50 °C, obtaining a white solid, hereafter named Sb(III)-POSS (yield = 81.7%).

3.2.2. Synthesis of Sb(V)-POSS

Sb(V)-POSS was prepared following the procedure previously described. In detail, 78.5 mL of $SbCl_5$ (0.66 mmol; ≥99.99%), carefully dissolved in 10 mL of anhydrous distilled THF (0.12 mol), was added with vigorous stirring to a solution of 0.500 g of hib-POSS (0.63 mmol) in 20 mL of anhydrous distilled THF (0.25 mol) in the presence of 544 mL of Et_3N (5.22 mmol). The reaction was carefully purged with nitrogen for 10 min. The temperature was increased to 50 °C and the mixture was stirred for 4 h. Afterwards, the reaction mixture was filtered to remove unreacted reagents and by-products. The filtered orange solution was evaporated in vacuo and the resulting oily sample was extracted in $CHCl_3$; an appropriate amount of Na_2SO_4 was then used to remove any traces of residual water. The extracted sample was filtered again and evaporated in vacuo, obtaining an oily orange compound, hereafter named Sb(V)-POSS (yield = 62.1%).

3.3. Analytical Methods

Elemental analyses were performed with an Ametek Spectro Genesis EOP Inductively Coupled Plasma Atomic Emission Spectrometer (ICP-AES) (Kleve, Germany) equipped with a cross-flow nebulizer with simultaneous spectrum capture in the 175–770 nm wavelength range. The compounds were mineralized in a mixture of HNO_3 70% and HF 48% at 373 K for 8 h, and then opportunely diluted in 1 wt.% HNO_3 solutions before analysis.

CHN elemental analyses were performed with an EA3000 CHN Elemental Analyzer (EuroVector, Milano, Italy). Acetanilide, purchased from EuroVector (Milano, Italy), was used as the calibration standard (C % = 71.089, H % = 6.711, N % = 10.363).

X-ray powder diffractograms (XRPDs) were collected on unoriented ground powders using a Bruker D8 Advance Powder Diffractometer (Karlsruhe, Germany), operating in Bragg–Brentano geometry with a Cu anode target equipped with a Ni filter (used as the X-ray source) and a Lynxeye XE-T high-resolution position-sensitive detector. Trio and Twin/Twin optics were mounted on the DaVinci Design modular XRD system. The X-ray tube of the instrument operated with a Cu-K$_{\alpha 1}$ monochromatic radiation (λ = 1.54062 Å), with the current intensity and operative electric potential difference set to 40 mA and 40 kV, respectively, and with automatic variable primary divergent slits and primary and secondary Soller slits of 2.5°. The X-ray profiles were recorded at room temperature in the 5°–50° 2θ range with a coupled 2θ–θ method, continuous PSD fast scan mode, time per step (rate or scan speed) of 0.100 s/step, and a 2θ step size (increment) of 0.01°, with automatic synchronization of the air scatter (or anti-scatter) knife and slits, and a fixed illumination sample set at 15 mm.

Fourier-transform infrared (FT-IR) spectra were collected using a Thermo Electron Corporation FT Nicolet 5700 Spectrometer (Waltham, MA, USA) in the range 4000–400 cm^{-1} with a resolution of 4 cm^{-1}. IR spectra of the solids mixed with potassium bromide (KBr, 0.5 wt.%) pellets were measured in absorbance mode at beam temperature (b.t.). All spectra were normalized to the intensity of the bending modes of isobutyl groups in the 1500–1200 cm^{-1} region.

Then, 1D ^1H (500 MHz), ^{29}Si-{^1H} (99.34 MHz), and 2D DOSY (diffusion-ordered spectroscopy) ^1H Nuclear Magnetic Resonance (NMR) spectra in solution were recorded at 300 K with a Bruker Advance III Spectrometer equipped with a wide bore 11.7 Tesla magnet. The 1D experiments were carried out by dissolving an appropriate amount of each sample (40 mg) in 600 µL of $CDCl_3$ and placing them in 5 mm NMR tubes for analyses. A ^1H-decoupling method was used in ^{29}Si experiments to enhance their signals. The 2D DOSY 1H experiments were performed by dissolving ~55–65 mg of each sample in 550 µL of $CDCl_3$. The spectrometer was equipped with a 5 mm double-resonance Z-gradient broadband probe, with the inner coil optimized for the observation of nuclei between ^{31}P and ^{15}N and for ^{19}F (BBFO), and a Bruker BVT-3000 unit for temperature control. All chemical shifts were reported using the δ [ppm] scale and were externally referenced to tetramethylsilane (TMS) at 0 ppm. Thermogravimetric analysis, TGA, was performed over a Perkin Elmer 7HT apparatus. Analyses were run under dry air with a heating temperature program of 5 °C min^{-1}, from 50 °C to 950 °C. Samples of ca. 10 mg were weighed.

4. Conclusions

In conclusion, we report the direct incorporation of pentavalent antimony ions into a polycondensed silsesquioxane network consisting of heptaisobutyl polyhedral oligomeric silsesquioxane cages as base units. The synthesis of the Sb(V)-POSS compound was accomplished through a corner-capping reaction carried out under mild experimental conditions. A completely condensed heptaisobutyl POSS bound to a single trivalent antimony ion was also prepared as a reference compound. A detailed investigation of the chemical properties of the samples was performed through a multi-technique approach. The ratio between POSS molecules and Sb(V) ions, estimated by CHN and ICP-AES analyses, suggests the average presence of three silsesquioxane cages linked to one Sb(V) ion. Infrared analyses confirmed the successful incorporation of the Sb(V) ions into the inorganic framework. X-ray powder diffraction studies, combined with 1D ^{1}H and ^{29}Si NMR spectra, further demonstrate the multimeric nature of the Sb(V)-POSS sample, consisting of an irregular organization of Sb(V)-linked POSS cores in the final structural network. Moreover, the molecular size of Sb(V)-POSS was found to be approx. two times greater than that of trisilanol heptaisobutyl POSS. Future studies will be focused on the application of these Sb-POSSs as flame-retardant additives for polymeric composites.

Author Contributions: Conceptualization, S.M., F.C. and E.B.; methodology, S.M. and F.C.; supervision: E.B.; formal analysis, S.M. and F.C.; investigation, S.M.; data curation, S.M.; writing—original draft preparation, S.M.; writing—review and editing, S.M., C.B., F.C. and E.B. All authors have read and agreed to the published version of the manuscript.

Funding: This research received no external funding.

Data Availability Statement: Not applicable.

Acknowledgments: The authors are fully grateful to Elena Perin for CHN elemental analyses and Chiara Zaccone for ICP-AES elemental analyses.

References

1. Pescarmona, P.P.; Maschmeyer, T. Review: Oligomeric Silsesquioxanes: Synthesis, Characterization and Selected Applications. *Aust. J. Chem.* **2001**, *54*, 583–596. [CrossRef]
2. Cordes, D.B.; Lickiss, P.D.; Rataboul, F. Recent Developments in the Chemistry of Cubic Polyhedral Oligosilsesquioxanes. *Chem. Rev.* **2010**, *110*, 2081–2173. [CrossRef] [PubMed]
3. Kalia, S.; Pielichowski, K. (Eds.) *Polymer/POSS Nanocomposites and Hybrid Materials: Preparation, Properties, Applications*; Springer Series on Polymer and Composite Materials; Springer International Publishing: Cham, Switzerland, 2018; ISBN 978-3-030-02326-3.
4. Laird, M.; Herrmann, N.; Ramsahye, N.; Totée, C.; Carcel, C.; Unno, M.; Bartlett, J.R.; Wong Chi Man, M. Large Polyhedral Oligomeric Silsesquioxane Cages: The Isolation of Functionalized POSS with an Unprecedented Si$_{18}$O$_{27}$ Core. *Angew. Chem. Int. Ed.* **2021**, *60*, 3022–3027. [CrossRef] [PubMed]
5. Quadrelli, E.A.; Basset, J.-M. On Silsesquioxanes' Accuracy as Molecular Models for Silica-Grafted Complexes in Heterogeneous Catalysis. *Coord. Chem. Rev.* **2010**, *254*, 707–728. [CrossRef]
6. Feher, F.J.; Newman, D.A.; Walzer, J.F. Silsesquioxanes as Models for Silica Surfaces. *J. Am. Chem. Soc.* **1989**, *111*, 1741–1748. [CrossRef]
7. Zhou, H.; Ye, Q.; Xu, J. Polyhedral Oligomeric Silsesquioxane-Based Hybrid Materials and Their Applications. *Mater. Chem. Front.* **2017**, *1*, 212–230. [CrossRef]
8. Dong, F.; Lu, L.; Ha, C.-S. Silsesquioxane-Containing Hybrid Nanomaterials: Fascinating Platforms for Advanced Applications. *Macromol. Chem. Phys.* **2019**, *220*, 1800324. [CrossRef]
9. Tunstall-Garcia, H.; Charles, B.L.; Evans, R.C. The Role of Polyhedral Oligomeric Silsesquioxanes in Optical Applications. *Adv. Photonics Res.* **2021**, *2*, 2000196. [CrossRef]
10. John, Ł.; Ejfler, J. A Brief Review on Selected Applications of Hybrid Materials Based on Functionalized Cage-like Silsesquioxanes. *Polymers* **2023**, *15*, 1452. [CrossRef]
11. Calabrese, C.; Aprile, C.; Gruttadauria, M.; Giacalone, F. POSS Nanostructures in Catalysis. *Catal. Sci. Technol.* **2020**, *10*, 7415–7447. [CrossRef]
12. Blanco, I.; Abate, L.; Bottino, F.A.; Bottino, P. Hepta Isobutyl Polyhedral Oligomeric Silsesquioxanes (Hib-POSS). *J. Therm. Anal. Calorim.* **2012**, *108*, 807–815. [CrossRef]
13. Ye, M.; Wu, Y.; Zhang, W.; Yang, R. Synthesis of Incompletely Caged Silsesquioxane (T7-POSS) Compounds via a Versatile Three-Step Approach. *Res. Chem. Intermed.* **2018**, *44*, 4277–4294. [CrossRef]
14. Imoto, H.; Ueda, Y.; Sato, Y.; Nakamura, M.; Mitamura, K.; Watase, S.; Naka, K. Corner-and Side-Opened Cage Silsesquioxanes: Structural Effects on the Materials Properties. *Eur. J. Inorg. Chem.* **2020**, *2020*, 737–742. [CrossRef]
15. Kaneko, Y.; Coughlin, E.B.; Gunji, T.; Itoh, M.; Matsukawa, K.; Naka, K. Silsesquioxanes: Recent Advancement and Novel Applications. *Int. J. Polym. Sci.* **2012**, *2012*, e453821. [CrossRef]
16. Lorenz, V.; Edelmann, F.T. Dimeric Silsesquioxanes and Metallasilsesquioxanes—En Route to Large, Welldefined Si-O-Assemblies. *Z. Anorg. Allg. Chem.* **2004**, *630*, 1147–1157. [CrossRef]
17. Ye, X.; Li, J.; Zhang, W.; Yang, R.; Li, J. Fabrication of Eco-Friendly and Multifunctional Sodium-Containing Polyhedral Oligomeric Silsesquioxane and Its Flame Retardancy on Epoxy Resin. *Compos. B Eng.* **2020**, *191*, 107961. [CrossRef]
18. Prigyai, N.; Chanmungkalakul, S.; Ervithayasuporn, V.; Yodsin, N.; Jungsuttiwong, S.; Takeda, N.; Unno, M.; Boonmak, J.; Kiatkamjornwong, S. Lithium-Templated Formation of Polyhedral Oligomeric Silsesquioxanes (POSS). *Inorg. Chem.* **2019**, *58*, 15110–15117. [CrossRef]

19. Gießmann, S.; Lorenz, V.; Liebing, P.; Hilfert, L.; Fischer, A.; Edelmann, F.T. Synthesis and Structural Study of New Metallasilsesquioxanes of Potassium and Uranium. *Dalton Trans.* **2017**, *46*, 2415–2419. [CrossRef]
20. Hanssen, R.W.J.M.; Meetsma, A.; van Santen, R.A.; Abbenhuis, H.C.L. Synthesis, Structural Characterization, and Transmetalation Reactions of a Tetranuclear Magnesium Silsesquioxane Complex. *Inorg. Chem.* **2001**, *40*, 4049–4052. [CrossRef]
21. Lorenz, V.; Fischer, A.; Edelmann, F.T. Silsesquioxane Chemistry, 6: The First Beryllium Silsesquioxane: Synthesis and Structure of $[Cy_7Si_7O_{12}BeLi]_2 \cdot 2THF$. *Inorg. Chem. Commun.* **2000**, *3*, 292–295. [CrossRef]
22. Levitsky, M.M.; Zubavichus, Y.V.; Korlyukov, A.A.; Khrustalev, V.N.; Shubina, E.S.; Bilyachenko, A.N. Silicon and Germanium-Based Sesquioxanes as Versatile Building Blocks for Cage Metallacomplexes. A Review. *J. Clust. Sci.* **2019**, *30*, 1283–1316. [CrossRef]
23. Kaźmierczak, J.; Kuciński, K.; Stachowiak, H.; Hreczycho, G. Introduction of Boron Functionalities into Silsesquioxanes: Novel Independent Methodologies. *Chem.—Eur. J.* **2018**, *24*, 2509–2514. [CrossRef]
24. Astakhov, G.S.; Levitsky, M.M.; Zubavichus, Y.V.; Khrustalev, V.N.; Titov, A.A.; Dorovatovskii, P.V.; Smol'yakov, A.F.; Shubina, E.S.; Kirillova, M.V.; Kirillov, A.M.; et al. Cu_6- and Cu_8-Cage Sil-and Germsesquioxanes: Synthetic and Structural Features, Oxidative Rearrangements, and Catalytic Activity. *Inorg. Chem.* **2021**, *60*, 8062–8074. [CrossRef] [PubMed]
25. Gerritsen, G.; Duchateau, R.; Yap, G.P.A. Boron, Aluminum, and Gallium Silsesquioxane Compounds, Homogeneous Models for Group 13 Element-Containing Silicates and Zeolites. *Organometallics* **2003**, *22*, 100–110. [CrossRef]
26. Besselink, R.; Venkatachalam, S.; van Wüllen, L.; ten Elshof, J.E. Incorporation of Niobium into Bridged Silsesquioxane Based Silica Networks. *J. Sol-Gel Sci. Technol.* **2014**, *70*, 473–481. [CrossRef]
27. Marchesi, S.; Carniato, F.; Boccaleri, E. Synthesis and Characterisation of a Novel Europium(III)-Containing Heptaisobutyl-POSS. *New J. Chem.* **2014**, *38*, 2480–2485. [CrossRef]
28. Willauer, A.R.; Dabrowska, A.M.; Scopelliti, R.; Mazzanti, M. Structure and Small Molecule Activation Reactivity of a Metallasilsesquioxane of Divalent Ytterbium. *Chem. Commun.* **2020**, *56*, 8936–8939. [CrossRef]
29. Lorenz, V.; Fischer, A.; Gießmann, S.; Gilje, J.W.; Gun'ko, Y.; Jacob, K.; Edelmann, F.T. Disiloxanediolates and Polyhedral Metallasilsesquioxanes of the Early Transition Metals and F-Elements. *Coord. Chem. Rev.* **2000**, *206*, 321–368. [CrossRef]
30. Marchesi, S.; Carniato, F.; Marchese, L.; Boccaleri, E. Luminescent Mesoporous Silica Built through Self-Assembly of Polyhedral Oligomeric Silsesquioxane and Europium(III) Ions. *ChemPlusChem* **2015**, *80*, 915–918. [CrossRef]
31. Kumar, B.; Kumar, A.P.; Bindu, P.; Mukherjee, A.; Patra, S. Red Light Emission of POSS Triol Chelated with Europium. *Asian J. Nanosci. Mater.* **2019**, *2*, 244–256. [CrossRef]
32. Marchesi, S.; Bisio, C.; Boccaleri, E.; Carniato, F. A Luminescent Polysilsesquioxane Obtained by Self-Condensation of Anionic Polyhedral Oligomeric Silsequioxanes (POSS) and Europium(III) Ions. *ChemPlusChem* **2020**, *85*, 176–182. [CrossRef]
33. Marchesi, S.; Bisio, C.; Carniato, F. Synthesis of Novel Luminescent Double-Decker Silsesquioxanes Based on Partially Condensed TetraSilanolPhenyl POSS and Tb^{3+}/Eu^{3+} Lanthanide Ions. *Processes* **2022**, *10*, 758. [CrossRef]
34. Marchesi, S.; Miletto, I.; Bisio, C.; Gianotti, E.; Marchese, L.; Carniato, F. Eu^{3+} and Tb^{3+} @ PSQ: Dual Luminescent Polyhedral Oligomeric Polysilsesquioxanes. *Materials* **2022**, *15*, 7996. [CrossRef] [PubMed]
35. Levitsky, M.M.; Yalymov, A.I.; Kulakova, A.N.; Petrov, A.A.; Bilyachenko, A.N. Cage-like Metallasilsesquioxanes in Catalysis: A Review. *J. Mol. Catal. A-Chem.* **2017**, *426*, 297–304. [CrossRef]
36. Lorenz, V.; Edelmann, F.T. Metallasilsesquioxanes. In *Advances in Organometallic Chemistry*; West, R., Hill, A.F., Stone, F.G.A., Eds.; Academic Press: Cambridge, MA, USA, 2005; Volume 53, pp. 101–153.
37. Levitsky, M.M.; Bilyachenko, A.N. Modern Concepts and Methods in the Chemistry of Polyhedral Metallasiloxanes. *Coord. Chem. Rev.* **2016**, *306*, 235–269. [CrossRef]
38. Pan, G. Polyhedral Oligomeric Silsesquioxane (POSS). In *Physical Properties of Polymers Handbook*; Mark, J.E., Ed.; Springer: New York, NY, USA, 2007; pp. 577–584. ISBN 978-0-387-69002-5.
39. Ward, A.J.; Masters, A.F.; Maschmeyer, T. Metallasilsesquioxanes: Molecular Analogues of Heterogeneous Catalysts. In *Applications of Polyhedral Oligomeric Silsesquioxanes*; Hartmann-Thompson, C., Ed.; Advances in Silicon Science; Springer: Dordrecht, The Netherlands, 2011; pp. 135–166. ISBN 978-90-481-3787-9.
40. Henig, J.; Tóth, É.; Engelmann, J.; Gottschalk, S.; Mayer, H.A. Macrocyclic Gd^{3+} Chelates Attached to a Silsesquioxane Core as Potential Magnetic Resonance Imaging Contrast Agents: Synthesis, Physicochemical Characterization, and Stability Studies. *Inorg. Chem.* **2010**, *49*, 6124–6138. [CrossRef]
41. Köytepe, S.; Demirel, M.H.; Gültek, A.; Seçkin, T. Metallo-Supramolecular Materials Based on Terpyridine-Functionalized Polyhedral Silsesquioxane. *Polym. Int.* **2014**, *63*, 778–787. [CrossRef]
42. Li, Y.; Dong, X.-H.; Zou, Y.; Wang, Z.; Yue, K.; Huang, M.; Liu, H.; Feng, X.; Lin, Z.; Zhang, W.; et al. Polyhedral Oligomeric Silsesquioxane Meets "Click" Chemistry: Rational Design and Facile Preparation of Functional Hybrid Materials. *Polymer* **2017**, *125*, 303–329. [CrossRef]
43. Murugavel, R.; Voigt, A.; Walawalkar, M.G.; Roesky, H.W. Hetero- and Metallasiloxanes Derived from Silanediols, Disilanols, Silanetriols, and Trisilanols. *Chem. Rev.* **1996**, *96*, 2205–2236. [CrossRef]
44. Feher, F.J.; Budzichowski, T.A. Heterosilsesquioxanes: Synthesis and Characterization of Group 15 Containing Polyhedral Oligosilsesquioxanes. *Organometallics* **1991**, *10*, 812–815. [CrossRef]

45. Feher, F.J.; Budzichowski, T.A.; Rahimian, K.; Ziller, J.W. Reactions of Incompletely-Condensed Silsesquioxanes with Pentamethylantimony: A New Synthesis of Metallasilsesquioxanes with Important Implications for the Chemistry of Silica Surfaces. *J. Am. Chem. Soc.* **1992**, *114*, 3859–3866. [CrossRef]
46. Zhang, W.; Camino, G.; Yang, R. Polymer/Polyhedral Oligomeric Silsesquioxane (POSS) Nanocomposites: An Overview of Fire Retardance. *Prog. Polym. Sci.* **2017**, *67*, 77–125. [CrossRef]
47. Glodek, T.; Boyd, S.; Mcaninch, I.; Lascala, J. Properties and Performance of Fire Resistant Eco-Composites Using Polyhedral Oligomeric Silsesquioxane (POSS) Fire Retardants. *Compos. Sci. Technol.* **2008**, *68*, 2994–3001. [CrossRef]
48. Kim, H.-J.; Kwon, Y.; Kim, C.K. Mechanical Property and Thermal Stability of Polyurethane Composites Reinforced with Polyhedral Oligomeric Silsesquioxanes and Inorganic Flame Retardant Filler. *J. Nanosci. Nanotechnol.* **2014**, *14*, 6048–6052. [CrossRef]
49. Palin, L.; Rombolà, G.; Milanesio, M.; Boccaleri, E. The Use of POSS-Based Nanoadditives for Cable-Grade PVC: Effects on Its Thermal Stability. *Polymers* **2019**, *11*, 1105. [CrossRef] [PubMed]
50. Weil, E.D.; Levchik, S.V. Commercial Flame Retardancy of Unsaturated Polyester and Vinyl Resins: Review. *J. Fire Sci.* **2004**, *22*, 293–303. [CrossRef]
51. Li, N.; Xia, Y.; Mao, Z.; Wang, L.; Guan, Y.; Zheng, A. Influence of Antimony Oxide on Flammability of Polypropylene/Intumescent Flame Retardant System. *Polym. Degrad. Stab.* **2012**, *97*, 1737–1744. [CrossRef]
52. Kim, H.; Kim, J.-S.; Jeong, W. Study on the Flame Retardancy and Hazard Evaluation of Poly(Acrylonitrile-Co-Vinylidene Chloride) Fibers by the Addition of Antimony-Based Flame Retardants. *Polymers* **2022**, *14*, 42. [CrossRef]
53. Zanetti, M.; Camino, G.; Canavese, D.; Morgan, A.B.; Lamelas, F.J.; Wilkie, C.A. Fire Retardant Halogen−Antimony−Clay Synergism in Polypropylene Layered Silicate Nanocomposites. *Chem. Mater.* **2002**, *14*, 189–193. [CrossRef]
54. Faryal, S.; Zafar, M.; Nazir, M.S.; Ali, Z.; Hussain, M.; Muhammad Imran, S. The Synergic Effect of Primary and Secondary Flame Retardants on the Improvement in the Flame Retardant and Mechanical Properties of Thermoplastic Polyurethane Nanocomposites. *Appl. Sci.* **2022**, *12*, 10866. [CrossRef]
55. Alphazan, T.; Díaz Álvarez, A.; Martin, F.; Grampeix, H.; Enyedi, V.; Martinez, E.; Rochat, N.; Veillerot, M.; Dewitte, M.; Nys, J.-P.; et al. Shallow Heavily Doped N++ Germanium by Organo-Antimony Monolayer Doping. *ACS Appl. Mater. Interfaces* **2017**, *9*, 20179–20187. [CrossRef] [PubMed]
56. Alphazan, T.; Florian, P.; Thieuleux, C. Ethoxy and Silsesquioxane Derivatives of Antimony as Dopant Precursors: Unravelling the Structure and Thermal Stability of Surface Species on SiO$_2$. *Phys. Chem. Chem. Phys.* **2017**, *19*, 8595–8601. [CrossRef] [PubMed]
57. Khouchaf, L.; Boulahya, K.; Das, P.P.; Nicolopoulos, S.; Kis, V.K.; Lábár, J.L. Study of the Microstructure of Amorphous Silica Nanostructures Using High-Resolution Electron Microscopy, Electron Energy Loss Spectroscopy, X-ray Powder Diffraction, and Electron Pair Distribution Function. *Materials* **2020**, *13*, 4393. [CrossRef] [PubMed]
58. Carniato, F.; Boccaleri, E.; Marchese, L.; Fina, A.; Tabuani, D.; Camino, G. Synthesis and Characterisation of Metal Isobutylsilsesquioxanes and Their Role as Inorganic–Organic Nanoadditives for Enhancing Polymer Thermal Stability. *Eur. J. Inorg. Chem.* **2007**, *2007*, 585–591. [CrossRef]
59. Marchesi, S.; Carniato, F.; Palin, L.; Boccaleri, E. POSS as Building-Blocks for the Preparation of Polysilsesquioxanes through an Innovative Synthetic Approach. *Dalton Trans.* **2015**, *44*, 2042–2046. [CrossRef] [PubMed]
60. Komagata, Y.; Iimura, T.; Shima, N.; Kudo, T. A Theoretical Study of the Insertion of Atoms and Ions into Titanosilsequioxane (Ti-POSS) in Comparison with POSS. *Int. J. Polym. Sci.* **2012**, *2012*, e391325. [CrossRef]
61. Jin, L.; Hong, C.; Li, X.; Sun, Z.; Feng, F.; Liu, H. Corner-Opening and Corner-Capping of Mono-Substituted T8 POSS: Product Distribution and Isomerization. *Chem. Commun.* **2022**, *58*, 1573–1576. [CrossRef]
62. Loman-Cortes, P.; Binte Huq, T.; Vivero-Escoto, J.L. Use of Polyhedral Oligomeric Silsesquioxane (POSS) in Drug Delivery, Photodynamic Therapy and Bioimaging. *Molecules* **2021**, *26*, 6453. [CrossRef]
63. Alphazan, T.; Mathey, L.; Schwarzwälder, M.; Lin, T.-H.; Rossini, A.J.; Wischert, R.; Enyedi, V.; Fontaine, H.; Veillerot, M.; Lesage, A.; et al. Monolayer Doping of Silicon through Grafting a Tailored Molecular Phosphorus Precursor onto Oxide-Passivated Silicon Surfaces. *Chem. Mater.* **2016**, *28*, 3634–3640. [CrossRef]
64. Carniato, F.; Boccaleri, E.; Marchese, L. A Versatile Route to Bifunctionalized Silsesquioxane (POSS): Synthesis and Characterisation of Ti-Containing Aminopropylisobutyl-POSS. *Dalton Trans.* **2007**, *1*, 36–39. [CrossRef]
65. Klein, R.A.; Marinus, C. Synthesis and use of metallized polyhedral oligomeric silsequioxane catalyst compositions. WO2015062759A1, 7 May 2015.
66. Evans, R. The Interpretation of Small Molecule Diffusion Coefficients: Quantitative Use of Diffusion-Ordered NMR Spectroscopy. *Prog. Nucl. Magn. Reson. Spectrosc.* **2020**, *117*, 33–69. [CrossRef] [PubMed]
67. Dill, K.A.; Bromberg, S. *Molecular Driving Forces: Statistical Thermodynamics in Chemistry and Biology*; Garland Science: New York, NY, USA, 2003; ISBN 978-0-8153-2051-7.
68. Einstein, A. Über Die von Der Molekularkinetischen Theorie Der Wärme Geforderte Bewegung von in Ruhenden Flüssigkeiten Suspendierten Teilchen. *Ann. Phys.* **1905**, *322*, 549–560. [CrossRef]

Article

Hydrazine Oxidation in Aqueous Solutions I: N_4H_6 Decomposition

Martin Breza [1,*] **and Alena Manova** [2]

[1]   Department of Physical Chemistry, Slovak Technical University, Radlinskeho 9, SK-81237 Bratislava, Slovakia
[2]   Department of Analytical Chemistry, Slovak Technical University, Radlinskeho 9, SK-81237 Bratislava, Slovakia; alena.manova@stuba.sk
*   Correspondence: martin.breza@stuba.sk

Abstract: A mixture of nonlabeled ($^{14}N_2H_4$) and ^{15}N labeled hydrazine ($^{15}N_2H_4$) in an aqueous solution is oxidized to $^{15}N_2$, $^{14}N_2$, and $^{14}N^{15}N$ molecules, indicating the intermediate existence of the $^{14}NH_2$-^{14}NH-^{15}NH-$^{15}NH_2$ with subsequent hydrogen transfers and splitting of side N-N bonds. The structures, thermodynamics and electron characteristics of various N_4H_6 molecules in aqueous solutions are investigated using theoretical treatment at the CCSD/cc-pVTZ level of theory to explain the crucial part of the hydrazine oxidation reaction. Most N_4H_6 structures in aqueous solutions are decomposed during geometry optimization. Splitting the bond between central nitrogen atoms is the most frequent method, but the breakaway of the side nitrogen is energetically the most preferred one. The N-N fissions are enabled by suitable hydrogen rearrangements. Gibbs free energy data indicate the dominant abundance of NH_3... N_2... NH_3 species. The side N atoms have very high negative charges, which should support hydrogen transfers in aqueous solutions. The only stable cyclo-$(NH)_4$...H_2 structure has a Gibbs energy that is too high and breaks the H_2 molecule. The remaining initial cyclic structures are split into hydrazine and $HN\equiv NH$ or $H_2N\equiv N$ species, and their relative abundance in aqueous solutions is vanishing.

Keywords: Coupled Cluster; geometry optimization; N-N bond splitting; QTAIM analysis; electron structure

Citation: Breza, M.; Manova, A. Hydrazine Oxidation in Aqueous Solutions I: N₄H₆ Decomposition. *Inorganics* 2023, *11*, 413. https://doi.org/10.3390/inorganics11100413

Academic Editors: Hicham Idriss, Roberto Nisticò, Torben R. Jensen, Luciano Carlos and Eleonora Aneggi

Received: 29 September 2023
Revised: 14 October 2023
Accepted: 17 October 2023
Published: 18 October 2023

1. Introduction

Hydrazine N_2H_4 is a colorless flammable liquid that is used in industry and agriculture due to its reducing properties. It is used as a corrosion inhibitor in boilers, as a rocket propellant, antioxidant, catalyst, and pesticide precursor. In boiler water, it serves as an oxygen scavenger that reacts with oxygen into nitrogen and water only, which does not cause corrosion of ferrous metals. Unreacted hydrazine can be decomposed into ammonia, which can be corrosive to copper and copper-containing alloys [1]. Thus, the knowledge of the exact mechanism of its oxidation is of practical importance so far.

Higginson and Sutton [2] studied the oxidation of ^{15}N-enriched hydrazine by an excess of various oxidizing agents in aqueous solutions. Mass spectroscopic analysis of the evolved nitrogen for 28, 29 and 30 mass-number abundance (i.e., incidence of $^{14}N_2$, $^{15}N^{14}N$ and $^{15}N_2$ molecules, respectively) has shown that the proportion of $^{15}N_2$ molecules decreased while that of $^{15}N^{14}N$ molecules increased depending on the oxidizing agent used. If the nitrogen produced by the reaction

$$2\,N_2H_4 \rightarrow N_2 + 2\,NH_3 \tag{1}$$

involves no N-N fission, the evolved N_2 molecule originates in the same N_2H_4 molecule, and therefore it must have the same distribution of ^{15}N isotopes as the hydrazine reactant. This implies that some of the nitrogen molecules are formed by a mechanism involving a N-N fission and the formation of nitrogen-containing radicals from two different hydrazine molecules as follows:

$$^{15}NH_2\text{-}^{15}NH_2 + {}^{14}NH_2\text{-}^{14}NH_2 \rightarrow {}^{15}NH_2\text{-}^{15}NH\bullet + \bullet^{14}NH\text{-}^{14}NH_2 \rightarrow {}^{15}NH_2\text{-}^{15}NH\text{-}^{14}NH\text{-}^{14}NH_2 \tag{2}$$

$$^{15}NH_2\text{-}^{15}NH\text{-}^{14}NH\text{-}^{14}NH_2 \rightarrow {}^{15}NH_3 + {}^{15}NH = {}^{14}N\text{-}^{14}NH_2 \rightarrow {}^{15}NH_3 + {}^{15}N^{14}N + {}^{14}NH_3 \tag{3}$$

$$^{15}NH_2\text{-}^{15}NH\text{-}^{14}NH\text{-}^{14}NH_2 \rightarrow {}^{15}NH_2\text{-}^{15}N = {}^{14}NH + {}^{14}NH_3 \rightarrow {}^{15}NH_3 + {}^{15}N^{14}N + {}^{14}NH_3 \tag{4}$$

Cahn and Powell [3] confirmed the randomized $^{15}N^{14}N$ composition obtained by one-electron oxidation of ^{15}N enriched hydrazines with a number of oxidizing agents unlike exclusively four-electron oxidizing agents (acid iodate, alkaline ferricyanide) that produced unrandomized N_2 molecules (all four hydrogen atoms must be removed from a single hydrazine molecule). Petek and Bruckenstein [4] observed that the electrooxidation of ^{15}N labelled hydrazine (96.7% enrichment) at the Pt electrode produced N_2 molecules with the ratio of $^{14}N^{15}N/^{15}N^{15}N = 0.07 \pm 0.01$ while in Ce(IV) solutions it was 0.9 ± 0.2. A ratio of both isotopic forms between these two limits was produced by simultaneous electrooxidation and homogeneous oxidation with electrogenerated Ce(IV).

A bright yellow substance, stable under $-178\,^\circ C$, is formed after thermal decomposition of hydrazine at high temperatures (~850 $^\circ F$) and low pressures (~0.5 mm Hg) in a flowing system [5]. The authors suppose that it is tetrazane N_4H_6.

Based on polarographic and voltammetric studies of hydrazine in alkali solutions, Karp and Meites [6] suggested its two-electron oxidation to diimide with subsequent dimerization and decomposition as follows

$$2\,N_2H_2 \rightarrow N_4H_4 \rightarrow NH_4^+ + N_3^- \tag{5}$$

The proposed mechanism is also capable of explaining the randomized $^{15}N^{14}N$ composition.

Ball [7] investigated the structure and some thermochemical properties of the cis- and trans-conformations of tetrazane $NH_2\text{-}NH\text{-}NH\text{-}NH_2$ using various-level ab initio methods. Unlike nearly planar trans-conformation, the cis-conformation should be denoted as a gauche structure (N-N-N-N dihedral angle of ca 90°).

The decomposition of hydrazine was studied at the CCSD(T)-F12a/aug-ccpVTZ// ωB97x-D3/6-311++G(3df,3pd) level of theory [8]. A comprehensive analysis of the N_4H_6 singlet potential energy surfaces was performed. Three stable isomers, $NH_2\text{-}NH\text{-}NH\text{-}NH_2$, $NH_2\text{-}NH\text{-}NH_2{=}NH$ and $NH_2\text{-}NH_2\text{-}N{=}NH_2$, and the transition states for H transfers between them were obtained as well. Stabilized $NH_2\text{-}NH\text{-}NH\text{-}NH_2$ formation becomes significant only at relatively high pressures and low temperatures due to its decomposition into $N_2H_3\bullet + N_2H_3\bullet$. No direct reaction between $NH_2\text{-}NH\text{-}NH\text{-}NH_2$ and $NH_2\text{-}NH_2\text{-}N{=}NH_2$ was found. NH_3 eliminations from $NH_2\text{-}NH\text{-}NH\text{-}NH_2$ and $NH_2\text{-}NH_2\text{-}N{=}NH_2$ are energetically preferred, but only $NH_2\text{-}NH_2\text{-}N{=}NH_2$ has relatively small activation energy for this reaction (see Table 1).

Table 1. Reaction, ΔE_r, and activation, E_a, energy data from elementary reactions on the N_4H_6 potential energy surface [8].

Reaction	ΔE_r (kJ/mol)	E_a (kJ/mol)
$N_2H_4 + H_2N{=}N \rightarrow NH_2\text{-}NH\text{-}NH\text{-}NH_2$	-103.6	50.6
$N_2H_4 + H_2N{=}N \rightarrow NH_2NH_2N{=}NH_2$	29.0	55.4
$NH_2\text{-}NH\text{-}NH\text{-}NH_2 \rightarrow NH_2NH{=}N + NH_3$	7.5	178.7
$NH_2\text{-}NH\text{-}NH\text{-}NH_2 \rightarrow NH_2\text{-}N{=}NH + NH_3$	-102.5	214.1
$NH_2\text{-}NH_2\text{-}N{=}NH_2 \rightarrow NH_2\text{-}N{=}NH + NH_3$	-245.1	38.7
$NH_2\text{-}NH\text{-}NH\text{-}NH_2 \rightarrow NH_2\text{-}NH\text{-}NH_2{=}NH$	151.1	158.6
$NH_2\text{-}NH\text{-}NH_2{=}NH \rightarrow NH_2\text{-}NH_2\text{-}N{=}NH_2$	-18.5	74.4
$N_2H_3\bullet + N_2H_3\bullet \rightarrow NH_2\text{-}NH\text{-}NH\text{-}NH_2$	-152.9	0.2
$NH_2\text{-}NH{=}NH\bullet + NH_2\bullet \rightarrow NH_2\text{-}NH\text{-}NH\text{-}NH_2$	208.9	0.2
$NH{=}NH_2\text{-}NH\bullet + NH2\bullet \rightarrow NH_2\text{-}NH\text{-}NH_2{=}NH$	682.7	0.2
$NH_2\text{-}N{=}NH_2\bullet + NH_2\bullet \rightarrow NH_2\text{-}NH_2\text{-}N{=}NH_2$	37.9	2.8

It is evident that the decomposition of N_4H_6 is crucial for hydrazine oxidation with subsequent $^{15}N^{14}N$ molecule formation. It depends on the suitable N_4H_6 site of N-N bond splitting. At first, the NH_2-NH-NH-NH_2 isomer is formed by the reaction

$$N_2H_3\bullet + N_2H_3\bullet \rightarrow NH_2\text{-}NH\text{-}NH\text{-}NH_2 \tag{6}$$

In the next steps, H transfers and possible N-N bond splitting may proceed. The main aim of this study is a quantum-chemical study of N_4H_6 isomers in aqueous solutions solely at the Coupled Cluster level of theory and to determine the sites of the possible N-N fission within them. The thermodynamic properties of the decomposition reaction products enable us to predict the possible formation of $^{15}N^{14}N$ molecules in real systems. The electronic structure of the optimized structures will also be discussed.

2. Results and Discussion

We consider possible linear isomers of N_4H_6 with an N1-N2-N3-N4 backbone and the composition of $N1H_m$-$N2H_n$-$N3H_p$-$N4H_q$, where subscripts m, n, p and q denote the number of H atoms bonded to individual Ni; i = 1→ 4, atoms, and m + n + p + q = 6. We started geometry optimizations from anti- and syn-conformations of N1-N2-N3-N4. The optimized structures usually correspond to gauche conformers, or some N-N bonds are split (see Table 2). If N1 and N2 correspond to ^{15}N atoms, while N3 and N4 correspond to the ^{14}N ones, then N1-N2 and N3-N4 fissions would lead to $^{15}N^{14}N$ molecules, unlike the N2-N3 fissions.

Table 2. N1-N2-N3-N4 dihedral angles (Θ_{1234}), absolute (G_{298}) and relative (ΔG_{298}) Gibbs free energies at 298.15 K for the optimized N_4H_6 structures obtained from the starting ones. The most stable structure is highlighted in bold. The different structures with the same notation are distinguished by additional letters a, b, c, or d.

Starting	Optimized	Θ_{1234} [°]	G_{298} [Hartree]	ΔG_{298} [kJ/mol]	Remarks
A2112	D2112a	168.3	−222.09177	0.00	
A2121	D2121a	−161.4	−222.04654	118.75	
A2211	E(22)(11)a	−33.7	−222.10760	−41.56	H_2N-NH_2 + HN=NH
A2202	A2202	−179.9	−222.04618	119.71	
A2220	E(22)(20)a	146.5	−222.07817	35.7	H_2N-NH_2 + H_2N=N
A1221	D1221	−168.5	−222.00357	231.58	
A3210	E(32)(10)	14.3	−222.04629	119.42	H_3N-NH_2 + HN=N
A3201	E(22)(11)b	−142.8	−222.10755	−41.43	H_2N-NH_2 + HN=NH, 1→3 H rearrangement
A3201	E(3)(201)	−26.5	−222.14890	−150.04	NH_3 + H_2N-N=NH
A3111	D2112b	75.6	−222.09311	−3.52	1→4 H rearrangement
A3120	E(31)(20)	−21.4	−222.03229	156.16	H_3N-NH + H_2N=N
A3102	E(3)(102)a	−177.1	−222.14926	−150.94	NH_3 + HN=N-NH_2
A3012	D3012a	88.8	−222.04832	114.07	
A3021	A3021	176.8	−221.99866	244.46	
A3003	**E(3)(00)(3)**	**60.4**	**−222.26295**	**−449.44**	$2NH_3$ + N_2
B2112	D2112c	72.0	−222.09665	−12.80	
B2121	D2121b	−65.0	−222.04530	122.02	
B2121	D2121c	−44.8	−222.04982	110.13	
B2211	E(22)(11)a	−33.7	−222.10760	−41.56	H_2N-NH_2 + HN=NH
B2202	D2202a	73.7	−222.05056	108.21	
B2220	E(22)(20)b	−32.9	−222.07820	35.64	H_2N-NH_2 + H_2N=N
B1221	F12)(21	75.8	−222.09760	−15.30	N2-N3 fission, N1-N4 bonding
B1221	F1)(22)(1	−33.3	−222.10756	−41.45	H_2N-NH_2 + HN=NH, N1-N4 bonding
B3210	E(22)(11)c	−34.0	−222.10754	−41.41	H_2N-NH_2 + HN=NH, 1→4 H rearrangement
B3201	D2202b	80.1	−222.04758	116.02	1→4 H rearrangement
B3201	E(3)(201)	−26.5	−222.14892	−150.04	NH_3 + H_2N-N=NH
B3111	D2112b	75.6	−222.09311	−3.52	1→4 H rearrangement
B3120	D2121d	68.8	−222.04651	118.82	1→4 H rearrangement
B3102	E(3)(102)b	−20.6	−222.14678	−144.42	NH_3 + HN=N-NH_2

Table 2. *Cont.*

Starting	Optimized	Θ_{1234} [°]	G_{298} [Hartree]	ΔG_{298} [kJ/mol]	Remarks
B3012	D3012b	−59.7	−222.04910	112.02	
B3021	D2022	−73.8	−222.05056	108.21	1→4 H rearrangement
B3003	**E(3)(00)(3)**	**60.4**	**−222.26235**	**−449.44**	2 NH_3 + N_2
C2211	E(22)(11)d	12.3	−222.10759	−41.45	$H_2N\text{-}NH_2$ + HN=NH
C2202	E(22)(02)	29.6	−222.07822	35.58	$H_2N\text{-}NH_2$ + N=NH_2
C2121	E1111	23.1	−221.99631	250.64	Cyclo-N_4H_4 + H_2

In the case of cyclo-N_4H_6 isomers we can use the same notation, but any N-N fission can lead to $^{15}N^{14}N$ molecules because of suitable H transfers within the cycle.

We introduce the notation Xmnpq for the individual systems under study, where X = A, B and C, and D stands for anti-, syn-, cyclic and gauche-structures and the indices m, n, p and q are explained above. X = E denotes structures with N-N fissions, i.e., consisting of two or three molecules after geometry optimization. X = F stands for structures with N2-N3 fissions and subsequent N1-N4 bond formations. The N-N fissions in E and F systems are denoted by round brackets where the mutually bonded N atoms are included in the same bracket couple. The different structures with the same Xmnpq notation can be distinguished by additional letters a, b, c, etc. For example, E(22)(11)a and E(22)(11)b denote two different structures composed of $H_2N\text{-}NH_2$ and HN=NH molecules.

The N_4H_6 structures under study are shown in Table 2 and are divided into three groups according to the initial N1-N2-N3-N4 conformations. The H atom rearrangements during geometry optimizations are less frequent in the anti-conformations (starting A structures) than in the syn-conformations (starting B structures). In both groups the probability of N-N fissions is approximately 50%, and N2-N3 fissions prevail. On the other hand, the N1-N2 fissions lead to energetically preferred products such as E(3)(201), E(3)(102) and especially E(3)(00)(3). In the B1221 syn-conformation the mutual interaction of N1 and N4 causes the formation of the N1-N4 bond and N2-N3 fission leading to the structure of $H_2N2\text{-}N1H\text{-}N4H\text{-}N3H_2$, i.e., F12)(21, in gauche conformation or decomposition to more stable HN1=N4H and $H_2N2\text{-}N3H_2$ species denoted as the F1)(22)(1 system.

The relative Gibbs free energies in Table 2 are related to the structure D2112a obtained by the reaction (6) in the first step. According to these data, the system E(3)(00)(3), which corresponds to $^{15}NH_3$, $^{14}NH_3$ and $^{15}N^{14}N$ molecules, is dominant among all N_4H_6 structures in aqueous solutions under normal conditions and the relative abundance of the remaining systems vanishes. In general, the decomposed E systems are more stable than the remaining structures (see Tables 2–4, Figures 1 and 2).

Table 3. Interatomic distances (in Å) in the optimized Amnpq and Dmnpq structures. The different structures with the same notation are distinguished by additional letters a, b, c, or d.

Structure	N1-N2	N2-N3	N3-N4	N1-H	N2-H	N3-H	N4-H
D1221	1.341	1.840	1.338	1.017	1.019 1.016	1.015 1.021	1.018
D2112a	1.423	1.467	1.431	1.012 1.018	1.014	1.016	1.011 1.014
D2112b	1.432	1.419	1.440	1.012 1.015	1.018	1.013	1.011 1.015
D2112c	1.424	1.428	1.437	1.013 1.017	1.016	1.014	1.011 1.015
D2121a	1.413	1.480	1.412	1.011 1.017	1.015	1.020 1.021	1.018
D2121b	1.423	1.467	1.417	1.010 1.013	1.018	1.017 1.020	1.019

Table 3. *Cont.*

Structure	N1-N2	N2-N3	N3-N4	N1-H	N2-H	N3-H	N4-H
D2121c	1.413	1.504	1.409	1.013 1.024	1.017	1.017 1.020	1.020
D2121d	1.423	1.467	1.415	1.010 1.013	1.018	1.016 1.021	1.018
A2202	1.427	1.454	1.443	1.016(2×)	1.021(2×)	-	1.013(2×)
D2202a	1.459	1.422	1.446	1.017(2×)	1.016 1.022	-	1.012 1.013
D2202b	1.464	1.418	1.446	1.016 1.018	1.017 1.021	-	1.012 1.014
D2022	1.459	1.421	1.447	1.017(2×)	-	1.016 1.022	1.012 1.013
A3021	1.463	1.452	1.433	1.016 1.024(2×)	-	1.020 1.025	1.019
D3012a	1.463	1.418	1.463	1.016 1.021(2×)	-	1.013	1.015 1.017
D3012b	1.493	1.395	1.485	1.014 1.021(2×)	-	1.021	1.014 1.017

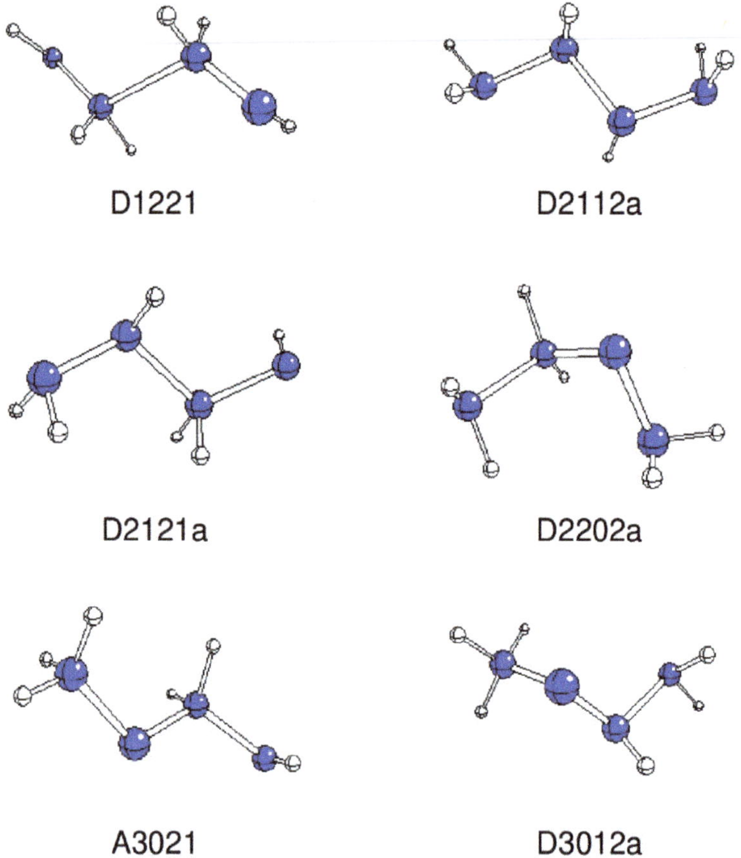

Figure 1. Optimized geometries of stable A and D structures (N—blue, H—white).

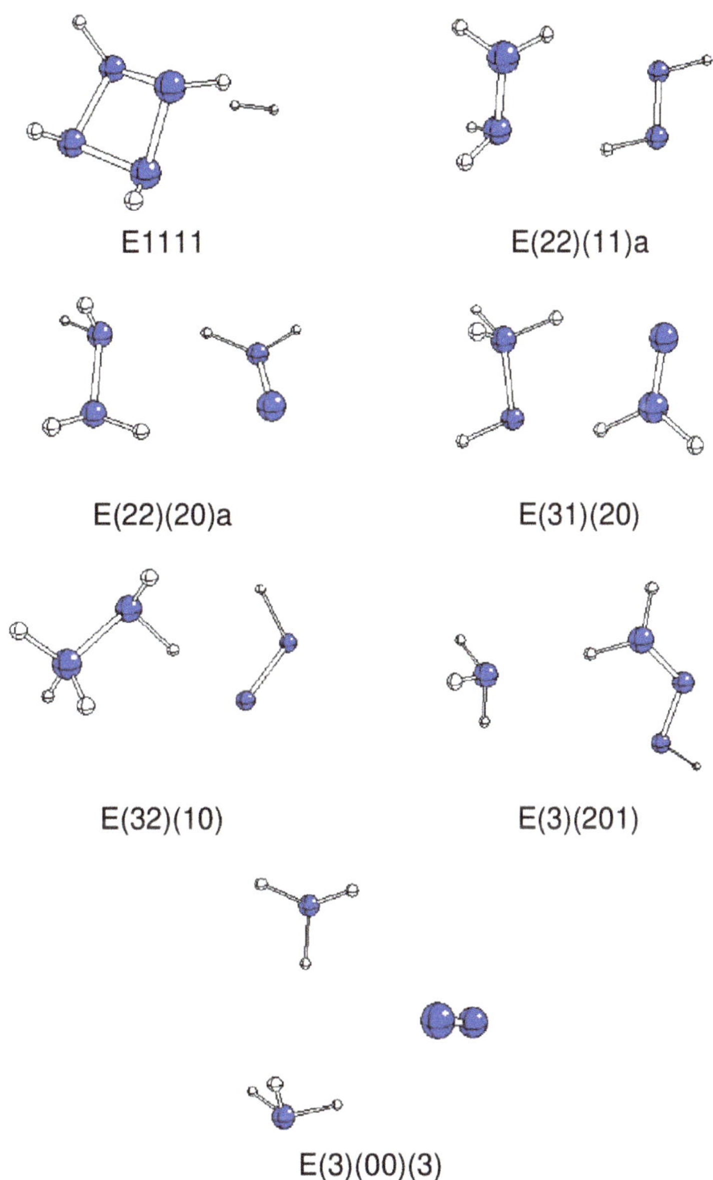

Figure 2. Optimized geometries of stable E systems (N—blue, H—white).

Table 4. Interatomic distances (in Å) in the optimized Emnpq and Fmnpq systems. The different structures with the same notation are distinguished by additional letters a, b, c, or d.

System	N1-N2	N2-N3	N3-N4	N1-H	N2-H	N3-H	N4-H
E1111 [a]	1.476	1.481	1.481	1.023	1.017	1.017	1.017
E(22)(11)a	1.446	3.118	1.245	1.012 1.014	1.011 1.014	1.030	1.027

Table 4. *Cont.*

System	N1-N2	N2-N3	N3-N4	N1-H	N2-H	N3-H	N4-H
E(22)(11)b	1.446	3.465	1.245	1.011 1.014	1.012 1.014	1.030	1.027
E(22)(11)c	1.445	3.292	1.245	1.011 1.014	1.012 1.014	1.027	1.030
E(22)(11)d	1.446	3.116	1.245	1.012 1.014	1.011 1.014	1.030	1.027
E(22)(20)a	1.446	3.271	1.225	1.011 1.014	1.013 1.014	1.028 1.034	-
E(22)(20)b	1.446	2.971	1.225	1.013 1.014	1.011 1.014	1.028 1.033	-
E(22)(02)	1.447	3.276	1.225	1.011 1.014	1.013 1.014	-	1.028 1.033
E(31)(20)	1.468	2.750	1.230	1.018(2×) 1.029	1.016	1.029 1.057	-
E(32)(10)	1.445	3.035	1.242	1.018(2×) 1.021	1.015 1.079	1.076	-
E(3)(201)	3.088	1.350	1.249	1.013(3×)	1.006 1.022	-	1.019
E(3)(102)a	3.117	1.243	1.365	1.013(3×)	1.026	-	1.008 1.014
E(3)(102)b	3.760	1.246	1.356	1.013(2×) 1.014	1.032	-	1.007 1.024
E(3)(00)(3)	3.636	1.096	3.711	1.014(3×)	-	-	1.013(3×)
F(11)(22)	1.245	3.291	1.446	1.030	1.027	1.012 1.014	1.011 1.014
F12)(21 [b]	1.430	3.017	1.424	1.012	1.012 1.018	1.012 1.017	1.017

Remarks: [a] N1-N4 bond length of 1.476 Å, [b] N1-N4 bond length of 1.432 Å.

During the geometry optimization of the starting cyclic C structures (Table 2), only the least stable cyclo-$(NH)_4$ structure, denoted as E1111, preserves its tetraatomic ring after removing an H_2 molecule. The remaining C structures split into hydrazine and HN=NH in E(22)(11)d or $H_2N=N$ in E(22)(02). The disadvantage of cyclic structure preservation is indicated by preferring the above-mentioned F structures after N1-N4 bonding within geometry optimization of the starting B1221 syn-conformation.

The bonding within N_4H_6 structures can be described by individual bond lengths d (Tables 3 and 4) as well as by the corresponding electron density ρ (Tables 5 and 6) and ellipticity ε (Tables 7 and 8) at their bond-critical points (BCP) [9]. Bond strengths decrease with bond lengths d and increase with their BCP electron densities ρ_{BCP}. Their double bond character in acyclic structures increases with their BCP ellipticities ε_{BCP}.

Table 5. BCP electron density (in e/Bohr^3) of N-N and N-H bonds in the optimized Amnpq and Dmnpq structures. The different structures with the same notation are distinguished by additional letters a, b, c, or d.

Structure	N1-N2	N2-N3	N3-N4	N1-H	N2-H	N3-H	N4-H
D1221	0.3705	0.1281	0.3734	0.3450	0.3483 0.3528	0.3467 0.3534	0.3448
D2112a	0.3156	0.2911	0.3092	0.3459 0.3515	0.3574	0.3551	0.3492 0.3518
D2112b	0.3092	0.3237	0.3036	0.3494 0.3508	0.3527	0.3561	0.3484 0.3514

Table 5. *Cont.*

Structure	N1-N2	N2-N3	N3-N4	N1-H	N2-H	N3-H	N4-H
D2112c	0.3149	0.3165	0.3050	0.3472 0.3513	0.3542	0.3552	0.3484 0.3515
D2121a	0.3212	0.2827	0.3080	0.3458 0.3516	0.3544	0.3520 0.3527	0.3415
D2121b	0.3149	0.2920	0.3074	0.3509 0.3517	0.3524	0.3521 0.3551	0.3404
D2121c	0.3224	0.2668	0.3133	0.3406 0.3505	0.3542	0.3508 0.3556	0.3404
D2121d	0.3149	0.2928	0.3086	0.3503 0.3521	0.3523	0.3509 0.3559	0.3417
A2202	0.3148	0.2864	0.2965	0.3478(2×)	0.3514(2×)	-	0.3504 0.3503
D2202a	0.2928	0.3098	0.2951	0.3469 0.3474	0.3499 0.3553	-	0.3506(2×)
D2202b	0.2898	0.3107	0.2956	0.3447 0.3478	0.3507 0.3549	-	0.3501 0.3510
D2022	0.2949	0.3102	0.2930	0.3505 0.3507	-	0.3498 0.3552	0.3470 0.3474
A3021	0.2759	0.2964	0.2956	0.3431 0.3437 0.3480	-	0.3499 0.3542	0.3397
D3012a	0.2761	0.3196	0.2874	0.3454 0.3473 0.3500	-	0.3564	0.3463 0.3476
D3012b	0.2562	0.3362	0.2736	0.3446 0.3450 0.3501	-	0.3488	0.3455 0.3497

Table 6. BCP electron density (in e/Bohr^3) of N-N and N-H bonds in the optimized Emnpq and Fmnpq systems. The different structures with the same notation are distinguished by additional letters a, b, c, or d.

System	N1-N2	N2-N3	N3-N4	N1-H	N2-H	N3-H	N4-H
E1111 [a]	0.2858	0.2828	0.2824	0.3506	0.3545	0.3566	0.3546
E(22)(11)a	0.2953	-	0.4863	0.3502 0.3529	0.3500 0.3529	0.3463	0.3483
E(22)(11)b	0.2953	-	0.4826	0.3500 0.3529	0.3502 0.3529	0.3462	0.3482
E(22)(11)c	0.2954	-	0.4863	0.3500 0.3529	0.3503 0.3529	0.3482	0.3462
E(22)(11)d	0.2953	-	0.4863	0.3502 0.3529	0.3500 0.3529	0.3463	0.3482
E(22)(20)a	0.2947	-	0.4970	0.3499 0.3529	0.3501 0.3518	0.3367 0.3422	-
E(22)(20)b	0.2945	-	0.4967	0.3501 0.3517	0.3499 0.3529	0.3367 0.3423	-
E(22)(02)	0.2945	-	0.4967	0.3500 0.3529	0.3501 0.3517	-	0.3367 0.3423
E(31)(20)	0.2696	-	0.4927	0.3382 0.3486 0.3493	0.3426	0.3123 0.3412	-
E(32)(10)	0.2929	-	0.4831	0.3471 0.3453 0.3474	0.2909 0.3485	0.3032	-

Table 6. *Cont.*

System	N1-N2	N2-N3	N3-N4	N1-H	N2-H	N3-H	N4-H
E(3)(201)	-	0.3794	0.4825	0.3434 0.3435 0.3436	0.3531 0.3368	-	0.3503
E(3)(102)a	-	0.4891	0.3669	0.3435(2×) 0.3436	0.3448	-	0.3457 0.3521
E(3)(102)b	-	0.4833	0.3719	0.3432 0.3435 0.3436	0.3375	-	0.3345 0.3523
E(3)(00)(3)	-	0.7140	-	0.3432 0.3433 0.3433	-	-	0.3435 0.3437 0.3442
F(11)(22)	0.4863	-	0.2954	0.3463	0.3482	0.3502 0.3529	0.3500 0.3529
F12)(21 [b]	0.3096	-	0.3150	0.3573	0.3451 0.3515	0.3470 0.3515	0.3522

Remarks: [a] N1-N4 BCP electron density of 0.2858 e/Bohr^3, [b] N1-N4 BCP electron density of 0.3134 e/Bohr^3.

Table 7. BCP ellipticity of N-N and N-H bonds in the optimized Amnpq and Dmnpq structures. The different structures with the same notation are distinguished by additional letters a, b, c, or d.

Structure	N1-N2	N2-N3	N3-N4	N1-H	N2-H	N3-H	N4-H
D1221	0.230	0.107	0.231	0.048	0.015 0.017	0.015 0.016	0.047
D2112a	0.003	0.149	0.024	0.045 0.050	0.041	0.036	0.046 0.051
D2112b	0.040	0.039	0.012	0.046 0.051	0.043	0.051	0.047 0.051
D2112c	0.027	0.046	0.015	0.044 0.047	0.047	0.050	0.046 0.050
D2121a	0.027	0.123	0.198	0.048 0.051	0.046	0.007 0.009	0.073
D2121b	0.035	0.070	0.182	0.045 0.049	0.034	0.011 0.015	0.074
D2121c	0.026	0.074	0.192	0.039 0.048	0.038	0.013(2×)	0.071
D2121d	0.025	0.069	0.178	0.045 0.050	0.033	0.012 0.013	0.073
A2202	0.060	0.302	0.089	0.036(2×)	0.008(2×)	-	0.055(2×)
D2202a	0.045	0.288	0.084	0.034 0.035	0.012 0.013	-	0.053(2×)
D2202b	0.086	0.301	0.087	0.039 0.041	0.010 0.082	-	0.052 0.053
D2022	0.084	0.288	0.046	0.052 0.053	-	0.012 0.013	0.034 0.035
A3021	0.268	0.222	0.169	0.106 0.108 0.005	-	0.006 0.009	0.079
D3012a	0.267	0.124	0.064	0.006 0.007 0.008	-	0.049	0.044 0.048
D3012b	0.248	0.113	0.123	0.004 0.005(2×)	-	0.051	0.038(2×)

Table 8. BCP ellipticity of N-N and N-H bonds in the optimized Emnpq and Fmnpq systems. The different structures with the same notation are distinguished by additional letters a, b, c, or d.

System	N1-N2	N2-N3	N3-N4	N1-H	N2-H	N3-H	N4-H
E1111 [a]	0.103	0.108	0.108	0.029	0.030	0.027	0.030
E(22)(11)a	0.008	-	0.189	0.047 0.049	0.046 0.050	0.004	0.004
E(22)(11)b	0.008	-	0.189	0.046 0.050	0.047 0.049	0.004	0.004
E(22)(11)c	0.008	-	0.189	0.046 0.050	0.047 0.049	0.004	0.004
E(22)(11)d	0.008	-	0.189	0.047 0.049	0.046 0.049	0.004	0.004
E(22)(20)a	0.008	-	0.021	0.046 0.049	0.047(2×)	0.035 0.038	-
E(22)(20)b	0.007	-	0.020	0.047(2×)	0.046 0.049	0.035 0.038	-
E(22)(02)	0.007	-	0.020	0.046 0.049	0.047(2×)	-	0.035 0.039
E(31)(20)	0.156	-	0.005	0.006 0.011 0.012	0.080	0.029 0.035	-
E(32)(10)	0.089	-	0.072	0.009 0.010(2×)	0.027 0.045	0.005	-
E(3)(201)	-	0.138	0.229	0.033(3×)	0.043 0.053	-	0.008
E(3)(102)a	-	0.218	0.118	0.326 0.327(2×)	0.005	-	0.047 0.051
E(3)(102)b	-	0.238	0.133	0.324(2×) 0.329	0.001	-	0.041 0.052
E(3)(00)(3)	-	0.000	-	0.033 0.034(2×)	-	-	0.033(3×)
F(11)(22)	0.189	-	0.008	0.004	0.004	0.047 0.049	0.046 0.050
F12)(21 [b]	0.012	-	0.031	0.054	0.046 0.051	0.045 0.047	0.049

Remarks: [a] N1-N4 BCP ellipticity of 0.103; [b] N1-N4 BCP ellipticity of 0.041.

The D1221 structure has an extremely long N2-N3 bond, and the remaining N-N bonds are shorter than the average N_4H_6 ones. The ρ_{BCP}(N2-N3)~0.1 e/Bohr^3 corresponds to a very weak bond, and the remaining N-N bonds are approximately three times stronger. The ε_{BCP}(N2-N3)~0.1 is relatively high, and the remaining double N-N bonds have a ca. two times higher ellipticity.

The D2112a-c structures differ in N1-N2-N3-N4 dihedral angles, and their bond length alternation decreases with non-planarity of their backbone. Their ρ_{BCP}(N-N) values vary by about ~0.3 e/Bohr^3, as in single N-N bonds. The ε_{BCP}(N2-N3) values decrease with non-planarity (~0.1 and less), while they are very small for the remaining N-N bonds, which correspond to single bonds.

Similarly, the D2121a-d structures differ in the N1-N2-N3-N4 dihedral angles, with the N2-N3 bond length being longer and weaker than the remaining ones'. The ρ_{BCP}(N-N) values that vary by about ~0.3 e/Bohr^3 correspond to single N-N bonds. The ε_{BCP}(N2-N3) values decrease with non-planarity (~0.1 and less); ε_{BCP}(N3-N4)~0.2 is typical for double bonds.

The N-N bond properties in the A2202, D2202a-b and D2022 structures (aside from reverse numbering of N atoms) vary with the N1-N2-N3-N4 dihedral angles. The N2-N3 bonds are the shortest in all these systems. The ρ_{BCP}(N-N) values that vary by about ~0.3 e/Bohr^3 are typical for single N-N bonds but the ε_{BCP}(N2-N3)~0.3 in all structures indicate the double-bond character of this bond.

In A3021 the N-N bond lengths decrease with the distance from N1, and the BCP ellipticity values indicate the same trend in decreasing double-bond character. However, the ρ_{BCP}(N-N) values of about 0.3 e/Bohr^3 correspond to single N-N bonds.

Analogous trends are observed for D3012a-b structures.

In E1111 with N-N bond lengths of ca. 1.5 Å and ρ_{BCP}(N-N)~0.3 e/Bohr^3 typical for single N-N bonds, the ε_{BCP}(N-N) values of 0.108 can be explained by mechanical strain in its four-membered ring rather than by its double-bond character.

The remaining E systems consist of two or three independent molecules, interacting through weak hydrogen bonds only, which can be treated independently of their parent E structures. The possible biradical character of E(32)(10) can be excluded on the basis of its atomic charges (see later) which indicate the existence of [NH$_3$-NH$_2$]$^+$ and [HN≡N]$^-$ charged species.

H$_2$N-NH$_2$ with an N-N distance of 1.45 Å, ρ_{BCP}(N-N) = 0.295 e/Bohr^3 and ε_{BCP}(N-N) = 0.008 in all E systems is typical for a single N-N bond.

HN=NH with an N-N distance of 1.245 Å, ρ_{BCP}(N-N) = 0.486 e/Bohr^3 and ε_{BCP}(N-N) = 0.189 in all E systems corresponds to the double N-N bond.

Its isomer H$_2$N=N has a N-N distance of 1.23 Å and ρ_{BCP}(N-N) = 0.497 e/Bohr^3 which correspond to the double N-N bond in contradiction with ε_{BCP}(N-N) = 0.020, which corresponds to single or triple bonds.

On the other hand, NH$_3$-NH has a N-N distance of 1.47 Å and ρ_{BCP}(N-N) = 0.27 e/Bohr^3, which corresponds to the single N-N bond in contrast to the high ε_{BCP}(N-N) value of 0.156.

The [NH$_3$-NH$_2$]$^+$ cation with an N-N distance of 1.446 Å, ρ_{BCP}(N-N) = 0.293 e/Bohr^3 and ε_{BCP}(N-N) = 0.089 corresponds to a single N-N bond.

Its counterpart [HN≡N]$^-$ has a N-N distance of 1.242 Å and ρ_{BCP}(N-N) = 0.48 e/Bohr^3 which correspond to the double N-N bond in contradiction with its too low ε_{BCP}(N-N) = 0.072.

N$_2$ has a N-N distance of 1.096 Å, ρ_{BCP}(N-N) = 0.714 e/Bohr^3 and ε_{BCP}(N-N) = 0.000, which is typical for the triple bond.

Finally, H$_2$N-N=NH with N-N distances of 1.36 and 1.24 Å, ρ_{BCP}(N-N) values of 0.37 and 0.48 e/Bohr^3 as well as ε_{BCP}(N-N) values of 0.23 and 0.13, respectively, probably correspond to nearly-double N-N bonds.

The F(11)(22) system is explained within the HN≡NH and H$_2$N-NH$_2$ structures above.

The F12)(21 structure H$_2$N2-N1H-N4H-N3H$_2$ (aside from different numbering of N atoms) corresponds to the D2112 structures explained above.

We have not discussed N-H bonding in the systems under study because the differences in their bond lengths and BCP electron densities are too small. However, their BCP electron densities are higher than those of N-N bonds except HN=NH, H$_2$N=N, [HN≡N]$^-$ and N$_2$. Increased ε_{BCP}(N-H) values can mostly be ascribed to the double-bond character of neighboring N-N bonds, except ε_{BCP}(N-H) = 0.3 in NH$_3$ molecules within the E(3)(102) systems.

The nitrogen atomic charges in the A and D structures (Table 9) on the N1 and N4 atoms are more negative (−0.65 to −0.82) than on the central N2 and N3 atoms (−0.35 to −0.50). Positive hydrogen atomic charges bonded to side N1 and N4 atoms increase with the number of bonded H atoms. The same trend holds for H atoms bonded to central N2 and N3 atoms which are more positive than the side hydrogens.

In the decomposed E systems (Table 10), negative N charges increase with the number of bonded H atoms. An analogous trend for positive H charges cannot be confirmed. Atomic charges are only slightly affected by hydrogen bonding. In the E(32)(10) system, the charges of its [NH$_3$-NH$_2$]$^+$ and [HN≡N]$^-$ subsystems are +0.97 and −0.68, respectively (the ideal charges are +1.00 and −1.00, respectively). The errors can be ascribed to numerical integration of electron density up to 0.001 e/Bohr^3 (instead of 0.000 e/Bohr^3). A significantly higher error of [HN≡N]$^-$ is caused by the higher diffusive character of the electron density of anionic species. When accounting for the errors in the electron density integration over atomic basins, the alternative biradical structure of the neutral E(32)(10) subsystems (the ideal charges of both species should be 0.00) seems to be less probable.

Table 9. Atomic charges of N and H (bonded to N in brackets) in the optimized Amnpq and Dmnpq structures. The asterisks denote the atoms also included in hydrogen bonds. The different structures with the same notation are distinguished by additional letters a, b, c, or d.

Structure	N1	N2	N3	N4	H(N1)	H(N2)	H(N3)	H(N4)
D1221	−0.657	−0.488	−0.485	−0.649	0.342	0.452 0.455	0.444 0.457	0.342
D2112a	−0.699	−0.347	−0.367	−0.706	0.379 0.392	0.391	0.382	0.391 0.404
D2112b	−0.691	−0.357	−0.354	−0.726	0.378 0.394	0.372	0.395	0.387 0.398
D2112c	−0.711	−0.354	−0.368	−0.729	0.377 0.389	0.382	0.396	0.389 0.401
D2121a	−0.700	−0.341	−0.398	−0.787	0.394 0.413	0.417	0.452(2×)	0.309
D2121b	−0.704	−0.361	−0.394	−0.811	0.400 0.416	0.405	0.470 0.560	0.302
D2121c	−0.709	−0.365	−0.412	−0.800 *	0.396 0.407 *	0.406	0.463 0.468	0.310
D2121d	−0.707	−0.361	−0.395	−0.809	0.402 0.420	0.408	0.458 0.471	0.304
A2202	−0.664	−0.388	−0.435	−0.750	0.418(2×)	0.450(2×)	-	0.362(2×)
D2202a	−0.712	−0.404	−0.430	−0.760	0.409 0.410	0.455 0.475	-	0.361 0.364
D2202b	−0.705	−0.397	−0.432	−0.737	0.407 0.411	0.459 0.465	-	0.357 0.367
D2022	−0.761	−0.430	−0.402	−0.711	0.361 0.365	-	0.455 0.475	0.409 0.410
A3021	−0.730	−0.368	−0.388	−0.824	0.460 0.461 0.496	-	0.403 0.423	0.286
D3012a	−0.732	−0.436	−0.390	−0.739	0.449(2×) 0.472	-	0.370	0.360 0.372
D3012b	−0.762	−0.417	−0.384	−0.754 *	0.444 0.466 * 0.473	-	0.345	0.376 0.378

Table 10. Atomic charges of N and H (bonded to N in bracket) in the optimized Emnpq and Fmnpq systems. Asterisks denote atoms also included in hydrogen bonds. The different structures with the same notation are distinguished by additional letters a, b, c, or d.

System	N1	N2	N3	N4	H(N1)	H(N2)	H(N3)	H(N4)
E1111	−0.345	−0.367 *	−0.373	−0.367	0.383	0.408	0.396 *	0.403
E(22)(11)a	−0.707	−0.727 *	−0.358	−0.348 *	0.380 0.392 *	0.385 0.393	0.409 *	0.380
E(22)(11)b	−0.727 *	−0.707	−0.360	−0.348 *	0.384 0.393	0.380 0.388 *	0.409 *	0.380
E(22)(11)c	−0.726 *	−0.706	−0.349h	−0.360	0.384 0.393	0.380 0.387 *	0.380	0.409h
E(22)(11)d	−0.706	−0.726 *	−0.359	−0.347 *	0.380 0.388 *	0.384 0.393	0.409 *	0.380
E(22)(20)a	−0.732 *	−0.714	−0.519	−0.271 *	0.380 0.393 *	0.387 0.395	0.417 0.460 *	-
E(22)(20)b	−0.713	−0.732 *	−0.517	−0.273 *	0.380 0.393 *	0.387 0.395	0.417 0.461 *	-

Table 10. *Cont.*

System	N1	N2	N3	N4	H(N1)	H(N2)	H(N3)	H(N4)
E(22)(02)	−0.714	−0.732 *	−0.272 *	−0.517	0.380 0.393 *	0.387 0.395	-	0.417 0.460 *
E(31)(20)	−0.751	−0.831 *	−0.543	−0.306 *	0.447 0.452 0.491 *	0.315	0.408 0.511 *	-
E(32)(10)	−0.718	−0.731	−0.426	−0.530 *	0.496 0.508(2×)	0.408 0.501 *	0.185	-
E(3)(201)	−1.079 *	−0.734	−0.035	−0.436	0.394(3×)	0.443 0.473 *	-	0.388
E(3)(102)a	−1.076 *	−0.454	−0.033	−0.686	0.394(3×)	0.428 *	-	0.429 0.445
E(3)(102)b	−1.084 *	−0.396	−0.030	−0.739	0.394 0.395 0.396	0.352	-	0.445 0.470 *
E(3)(00)(3)	−1.077 *	0.076 *	−0.049	−1.059	0.382 0.382 * 0.384	-	-	0.373 0.380 * 0.386
F(11)(22)	−0.359	−0.347 *	−0.706	−0.725 *	0.409 *	0.380	0.380 0.388 *	0.393 0.394
F12)(21	−0.356	−0.722	−0.702	−0.368	0.403	0.377 0.392	0.371 0.395	0.381

3. Method

Geometry optimizations for various isomers of neutral N_4H_6 molecules were performed at the CCSD (Coupled Cluster using Single and Double substitutions from the Hartree-Fock determinant) [10] level of theory and cc-pVTZ basis sets [11]. The effects of the aqueous solution were taken into account within the SMD (Solvation Model based on the solute electron Density) solvation model [12]. The optimized structures were tested by vibrational analysis for the absence of imaginary vibrations. Gaussian16 (Revision B.01) software [13] was used for all quantum-chemical calculations.

The electron structures of the systems under study were evaluated in terms of Quantum Theory of Atoms-in-Molecules (QTAIM) [9] using AIM2000 (Version 1.0) software [14]. The bond strengths were compared according to the electron densities ρ at the bond-critical points (BCP). The BCP bond ellipticities ε_{BCP} were evaluated as

$$\varepsilon_{BCP} = \lambda_1/\lambda_2 - 1 \tag{7}$$

where λ_i are the eigenvalues of the Hessian of the BCP electron density within the sequence $\lambda_1 < \lambda_2 < 0 < \lambda_3$. Atomic charges were obtained by integration over atomic basins up to 0.001 e/Bohr^3.

Visualization and geometry modification were performed using MOLDRAW (Release 2.0) software (https://www.moldraw.software.informer.com, accessed on 9 September 2019) [15].

4. Conclusions

We have shown that most N_4H_6 structures in aqueous solutions are decomposed during geometry optimization. Splitting the bond between central nitrogen atoms is the most frequent method, but the breakaway of the side nitrogen is energetically the most preferred one. The N-N fissions are enabled by suitable hydrogen rearrangements. The initial H_2N-NH-NH-NH_2 structure (D2112) has a very weak central N-N bond, which explains the high degree of reversibility for the reaction (6). The most stable system NH_3...N_2...NH_3 (E(3)(00)(3) system) might be obtained by transfers of both H atoms bonded with central nitrogens to the side N atoms. According to [8], such double H transfer was not found by quantum-chemical calculations in vacuo, and so must be decomposed into several steps and this instantaneous decomposition should be slowed down. Furthermore, our calculations show that the transfer of the third H atom to the side nitrogen is very energetically disadvantageous, as indicated by the Gibbs energies of the structures NH_3-$N=NH_2$-NH and NH_3-N=NH-NH_2 (A3021 and D3012, respectively, see Table 2). In aqueous solutions, H atom transfers can be mediated by H_2O, H_3O^+ and/or OH^- species. We have shown that side N atoms have very high negative charges that should support such hydrogen transfers.

The experimentally observed formation of $^{15}N^{14}N$ molecules [1–4] is enabled by side N-N fissions. We have shown that the Gibbs free energy data (Table 2) indicate the dominant abundance of the NH_3... N_2... NH_3 species (E(3)(00)(3) system) in aqueous solutions, which explains the mentioned observations.

The $^{15}N^{14}N$ molecules can also be created by the decomposition of cyclic N_4H_6 structures. We have shown the high instability of such species. The only stable cyclo-$(NH)_4$...H_2 structure (E1111) has a too-high Gibbs energy and breaks the H_2 molecule instead. The remaining initial cyclic structures are split into hydrazine and $HN\equiv NH$ (E(22)(11)d) or $H_2N\equiv N$ species (E(22)(02), see Table 2), and their relative abundance in aqueous solutions vanishes.

We can deduce from the QTAIM analysis of our systems that single, double and triple N-N bonds exhibit BCP electron densities of ca. 0.2, 0.5 and 0.7 e/$Bohr^3$ with BCP ellipticities of ca 0, 0.2 and 0, respectively. The bonds in the N_4H_6 structures often exhibit significant deviations from these values.

Our study did not solve all of the problems related to hydrazine oxidation in aqueous solutions. The role of various water forms and the corresponding transition states should also be investigated. The transition states can possibly be of extremely high-energy. Thus, the thermodynamic stability of the products means less if their formation is kinetically hindered. Moreover, directly accounting for the solvent molecules is required. An alternative reaction pathway through N_4H_4 [6] according to reaction (5) is worth studying as well. Further theoretical studies in these fields are desirable.

Author Contributions: Methodology, software, investigation, writing—original draft preparation, writing—review and editing, M.B.; conceptualization, supervision, project administration, funding acquisition, A.M. All authors have read and agreed to the published version of the manuscript.

Funding: This publication was supported by the Competence Center for SMART Technologies for Electronics and Informatics Systems and Services under the project no. ITMS 26240220072.

Institutional Review Board Statement: Not applicable.

Informed Consent Statement: Not applicable.

Data Availability Statement: All necessary research data are presented in the article.

Acknowledgments: M.B. thanks the HPC center at the Slovak University of Technology in Bratislava, which is a part of the Slovak Infrastructure of High Performance Computing (SIVVP Project No. 26230120002, funded by the European Region Development Funds), for computing facilities.

Conflicts of Interest: The authors declare no conflict of interest.

References

1. Lauko, L.; Hudec, R.; Lenghartova, K.; Manova, A.; Cacho, F.; Beinrohr, E. Simple Electrochemical Determination of Hydrazine in Water. *Pol. J. Environ. Stud.* **2015**, *24*, 1659–1666. [CrossRef] [PubMed]
2. Higginson, W.C.E.; Sutton, D. The Oxidation of Hydrazine in Aqueous Solution. Part II. The Use of ^{15}N as a Tracer in the Oxidation of Hydrazine. *J. Chem. Soc.* **1953**, 1402–1406. [CrossRef]
3. Cahn, J.W.; Powell, R.E. Oxidation of Hydrazine in Solution. *J. Am. Chem. Soc.* **1954**, *76*, 2568–2572. [CrossRef]
4. Petek, M.; Bruckenstein, S. An Isotopic Labeling Investigation of the Mechanism of the Electrooxidation of Hydrazine at Platinum. An Electrochemical Mass Spectrometric Study. *Electroanal. Chem. Interrac. Electrochem.* **1973**, *47*, 329–333. [CrossRef]
5. Rice, F.O.; Sherber, F. The Hydrazino Radical and Tetrazane. *J. Am. Chem. Soc.* **1955**, *77*, 291–293. [CrossRef]
6. Karp, S.; Meites, L. The Voltammetric Characteristics and Mechanism of Electrooxidation of Hydrazine. *J. Am. Chem. Soc.* **1962**, *84*, 906–912. [CrossRef]

7.   Ball, D.W. Tetrazane: Hartree-Fock, Gaussian-2 and -3, and Complete Basis Set Predictions of Some Thermochemical Properties of N_4H_6. *J. Phys. Chem. A* **2001**, *105*, 465–470. [CrossRef]

8.   Dana, A.G.; Moore, K.B., III; Jasper, A.W.; Green, W.H. Large Intermediates in Hydrazine Decomposition: A Theoretical Study of the N_3H_5 and N_4H_6 Potential Energy Surfaces. *J. Phys. Chem. A* **2019**, *123*, 4679–4692. [CrossRef] [PubMed]

9.   Bader, R.F.W. *Atoms in Molecules: A Quantum Theory*; Clarendon Press: Oxford, UK, 1990; ISBN 9780198558651.

10.   Scuseria, G.E.; Janssen, C.L.; Schaefer, H.F., III. An efficient reformulation of the closed-shell coupled cluster single and double excitation (CCSD) equations. *J. Chem. Phys.* **1988**, *89*, 7382–7387. [CrossRef]

11.   Dunning, T.H., Jr. Gaussian basis sets for use in correlated molecular calculations. I. The atoms boron through neon and hydrogen. *J. Chem. Phys.* **1989**, *90*, 1007–1023. [CrossRef]

12.   Marenich, A.V.; Cramer, C.J.; Truhlar, D.G. Universal solvation model based on solute electron density and a continuum model of the solvent defined by the bulk dielectric constant and atomic surface tensions. *J. Phys. Chem. B* **2009**, *113*, 6378–6396. [CrossRef] [PubMed]

13.   Frisch, G.W.; Trucks, M.J.; Schlegel, H.B.; Scuseria, G.E.; Robb, M.A.; Cheeseman, J.R.; Scalmani, G.; Barone, V.; Petersson, G.A.; Nakatsuji, H.; et al. *Gaussian 16, Revision B.01*; Gaussian, Inc.: Wallingford, CT, USA, 2016.

14.   Biegler-König, F.; Schönbohm, J.; Bayles, D. AIM2000—A Program to Analyze and Visualize Atoms in Molecules. *J. Comput. Chem.* **2001**, *22*, 545–559.

15.   Ugliengo, P. MOLDRAW: A Program to Display and Manipulate Molecular and Crystal Structures, University Torino, Torino. 2012. Available online: https://www.moldraw.software.informer.com (accessed on 9 September 2019).

MDPI

Article

Photocatalytic Degradation of Emerging Contaminants with N-Doped TiO$_2$ Using Simulated Sunlight in Real Water Matrices

Elisa Gaggero, Arianna Giovagnoni, Alessia Zollo, Paola Calza * and Maria Cristina Paganini

Department of Chemistry, University of Turin, Via Pietro Giuria 5/7, 10126 Turin, Italy;
elisa.gaggero@unito.it (E.G.); arianna.giovagnoni@edu.unito.it (A.G.); alessia.zollo@unito.it (A.Z.);
mariacristina.paganini@unito.it (M.C.P.)
* Correspondence: paola.calza@unito.it

Abstract: In the present work, the photodegradation performances of N-doped TiO$_2$ photocatalysts with enhanced absorption of visible light were exploited for the abatement of some representative contaminants of emerging concern (CECs). Pristine TiO$_2$ and N-TiO$_2$ were synthesized using hydrothermal (HT) and sol–gel (SG) routes, they were characterized using XRD and UV-Vis spectroscopy, and their band gaps were determined via analysis in diffuse reflectance. Their photodegradation efficiency was tested on a mixture of recalcitrant organic pollutants, namely, benzotriazole, diclofenac, sulfamethoxazole, and bisphenol A, using a solar simulator lamp with two different cut-off filters ($\lambda > 340$ nm and $\lambda > 400$ nm). The evaluation of the photocatalytic performances was initially carried out in spiked ultrapure water and subsequently in aqueous matrices of increasing complexity such as Po River water and water coming from an aquaculture plant. The exclusive utilization of visible light ($\lambda > 400$ nm) highlighted the advantage of introducing the dopant into the TiO$_2$ photocatalyst since this modification allows for the material to be responsive to visible light, which is not sufficient in the case of pristine TiO$_2$ and the higher efficiency of materials obtained via the sol–gel route. Thanks to the doping, improved performance was obtained in both ultrapure water and real water matrices, indicating the potential of the doped material for future applications in the field.

Keywords: N-doped TiO$_2$; photocatalysis; contaminants of emerging concern

Citation: Gaggero, E.; Giovagnoni, A.; Zollo, A.; Calza, P.; Paganini, M.C. Photocatalytic Degradation of Emerging Contaminants with N-Doped TiO$_2$ Using Simulated Sunlight in Real Water Matrices. *Inorganics* **2023**, *11*, 439. https://doi.org/10.3390/inorganics11110439

Academic Editor: Francis Verpoort

Received: 6 October 2023
Revised: 7 November 2023
Accepted: 15 November 2023
Published: 17 November 2023

1. Introduction

Freshwater, a limited but renewable resource, is essential for life on Earth. The water cycle replenishes it, but overconsumption can harm ecosystems, and global per capita renewable freshwater resources are declining due to economic development, overexploitation of water resources, pollution, anthropogenic activities, and related climate change. Among the various sources of pollution, the ones with the most significant impact on natural water sources are the agricultural and industrial sources, especially in developing countries [1]. Industrial and domestic wastewater discharges have led to an increase in freshwater pollution and a decrease in clean water resources, particularly in relation to rivers. Rivers can contain microorganisms, heavy metals, and dangerous chemicals, rendering the water unsuitable for direct use and elevating the costs of purification processes [2]. Pollution takes various forms, from aesthetically unpleasant water due to untreated wastewater to organic materials in wastewater causing oxygen depletion and damages to aquatic life [3]. The European Union has identified over a hundred thousand chemical substances as micropollutants and generates continuously updated lists of contaminants of emerging concern (CECs) and persistent organic pollutants (POPs). Micro-pollutants encompass items like pharmaceuticals, numerous pesticides, metals, and phthalates [4]. Many commercially available drugs and their byproducts fall under the category of CECs. There has been a global ban on persistent organic pollutants (POPs) since 2004, backed by 179 countries since micro-pollutants are recognized for their non-biodegradability, high toxicity, potential for bioaccumulation due to their persistent nature, and properties such as genotoxicity, mutagenicity, and oestrogenicity [5,6]. Common micro-pollutants found in groundwater include substances like triclosan, sulfamethoxazole, carbamazepine, and specific pharmaceuticals [7]. Recently, there has been a growing focus of scientific studies on emerging pollutants, leading to the formation of research groups aiming to collect and share detailed information. Their objective is to create management strategies for environmental protection and implement measures to limit pollutants in water environments, alongside technological advancements in water treatment. This interest also arises partly from the lack of regulation for pharmaceutical products and their metabolic byproducts, even if they will likely be subject to future regulations due to their potential impact on both ecosystems and human health. The main concerns lie in the diversity of compounds falling under this category and their proven harmful effects, even at low concentrations. Analytically identifying and removing these substances using conventional methods is challenging since emerging pollutants cover a broad range of compounds, including pharmaceuticals, hormones, pesticides, flame retardants, and fragrances [8]. Addressing their impact may not be simple also because of the difficulty in removing these molecules with conventional techniques. This makes the development of innovative, environmentally friendly, and low-cost systems for their removal essential. In the present work, we aimed at exploiting the photodegradation performances of N-doped TiO_2 photocatalysts with enhanced absorption of visible light for the abatement of some CECs, namely, phenol, bisphenol A, sulfamethoxazole, diclofenac, and benzotriazole, representative of different classes [9–13].

One of the strategies studied and used to degrade emerging pollutants and organic contaminants in water is the employment of visible-light-activated photocatalysts [14–16]. This study focused on four photocatalysts: TiO_2 synthesized through sol–gel and hydrothermal methods, and nitrogen-doped TiO. Among the photocatalysts, TiO_2 is widely studied due to its stability and affordability, though its high band gap limits its efficiency [17,18]. When titanium dioxide absorbs photons compatible with its band gap, it produces hole-electron pairs that allow for the oxidation of water and organic matter and the reduction of oxygen and its reactive species that lead to the formation of radicals that contribute to the further abatement of the contaminants. However, its limited response to visible light restricts its potential; so, to enhance its efficiency, doping with transition metals or non-metals like nitrogen is used [19]. In order to overcome this limitation, nitrogen doping was used. Currently, a substantial body of research results demonstrates that N-doped TiO_2 and N co-doping with transition elements [20–22] exhibit higher photoactivity in visible or solar light compared to the conventional TiO_2, which serves as the true benchmark for photocatalytic phenomena under the same conditions. However, there is a limited number of experimental studies dedicated to comprehending the underlying reasons for this heightened photoactivity and devising strategies to enhance the photocatalyst's performance (as indicated in references [23–30]). Consequently, it remains uncertain whether this novel material will ultimately be recognized as a significant breakthrough in the realm of photocatalytic applications. To overcome this challenge, it is imperative to devote more attention to both the electronic properties of N-TiO_2 and the mechanisms governing the dynamics of charge carriers during irradiation. Furthermore, such efforts should be aimed at tailoring appropriate modifications of the material to influence its properties. There is now a general consensus on the electronic structure of N-TiO_2 as most of the authors agree on the presence of N 2p intraband gap states some tenths of electronvolts over the valence band (VB) limit. There are currently two distinct categories of photoactive centers that have been identified. The first type arises from the substitution of nitrogen (where nitrogen replaces oxygen in certain lattice positions) and is formed through high-temperature nitridation processes of TiO. The second type, which is more prevalent and occurs in most wet chemistry methods used to prepare the material, involves interstitial nitrogen chemically bonded to a lattice oxygen ion, resulting in the formation of a type of NO group within the bulk of the solid. This species is responsible for the optical absorption in the visible range of N doped titania [31]. Although there is consensus regarding the electronic structure of N-TiO_2, there remains a lack of comprehensive understanding concerning how the system behaves in terms of charge dynamics when exposed to visible light. It is of great significance, for instance, to conclusively determine whether visible photons have the capability to induce charge separation, leading to the migration of both electrons and holes at the surface. The presence of electrons and holes at the surface is indeed crucial for initiating the redox processes characteristic of photocatalysis. While the dynamics of charge carriers in photocatalytic systems have primarily been studied using time-resolved spectroscopic techniques in the case of pure TiO_2, there are relatively few reports available for $N\text{-}TiO_2$ (as mentioned in references [32,33]). In these recent investigations, it has been suggested that the efficiency of charge separation in $N\text{-}TiO_2$ under visible light is, as anticipated, lower than that observed under UV light [34,35].

We assessed the efficiency of the synthetized photocatalysts in degrading persistent organic pollutants using a solar simulator lamp equipped with two distinct cut-off filters ($\lambda > 340$ nm and $\lambda > 400$ nm). The evaluation of their photocatalytic effectiveness commenced with pure-water samples spiked with pollutants and progressed to more challenging aqueous environments like Po River water and water from an aquaculture farm.

2. Results and Discussion

2.1. XRD and UV-VIS Characterization

The synthesized materials were characterized through XRD and Diffuse Reflectance UV Vis analysis.

Figure 1 shows the X-ray diffraction patterns for the photocatalysts obtained using sol–gel and hydrothermal methodologies. As expected, the XRD diffractograms of the nitrogen-doped sample are almost completely overlapping the spectrum of the reference, TiO_2. The patterns of the two kinds of syntheses present different features. The sol–gel process leads to well crystallized materials with a relatively large size of the crystallites. In the case of pure TiO_2, some traces of rutile are also present.

In both cases (SG and HT), the diffraction pattern shows a predominance of the anatase phase. This is demonstrated by the presence of peaks at $25.33°$, $37.96°$, $48.04°$, $54.37°$, $55.29°$, and $62.75°$ that correspond to the planes of the crystal lattice (101), (004), (200), (105), (211), and (204), respectively [36]. Samples obtained with hydrothermal synthesis also present two peaks that could be assigned to the brookite phase. In general, samples obtained via the hydrothermal method seem to be less crystalline and show smaller crystallites sizes. The presence of Nitrogen is not revealed with this technique, and this is justified by the fact that elements such as N are not visible in X-rays [36]; moreover, it has been demonstrated elsewhere that N species occupy an interstitial position in the lattice of TiO_2 and do not form a periodic phase inside the matrix [25].

The absorption spectra of the synthesized materials are shown in Figure 2. For the N-doped samples, the absorption due to electronic transitions from a valence band to a conduction band occurring in the UV region and typical of TiO_2 is modified by the onset of a relatively broad absorption band in the visible region whose intensity and width depend on the type of sample. These bands are similar to others reported in the literature [37] for nitrogen-doped TiO_2 and are usually associated with nitrogen insertion in the oxide. It is evident that the N doped sample synthetized via a sol–gel presents the deepest shoulder in the visible region.

From the study on diffuse reflectance, it is possible to determine the band gaps (Table 1) of the materials and highlight any differences due to the presence of the dopant.

The band gap values obtained from processing the Kubelka–Munk function of the spectra using the Tauc plot method highlight that the introduction of nitrogen does not significantly change the band gap value, which remains in the range of 3.34–3.38 eV, around the reference value found in the literature. In fact, the band gap for anatase is 3.2 eV, requiring irradiation with a wavelength $\lambda < 387$ nm [38].

Table 1. Band gap (eV) values determined from absorption spectra for different synthesized materials.

Sample	Band Gap (eV)
TiO_2-SG	3.38
N-TiO_2-SG	3.37
TiO_2-HT	3.37
N-TiO_2-HT	3.34

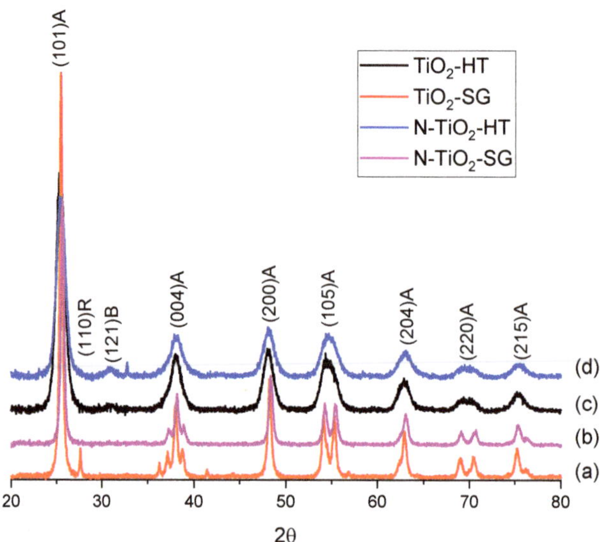

Figure 1. XRD diffractograms for materials obtained through sol–gel synthesis. (a) TiO_2-SG, (b) N-TiO_2-SG and hydrothermal synthesis, (c) TiO_2-HT, (d) N-TiO_2-HT. (Together with the prominent anatase phase are also present traces of rutile and brookite phase; residual NH_4Cl from the synthesis) [39,40].

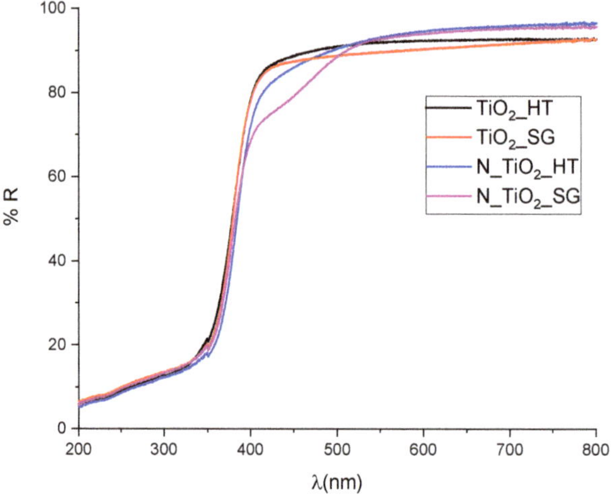

Figure 2. DR-UV-Vis spectra for the materials obtained through hydrothermal synthesis TiO_2-HT and N-TiO_2-HT, and sol–gel syntheses TiO_2-SG and N-TiO_2-SG.

2.2. Photocatalytic Degradation of Organic Pollutants

Heterogeneous photocatalysis tests in solution were conducted to determine the photocatalytic activity of the synthesized TiO_2-based materials. Preliminarily, dark adsorption toward the considered molecules and direct photolysis experiments were carried out with all the materials. Both contributions were found to be negligible over the examined time window. Figure 3 shows the degradation trends of 10 ppm phenol using the synthesized materials under irradiation with a 340 nm cut-off. TiO_2-based materials doped with nitrogen exhibited excellent photocatalytic performance, leading to complete degradation of phenol within 2 h. It is further inferred that the performance improvement resulting from nitrogen doping is more evident for materials obtained via sol–gel synthesis; this fact could be related to the greater amount of crystalline materials.

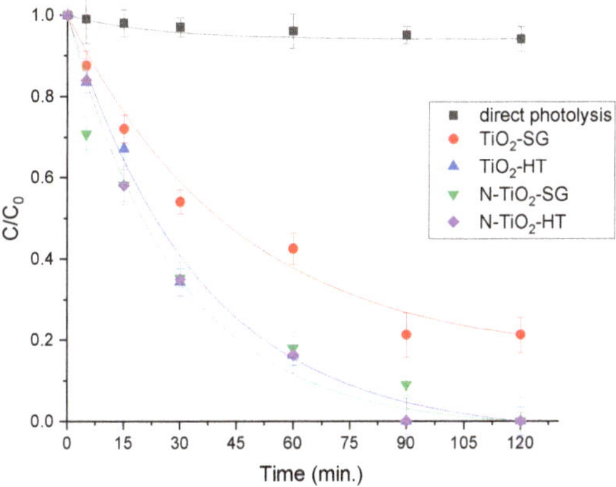

Figure 3. Phenol degradation rate using prepared photocatalyst ($\lambda > 340$ nm).

In order to determine whether nitrogen doping can provide better photocatalytic activity using only visible light, at which the activity of pristine TiO_2 is insufficient because of its high band gap, experiments were repeated by irradiating with a $\lambda > 400$ nm cut-off. Although the nitrogen-doped materials present the same band gap of the pure titania, the DR UV vis spectra highlighted the presence of an absorption shoulder in the region of the visible light. We know from the literature [31] that this shoulder has been associated with intraband gap states generated from the presence of nitrogen in the lattice of TiO. These states can act as steps for the electrons, so the irradiation with visible light provides enough energy to promote electron migration from the VB to these states and from these states to the CB. Being isolated states, the total efficiency of the process should be expected to be much lower than the classic process from VB to CB.

The kinetic constants of the degradations were calculated by considering the expression of the pseudo-first-order reaction (Equation (1)):

$$\ln \frac{C_t}{C_0} = -kt \tag{1}$$

where C_t and C_0 are the concentrations at time zero and at any time t (min), respectively, and k (min^{-1}) is the rate constant of the equation.

The pseudo first-order kinetic constants and abatement percentage after six hours are reported in Figure 4.

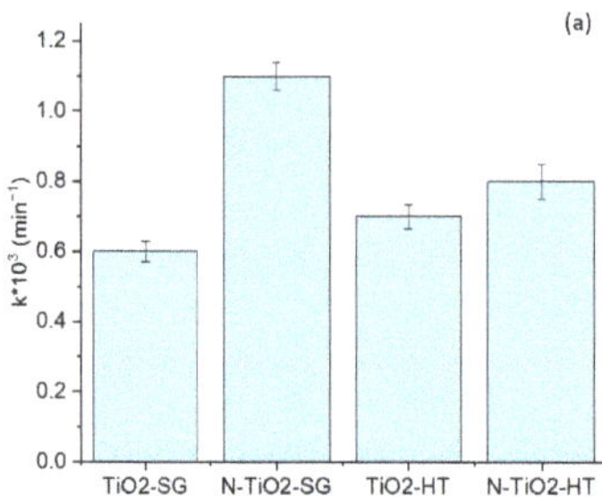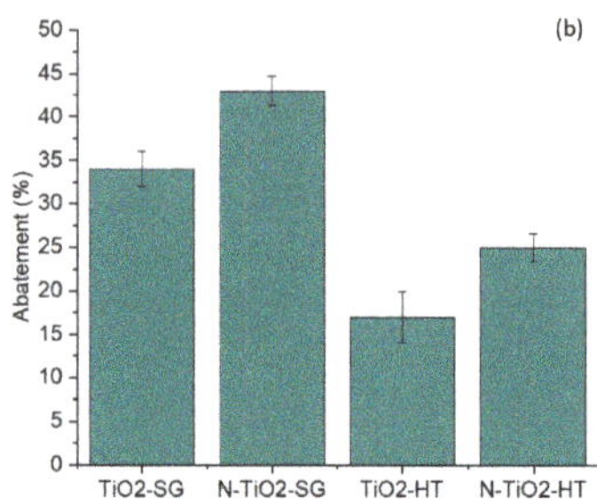

Figure 4. Pseudo-first order kinetic constants (**a**) and abatement percentage (**b**) after six hours obtained for phenol degradation ($\lambda > 400$ nm).

In this case, as expected, lower performances are achieved than using $\lambda > 340$ nm. In fact, after 90 min, the maximum abatement for N-TiO$_2$-HT reached 25% compared to the 100% achieved in the previous case. Interestingly, there is an improvement in the photocatalytic performance of materials obtained via the sol–gel route compared to those obtained via hydrothermal synthesis, presenting an opposite trend in respect to the tests previously carried out. After preliminary tests using phenol as a target molecule, photocatalytic experiments were carried out on a mixture of CECs (benzotriazole, bisphenol A, diclofenac, and sulfamethoxazole) prepared in three matrices of increasing complexity: Milli-Q water, Po river water, and water collected in a koi-carps aquaculture farm. For the actual waters used as matrices, both pH and NPOC values were recorded. The pH was 8.02 and 8.60 for Po River water and water from carp aquaculture, respectively. The NPOC values measured were 1.51 mg/L and 14.2 mg/L for Po River and aquaculture water, respectively.

The activity of all materials was first tested using the $\lambda > 340$ nm cut-off. Observing the degradation curves reported in Figures S1–S4, it can be pointed out that in the case of hydrothermally obtained materials, abatement around 100% is achieved for all matrices, with no substantial differences between Milli-Q and real waters. Using N-TiO$_2$-HT, an abatement greater than 75% is achieved for all contaminants already after the first 60 min and almost-complete abatement in all cases evaluated at the end of the 2-hour experiment. In comparison with the reference, the introduction of nitrogen as a dopant causes a slight increase in the pseudo first-order kinetic constants when real waters are used as matrices.

On the other hand, the performance of the sol–gel-derived materials using a filter at $\lambda > 340$ nm is inferior to that of hydrothermal-synthetized photocatalysts, particularly when experiments are carried out in aquaculture water; moreover, the introduction of the nitrogen does not appear to positively affect the rate of degradation.

As can be observed in Figure 5, in general, considering pristine materials, a decrease in kinetic constants is registered when moving from Milli-Q water to real water matrices even if the diminution is less pronounced for doped materials. This could be due to the higher content of organic matter and to the presence of naturally occurring inorganic species that can act as scavengers during the oxidation processes, for example, carbonate and bicarbonate ions that can act as scavengers for the hydroxyl radical, reducing the contaminant removal efficiency [41].

With the aim of verifying the performance improvement of the doped material compared to the reference using visible light only, the experiments were repeated using a lamp cut-off that excluded UV-A. In these operating conditions, the advantage of using the doped material is evident from the degradation curves and the abatement values achieved; therefore, using only visible wavelengths reveals the advantage of introducing the dopant inside a photocatalyst such as TiO$_2$, which otherwise works mainly in UV-A. Moreover, sol–gel-synthetized photocatalysts exhibit a noticeable increase in degradative capacity compared to hydrothermally obtained materials, confirming what was observed for phenol, and nitrogen-doped sol–gel photocatalyst shows better performance than the reference and hydrothermal photocatalysts for all CECs monitored and for all matrices studied. The differences between the sol–gel-doped and pure materials are most pronounced for the Milli-Q water matrix and are confirmed by the results obtained in real water matrices. In Milli-Q water, using N-TiO$_2$-SG, the complete abatement is achieved for all contaminants within two hours, while for pristine material the degradation rate ranges from 20 to 51% depending on the monitored pollutant. As regards the Po River water matrix, the abatement achieved is between 54% and 68% within two hours for the doped material, whereas it does not exceed 48% for the pristine material. Finally, for the aquaculture water matrix, N-TiO$_2$-SG leads to an abatement of at least 70% for all contaminants while it does not exceed 56% for the pristine material. This trend is confirmed by the pseudo first-order kinetic constants' values for the photocatalytic processes, which are always higher than those of the pristine material, in some cases even by an order of magnitude, as shown in Figure 6. Hydrothermal-derived materials also showed improved performance thanks to nitrogen doping, but the degradation efficiency was far lower than that of the sol–gel materials, especially for benzotriazole and bisphenol A. The reason of this difference in the abatement efficiency between the materials obtained with the two synthetic methods can be ascribed to the crystallinity of the prepared samples (much higher in the case of the sol–gel groups than in the case of hydrothermal samples). This explanation can be exhaustive in the case of the pure titania: regarding the abatement activity of N doped samples, it is evident that the SG material is much more active than the HT one. The reason is attributable to the presence of nitrogen in the lattice of titania; in particular, the amount of nitrogen in an interstitial position in the matrix of the SG samples generates enough intraband gap states to allow for electronic migration from VB to these states and from these states to CB, being the energies involved in the range of visible frequencies. In the case of HT doped samples, the amount of nitrogen is not enough to promote a sufficient electronic flow, so it shows a lower photocatalytic activity.

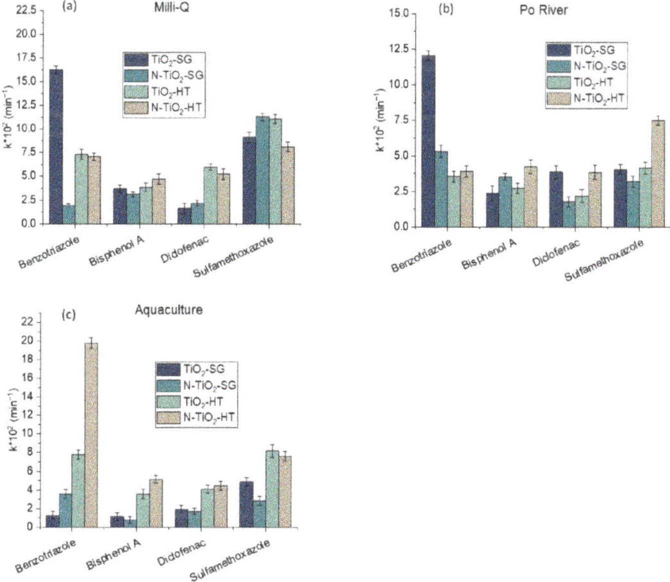

Figure 5. Pseudo first-order kinetic constants for the abatement of the mixture of CECs using synthetized photocatalyst, irradiation with λ > 340 nm cut-off, and the three matrices: (**a**) Milli-Q water, (**b**) Po River water, (**c**), aquaculture water.

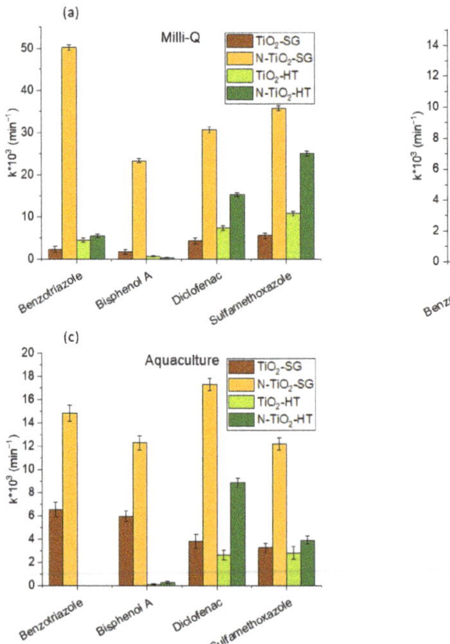
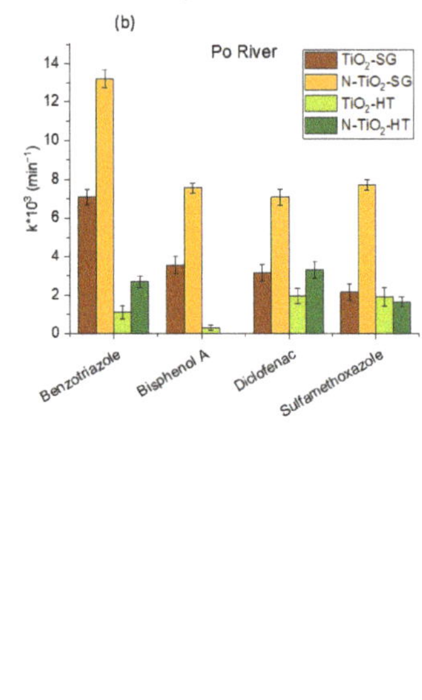

Figure 6. Pseudo first-order kinetic constants for the abatement of the mixture of CECs using synthetized photocatalyst, irradiation with λ > 400 nm cut-off, and the three matrices: (**a**) Milli-Q water, (**b**) Po river water, (**c**), aquaculture water.

In order to have a broader understanding of the degradation mechanism of the considered pollutants by nitrogen-doped materials, experiments were repeated with irradiation at wavelengths longer than 400 nm using different scavengers. In particular, tert-butyl alcohol, p-benzoquinone, ethylenediaminetetraacetic acid (EDTA), and 2,2,6,6-tetramethylpiperidine (TEMP) were separately added as a scavenger of hydroxyl radicals, superoxide radicals, holes, and singlet oxygen, respectively. Results are reported in Figures S5 and S6 and evidence that depending on the emerging contaminant examined, the reactive species involved in degradation are different, while no substantial dissimilarities are observed between the two materials obtained via the different synthesis routes. In particular, benzotriazole is degraded exclusively by superoxide radicals, while in the degradation of bisphenol A and diclofenac, several reactive species such as superoxide radicals, holes, and singlet oxygen are involved. As for sulfamethoxazole, singlet oxygen, holes, and minimally hydroxyl radicals are involved.

For experiments performed using sol–gel materials on the CECs mixture in Milli-Q water irradiated with a cut-off at 400 nm, the decrease in total organic carbon over time was measured, as shown in Figure 7. Its decrease reflects the degree of mineralization at the end of the photocatalytic process, which is greater for nitrogen-doped materials.

The reason for the remarkable increase in efficiency in the photocatalytic degradation of pollutants for N-doped materials cannot be attributed to a reduction in the band gap because, as demonstrated by the calculated band gap values (Table 1), there is no significant change due to the introduction of the dopant, but rather due to the reduced recombination of photo-generated hole-electron pairs on the material surface; in fact, defects due to oxygen vacancies produced by doping with nitrogen generate intermediate band levels and additional electronic states that can trap electrons and promote absorption into the visible range. It is widely recognized that charge-carrier separation plays a key role in a photocatalytic process, and this, in the present case, results in the increased degradation efficiency of N-TiO$_2$-SG using the visible spectrum due to the generation of intraband gap states produced by the interstitial nitrogen species present both on the surface and in the bulk of the material [42].

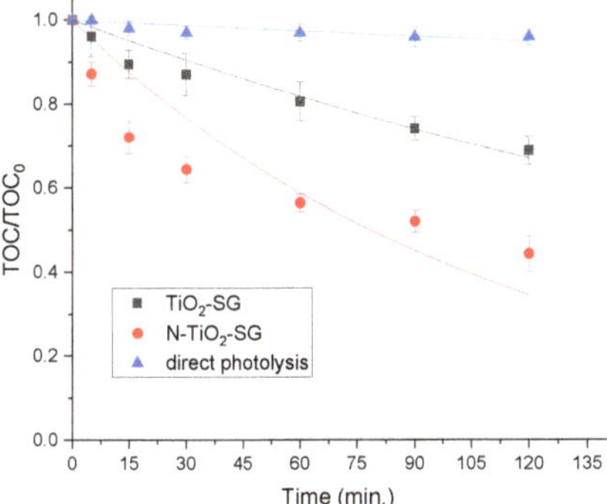

Figure 7. TOC abatement of CECs mixture in Milli-Q water using sol–gel materials and irradiation λ > 400 nm.

3. Materials and Methods

Titanium isopropoxide (97%), 2-propanol (≥99.5%), ammonium chloride (≥99.5%), potassium permanganate (≥99%), fumaric acid (≥99%), maleic acid (≥99%), phenol (≥99%), bisphenol A (≥99%), benzotriazole (99%), sulfamethoxazole (analytical standard), and diclofenac sodium salt (pure) were purchased by Sigma Aldrich. All solutions were prepared with ultrapure water Millipore Milli-Q™ (TOC < 2 ppb, conductivity ≥ 18 MΩ cm).

3.1. Synthesis

Four different samples were synthesized: TiO$_2$ from hydrothermal synthesis (TiO$_2$-HT), TiO$_2$ from sol–gel synthesis (TiO$_2$-SG), N-TiO$_2$ from hydrothermal synthesis (N-TiO$_2$-HT), and N-TiO$_2$ from sol–gel synthesis (N-TiO$_2$-SG).

3.1.1. Synthesis of TiO$_2$ Using Sol–Gel

Precursors were mixed in a 1:1:1 ratio in a 50 mL beaker. Then, 8 mL titanium isopropoxide, 8 mL 2-propanol, and 8 mL deionized water were added [43]. As soon as the water was added under stirring, the gelling process was immediate. After a few minutes of stirring with a magnetic stirrer bar, the beaker was covered and allowed to stand for at least 24 h to ensure cross-linking and thus the correct execution of the sol–gel process. The catalyst obtained was TiO$_2$ with numerous impurities due to solvent, reagent, and carbon residues. A drying step of at least 4 h in an oven at 70 °C was carried out to remove the most volatile species. Then, the sample was placed in a muffle furnace at 450 °C for 1 h with a ramp of 5 °C/min to remove carbon residues and achieve complete calcination [44].

3.1.2. Synthesis of TiO$_2$ Using Hydrothermal Method

The hydrothermal method implies shorter times of reaction than the sol–gel method due to the high-pressure and -temperature conditions achieved. The same solution described in the previous paragraph was placed in a Teflon beaker and inserted into an autoclave. The system was placed in the oven for 16 h at 175 °C and then cooled down to ambient temperature. The sample thus collected was washed with deionized water, centrifugated, and dried in the oven at 70 °C for 4 h.

3.1.3. Synthesis of N-TiO$_2$ Using Sol–Gel Method

Two solutions were prepared: solution A or the mixture of 8 mL of isopropoxide and 8 mL of 2-propanol (1:1 ratio) and solution B that consisted of 0.28 g of ammonium chloride in 8 mL of water [43]. Once the two solutions were mixed, the gel was allowed to stand for at least 24 h and then dried in the oven at 70 °C for at least 4 h. After that, it was calcined at 450 °C for 1 h in a muffle furnace, exploiting a heating ramp of 5 °C/min.

3.1.4. Synthesis of N-TiO$_2$ Using Hydrothermal Method

Two solutions were prepared to obtain the desired product: for solution A, 4 mL of isopropoxide and 4 mL of 2-propanol (1:1 ratio) were mixed [43], while for solution B, 0.14 g of NH$_4$Cl were added to 4 mL of deionized water. The two solutions were mixed in a Teflon beaker, inserted into an autoclave, and placed in the oven at 175 °C for 16 h. The sample was cooled to room temperature, washed with deionized water, centrifugated, and dried in the oven at 70 °C for 4 h.

3.2. Characterization

All synthesized materials were characterized through powder X-ray diffractometry (PXRD) to obtain phase identification and gain information about the crystallinity of the materials. Diffractograms were recorded using the PANalytical PW3040/60 X'Pert Pro MPD instrument, with Bragg–Brentano geometry, equipped with an X-ray generator supplied with a voltage of 45 kV and a current of 40 mA and an X'Celerator scintillation detector. Powder samples were exposed to X Cu Kα radiation (λ = 0.154056 nm) generated by a ceramic tube with a copper anode, and diffracted beams were collected at 2θ values between 0 and 180°. The instrument was interfaced to a PC configured with the X'Pert Data Collector and X'Pert HighScore software version 1.0c for recording and interpretation of diffraction data. Diffused Reflectance UV-vis spectroscopy was utilized for the optical characterization of the samples. The optical spectra were captured using a Varian Cary 5000 spectrophotometer (Agilent, CA, USA) and analyzed with the Carywin-UV/scan software version 4.1 (Agilent, CA, USA). A reference sample of PTFE with 100% reflectance was employed. The optical band gap energies were determined using the Tauc plot method applied to the acquired spectra. The energy gap is directly linked to the absorption coefficient near the absorption edge, and its magnitude is contingent upon the nature of the transition (whether direct or indirectly allowed) [45].

3.3. Photocatalytic Activity

The photocatalytic activity of the synthesized materials was studied through heterogeneous photocatalysis tests for the degradation of organic pollutants in aqueous solutions. Tests were carried out in Pyrex glass cells of a 5 mL capacity, equipped with a side opening with a screw cap and with a magnetic stirring bar. Substrate solutions and suspensions of each catalyst were prepared in Milli-Q water. The suspensions were put in ultrasonic bath for 30 min before use to make them as homogeneous as possible. The solution containing the organic target and the catalyst suspension was transferred to the cells at a ratio of 1:1. The samples were irradiated for different times (from 0 to 120 min) and then filtered to remove the catalyst and stop the degradation reactions. The filters used were Millipore millex-LCR Hydrophilic PTFE 0.45 μm.

An initial evaluation of the photocatalytic activity of the materials was carried out by monitoring the removal of a 10 mg/L phenol solution in Milli-Q water using a concentration of photocatalyst equal to 1 g/L. Photodegradation experiments were conducted inside a Solarbox capable of simulating the spectrum of solar radiation. The Solarbox system, manufactured by CO.FO.ME.GRA., was equipped with a 1500 W Xenon lamp with an irradiance of 19 W/m^2 in the range 295–400 nm. Two different filters were used to select the portion of radiation of interest: one for λ > 400 nm, selecting the visible range and excluding UV, and one for λ > 340 nm, selecting UV-A and visible.

Furthermore, to gain a deeper understanding of the degradation mechanism of CECs by synthetized photocatalysts, additional experiments at λ > 400 nm were conducted, employing different ROS scavengers. Tert-butyl alcohol, p-benzoquinone, ethylenediaminetetraacetic acid (EDTA), and 2,2,6,6-tetramethylpiperidine (TEMP) were separately added as a scavenger of hydroxyl radicals, superoxide radicals, holes, and singlet oxygen, respectively.

Analyses of organic substrates were performed with an HPLC-UV instrument equipped with a Rheodyne manual injector, two Merck Hitachi L-6200 pumps, a lichrospher 100 RP C18 column (125 × 4 mm, part-cell diameter 5 μm), and a Merk Hitachi L-4200 single-channel UV-visible detector. The instrument was interfaced to a PC configured with D-7000 HPLC System Manager software version 3.0 for chromatogram monitoring and processing. Then, 10^{-2} M orthophosphoric acid buffer (pH 2.8) (A) and acetonitrile (B) were used as eluents for all analyses.

Analyses of phenol photodegradation samples in Milli-Q matrix were performed in isocratic mode with 70:30 A/B at 1 mL/min flow, λ = 220 nm. After the preliminary experiments on phenol, the materials were tested on a mixture of four emerging contaminants, namely, benzotriazole, bisphenol A, diclofenac, and sulfamethoxazole at 4 mg/L each. Milli-Q water, Po River water (sampled on 30 September 2022 in Turin, Valentino Park), and water sampled in an aquaculture plant were used as matrices. The analyses were carried out using the following gradient at 1 mL/min flow: 75:25 A/B for 3 min; 75:25 to 65:35 over 4 min; 65:35 to 40:60 over 8 min; 40:60 for 5 min; 40:60 to 75:25 over 0.5 min; 75:25 for 5 min. The UV detector was set at 220 nm, and the retention times were 2.96, 5.35, 14.08, and 18.20 min for benzotriazole, sulfamethoxazole, bisphenol A, and diclofenac, respectively.

A Shimadzu Total Organic Carbon (TOC) analyzer, model TOC-V CSH, based on catalytic oxidation of carbon on Pt at 680 °C, was used to determine the TOC, so the degree of mineralization of samples containing emerging contaminant mixture after photocatalytic treatment and the Non-purgeable Organic Carbon (NPOC) of real waters were used as matrices.

4. Conclusions

In this study, the synthesis of new TiO_2-based materials was performed via the hydrothermal and sol–gel routes with the aim of obtaining a doped material efficient in degrading a mixture of emerging contaminants, namely, Benzotriazole, Bisphenol A, Diclofenac, and Sulfamethoxazole, exploiting the use of visible light. The materials obtained were characterized using powder X-ray diffractometry and diffused reflectance UV-vis spectroscopy and tested for the degradation of phenol and mixture of CECs by first irradiating with a lamp equipped with a λ > 340 nm cut-off and then using only the visible spectrum (λ > 400 nm) and matrices of increasing complexity. Based on the results obtained, it can be inferred that when utilizing an irradiation source that encompasses the UV-A component of the solar spectrum, the superiority of the use of doped materials compared to their pristine counterparts is not readily evident, and there is no substantial distinction attributable to the two distinct synthesis approaches. The most evident difference in the materials obtained with the two synthesis approaches is the crystallinity, which is much higher for samples obtained via the sol–gel process. Further analysis will be performed to obtain more information and to better understand the reason for this difference. Conversely, experiments conducted solely with the visible portion of light exhibit a noticeable advantage in the use of nitrogen-doped materials and a heightened efficiency of the photocatalysts produced through the sol–gel method, both in ultrapure and real water. N-TiO_2-SG emerges as a promising photocatalytic material when exposed to light with a wavelength greater than 400 nm. This is achieved through the introduction of nitrogen that leads to the generation of oxygen vacancies and, more importantly, to the formation of intraband gap states that can play a crucial role in the double-step excitation of the light. These defects give rise to intermediate band levels and extra electronic states, which have the capacity to capture electrons and promote the absorption of visible light, making N-TiO_2-SG a viable choice for treating contaminated water. In fact, it can be effortlessly synthesized, offering a cost-effective and eco-friendly approach that relies solely on visible light.

Supplementary Materials: The following supporting information can be downloaded at: https://www.mdpi.com/article/10.3390/inorganics11110439/s1, Figure S1: Degradation curves of benzotriazole using the synthetized photocatalyst in different matrices (a) Milli-Q, (b) Po River, (c) aquaculture water irradiating with lamp equipped with $\lambda > 340$ nm cut-off. Figure S2. Degradation curves of bisphenol A using the synthetized photocatalyst in different matrices (a) Milli-Q, (b) Po River, (c) aquaculture water irradiating with lamp equipped with $\lambda > 340$ nm cut-off. Figure S3. Degradation curves of sulfamethoxazole using the synthetized photocatalyst in different matrices (a) Milli-Q, (b) Po River, (c) aquaculture water irradiating with lamp equipped with $\lambda > 340$ nm cut-off. Figure S4. Degradation curves of diclofenac using the synthetized photocatalyst in different matrices (a) Milli-Q, (b) Po River, (c) aquaculture water irradiating with lamp equipped with $\lambda > 340$ nm cut-off. Figure S5. The degradation percentage of CECs obtained using N-TiO$_2$ HT photocatalyst after two hours of irradiation with $\lambda > 400$ nm in the presence of t-butyl alcohol, p-benzoquinone, EDTA, and TEMP as scavengers of hydroxyl radicals, superoxide radicals, holes, and singlet oxygen, respectively. Figure S6. The degradation percentage of CECs obtained using N-TiO$_2$ SG photocatalyst after two hours of irradiation with $\lambda > 400$ nm in the presence of t-butyl alcohol, p-benzoquinone, EDTA, and TEMP as scavengers of hydroxyl radicals, superoxide radicals, holes, and singlet oxygen, respectively.

Author Contributions: E.G.: investigation, data curation, writing—original draft, editing; A.G.: investigation; A.Z.: data curation, editing; M.C.P.: conceptualization, writing—review and editing; P.C.: funding acquisition, conceptualization, supervision, writing—review and editing. All authors have read and agreed to the published version of the manuscript.

Funding: This research was funded by funding from the European Union's Horizon 2020 Research and Innovation Programme under the MarieSkłodowska-Curie Grant Agreement No101007578 (SusWater).

Data Availability Statement: Data, associated metadata, and calculation tools are available from the corresponding author.

Conflicts of Interest: The authors declare no conflict of interest.

References

1. Gleick, P.H. Global Freshwater Resources: Soft-Path Solutions for the 21st Century. *Science* **2003**, *302*, 1524–1528. [CrossRef]
2. Konar, M.; Evans, T.P.; Levy, M.; Scott, C.A.; Troy, T.J.; Vörösmarty, C.J.; Sivapalan, M. Water resources sustainability in a globalizing world: Who uses the water? *Hydrol. Process.* **2016**, *30*, 3330–3336. [CrossRef]
3. Chaudhry, F.N.; Malik, M. Factors affecting water pollution: A review. *J. Ecosyst. Ecography* **2017**, *7*, 225–231.
4. Kumar, N.M.; Sudha, M.C.; Damodharam, T.; Varjani, S. Chapter 3-Micro-pollutants in surface water: Impacts on the aquatic environment and treatment technologies. In *Current Developments in Biotechnology and Bioengineering*; Varjani, S., Pandey, A., Tyagi, R.D., Ngo, H.H., Larroche, C., Eds.; Elsevier: Amsterdam, The Netherlands, 2020; pp. 41–62. [CrossRef]
5. Ashraf, M.A. *Persistent Organic Pollutants (POPs): A Global Issue, a Global Challenge*; Springer: Berlin, Germany, 2017; pp. 4223–4227.
6. Lofrano, G.; Libralato, G.; Meric, S.; Vaiano, V.; Sacco, O.; Venditto, V.; Guida, M.; Carotenuto, M. 1-Occurrence and potential risks of emerging contaminants in water. In *Visible Light Active Structured Photocatalysts for the Removal of Emerging Contaminants*; Sacco, O., Vaiano, V., Eds.; Elsevier: Amsterdam, The Netherlands, 2020; pp. 1–25.
7. Goswami, L.; Kumar, R.V.; Borah, S.N.; Manikandan, N.A.; Pakshirajan, K.; Pugazhenthi, G. Membrane bioreactor and integrated membrane bioreactor systems for micropollutant removal from wastewater: A review. *J. Water Process Eng.* **2018**, *26*, 314–328. [CrossRef]
8. Salimi, M.; Esrafili, A.; Gholami, M.; Jafari, A.J.; Kalantary, R.R.; Farzadkia, M.; Kermani, M.; Sobhi, H.R. Contaminants of emerging concern: A review of new approach in AOP technologies. *Environ. Monit. Assess.* **2017**, *189*, 414. [CrossRef] [PubMed]
9. Said, K.A.M.; Ismail, A.F.; Karim, Z.A.; Abdullah, M.S.; Hafeez, A. A review of technologies for the phenolic compounds recovery and phenol removal from wastewater. *Process Saf. Environ. Prot.* **2021**, *151*, 257–289. [CrossRef]
10. Abraham, A.; Chakraborty, P. A review on sources and health impacts of bisphenol A. *Rev. Environ. Health* **2020**, *35*, 201–210. [CrossRef] [PubMed]
11. Datta, A.; Alpana, A.; Shrikant, M.; Pratyush, K.; Abhibnav, B.; Ruchita, T. Review on synthetic study of benzotriazole. *GSC Biol. Pharm. Sci.* **2020**, *11*, 215–225. [CrossRef]
12. Alessandretti, I.; Rigueto, C.V.T.; Nazari, M.T.; Rosseto, M.; Dettmer, A. Removal of diclofenac from wastewater: A comprehensive review of detection, characteristics and tertiary treatment techniques. *J. Environ. Chem. Eng.* **2021**, *9*, 106743. [CrossRef]
13. Wang, J.; Wang, S. Microbial degradation of sulfamethoxazole in the environment. *Appl. Microbiol. Biotechnol.* **2018**, *102*, 3573–3582. [CrossRef]
14. Ren, L.; Zhou, W.; Sun, B.; Li, H.; Qiao, P.; Xu, Y.; Wu, J.; Lin, K.; Fu, H. Defects-engineering of magnetic γ-Fe$_2$O$_3$ ultrathin nanosheets/mesoporous black TiO$_2$ hollow sphere heterojunctions for efficient charge separation and the solar-driven photocatalytic mechanism of tetracycline degradation. *Appl. Catal. B Environ.* **2019**, *240*, 319–328. [CrossRef]

15. Rahman, M.U.; Qazi, U.Y.; Hussain, T.; Nadeem, N.; Zahid, M.; Bhatti, H.N.; Shahid, I. Solar driven photocatalytic degradation potential of novel graphitic carbon nitride based nano zero-valent iron doped bismuth ferrite ternary composite. *Opt. Mater.* **2021**, *120*, 111408. [CrossRef]
16. Hao, Q.; Liu, Y.; Chen, T.; Guo, Q.; Wei, W.; Ni, B. Bi_2O_3@ carbon nanocomposites for solar-driven photocatalytic degradation of chlorophenols. *ACS Appl. Nano Mater.* **2019**, *2*, 2308–2316. [CrossRef]
17. Lee, S.-Y.; Park, S.-J. TiO2 photocatalyst for water treatment applications. *J. Ind. Eng. Chem.* **2013**, *19*, 1761–1769. [CrossRef]
18. Gomes, J.; Lincho, J.; Domingues, E.; Quinta-Ferreira, R.M.; Martins, R.C. $N–TiO_2$ Photocatalysts: A Review of Their Characteristics and Capacity for Emerging Contaminants Removal. *Water* **2019**, *11*, 373. [CrossRef]
19. Natarajan, T.S.; Mozhiarasi, V.; Tayade, R.J. Nitrogen doped titanium dioxide ($N-TiO_2$): Synopsis of synthesis methodologies, doping mechanisms, property evaluation and visible light photocatalytic applications. *Photochem* **2021**, *1*, 371–410. [CrossRef]
20. Varma, R.; Yadav, M.; Tiwari, K.; Makani, N.; Gupta, S.; Kothari, D.C.; Miotello, A.; Patel, N. Roles of Vanadium and Nitrogen in Photocatalytic Activity of VN-Codoped TiO_2 Photocatalyst. *Photochem. Photobiol.* **2018**, *94*, 955–964. [CrossRef] [PubMed]
21. El Koura, Z.; Cazzanelli, M.; Bazzanella, N.; Patel, N.; Fernandes, R.; Arnaoutakis, G.E.; Gakamsky, A.; Dick, A.; Quaranta, A.; Miotello, A. Synthesis and Characterization of Cu and N Codoped RF-Sputtered TiO_2 Films: Photoluminescence Dynamics of Charge Carriers Relevant for Water Splitting. *J. Phys. Chem. C* **2016**, *120*, 12042–12050. [CrossRef]
22. Bazzanella, N.; Bajpai, O.P.; Fendrich, M.; Guella, G.; Miotello, A.; Orlandi, M. Ciprofloxacin degradation with a defective TiO2-x nanomaterial under sunlight. *MRS Commun.* **2023**, *13*, 1252–1259. [CrossRef]
23. Nakamura, R.; Tanaka, T.; Nakato, Y. Mechanism for Visible Light Responses in Anodic Photocurrents at N-Doped TiO_2 Film Electrodes. *J. Phys. Chem. B* **2004**, *108*, 10617–10620. [CrossRef]
24. Li, D.; Haneda, H.; Hishita, S.; Ohashi, N. Visible-light-driven N− F− codoped TiO_2 photocatalysts. 2. Optical characterization, photocatalysis, and potential application to air purification. *Chem. Mater.* **2005**, *17*, 2596–2602. [CrossRef]
25. Li, D.; Ohashi, N.; Hishita, S.; Kolodiazhnyi, T.; Haneda, H. Origin of visible-light-driven photocatalysis: A comparative study on N/F-doped and N–F-codoped TiO_2 powders by means of experimental characterizations and theoretical calculations. *J. Solid State Chem.* **2005**, *178*, 3293–3302. [CrossRef]
26. Lin, Z.; Orlov, A.; Lambert, R.M.; Payne, M.C. New Insights into the Origin of Visible Light Photocatalytic Activity of Nitrogen-Doped and Oxygen-Deficient Anatase TiO_2. *J. Phys. Chem. B* **2005**, *109*, 20948–20952. [CrossRef] [PubMed]
27. Nakano, Y.; Morikawa, T.; Ohwaki, T.; Taga, Y. Band-gap narrowing of TiO_2 films induced by N-doping. *Phys. B Condens. Matter* **2006**, *376–377*, 823–826. [CrossRef]
28. Emeline, A.; Kuzmin, G.; Serpone, N. Wavelength-dependent photostimulated adsorption of molecular O2 and H2 on second generation titania photocatalysts: The case of the visible-light-active N-doped TiO_2 system. *Chem. Phys. Lett.* **2008**, *454*, 279–283. [CrossRef]
29. Emeline, A.V.; Sheremetyeva, N.V.; Khomchenko, N.V.; Ryabchuk, V.K.; Serpone, N. Photoinduced Formation of Defects and Nitrogen Stabilization of Color Centers in N-Doped Titanium Dioxide. *J. Phys. Chem. C* **2007**, *111*, 11456–11462. [CrossRef]
30. Ohsawa, T.; Lyubinetsky, I.; Du, Y.; Henderson, M.A.; Shutthanandan, V.; Chambers, S.A. Crystallographic dependence of visible-light photoactivity in epitaxial $TiO_{2−x} N_x$ anatase and rutile. *Phys. Rev.* **2009**, *79*, 085401. [CrossRef]
31. Barolo, G.; Livraghi, S.; Chiesa, M.; Paganini, M.C.; Giamello, E. Mechanism of the Photoactivity under Visible Light of N-Doped Titanium Dioxide. Charge Carriers Migration in Irradiated N-TiO_2 Investigated by Electron Paramagnetic Resonance. *J. Phys. Chem. C* **2012**, *116*, 20887–20894. [CrossRef]
32. Katoh, R.; Furube, A.; Yamanaka, K.-I.; Morikawa, T. Charge Separation and Trapping in N-Doped TiO_2 Photocatalysts: A Time-Resolved Microwave Conductivity Study. *J. Phys. Chem. Lett.* **2010**, *1*, 3261–3265. [CrossRef]
33. Yamanaka, K.-i.; Morikawa, T. Charge-carrier dynamics in nitrogen-doped TiO_2 powder studied by femtosecond time-resolved diffuse reflectance spectroscopy. *J. Phys. Chem. C* **2012**, *116*, 1286–1292. [CrossRef]
34. Zhang, Q.; Xiao, Y.; Yang, L.; Wen, Y.; Xiong, Z.; Lei, L.; Wang, L.; Zeng, Q. Branched core-shell a-TiO_2@ N-TiO_2 nanospheres with gradient-doped N for highly efficient photocatalytic applications. *Chin. Chem. Lett.* **2023**, *34*, 107628. [CrossRef]
35. Jin, Y.; Zhang, S.; Xu, H.; Ma, C.; Sun, J.; Li, H.; Pei, H. Application of N-TiO_2 for visible-light photocatalytic degradation of Cylindrospermopsis raciborskii—More difficult than that for photodegradation of Microcystis aeruginosa? *Environ. Pollut.* **2019**, *245*, 642–650. [CrossRef] [PubMed]
36. Feng, J.; Bao, W.; Li, L.; Cheng, H.; Huang, W.; Kong, H.; Li, Y. The synergistic effect of nitrogen-doped titanium dioxide/mercaptobenzoic acid/silver nanocomplexes for surface-enhanced Raman scattering. *J. Nanopart. Res.* **2018**, *20*, 82. [CrossRef]
37. Sato, S. Photocatalytic activity of NOx-doped TiO_2 in the visible light region. *Chem. Phys. Lett.* **1986**, *123*, 126–128. [CrossRef]
38. Asahi, R.; Morikawa, T.; Ohwaki, T.; Aoki, K.; Taga, Y. Visible-Light Photocatalysis in Nitrogen-Doped Titanium Oxides. *Science* **2001**, *293*, 269–271. [CrossRef] [PubMed]
39. Swanson, H.E. *Standard X-ray Diffraction Powder Patterns*; US Department of Commerce, National Bureau of Standards: Washington, DC, USA, 1971; Volume 25.
40. Hanawalt, J.D.; Rinn, H.W.; Frevel, L.K. Chemical Analysis by X-ray Diffraction. *Ind. Eng. Chem. Anal. Ed.* **1938**, *10*, 457–512. [CrossRef]
41. Fernandes, E.; Martins, R.C.; Gomes, J. Photocatalytic ozonation of parabens mixture using 10% N-TiO2 and the effect of water matrix. *Sci. Total. Environ.* **2020**, *718*, 137321. [CrossRef]

42. Li, Y.; Peng, Y.-K.; Hu, L.; Zheng, J.; Prabhakaran, D.; Wu, S.; Puchtler, T.J.; Li, M.; Wong, K.-Y.; Taylor, R.A.; et al. Photocatalytic water splitting by N-TiO$_2$ on MgO (111) with exceptional quantum efficiencies at elevated temperatures. *Nat. Commun.* **2019**, *10*, 1–9. [CrossRef]
43. Livraghi, S.; Paganini, M.C.; Giamello, E.; Selloni, A.; Di Valentin, C.; Pacchioni, G. Origin of Photoactivity of Nitrogen-Doped Titanium Dioxide under Visible Light. *J. Am. Chem. Soc.* **2006**, *128*, 15666–15671. [CrossRef]
44. Di Valentin, C.; Pacchioni, G.; Selloni, A.; Livraghi, S.; Giamello, E. Characterization of Paramagnetic Species in N-Doped TiO$_2$ Powders by EPR Spectroscopy and DFT Calculations. *J. Phys. Chem. B* **2005**, *109*, 11414–11419. [CrossRef]
45. Tauc, J.; Scott, T.A. The optical properties of solids. *Phys. Today* **1967**, *20*, 105–107. [CrossRef]

Article

Synthesis of Polystyrene@TiO$_2$ Core–Shell Particles and Their Photocatalytic Activity for the Decomposition of Methylene Blue

Naoki Toyama [1,*], Tatsuya Takahashi [2], Norifumi Terui [2] and Shigeki Furukawa [1]

[1] Department of Sustainable Engineering, College of Industrial Technology, Nihon University, 1-2-1, Izuni-cho, Narashino 275-8575, Chiba, Japan; furukawa.shigeki@nihon-u.ac.jp
[2] Department of Engineering for Future Innovation, National Institute of Technology, Ichinoseki College, Takanashi, Hagisho, Ichinoseki 021-8511, Iwate, Japan; t.tatsuya21806@gmail.com (T.T.); terui@ichinoseki.ac.jp (N.T.)
* Correspondence: toyama.naoki@nihon-u.ac.jp

Abstract: In this study, we investigated the preparation conditions of polystyrene (PS)@TiO$_2$ core–shell particles and their photocatalytic activity during the decomposition of methylene blue (MB). TiO$_2$ shells were formed on the surfaces of PS particles using the sol–gel method. Homogeneous PS@TiO$_2$ core–shell particles were obtained using an aqueous NH$_3$ solution as the promoter of the sol–gel reaction and stirred at room temperature. This investigation revealed that the temperature and amount of the sol–gel reaction promoter influenced the morphology of the PS@TiO$_2$ core–shell particles. The TiO$_2$ shell thickness of the PS@TiO$_2$ core–shell particles was approximately 5 nm, as observed using transmission electron microscopy. Additionally, Ti elements were detected on the surfaces of the PS@TiO$_2$ core–shell particles using energy-dispersive X-ray spectroscopy analysis. The PS@TiO$_2$ core–shell particles were used in MB decomposition to evaluate their photocatalytic activities. For comparison, we utilized commercial P25 and TiO$_2$ particles prepared using the sol–gel method. The results showed that the PS@TiO$_2$ core–shell particles exhibited higher activity than that of the compared samples.

Keywords: titania; polystyrene; core–shell; methylene blue; photocatalytic activity

Citation: Toyama, N.; Takahashi, T.; Terui, N.; Furukawa, S. Synthesis of Polystyrene@TiO$_2$ Core–Shell Particles and Their Photocatalytic Activity for the Decomposition of Methylene Blue. *Inorganics* **2023**, *11*, 343. https://doi.org/10.3390/inorganics11080343

Academic Editors: Roberto Nisticò, Torben R. Jensen, Luciano Carlos, Hicham Idriss, Eleonora Aneggi and Antonino Gulino

Received: 31 May 2023
Revised: 14 August 2023
Accepted: 17 August 2023
Published: 21 August 2023

1. Introduction

As environmental pollution continues to worsen, technologies and materials that mitigate this issue have become widely researched topics [1,2]. Organic dyes primarily released from industrial plants are among such pollutants that form part of effluents in various water bodies [3,4]. Dyes such as methylene blue (MB) are toxic and nonbiodegradable, and they remain in the water for long periods of time [5]. Therefore, developing methods for removing dyes from water bodies is crucial.

Many approaches have been reported for dye removal, such as flocculation, chemical oxidation, membrane filtration, chemical coagulation, photochemical degradation, and biological degradation [4–6]. Notably, photocatalysis is an effective method for dye degradation because of its advantages, such as its eco-friendliness, high efficiency, low cost, and reusability [7,8]. Recently, hydrogen evolution was promoted in the presence of photocatalysts such as metal compound/g-C$_3$N$_4$ systems and flower-like carbon quantum dots/BiOBr composites [9–13].

Titanium dioxide (TiO$_2$) is well known for being an excellent photocatalyst. This oxide has attracted attention for use in pollutant removal because it offers high conversion efficiency, good chemical stability, and low cost [14–16]. The sol–gel reaction has been reported as a synthetic method for generating TiO$_2$ particles [17,18]. In general, TiO$_2$ particles with a small particle size can efficiently act as a photocatalyst; however, these tend to agglomerate due to their high energy and reduce the number of available active sites [19]. To solve this problem, TiO_2 particles have been immobilized on polymer substrates such as polymers, including polyethlene [20,21] and poly (dimethylsiloxane) beads, [22] and so on [23]. Moreover, it is difficult to obtain TiO_2 particles with uniform size due to the fast sol–gel reaction rates exhibited by Ti precursors [24,25]. Therefore, the challenge is to immobilize TiO_2 particles uniformly on the surfaces of substrates. Additionally, many complex processes are required in order to obtain substrate-supported TiO_2 particles due their long synthesis time [26–28]. We propose a facile method for the immobilization of TiO_2 particles on the surfaces of polystyrene (PS) particles.

Monodisperse PS particles with a positive charge were prepared using emulsifier-free emulsion polymerization using 2′2-azodiisobutyramidine dihydrochloride as the initiator and poly (vinylpyrrolidone) as the stabilizer [29–31]. Previous studies found that SiO_2-Al_2O_3 nanoparticles could be collected on the surfaces of PS particles. In this method, the SiO_2-Al_2O_3 nanoparticles were prepared using the sol–gel method, and the nanoparticles with negative charges were attracted to the surfaces of the PS particles because of their positive charges, thus forming SiO_2-Al_2O_3 shells [32]. With this knowledge in mind, we attempted to collect TiO_2 particles on the surfaces of PS particles using this simple method.

Herein, we report PS@TiO_2 core–shell particles synthesized using the sol–gel method. In this method, TiO_2 shells were formed on the surfaces of PS particles. Homogeneous spherical particles were obtained by adjusting the reaction temperature and amount of the promoter aqueous NH_3 solution. A schematic illustration of the synthesis of the PS@TiO_2 core–shell particles is shown in Figure 1. Experimental observations confirmed that the diameters of PS@TiO_2 core–shell particles were approximately 220 nm. Furthermore, TiO_2 was confirmed to be present on the surfaces of the PS@TiO_2 core–shell particles. The photocatalytic activity of the PS@TiO_2 core–shell particles was evaluated using MB decomposition. The PS@TiO_2 core–shell particles showed better performance than that of the commercial P25.

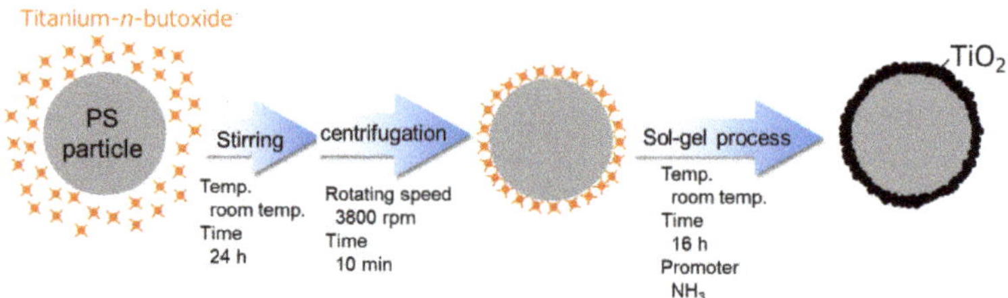

Figure 1. Schematic illustration of the formation process of PS@TiO_2 core–shell particles.

2. Results and Discussion

First, we investigated the influence of preparation conditions such as the preparation temperature and amount of aqueous NH_3 solution on the morphology of PS@TiO_2 core–shell particles. Figure 2a–c show scanning electron microscopy (SEM) images of PS@TiO_2 core–shell particles prepared at various preparation temperatures and with varying amounts of aqueous NH_3 solution. Uniform spherical particles were prepared using 300 µL of aqueous NH_3 solution at room temperature (Figure 2a), whereas spherical particles and particles with irregular shapes were prepared using 300 µL of aqueous NH_3 solution at 323 K (Figure 2b). Meanwhile, the rough layers, including spherical particles on the surface, were observed in SEM images of Figure 2c when prepared using 1000 µL of aqueous NH_3 solution at room temperature. Using this process, the TiO_2 nanoparticles prepared via the sol–gel reaction were collected on the surfaces of PS particles. Reportedly, the sol–gel reaction rate enhances higher temperatures and increases the amounts of promoters such as NH_3 [33]. Consistent with the above reports, a faster sol–gel reaction rate promoted TiO_2 particle growth, leading to the formation of particles with irregular shapes and rough layers, as shown in Figure 2b,c. Moreover, we confirmed the presence of TiO_2

on the surfaces of uniform PS@TiO$_2$ core–shell particles prepared using 300 μL of aqueous NH$_3$ solution at room temperature through energy-dispersive X-ray spectroscopy (EDX) measurements. Figure 3 shows elemental mapping images of O, Ti, and C for the prepared PS@TiO$_2$ core–shell particles. O and Ti were observed in the overall images, and the ratio of O:Ti was approximately 1:1. Additionally, the areas of C elemental mapping overlapped the areas of Ti and O elemental mapping. This result suggests that very thin TiO$_2$ shells can be formed on the surfaces of PS particles. Meanwhile, we measured the specific surface area of the PS@TiO$_2$ core–shell particles using the Brunauer–Emmett–Teller (BET) method. The results showed that the specific surface area of the PS@TiO$_2$ core–shell particles was 125 m^2 g^{-1}.

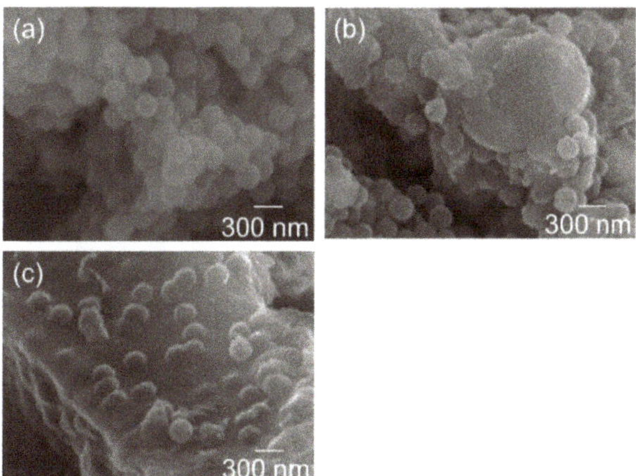

Figure 2. SEM images of PS@TiO$_2$ core–shell particles prepared at (**a**) room temperature, (**b**) 323 K using 300 μL of aqueous NH$_3$ solution, and (**c**) room temperature using 1000 μL of aqueous NH$_3$ solution.

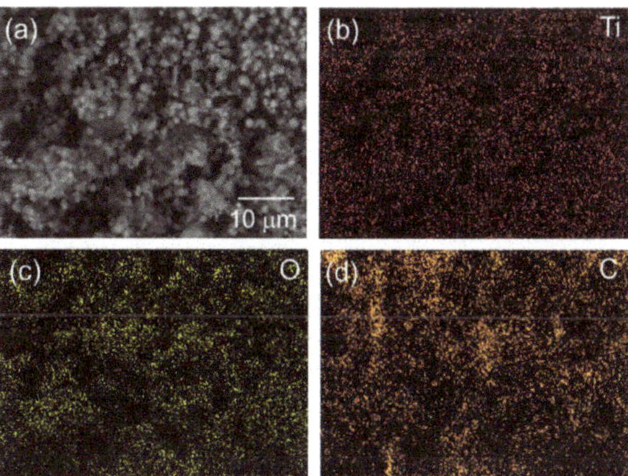

Figure 3. SEM/EDX images of (**a**) PS@TiO$_2$ core–shell particles and element mappings of (**b**) Ti, (**c**) O, and (**d**) C.

Next, the morphology of PS@TiO$_2$ core–shell particles were observed in detail using transmission electron microscopy (TEM). For comparison, we used TiO$_2$ particles prepared using sol–gel reactions and commercial P25. Figure 4a shows a TEM image of PS particles. In this image, the diameters of the PS particles are approximately 280 nm. Figure 4b,c show TEM images of the PS@TiO$_2$ core–shell particles. The obtained images show that the diameters of the spherical particles were approximately 290 nm. The presence of the TiO_2 shell was confirmed by the different contrasts between core and shell. As shown in Figure 4a–c, the shell thickness of the PS@TiO_2 core–shell particles could be ~5 nm. This result is also consistent with the results of the SEM/energy-dispersive X-ray spectroscopy (EDX) element mapping images, which show the detection of C (Figure 3). Meanwhile, the presence of prepared TiO_2 particles and P25 was also confirmed using the TEM images. In these results, the prepared TiO_2 is presented in random shapes (Figure 4d), whereas crystal particles with diameters of 20–50 nm are observed in the TEM image of P25 (Figure 4e).

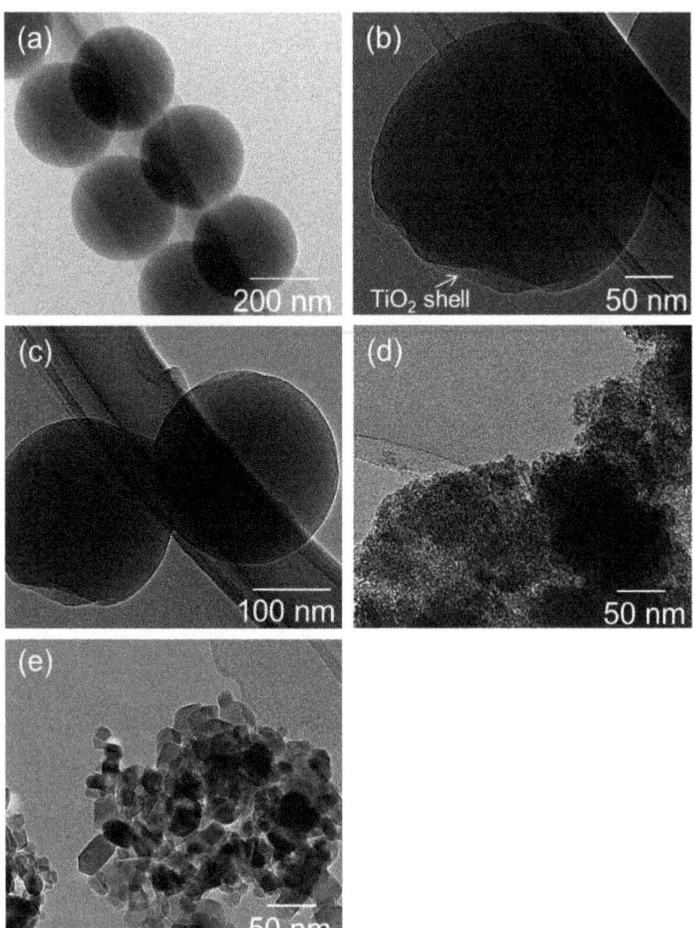

Figure 4. TEM images of (**a**) PS particles, (**b**,**c**) PS@TiO_2 core–shell particles, (**d**) prepared TiO_2, and (**e**) commercial P25.

The crystallinity of the samples was measured using powder X-ray diffraction (XRD). Figure 5 shows the XRD patterns of the PS@TiO_2 core–shell particles, the prepared TiO_2, and the commercial P25. The PS@TiO_2 core–shell particles showed a broad diffraction peak, whereas the prepared TiO_2 showed no peak, indicating that both samples were composed of an amorphous phase. Meanwhile, the commercial P25 showed sharp peaks, indicating that it was composed of rutile and anatase phases [34].

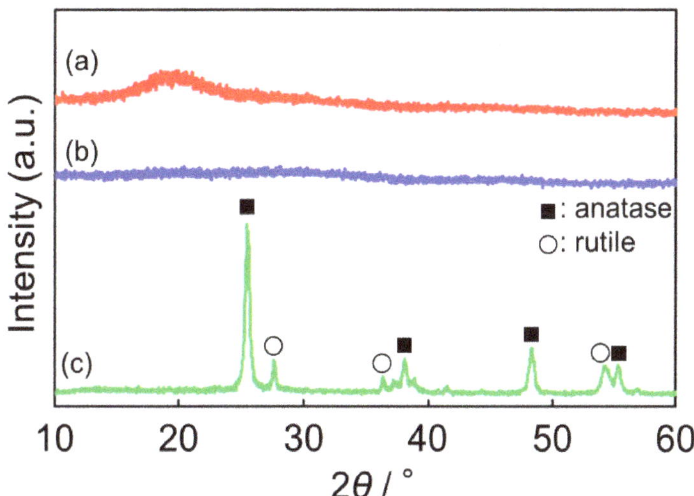

Figure 5. XRD patterns of (**a**) the PS@TiO$_2$ core–shell particles, (**b**) the prepared TiO$_2$, and (**c**) the commercial P25.

Finally, the photocatalytic activities of the PS@TiO$_2$ core–shell particles, the prepared TiO$_2$, and the commercial P25 were evaluated through MB decomposition. The main peak of MB was observed at 664 nm. The UV visible (UV-Vis) absorption spectra of the PS@TiO$_2$ core–shell particles, the prepared TiO$_2$, the commercial P25, and the PS particles monitored at different times are shown in Figure 6a–d. As seen in these results, the intensity of the absorption peak gradually decreased with the reaction time in the presence of the PS@TiO$_2$ core–shell particles, the prepared TiO$_2$, and the commercial P25. However, the intensity of the absorption peak was not changed in the presence of the PS particles. Figure 6e shows the reaction time dependence of MB decomposition in the presence of these samples with UV irradiation. The relative intensity of the absorbance at 664 nm was used to calculate the value of C/C_0, where C_0 and C are the MB concentrations of the aqueous MB solution at the beginning and reaction time, respectively. In the presence of the PS@TiO$_2$ core–shell particles and the prepared TiO$_2$, 84% and 38% of MB degraded within 3 min. Meanwhile, 84% and 60% of MB degraded within 24 min, indicating that both samples did not promote MB decomposition after 4 min. In the case of the commercial P25, the intensity of the main peak gradually decreased with the increase in reaction time, and 58% of MB degraded after 24 min. On the other hand, the PS particles showed no activity. These results indicate that the PS@TiO$_2$ core–shell particles demonstrated a faster reaction rate than those of the prepared TiO$_2$ and the commercial P25.

The PS@TiO$_2$ core–shell particles were used in MB decomposition with and without UV irradiation to evaluate their photocatalytic activity. Figure 7a shows the UV-Vis absorption spectra of the reaction solutions with the PS@TiO$_2$ core–shell particles without UV irradiation. The intensity of the absorption peak at 664 nm was unchanged without UV irradiation. Figure 7b shows the reaction time dependence in the presence of the PS@TiO$_2$ core–shell particles with and without UV irradiation. With UV irradiation, 87% of MB degraded after 24 min, as described in the above results, whereas without UV irradiation, MB decomposition was not promoted.

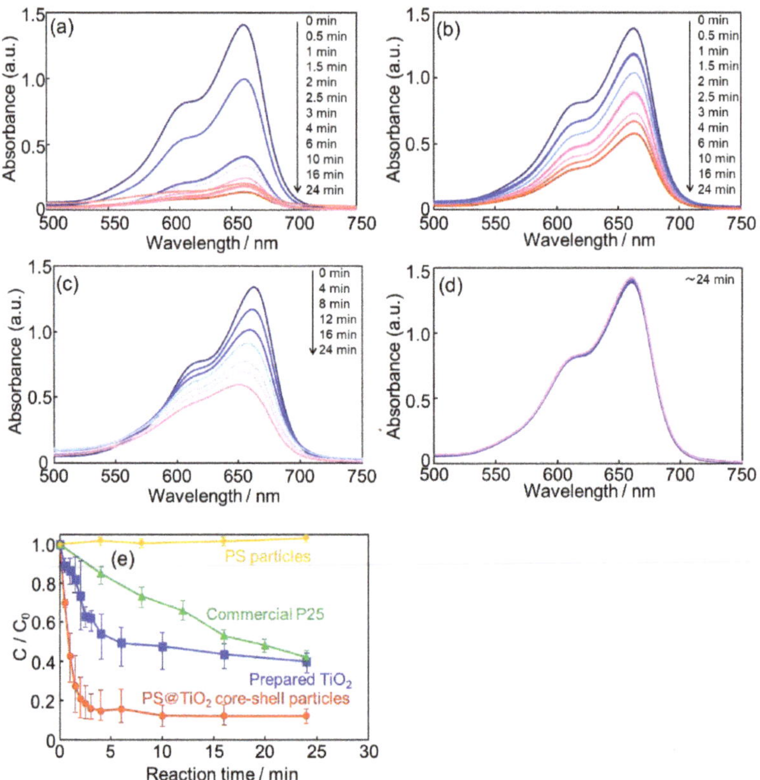

Figure 6. UV-Vis absorption spectra of the reaction solutions with (**a**) PS@TiO$_2$ core–shell particles, (**b**) prepared TiO$_2$, (**c**) commercial P25, and (**d**) PS particles. (**e**) Relationship between C/C_0 and the reaction time for MB decomposition.

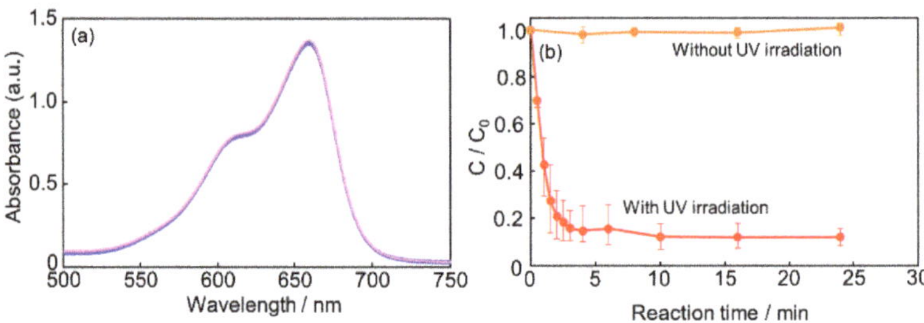

Figure 7. UV-Vis absorption spectra of the reaction solutions with (**a**) PS@TiO$_2$ core–shell particles without UV irradiation. (**b**) Relationship between C/C_0 and the reaction time for MB decomposition in the presence of the PS@TiO$_2$ core–shell particles with and without UV irradiation.

Herein, we discussed the differences in the photocatalytic activities of the PS@TiO$_2$ core–shell particles, the prepared TiO$_2$, and the commercial P25. First, we noted that PS@TiO$_2$ core–shell particles exhibited the highest MB decomposition activity. In this activity evaluation, the masses of the samples were equal, whereas the masses of TiO$_2$ were lower than the practical masses of the samples because PS@TiO$_2$ core–shell particles contained PS particles as the core. Nevertheless, the PS@TiO$_2$ core–shell particles exhibited the highest MB decomposition activity among all samples. This may have been because

TiO$_2$ shells with a shell thickness of about 5 nm were uniformly coated on surface of PS particles leading to efficient activity promotion. Therefore, the PS@TiO$_2$ core–shell particles, the prepared TiO$_2$, and the PS particles were investigated via photoluminescence (PL) measurements. Figure 8 shows the PL spectra of the PS@TiO$_2$ core–shell particles, the prepared TiO$_2$, and the PS particles. From this result, a broad peak was observed at around 390 nm for the PS@TiO$_2$ core–shell particles and the prepared TiO$_2$, while no peak was observed for the PS particles. In addition, the PS@TiO$_2$ core–shell particles exhibited a higher peak intensity than the prepared TiO$_2$. This peak was attributed to band-to-band transition [35]. Previous studies have reported that a peak was not observed for amorphous TiO$_2$ because of the lack of fluorescence [36], whereas for the TiO$_2$ with anatase crystalline phases, a peak was observed [35]. It may be that the TiO$_2$ shells on the surface of the PS@TiO$_2$ core–shell particles formed very tiny crystal particles with poor crystalline phases. The broad peaks observed at 20° in the XRD results may imply the existence of poor crystalline phases. In other reports, the peak intensity of the amorphous TiO$_2$ coating on the surface of ZnO particles increased compared with that of ZnO particles, indicating that the fine control of TiO$_2$ shell thickness on the surface of ZnO nanoparticles was very important for the enhancement of PL intensity [37]. Considering this, the crystallinity and thickness of the TiO$_2$ shell of PS@TiO$_2$ core–shell particles may play an important role in MB decomposition.

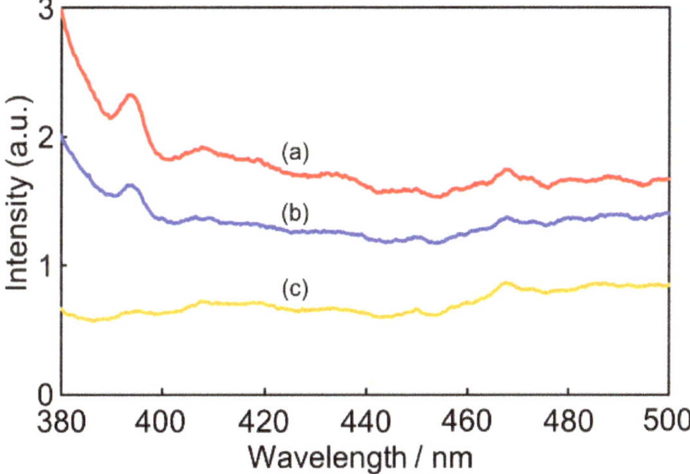

Figure 8. PL spectra of (a) PS@TiO$_2$ core–shell particles, (b) prepared TiO$_2$, and (c) PS particles.

Next, a comparison of the MB decomposition obtained in this study with those of other previously reported studies [38–42] is listed in Table 1. It was found that PS@TiO$_2$ core–shell particles show higher activity than most TiO$_2$-polymers and their related catalysts. However, the MB degradation rate per dosage of the PS@TiO$_2$ core–shell particles (0.379% min^{-1}mg^{-1}) was lower than that of TiO$_2$/graphene porous composites (0.64% min^{-1}mg^{-1}). This result might be ascribed to the different crystallinity of TiO$_2$. The TiO$_2$ shells of the PS@TiO$_2$ core–shell particles were probably amorphous phases, as shown in Figure 5a. On the other hand, the TiO$_2$ on the graphene porous composites was an anatase phase [42]. Indeed, the band gap of the amorphous TiO$_2$ is higher than that of the rutile and anatase TiO$_2$ [43]. Moreover, there may also be a difference in the mechanism of photocatalysis.

Table 1. Comparison of the MB decomposition of TiO_2-polymer photocatalysts in previous reports.

Samples	Light Source	MB Concentration	Dosage/mg	Degradation Rate/% min^{-1}	References
PS@TiO_2 core–shell particles	UV lamp (150 mW cm^{-2})	7 ppm (1.9×10^{-5} M)	10	3.67	This study
TiO_2 firm/ABS	Xenon lamp (300 W)	1.0×10^{-6} M	No data (3×3 cm)	0.67	38
PU/aTiO_2 Hybrid (10 wt%)	UV-A irradiation	10 ppm	No data	3.19	39
CAT-1 [*1]	Xenon lamp (300 W)	50 ppm	20	0.29	40
TiO_2/graphene porous composite	Xenon lamp (50 W)	10 ppm	1	0.64	41
TiO_2/porous carbon nanofibers	Xenon lamp (800 W)	10 ppm	50	1.88	42

[*1] CAT-1: TiO_2-MWCNT-C 1 hybrid aerogel.

In addition, we discussed the differences in reaction rates between the PS@TiO_2 core–shell particles and the commercial P25. The reaction rates of MB decomposition in the presence of the PS@TiO_2 core–shell particles and the commercial P25 from 0 to 4 min were 21.3% min^{-1} and 3.7% min^{-1}, respectively. Meanwhile, the reaction rates of MB decomposition from 4 to 24 min were 0.13% min^{-1} and 2.1% min^{-1}, respectively. The significantly faster reaction rate of the PS@TiO_2 core–shell particles might be related to the zeta potential of the samples. TiO_2 samples such as P25 reportedly have a negatively charged surface at a pH value of greater than 7 [44]. Meanwhile, the PS@TiO_2 core–shell particles have a negatively charged surface at a pH value of more than 4–5 [45,46]. Because the pH of the reaction solution was 6 in this study, the surface of the PS@TiO_2 core–shell particles was negatively charged, whereas the surface of the commercial P25 was positively charged. MB which have cationic ions were electrochemically attracted to the surface of the PS@TiO_2 core–shell particles, leading to an increase in the initial reaction rate. Meanwhile, the MB was electrochemically repelled from the surface of the commercial P25. The photocatalytic activity decreased after 4 min, probably because the rate of MB adsorption on the surface was faster than the rate of decomposition induced by the photocatalyst, and the MB covered the TiO_2 layer as it prevented the blocking of UV irradiation. In order to solve this problem, we suggest that TiO_2 hollow spheres are prepared via calcination of the PS@TiO_2 core–shell particles. The hollow spheres have high specific surface areas because of the void spaces, leading to an increase in contact frequency with the MB. We found that the PS particles can be decomposed via calcination at 723–873 K [47]. The crystallinity of the TiO_2 simultaneously grows during calcination. Based on this knowledge, we expect that TiO_2 hollow spheres with crystalline shells can efficiently act as photocatalysts.

3. Materials and Methods

3.1. Synthesis of PS@TiO_2 Core–Shell Particles and TiO_2 particles

PS particles were prepared with reference to a previous study [29]. The collected PS was washed with ethanol (10 mL, Kanto Chemical Co., Tokyo, Japan, >99.5%) via centrifugation at 3800 rpm for 10 min, twice. The final PS contents were dispersed in ethanol (15 mL). This PS suspension (10 g), titanium-n-butoxide (1000 μL, Kanto Chemical Co., >98.0%), and ethanol (40 mL) were stirred at room temperature for 24 h. The obtained suspension was centrifuged at 3800 rpm for 10 min and washed with ethanol to obtain the precursor of the PS@TiO_2 core–shell particles. The precursor was spread in ethanol (20 mL). After that, ethanol and an aqueous NH_3 solution (300 or 1000 μL, Kanto Chemical Co., 28.0–30.0% NH_3) were added to the precursor suspension, which was then stirred at room temperature or 323 K for 17 h. Then, the suspension was centrifuged at 3800 rpm for 10 min, and the collected white contents were dried in a desiccator overnight to obtain the PS@TiO_2 core–shell particles. For comparison, we utilized commercial P25 (Kanto Chemical Co.)

and TiO_2 particles prepared using the sol–gel method. The TiO_2 particles were prepared as follows: The titanium-*n*-butoxide (1000 μL), aqueous NH_3 solution (300 μL), and ethanol were stirred at room temperature for 1 h. After that, the white suspension was separated via filtration. The collected white powders were dried in a desiccator overnight to obtain the TiO_2 particles.

3.2. Characterization

The morphologies of the samples were observed via TEM (JEM 2010F, JEOL), operating at an accelerating voltage of 200 kV, and SEM (SU3500, JEOL), operating at an accelerating voltage of 15 kV. The chemical compositions and mapping images of the samples were determined using EDX spectroscopy. The specific surface areas of the samples were measured using N_2 sorption at 77 K using the BET method (Belsorp-mini, MicrotracBEL). The crystallites of the samples were determined using XRD (MultiFlex X-ray diffractometer, Rigaku) at 30 kV and 15 mA of $CuK\alpha$ radiation. PL measurement was performed to evaluate the samples using a fluorospectrophotometer with a Xe lamp light source (HITACHI, F-4500). The measurement excitation wavelength was set to 365 nm. All measurements of the sample were carried out at room temperature.

3.3. Photocatalytic Activity in MB Decomposition

The photocatalytic activities of the $PS@TiO_2$ core–shell particles, the prepared TiO_2, the commercial P25, and the PS particles were evaluated using MB decomposition. Samples (10 mg) were added to 7 ppm aqueous MB solutions (100 mL). Thereafter, this suspension was stirred at 30 min in a black box (40 cm $\times$ 40 cm $\times$ 40 cm). Then, the photocatalytic reaction was started after irradiation with a UV lamp (365 nm) with the irradiation power (150 mW/cm^2) in the black box. A total of 3 mL of the mixture was withdrawn, and the pure aqueous MB solution was separated from the suspension by centrifugation at 3800 rpm for 5 min. The adsorption spectra of these solutions were recorded on spectrophotometer (UV-1800, Shimazu).

4. Conclusions

We synthesized $PS@TiO_2$ core–shell particles and demonstrated their photocatalytic activity in MB decomposition. In this method, TiO_2 nanoparticles prepared using the sol–gel method were collected on the surfaces of PS particles. Homogeneous $PS@TiO_2$ core–shell particles were obtained using 300 μL of an aqueous NH_3 solution and stirred at room temperature. TEM images showed that the TiO_2 shell thickness on the surfaces of the PS particles was approximately 5 nm. Additionally, Ti elements were confirmed to be present on the surfaces of the $PS@TiO_2$ core–shell particles using EDX analysis. The photocatalytic activities of the $PS@TiO_2$ core–shell particles were evaluated using MB decomposition. For comparison, we utilized TiO_2, prepared using the sol–gel method, and commercial P25. The MB decompositions of the $PS@TiO_2$ core–shell particles, the prepared TiO_2, and the commercial P25 after 24 min were 87%, 60%, and 58%, respectively. The crystallinity and shell thickness of TiO_2 on the surface of PS particles might play important role in MB decomposition. On the other hand, the reaction rates of the $PS@TiO_2$ core–shell particles between 0 and 4 min (21.3% min^{-1}) were faster than those of the commercial P25 (3.7% min^{-1}). Meanwhile, the rate of MB decomposition from 4 to 24 min was likely promoted in the presence of the $PS@TiO_2$ core–shell particles, probably because the surface of the $PS@TiO_2$ core–shell particles may have adsorbed the MB and its related decomposition products, leading to a decrease in photocatalytic activity. As one of the approaches to solve this problem, $PS@TiO_2$ core–shell particles can be synthesized via calcination to obtain TiO_2 hollow spheres.

Author Contributions: Conceptualization: N.T. (Naoki Toyama); formal analysis: N.T. (Naoki Toyama) and T.T.; investigation: N.T. (Naoki Toyama), T.T., N.T. (Norifumi Terui), and S.F.; writing—original draft preparation: N.T. (Naoki Toyama); writing—review and editing: N.T. (Norifumi Terui) and S.F.; supervision: N.T. (Norifumi Terui) and S.F. All authors have read and agreed to the published version of the manuscript.

Funding: This research received no external funding.

Data Availability Statement: Not applicable.

Acknowledgments: This work was supported by the Nanotechnology Platform of the University of Tokyo, supported by "Nanotechnology Platform" of the Ministry of Education, Culture, Sports, Science and Technology (MEXT), Japan (grant number A-21-UT-0367). We are grateful to Oshikawa (University of Tokyo) for the TEM measurements. We also thank Teshima, Sato, and Chiba (South Iwate Research Center of Technology) for the SEM measurements.

Conflicts of Interest: The authors declare no conflict of interest.

References

1. Bastakoti, B.P.; Kuila, D.; Salomon, C.; Konarova, M.; Eguchi, M.; Na, J.; Yamauchi, Y. Metal–incorporated mesoporous oxides: Synthesis and applications. *J. Hazard. Mater.* **2021**, *401*, 123348. [CrossRef]
2. Olatidoye, O.; Thomas, D.; Bastakoti, B.P. Facile synthesis of a mesoporous TiO_2 film templated by a block copolymer for photocatalytic applications. *New J. Chem.* **2021**, *45*, 15761. [CrossRef]
3. Mahanta, U.; Khandelwal, M.; Deshpande, A.S. $TiO_2@SiO_2$ nanoparticles for methylene blue removal and photocatalytic degradation under natural sunlight and low–power UV light. *Appl. Sur. Sci.* **2022**, *576*, 151745. [CrossRef]
4. Priya, R.; Stanly, S.; Kavitharani, K.; Mohammad, F.; Sagadevan, S. Highly effective photocatalytic degradation of methylene blue using PrO_2–MgO nanocomposites under UV light. *Optik* **2020**, *206*, 164318.
5. Din, M.I.; Khalid, R.; Najeeb, J.; Hussain, Z. Fundamentals and photocatalysis of methylene blue dye using various nanocatalytic assemblies—A critical review. *J. Clean. Prod.* **2021**, *298*, 126567. [CrossRef]
6. Kumar, S.; Ahlawat, W.; Bhanjana, G.; Heydarifard, S.; Nazhad, M.M.; Dilbaghi, N. Nanotechnology–based water treatment strategies. *J. Nanosci. Nanotechnol.* **2014**, *14*, 1838–1858. [CrossRef] [PubMed]
7. Gaya, U.I.; Abdullah, A.H. Heterogeneous photocatalytic degradation of organic contaminants over titanium dioxide: A review of fundamentals, progress and problems. *J. Photochem. Photobiol. C Photochem. Rev.* **2008**, *9*, 1–12. [CrossRef]
8. Che Ramli, Z.A.; Asim, N.; Isahak, W.N.R.W.; Emdadi, Z.; Ahmad–Ludin, N.; Yarmo, M.A.; Sopian, K. Photocatalytic degradation of methylene blue under UV light irradiation on prepared carbonaceous TiO_2. *Sci. World J.* **2014**, *2014*, 13–15. [CrossRef]
9. Shen, R.; Hao, L.; Chen, Q.; Zheng, Q.; Zhang, P.; Li, X. P-Doped g-C_3N_4 Nanosheets with Highly Dispersed $Co_{0.2}Ni_{1.6}Fe_{0.2}P$ Cocatalyst for Efficient Photocatalytic Hydrogen Evolution. *Acta Phys.-Chim. Sin.* **2022**, *38*, 2110014.
10. Lei, Z.; Ma, X.; Hu, X.; Fan, J.; Liu, E. Enhancement of photocatalytic H_2-evolution kinetics through the dual cocatalyst activity of Ni_2P-NiS-decorated g-C_3N_4 heterojunctions. *Acta Phys.-Chim. Sin.* **2022**, *38*, 2110049.
11. Lu, N.; Jing, X.; Xu, Y.; Lu, W.; Liu, K.; Zhang, Z. Effective cascade modulation of charge-carrier kinetics in the well-designed multi-component nanofiber system for highly-efficient photocatalytic hydrogen generation. *Acta Phys.-Chim. Sin.* **2023**, *39*, 2207045.
12. Lu, N.; Jing, X.; Zhang, J.; Zhang, P.; Qiao, Q.; Zhang, Z. Photo-assisted self-assembly synthesis of all 2D-layered heterojunction photocatalysts with long-range spatial separation of charge-carriers toward photocatalytic redox reactions. *Chem. Eng. J.* **2022**, *431*, 134001. [CrossRef]
13. Yan, X.; Wang, B.; Ji, M.; Jiang, Q.; Liu, G.; Liu, P.; Yin, S.; Li, H.; Xia, J. In-situ synthesis of CQDs/BiOBr material via mechanical ball milling with enhanced photocatalytic performances. *Chin. J. Struct. Chem.* **2022**, *41*, 2208044–2208051.
14. Hashimoto, K.; Irie, H.; Fujishima, A. TiO_2 Photocatalysis: A Historical Overview and Future Prospects. *Jpn. J. Appl. Phys.* **2005**, *44*, 8269. [CrossRef]
15. Bahnemann, D. Photocatalytic water treatment: Solar energy applications. *Sol. Energy* **2004**, *77*, 445–459. [CrossRef]
16. Ertuş, E.B.; Vakifahmetoglu, C.; Öztür, A. Enhanced methylene blue removal efficiency of TiO_2 embedded porous glass. *J. Eur. Ceram. Soc.* **2021**, *41*, 1530–1536. [CrossRef]
17. Legrand-Buscema, C.; Malibert, C.; Bach, S. Elaboration and characterization of thin films of TiO_2 prepared by sol–gel process. *Thin Solid Films* **2002**, *418*, 79–84. [CrossRef]
18. Lopez, T.; Sanchez, E.; Bosch, P.; Meas, Y.; Gomez, R. FTIR and UV-Vis (diffuse reflectance) spectroscopic characterization of TiO_2 sol-gel. *Mater. Chem. Phys.* **1992**, *32*, 141–152. [CrossRef]
19. Schneider, J.; Matsuoka, M.; Takeuchi, M.; Zhang, J.; Horiuchi, Y.; Anpo, M.; Bahneman, D.W. Understanding TiO_2 Photocatalysis: Mechanisms and Materials. *Chem. Rev.* **2014**, *114*, 9919–9986. [CrossRef] [PubMed]
20. Kasanen, J.; Salstela, J.; Suvanto, M.; Pakkanen, T.T. Photocatalytic degradation of methylene blue in water solution by multilayer TiO_2 coating on HDPE. *Appl. Sur. Sci.* **2011**, *258*, 1738–1743. [CrossRef]
21. Abd El-Rehim, H.A.; Hegazy, E.-S.A.; Diaa, D.A. Photo-catalytic degradation of Metanil Yellow dye using TiO_2 immobilized into polyvinyl alcohol/acrylic acid microgels prepared by ionizing radiation. *React. Funct. Polym.* **2012**, *72*, 823–831. [CrossRef]

22. Bertram, J.R.; Nee, M.J. A buoyant, microstructured polymer substrate for photocatalytic degradation applications. *Catalysis* **2018**, *10*, 482. [CrossRef]
23. Shigh, S.; Mahalingam, H.; Kumar Singh, P. Polymer-supported titanium dioxide photocatalysts for environmental remediation: A review. *Appl. Catal. A* **2013**, *462–463*, 178–195.
24. Mahshid, S.; Askari, M.; Sasani Ghamsari, M.; Afshar, M.; Lahuti, S. Mixed-phase TiO_2 nanoparticles preparation using sol–gel method. *J. Alloys Compd.* **2009**, *478*, 586–589. [CrossRef]
25. Luo, M.-L.; Tang, W.; Zhao, J.-Q.; Pu, C.-S. Hydrophilic modification of poly(ether sulfone) used TiO_2 nanoparticles by a sol–gel process. *J. Mater. Proc. Technol.* **2006**, *172*, 431–436. [CrossRef]
26. Wang, X.-J.; Li, F.-T.; Hao, Y.-J.; Liu, S.-J.; Yang, M.-L. TiO_2/SBA–15 composites prepared using H_2TiO_3 by hydrothermal method and its photocatalytic activity. *Mater. Lett.* **2013**, *99*, 38–41. [CrossRef]
27. Jiang, C.; Lee, K.Y.; Parlett, C.M.A.; Bayazit, M.K.; Lau, C.C.; Ruan, Q.; Moniz, S.J.A.; Lee, A.F.; Tang, J. Size–controlled TiO_2 nanoparticles on porous hosts for enhanced photocatalytic hydrogen production. *Appl. Catal. A* **2016**, *521*, 113–139. [CrossRef]
28. Wu, F.; Liu, W.; Qiu, J.; Li, J.; Zhou, W.; Fang, Y.; Zhang, S.; Li, X. Enhanced photocatalytic degradation and adsorption of methylene blue via TiO_2 nanocrystals supported on graphene–like bamboo charcoal. *Appl. Sur. Sci.* **2015**, *358*, 425–435. [CrossRef]
29. Deng, Z.; Chen, M.; Zhou, S.; You, B.; Wu, L. A Novel Method for the Fabrication of Monodisperse Hollow Silica Spheres. *Langmuir* **2006**, *22*, 6403–6407. [CrossRef]
30. Song, X.; Gao, L. Fabrication of Hollow Hybrid Microspheres Coated with Silica/Titania via Sol−Gel Process and Enhanced Photocatalytic Activities. *J. Phys. Chem. C* **2007**, *111*, 8180–8187. [CrossRef]
31. Leng, W.; Chen, M.; Zhou, S.; Wu, L. Capillary Force Induced Formation of Monodisperse Polystyrene/Silica Organic−Inorganic Hybrid Hollow Spheres. *Langmuir* **2010**, *26*, 14271–14275. [CrossRef]
32. Toyama, N.; Ohki, S.; Tansho, M.; Shimizu, T.; Umegaki, T.; Kojima, Y. Influence of alcohol solvents on morphology of hollow silica–alumina composite spheres and their activity for hydrolytic dehydrogenation of ammonia borane. *J. Sol-Gel Sci. Technol.* **2017**, *82*, 92–100. [CrossRef]
33. Rao, K.S.; El–Hami, K.; Kodaki, T.; Matsushige, K.; Makino, K. A novel method for synthesis of silica nanoparticles. *J. Colloid Inter. Sci.* **2005**, *289*, 125–131. [CrossRef]
34. Markowska–Szczupak, A.; Wang, K.; Rokicka, P.; Endo, M.; Wei, Z.; Ohtani, B.; Morawski, A.W.; Kowalska, E. The effect of anatase and rutile crystallites isolated from titania P25 photocatalyst on growth of selected mould fungi. *J. Photochem. Photobiol.* **2015**, *151*, 54–62. [CrossRef] [PubMed]
35. Zou, J.; Gao, J.; Xie, F. Anamorphous TiO_2 sol sensitized with H_2O_2 with the enhancement of photocatalytic activity. *J. Alloys Compd.* **2010**, *497*, 420–427. [CrossRef]
36. Wang, Z.; Zhang, F.; Yang, Y.; Xue, B.; Cui, J.; Guan, N. Facile Postsynthesis of Visible-Light-Sensitive Titanium Dioxide/Mesoporous SBA-15. *Chem. Mater.* **2007**, *19*, 3286–3293. [CrossRef]
37. Liaoa, M.-H.; Hsu, C.-H.; Chen, D.-H. Preparation and properties of amorphous titania-coated zinc oxide nanoparticles. *J. Solid State Chem.* **2006**, *179*, 2020–2026. [CrossRef]
38. Yang, J.-H.; Han, Y.-S.; Choy, J.-H. TiO_2 thin-films on polymer substrates and their photocatalytic activity. *Thin Solid Films* **2006**, *495*, 266–271. [CrossRef]
39. Metanawin, T.; Panutumrong, P.; Metanawin, S. Synthesis of polyurethane/TiO_2 hybrid with high encapsulation efficiency using one-step miniemulsion polymerization for methylene blue degradation and its antibacterial applications. *ChemistrySelect* **2023**, *8*, e202204522. [CrossRef]
40. Salehi Taleghani, M.; Salman Tabrizi, N.; Sangpour, P. Enhanced visible-light photocatalytic activity of titanium dioxide doped CNT-C aerogel. *Chem. Eng. Res. Design* **2022**, *179*, 162–174. [CrossRef]
41. Yang, Y.; Xu, L.; Wang, H.; Wang, W.; Zhang, L. TiO_2/graphene porous composite and its photocatalytic degradation of methylene blue. *Mater. Design* **2016**, *108*, 632–639. [CrossRef]
42. Li, X.; Lin, H.; Chen, X.; Niu, H.; Zhang, T.; Liu, J. Fabrication of TiO_2/porous carbon nanofibers with superior visible photocatalytic activity. *New J. Chem.* **2015**, *39*, 7863–7872. [CrossRef]
43. Kaur, K.; Singh, C.V. Amorphous TiO_2 as a photocatalyst for hydrogen production: A DFT study of structural and electronic properties. *Energy Procedia* **2012**, *29*, 291–299. [CrossRef]
44. Zhao, J.; Hidaka, H.; Takamura, A.; Pelizzetti, E.; Serpone, N. Photodegradation of surfactants. 11. zeta.–Potential measurements in the photocatalytic oxidation of surfactants in aqueous titania dispersions. *Langmuir* **1993**, *9*, 1646–1650. [CrossRef]
45. Imhof, A. Preparation and characterization of titania–coated polystyrene spheres and hollow titania shells. *Langmuir* **2001**, *17*, 3579–3585. [CrossRef]
46. Jang, I.B.; Sung, J.H.; Choi, H.J.; Chin, I. Synthesis and characterization of titania coated polystyrene core–shell spheres for electronic ink. *Synth. Met.* **2005**, *152*, 9–12. [CrossRef]
47. Toyama, N.; Umegaki, T.; Kojima, Y. Fabrication of hollow silica–alumina composite spheres and their activity for hydrolytic dehydrogenation of ammonia borane. *Int. J. Hydrogen Energy* **2014**, *39*, 17136–17143. [CrossRef]

 *inorganics*

Article

Method for Decontamination of Toxic Aluminochrome Catalyst Sludge by Reduction of Hexavalent Chromium

Igor Pyagay, Olga Zubkova *, Margarita Zubakina * and Viktor Sizyakov

Scientific Center «Problems of Mineral Mineral and Technogenic Resources», Saint-Petersburg Mining University, 21st Line of V. I., 2, 199106 St. Petersburg, Russia; pyagay_in@pers.spmi.ru (I.P.); siziakov_vm@pers.spmi.ru (V.S.)
* Correspondence: churkina_os@pers.spmi.ru (O.Z.); s212546@stud.spmi.ru (M.Z.)

Abstract: The article is devoted to the neutralization of the harmful effects of aluminochrome catalyst sludge. Catalyst sludge is a waste product from petrochemical production and poses a serious threat to the environment and humans because of the toxic hexavalent chromium it contains. The emissions of Russian petrochemical enterprises' alumochrome sludge is 10,000–12,000 tons per year. In this paper, research related to the possibility of reducing the harmful effects of sludge by converting hexavalent chromium to a less dangerous trivalent state is presented. The reduction of hexavalent chromium was carried out with different reagents: Na_2SO_3, $FeSO_4$, $Na_2S_2O_3$, and $Na_2S_2O_5$. Then, a comparative analysis was carried out, and sodium metabisulfite was chosen as the most preferred reagent. The peculiarity of the reducing method was carrying out the reaction in a neutral medium, pH = 7.0. The reduction was carried out in the temperature range of 60–85 °C and under standard conditions. The maximum recovery efficiency of chromium from the catalyst sludge (100%) was achieved at 85 °C and 10 min. This method did not involve the use of concentrated sulfuric acid, as in a number of common techniques, or additional reagents for the precipitation of chromium in the form of hydroxide.

Keywords: catalyst sludge; hexavalent chromium; reduction; neutralization; deposition

Citation: Pyagay, I.; Zubkova, O.; Zubakina, M.; Sizyakov, V. Method for Decontamination of Toxic Aluminochrome Catalyst Sludge by Reduction of Hexavalent Chromium. *Inorganics* **2023**, *11*, 284. https://doi.org/10.3390/inorganics11070284

Academic Editors: Roberto Nisticò, Torben R. Jensen, Luciano Carlos, Hicham Idriss and Eleonora Aneggi

Received: 29 March 2023
Revised: 22 June 2023
Accepted: 26 June 2023
Published: 30 June 2023

1. Introduction

Alumochrome catalyst is used in the petrochemical industry, in the process of the dehydrogenation of lower paraffins to C3–C5 olefins, which are used in the production of synthetic rubbers, plastic masses, and high-active additives for fuels [1–3]. The total volume of olefins produced by dehydrogenation methods in Russia is 600,000–700,000 tons per year [4].

The catalyst is an Al_2O_3-Cr_2O_3-SiO_2 system obtained by the thermal activation of gibbsite [5]. Russian companies use catalysts of IM-2201 and KDM grades. The catalyst undergoes irreversible deactivation due to changes in the state of the active component, as well as being destroyed by mechanical action. The catalyst fragments are carried out of the reactor zone in the form of dust, which makes it necessary to carry out wet scrubber cleaning [6,7]. As a result, watered-down catalyst sludge, which the U.S. Environmental Protection Agency (EPA) classifies as a Group A carcinogen for humans, is removed from the plant's reaction zone [8].

This sludge poses a serious threat because it contains trivalent and hexavalent chromium ions. The most stable oxidation degrees of chromium are trivalent and hexavalent. Hexavalent chromium is soluble in water in the entire pH range, while trivalent chromium can be dissolved or precipitated as chromium hydroxide in mildly acidic and alkaline environments of pH 4–8 [8]. Cr(VI) is the strongest carcinogen and mutagen; it is about 1000 times more toxic than Cr(III) [9,10]. Chromium is considered one of the top 20 pollutants on Superfund's list of priority hazardous substances over the past 15 years [11]. From this, we can conclude that all industrial waste containing hexavalent chromium must be thoroughly recycled.

Alumochrome catalyst sludge is not recycled and is stored in special landfills, which leads to the accumulation of large-scale hazardous waste [12–14]. The storage of highly toxic sludge is costly and creates environmental problems, such as the pollution of sewage, groundwater, and air space, and has a detrimental effect on the health of living organisms [15–17]. In this regard, the development of technology to reduce the toxicity of waste by removing hexavalent chromium, with subsequent recycling, is relevant [18,19].

Since the catalyst sludge has an initial pH = 7, the reduction of Cr(VI) to Cr(III) was carried out in a neutral medium rather than in a strongly acidic one, as previously used, which is a novelty. Moreover, sodium metabisulfite was used as a reducing agent, which was not previously used for the reduction of chromium. The method proposed in this work to recover hexavalent chromium from catalyst sludge can be used at petrochemical plants to reduce the environmental load.

2. Literature Review

To date, there are various methods for cleaning solutions of hexavalent chromium, such as extraction, sorption, and bacterial, and using inorganic and organic reducing agents. Over the past few years, methods of removing chromium from waste solutions from various industries have become increasingly diverse.

The extraction method for extracting hexavalent chromium from wastewater is discussed in an article by Ziwen Ying et al., which compared nine amide extractants with different alkyl groups and six diluents with different polarity [20].

Liping Dai and Anil Kumar considered liquid membranes as extractants [21,22]. Liu and Xiaoyun Wu et al. constructed a bipolar membrane electrodialysis system (BMES) for the simultaneous extraction of Cr(III) and Cr(VI) from chromium slurry in the form of Na_2CrO_4 [23,24].

Ion exchange is widely used for the wastewater treatment of Cr(VI) ions, and various types of resins, zeolites, and natural sorbents are used as sorbents. In a study by Zhenxiong Ye et al., a process combining ion exchange and reduction–deposition, based on pyridine resin SiPyR-N4 deposited on a silicon dioxide carrier, with a Cr (VI) removal efficiency of 99.3% from the solution [25], was proposed. Jiayu Lu et al. used a binary solution of sapind saponin and the surfactant cetyltrimethylammonium bromide to remove Cr (VI) and recorded 94% removal [26].

Biochar is used as an adsorbent to remove pollutants from industrial wastewater. Studies by Zixi Fan et al. and Jianhua Qu et al. investigated several types of activated and deactivated biochar to effectively remove Cr (VI) from polluted industrial wastewater [27,28]. For the adsorption of chromium from solutions in the works of Paulina Janik et al. and Changwoo Kim et al., graphene oxide was modified with various aminosilanes containing one, two, or three nitrogen atoms in the molecule [29,30]. In a study by Xufan Zhang et al., magnetic titanomagnegemite (Fe_2TiO_5) was sulfated with H_2S gas to increase the removal efficiency of Cr (VI) [31].

The electrochemical reduction (ECR) of Cr (VI) to Cr (III) was extensively described in the removal of Cr (VI) from contaminated water. In a research study, Fubing Yao et al. used a single-chamber cell with a titanium anode for the indirect reduction of Cr (VI) and the deposition of Cr (III) [32].

Electroflotation (EF) is effective for removing various types of organic and inorganic pollutants. A. V. Kolesnikov et al. studied the EF process in the removal of hydroxide Ni(II), Cr(III), and other metals from the solution, using sodium sulfate as the background electrolyte [33].

The use of nanomaterials, which are considered to be materials smaller than 100 nm, is gaining popularity. The use of some inorganic nanoparticles as reducing agents or catalysts can significantly increase the reduction rate of Cr (VI) [30]. In the works of Miroslav Brumovský et al. and Qianqian Shao et al., the exploitation of sulfidized iron nanoparticles to increase the reactivity and selectivity of nZVI, with respect to the target pollutant chromium [34,35], was considered. Some of the nanomaterials used were bifunctional MOF/titanate nanotube composites [36], chitosan-grafted graphene oxide (CS-GO) nanocomposite [37], and graphene/SiO_2 nanocomposites @polypyrrole [38].

Over the past few years, methods of removing chromium from solutions by biological waste from various industries have been gaining popularity. Living organisms such as bacteria, yeast, fungi, algae, and plants proved to be effective bioremediation agents. Biosorption is another phenomenon in which a product of biological origin and byproducts are used as an adsorbent, and the process of adsorption is called biosorption. In the works of Golovina V.V. and Bagauv A.I. [39,40], chromium was extracted from solutions by logging wastes, namely bark and sawdust. Cherdchoo W. et al. conducted studies on the adsorption of chromium on ground coffee and mixed tea waste [41]. Vilardi G. et al. studied adsorption on biomass from olive seeds [42]. Bacterial methods of chromium extraction were described in the works of Zorkina O.V. and Singh P. [43,44].

All of these methods have a number of drawbacks. Electrochemical processes are efficient but require a high energy supply and skilled labor. The adsorption process is also efficient, but, after several cycles, the material removal potential decreases. The bioremediation process is environmentally friendly, but it takes a long time to complete [44,45]. The choice of a suitable purification technology depends on the initial concentration of heavy metals, the operating costs, and the characteristics of the solutions containing chromium ions.

The reagent method (recovery method) is widespread because of the minimum cost and relative simplicity of its technological process, due to which the maximum extraction in a wide range of parameters of chromium-containing waste is provided [46]. The reduction of hexavalent chromium to trivalent chromium by inorganic reducing agents is carried out in an acidic environment, followed by the precipitation of metal ions in the form of hydroxides in an alkaline environment [47].

The most commonly used reagents to convert Cr (VI) to Cr (III) are the sodium salts of sulfuric acid: sulfite (Na_2SO_3), bisulfite ($NaHSO_3$), thiosulfate ($Na_2S_2O_5$), iron sulfate ($FeSO_4$), iron chips, and organic reducing agents. The reductions proceed according to the following reactions (1)–(4) [47–49]:

Reaction with sodium sulfite:

$$H_2Cr_2O_7 + 3Na_2SO_3 + 3H_2SO_4 \rightarrow Cr_2(SO_4)_3 + 3Na_2SO_4 + 4H_2O \tag{1}$$

Reaction with sodium thiosulfate:

$$H_2Cr_2O_7 + 2Na_2S_2O_3 + H_2SO_4 \rightarrow Cr_2(SO_4)_3 + 2Na_2SO_4 + 2H_2O \tag{2}$$

Reaction with ferrous sulfate:

$$H_2Cr_2O_7 + 6FeSO_4 + 6H_2SO_4 \rightarrow Cr_2(SO_4)_3 + 3Fe_2(SO_4)_3 + 7H_2O \tag{3}$$

Reaction with iron shavings:

$$H_2Cr_2O_7 + 2Fe + 6H_2SO_4 \rightarrow Cr_2(SO_4)_3 + Fe_2(SO_4)_3 + 7H_2O \tag{4}$$

The above reactions occur in the strongly acidic region, at pH = 2.5. Chromium is converted to the trivalent form as sulfate; the reagents used to precipitate the trivalent chromium as $Cr(OH)_3$ are $Ca(OH)_2$, Na_2CO_3, and $NaOH$. This reaction must be closely monitored, since chromium hydroxide completely dissolves when the pH is increased above 8–9 [8,48].

Such a purification scheme has a number of disadvantages, such as the use of concentrated sulfuric acid to acidify solutions and reagents for chromium precipitation in narrow pH ranges.

3. Results and Discussion

Alumina catalyst sludge, which was formed during isobutane dehydrogenation and is a green-colored slurry, was used as an object of study.

For analysis, the slurry had to be separated into liquid and solid phases, and their qualitative and quantitative compositions had to be determined. The filtration process was carried out using a vacuum.

The solid phase has a gray-green color, and the liquid phase has a bright yellow color (Figure 1).

Figure 1. Liquid and solid phases of catalyst slurry.

In the course of this research, it was revealed that trivalent chromium in the liquid phase of the catalyst sludge was absent, and no precipitate was formed when precipitating with a concentrated ammonia solution. The content of the hexavalent chromium in the solution is $1.2\ g/dm^3$.

The solid phase of the catalyst sludge is a finely dispersed powder with the presence of amorphous agglomerates ranging in size from 1 to 20 microns, as shown in Figure 2.

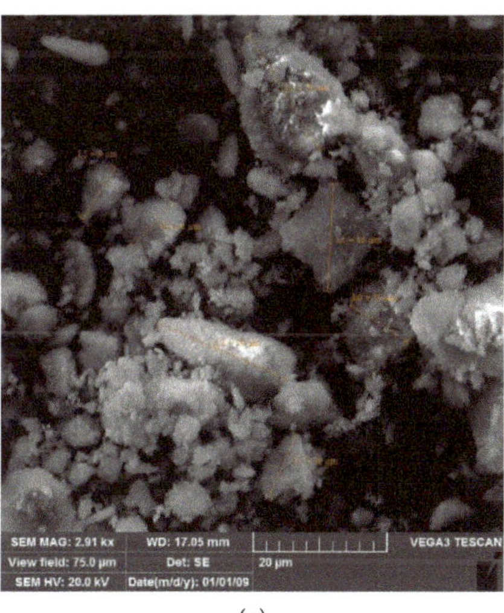

(**a**)

(**b**)

Figure 2. SEM of the solid phase of the catalyst slurry: (**a**) before reduction; (**b**) after reduction.

The process of the reduction of hexavalent chromium with sodium metabisulfite in the pulp had no effect on the morphological structure of the solid phase.

The solid phase of the catalyst slurry is a finely dispersed powder with the presence of amorphous agglomerates ranging in size from 1 to 20 microns.

The results of diffractograms performed using X-ray powder diffraction are shown in Figure 3. According to X-ray diffraction analysis, we can see from the crystalline phases in the sample there is a mix of alpha aluminum oxide and delta aluminum oxide, as well as roughly dispersed crystallites of the solid solution Cr_2O_3-Al_2O_3 (chromaluminum oxide). According to X-ray analysis, we can say that there is a high proportion of the amorphous phase. The results of infrared spectrometry (ALPHA II, Bruker, Billerica, MA, USA) showed nothing but the presence of aluminum alpha-oxide (Figure 4).

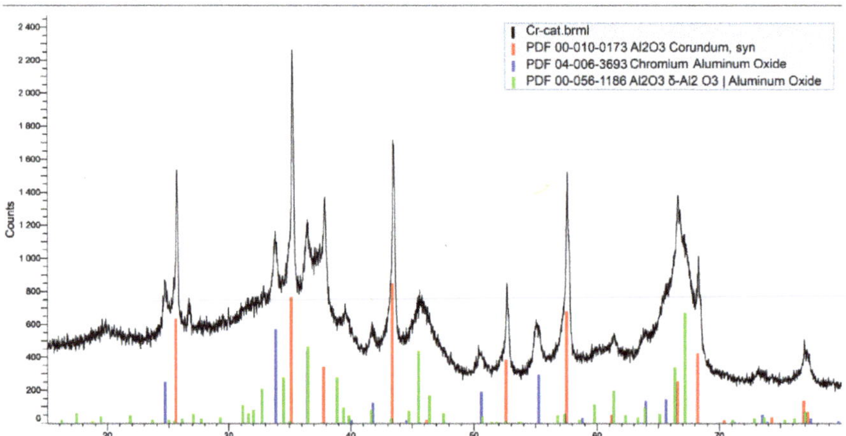

Figure 3. XRF of spent chromium catalyst.

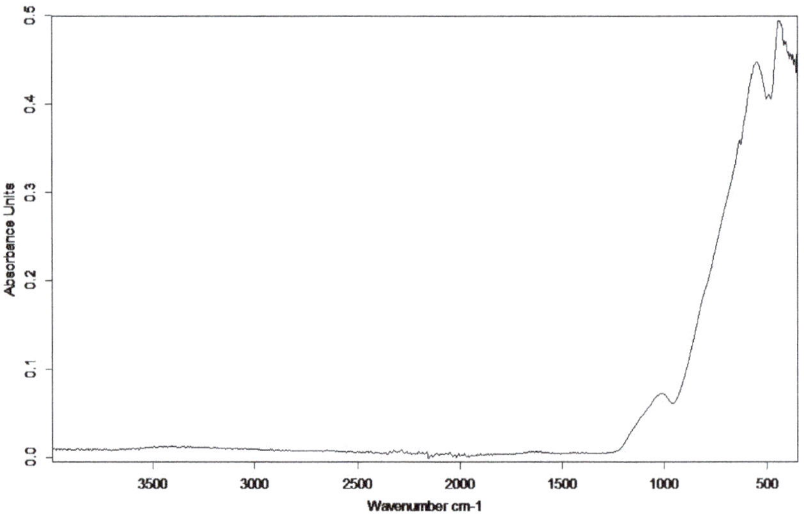

Figure 4. IR spectrum of the solid phase of catalyst sludge.

The composition of the solid phase of the catalyst slurry before and after the separation of water-soluble chromates is presented in Table 1.

Table 1. Chemical composition of the solid phase of catalyst sludge.

Name of Specimen and Research Instrument	Concentration, % wt.								
	Al_2O_3	SiO_2	K_2O	CaO	TiO_2	Cr_2O_3	Fe_2O_3	ZnO	ZrO_2
Solid phase of catalyst sludge (initial) (Shimadzu)	72.46	11.61	2.49	0.18	0.23	11.74	0.21	0.01	0.23
Solid phase of catalyst sludge (initial) (ISP MS)	70.20	8.8	1.90	0.30	0.23	13.5	0.31	-	-
Solid phase of catalyst sludge (initial) (microanalysis: Tescan Vega 3)	71.8	10.08	2.94	0.35	-	13.6	0.27	-	-

Studies to identify the free chromates on the surface of the solid phase of the catalyst showed their absence.

The results showed that all trivalent chromium is contained in the solid phase and all hexavalent in the liquid phase, which means that only the liquid phase of the catalyst slurry is subjected to the reduction process.

3.1. Recovery of Hexavalent Chromium in the Liquid Phase of Catalyst Sludge

All experiments with the initial liquid phase of the catalyst slurry were performed without acidification with sulfuric acid for all the selected reducing agents. The reduction of hexavalent chromium was carried out in the temperature range of 50 °C–85 °C.

Hexavalent chromium in the liquid phase, at pH = 7, is in the form of chromates, as chromate ions prevail at pH > 6.5, according to the literature [50,51]. The conditions of the liquid-phase reduction process, and the results are shown in Table 1.

The interaction of chromates with sodium sulfite in a neutral medium follows the reaction (5).

$$2CrO_4{}^{2-} + 3Na_2SO_3 + 5H_2O \rightarrow 2Cr(OH)_3 + 3Na_2SO_4 + 4OH^- \tag{5}$$

The reaction produces a precipitate of amorphous chromium hydroxide with a gray-green color.

The quantity of sodium sulfite added to the liquid phase of the catalyst slurry was chosen based on earlier studies on model solutions. The mass ratio of sodium sulfite to the mass of the liquid phase of the catalyst slurry ranged from 0.0019 to 0.0060. Based on the literature data, sodium sulfite was added in excess, two or more times the calculated stoichiometric value [52,53]. In the practical implementation of the above interaction, it was found that the excess sodium sulfite remaining after the reaction was hydrolyzed, with a characteristic odor of sulfur sulfide (SO_2). The resulting mother liquor was highly alkaline, with pH = 11.7–12.5. Already with a recovery time of 30 min or more, the concentration of chromium ions in the liquid phase was less than 0.005 mg/L, which is less than the limit that the USEPA set for industrial wastewater discharges [8].

In order to reduce the formation of gas release and alkaline environment reduction, reactions were carried out with other reagents.

The interaction of hexavalent chromium with iron sulfate in a neutral medium follows the reaction (6).

$$CrO_4{}^{2-} + 3FeSO_4 + 4H_2O \rightarrow 2FeOHSO_4 + Cr(OH)_3 + Fe(OH)_3 + K_2SO_4 \tag{6}$$

Iron (III) hydroxide is a brown precipitate.

During the reduction with iron sulfate, in addition to chromium hydroxide, iron hydroxide precipitated out, which was an undesirable effect, since these precipitates must be separated later. Moreover, ferrous sulfate had worse chromium extraction results than sodium sulfite, so this reducing agent had to be abandoned.

Sodium thiosulfate showed a negative result when reducing Cr (VI) to Cr (III) in a neutral medium. The concentration of Cr (VI) in the solution in the temperature range of 50–75 °C did not change, so further studies with this reagent were discontinued.

In this series of experiments, it was assumed that reactions (7) and (8) proceed under these conditions under reduction. The reactions are presented in shrunken ionic form:

$$3S_2O_5{}^{2-}_3 + 4CrO_4{}^{2-} + 7H_2O \rightarrow 6SO_4{}^{2-} + 4Cr(OH)_3 \downarrow + 2OH^- \tag{7}$$

$$3S_2O_5{}^{2-}_3 + 2Cr_2O_7{}^{2-} + 5H_2O \rightarrow 6SO_4{}^{2-} + 3Cr(OH)_3 \downarrow + Cr^{3+} + OH^- \tag{8}$$

Since chromate ions predominate in a neutral medium [50,51], the Gibbs energy of reaction (7) was calculated under standard conditions, as follows: $\Delta_r G^0_{298} = -613.96$ kJ/mol. This indicates that this reaction is thermodynamically possible and runs spontaneously.

The application of sodium metabisulfite as a reductant of hexavalent chromium from the liquid phase of catalyst sludge in a neutral medium (pH = 7) leads to the transfer of up to 100% Cr (VI) to Cr (III), with the precipitation in the form of hydroxide, so this reagent was chosen for the reduction of chromium in the catalyst sludge pulp. Table 2 shows the conditions of the liquid-phase reduction process and the results of reduction.

Table 2. The conditions of the liquid-phase reduction process and the results.

Experiment Number	Ratio of the Reducing Agent's Mass to the Mass of the Liquid Phase of the Catalyst Slurry, g/g	Name of Reducing Agent	Process Temperature, °C	Process Time, min	Cr(VI) Removal Efficiency, %	Concentration of Cr^{+6} in the Liquid Phase before Reduction, g/L	Concentration of Cr^{+6} in the Liquid Phase after Recovery, g/L	pH of the Liquid Phase before Recovery	pH of the Liquid Phase after Recovery
			Reduction with sodium sulfite (neutral medium)						
1			50		20		1.113		
2			75	30	98		Менее 0.005		
3			85		100		0		
4			50		68		0.390		
5	0.0060		75	60	98		Менее 0.005		
6			85		100		0		
7		Na_2SO_3	50		78	1.205	0.265	7	11.7–12.5
8			75	120	100		0		
9			85		100		0		
10			50		7		1.105		
11	0.0037		75	60	12		0.64		
12			85		46		0.65		
13	0.0019		75		10		0.89		
			Recovery with ferrous sulfate (neutral medium)						
14			50		31.10		0.375		
15	0.0060	$FeSO_4$	75	60	33.61	1.205	0.405	7	2.5–3.0
16			85		30.29		0.365		
			Recovery with sodium thiosulfate (neutral medium)						
17			50				1.160		
18	0.0060	$Na_2S_2O_3$	75	60	No sludge	1.205	1.070	7	8.8
19			85				1.085		
			Reduction with sodium metabisulphite (neutral medium)						
20			50	60	47		0.64		
21	0.0035		85	65	52		0.58		
22		$Na_2S_2O_5$	85	120	55	1.205	0.54	7	8.8
23			50	420	57		0.52		
24	0.0070		50	15	0		1.20		
25			85	60	100		0		

During storage of the reduced liquid phase, we observed that the precipitation of chromium hydroxide continues with time, and the solution becomes lighter. In this connection, a series of experiments on the incubation of the liquid phase of the catalyst sludge in the presence of sodium metabisulfite at 20 °C and with periodic stirring every 12 h was carried out. The results of the analyses are presented in Table 3.

Table 3. Changes in the parameters of the liquid phase of the catalyst slurry when incubated in the presence of sodium metabisulfite at 20 °C.

Duration of Experiment, Days	Concentration of Total Chromium in the Liquid Phase, g/L
0	1.21
2	1.10
3	0.92
4	0.89
5	0.86

Under this condition of the experiment, on the second day, the concentration of the chromium in the liquid phase of the catalyst sludge dropped by 18%. During reduction, the sediment, which over time thickened and separated from the liquid phase, also precipitated. During the next 120 h, the precipitation sludge considerably slowed down.

3.2. Recovery of Hexavalent Chromium in Catalyst Slurry

In this part of the work, studies were conducted on the recovery of hexavalent chromium from the liquid phase of the catalyst slurry without separating the solid phase. In the course of this study, the optimum ratio of the reducing agent mass ($Na_2S_2O_5$) to the slurry mass, g/g = 0.005, was found. The results of the study of the reduction kinetics are shown in Figure 5.

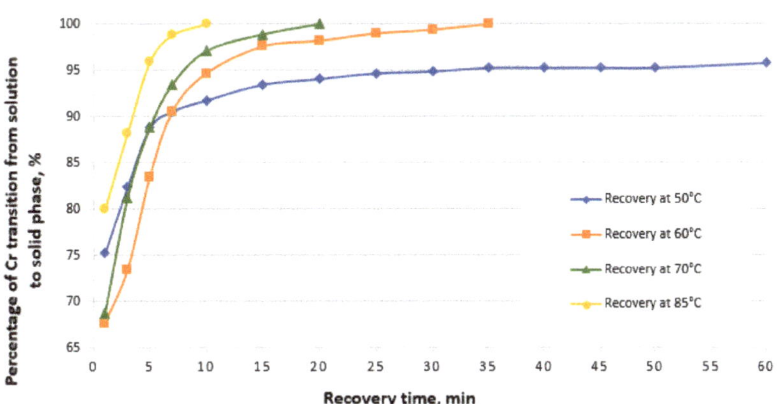

Figure 5. Kinetics of reduction of hexavalent chromium to trivalent chromium in catalyst slurry (ratio of reducing agent mass to slurry mass: g/g = 0.005).

Analysis of the mother liquor after pulp filtration showed that the recovery of hexavalent chromium from the catalyst slurry was possible at temperatures above 60 °C. At a temperature of 85 °C and a time of 10 min, the degree of conversion of chromium to the trivalent state reached 100%, which was more technologically optimal for the reduction process. The pH of the solution insignificantly increased and was 8.8.

Based on the results of the completed work, the following results can be summarized:

1. Sodium thiosulfate ($Na_2S_2O_3$) showed a negative result for the reduction of Cr (VI) to Cr (III) in a neutral medium. The concentration of Cr (VI) in the solution in the temperature range of 50–85 °C did not change.
2. Iron sulfate (anhydrous) in the reduction of Cr (VI) to Cr (III) in a neutral medium showed an average result, with an efficiency of about 30%. The reduction process proceeded without the use of additional precipitators, since the reduced chromium already precipitated out as $Cr(OH)_3$ in the process of reduction. Moreover, during the reduction with ferrous sulfate, in addition to chromium hydroxide, iron hydroxide precipitated out, which was an undesirable effect, since these precipitates must be separated later.
3. The reduction of Cr (VI) to Cr (III) in a neutral medium in the presence of sodium sulfite proceeded with 100% efficiency without the use of additional precipitators. However, during the reduction with sodium sulfite, the solutions acquired a strongly alkaline environment, which should be neutralized. In addition, this interaction produced a byproduct in the form of sulfur dioxide, which must be captured and disposed of.
4. The recovery of the liquid phase of the catalyst sludge with sodium metabisulfite ($Na_2S_2O_5$) completely proceeded without the formation of byproducts, as with the other reagents previously described. Its effect in a neutral environment led to a conversion of up to 100% of hexavalent chromium to the trivalent state, with the formation of a precipitate in the form of chromium hydroxide, without the use of additional reagents. Therefore, sodium metabisulfite was chosen for the reduction of hexavalent chromium in the catalyst slurry.
5. The experiments on the conversion of hexavalent chromium to the trivalent state in the catalyst slurry showed that the reduction process in the presence of sodium metabisulfite $Na_2S_2O_5$ can be directly carried out without the separation of the solid and liquid phases. Complete reduction took place at 85 °C after 10 min of interaction with the reducing agent.

4. Materials and Methods

This study assumed the possibility of using the reaction of effective reduction of Cr (VI) to Cr (III) by reducing reagents Na_2SO_3, $FeSO_4$, $Na_2S_2O_3$, and $Na_2S_2O_5$ not only from the liquid phase but also from the slurry. After the reduction reactions, the resulting precipitate was separated by filtration. The mother liquor was analyzed using X-ray fluorescence spectroscopy and the solid phase using an electron microscope equipped with an energy-dispersive analyzer.

The stages of the study included:

— separation of the liquid phase of the slurry from the solid phase;
— analysis of the solid and liquid phases before reduction;
— slurry liquid-phase recovery;
— filtration of liquid phase after recovery;
— analysis of liquid phase after recovery;
— selection of the most effective reagent for recovery of the liquid phase (Na_2SO_3, $FeSO_4$, $Na_2S_2O_3$, and $Na_2S_2O_5$);
— pulp reduction carried out with the selected reagent;
— determination of the chemical composition of the chromium content in the liquid and solid phases.

4.1. Separation of the Liquid Phase of the Pulp

Separation from the solid was carried out by filtration with a polyester filter cloth under a vacuum of 100 kPa. After filtration, the solid phase was washed with distilled water and dried until all moisture was removed.

4.2. Analysis of the Obtained Phases

Qualitative and quantitative analyses of the mother liquor after filtration were performed on a Shimadzu EDX-7000 energy dispersive spectrometer [54].

When a sample is irradiated with X-rays, the atoms in the sample emit fluorescent X-rays. Atoms of each element emit their (characteristic) radiation, which has a wavelength and energy strictly defined for the element. By recording the spectrum, the qualitative elemental composition of the sample is determined. By measuring the intensity of radiation of different wavelengths or energies, a conclusion can be made about the quantitative content of each element.

The EDX-7000 works based on a thermoelectrically cooled silicon drift detector with a chromium detection limit of 1 ppm. Quantitative sample analysis was performed by the calibration curve method. For this purpose, serially diluted standard samples were analyzed, and a fluorescence intensity vs. chromium content curve was plotted, which was used to quantify the element in unknown samples.

To determine the content of trivalent chromium in the solution, we used the gravimetric method, which is based on precipitation of trivalent chromium present in the liquid phase with a concentrated solution of ammonia [52].

Determination of the chemical composition of the solid phase of catalyst sludge was carried out using X-ray fluorescence analysis. Determination of morphology of solid phase of catalyst sludge was carried out using scanning electron microscopy and using microanalysis on a Tescan Vega 3 device with enlarging voltage of 20 kV. X-ray structural analysis of the solid phase using powder X-ray diffraction on Bruker D2 PHASER and infrared spectrometry (ALPHA II, Bruker) was also carried out.

Free chromates on the surface of the liquid phase of the catalyst were determined by boiling the solid phase of the catalyst in distilled water for 15, 30, and 60 min, and then the resulting solution was analyzed by X-ray fluorescence.

4.3. Recovering the Liquid Phase of the Slurry

The initial ratio of the reducing agent to the liquid phase was chosen based on analysis of the literature data [53,55], and then our own experiments were conducted with a decrease and increase in the dosage of the reducing agent. All reduction reactions were performed in a closed reactor at a stirring speed of 300 rpm. The weight of the samples was 100 g. The reduction process was carried out in the temperature range of 50–85 °C for 15–60 min and at room temperature with an interaction time of up to several days. During the reactions, the solution was discolored and the reduced chromium precipitated as $Cr(OH)_3$ hydroxide, which was consistent with the literature data of this study. The dynamics of pH change before and after the reaction was carried out using a pH meter.

4.4. Phase Analysis after Recovery

The cooled solution was filtered under vacuum at 100 kPa. The liquid phase was analyzed for residual hexavalent chromium content using X-ray fluorescence spectroscopy on a Shimadzu device. The chemical composition of the liquid phase was studied on a Shimadzu spectrometer, and the morphology of the solid phase was studied on a Tescan Vega 3 electron microscope.

4.5. Conducting Pulp Recovery with the Selected Reagent

The pulp was reconstituted in a closed reactor at 300 rpm agitation. The mass of the samples was 500 g. The process was carried out in the temperature range of 50–85 °C. A pulse of 5 mL was taken at regular intervals and filtered. The chemical composition of the liquid phase was studied on a Shimadzu EDX-7000 spectrometer, and the morphology of the solid phase was studied on a Tescan Vega 3 electron microscope.

5. Conclusions

According to the results of this study, we can conclude that the reduction of hexavalent chromium to trivalent chromium is possible in a neutral environment, rather than in a strongly acidic environment, as suggested in previously published works. Of all the studied reagents, sodium metabisulfite showed the most effective reduction of chromium, and it can be used to precipitate chromium in the form of hydroxide, both from the liquid phase and as part of the pulp of the catalyst slurry.

The advantage of the hexavalent chromium reduction process in a neutral medium is the formation of a precipitate without the use of concentrated sulfuric acid and additional precipitant reagents. Moreover, this method allows for the refusal of equipment designed for aggressive environments, which significantly reduces the economic costs.

Author Contributions: Conceptualization, V.S.; methodology, I.P.; formal analysis, writing—reviewing and editing, O.Z.; selection of methods and technological parameters of experiments, visualization, writing—original draft preparation, M.Z. All authors have read and agreed to the published version of the manuscript.

Funding: This research received no external funding.

Data Availability Statement: The data presented in this study are available on request from the corresponding author. The data falls within the scope of intellectual property: patent of the Russian Federation 2796659.

Acknowledgments: The authors are grateful to V.G. Povarov, the leading researcher of the Scientific Center "Ecosystem" of St. Petersburg Mining University, for his assistance in conducting the research to determine the chemical composition of the samples obtained by the XRF method.

Conflicts of Interest: The authors declare no conflict of interest.

References

1. Bekmukhamedov, G.E. Modified Silicon Dioxide Aluminochrome Catalyst for Isobutane Dehydrogenation. Kazan (Volga Region) Federal University, Kazan, Russia, 2015.
2. Kudinova, A.A.; Poltoratckaya, M.E.; Gabdulkhakov, R.R.; Litvinova, T.E.; Rudko, V.A. Parameters influence establishment of the petroleum coke genesis on the structure and properties of a highly porous carbon material obtained by activation of KOH. *J. Porous Mater.* **2022**, *29*, 1599–1616. [CrossRef]
3. Nasifullina, A.I.; Starkov, M.K.; Gabdulkhakov, R.R.; Rudko, V.A. Petroleum coking additive-raw material component for metallurgical coke production. Part 2. Experimental studies of obtaining a petroleum coking additive. *CIS Iron Steel Rev.* **2022**, *24*, 9–16. [CrossRef]
4. Pakhomov, N.A.; Kashkin, V.N.; Molchanov, V.V.; Noskov, A.S. Dehydrogenation of C2-C4 paraffins on Cr_2O_3/Al_2O_3 catalysts. *Gas Chem.* **2008**, *154*, 66–69.
5. Bekmukhamedov, G.E.; Morozov, V.I.; Tuktarov, R.R.; Bukharov, M.S.; Egorova, S.R.; Lamberov, A.A.; Yakhvarov, D.G. Electronic interaction between Cr(III) ions in chromia-alumina catalysts for light alkane dehydrogenation. *J. Phys. Chem. Solids* **2022**, *167*, 110778. [CrossRef]
6. Krylov, O.V. *Heterogeneous Catalysis: Textbook for Universities*; Academkniga: Moscow, Russia, 2004; p. 679.
7. Konoplin, R.; Kondrasheva, N. Difficulties in the industrial introduction of new effective hydrodesulfurization catalysts in the Russian Federation. *E3S Web Conf.* **2021**, *266*, 02016. [CrossRef]
8. Dhal, B.; Thatoi, H.N.; Das, N.N.; Pandeya, B.D. Chemical and microbial remediation of hexavalent chromium from contaminated soil and mining/metallurgical solid waste: A review. *J. Hazard. Mater.* **2013**, *250–251*, 272–291. [CrossRef]
9. Kamaludeen, S.P.B.; Megharaj, M.; Juhasz, A.L.; Sethunathan, N.; Naidu, R. Chromium-Microorganism Interactions in Soils: Remediation Implications. *Rev. Environ. Contam. Toxicol.* **2003**, *178*, 93–164. [CrossRef]
10. Chernobrovin, V.P. State of the chrome industry in Russia. *Bull. SUSU Metall. Ser.* **2017**, *17*, 44–48.
11. Chrysochoou, M.; Johnston, C.P. Reduction of chromium (VI) in saturated zone sediments by calcium polysulfide and nanoscale zerovalent iron derived from green tea extract. *State Art Pract. Geotech. Eng.* **2012**, *2012*, 3959–3967.
12. Lebedev, A.B.; Shuiskaya, V.S. Influence of composition and cooling rate of alumocalcium slag on its crumblability. *Izv. Ferr. Met.* **2022**, *65*, 806–813. [CrossRef]
13. Lurie, Y.Y. *Chemical Analysis of Industrial Wastewater*; Chemistry: Moscow, Russia, 1974; 448p.
14. Lebedev, A.B.; Musinova, P.V. Formation of the strength of pelletized multiphase dicalcium silicate sinter. *Chernye Met.* **2022**, 40–46. [CrossRef]
15. Shrivastava, R.; Upreti, R.K.; Chaturvedi, U.C. Various cells of the immune system and intestine differ in their capacity to reduce hexavalent chromium. *FEMS Immunol. Med. Microbiol.* **2003**, *38*, 65–70. [CrossRef] [PubMed]

16. Yemane, M.; Chandravanshi, B.S.; Wondimu, T. Levels of essential and nonessential metals in leaves of the tea plant (*Camellia sinensis* L.) and soil of Wushwush farms, Ethiopia. *Food Chem.* **2008**, *107*, 1236–1243.

17. Pécou, E.; Maass, A.; Remenik, D.; Briche, J.; Gonzalez, M. A mathematical model for copper homeostasis in Enterococcus hirae. *Math. Biosci.* **2006**, *203*, 222–239. [CrossRef]

18. Petrov, G.V.; Schneerson, J.M.; Andreev, Y.V. Extraction of platinum metals during processing of chromite ores from dunite massifs. *Notes Min. Inst.* **2018**, *231*, 281–286.

19. Petrov, G.V.; Graver, T.N.; Tikhonov, O.N.; Belenkiy, A.M.; Boduen, A.Y. Study of Russian chromite platinum-metal massifs. *Zap. Mines Inst.* **2006**, *169*, 173–177.

20. Ying, Z.; Ren, X.; Li, J.; Wu, G.; Wei, Q. Recovery of chromium(VI) in wastewater using solvent extraction with amide. *Hydrometallurgy* **2020**, *196*, 105440. [CrossRef]

21. Dai, L.; Ding, J.; Liu, Y.; Wu, X.; Chen, L.; Chen, R.; Van der Bruggen, B. Recovery of Cr(VI) and removal of cationic metals from chromium slag using a modified bipolar membrane system. *J. Membr. Sci.* **2021**, *639*, 119772. [CrossRef]

22. Kumar, A.; Thakur, A.; Panesar, P. Extraction of hexavalent chromium by environmentally benign green emulsion liquid membrane using tridodecyamine as an extractant. *J. Ind. Eng. Chem.* **2019**, *70*, 394–401. [CrossRef]

23. Liu, Y.; Zhu, H.; Zhang, M.; Chen, R.; Chen, X.; Zheng, X.; Jin, Y. Cr(VI) recovery from chromite ore processing residual using an enhanced electrokinetic process by bipolar membranes. *J. Membr. Sci.* **2018**, *566*, 190–196. [CrossRef]

24. Wu, X.; Zhu, H.; Liu, Y.; Chen, R.; Qian, Q.; Van der Bruggen, B. Cr(III) recovery in form of Na_2CrO_4 from aqueous solution using improved bipolar membrane electrodialysis. *J. Membr. Sci.* **2020**, *604*, 118097. [CrossRef]

25. Ye, Z.; Yin, X.; Chen, L.; He, X.; Lin, Z.; Liu, C.; Ning, S.; Wang, X.; Wei, Y. An integrated process for removal and recovery of Cr(VI) from electroplating wastewater by ion exchange and reduction–precipitation based on a silica-supported pyridine resin. *J. Clean. Prod.* **2019**, *236*, 117631. [CrossRef]

26. Lu, J.; Liu, Z.; Wu, Z.; Liu, W.; Yang, C. Synergistic effects of binary surfactant mixtures in the removal of Cr(VI) from its aqueous solution by foam fractionation. *Sep. Purif. Technol.* **2020**, *237*, 116346. [CrossRef]

27. Fan, Z.; Zhang, Q.; Gao, B.; Li, M.; Liu, C.; Qiu, Y. Removal of hexavalent chromium by biochar supported nZVI composite: Batch and fixed-bed column evaluations, mechanisms, and secondary contamination prevention. *Chemosphere* **2019**, *217*, 85–94. [CrossRef] [PubMed]

28. Qu, J.; Wang, Y.; Tian, X.; Jiang, Z.; Deng, F.; Tao, Y.; Jiang, Q.; Wang, L.; Zhang, Y. KOH-activated porous biochar with high specific surface area for adsorptive removal of chromium (VI) and naphthalene from water: Affecting factors, mechanisms and reusability exploration. *J. Hazard. Mater.* **2020**, *401*, 123292. [CrossRef] [PubMed]

29. Janik, P.; Zawisza, B.; Talik, E.; Sitko, R. Selective adsorption and determination of hexavalent chromium ions using graphene oxide modified with amino silanes. *Microchim. Acta* **2018**, *185*, 1–8. [CrossRef]

30. Kim, C.; An, S.; Lee, J.; Ghosh, A.; Zhong, M.; Fortner, J.D. Photoactive Polyethylenimine-Coated Graphene Oxide Composites for Enhanced Cr(VI) Reduction and Recovery. *ACS Appl. Mater. Interfaces* **2021**, *13*, 28027–28035. [CrossRef]

31. Zhang, X.; Yang, Z.; Mei, J.; Hu, Q.; Chang, S.; Hong, Q.; Yang, S. Outstanding performance of sulfurated titanomaghemite (Fe_2TiO_5) for hexavalent chromium removal: Sulfuration promotion mechanism and its application in chromium resource recovery. *Chemosphere* **2022**, *287*, 132360. [CrossRef] [PubMed]

32. Nasrollahzadeh, M.; Bidgoli, N.S.S.; Issaabadi, Z.; Ghavamifar, Z.; Baran, T.; Luque, R. *Hibiscus rosasinensis* L. aqueous extract-assisted valorization of lignin: Preparation of magnetically reusable Pd NPs@Fe_3O_4-lignin for Cr(VI) reduction and Suzuki-Miyaura reaction in eco-friendly media. *Int. J. Biol. Macromol.* **2020**, *148*, 265–275. [CrossRef]

33. Fang, W.; Jiang, X.; Luo, H.; Geng, J. Synthesis of graphene/SiO_2@polypyrrole nanocomposites and their application for Cr(VI) removal in aqueous solution. *Chemosphere* **2018**, *197*, 594–602. [CrossRef]

34. Wang, T.; Liu, Y.; Wang, J.; Wang, X.; Liu, B.; Wang, Y. In-situ remediation of hexavalent chromium contaminated groundwater and saturated soil using stabilized iron sulfide nanoparticles. *J. Environ. Manag.* **2019**, *231*, 679–686. [CrossRef] [PubMed]

35. Aigbe, U.O.; Osibote, O.A. A review of hexavalent chromium removal from aqueous solutions by sorption technique using nanomaterials. *J. Environ. Chem. Eng.* **2020**, *8*, 104503. [CrossRef]

36. Shao, Q.; Xu, C.; Wang, Y.; Huang, S.; Zhang, B.; Huang, L.; Fan, D.; Tratnyek, P.G. Dynamic interactions between sulfidated zerovalent iron and dissolved oxygen: Mechanistic insights for enhanced chromate removal. *Water Res.* **2018**, *135*, 322–330. [CrossRef] [PubMed]

37. Azeez, N.A.; Dash, S.S.; Gummadi, S.N.; Deepa, V.S. Nano-remediation of toxic heavy metal contamination: Hexavalent chromium [Cr(VI)]. *Chemosphere* **2021**, *266*, 129204. [CrossRef] [PubMed]

38. Chen, X.; Fan, G.; Zhu, X.; Li, H.; Li, Y.; Li, H.; Xu, X. The remediation of hexavalent chromium-contaminated soil by nanoscale zero-valent iron supported on sludge-based biochar. *J. Soils Sediments* **2023**, *23*, 1607–1616. [CrossRef]

39. Golovina, V.V.; Eremina, A.O.; Sobolev, A.A.; Chesnokov, N.V. Extraction of chromium from aqueous solutions by porous materials based on logging waste of local wood raw materials (bark and wood chips). *J. Sib. Fed. Univ. Ser. Chem.* **2017**, *10*, 186–205. [CrossRef]

40. Bagauva, A.I.; Stepanova, S.V.; Shaikhiev, I.G. Study of extracts from wood waste (sawdust oak bark) to remove chromium (VI) ions from the model solutions. *Bull. Kazan Technol. Univ.* **2011**, *14*, 74–79.

41. Cherdchoo, W.; Nithettham, S.; Charoenpanich, J. Removal of Cr(VI) from synthetic wastewater by adsorption onto coffee ground and mixed waste tea. *Chemosphere* **2019**, *221*, 758–767. [CrossRef]

42. Vilardi, G.; Ochando-Pulido, J.M.; Verdone, N.; Stoller, M.; Di Palma, L. On the removal of hexavalent chromium by olive stones coated by iron-based nanoparticles: Equilibrium study and chromium recovery. *J. Clean. Prod.* **2018**, *190*, 200–210. [CrossRef]
43. Zorkina, O.V. Methods for converting hexavalent chromium to trivalent form by organic reducing agent. *Proc. Penza State Pedagog. Univ. Named V.G. Belinsky* **2011**, *25*, 690–696.
44. Singh, P.; Itankar, N.; Patil, Y. Biomanagement of hexavalent chromium: Current trends and promising perspectives. *J. Environ. Manag.* **2020**, *279*, 111547. [CrossRef]
45. Yogeshwaran, V.; Priya, A.K. Removal of Hexavalent Chromium (Cr(VI)) Using Different Natural Adsorbents—A Review. *J. Chromatogr. Sep. Tech.* **2017**, *8*, 2–6.
46. Petrov, G.V.; Kalashnikova, M.I.; Fokina, S.B. Laws of behavior of selenium and chromium in oxidation-reduction processes during hydrometallurgical processing of solid-phase products of rhenium extraction. *Proc. Min. Univ.* **2016**, *220*, 601–606.
47. Tsybulskaya, O.N.; Ksenik, T.V.; Kisel, A.A.; Yudakov, A.A.; Perfiliev, A.V.; Chirikov, A.Y. Decontamination of chromium-containing electroplating waste. *New Technol.* **2015**, *4*, 104–112.
48. Karataev, O.R.; Kudryavtseva, E.S.; Mingazetdinov, I.H. Wastewater treatment from hexavalent chromium ions. *Bull. Kazan Technol. Univ.* **2014**, *2*, 52–54.
49. Klimova, O.V. Processes and Apparatus Design of Wastewater Treatment from Chromium (VI) Ions by Carbon Adsorbents. Ph.D. Thesis, Irkutsk National Research Technical University, Irkutsk, Russia, 2015.
50. Cespón-Romero, R.; Yebra-Biurrun, M.; Bermejo-Barrera, M. Preconcentration and speciation of chromium by the determination of total chromium and chromium(III) in natural waters by flame atomic absorption spectrometry with a chelating ion-exchange flow injection system. *Anal. Chim. Acta* **1996**, *327*, 37–45. [CrossRef]
51. Torkmahalleh, M.A.; Lin, L.; Holsen, T.M.; Rasmussen, D.H.; Hopke, P.K. The Impact of Deliquescence and pH on Cr Speciation in Ambient PM Samples. *Aerosol Sci. Technol.* **2021**, *46*, 690–696. [CrossRef]
52. Lee, T.J.; Kim, H.J. Interfering elements on determination of hexavalent chromium in papermaterials with UV-vis spectrophotometry. *Nord. Pulp Pap. Res. J.* **2021**, *37*, 130–137. [CrossRef]
53. Korostelev, P.P. Titrimetric and Gravimetric Analysis in Metallurgy: Handbook. 1985. Available online: https://www.chem.msu.ru/rus/teaching/analyt/korostelev/all.pdf (accessed on 28 March 2023).
54. Elanieva, S.I. Physico-chemical methods of reducing the aggressiveness of spent electrolytes by transferring CR (VI) into CR (III) Proceedings of PSPU. *Sect. Young Sci.* **2008**, *6*, 174–178.
55. Barrera-Díaz, C.E.; Lugo-Lugo, V.; Bilyeu, B. A review of chemical, electrochemical and biological methods for aqueous Cr(VI) reduction. *J. Hazard. Mater.* **2012**, *223–224*, 1–12. [CrossRef]

Article

Hydrogen Incorporation in $Ru_xTi_{1-x}O_2$ Mixed Oxides Promotes Total Oxidation of Propane

Wei Wang [1,2], Yu Wang [1,2], Phillip Timmer [2], Alexander Spriewald-Luciano [2], Tim Weber [2], Lorena Glatthaar [2], Yun Guo [1,*], Bernd M. Smarsly [2,*] and Herbert Over [2,*]

[1] Key Laboratory for Advanced Materials, Research Institute of Industrial Catalysis, School of Chemistry and Molecular Engineering, East China University of Science and Technology, Shanghai 200237, China

[2] Institute of Physical Chemistry, Justus Liebig University, Heinrich-Buff-Ring 17, 35392 Giessen, Germany

* Correspondence: yunguo@ecust.edu.cn (Y.G.); bernd.smarsly@phys.chemie.uni-giessen.de (B.M.S.); herbert.over@phys.chemie.uni-giessen.de (H.O.)

Abstract: A rational synthetic approach is introduced to enable hydrogen insertion into oxides by forming a solid solution of a reducible oxide with a less reducible oxide as exemplified with RuO_2 and TiO_2 (Ru_x, a mixture of x% RuO_2 with $(100-x)$% TiO_2). Hydrogen exposure at 250 °C to Ru_x (Ru_x_250R) results in substantial hydrogen incorporation accompanied by lattice strain that in turn induces pronounced activity variations. Here, we demonstrate that hydrogen incorporation in mixed oxides promotes the oxidation catalysis of propane combustion with Ru_60_250R being the catalytically most active catalyst.

Keywords: catalyst promotor; mixed oxides; hydrogenated $Ru_xTi_{1-x}O_2$; propane combustion; hydrogen-induced variation in the activity

Citation: Wang, W.; Wang, Y.; Timmer, P.; Spriewald-Luciano, A.; Weber, T.; Glatthaar, L.; Guo, Y.; Smarsly, B.M.; Over, H. Hydrogen Incorporation in $Ru_xTi_{1-x}O_2$ Mixed Oxides Promotes Total Oxidation of Propane. *Inorganics* 2023, *11*, 330. https://doi.org/10.3390/inorganics11080330

Academic Editors: Torben R. Jensen, Roberto Nisticò, Luciano Carlos, Hicham Idriss and Eleonora Aneggi

Received: 14 June 2023
Revised: 31 July 2023
Accepted: 4 August 2023
Published: 7 August 2023

1. Introduction

Strain-induced changes of the catalytic activity of transition metal compounds in thermal catalysis were predicted by theory [1] and attributed to a shift of the metal d band center with strain [2], so that strain engineering has been considered a promising way to tune activity. Strain can be introduced to the catalyst's material in various ways including epitaxial film growth [3], doping such as Li insertion [4], alloying–dealloying [5], formation of core-shell particles [6] and nano-structuring [7]. In fact, strain engineering has turned out to be an important tool to improve activity in electrocatalysis, most notably for water electrolysis [8–12]. Quite in contrast, in thermal catalysis strain engineering is less often encountered [6,7,13,14], due presumably to missing stability of strain-engineered materials at higher reaction temperatures.

Hydrogenation of the catalyst material might be a convenient way to introduce strain into the host lattice and thereby tune the activity of a catalyst as long as the catalyst material is able to incorporate a sufficient amount of hydrogen into the lattice. For instance, Pd-based catalysts [15,16] were reported to form both absorbed and adsorbed hydrogen that may play an important role in hydrogenation catalysis. Hydrogen interaction with oxides is more intricate than with metals [17] since reducible oxides can face stability problems due to partial or even total reduction up to the metal phase. Exposing hydrogen to oxide surfaces frequently forms surface hydroxyl groups, but it can also incorporate hydrogen into the bulk oxide. Hydrogen exposure to CeO_2 was reported to form of hydride species H^- in bulk CeO_2 [18–22], a process that is facilitated by oxygen vacancies. The formation of hydride species H^- in CeO_2 may explain its remarkable catalytic performance in the partial hydrogenation of alkynes to alkenes [23–25].

Recently, we reported that H_2 exposure at 250 °C to mixed $Ru_{0.3}Ti_{0.7}O_2$ is able to incorporate about 20 mol% hydrogen into the rutile lattice, thereby altering slightly the lattice parameters [26]. This was considered a remarkable finding since RuO_2 is not stable under such conditions and transforms readily to metallic Ru, while TiO_2 is not able to incorporate hydrogen into the lattice, at least not at 250 °C. Hydrogenation of $Ru_{0.3}Ti_{0.7}O_2$ was shown to increase substantially the catalytic oxidation activity in the total oxidation of propane and HCl oxidation reaction.

In the present study, we systematically vary the composition x of the mixed oxide $Ru_xTi_{1-x}O_2$ (Ru_x). For various compositions ranging from $x = 0.2$ to $x = 1.0$ in steps of 0.1, we explore the hydrogenation behavior at 250 °C and compare the catalytic activity of $Ru_xTi_{1-x}O_2$ with that of the corresponding hydrogenated catalysts in the total oxidation of propane. Without hydrogen treatment, Ru_x reveals a strict composition−activity correlation of activity in that the higher the Ru concentration the higher the activity; highest activity is achieved with Ru_100. However, when treating Ru_x with hydrogen at 250 °C for 3 h (Ru_x_250R), the highest activity is encountered for compositions where both Ru and Ti have similar concentrations. Ru_60_250R turns out to be the most active propane combustion catalyst exceeding even the activity of Ru_100. The hydrogen-induced activity variation is tentatively attributed to hydrogen-induced strain in the mixed oxide $Ru_xTi_{1-x}O_2$.

2. Experimental Results

2.1. Characterization of the Fresh Ruthenium−Titanium Mixed Oxide Samples

Figure 1a summarizes the X-ray diffraction (XRD) patterns of freshly prepared ruthenium−titanium mixed oxide catalysts Ru_x with different nominal Ru concentrations, x; the diffraction pattern of pure commercial rutile-TiO_2 is overlaid for comparison reasons. The XRD pattern of Ru_100 contains reflections from both a rutile structure and metallic Ru (hcp structure). With the addition of titanium to RuO_2, the rutile structure is preserved, and the (110) and the (101) reflections continuously shift towards the reflection of pure rutile TiO_2. Above a Ti concentration of 20 mol% no reflections from metallic Ru are discernible. In addition, the rutile related diffraction peaks split into two components with increasing Ti concentration.

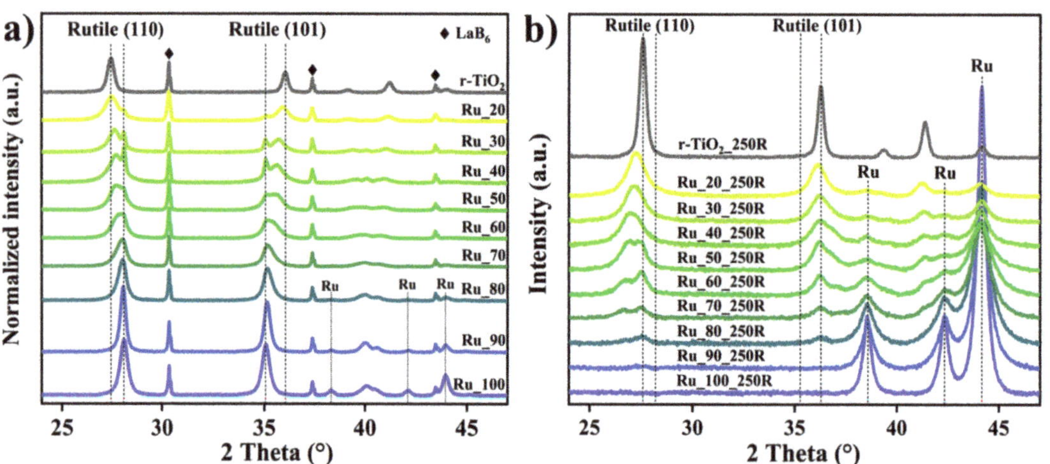

Figure 1. (**a**) X-ray diffraction (XRD) patterns of the ruthenium−titanium mixed oxide catalysts Ru_x; the composition x of Ru ranges from 20 mol% to 100 mol%. Dashed lines indicate the position of pure rutile-TiO_2 and RuO_2. (**b**) XRD patterns of Ru_x_250R catalysts treated in 4 vol% H_2/N_2 at 250 °C for various compositions x ranging from 20% to 100%.

The sharp reflections at 28.02° from rutile (110) and 35.07° from rutile (101) do not vary with the Ti concentration and therefore are assigned to the pure RuO_2 phase. The position of the broader component in XRD shifts continuously towards rutile TiO_2 and is hence ascribed to solid solution $Ru_xTi_{1-x}O_2$. The coexistence of pure RuO_2 and mixed $Ru_xTi_{1-x}O_2$ in a wide range of compositions evidences a miscibility gap consistent with previous findings based on a similar polymer-assisted preparation method [27] and in agreement with a recent DFT study [28].

In Figure S1, we present the peak deconvolution of the rutile (110) reflection and that of the (101) reflection. For high Ru-content samples like Ru_90, Ru_80 and Ru_70, the (110)/(101) peaks are found to be asymmetric, which clearly points toward a phase separation. For low Ru-content samples this phase separation is pronounced, with two separated reflections corresponding to the RuO_2 phase and $Ru_xTi_{1-x}O_2$ oxide phase, respectively. Therefore, we assume that a pure RuO_2 phase exists in the full composition range of Ru–Ti mixed oxides. In the decomposition of the diffraction patterns, we fix the peak position of pure RuO_2 and assume that the mixed $Ru_xTi_{1-x}O_2$ oxide phase crystallizes in the rutile structure. As a main result of the decomposition, the peak position of Ru–Ti solid solution turns out to linearly shift to lower angles with the increasing Ti concentration (cf. Figure S2), in accordance with Vegard's law [29]. The deconvolution analysis in the present study emphasizes that the prepared samples are not phase pure but facing a miscibility gap.

Together with the analysis of the rutile (101) reflection, we can derive the unit cell parameters a/b and c of the mixed oxide $Ru_xTi_{1-x}O_2$ as a function of the composition x that are summarized in Figure 2. The linear shift in a/b and c with the nominal composition indicates the fulfillment of Vegard's law, thus corroborating the formation of a solid solution $Ru_xTi_{1-x}O_2$ with nominal composition x. Besides, the calculated unit cell volumes for the Ru_x samples (cf. Figure S3) do not vary significantly with the nominal composition x. Note that phase separation becomes more severe for the Ru_20 sample with its solid solution phase starting to deviate from Vegard's law.

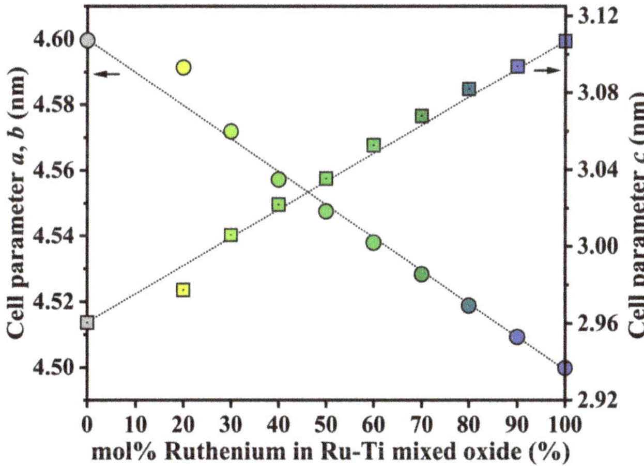

Figure 2. Lattice parameters of rutile $Ti_{1-x}Ru_xO_2$ solid solution as a function of the nominal composition x given in mol%, as derived from the peak deconvolution of rutile (110) and rutile (101).

As seen from Figure S1, the FWHM of the RuO_2 peak decreases, while that of the mixed oxide increases with Ti concentration. Utilizing the Scherrer equation, this observation translates to an increase of the crystallite size of RuO_2, while that of the mixed oxide decreases with increasing Ti concentration (until 50% Ti) (cf. Table 1). In order to consider also the micro-strain in the calculation of the crystallite size, we use the Williamson–Hall method [30], whose results are summarized in Figure S4 and Table S1; the Williamson–Hall plots of RuO_2 and the Ru–Ti solid solution phase are exemplified for the Ru_60 sample shown in Figure S5. Since the micro-strain values $\Delta\varepsilon$ ($\leq$0.004) of the powder materials are relatively low compared to those described in the literature [31], the values derived from the Scherrer equation are practically not affected by the micro-strain.

Table 1. XRD derived data of Ru–Ti mixed oxide catalysts.

Catalysts	Cell Parameter a/b (nm) [a]	Cell Parameter c (nm) [b]	Grain Size (RuO$_2$) (nm) [c]	Grain Size (Ru–Ti) (nm) [c]	Ru–Ti/(RuO$_2$ + Ru–Ti) (%) [d]
Ru_100	4.500	3.107	18 ± 0.5	-	0
Ru_90	4.509	3.094	27 ± 1	25 ± 2	81.2
Ru_80	4.519	3.082	32 ± 0.5	15 ± 0.3	89.9
Ru_70	4.528	3.068	33 ± 1	12 ± 0.2	90.9
Ru_60	4.538	3.053	36 ± 1	10 ± 0.5	93.7
Ru_50	4.548	3.035	46 ± 8	9 ± 0.5	96.4
Ru_40	4.558	3.022	45 ± 6	10 ± 0.5	88.8
Ru_30	4.572	3.006	43 ± 5	12 ± 2	81.9
Ru_20	4.591	2.977	28 ± 6	16 ± 1	92.5

a: Calculated by rutile (110) reflection of the Ru–Ti solid solution phase. *b*: Calculated based on obtained a/b value and rutile (101)/(101) reflections of the Ru–Ti solid solution phase. *c*: Determined by Scherrer equation from the (110)/(101) reflections of the RuO$_2$ phase after peak deconvolution. *d*: Determined by peak deconvolution of the (110)/(101) reflections.

From XRD we gain the following structural information of the Ru–Ti mixed oxides: RuO$_2$ nanoparticles and Ru–Ti solid solutions co-exist throughout the entire composition range, while metallic ruthenium is eliminated when more than 20 mol% titanium is incorporated. The Ru–Ti solid solutions fulfill Vegard's law, thus evidencing that ruthenium and titanium form a solid solution despite phase separation.

While XRD is a bulk characterization method, X-ray photoelectron spectroscopy (XPS) can determine the surface/near-surface compositions of Ru_x catalysts with varying nominal x values (as presented in Figure S6, fitting parameters are compiled in Table S2). From the Ru 3d XP spectra, Ti-rich samples (Ru_40, Ru_30 and Ru_20) show a weak peak at ~288.0 eV whose origin may be attributed to the formed carbonates (O=C=O) from exposure to air or from residual carbon of preparation. It is evident that ruthenium is in the Ru^{4+} oxidation state in each sample Ru_x; this assignment is corroborated by the pronounced satellite features [32]. No metallic ruthenium is detected on the catalyst surface, proving that the metallic ruthenium species (see in Figure 1a) is encapsulated by the oxide. Buried metallic ruthenium is not expected to participate in the catalytic reaction.

Additionally, the binding energies of Ru3d$_{5/2}$ and Ru3d$_{3/2}$ do not vary with the composition, while, surprisingly, the Ru satellite features monotonically shift to lower binding energies with lower ruthenium concentration (see in Figure 3), from 282.69 eV for Ru_100 to 282.34 eV for Ru_30. As the satellite feature is attributed to surface plasmon excitation [33], this shift is correlated to a reduced valence electron density in Ru$_x$Ti$_{1-x}$O$_2$.

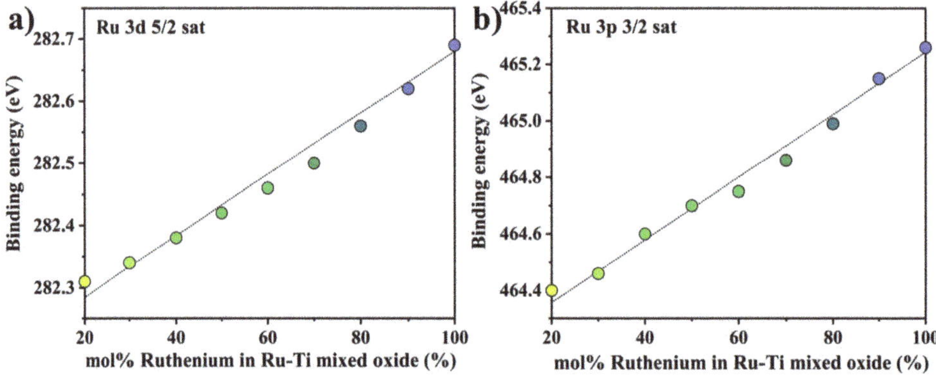

Figure 3. Shift of binding energies of (**a**) Ru 3d satellite features. (**b**) Ru 3p satellite features derived from peak deconvolution of XPS data when varying the composition *x* of Ru_x.

The O 1s spectrum in Figure S7 depicts two chemical states of near surface oxygen: one is the O^{2-} from the lattice oxygen at ~529 eV, and the other, a shoulder peak at higher binding energy (around 531.8 eV), which is ascribed to the surface OH groups/oxygenated carbon [34]. The integral ratio of OH to the entire surface oxygen is continuously decreasing on lowering the ruthenium concentration, from 39.30% (for Ru_100) to 2.50% (for Ru_50), thus indicating that the incorporation of titanium reduces greatly the amount of surface OH on the catalyst surface.

Since the Ru 3p and Ti 2p XPS features are in the same binding energy region, the surface composition of ruthenium and titanium can be quantified by peak deconvolution. As presented in Table S2 and Figure S8, the Ru 3p region also exhibits satellite features, which behave identically to those in the Ru 3d region; the shifts to lower binding energy further evidence that doping of titanium leads to lower electron density. As compiled in Table 2, the actual surface concentrations deviate quite substantially above a nominal Ru concentration of 60%, while below 50 mol% Ru the surface composition agrees well with the nominal composition.

Table 2. Compositional information from the surface and bulk region of ruthenium−titanium mixed oxide catalysts.

Catalysts	S_{BET} (m²/g)	Ru/(Ti + Ru)(mol%) [a]	Ru/(Ti + Ru)(mol%) [b]	(OH + Oxygenated Carbon)/O 1s Total [b]
Ru_100	9	100	100	39.90
Ru_90	6	89.3	70.65	14.67
Ru_80	10	78.8	59.81	5.42
Ru_70	14	68.1	60.53	4.85
Ru_60	19	56.7	54.4	4.42
Ru_50	26	46.8	47.74	2.50
Ru_40	32	40.9	37.96	4.69
Ru_30	31	30.0	29.08	3.38
Ru_20	34	17.4	17.61	4.11
Ru_60_250R	19	56.2	52.12	4.43

a: Calculated from SEM-EDS mapping. *b*: Calculated from XPS.

From these XPS results one may ask whether this deviation of the surface composition from the nominal ones originates from an insufficient control of the composition during synthesis. In order to settle this question, we applied dispersive X-ray spectroscopy (EDS)-scanning electron microscopy (SEM) to gain compositional information of all samples. Since the detection depth from EDS reaches several micrometers, we can consider it as a bulk characterization method. The average composition quantified by EDS summarized in Table 2 indicates that, in the full range ($20 \leq x \leq 100$), our synthesized catalysts agree well with the nominal compositions x, thus excluding the possibilities that the deviation of surface concentration is caused by uncertainties in the synthesis procedure.

Besides, with SEM and TEM we also investigate the morphologies of our catalysts. The Ru_100, SEM images in Figure S9 reveal a dense and rough surface with additional macropores, round small particles agglomerate like a sponge. The particle size gradually becomes smaller with increasing titanium concentration until $1 - x = 50\%$. This is in agreement with the crystallite size derived from XRD data, indicating that a proper amount of titanium will efficiently decrease the crystallite size of Ru–Ti mixed oxides. Moreover, TEM micrographs confirm that the particle size of mixed oxides decreases with the addition of titanium. Table 2 summarizes also the Krypton physisorption experiments of all as-prepared samples. The Brunauer−Emmett−Teller (BET) surface area varies quite substantially, changing from 9 m²/g (for Ru_100) to 34 m²/g (for Ru_20).

2.2. Hydrogen-Induced Changes of Ruthenium−Titanium Mixed Oxides

We focus on the hydrogen-induced change of the Ru_x samples. The hydrogen treatment consists of an exposure of 4% H_2/N_2 for 3 h at 250 °C. The specific reduction temperature of 250 °C has been shown to result in a large hydrogen uptake for the case of Ru_30 [26]. As shown in Figure 1b, Ru_00, e.g., pure rutile-TiO$_2$, is not affected by this hydrogen treatment, while Ru_100 is fully transformed to metallic Ru phase, consistent with our previous study [26].

The diffraction peak of RuO$_2$ at 2θ = 28.02° disappears in Figure 1b when exposed to 4% H$_2$/N$_2$ at 250 °C, while the intensity of Ru metal-related diffraction increases. As expected, the metallic Ru diffraction intensity is higher the higher is the Ru content x of Ru_x. This means that the RuO$_2$ particles are easily reduced to metallic ruthenium at a temperature of 250 °C, consistent with previously published studies [35,36].

Quite in contrast, the rutile diffraction peaks of the mixed oxide phases (110) and (101) persist and shift only in position to the lower and higher angles, respectively, upon exposure to 4% H$_2$/N$_2$ at 250 °C. The degree of peak shift depends on the Ru concentration. Close to 50 mol% Ru, the observed shift is the highest. Actually, the shifted diffraction peak of rutile consists of two components, most likely evidencing different hydrogen concentrations in the mixed oxide crystallites. The bulk composition of Ru_x (cf. Table 2) does not change when exposed to H$_2$ at 250 °C (Ru_x_250R). As discussed recently [26], these shifts of the reflections are caused by the incorporation of hydrogen and not by reduction of the metal ions.

The Ru_x_250R samples are subsequently subject to a mild re-oxidation treatment that is conducted at 300 °C and the XRD patterns are presented in Figure S10a, and the changes of macrostrain (position) and micro-strain (FWHM) among the initial, reduced and re-oxidized samples as exemplified by Ru_20, Ru_40, Ru_60 and Ru_80 are compiled in Figure S10b. The rutile structure of Ru_x is restored after re-oxidation treatment, while most but not all of the metallic Ru transforms back to RuO$_2$. For the case of Ru_30_250R, it was shown that full recovery of Ru_30 requires oxidation temperatures of 400 °C [26].

With thermogravimetric-mass spectrometry (TG-MS) the amount of incorporated hydrogen can be quantified by the integrated water signal that is produced by reacting incorporated hydrogen with oxygen during heating of the sample in ambient air. We exemplify this experiment with Ru_60_250R since the catalytic activity of this sample is thoroughly tested. A nitrogen-treated Ru_60_250N sample serves as reference. As summarized in Figure 4, a small mass signal for H$_2$O (m/z = 18) is evident at 80 °C for Ru_60_250N, while, for Ru_60_250R, a broad and strong water signal appears. This water peak of Ru_60_250R contains actually two components, one is associated with water desorption (90 °C) and the other is related to the oxidation of incorporated hydrogen (maximum at 160 °C). Employing a deconvolution procedure, as indicated in Figure 4, the molar fraction of inserted hydrogen is determined to be 35 mol% based on the integrated water difference area of Ru_60_250R and Ru_60_250N. Recently, the molar fraction of inserted hydrogen for Ru_30_250R was found to be 18 mol% [26]. Another hydrogenation experiment for Ru_40_250R (reference Ru_40_250N) is shown in Figure S11 and yields 23 mol% of inserted hydrogen in Ru_40_250R. Given that Ru_100 and pure TiO$_2$ rutile cannot incorporate any hydrogen, the obtained amount of incorporated hydrogen among different Ru_x_250R catalysts (Figure S12) reveals a "volcano" type of the H-fraction with the increase of the Ru concentration in the Ru–Ti mixed oxides. The maximum amount of incorporated hydrogen is encountered at the Ru concentration of 60%.

Figure 5 compares the Ru3d XP spectrum of Ru_60 with those of the hydrogen-treated Ru_60_250R sample and the re-oxidized one, Ru_60_250R_300O. Three Ru components, namely metallic Ru, Ru^{4+}, and the satellite of Ru^{4+} and two carbon species are considered to fit the spectra. The Ru_60 sample reveals only the Ru^{4+} component (red) together with the corresponding satellite feature (blue), consistent with the corresponding XRD (pattern Figure 1a) that is composed only of diffraction peaks of the mixed oxide Ru$_{0.6}$Ti$_{0.4}$O$_2$ and pure RuO$_2$. Upon hydrogen exposure at 250 °C, a strong metallic Ru peak becomes apparent in the Ru 3d spectrum. The metallic Ru component comes from the reduction of RuO$_2$ towards metallic Ru, as indicated by XRD, while the Ru^{4+} component originates from Ru in Ru$_{0.6}$Ti$_{0.4}$O$_2$. Upon re-oxidation of Ru_60_250R at 300 °C (Ru_60_250R_300O), most of metallic component transforms back to Ru^{4+}. We conclude from these experiments that the Ru^{4+} oxidation state $Ru_{0.6}Ti_{0.4}O_2$ is preserved, regardless of the applied treatment (hydrogenation, re-oxidation).

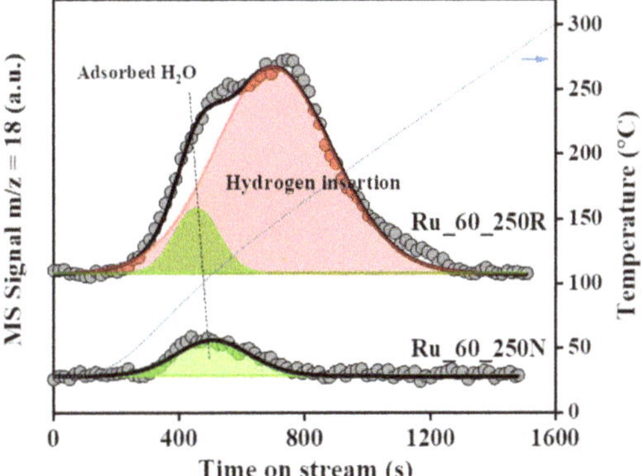

Figure 4. Peak deconvolution of H_2O signal (m/z = 18) of Ru_60_250R and Ru_60_250N from TG-MS analysis.

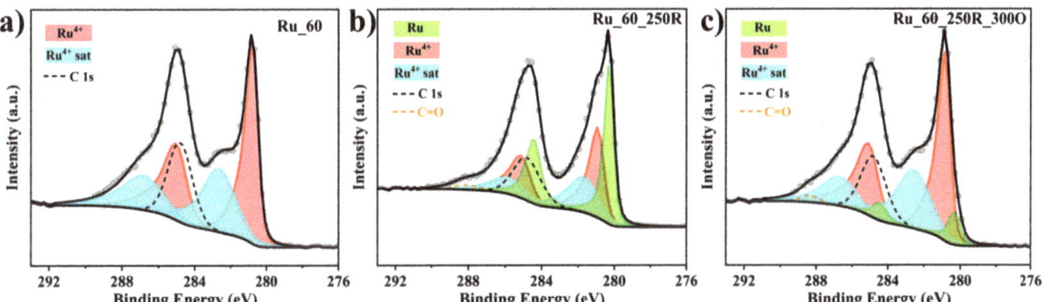

Figure 5. XP spectra of Ru_60 (**a**) in comparison with hydrogen treated sample at 250 °C. (**b**) Ru_60_250R and re-oxidized sample at 300 °C. (**c**) Ru_60_250R_300O. Using the CasaXPS software, the Ru3d spectra are decomposed into five components: Ru^{4+} (red), satellite Ru^{4+} (blue), metallic Ru (green) and two C1s species.

Hydrogen reduction of Ru from $Ru_{0.6}Ti_{0.4}O_2$ can, however, be excluded for the following reasons. The reduction treatment at 250 °C preserves the rutile structure, albeit with low intensity (cf. Figure 6). Upon re-oxidation at 450 °C (Ru_60_250R_450O), however, the rutile diffraction peaks of Ru_60 are practically restored. In particular, rutile diffraction peaks shift back to those positions of Ru_60 (with identical intensity), thus evidencing that the mol% of Ru in the mixed rutile structure has been preserved (Vegard's law).

From HRTEM and element mapping (cf. Figure S13), it is evident that Ru_60 and Ru_60_250R consist mainly of mixed $Ru_xTi_{1-x}O_2$ oxide whose composition has not changed. For Ru_60 larger RuO_2 particles are discernible.

In addition to the Ru 3d spectra (cf. Figure 5), Ti 2p XP spectra are compiled in Figure S14 for Ru_60, the hydrogen treated Ru_60_250R sample and the re-oxidized one, Ru_60_250R_300O. All Ti 2p spectra show only Ti^{4+}, and there is no indication of Ti^{3+}. The overall near-surface composition of Ru_60_250R is collected in Table 2. Corresponding O1s spectra in Figure S15 exhibit two components, one related to O^{2-} and the other assigned to OH/carbonate species. The OH/carbonate feature does not vary when Ru_60 is exposed to hydrogen or is re-oxidized, suggesting that incorporated H does not change the concentration of OH species or carbonate species.

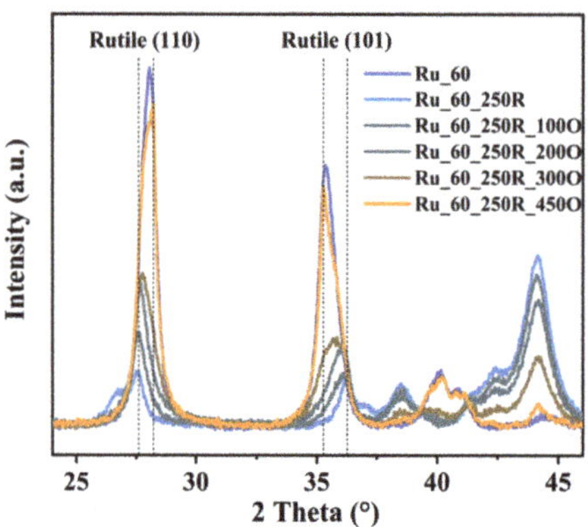

Figure 6. XRD patterns of Ru_60_250R mildly re-oxidized at various temperatures in air. For comparison, Ru_60 and Ru_60_250R are also included.

From all these experiments we infer that the reduction process results in the incorporation of hydrogen within in the mixed oxide and both Ru and Ti in the mixed Ru–Ti oxide phase remain in the 4+ oxidation state.

2.3. Catalytic Tests: Propane Combustion

In the following, we conducted catalytic tests of Ru_x and Ru_x_250R for the total oxidation of propane, serving here as a model reaction. The full set of light-off curves is presented in the ESI (cf. Figure S16a,b). Except for Ru_100, the hydrogenated Ru_x_250R sample is more active than Ru_x. The temperature differences for T_{90} is collected in Figure S16c (T_{90} is the temperature where 90% conversion is realized). From these conversion data, one can recognize that the Ru_x with x close to 60% exhibits highest activity after hydrogenation, that is even higher than that of Ru_100.

In Figure 7 we exemplify light-off curves for propane combustion of Ru_60, Ru100 and Ru_20 before and after hydrogenation. Ru_100 reveals the highest activity among the non-hydrogenated samples with T_{90} = 182 °C. Upon hydrogenation, the light-off curve of Ru_100_250R shifts to higher temperatures (T_{90} = 194 °C). Ru_20 is significantly less active than Ru_100; T_{90} = 231 °C; upon hydrogenation, T_{90} decreases to 211 °C for Ru_20_250R. Ru_60 reaches 90% conversion at 203 °C, while 90% conversion is achieved at 168 °C for Ru_60_250R, i.e., 35 °C lower than for Ru_20 and even 14 °C lower than that for Ru_100.

Since the inserted H in $Ru_xTi_{1-x}O_2$ is a labile species that leaves the sample already at about 100 °C by water formation under ambient atmosphere [26], we performed an additional experiment, where the Ru_60 sample is in situ hydrogenated at 250 °C in the reactor; this procedure allows us to keep the hydrogenation level high in the mixed oxide sample Ru_60. For the catalytic test of propane oxidation, we choose a reaction temperature of 150 °C to avoid full conversion (cf. Figure 8).

Figure 8 indicates that the activity of Ru_60_250R at 150 °C in the first cycle is quite high with a STY value of 5.2 $mol_{(CO_2)} \cdot kg^{-1}_{(Cat)} \cdot h^{-1}$, and it declines in the second cycle to 2.4 $mol_{(CO_2)} \cdot kg^{-1}_{(Cat)} \cdot h^{-1}$ likely due to the removal of incorporated hydrogen [26]. Conducting an in situ hydrogen treatment at 250 °C of Ru_60_250R leads to a re-activation of the catalyst with a steady state STY of 6.1 $mol_{(CO_2)} \cdot kg^{-1}_{(Cat)} \cdot h^{-1}$ that is even higher than the initial activity during the first cycle. Hydrogen exposure at higher temperature will lead to lower activity of the Ru_60_400R sample (STY = 3.1 $mol_{(CO_2)} \cdot kg^{-1}_{(Cat)} \cdot h^{-1}$) (cf. Figure S17). After in situ hydrogen treatment at 150 °C, Ru_60_400R is reactivated with a higher STY value of 4.7 $mol_{(CO_2)} \cdot kg^{-1}_{(Cat)} \cdot h^{-1}$, that is lower than that of Ru_60_250R (cf. Figure S17). Altogether, hydrogen treatment at 250 °C seems to optimize the promotional effect of hydrogen in the total oxidation of propane.

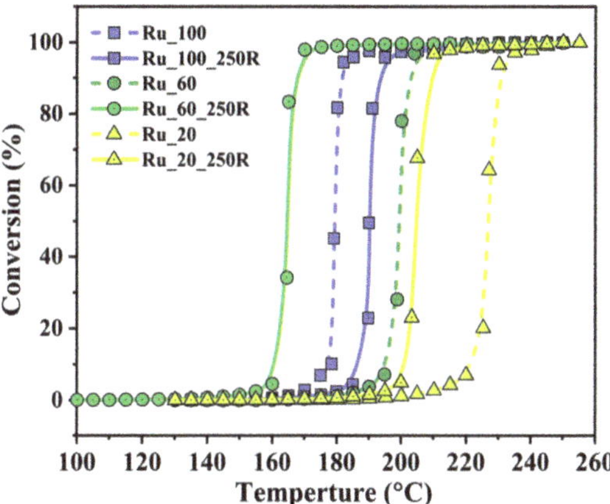

Figure 7. Selection of full conversion curves of catalytic propane combustion over Ru_x and Ru_x_250R (x = 100%, 60% and 20%) as a function of reaction temperature, when cycling the reaction temperature from 30 °C to 250 °C. The full set of conversion curves can be found in Figure S16.

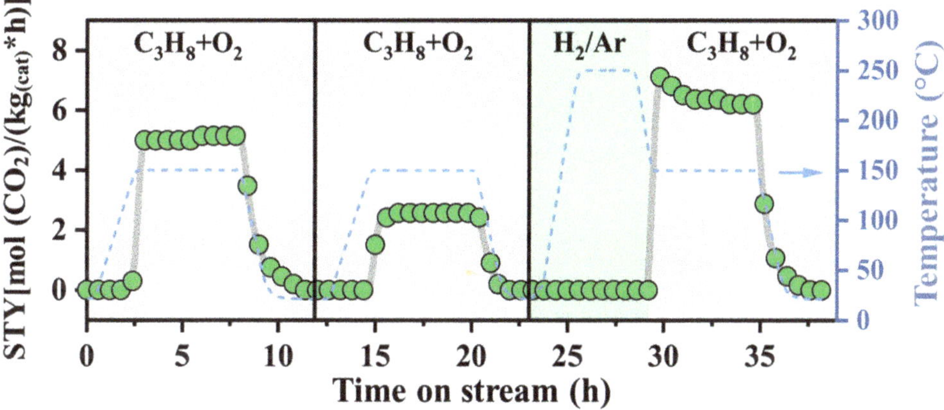

Figure 8. STY as a function of reaction time on catalytic propane oxidation over Ru_60_250R when cycling the reaction temperature from 30 °C to 150 °C (blue dotted line). The gray background represents total C_3H_8 oxidation conditions: 1 vol% C_3H_8, 5 vol% O_2, balanced by N_2; total volume flow: 100 sccm/min, temperature ramp: 1 K/min. The green background represents the gas mixture during heating and cooling stage: 4% H_2/Ar, total volume flow: 50 sccm/min. When reaching 150 °C, the gas composition is switched to the reaction mixture (gray background).

3. Discussion

3.1. Formation of Ru_x and Ru_x_250R

A mixed ruthenium–titanium oxide material with nominal varying concentration x of Ru (Ru_x) is successfully prepared by a conventional sol-gel method. Below x = 70%, Ru_x consists of pure RuO_2 and a mixed oxide $Ru_xTi_{1-x}O_2$. For higher concentration of Ru, in addition, a metallic Ru is formed. The metallic Ru phase is caused by the preparation method. For sol-gel synthesis, O_2 cannot penetrate into the polymer network when metal cations nucleate during the calcination stage, thus causing a net reducing environment for the growing particle where the metal cations Ru^{3+} is reduced to Ru^0 [34]. From EDS-SEM and XPS, the bulk and surface compositions for x < 50% are similar and close to the nominal composition of Ru_x, while for x > 50%, the surface composition of Ru is significantly lower than the bulk concentration.

For Ru_x we see a clear linear correlation of the lattice parameters of $Ru_xTi_{1-x}O_2$ with the composition x of ruthenium. The cell parameters a/b of RuO_2 are smaller than those of TiO_2 while the c value of RuO_2 is larger than that of TiO_2, therefore for a mixed oxide, a/b and c cell parameters of the Ru–Ti phase indicate an "anti-symbatic" behavior (Figure 2). In addition, there is a clear linear correlation of the binding energy of the satellite peak in Ru3d with the composition x of ruthenium. (cf. Figure 3) With increasing Ru concentration, the energy of the surface plasmon increases, consistent with the surface plasmon energy being expected to increase in energy with increasing electron density [33].

From XPS, we conclude that both Ti and Ru in Ru_x for $x \leq 70\%$ are in the 4+ oxidation state independent of the composition x. Metallic Ru, if present, is buried, as evidenced by XPS, and does therefore not participate in the catalytic reaction.

Hydrogen exposure to Ru_x at 250 °C (Ru_x_250R) leads to hydrogen incorporation in Ru_x for Ru concentration $x = 20\%$ up to 80%. (cf. Figure 1) The extremes Ru_00 (e.g., pure rutile-TiO_2) and Ru_100 are not able to incorporate hydrogen: TiO_2 is not affected at all by hydrogenation at 250 °C, as reduction to Magneli phases requires much higher temperatures [37,38]. Ru_100 is completely transformed to metallic Ru, and with increasing Ru concentration, more metallic Ru is formed upon hydrogen exposure at 250 °C. However, 18, 23 and 35mol% of hydrogen are inserted in Ru_30_250R [26], Ru_40_250R, (Figure S11) and R_60_250R, (Figure 4), respectively. From Ru 3d XPS data, there is no indication of an oxidation state of ruthenium other than Ru^0 and Ru^{4+}. The rutile structure of mixed oxide $Ru_xTi_{1-x}O_2$ is maintained upon H_2 exposure at 250 °C, suggesting that the Ru^{4+} can be preserved in rutile structure when mixed with less reducible TiO_2. Therefore, the peak shifts in XRD (Figure 1b) after reduction treatment are caused by the incorporation of hydrogen in the rutile structure and not by the reduction of the metal ions, which is consistent with literature [26]. The only change observed for mixed $Ru_xTi_{1-x}O_2$ is a hydrogen-induced shift in the lattice parameters (cf. Figure 2), i.e., the introduction of hydrogen-induced strain into the mixed oxide lattice. Hydrogen absorption in mixed oxide $Ru_xTi_{1-x}O$ manifests therefore a synergy effect in that Ru enables the activation of H_2 while Ti stabilizes Ru^{4+} against reduction to metallic Ru.

The possible types of inserted hydrogen could be proton, neutral H or hydride H^- species. Based on the XPS data, there are no changes in the oxidation states of Ru^{4+} and Ti^{4+} in the rutile phase (as shown in Figure 5 and Figures S14 and S15). Moreover, from the O 1s spectrum, there is no obvious increase of the OH-related signals, so that we exclude the existence of protons and neutral H in the mixed oxide lattice. Instead, we favor that the incorporated hydrogen is a hydride species $H^{-\delta}$, which is reconciled with the energetic shift of the Ru satellite peak in the Ru3d spectrum and that is correlated with a reduced electron density for exciting the surface plasmon. Obviously, some of the delocalized electrons are localized at $H^{-\delta}$.

Ru_60 can almost be recovered from Ru_60_250R by mild re-oxidation at 450 °C. Re-oxidation at 100 °C starts to remove H-induced changes in strain, both macro- and micro-strain, as reflected by changes in diffraction peak position and FWHM. This behavior was also observed in a previous study of Ru_30, and it was traced to the removal of absorbed hydrogen via water formation [26].

3.2. Improved Oxidation Catalysis of Ru_x_250R in Comparison to Ru_x

The light-off curves of Ru_x and hydrogen inserted Ru_x catalysts are measured for the total oxidation of propane (cf. Figure 7 and Figure S16). These experiments provide compelling evidence that H insertion is beneficial for the oxidation catalysis of propane, with the optimum catalyst being identified with Ru_60_250R. For Ru_100, hydrogenation at 250 °C leads to lower propane combustion activity.

Hydrogen exposure at 250 °C produces a labile hydrogen species in the mixed oxide lattice that can readily be removed by increasing the temperature in the ambient atmosphere (cf. Figure 6). Already a mild re-oxidation at 100 °C restores part of the hydrogen-induced strain in the mixed oxide. From activity experiments in Figure 8, we can conclude that hydrogen is lost during temperature ramping in the first cycle, although the catalyst runs stably at the final reaction temperature of 150 °C. In order to stabilize inserted H during reaction, the surface needs to be continuously supplied by hydrogen from propane activation. Propane dissociation is activated and needs a temperature of at least 100 °C [39,40]. Therefore, during the heating and cooling ramp in the temperature window of 50–100 °C, the catalyst is exposed to a reaction mixture that is not able to supply the surface with hydrogen. The temperature range of 50–100 °C is, however, high enough to consume inserted hydrogen by water formation due to oxygen in the feed. From this discussion, it is clear that higher activities are expected when the Ru_60_250R catalyst is in situ hydrogen-treated in the reactor during temperature ramping. The activity of propane oxidation (cf. Figure 8) is indeed significantly higher and stable when the catalyst is heated under a hydrogen atmosphere first to the specific reaction temperature and then the gas atmosphere is switched to the actual reaction mixtures. Propane decomposition at high temperature can supply the catalyst surface constantly with hydrogen. Therefore, dissolved hydrogen does not experience a driving force to diffuse towards the surface, thus stabilizing the inserted hydrogen in the lattice of the mixed oxide.

We can safely assume that inserted hydrogen does not take place in the catalytic cycle, since we focus here only on the catalytic propane oxidation reaction. The incorporated hydrogen stabilized by the presence of hydrogen at the surface is, however, the key factor to promote the oxidation catalysis. We expect, however, that for catalytic hydrogenation reaction, the beneficial effect of inserted H may be even more pronounced since inserted hydrogen is able participate in the catalytic cycle.

Hydrogen incorporation is accompanied by the development of macro- and micro-strain of Ru_x, as evidenced by the shift and broadening of the rutile diffraction maxima in the powder XRD (cf. Figure 1), while, after mild re-oxidation at 300 °C, the macro- and micro-strain of Ru_x_250R is largely removed, as reflected by the recovery of the peak position and FWHM (cf. Figure S10). Whether H insertion itself or the induced changes of lattice parameters leads to electronic modifications cannot be disentangled at this point and needs to be scrutinized by future ab initio studies. In any case, the electronic structure is affected by H insertion, as corroborated in the observed binding energy shift of the satellite feature in Ru3d (cf. Figure 3). The increased activity can therefore be related to the altered electronic structure of $Ru_xTi_{1-x}O_2$. A direct correlation of strain with activity is, however, not evident for Ru_x_250R and needs further studies.

4. Materials and Methods

4.1. Materials

Ruthenium (III) trichloride hydrate ($RuCl_3 \cdot xH_2O$, Ruthenium content: 40.00–49.00%, ReagentPlus®), anhydrous citric acid ($C_6H_8O_7$, ≥99.5%), ammonia solution (NH_4OH, 25%) and titanium butoxide ($C_{16}H_{36}O_4Ti$, 97%) are purchased from Sigma-Aldrich (St. Louis, MO, USA) and used without further purification.

4.2. Catalysts Preparation

Ru–Ti mixed oxide materials with varying compositions are prepared by a conventional citric acid assisted sol-gel method (cf. Figure 9) and are denoted as Ru_x, where x represents the nominal molar percentage of ruthenium varying from 20% to 100%. For the procedural synthesis of Ru_60: 0.024 mol citric acid (anhydrous) is dissolved into 50 mL deionized water, and the solution is stirred and kept at 60 °C. Then, 5 mL anhydrous ethanol containing 0.8 mmol titanium butoxide is quickly injected into the citric acid solution. After thorough mixing, 0.0012 mol $RuCl_3 \cdot xH_2O$ is added to the solution and stirred for another 30 min at 60 °C. The full complexation of the ruthenium and titanium cations is accomplished by slowly heating the mixture to 80 °C, afterwards, the aqueous ammonia solution of 2 mol/L is added dropwise to adjust the pH of the solution to ~6. Finally, the dark brown but transparent solution is evaporated at 90 °C and the obtained dark green gel is dried at 120 °C for 12 h. Subsequently, the dark foamy material is carefully ground before calcination at 450 °C for 4 h in static air at 2 °C/min, and the product is ground again for further catalytic tests and characterizations. Note that, contrary to the commonly used version of the Pechini method, here, no glycol is added. The preparation of the other Ru_x catalysts follows the same procedure but adjusting the molar ratio of the ruthenium and titanium precursors.

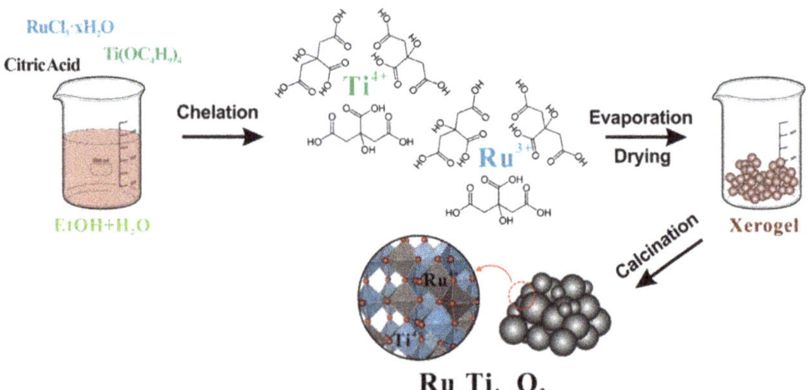

Figure 9. Illustration of the citric acid assisted sol-gel method to prepare mixed oxides of RuO$_2$ and TiO$_2$ with varying concentration x of Ru: Ru_x.

Hydrogen insertion experiments are performed as follows: 50 mg of fresh Ru_x powder material is treated at various temperatures under 4 vol% H$_2$/N$_2$ for 3 h, the total flow rate being set to 50 sccm/min. After hydrogen treatment, the catalyst is cooled down to room temperature under the same atmosphere. The obtained catalyst is referred to as Ru_x_yR, where x represents the nominal molar percentage of ruthenium, y represents the reduction temperature. The re-oxidation experiment follows the same procedure while the gas atmosphere is changed to dry air. The catalyst is now denoted as Ru_x_yR_zO, where z stands for the re-oxidation temperature. For comparison, a blank experiment is also conducted by calcining the Ru_x under pure N$_2$ and the catalyst is referred to as Ru_x_yN.

4.3. Catalysts Characterization

Powder XRD patterns are recorded on a Panalytical Empyrean diffractometer (Malvern, UK) equipped with a Cu K$_\alpha$ radiation (40 kV, 40 mA). The correction of the 2θ shift that may originate from different positions of the sample holder is assisted by mixing LaB$_6$ standard powder (NIST) with all the catalysts before measurement. The Scherrer equation is applied to calculate the crystallite size.

Kr physisorption experiments are performed at T = 77 K with an Autosorb 6 instrument (Quantachrome, Ostfildern, Germany), all the catalysts are pre-treated in vacuum for 12 h at 100 °C. The specific surface area is calculated by the BET method. Note that we deliberately used Kr, instead of nitrogen or argon, because the materials we anticipated to exhibit small surface areas, and Kr has a higher sensitivity in this respect.

Scanning transmission electron microscopy (STEM) micrographs are recorded on a ThermoFisher Talos F200X electron microscope (Waltham, MA, USA). SEM images are obtained on a Gemini SEM 560 instrument (Carl Zeiss MicroImaging GmbH, Göttingen, Germany). The average composition of the catalyst is quantified by energy dispersive X-ray spectroscopy (EDS).

XPS spectra are acquired on a PHI VersaProbe II instrument (Feldkirchen, Germany) equipped with a monochromatized Al-Kα line, the photon energy is 1486.6 eV. XPS spectrum data analysis is performed by using a CasaXPS software (Version 2.3.17), and the standard Carbon 1s (at 284.8 eV) is used to re-determine all the binding energy. After calcination, the Ru_x samples do not show any residual chlorine in the XP overview spectra.

A TG-MS experiment is conducted on a STA 409PC thermoscale (Netzsch, Selb, Germany) analyzer coupled with a QMG421 quadrupole mass spectrometer (MS) from Balzers (Balzers, Liechtenstein) with an ionization energy of 70 eV. The catalyst is heated under dry air (30 sccm/min) from 25 to 500 °C, while the heating rate is 10 °C/min. The detailed procedure for calculating the amount of inserted hydrogen is as follows: dry air is applied (30 mL/min) during the TG-MS experiment. Since the flow rate/MS signal of N_2 is constant in air (78.1%), we can use the concentration of N_2 as reference to obtain the flow rate of H_2O (gaseous) by dividing the MS signal of H_2O by N_2. After having the ratio of H_2O/N_2, the total H_2O volume is determined by integration, from which the molar amount of produced H_2O and hence the molar amount of inserted hydrogen are derived; note the sample storage conditions of Ru_x_250R and Ru_x_250N are identical to guarantee accurate data processing.

4.4. Catalytic Tests

The catalytic performance of propane oxidation on the mixed Ru–Ti oxide catalysts is evaluated in a home-made quartz reactor (inner diameter = 6 mm). The feed gas contains 1 vol.% C_3H_8 (purity: 3.5), 5 vol.% O_2 (purity: 4.8) and 94 vol.% N_2 (purity: 4.8) and is admitted to reactor with a total mass flow rate of 100 cm^3 STP min^{-1} (sccm). During catalytic measurement, the temperature of a mixture consisting of 20 mg of catalyst and 40 mg of quartz sand is programmed from 25 °C to 250 °C with a heating rate of 1 °C/min. The corresponding weight hourly space velocity is 345000 $mL·g^{-1}·h^{-1}$. For product analysis, a nondispersive infrared sensor is coupled downstream to detect the volumetric concentration of CO/CO_2 and C_3H_8. The conversion of propane (%) is determined based on the following equation:

$$X_{C_3H_8} = \frac{c(CO_2)}{c_{max}(CO_2)} \times 100\% \tag{1}$$

where $c(CO_2)$ is the real-time concentration of CO_2 in the outlet gas and $c_{max}(CO_2)$ is the steady-state concentration of CO_2 when full conversion is achieved. Propane conversion calculated by the change of the propane concentration is simultaneously conducted to countercheck data accuracy. During the whole temperature range, there is no CO detected and the concentration of CO_2 at full conversion state is virtually three times the inlet propane concentration, thus evidencing that the carbon mass is balanced and no other byproduct is formed. Finally, we use space time yield ($mol_{(CO_2)}·kg_{(Cat)}^{-1}·h^{-1}$, molar amount of CO_2 per kilogram catalyst and hour, STY) to quantify the activity of the catalyst in total propane oxidation reaction.

5. Conclusions

A rational synthesis approach is introduced to favor hydrogen incorporation in the oxide lattice by mixing a reducible oxide with a less reducible oxide, as exemplified with the solid solution of RuO_2 and rutile TiO_2. Neither RuO_2 nor rutile-TiO_2 is able to incorporate hydrogen into the lattice just by hydrogen exposure at elevated temperatures: rutile-TiO_2 is not affected at all, while Ru_100 is fully reduced to metallic ruthenium. We show that mixed $Ru_xTi_{1-x}O_2$ is stable against H_2 exposure at 250 °C for compositions $0.2 < x < 0.8$ and hydrogen can be incorporated into the lattice. Hydrogen incorporation in mixed oxide $Ru_xTi_{1-x}O_2$ reveals a synergy effect in that Ru enables the activation of H_2, while Ti stabilizes the oxidation state of Ru (Ru^{4+}) against reduction to metallic Ru.

Hydrogen insertion into the rutile lattice of $Ru_xTi_{1-x}O_2$ is accompanied by a change of the lattice constants (XRD) and increased micro-strain. Hydrogen insertion affects directly or indirectly via macro- and micro-strain the electronic structure of $Ru_xTi_{1-x}O_2$ that in turn is expected to be responsible for the improved catalytic activity, not only for oxidation catalysis as exemplified with the propane oxidation, but may be equally beneficial for the selective hydrogenation and oxidation of other organic compounds. For propane combustion, we show that the activity of Ru_x is significantly increased by H_2 exposure at 250 °C. The optimum catalyst is identified with Ru_60_250R, whose activity is substantially higher than that of Ru_100.

Hydrogen treatment can also be conducted in situ by H_2 exposure during temperature ramping and switching to the reaction mixture when the reaction temperature is reached, thus providing an additional parameter to tune the catalytic performance of a mixed oxide catalyst in the reactor. This approach is of general interest in catalysis research and inorganic chemistry to fine-tune properties of (mixed) oxides and may therefore open exciting perspectives for tuning the catalytic activity of mixed oxide catalysts, not only in thermal catalysis but also in electrocatalysis of acidic water splitting at the anode side.

Supplementary Materials: The following supporting information can be downloaded at: https://www.mdpi.com/article/10.3390/inorganics11080330/s1, Figure S1: Decomposition of the (110) and (101) reflection of Ru_x; Figure S2: Peak shift of rutile (110) and (101); Figure S3: Calculated cell volumes of the mixed oxide RuxTi1−xO2 phase; Table S1: Calculation of grain size and micro-strain of Ru–Ti mixed oxide catalysts by the Williamson–Hall method; Figure S4: Calculated crystallite size of RuO2 phase and Ru–Ti solid solution phase; Figure S5: Williamson–Hall plot of RuO2 phase and Ru–Ti solid solution phase as exemplified by Ru_60 sample; Table S2: Optimized fitting parameters for the XPS data deconvolution; Figure S6: Ru 3d XP spectra of Ru_x catalysts; Figure S7: O 1s spectra of Ru_x catalysts; Figure S8: Ru3p and Ti2p spectra of Ru_x catalysts; Figure S9: SEM micrographs for the various Ru_x samples; Figure S10: XRD patterns of Ru_x_250R samples re-oxidized at 300 °C; Figure S11: H2O signal from TG-MS analysis; Figure S12: The calculated amount of incorporated hydrogen when varying the composition x of Ru_x; Figure S13: HAADF-STEM images/element mapping of Ru_60 sample; Figure S14: Ti 2p XP spectra of Ru_60 sample; Figure S15: O 1s XP spectra of Ru_60sample; Figure S16: Light-off curves of catalytic propane combustion; Figure S17: STY as a function of reaction time over Ru_60_400R at 150 °C.

Author Contributions: Conceptualization, methodology, writing—original draft, W.W.; data curation, investigation, Y.W.; methodology, P.T. and A.S.-L.; investigation, T.W. and L.G.; supervision, resources, Y.G. and B.M.S.; conceptualization, supervision, writing—review and editing, H.O. All authors have read and agreed to the published version of the manuscript.

Funding: This project was supported financially by National Key Research and Development Program of China (2022YFB3504200), the National Natural Science Foundation of China (U21A20326, 21976057, 21922602 and 21673072), the fund of the National Engineering Laboratory for Mobile Source Emission Control Technology (NELMS2020A05) and the Fundamental Research Funds for the Central Universities.

Data Availability Statement: Not applicable.

Acknowledgments: W.W. gratefully acknowledges the China Scholarship Council for the Joint-Ph.D. program between the China Scholarship Council and the Institute of Physical Chemistry of the Justus-Liebig-University Giessen. We acknowledge support from the Center for Materials Research at the JLU. H.O. and L.G. acknowledge funding by the German Research Foundation (DFG, Deutsche Forschungsgemeinschaft—493681475).

Conflicts of Interest: The authors declare that they have no known competing financial interests or personal relationships that could have appeared to influence the work reported in this paper.

References

1. Mavrikakis, M.; Hammer, B.; Nørskov, J.K. Effect of Strain on the Reactivity of Metal Surfaces. *Phys. Rev. Lett.* **1998**, *81*, 2819–2822. [CrossRef]
2. Hammer, B.; Nørskov, J.K. *Chemisorption and Reactivity on Supported Clusters and Thin Films*; Kluwer Academic: Dordrecht, The Netherlands, 1997; pp. 285–351.
3. Buvat, G.; Eslamibidgoli, M.J.; Youssef, A.H.; Garbarino, S.; Ruediger, A.; Eikerling, M.; Guay, D. Effect of IrO_6 Octahedron Distortion on the OER Activity at (100) IrO_2 Thin Film. *ACS Catal.* **2020**, *10*, 806–817. [CrossRef]
4. Wang, H.; Xu, S.; Tsai, C.; Li, Y.; Liu, C.; Zhao, J.; Liu, Y.; Yuan, H.; Abild-Pedersen, F.; Prinz, F.B.; et al. Direct and Continuous Strain Control of Catalysts with Tunable Battery Electrode Materials. *Science* **2016**, *354*, 1031–1036. [CrossRef]
5. Strasser, P.; Kühl, S. Dealloyed Pt-based Core-shell Oxygen Reduction Electrocatalysts. *Nano Energy* **2016**, *29*, 166–177. [CrossRef]

6. Gawande, M.B.; Goswami, A.; Asefa, T.; Guo, H.; Biradar, A.V.; Peng, D.-L.; Zboril, R.; Varma, R.S. Core-shell Nanoparticles: Synthesis and Applications in Catalysis and Electrocatalysis. *Chem. Soc. Rev.* **2015**, *44*, 7540–7590. [CrossRef] [PubMed]

7. Zhang, S.; Zhang, X.; Jiang, G.; Zhu, H.; Guo, S.; Su, D.; Lu, G.; Sun, S. Tuning Nanoparticle Structure and Surface Strain for Catalysis Optimization. *J. Am. Chem. Soc.* **2014**, *136*, 7734–7739. [CrossRef]

8. Kibler, L.A.; El-Aziz, A.M.; Hoyer, R.; Kolb, D.M. Tuning Reaction Rates by Lateral Strain in a Palladium Monolayer. *Angew. Chem. Int. Ed.* **2005**, *44*, 2080–2084. [CrossRef] [PubMed]

9. Strasser, P.; Koh, S.; Anniyev, T.; Greeley, J.; More, K.; Yu, C.; Liu, Z.; Kaya, S.; Nordlund, D.; Ogasawara, H.; et al. Lattice-strain Control of the Activity in Dealloyed Core-shell Fuel Cell Catalysts. *Nat. Chem.* **2010**, *2*, 454–460. [CrossRef]

10. Xia, Z.; Guo, S. Strain Engineering of Metal-based Nanomaterials for Energy Electrocatalysis. *Chem. Soc. Rev.* **2019**, *48*, 3265–3278. [CrossRef]

11. Wang, L.; Zeng, Z.; Gao, W.; Maxson, T.; Raciti, D.; Giroux, M.; Pan, X.; Wang, C.; Greeley, J. Tunable Intrinsic Strain in Two-dimensional Transition Metal Electrocatalysts. *Science* **2019**, *363*, 870–874. [CrossRef]

12. You, B.; Tang, M.T.; Tsai, C.; Abild-Pedersen, F.; Zheng, X.; Li, H. Enhancing Electrocatalytic Water Splitting by Strain Engineering. *Adv. Mater.* **2019**, *31*, 1807001. [CrossRef] [PubMed]

13. Alayoglu, S.; Nilekar, A.U.; Mavrikakis, M.; Eichhorn, B. Ru-Pt Core-shell Nanoparticles for Preferential Oxidation of Carbon Monoxide in Hydrogen. *Nat. Mater.* **2008**, *7*, 333–338. [CrossRef] [PubMed]

14. Schlapka, A.; Lischka, M.; Groß, A.; Käsberger, U.; Jakob, P. Surface Strain versus Substrate Interaction in Heteroepitaxial Metal Layers: Pt on Ru(0001). *Phys. Rev. Lett.* **2003**, *91*, 016101. [CrossRef]

15. Teschner, D.; Borsodi, J.; Wootsch, A.; Révay, Z.; Hävecker, M.; Knop-Gericke, A.; Jackson, S.D.; Schlögl, R. The Roles of Subsurface Carbon and Hydrogen in Palladium-Catalyzed Alkyne Hydrogenation. *Science* **2008**, *320*, 86–89. [CrossRef] [PubMed]

16. Wilde, M.; Fukutani, K.; Ludwig, W.; Brandt, B.; Fischer, J.-H.; Schauermann, S.; Freund, H.-J. Influence of Carbon Deposition on the Hydrogen Distribution in Pd Nanoparticles and Their Reactivity in Olefin Hydrogenation. *Angew. Chem. Int. Ed.* **2008**, *47*, 9289–9293. [CrossRef] [PubMed]

17. Copéret, C.; Estes, D.P.; Larmier, K.; Searles, K. Isolated Surface Hydrides: Formation, Structure, and Reactivity. *Chem. Rev.* **2016**, *116*, 8463–8505. [CrossRef]

18. Wu, Z.; Cheng, Y.; Tao, F.; Daemen, L.; Foo, G.S.; Nguyen, L.; Zhang, X.; Beste, A.; Ramirez-Cuesta, A.J. Direct Neutron Spectroscopy Observation of Cerium Hydride Species on a Cerium Oxide Catalyst. *J. Am. Chem. Soc.* **2017**, *139*, 9721–9727. [CrossRef]

19. Werner, K.; Weng, X.; Calaza, F.; Sterrer, M.; Kropp, T.; Paier, J.; Sauer, J.; Wilde, M.; Fukutani, K.; Shaikhutdinov, S.; et al. Toward an Understanding of Selective Alkyne Hydrogenation on Ceria: On the Impact of O Vacancies on H_2 Interaction with CeO_2(111). *J. Am. Chem. Soc.* **2017**, *139*, 17608–17616. [CrossRef]

20. Cao, T.; You, R.; Zhang, X.; Chen, S.; Li, D.; Zhang, Z.; Huang, W. An in situ DRIFTS Mechanistic Study of CeO_2-catalyzed Acetylene Semihydrogenation Reaction. *Phys. Chem. Chem. Phys.* **2018**, *20*, 9659–9670. [CrossRef]

21. Cheng, H.; Wen, M.; Ma, X.; Kuwahara, Y.; Mori, K.; Dai, Y.; Huang, B.; Yamashita, H. Hydrogen Doped Metal Oxide Semiconductors with Exceptional and Tunable Localized Surface Plasmon Resonances. *J. Am. Chem. Soc.* **2016**, *138*, 9316–9324. [CrossRef]

22. Li, Z.; Werner, K.; Qian, K.; You, R.; Płucienik, A.; Jia, A.; Wu, L.; Zhang, L.; Pan, H.; Kuhlenbeck, H.; et al. Oxidation of Reduced Ceria by Incorporation of Hydrogen. *Angew. Chem. Int. Ed.* **2019**, *58*, 14686–14693. [CrossRef]

23. Vilé, G.; Bridier, B.; Wichert, J.; Pérez-Ramírez, J. Ceria in Hydrogenation Catalysis: High Selectivity in the Conversion of Alkynes to Olefins. *Angew. Chem. Int. Ed.* **2012**, *51*, 8620–8623. [CrossRef]

24. Vilé, G.; Colussi, S.; Krumeich, F.; Trovarelli, A.; Pérez-Ramírez, J. Opposite Face Sensitivity of CeO_2 in Hydrogenation and Oxidation Catalysis. *Angew. Chem. Int. Ed.* **2014**, *53*, 12069–12072. [CrossRef]

25. Carrasco, J.; Vilé, G.; Fernández-Torre, D.; Pérez, J.; Pérez-Ramírez, R.; Ganduglia-Pirovano, M.V. Molecular-Level Understanding of CeO_2 as a Catalyst for Partial Alkyne Hydrogenation. *J. Phys. Chem. C* **2014**, *118*, 5352–5360. [CrossRef]

26. Wang, W.; Timmer, P.; Luciano, A.S.; Wang, Y.; Weber, T.; Glatthaar, L.; Guo, Y.; Smarsly, B.M.; Over, H. Inserted Hydrogen Promotes Oxidation Catalysis of Mixed $Ru_{0.3}Ti_{0.7}O_2$ as Exemplified with Total Propane Oxidation and the HCl Oxidation Reaction. *Catal. Sci. Technol.* **2023**, *13*, 1395–1408. [CrossRef]

27. Colomer, M.T.; Jurado, J.R. Structural, Microstructural, and Electrical Transport Properties of TiO_2-RuO_2 Ceramic Materials Obtained by Polymeric Sol-Gel Route. *Chem. Mater.* **2000**, *12*, 923–930. [CrossRef]

28. Wang, X.; Shao, Y.; Liu, X.; Tang, D.; Wu, B.; Tang, Z.; Wang, X.; Lin, W. Phase Stability and Phase Structure of Ru–Ti–O Complex Oxide Electrocatalyst. *J. Am. Ceram. Soc.* **2015**, *98*, 1915–1924. [CrossRef]

29. Ashcroft, N.; Denton, A. Vegard's Law. *Phys. Rev. A* **1991**, *43*, 3161–3164.

30. Özkan, E.; Cop, P.; Benfer, F.; Hofmann, A.; Votsmeier, M.; Guerra, J.M.; Giar, M.; Heiliger, C.; Over, H.; Smarsly, B.M. Rational Synthesis Concept for Cerium Oxide Nanoparticles: On the Impact of Particle Size on the Oxygen Storage Capacity. *J. Phys. Chem. C* **2020**, *124*, 8736–8748. [CrossRef]

31. Sivakami, R.; Dhanuskodi, S.; Karvembu, R. Estimation of Lattice Strain in Nanocrystalline RuO_2 by Williamson-Hall and Size-strain Plot Methods. *Spectrochim. Acta A Mol. Biomol. Spectrosc.* **2016**, *152*, 43–50. [CrossRef]

32. Over, H.; Muhler, M. Catalytic CO Oxidation over Ruthenium—Bridging the Pressure Gap. *Prog. Surf. Sci.* **2003**, *72*, 3–17. [CrossRef]

33. Over, H.; Seitsonen, A.P.; Lundgren, E.; Smedh, M.; Andersen, J.N. On the Origin of the Ru-3d$_{5/2}$ Satellite Feature from RuO$_2$(110). *Surf. Sci.* **2002**, *504*, L196–L200. [CrossRef]
34. Khalid, O.; Weber, T.; Drazic, G.; Djerdj, I.; Over, H. Mixed Ru$_x$Ir$_{1-x}$O$_2$ Oxide Catalyst with Well-Defined and Varying Composition Applied to CO Oxidation. *J. Phys. Chem. C* **2020**, *124*, 18670–18683. [CrossRef]
35. Assmann, J.; Narkhede, V.; Khodeir, L.; Löffler, E.; Hinrichsen, O.; Birkner, A.; Over, H.; Muhler, M. On the Nature of the Active State of Supported Ruthenium Catalysts Used for the Oxidation of Carbon Monoxide: Steady-state and Transient Kinetics Combined with in Situ Infrared Spectroscopy. *J. Phys. Chem. B* **2004**, *108*, 14634–14642. [CrossRef]
36. Wang, Z.; Khalid, O.; Wang, W.; Wang, Y.; Weber, T.; Luciano, A.S.; Zhan, W.; Smarsly, B.M.; Over, H. Comparison Study of the Effect of CeO$_2$-based Carrier Materials on the Total Oxidation of CO, Methane, and Propane over RuO$_2$. *Catal. Sci. Technol.* **2021**, *11*, 6839–6853. [CrossRef]
37. Walsh, F.; Wills, R. The Continuing Development of Magnéli Phase Titanium Sub-Oxides and Ebonex® Electrodes. *Electrochim. Acta* **2010**, *55*, 6342–6351. [CrossRef]
38. Malik, H.; Sarkar, S.; Mohanty, S.; Carlson, K. Modeling and Synthesis of Magneli Phases in Ordered Titanium Oxide Nanotubes with Preserved Morphology. *Sci. Rep.* **2020**, *10*, 8050. [CrossRef]
39. Wang, Z.; Huang, Z.; Brosnahan, J.T.; Zhang, S.; Guo, Y.; Guo, Y.; Wang, L.; Wang, Y.; Zhan, W. Ru/CeO$_2$ Catalyst with Optimized CeO$_2$ Support Morphology and Surface Facets for Propane Combustion. *Environ. Sci. Technol.* **2019**, *53*, 5349–5358. [CrossRef]
40. Wu, J.; Chen, B.; Yan, J.; Zheng, X.; Wang, X.; Deng, W.; Dai, Q. Ultra-active Ru Supported on CeO$_2$ Nanosheets for Catalytic Combustion of Propane: Experimental Insights into Interfacial Active Sites. *J. Chem. Eng.* **2022**, *438*, 135501. [CrossRef]

Article

Superoxide Radical Formed on the TiO$_2$ Surface Produced from Ti(OiPr)$_4$ Exposed to H$_2$O$_2$/KOH

Rimma I. Samoilova [1] and Sergei A. Dikanov [2,*]

[1] Voevodsky Institute of Chemical Kinetics and Combustion, Russian Academy of Sciences, Novosibirsk 630090, Russia; samoilov@kinetics.nsc.ru

[2] Department of Veterinary Clinical Medicine, University of Illinois at Urbana-Champaign, Urbana, IL 61801, USA

* Correspondence: dikanov@illinois.edu

Abstract: In this study, the superoxide radical O$_2^{\bullet-}$ formed by treating Ti(OR)$_4$ (R = iPr, nBu) with H$_2$O$_2$ in the presence of KOH was detected in the EPR spectra. The g-tensor of this radical differs from the typical values reported for a superoxide on various TiO$_2$ surfaces. On the other hand, similar g-tensor components g$_{||(zz}$ = 2.10 ± 0.01, g$_\perp$ = 2.005 ± 0.003 assigned to the O$_2^{\bullet-}$ were previously observed for radicals in aqueous solutions in the presence of K$_2$O, alkaline solutions of DMSO, and water/DMSO mixtures. A common factor in all these systems is the presence of alkali ions. However, there was no structural support for the possible interaction of alkali ions with a superoxide in these systems. The use of multifrequency pulsed EPR techniques in this work revealed the stabilization of the O$_2^{\bullet-}$ near the K$^+$ ion and its involvement in a strong hydrogen bond with the surface. These findings are consistent with the features previously reported for superoxides on a Na pre-covered MgO surface. Interactions with a closely located ^{23}Na and a strongly coupled ^{1}H proton were also seen in the HYSCORE spectra but assigned to two different superoxides with various g$_{zz}$ values presented in the sample.

Keywords: TiO$_2$; hydrogen peroxide H$_2$O$_2$; KOH; superoxide radical; pulsed EPR; hyperfine coupling

Citation: Samoilova, R.I.; Dikanov, S.A. Superoxide Radical Formed on the TiO$_2$ Surface Produced from Ti(OiPr)$_4$ Exposed to H$_2$O$_2$/KOH. *Inorganics* **2023**, *11*, 274. https://doi.org/10.3390/inorganics11070274

Academic Editors: Hicham Idriss, Eleonora Aneggi, Richard Walton, Roberto Nisticò, Torben R. Jensen and Luciano Carlos

Received: 8 May 2023
Revised: 11 June 2023
Accepted: 24 June 2023
Published: 27 June 2023

1. Introduction

Superoxide anion (O$_2^{\bullet-}$) is a radical formed after the one-electron reduction of dioxygen O$_2$, in different chemical processes [1]. In reactions with organic compounds, it can behave as a base, a nucleophile, and an oxidizing or reducing agent [2,3]. O$_2^{\bullet-}$ is paramagnetic, which has allowed for broad applications of multifrequency, continuous wave (CW) EPR spectroscopy for its studies [4–8].

The treatment of an oxide with a solution of hydrogen peroxide (H$_2$O$_2$), followed by drying the obtained solid under a vacuum has been also employed for generations of the superoxide radicals [9]. A product with matrix-bound O$_2^{\bullet-}$, produced by treating Ti(OR)$_4$ (R = iPr, nBu) with H$_2$O$_2$, was described and used as a selective heterogeneous catalyst for the oxidation of organic compounds [10]. It is effective at room temperatures and with various solvents including water.

Detailed studies of the catalyst using various experimental methods have shown that O$_2^{\bullet-}$ is responsible for the reaction, and its exceptional stability results from a stabilization near Ti^{4+} on the TiO$_2$ surface with a contribution of H$_2$O molecules and/or OH groups [10–12]. However, the nature and strength of O$_2^{\bullet-}$ interactions with the surrounding molecules was not characterized.

In our previous work, paramagnetic O$_2^{\bullet-}$ intermediates formed during the decomposition of H$_2$O$_2$ on the TiO$_2$ surface have been studied employing X- and Q-band CW and pulsed EPR spectroscopy. Exploiting high-resolution pulsed EPR techniques, i.e., 1D and 2D ESEEM (Electron Spin Echo Envelope Modulation) and ENDOR (Electron-Nuclear Double Resonance), weak interactions between the superoxide unpaired electron and the surrounding protons were quantitatively characterized. This enabled us to modify the model of the $O_2^{\bullet-}$ with its surrounding environment on the TiO_2 surface [13]. In this work, we found that the superoxide radical with different EPR spectroscopic characteristics is formed in reaction with $Ti(OR)_4$ (R = iPr, nBu) and H_2O_2 in the presence of a KOH solution. The application of pulsed EPR techniques has led to our finding that the stabilization of this $O_2^{\bullet-}$ is due to proximity to K^+ ion and its involvement in a strong hydrogen-bonding interaction with the surface.

2. Materials and Methods

2.1. Preparation of the Catalysts

The $TiO_2/O_2^{\bullet-}$ catalyst studied in our previous work [13] was prepared from $Ti(OiPr)_4$ exposed to H_2O_2 following the method described in [10]. The dried powder obtained at the end of this procedure was transferred in quartz X- and Q-band EPR tubes, degassed, sealed, and used in the EPR experiments described in [13] (further called "sample **I**" in this article). It is known that radiolytically or photochemically generated superoxide reacts with tyrosine, forming phenoxyl radicals of tyrosine [14,15]. We aimed to test the appearance of this species in a similar reaction using tyrosine with superoxide on the surface without prior irradiation. The solubility of tyrosine in alcohol is highly pH-dependent [16]. Initially, we tried to initiate a reaction of tyrosine with superoxide by adding tyrosine to the TiO_2 dispersion powder from an alcohol solution or from its mixture with low concentration KOH. EPR spectra of these samples show a rhombic signal consistent with the spectrum of the $O_2^{\bullet-}$ radical found in sample **I**. On the contrary, a signal with an axial g-tensor and increased g_{zz} was observed in samples with a higher KOH concentration (pH > 10). Similar results were obtained upon joint addition of peroxide and tyrosine in an alkaline alcohol solution to $Ti(OCH(CH_3)_2)_4$. The presence of an axial EPR signal was also confirmed in the control experiment by adding KOH in methanol without tyrosine to the $TiO_2/O_2^{\bullet-}$ catalyst (sample **II**).

Earlier, in aqueous solutions and mixtures containing alkali metals (M), radicals with similar EPR characteristics attributed to $O_2^{\bullet-}$ were found, which suggests a special role of M^+ ions in these samples for the radical stabilization. However, no structural EPR support was provided for this hypothesis. Therefore, this paper describes the pulsed EPR characterization of the superoxide radical and its environment in the type **II** samples containing K^+ ions.

2.2. EPR Measurements

The CW EPR, two-dimensional, four-pulse hyperfine sublevel correlation (HYSCORE, $\pi/2 - \tau - \pi/2 - t1 - \pi - t2 - \pi/2 - \tau -$ echo) [17], and Davies pulsed ENDOR ($\pi - t - \pi/2 - \tau - \pi - \tau -$ echo) [18]) experiments were performed as previously described elsewhere [13]. The Bruker WIN-EPR software was used for spectral processing.

3. Results and Discussions

3.1. EPR Spectra of Dried TiO_2 + H_2O_2

X- and Q-band EPR spectra of a dried sample of TiO_2 (solution of $Ti(OiPr)_4$ treated with H_2O_2, sample **I**) were reported in [13]. They show a signal with rhombic g-tensor produced by the decomposition of H_2O_2 on the TiO_2 surface (Figure S1). The g-tensor principal components (2.024, 2.009, and 2.003) determined from the Q-band spectrum supports the formation of a stable superoxide radical $O_2^{\bullet-}$ [10], because they are in line with values usually reported for the superoxide on various TiO_2 surfaces ($O_2^{\bullet-}$-Ti^{4+}) (see Table S1).

3.2. EPR Spectra in the Presence of KOH

Figures 1 and S2 show Q- and X-band EPR spectra of the radical formed in sample **II**. The spectra were obtained as a field-swept two-pulse Electron Spin Echo (ESE) signal and its calculated derivative. In contrast to sample **I**, the shapes of these spectra possess a typical axial g-tensor anisotropy with principal values $g_{||(zz)} = 2.10$, $g_{\perp(x,y)} = 2.002$, and a total width of ~20 mT in the X-band. We used the field-sweep ESE because a broad $g_{||(zz)}$ feature was not clearly resolved in CW EPR spectra. The EPR signals with similar g-tensor components $g_{||(zz)} = 2.10 \pm 0.01$ and $g_{\perp} = 2.005 \pm 0.003$ assigned to the superoxide radical have previously been observed in water solutions with the presence of K_2O, alkaline DMSO solutions, and water/DMSO mixtures (Table 1). One can note that the common factor in all these systems is the presence of alkaline ions. Earlier studies have found a good correlation of the superoxide g_{zz} value with the oxidation state of the nearest metal cation [5,19]. Particularly, a comparison of our data with the reported empirical dependence [5] indicates that $g_{zz} = 2.0227$ for the superoxide in sample **I** is consistent with its suggested location near Ti^{4+}. In contrast, $g_{zz} = 2.10$ is within the region reported for M^+ and may display the radical location near an alkaline cation in sample **II** as well as in the compounds shown in Table 1.

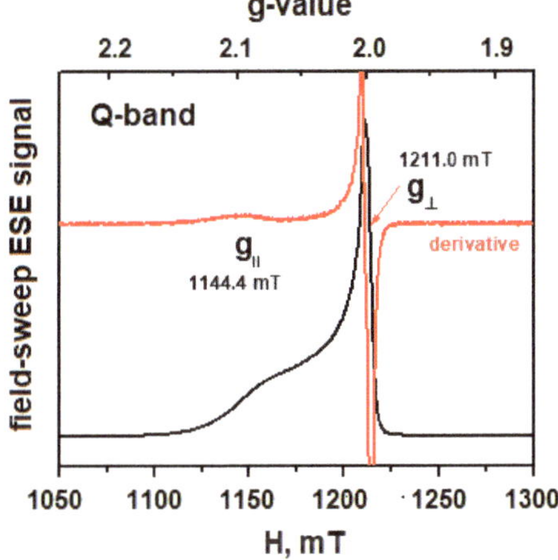

Figure 1. Q-band two-pulse ESE field-sweep spectrum and its first derivative of the radical formed in sample **II**. Field-sweep ESE generates a spectrum similar in shape to the absorption spectrum produced on the integration of the normal continuous-wave EPR derivative. Microwave frequency is 33.9523 GHz, length of $\pi/2$ pulse is 100 ns, time τ between first and second pulses is 400 ns, and temperature is 15 K.

Table 1. Systems where the EPR signal with axial g-tensor was assigned to a superoxide.

Matrix	$g_{||}(g_{zz})$	$g_{\perp}(g_{x,y})$	Reference
H_2O ice (K_2O)	2.110		[20]
D_2O ice (0.1 mM K_2O)	2.110	2.002	[21]
KO_2 in DMSO/H_2O in the presence of ubiquinone-10	2.108	2.004	[22]
Alkaline DMSO	2.098	2.005	[23,24]
10 µL of 0.5M NaOH/mL of DMSO	2.089	2.007	[25]
$TiO_2 + H_2O_2 + KOH$	2.10	2.002	This work

3.3. Pulsed EPR Characterization of the Radical

Interactions between the $O_2^{\bullet-}$ species and its environment in samples **I** and **II** were probed using HYSCORE and pulsed ENDOR techniques. Our previous HYSCORE studies of the superoxide in sample **I** have found only weakly coupled protons, with the anisotropic couplings T not exceeding ~2 MHz (Supplementary Materials, Section S1) [13]. These protons produce cross-features located along an antidiagonal crossing the diagonal of the (++) quadrant at the (ν_{1H}, ν_{1H}) point, where ν_{1H} is the proton Zeeman frequency in the applied magnetic field (Figure S3).

In contrast, the X-band HYSCORE spectrum of sample **II** (Figure 2) is dominated by a pair of cross-ridges **1**$_H$ that significantly deviated from the (ν_{1H}, ν_{1H}) antidiagonal. This indicates the presence of proton(s) with a substantially stronger anisotropic hyperfine interaction [22] than the protons contributing to the spectra of sample **I** (Figure S3). One can also note that the ^{1}H spectrum in Figure 2 clearly shows additional features **1'**$_H$ that also deviated from the ^{1}H antidiagonal but oriented in the opposite matter relative to this line. In the course of the analysis described below, we provide evidence that the **1**$_H$ and **1'**$_H$ lines are parts of the same cross-feature located on opposite sides of the diagonal of the (++) quadrant. Lines from weaker coupled protons, elongated along the antidiagonal (ν_{1H}, ν_{1H}), are also present in the spectra of sample **II** but possess a lower intensity.

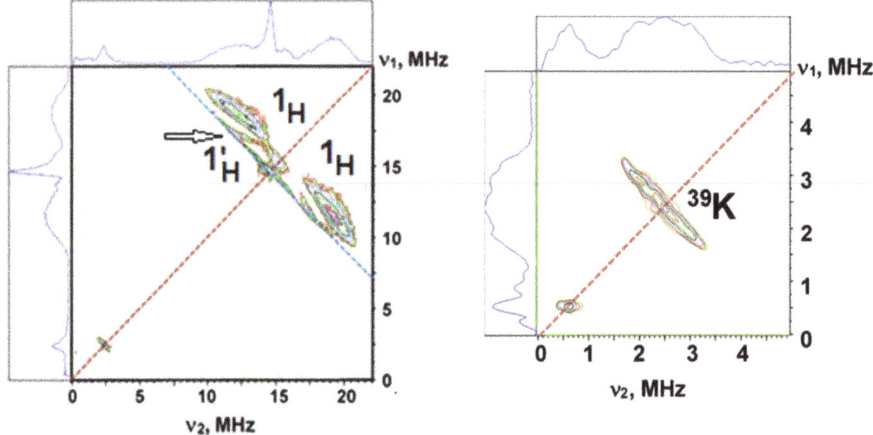

Figure 2. Contour representation of the X- (**left**) and Q-band (**right**) HYSCORE spectra of the superoxide radical in sample **II**. The time τ, between the first and the second microwave pulses, was 136 ns (X) and 200 ns (Q). The spectra were obtained by FT of the 2D time domain patterns containing 256×256 points with a 16 ns step in t_1 and t_2, which are the intervals between the second and the third microwave pulses, and the third and the fourth microwave pulses, respectively. The microwave frequency was 9.671 GHz (X) and 33.9915 GHz (Q), and the magnetic field was set to 343 mT (X) and 12,125 T (Q), and the temperature was 15 K.

The Q-band HYSCORE spectrum of sample **II** (Figure 2, right) shows an extended straight ridge around a diagonal frequency of 2.4 MHz with a total length of ~2 MHz. This line is produced by the interaction with ^{39}K (nuclear spin $I = 3/2$) possessing a Zeeman frequency of 2.409 MHz in the applied magnetic field 1212.5 mT. The natural abundance of ^{39}K is 93.26%. The second stable isotope ^{41}K has the same nuclear spin and natural abundance of 6.73%. The Zeeman frequency of ^{41}K in the specified field is 1.32 MHz. However, the spectrum in Figure 2 does not contain any features near this frequency on the diagonal line.

4. Discussions

4.1. Analysis of the ^{1}H HYSCORE Spectra

Quantitative analysis of ^{1}H cross-ridges from the HYSCORE spectra of the $O_2^{\bullet-}$ in samples **I** and **II**, based on linear regression of contour line shapes in ν_1^2 vs. ν_2^2 coordinates [26], gives isotropic and anisotropic components of hyperfine tensors in axial approximation for protons interacting with the electron spin of the superoxide. Detailed explanations and results of the analysis are provided in the Supplementary Materials, Section S1.

In particular, a representation of the cross-ridges 1_H from the spectra of sample **I** prepared with H_2O_2 and D_2O_2 in coordinates $\nu_1{}^2$ vs. $\nu_2{}^2$ gives anisotropic hyperfine coupling $T = 2.0 \pm 0.2$ MHz for the contributing proton(s) (Figure S4 and Table S2) [13]. The estimated value of $T \sim 2$ MHz is supported by the negligible deviation of the cross-ridges from the antidiagonal in the experimental spectra. A similar analysis of the cross-ridges 1_H and $1'_H$ with the visible deviation from the antidiagonal (Figure S5) in sample **II** provides the value of anisotropic coupling of $T \sim 7.0$ MHz, which significantly exceeds the hyperfine coupling of $T \sim 2$ MHz for protons interacting with $O_2{}^{\bullet-}$ in sample **I**.

4.2. Evaluation of the Proton Anisotropic Hyperfine Couplings

The anisotropic couplings of $T \sim 2$ MHz and ~7 MHz for the superoxide-proton interactions in samples **I** and **II** indicate different relative locations of the O–O molecule and protons. The value of ~7 MHz is closer to previously reported ^{1}H hyperfine couplings of $T \sim 10$ MHz and 9.8 MHz for a species with similar g-tensor generated on from KO_2 reacting with water in a $H_2O/DMSO$ mixture in the presence of ubiquinone-10 [22] and a Na pre-covered MgO surface [27], respectively. The anisotropic hyperfine tensor for the proton located near the $O_2{}^{\bullet-}$ is the result of a magnetic dipole–dipole interaction with an unpaired π spin density distributed approximately equally over two oxygens. The tensor depends on a proton position relative to the O–O bond and is generally rhombic [28]. When the interaction between the electron and the proton spins is described by the point dipole approximation the anisotropic parameter is defined by the expression $T = \frac{79}{r^3}$ (MHz) [13], where r is the distance between spins.

The dipole–dipole interaction is described by an axially symmetric tensor with diagonal principal values

$$T = [T_{xx}, T_{yy}, T_{zz}] = T\,[-1, -1, 2] \tag{1}$$

in the principal axes coordinate, with the **z** axis directed along the $\vec{r}$ direction.

A proton near the superoxide oxygens O_1 and O_2, carrying unpaired spin densities ρ_1 and ρ_2, experiences a local magnetic field, which is a vector sum of two contributions depending on the O_1–H (r_1) and the O_2–H (r_2) distances (Figure S9). The principal values of the rhombic hyperfine tensor in this case are [28]

$$T = [1/2\,(T_1 + T_2 - 3\delta),\ -(T_1 + T_2),\ \frac{1}{2}(T_1 + T_2 + 3\delta)] \tag{2}$$

where $\delta = [T_1{}^2 + T_2{}^2 + 2T_1T_2\cos(2\alpha + 2\beta)]^{1/2}$, $T_1 = 79\rho_1/r_1{}^3$, $T_2 = 79\rho_2/r_2{}^3$. Equation (2) transforms to the traditional axial form

$$T = [-(T_1 + T_2),\ -(T_1 + T_2),\ 2(T_1 + T_2)] \tag{3}$$

for the proton located on the O_1—O_2 line with $\beta = 180°$ and $\delta = T_1 + T_2$.

The relations between the sides and angles of the triangle HO_1O_2 (Figure S8)

$$r_2^2 = [r_1^2 + r_{O-O}^2 - 2r_1 r_{O-O}\cos\beta]^{1/2} \text{ and } \alpha = \arcsin[\frac{r_1}{r_2}\sin\beta] \tag{4}$$

allow us to define the tensor components based on one distance (r_1, O_1–H distance) and one angle (β, angle between H–O_1 and O_1–O_2).

It has been shown that the unpaired spin density is almost equally ($\rho_{1,2} \sim 0.5$) distributed on the $2p_\pi{}^x$ of each oxygen in superoxide radicals studied in a solution [29], on a MgO surface [30], and generated in TiAlPO-5 [31]. The reported O–O distance in the superoxide varies between 1.32–1.35 Å [29,32–34] and increases under the influence of hydrogen bonds [29,32].

For a direct comparison with the HYSCORE determined values of $T = 2$ [13], 7 (this work) and 10 MHz [22,27], we calculated the term $T_1 + T_2$ from Equation (2), which lacks the rhombic term 3δ and is equal to $T = 39.5 \left[\frac{1}{r_1^3} + \frac{1}{r_2^3}\right]$ for $\rho_1 \approx \rho_2 \approx 0.5$. This term is shown in the form of contours, where each point defines r_1 and β with the selected T (2, 7, or 10 MHz) (Figure 3), i.e., as a function of the O_1–H distance and the angle β between the H–O_1 and O_1–O_2 directions. Calculated graphs show that $T = 2$ MHz corresponds to the H–O_1 distance 2.97–3.4 Å for the angles $\beta < 180°$. On the contrary, this distance is ~1.0–1.2 Å less for $T = 7$ or 10 MHz (Table 2).

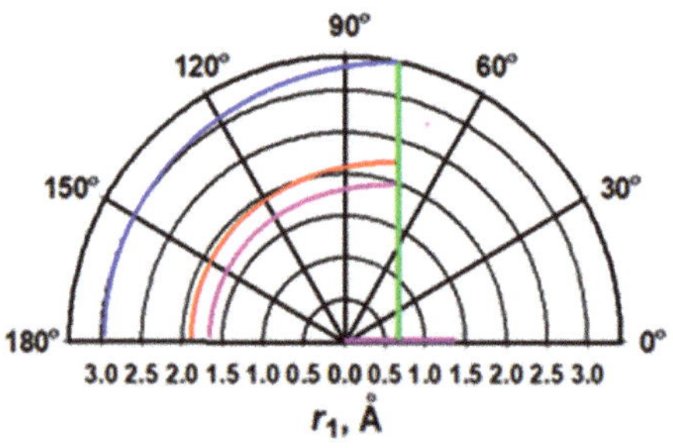

Figure 3. Polar graphs (r_1, β) with representative contours $T = 39.5 \left[\frac{1}{r_1^3} + \frac{1}{r_2^3}\right]$ equal to 2.0 (blue), 7.0 (red), and 10 (pink) MHz calculated for a model with one superoxide oxygen located at the (0,0) point of the coordinates and a second oxygen at the point (1.33 Å, 0°). O_1–O_2 distance equals to 1.33 Å marked in purple along the line with $\beta = 0°$. Green line normal to r_1 axis corresponds to the middle of the O_1–O_2 distance. Adapted by permission from Copyright Clearance Center: Springer Nature, Samoilova et al. [13]. Copyright 2022.

Table 2. Distances of a proton location relative to a superoxide radical at different values of the anisotropic hyperfine coupling.

T, MHz	r_1 (H–O_1), Å	β, Degree	r_1 at $\beta = 180°$, Å	r_1 at $\beta = 120°$, Å	References [a]
2	2.97–3.4	180–75	2.97	3.08	[13]
7	1.89–2.24	180–73	1.89	1.95	This work
10	1.67–1.99	180–70	1.67	1.71	[22,27]

[a] References, where parameters T from first column were experimentally determined. Adapted by permission from Copyright Clearance Center: Springer Nature, Samoilova et al. [13]. Copyright 2022.

Available models of a hydration shell for the aqueous $O_2^{\bullet-}$ include four water molecules, with two waters forming hydrogen bonds with each oxygen atom (Figure S9). The hydrogen bond lengths vary between 1.72–1.94 Å [30,35,36]. The lower limit of the H–O_1 distance 1.89 Å obtained for $T = 7$ MHz is still within the interval shown above.

4.3. Interaction with ^{39}K Nucleus

^{39}K possesses the nuclear spin $I = 3/2$ with a natural abundance of 93.26%. The orientation disordered HYSCORE spectra of the $S = 1/2$ and $I = 3/2$ system are influenced by hyperfine and nuclear quadrupole interactions. Both of them are anisotropic. Available data about the nuclear quadrupole coupling constant in ^{39}K$^+$ state indicate that it is quite small [37,38]. Model simulations of the HYSCORE spectra from $I = 3/2$ nuclei with hyperfine and nuclear quadrupole tensors satisfying the conditions [$\nu_I > A_{ZZ} > Q_{ZZ}$] have shown that the hyperfine interaction creates cross-ridges normal to the diagonal line in the (++) quadrant, whereas the nuclear quadrupole interaction produces an additional splitting of these cross-features in the direction parallel to the diagonal [39]. This simple manual is helpful for the qualitative analysis of the observed spectrum from ^{39}K (Figure 2).

The spectrum consists of a straight segment normal to the diagonal of the (++) quadrant and is located symmetrically relative to the (ν_{39K}, ν_{39K}) diagonal point. The length of the ridge is about ~2 MHz along each coordinate. A projection of the spectrum on each coordinate and stacked presentation of the spectrum shows a weakly resolved triplet structure of the ridge with the hyperfine splitting A ~ 1 MHz between cross-peaks with permutated coordinates of ~2 and 3 MHz. The spectrum does not show any additional splitting of the ridge along the diagonal, assuming the small value of the quadrupole coupling constant of ^{39}K [39].

The crystal structure of an α-potassium superoxide shows an octahedral environment of the superoxide ion near K$^+$ ions with smallest contact distances of 2.71 Å in the direction parallel to the O–O bond and another 2.92 Å away, in the direction approximately normal to the O–O bond [40]. The calculated value of the principal component ($T_1 + T_2$) of the hyperfine tensor defined by Equation (2) for the two indicated locations of ^{39}K$^+$ using formulae $T = (3.7/2)[1/r_1{}^3 + 1/r_2{}^3]$ is equal to 0.12 and 0.148 MHz, respectively. The anisotropic width of the single-quantum transitions in the powder spectrum $3T/2$ does not explain the line splitting ~1 MHz from ^{39}K, as detected in the HYSCORE spectrum. Consequently, there remains only one source resulting in the observed line shape—the isotropic hyperfine interaction. However, the simulation of spectra with parameters a ~ 1 MHz and T ~ 0.15 MHz did not reproduce the presence of a spectral intensity around the diagonal between two peaks with a ~1 MHz splitting. Additional ideas about hyperfine interactions between ^{39}K and the $O_2{}^{\bullet-}$ can be obtained by taking into account the available data on hyperfine couplings between a superoxide and ^{23}Na or ^{133}Cs on the MgO surface.

4.4. Comparison with Superoxide on a Na or Cs Pre-Covered MgO Surface

The EPR spectrum of superoxide species formed on a Na pre-covered MgO surface [27] shows a formation of two species with g_{zz} values equal to 2.091 and 2.14. The $g_{x,y}$ components of both species are close to g = 2 and produce a single intensive line with a width of 1.5–2.0 mT in the spectrum. The species with g_{zz} = 2.091 have been assigned to superoxide ions on Mg^{2+} matrix sites. The value g_{zz} = 2.14 is within the range typical for superoxide anions stabilized on monovalent cations; thus, the corresponding EPR signal was designated as a surface $O_2{}^{\bullet-}$ − Na$^+$ adduct.

The HYSCORE spectra collected in the $g_{x,y}$ area of the Na/MgO sample contain cross-features from the ^{23}Na and ^{1}H nuclei (Figure S10). Two of them belong to ^{23}Na with strong and weak hyperfine couplings ~17 MHz and ~<3 MHz, respectively. The ^{1}H spectrum consists of two extended ridges with a clearly visible deviation from the antidiagonal crossing (ν_{1H}, ν_{1H}) point of the diagonal, indicating a strong anisotropic interaction between the superoxide and a proton. Computer simulations of the spectrum have provided hyperfine tensors $a = -15 \pm 2$ MHz, $T = (-0.1, 3.1, -3.0,)$ MHz (± 0.5 MHz) for the strongly coupled ^{23}Na from the $O_2{}^{\bullet-}$–Na$^+$ adduct and $a = -5.0 \pm 0.5$ MHz, $T = (-9.8, 19.6, -9.8)$ MHz (± 0.1 MHz) for the proton that produced extended cross-ridges. It was suggested that this proton belongs to the superoxide stabilized on the Mg^{2+} matrix site in the proximity of a surface OH$^-$. Our analysis (Figure 3) shows that the ^{1}H coupling of T ~ 10 MHz indicates the formation of an H-bond between the superoxide and proton with an O–H distance of ~1.7–2.0 Å. Furthermore, the authors have proposed that the line around the diagonal (ν_{Na}, ν_{Na}) point is produced by remote ^{23}Na nuclei that are randomly distributed between 3.5 Å and 4.5 Å away (i.e., $0.5 > |T| > 0.2$ MHz in the point dipole approximation), and the species of both g_{zz} contribute to this feature [27].

To compare the experimental and calculated spectra, this feature was calculated using a single hyperfine tensor $a = -2.5$ MHz, $T = (-0.5, -0.5, 1.0)$ MHz (Figure S10). On the other hand, we estimated $T_{max} = (20.9/2)[1/r_1{}^3 + 1/r_2{}^3]$ for two models of the ^{23}Na location relative to the superoxide (O–O length 1.33 Å) using a Na–O distance of 2.38–2.39 Å reported for the orthorhombic structure of sodium superoxide [41,42]. The corresponding values are T_{max} ~ 1 MHz and 1.5 MHz for the ^{23}Na locations on the line extending the O–O bond and normal to the middle of the O–O bond that is consistent with the value of

$T_{max}/2$ used for calculating the spectrum. This estimate shows that the signal assigned to randomly distributed ^{23}Na nuclei at distances in the range between 3.5 Å and 4.5 Å can be produced just by one nucleus located at the Na–O distance found in the crystal structures. More convincing conclusions about the nature of the signal from weakly coupled Na could be supported by the relative intensities of two sodium signals which are not available from the published spectra.

The increased hyperfine parameters for strongly coupled ^{23}Na is explained by the formation of the ionic sodium superoxide $Na^+O_2^-$ as symmetrical triangular molecules with an interatomic distance of 1.96 Å that was deduced from vibrational spectra [43]. One can recalculate the characteristic hyperfine parameters of the $^{23}Na^+O_2^{\bullet-}$ species obtained in this work for a superoxide interacting with a nucleus of $^{39}K^+$. The ratio of ^{23}Na/^{39}K magnetic moments is 5.56. Then, the parameters $|a|$ =15 MHz and $|T|$ = 3 MHz found for the ^{23}NaO$_2$ species will give formal values of a = 2.7 MHz and T = 0.54 MHz for ^{39}K. The ratio of the atomic isotropic hyperfine constants 927.1 MHz (^{23}Na, 3s) and 228.5 MHz (^{39}K, 4s) [44] calculated for unit spin density is 4.06. It leads to a decreasing a value of 0.67 MHz. Another possible factor that may reduce the hyperfine parameters of ^{39}K compared to ^{23}Na is the larger value of its ionic radius (1.02 Å for ^{23}Na and 1.38 Å for ^{29}K). Simulations of the ^{39}K HYSCORE spectra with estimated parameters a = 0.6–0.7 MHz and T = 0.6 MHz confirm the increased total length of the cross-ridge and its line shape without well pronounced maxima (Figure S11). Thus, the recalculation analysis assuming similar structural motifs of the MO$_2$ species predicts significantly lower hyperfine couplings for ^{39}K. However, the value of the anisotropy parameter is greater than the calculated ~0.15 MHz using the crystallographic structure. That increase provides an extended line shape from ^{39}K, which is consistent with the experimental HYSCORE spectra.

DFT calculations of the hyperfine parameters for $O_2^{\bullet-}$ in Na/MgO were performed for different elements of the MgO surface [27]. Based on the analysis of the full set of **g** and **A** tensors, the NaO$_2$ species formed on an edge site of the MgO surface, where O$_2$ is simultaneously bound to Na and to MgO, give the best description of the large g_{zz} and the ^{23}Na hyperfine interaction observed in the EPR experiments. On the other hand, the DFT analysis of the ^{1}H tensor determined from the HYSCORE spectrum of $O_2^{\bullet-}$ in Na/MgO has not been carried out even for the proposed $O_2^{\bullet-}$/HMgO center, although this tensor possesses an a_{iso} = −5 MHz and T_{max} = 19.6 MHz, which significantly exceeds the a_{iso} ~ 0 MHz and T_{max} ~ 10 MHz reported for $O_2^{\bullet-}$ at the surface of MgO [45].

One can note that the exchangeable proton with T_{max} ~ 20 MHz was found in the HYSCORE spectra of the superoxide in the KO$_2$/DMSO/H$_2$O mixture in the presence of ubiquinone-10 [24]. However, the X-band HYSCORE spectra obtained in this work were not suitable for the detection of ^{39}K signals due to a low Zeeman frequency in this band.

Other experimental examples relevant to this work are EPR studies of ^{23}NaO$_2$, ^{39}KO$_2$, ^{87}RbO$_2$, and ^{133}CsO$_2$ in rare gas matrices [46] and superoxides on a ^{133}Cs/MgO surface [47]. Similar to Table 1, the g_{zz} component of these species varies between 2.10–2.12, and the alkali metals ^{23}Na and ^{133}Cs produce a resolved hyperfine splitting of the g_{zz} and $g_{xx,yy}$ components presented in Table 3.

Table 3. EPR parameters of MO$_2$ species in a rare gas and on a MgO surface.

Sample	g_{zz}	T_{11}	T_{22}	T_{33}	a_{iso}	A_0	$\rho \times \bullet 10^{-3}$	Ref. [a]
^{23}NaO$_2$	2.111	−2.8	1.68	−1.12	9	927.1	9.7	[46]
$O_2^{\bullet-}$ on Na/MgO	2.14	−0.1	3.1	−3.0	−15	927.1	16.2	[27]
^{133}CsO$_2$	2.107	−0.56	0.84	−0.28	14	2464	5.7	[46]
$O_2^{\bullet-}$ on Cs/MgO	2.120	5.1	2.6	−7.7	29.6	2464	12.0	[47]

[a] T_{ii} components of an anisotropic tensor, isotropic constant a_{iso}, and atomic isotropic hyperfine constant A_0 are in MHz.

A comparison with rare-gas matrix-trapped MO_2 molecules shows that the surface-stabilized complexes are characterized by larger a_{iso} and g_{zz} parameters. The differences between isolated MO_2 molecules and surface-adsorbed NaO_2/MgO species have been examined with the help of DFT calculations, which illustrated the role of the matrix in the stabilization of the superoxide with particular magnetic characteristics [27]. A set of surface sites was compatible with the observed experimental results, which are characterized by a mutual interaction between the superoxide anion and the Mg^{2+} matrix ions and adsorbed Na^+ species. Particularly, in the case of the surface-stabilized NaO_2 complex, the unpaired electron is localized in a π^* orbital lying in the O_2–Na plane, whereas the π^* orbital hosting the unpaired electron is found to be perpendicular to the M(OO) plane for matrix-trapped NaO_2 in the rare gas. This difference will influence the hyperfine interaction with the alkaline atom nucleus [46].

5. Conclusions

Our experiments with the $TO_2 \ldots O_2^{\bullet-}$ catalyst obtained with the joint addition of peroxide and a KOH solution of alcohol show the formation of the superoxide radical with the atypical $g_{zz} = 2.10$. A similar EPR signal was previously observed in various systems containing alkaline ions that allowed us to suggest the special role of K^+ ions in the $O_2^{\bullet-}$ stabilization. However, ^{39}K nucleus(i) did not produce any resolved features in the reported EPR spectra. In this work, we applied 2D HYSCORE spectroscopy to characterize hyperfine interactions between the $O_2^{\bullet-}$ and the ^{39}K and ^1H nuclei in its environment. Q-band HYSCORE spectra have shown the presence of ^{39}K near the superoxide. A comparison of the magnetic characteristics and electronic configuration defining the isotropic coupling of ^{39}K with ^{23}Na in Na/MgO [27] allowed us to predict significantly smaller ^{39}K hyperfine couplings for the similar structural MO_2 motifs. The estimated values give a reasonable agreement between the calculated and experimental ^{39}K HYSCORE spectra. Another finding of this superoxide is the existence of a strongly coupled ^1H with the anisotropic coupling $T \sim 7$ MHz, which suggests the formation of an H-bond with an O–H distance of <2Å. So far, an interaction with the closely located alkaline ion and the strongly coupled proton $T \sim 10$ MHz has been reported for the $O_2^{\bullet-}$ on Na/MgO only [27]. However, two different $O_2^{\bullet-}$ species with $g_{zz} = 2.014$ and 2.091 were present in this sample, and strongly coupled ^{23}Na with $a = 15$ MHz and the ^1H with anisotropic coupling $T = 9.8$ MHz were assigned to different species, $- O_2^{\bullet-}/NaMgO$ and $O_2^{\bullet-}/HMgO$, respectively. Only one type of $O_2^{\bullet-}/KTiO_2$ species with $g_{zz} = 2.10$ was found in our work. This means that signals from ^{39}K and a strongly coupled proton with $T \sim 7$ MHz in the HYSCORE spectra are produced by interactions with this superoxide and should be considered as the elements of its structure. Therefore, independent data to support the hypothesis that the presence of a stable superoxide radical in systems containing alkali metals requires the simultaneous proximity of M^+ ion(s) and the formation of a hydrogen bond with its environment are still needed.

Supplementary Materials: The following supporting information can be downloaded at: https://www.mdpi.com/article/10.3390/inorganics11070274/s1, Figure S1: X- and Q-band EPR spectra of the superoxide radical in sample **I**; Figure S2: X-band two-pulse ESE field-swept spectrum and its first derivative of the TiO_2 surface produced from $Ti(OiPr)_4$ exposed to H_2O_2/KOH; Figure S3: Contour representation of the HYSCORE spectra of the superoxide radical in the sample **I** (a) and in the similar sample prepared using D_2O_2 (b); Figure S4: Plots of cross-ridges 1_H from HYSCORE spectra of the superoxide radical in sample **I** in the $(\nu_1)^2$ vs. $(\nu_2)^2$ coordinate system; Figure S5: Plots of cross-ridges 1_H and $1_{H'}$ from HYSCORE spectrum of the superoxide radical in sample **II** in the $(\nu_1)^2$ vs. $(\nu_2)^2$ coordinate system; Figure S6: Q-band field-sweep 2-pulse ESE spectrum (a) and Q-band Davies ENDOR spectra (b) of the superoxide radical in sample **I**; Figure S7: Q-band Davies ENDOR spectrum of the superoxide radical in sample **II**; Figure S8: Definition of the distances and angles describing the location of the proton relative to O_1 and O_2 of the superoxide radical; Figure S9: Structural model of $[O_2(H_2O)_4]^-$; Figure S10: Experimental and simulated HYSCORE spectra of the $O_2^{\bullet-}$ species on Na/MgO; Figure S11: Experimental and simulated ^{39}K HYSCORE spectra; Table S1: g-tensors assigned to the $O_2^{\bullet-}$ radical in different TiO_2 samples; Table S2: ^1H hyperfine tensor parameters determined from linear regressions of the cross-ridges; Section S1: Square frequency fitting of the ^1H HYSCORE spectra and its comparison with the pulsed ENDOR data [4,10,13,22,26,27,29,31,35,36,48–56].

Author Contributions: Conceptualization, R.I.S. and S.A.D.; methodology, R.I.S. and S.A.D.; validation, R.I.S. and S.A.D.; resources, S.A.D.; writing—original draft preparation, S.A.D.; writing—review and editing, S.A.D. and R.I.S.; project administration, S.A.D.; funding acquisition, S.A.D. All authors have read and agreed to the published version of the manuscript.

Funding: This work was supported by Grant DE-FG02-08ER15960 (S.A.D.) from the Chemical Sciences, Geosciences and Biosciences Division, the Office of Basic Energy Sciences, and the Office of Sciences, U.S. Department of Energy.

Institutional Review Board Statement: Not applicable.

Informed Consent Statement: Not applicable.

Data Availability Statement: All data and information recorded or analyzed throughout this study are included in this paper.

Acknowledgments: The authors are grateful to Andrei Astashkin (University of Arizona) for very useful discussions and Quan Lam for a critical reading of the manuscript.

Conflicts of Interest: The authors declare no conflict of interest.

References

1. Hayyan, M.; Hashim, M.A.; AlNashef, I.M. Superoxide Ion: Generation and Chemical Implications. *Chem. Rev.* **2016**, *116*, 3029–3085. [CrossRef]
2. Afanas'ev, I.B. The Oxygen Radical-anion O_2^- in Chemical and Biochemical Processes. *Russ. Chem. Rev.* **1979**, *48*, 527–549. [CrossRef]
3. Sawyer, D.T.; Valentine, J.S. How Super Is Superoxide? *Acc. Chem. Res.* **1981**, *14*, 393–400. [CrossRef]
4. Anpo, M.; Che, M.; Fubini, B.; Garrone, E.; Giamello, E.; Paganini, M.C. Generation of Superoxide Ions at Oxide Surfaces. *Top. Catal.* **1999**, *8*, 189–198. [CrossRef]
5. Che, M.; Tench, A.J. Characterization and Reactivity of Molecular Oxygen Species on Oxide Surfaces. *Adv. Catal.* **1983**, *32*, 1–148.
6. Kokorin, A.I. Electron Spin Resonance of Nanostructured oxide semiconductors. In *Chemical Physics of Nanostructured Semiconductors*; Kokorin, A.I., Bahnemann, D.W., Eds.; CRC Press: Boca Raton, FL, USA, 2003; Chapter 8, pp. 203–263.
7. Chiesa, M.; Giamello, E.; Che, M. EPR Characterization and Reactivity of Surface-Localized Inorganic Radicals and Radical Ions. *Chem. Rev.* **2010**, *110*, 1320–1347. [CrossRef]
8. Sobańska, K.; Krasowska, A.; Mazur, T.; Podolska-Serafin, K.; Pietrzyk, P.; Sojka, Z. Diagnostic Features of EPR Spectra of Superoxide Intermediates on Catalytic Surfaces and Molecular Interpretation of Their g and A Tensors. *Top. Catal.* **2015**, *58*, 796–810. [CrossRef]
9. Giamello, E.; Rumori, P.; Geobaldo, F.; Fubini, B.; Paganini, M.C. The Interaction between Hydrogen Peroxide and Metal Oxides: EPR Investigations. *Appl. Magn. Res.* **1996**, *10*, 173–192. [CrossRef]
10. Dewkar, G.K.; Nikalje, M.D.; Sayyed, I.A.; Paraskar, A.S.; Jagtap, H.S.; Sudalai, A. An Exceptionally Stable Ti Superoxide Radical Ion: A Novel Heterogeneous Catalyst for the Direct Conversion of Aromatic Primary Amines to Nitro Compounds. *Angew. Chem. Int. Ed.* **2001**, *40*, 405–408. [CrossRef]
11. Dewkar, G.K.; Shaikh, T.M.; Pardhy, S.; Kulkarni, S.S.; Sudalai, A. Titanium superoxide catalyzed selective oxidation of phenols to p-quinones with aq. H_2O_2. *Indian J. Chem. B* **2005**, *44*, 1530–1532. [CrossRef]
12. Reddy, R.S.; Shaikh, T.M.; Rawat, V.; Karabal, P.U.; Dewkar, G.; Suryavanshi, G.; Sudalai, A. A Novel Synthesis and Characterization of Titanium Superoxide and its Application in Organic Oxidative Processes. *Catal. Surv. Asia* **2010**, *14*, 21–32. [CrossRef]
13. Samoilova, R.I.; Dikanov, S.A. Local environment of superoxide radical formed on the TiO_2 surface produced from $Ti(OiPr)_4$ exposed to H_2O_2. *Appl. Magn. Reson.* **2022**, *53*, 1089–1104. [CrossRef]
14. Jin, F.; Leitich, J.; von Sonntag, C. The superoxide radical reacts with tyrosine-derived phenoxyl radicals by addition rather than by electron transfer. *J. Chem. Soc. Perkin. Trans. II* **1993**, 1583–1588. [CrossRef]
15. Schunemann, V.; Lendzian, F.; Jung, C.; Contzen, J.; Barra, A.-L.; Sligar, S.G.; Trautwein, A.X. Tyrosine Radical Formation in the Reaction of Wild Type and Mutant Cytochrome $P450_{cam}$ with Peroxy. *J. Biol. Chem.* **2004**, *279*, 10919–10930. [CrossRef] [PubMed]
16. Hitchcock, D.I. The solubility of tyrosine in acid and alkali. *J. Gen. Physiol.* **1924**, *6*, 747–757. [CrossRef] [PubMed]
17. Höfer, P.; Grupp, A.; Nebenführ, H.; Mehring, M. Hyperfine Sublevel Correlation (HYSCORE) spectroscopy: A 2D ESR investigation of the squaric acid radical. *Chem. Phys. Lett.* **1986**, *132*, 279–282. [CrossRef]
18. Davies, E.R. A New Pulse Endor Technique. *Phys. Lett. A* **1974**, *47*, 1–2. [CrossRef]
19. Lunsford, J.H. ESR of Adsorbed Oxygen Species. *Catal. Rev.* **1973**, *8*, 135–157. [CrossRef]

20. Symons, M.C.R.; Eastland, G.W.; Denny, L.R. Effect of Solvation on the Electron Spin Resonance Spectrum of the Superoxide ion. *J. Chem. Soc. Faraday Trans. I* **1980**, *76*, 1868–1874. [CrossRef]
21. Narayana, P.A.; Suryanarayana, D.; Kevan, L. Electron Spin-Echo Studies of the Solvation Structure of Superoxide ion (O_2^-) in Water. *J. Am. Chem. Soc.* **1982**, *104*, 3552–3555. [CrossRef]
22. Samoilova, R.I.; Crofts, A.R.; Dikanov, S.A. Reaction of Superoxide Radical with Quinone Molecules. *J. Phys. Chem. A* **2011**, *115*, 11589–11593. [CrossRef]
23. Haseloff, R.; Ebert, B.; Damerau, W. Superoxide generation in alkaline dimethyl sulfoxide. *Anal. Chim. Acta.* **1989**, *218*, 179–184. [CrossRef]
24. Krager, K. *Superoxide in Aprotic Solvents*; The University of Iowa: Iowa City, IA, USA, 2003.
25. Hyland, K.; Auclair, C. The Formation of Superoxide Radical Anions by a Reaction Between O_2, OH^- and Dimethyl Sulfoxide. *Biochem. Biophys. Res. Comm.* **1981**, *102*, 531–537. [CrossRef] [PubMed]
26. Dikanov, S.A.; Bowman, M.K. Cross-peak lineshape of two-dimensional ESEEM spectra in disordered $S=1/2$, $I=1/2$ spin system. *J. Magn. Reson., Ser. A* **1995**, *116*, 125–128. [CrossRef]
27. Napoli, F.; Chiesa, M.; Giamello, E.; Preda, G.; Di Valentin, C.; Pacchioni, G. Formation of Superoxo Species by Interaction of O_2 with Na Atoms Deposited on MgO Powders: A Combined Continuous-Wave EPR (CW-EPR), Hyperfine Sublevel Correlation (HYSCORE) and DFT Study. *Chem. Eur. J.* **2010**, *16*, 6776–6785. [CrossRef]
28. Randall, D.W.; Gelasco, A.; Caudle, M.T.; Pecoraro, V.L.; Britt, R.D. ESE-ENDOR and ESEEM Characterization of Water and Methanol Ligation to a Dinuclear Mn(III)Mn(IV) Complex. *J. Am. Chem. Soc.* **1997**, *119*, 4481–4491. [CrossRef]
29. Janik, I.; Tripathi, G.N.R. The nature of the superoxide radical anion in water. *J. Chem. Phys.* **2013**, *139*, 014302. [CrossRef] [PubMed]
30. Chiesa, M.; Giamello, E.; Paganini, M.C.; Sojka, Z.; Murphy, D.M. Continuous Wave Electron Paramagnetic Resonance Investigation of the Hyperfine Structure of $^{17}O_2^-$ Adsorbed on the MgO Surface. *J. Chem. Phys.* **2002**, *116*, 4266–4274. [CrossRef]
31. Maurelli, S.; Vishnuvarthan, M.; Berlier, G.; Chiesa, M. NH_3 and O_2 interaction with tetrahedral Ti^{3+} ions isomorphously substituted in the framework of TiAlPO-5. A combined pulse EPR, pulse ENDOR, UV-Vis and FT-IR study. *Phys. Chem. Chem. Phys.* **2012**, *14*, 987–995. [CrossRef] [PubMed]
32. Dietzel, P.D.C.; Kremer, R.K.; Jansen, M. Tetraorganylammonium Superoxide Compounds: Close to Unperturbed Superoxide Ions in the Solid State. *J. Am. Chem. Soc.* **2004**, *126*, 4689–4696. [CrossRef]
33. Seyeda, H.; Jansen, M. A novel access to ionic superoxides and the first accurate determination of the bond distance in O_2^-. *J. Chem. Soc. Dalton Trans.* **1998**, *6*, 875–876. [CrossRef]
34. Halverson, F. Comments on Potassium Superoxide Structure. *J. Phys. Chem. Solids* **1962**, *23*, 207–214. [CrossRef]
35. Antonchenko, V.Y.; Kryachko, E.S. Interaction of Superoxide O_2^- with Water Hexamer Clusters. *Chem. Phys.* **2006**, *327*, 485–493. [CrossRef]
36. Antonchenko, V.Y.; Kryachko, E.S. Structural, Energetic, and Spectroscopic Features of Lower Energy Complexes of Superoxide Hydrates $O_2^-(H_2O)_{1-4}$. *J. Phys. Chem. A* **2005**, *109*, 3052–3059. [CrossRef] [PubMed]
37. Barkhuijsen, H.; de Beer, R.; Deutz, A.F.; van Ormondt, D.; Völkel, G. Observation of Potassium Hyperfine Interactions in X-irradiated KH_2AsO_4 through the Method of Electron Spin Echo Envelope Modulation. *Solid State Commun.* **1984**, *49*, 679–684. [CrossRef]
38. Nellutla, S.; Morley, G.W.; van Tol, J.; Pati, M.; Dalal, N.S. Electron Spin Relaxation and ^{39}K Pulsed ENDOR Studies on Cr^{5+}-doped K_3NbO_8 at 9.7 and 240 GHz. *Phys. Rev. B* **2008**, *78*, 054426. [CrossRef]
39. Gutjahr, M.; Böttcher, R.; Pöppl, A. Analysis of Correlation Patterns in Hyperfine Sublevel Correlation Spectroscopy of $S = 1/2$, $I = 3/2$ Systems. *Appl. Magn. Reson.* **2002**, *22*, 401–414. [CrossRef]
40. Abrahams, S.C.; Kalnajs, J. The Crystal Structure of α-Potassium Superoxide. *Acta Crystallogr.* **1955**, *8*, 503–506. [CrossRef]
41. Carter, G.F.; Templeton, D.H. Polymorphism of Sodium Superoxide. *J. Am. Chem. Soc.* **1953**, *75*, 5247–5249. [CrossRef]
42. Deng, N.; Yang, G.; Wang, W.; Qiu, Y. Structural Transitions and Electronic Properties of Sodium Superoxide at High Pressures. *RSC Adv.* **2016**, *6*, 67910–67915. [CrossRef]
43. Smardzewski, R.R.; Andrews, L. Raman Spectra of the Products of Na and K Atom Argon Matrix Reactions with O_2 Molecules. *J. Chem. Phys.* **1972**, *57*, 1327–1333. [CrossRef]
44. Morton, J.R.; Preston, K.F. Atomic Parameters for Paramagnetic Resonance Data. *J. Magn. Reson.* **1978**, *30*, 577–583. [CrossRef]
45. Giamello, E.; Ugliengo, P.; Garrone, E.; Che, M.; Tench, A.J. Experimental Evidence for the Hyperfine Interaction between a Surface Superoxide Species on MgO and a Neighbouring Hydroxylic Proton. *J. Chem. Soc. Faraday Trans. I* **1989**, *85*, 3987–3994. [CrossRef]
46. Lindsay, D.M.; Herschbach, D.R.; Kwiram, A.L. ESR of Matrix Isolated Alkali Superoxides. *Chem. Phys. Lett.* **1974**, *25*, 175–181. [CrossRef]
47. Chiesa, M.; Paganini, M.C.; Giamello, E.; Murphy, D.M. Partial Ionization of Cesium Atoms at Point Defects over Polycrystalline Magnesium Oxide. *J. Phys. Chem. B* **2001**, *105*, 10457–10460. [CrossRef]
48. Antcliff, K.L.; Murphy, D.M.; Griffithsa, E.; Giamello, E. The interaction of H_2O_2 with exchanged titanium oxide systems (TS-1, TiO_2, [Ti]-APO-5, Ti-ZSM-5). *Phys. Chem. Chem. Phys.* **2003**, *5*, 4306–4316. [CrossRef]
49. Ramaswamy, V.; Awati, P.; Ramaswamy, A.V. Epoxidation of indene and cyclooctene on nanocrystalline anatase titania catalyst. *Top. Catal.* **2006**, *38*, 251–259. [CrossRef]

50. Tengvall, P.; Lundström, I.; Sjöqvist, L.; Elwing, H.; Bjursten, L.M. Titanium-hydrogen peroxide interaction: Model studies of the influence of the inflammatory response on titanium implants. *Biomaterials* **1989**, *10*, 166–175. [CrossRef]
51. Green, J.; Carter, E.; Murphy, D.M. Interaction of molecular oxygen with oxygen vacancies on reduced TiO_2: Site specific blocking by probe molecules. *Chem. Phys. Lett.* **2009**, *477*, 340–344. [CrossRef]
52. Carter, E.; Carley, A.F.; Murphy, D.M. Evidence for O^{2-} Radical Stabilization at Surface Oxygen Vacancies on Polycrystalline TiO_2. *J. Phys. Chem. C* **2007**, *111*, 10630–10638. [CrossRef]
53. Liu, F.; Feng, N.; Wang, Q.; Xu, J.; Qi, G.; Wang, W.; Deng, F. Transfer Channel of Photoinduced Holes on a TiO_2 Surface As Revealed by Solid-State Nuclear Magnetic Resonance and Electron Spin Resonance Spectroscopy. *J. Am. Chem. Soc.* **2017**, *139*, 10020–10028. [CrossRef] [PubMed]
54. Rahemi, V.; Trashin, S.; Hafideddine, Z.; Meynen, V.; Van Doorslaer, S.; De Wael, K. Enzymatic sensor for phenols based on titanium dioxide generating surface confined ROS after treatment with H_2O_2. *Sens. Actuators B Chem.* **2019**, *283*, 343–348. [CrossRef]
55. Yu, J.; Chen, J.; Li, C.; Wang, X.; Zhang, B.; Ding, H. ESR Signal of Superoxide Radical Anion Adsorbed on TiO_2 Generated at Room Temperature. *J. Phys. Chem. B* **2004**, *108*, 2781–2783. [CrossRef]
56. Dikanov, S.A.; Tyryshkin, A.M.; Bowman, M.K. Intensity of cross-peaks in HYSCORE spectra of $S = 1/2$, $I = 1/2$ spin systems. *J. Magn. Reson.* **2000**, *144*, 228–242. [CrossRef] [PubMed]

Communication

Binder-Free CoMn$_2$O$_4$ Nanoflower Particles/Graphene/Carbon Nanotube Composite Film for a High-Performance Lithium-Ion Battery

Xin Tong [1,*,†], Bo Yang [1,†], Fei Li [2], Manqi Gu [1], Xinxing Zhan [1], Juan Tian [1], Shengyun Huang [3,*] and Gang Wang [2,*]

[1] School of Chemistry and Material Science, Guizhou Normal University, Guiyang 550001, China
[2] State Key Laboratory of Photon-Technology in Western China Energy, Institute of Photonics & Photon-Technology, Northwest University, Xi'an 710127, China
[3] Ganjiang Innovation Academy, Chinese Academy of Sciences, Ganzhou 341000, China
*   Correspondence: tongxin@gznu.edu.cn (X.T.); shyhuang@gia.cas.cn (S.H.); gangwang@nwu.edu.cn (G.W.)
†   These authors contributed equally to this work.

Abstract: Manganese-based bimetallic oxides show a high theoretical specific capacity, making them a potential next-generation lithium-ion battery anode material. However, as with metal oxide anode materials, aggregation, volume expansion, and poor conductivity are the main obstacles. In this manuscript, flexible CoMn$_2$O$_4$/graphene/carbon nanotube films were successfully prepared through a facile filtration strategy and a subsequent thermal treatment process. When used as anodes for lithium batteries, these films can be pressed onto nickel foam without other conductive additives and binders, which simplifies the manufacturing process. When used as an anode in the lithium-ion battery, CoMn$_2$O$_4$/GR/CNT film exhibits a high discharge capacity of 881 mAh g^{-1} after 55 cycles. This value is ~2 times higher than the discharge capacity of CoMn$_2$O$_4$. The three-dimensional GR/CNT carrier effectively dispersed CoMn$_2$O$_4$, preventing its aggregation and alleviating the problem of volume expansion.

Keywords: lithium-ion battery; anode materials; CoMn$_2$O$_4$ nanoflower; graphene; film

Citation: Tong, X.; Yang, B.; Li, F.; Gu, M.; Zhan, X.; Tian, J.; Huang, S.; Wang, G. Binder-Free CoMn$_2$O$_4$ Nanoflower Particles/Graphene/ Carbon Nanotube Composite Film for a High-Performance Lithium-Ion Battery. *Inorganics* **2023**, *11*, 314. https://doi.org/10.3390/ inorganics11080314

Academic Editors: Roberto Nisticò, Torben R. Jensen, Luciano Carlos, Hicham Idriss and Eleonora Aneggi

Received: 17 May 2023
Revised: 19 July 2023
Accepted: 23 July 2023
Published: 25 July 2023

1. Introduction

The development and progress of society have made people heavily dependent on fossil fuels. However, this dependence has led to energy depletion and serious environmental pollution problems. Exploring various devices for efficient energy storage and conversion is crucial to enable the effective use of energy. In recent years, lithium-ion batteries (LIBs) have received extensive attention due to their large capacity, high power density, and good cycle stability. LIBs have emerged as critical power sources for a wide range of applications, including portable electronics and electric vehicles [1]. As demand for these devices continues to grow, it is crucial to improve their energy density, rapid charge capability, power density, as well as their durability [2–4]. Research on lithium-ion batteries has focused on developing electrode materials with higher capacity, greater safety, and lighter weight. Unfortunately, the current anode materials, such as graphite and carbon-based electrode materials, are unable to meet these requirements due to their poor rate capability and low theoretical capacity [5,6]. To overcome this technological bottleneck, high-performance anode materials are required.

Numerous micro-/nano-structured metal oxides have garnered significant attention as potential anode materials due to their high energy density. In recent decades, transition metal oxides (TMOs) have gained significant attention due to their structural diversity and unique electron transport properties that result from the properties of the outer d electrons. They offer promising applications in various fields such as spintronics and thermoelectrics [7–9]. As research continues, complex transition metal oxides, with a general formula of $X_xY_{3-x}O_4$ (X, Y = Fe, Co, Ni, Zn, Mn, etc.), have emerged as superior alternatives. Binary transition metal oxides typically adopt a spinel structure. By incorporating additional transition metals, more active sites can be introduced and the crystal structure and valence state of any metal element can be optimized. The synergistic effect of the two metal oxides during charging and discharging, along with the attractive interactions between different cations, can facilitate a greater number of redox reactions. Consequently, binary transition metal oxides exhibit higher reversible storage capacity and enhanced electronic conductivity compared with single metal oxides [10]. This excellent electrochemical activity positions them with great potential for energy storage and conversion applications [11]. Among various transition metal oxide materials, manganese-based oxides have garnered considerable attention due to their lower operating voltage and higher energy density. For instance, Yu et al. employed a simple solvothermal and thermal treatment process to grow hierarchical $CoM_{3-x}O_4$ arrays/nanostructures on stainless steel substrates [12]. By adjusting the volume ratio of solvent, $CoMn_2O_4$ nanowires and $MnCo_2O_4$ nanosheets were synthesized. The layered structure of these materials facilitates contact between the electroactive surface and the electrolyte, resulting in excellent electrochemical properties and cycling stability. In particular, the $CoMn_2O_4$ material, with its combination of cobalt's high oxidation potential and manganese's high capacity and excellent electron transport properties, exhibits great promise as an anode material for lithium-ion batteries [13]. However, their commercial application is hindered by inferior cycling stability and poor rapid charge capability, which are possibly due to sluggish electronic/ionic diffusion and unsatisfactory structural stability.

To address these issues, many strategies have been attempted. One useful method is to build different nanostructures from transition metal oxides using various strategies such as hierarchical mesoporous microspheres [14], hollow nanospheres [15], yolk–shell microspheres [16], hollow nanofibers [17], and bubble-like structures [18]. Hu et al. prepared hierarchical mesoporous $CoMn_2O_4$ microspheres using a facile solvothermal carbon templating method [14]. This material exhibited excellent cycling stability and high discharge capacity due to its unique hierarchical structure, which enhances electrolyte diffusion, shortens the Li^+ ion diffusion length, and accommodates volume-change-induced strain during cycling. Zhou et al. reported the synthesis of double-shell $CoMn_2O_4$ hollow microcubes using a co-precipitation and thermal annealing method [19]. The unique structure allowed the synthesized $CoMn_2O_4$ material to have a reversible capacity of 830 mA h g^{-1} at a current density of 200 Ag^{-1}. A discharge capacity of 406 mA h g^{-1} was still maintained after 50 cycles at 800 mA g^{-1}, demonstrating excellent cycling performance. Another important strategy is to combine TMOs with conductive carbon materials, especially graphene [20], nitrogen-doped graphene [21], and carbon hollow spheres [22]. The introduction of carbon materials not only provides high electrical conductivity but also serves as a support material to alleviate the volume expansion of the metal oxide for energy application. Cai et al. have synthesized a composite material consisting of $CoMn_2O_4$ nanoparticles anchored on reduced graphene oxide (rGO) sheets [23]. The rGO can prevent the aggregation of $CoMn_2O_4$ nanoparticles, which could otherwise reduce the active surface area of the cathode material and limit its performance. Furthermore, the flexible and porous structure of rGO can help to build a suitable flexible carbon matrix [24], allowing the material to accommodate the volume change that occurs during the charging and discharging of the battery, while also improving the electrical conductivity of the cathode material. This is important for maintaining the stability and performance of the battery over repeated charge–discharge cycles. To explore high-performance transition metal oxide electrodes, the rational design of the material's structure is essential. In the preparation of electrodes, the nature of the binder directly affects the electrochemical properties of transition metal oxides. Conventional electrodes are used by mixing electroactive materials with an insulating polymer binder and coating them on the collector. The ratio and distribution of the binder directly impact performance. Compositing transition metals with conductive materials avoids the use of binders. With this structure, the weight is reduced while exposing more active sites and the electrode can directly contact the electrolyte better, exhibiting satisfactory lithium storage performance [25,26].

In this manuscript, the $CoMn_2O_4$ nanoflower particles are synthesized by hydrothermal method. Then one-dimensional carbon nanotube (CNT) and two-dimensional graphene (GR) are used to build a three-dimensional network structure of GR/CNT to support the $CoMn_2O_4$ nanoflower particles. The $CoMn_2O_4$/GR/CNT composite film not only has the potential to improve the overall conductivity of the material but also effectively alleviate volume expansion and aggregation problems.

2. Experimental Section

2.1. Preparation of $CoMn_2O_4$

A total of 0.33 g of polyvinylpyrrolidone, 1.5 mmol of $Co(CH_3COO)_2$, and 3 mmol of $Mn(CH_3COO)_2$ were sequentially added into a mixed solution consisting of 90 mL of ethylene glycol and 1.5 mL of deionized water. After moderate stirring, the resulting homogeneous solution was transferred into two 50 mL hydrothermal autoclaves, heated to 180 °C, and maintained for 12 h [27]. Upon cooling to room temperature naturally, the product was obtained by washing with ethanol through centrifugation and drying in a vacuum at 60 °C for 12 h. Finally, the dried powder was heated from room temperature to 500 °C at a rate of 1 °C/min under ambient atmosphere and calcined at this temperature for 4 h. The obtained black powder is referred to as $CoMn_2O_4$.

2.2. Fabrication of $CoMn_2O_4$/GR/CNT Composite Film

The preparation of graphene oxide (GO) was achieved using an improved Hummers method [28] through a three-step process: low-temperature oxidation, high-temperature oxidation, and hydrolysis. Firstly, 1 g of graphite powder provided by the China National Pharmaceutical Group Co. was added to a mixture of 92 mL of concentrated sulfuric acid (95–98%) and 24 mL of nitric acid (65%) in a 500 mL flask placed in an ice bath. The suspended solution was vigorously stirred at 0 °C for 1h, constituting the low-temperature oxidation process. Next, to further enhance the oxidation degree of the graphite, the mixture was heated to 80 °C for 30 min, representing the high-temperature oxidation step. Following that, 92 mL of deionized water was slowly added and the suspension was stirred at 80 °C for an additional 30 min, concluding the hydrolysis process. Afterward, to reduce the remaining permanganate, 5 mL of H_2O_2 (30%) was added to the solution. Upon completion of the reaction, the resulting mixture was washed with a 5% solution of HCl and water until no sulfate was detected with $BaCl_2$. Finally, the GO powders were dried at 80 °C for 24 h, completing the post-treatment steps of drying and washing. The GO (20 mg) and the carbon nanotube (CNT) (10 mg) provided by Zhongke Times Nanomaterials Co., Ltd. were dispersed in 10 mL of deionized water by ultrasonication. The resulting suspension was stirred continuously for 12 h on a magnetic stirrer after adding 20 mg of the as-prepared $CoMn_2O_4$. The suspension was then transferred to a vacuum filtration apparatus (1000 mL) and continuously filtered for 4 h. The resulting thin film was transferred to a petri dish and briefly soaked in acetone solution. The synthesized thin film was then quickly peeled off from the microporous membrane using forceps. Finally, the peeled thin film was calcined at 220 °C for 2 h and the resulting $CoMn_2O_4$/GR/CNT composite film was synthesized.

2.3. Characterization

The surface composition of the samples was analyzed by X-ray diffraction (XRD) patterns using a Bruker D8 ADVANCE X-ray diffractometer equipped with a Cu-Kα radiation source and scanning ranges of 2–10° and 10–70°. The scanning electron microscopy (SEM) images were obtained using an FEI Quanta 400 ESEM-FEG instrument with an accelerating voltage of 20–40 kV.

2.4. Electrochemical Measurements

Foam nickel was used as the current collector, and the prepared thin film material was cut into appropriate sizes and pressed into electrode plates using a powder press machine at 20 MPa. The loading amount of the active substance on each electrode plate was approximately 2 mg cm^{-2}. The electrode plates were used as the working electrode, lithium foil as the counter and reference electrode, and polypropylene porous membrane (Celgard2400) as the separator. The electrolyte was 1 M LiPF$_6$/ethylene carbonate–dimethyl carbonate–methyl carbonate (EC-DMC-EMC) with a volume ratio of 1:1:1. The energy storage characteristics of the battery materials, such as specific capacity, rate performance, cycle life, and Coulomb efficiency, were mainly evaluated through charge and discharge cycle tests. The battery testing system used in this manuscript was produced by Wuhan LAND Electronic Technology Co., Ltd. The test conditions were a constant temperature of 30 °C, a voltage range of 0–3.0 V, and a current density of 50–1600 mA/g to complete the charge and discharge cycle tests. Cyclic voltammetry (CV) was performed using the CS310 electrochemical workstation produced by Wuhan KEJIA Technology Co., Ltd, Wuhan, China. The CV test conditions were a voltage range of 0–3.0 V and a scan rate of 0.2 mV/s.

3. Results and Discussion

Figure 1 illustrates the fabrication process of the CoMn$_2$O$_4$/GR/CNT composite film. In a typical experiment, CoMn$_2$O$_4$ nanoflower particles were initially synthesized using a hydrothermal method. This process involved the controlled reaction between specific precursor materials under suitable temperature and pressure conditions, resulting in the formation of the CoMn$_2$O$_4$ nanoflowers. Next, two-dimensional graphite oxide (GO) and one-dimensional carbon nanotube (CNT) were dispersed in deionized water and mixed with the CoMn$_2$O$_4$ nanoflowers. The mixture was stirred for 12 h to ensure a homogeneous distribution of the nanomaterials and promote the interaction between the components. The resulting suspension, containing the CoMn$_2$O$_4$ nanoflowers, GO, and CNT, was then subjected to vacuum filtration. This process involved filtering the suspension through a porous membrane, resulting in the formation of a thin brown film on the membrane surface. Subsequently, the film was calcined at 220 °C for 2 h to remove any residual oxygen group and promote the formation of strong interfacial bonds between the components. The calcination process led to the formation of the final black product, which represents the CoMn$_2$O$_4$/GR/CNT composite film. The integration of the CoMn$_2$O$_4$ nanoflower particles within the three-dimensional network structure of GR and CNT provides several advantages. Firstly, the GR/CNT network offers a large surface area, allowing for enhanced interaction and efficient utilization of the active materials. Secondly, the presence of CNT and GR improves the overall electrical conductivity of the composite, facilitating rapid charge and ion transport within the electrode material.

To analyze the structure of the CoMn$_2$O$_4$/GR/CNT film and evaluate the impact of the incorporation of GR/CNT on CoMn$_2$O$_4$, SEM characterization was conducted. Figure 2a,b present SEM images of the carbon nanotube (CNT) and graphene (GR). The SEM image of the CNT (Figure 2a) reveals a distinctive tubular structure with a curved morphology. The CNT exhibit a range of orientations, intertwining and interconnecting with each other. Moreover, in Figure 2b, the graphene displays a layered two-dimensional arrangement, demonstrating a stacked structure. The graphene layers are stacked on top of each other, creating a layered configuration with well-defined boundaries between the individual graphene sheets. Figure 2c,d show SEM images of pure CoMn$_2$O$_4$ and CoMn$_2$O$_4$/GR/CNT composite films at different magnifications. It can be observed from Figure 2c,d that without the introduction of GR/CNT, CoMn$_2$O$_4$ aggregates severely. From the high-magnification SEM image in Figure 2c, it can be seen that the prepared CoMn$_2$O$_4$ is composed of porous nanosheets, forming nanoflowers that have diameters of about 300 nm. Figure 2e,f show SEM images of CoMn$_2$O$_4$/GR/CNT composite films. The material with a sheet-like structure is graphene, whereas the material with a curved and fluffy structure is the carbon nanotube (CNT). The graphene sheets formed a three-dimensional framework that supported the $CoMn_2O_4$ particles, whereas the long and curved carbon nanotube filled the gaps between the graphene sheets and between $CoMn_2O_4$ particles and graphene sheets, facilitating connectivity between the $CoMn_2O_4$ and graphene. It can be seen that after vacuum filtration and heat treatment, the shape of $CoMn_2O_4$ nanoflowers did not change, and they were uniformly embedded in the GR/CNT composite film. The existence of the 3D structure of GR/CNT effectively mitigated the phenomenon of $CoMn_2O_4$ nanoflower particle aggregation compared with Figure 2e. The CNT in the film can effectively compensate for the defect of poor conductivity in the GR layer.

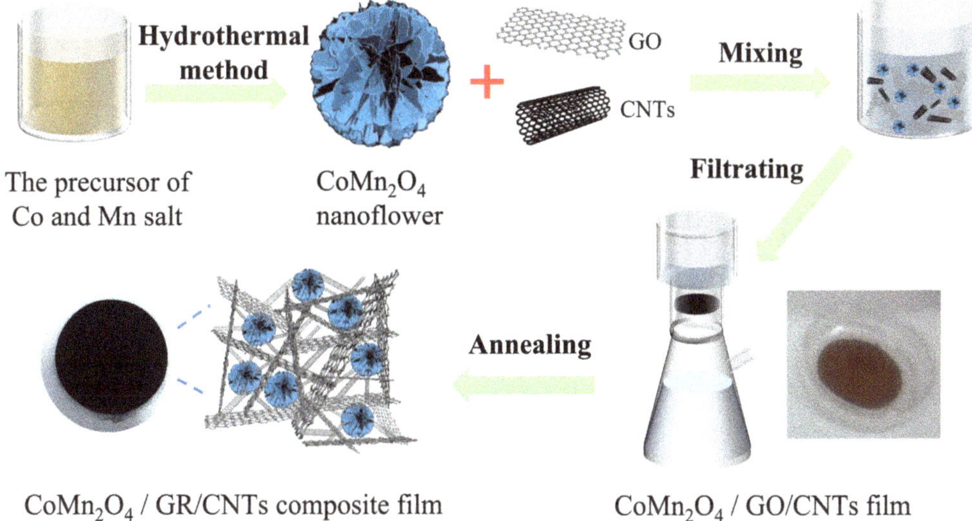

Figure 1. Schematic diagram of $CoMn_2O_4$ nanoflower particles/graphene/carbon nanotube composite film.

To determine the phase composition of the as-prepared materials, X-ray diffraction (XRD) analysis were carried out on the as-synthesized GR, $CoMn_2O_4$ microflowers, and the $CoMn_2O_4$/GR/CNT composite film. Figure 3a shows the low-angle X-ray diffraction (XRD) patterns of the as-prepared materials within the range of 2–10°. Upon analysis of the XRD patterns, it was observed that the GR exhibits a relatively featureless pattern in this low-angle range, with no distinct peaks observed. This indicates the amorphous nature or very fine crystalline structure of the graphene. In contrast, the $CoMn_2O_4$ nanoflower particles exhibit a characteristic peak at approximately 2.46°, suggesting the presence of a crystalline phase in the material. Interestingly, in the XRD pattern of the $CoMn_2O_4$/GR/CNT composite film, a peak is observed at the same position as the $CoMn_2O_4$ nanoflower particles. However, this peak appears to be slightly broadened compared with the pure $CoMn_2O_4$, indicating a potential modification in the crystalline structure or an influence from the presence of GR and CNT. From the wide-angle XRD diffractogram of GR in Figure 3b, it is evident that a diffraction peak is observed around 25°, which corresponds to the (002) crystal plane of graphene. This peak signifies that GR possesses an amorphous structure with no prominent crystalline peaks observed within the low-angle range of 2–10°. The pure $CoMn_2O_4$ nanoflowers (represented by the green line) showed diffraction peaks at $2\theta = 18.2°, 29.3°, 31.2°, 32.9°, 36.4°, 4.8°, 59.0°$, and 60.7°, corresponding to the (111), (202), (220), (113), (311), (400), (511), and (404) crystal faces, respectively, consistent with the PDF card (JCPDS 23-1237), with no other impurity peaks appearing, indicating the high purity of $CoMn_2O_4$. The brown line in Figure 3 represents the $CoMn_2O_4$/GR/CNT film. In the $CoMn_2O_4$/GR/CNT composites, some of the XRD peaks have become broader and less distinct. This could be attributed to the introduction of graphene and carbon nanotube that could have caused some lattice distortion and reduced the crystallinity of the $CoMn_2O_4$ nanoparticles. However, the main diffraction peaks of $CoMn_2O_4$ were still identifiable, indicating that the introduction of GR/CNT did not significantly affect the crystal structure of $CoMn_2O_4$.

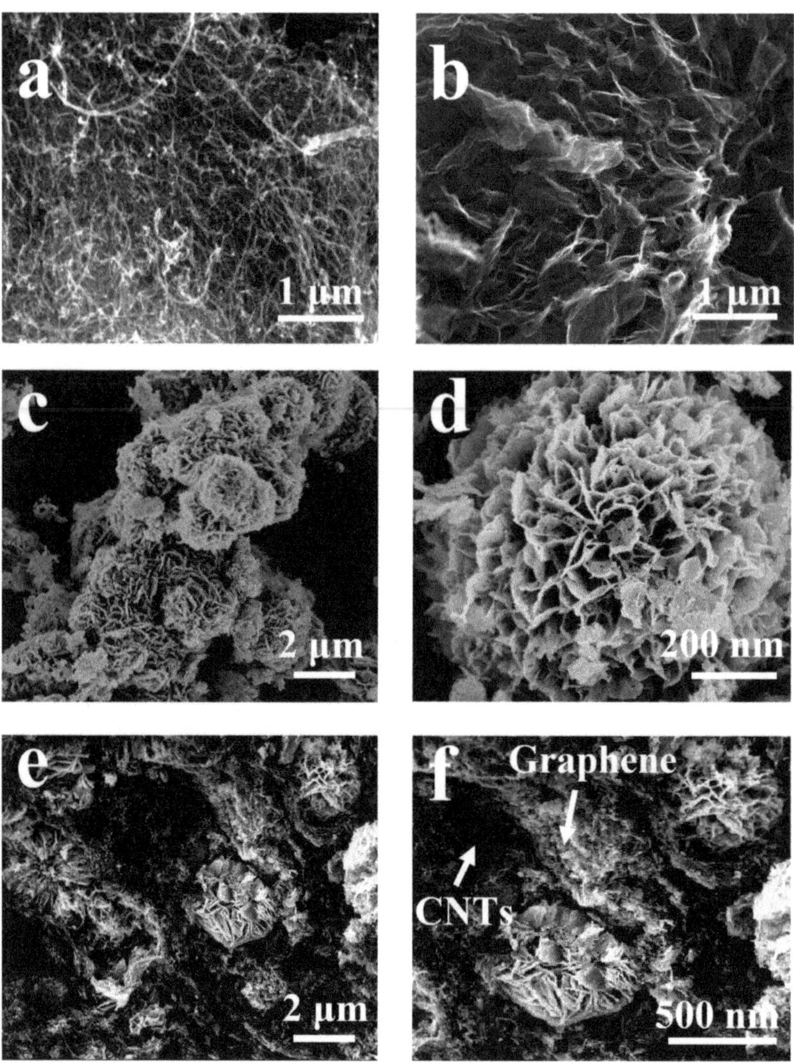

Figure 2. The SEM images of as-prepared (**a,b**) CNT and GR, (**c,d**) $CoMn_2O_4$ nanoflower particles, and (**e,f**) $CoMn_2O_4$/GR/CNT composite film.

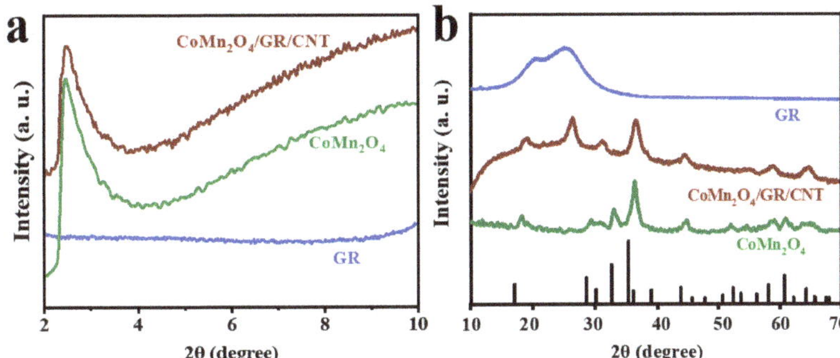

Figure 3. (**a**) Low-angle and (**b**) wide-angle X-ray diffraction (XRD) patterns of as-prepared GR, $CoMn_2O_4$ nanoflower particles, and $CoMn_2O_4$/GR/CNT composite film.

The cyclic voltammetry (CV) technique was used to study the electrochemical behavior of the $CoMn_2O_4$/GR/CNT composites. Figure 4a illustrates the cyclic voltammetry (CV) curves of the composite film electrode during the first three cycles. The CV curve of the initial cycle exhibits a distinct difference compared with the subsequent two cycles, particularly in the reduction curve. During the initial lithiation process, a broad reduction peak appears at 1.25 V, which is mainly attributed to the reduction of Co^{3+} to Co^{2+}. The sharp peak at 0.4 V is attributed to the transformation of Co^{2+} and Mn^{2+} to metal Co and Mn. The two peaks at 1.65 V and 2.0 V in the delithiation process are attributed to the oxidation of metal Co and Mn to Co^{2+} and Mn^{2+}. In the subsequent cycles, the peaks at 0.45/1.65 V and 1.1 V/2.0 V represent the repetitive reduction/oxidation of MnO and CoO, respectively [14]. Figure 4b shows the first three charge/discharge curves of the $CoMn_2O_4$/GR/CNT composite electrode at a current density of 100 mA g^{-1}. It can be seen from the figure that the initial specific capacities of $CoMn_2O_4$/GR/CNT for charge and discharge are 1164/2148 mAh g^{-1}, respectively. Due to the irreversible formation of SEI film during the first charge/discharge cycle, a part of Li^+ is lost, resulting in a low coulombic efficiency of 54%.

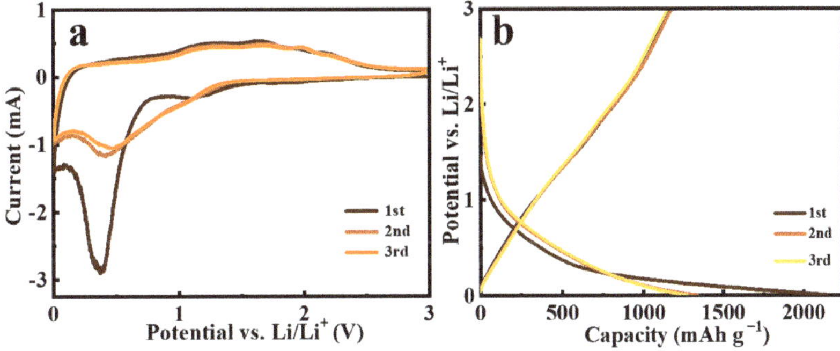

Figure 4. Electrochemical performance of $CoMn_2O_4$/GR/CNT film for LIBs at room temperature (25 °C). (**a**) Cyclic voltammograms between 0.0 V and 3.0 V at a scan rate of 0.1 mV s^{-1} for the initial three cycles. (**b**) Charge–discharge voltage profiles at a current density of 0.1 A g^{-1} for the initial three cycles.

A galvanostatic charge–discharge test was conducted to further evaluate the electrochemical performance of the $CoMn_2O_4$ and $CoMn_2O_4$/GR/CNT composite film electrodes. The cycling performance of $CoMn_2O_4$ and $CoMn_2O_4$/GR/CNT at a current density of 100 mA g^{-1} for the first 55 cycles is presented in Figure 5a. The initial discharge specific capacities of the $CoMn_2O_4$ and $CoMn_2O_4$/GR/CNT films were 1514 and

2148 mAh g^{-1}, respectively. Although the difference in specific capacity between $CoMn_2O_4$ and $CoMn_2O_4$/GR/CNT films was not significant during the first few cycles, the discharge specific capacity of the $CoMn_2O_4$ electrode decreased rapidly with increasing cycle number. After 55 cycles, the discharge specific capacity of the $CoMn_2O_4$ electrode was only 457 mAh g^{-1}, which was about half of that of the $CoMn_2O_4$/GR/CNT material (881 mAh g^{-1}) under the same testing conditions. The discharge specific capacity of $CoMn_2O_4$GR/CNT composites decreases very slowly and reaches a stable plateau after about 15 cycles. The enhanced electrochemical performance can be attributed to the synergistic effect of the three-dimensional network structure of GR/CNT and the highly dispersed $CoMn_2O_4$ nanoflowers, which provides a large surface area and efficient electron and ion transport pathways. Furthermore, the introduction of GR/CNT effectively alleviates the volume expansion and agglomeration of $CoMn_2O_4$ during the charge–discharge process, maintaining the integrity and activity of $CoMn_2O_4$ during the stability test.

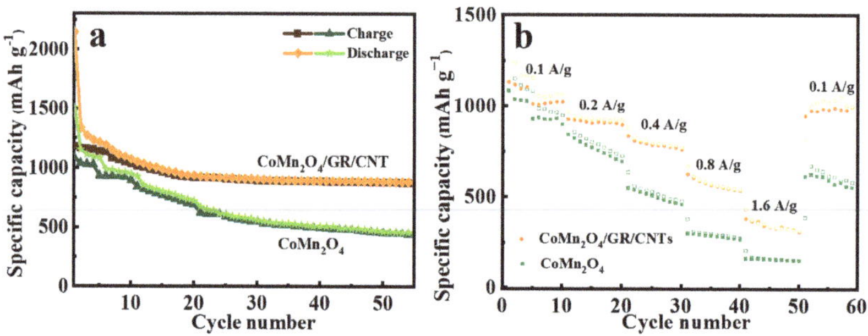

Figure 5. Electrochemical performance of $CoMn_2O_4$ and $CoMn_2O_4$/GR/CNT film for LIBs at room temperature (25 °C). (**a**) Cycling performance. (**b**) Rate capability at various current rates.

Figure 5b displays the rate performance curves of the $CoMn_2O_4$ and $CoMn_2O_4$/GR/CNT electrodes tested at current densities of 100, 200, 400, 800, and 1600 mA g^{-1}. In the tested conditions with current densities of 100, 200, 400, 800, and 1600 mA g^{-1}, the average discharge specific capacities of $CoMn_2O_4$ were found to be 1082, 800, 528, 296, and 161 mA g^{-1}, respectively. However, when the current density was suddenly reduced to 100 mA g^{-1}, the discharge specific capacity of the $CoMn_2O_4$ electrode could only recover to 616 mA g^{-1}. In contrast, under the same testing conditions, the average discharge specific capacities of the $CoMn_2O_4$/GR/CNT composite films were 1155, 923, 796, 572, and 339 mA g^{-1}, and when the current density was reduced to 100 mA g^{-1} again, the discharge specific capacity could be restored to the original level, reaching 1020 mA g^{-1}. The rate performance of the unmodified $CoMn_2O_4$ material is not satisfactory and its capacity decay is relatively rapid, especially after experiencing high current density charge-discharge. Furthermore, at any given current density, the specific capacity of the $CoMn_2O_4$/GR/CNT electrode is higher than that of the $CoMn_2O_4$ electrode, indicating that the introduction of GR/CNT has led to the better rate performance of the $CoMn_2O_4$/GR/CNT electrode. This suggests that the introduction of GR/CNT can provide a conductive network and increase the electron transfer rate, which can effectively alleviate the polarization effect and improve the rate performance of the $CoMn_2O_4$/GR/CNT electrode.

The $CoMn_2O_4$/GR/CNT composite film has shown improved performance in lithium-ion batteries (LIBs); this is mainly attributed to its unique structure. As shown in Figure 6, the $CoMn_2O_4$ nanoparticles are embedded in the three-dimensional network structure of the GR/CNT matrix, forming a continuous and porous conductive network. The GR sheets and CNT can not only provide good electrical conductivity but also prevent the aggregation of $CoMn_2O_4$ nanoparticles during the charge–discharge process, which can effectively improve the structural stability of the electrode and enhance the rate performance. Moreover, the $CoMn_2O_4$ particles in the composite film are distributed uniformly on the surface of the graphene and carbon nanotube, which increases the electrode's specific surface area and enhances the interaction between the electrode and electrolyte. This enhanced interaction can facilitate the diffusion of lithium ions and improve the electrode's electrochemical performance. In addition, the composite film also shows improved rate capability due to the enhanced conductivity and the reduced diffusion length of lithium ions in the electrode material. The conductive network provided by the GR and CNT can facilitate the movement of Li^+ ions and electrons in the electrode material, whereas the smaller diffusion length of lithium ions in the $CoMn_2O_4$ particles can improve the rate capability of the electrode.

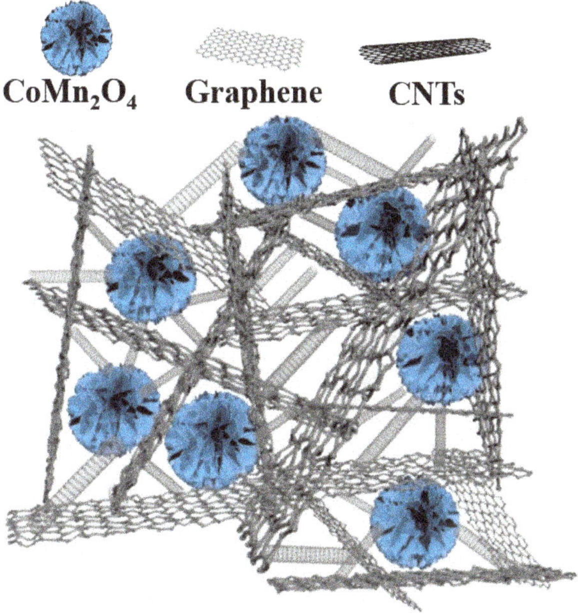

Figure 6. Schematic illustration of $CoMn_2O_4$/graphene/carbon nanotube composite film-based anode.

4. Conclusions

In summary, the $CoMn_2O_4$/graphene/carbon nanotube composite film was successfully prepared via a facile vacuum filtrating method and demonstrated improved performance as an anode material for lithium-ion batteries. The $CoMn_2O_4$/GR/CNT composite film exhibited a high initial discharge capacity of 2148 mAh g^{-1} and superior cycle stability (881 mAh g^{-1} after 55 cycles) compared with the unmodified $CoMn_2O_4$ film. The composite film also demonstrated better rate performance, with a higher average discharge capacity at various current densities. The results suggest that the $CoMn_2O_4$/graphene/carbon nanotube composite film is a promising anode material candidate for high-performance lithium-ion batteries. The addition of graphene and carbon nanotube to the $CoMn_2O_4$ film improved its conductivity and provided structural support, which effectively reduced volume expansion and aggregation during the charge and discharge process, thus maintaining the integrity and activity of the $CoMn_2O_4$ material. This study highlights the importance of designing and developing advanced materials for energy storage applications and the potential of using graphene and carbon nanotube as effective additives to improve the performance of metal-oxide-based anode materials in lithium-ion batteries.

In addition, the current electrochemical characterizations in this study were at room temperature; the performance of electrode materials in lithium-ion batteries at high temperatures is a critical aspect to consider. It is widely recognized that high temperatures can accelerate undesired side reactions such as electrolyte decomposition, electrode degradation, and increased interfacial resistance. These factors can lead to reduced capacity, increased irreversible capacity loss, accelerated capacity fading, and decreased cycling stability of the battery. Moreover, the higher operating temperatures can also affect the structural integrity of the electrode material, leading to morphological changes, phase transformations, and altered lithium-ion diffusion pathways. These effects can further influence the overall performance and lifespan of the battery. In the future, the behavior of the materials at high temperatures should be carefully investigated and evaluated to ensure the reliability and safety of lithium-ion batteries in real-world applications.

Author Contributions: Conceptualization, G.W. and X.T.; methodology, X.Z.; software, F.L. and B.Y.; validation, M.G., J.T., S.H. and G.W.; formal analysis, X.Z.; investigation, G.W. and X.T.; resources, G.W.; data curation, F.L. and B.Y.; writing—original draft preparation, F.L. and B.Y.; writing—review and editing, S.H., G.W. and X.T.; visualization, X.T.; supervision, X.T. and G.W.; project administration, X.T. and G.W.; funding acquisition, S.H., G.W. and X.T. All authors have read and agreed to the published version of the manuscript.

Funding: This study was accomplished with financial support by the Youth Science and Technology Talent Growth Program of Guizhou Provincial Education Department (no. QianjiaoheKY [2021]295), Youth Innovation Promotion Association of Chinese Academy of Sciences (2023341).

Data Availability Statement: All raw data in this study can be provided by the corresponding authors on request.

Conflicts of Interest: The authors declare no conflict of interest.

References

1. Wang, K.; Hua, W.; Huang, X.; Stenzel, D.; Wang, J.; Ding, Z.; Cui, Y.; Wang, Q.; Ehrenberg, H.; Breitung, B.; et al. Synergy of cations in high entropy oxide lithium ion battery anode. *Nat. Commun.* **2023**, *14*, 1487. [CrossRef]
2. Molaiyan, P.; Dos Reis, G.S.; Karuppiah, D.; Subramaniyam, C.M.; García-Alvarado, F.; Lassi, U. Recent Progress in Biomass-Derived Carbon Materials for Li-Ion and Na-Ion Batteries—A Review. *Batteries* **2023**, *9*, 116. [CrossRef]
3. Pavlovskii, A.A.; Pushnitsa, K.; Kosenko, A.; Novikov, P.; Popovich, A.A. Organic Anode Materials for Lithium-Ion Batteries: Recent Progress and Challenges. *Materials* **2022**, *16*, 177. [CrossRef] [PubMed]
4. Deng, F.; Zhang, Y.; Yu, Y. Conductive Metal–Organic Frameworks for Rechargeable Lithium Batteries. *Batteries* **2023**, *9*, 109. [CrossRef]
5. Nzereogu, P.U.; Omah, A.D.; Ezema, F.I.; Iwuoha, E.I.; Nwanya, A.C. Anode materials for lithium-ion batteries: A review. *Appl. Surf. Sci. Adv.* **2022**, *9*, 100233. [CrossRef]
6. Wang, C.; Yang, C.; Zheng, Z. Toward Practical High-Energy and High-Power Lithium Battery Anodes: Present and Future. *Adv. Sci.* **2022**, *9*, e2105213. [CrossRef] [PubMed]
7. Zhao, Y.; Li, X.; Yan, B.; Xiong, D.; Li, D.; Lawes, S.; Sun, X. Recent Developments and Understanding of Novel Mixed Transition-Metal Oxides as Anodes in Lithium Ion Batteries. *Adv. Energy Mater.* **2016**, *6*, 1502175. [CrossRef]
8. Roy, K.; Mishra, A.; Nayak, S.; Gupta, P.; Jena, B.B.; Bedanta, S. Spin to Charge Conversion in Semiconducting Antiferromagnetic Co_3O_4. *ACS Appl. Electron. Mater.* **2023**, *5*, 1575–1580. [CrossRef]
9. Ishibe, T.; Maeda, Y.; Terada, T.; Naruse, N.; Mera, Y.; Kobayashi, E.; Nakamura, Y. Resistive switching memory performance in oxide hetero-nanocrystals with well-controlled interfaces. *Sci. Technol. Adv. Mater.* **2020**, *21*, 195–204. [CrossRef]
10. Zhu, J.; Ding, Y.; Ma, Z.; Tang, W.; Chen, X.; Lu, Y. Recent Progress on Nanostructured Transition Metal Oxides As Anode Materials for Lithium-Ion Batteries. *J. Electron. Mater.* **2022**, *51*, 3391–3417. [CrossRef]
11. Manthiram, A. A reflection on lithium-ion battery cathode chemistry. *Nat. Commun.* **2020**, *11*, 1550. [CrossRef]
12. Yu, L.; Zhang, L.; Wu, H.B.; Zhang, G.; Lou, X.W. Controlled synthesis of hierarchical $Co_xMn_{3-x}O_4$ array micro-/nanostructures with tunable morphology and composition as integrated electrodes for lithium-ion batteries. *Energy Environ. Sci.* **2013**, *6*, 2664–2671. [CrossRef]
13. Deng, Y.; Wan, L.; Xie, Y.; Qin, X.; Chen, G. Recent advances in Mn-based oxides as anode materials for lithium ion batteries. *RSC Adv.* **2014**, *4*, 23914–23935. [CrossRef]
14. Hu, L.; Zhong, H.; Zheng, X.; Huang, Y.; Zhang, P.; Chen, Q. $CoMn_2O_4$ spinel hierarchical microspheres assembled with porous nanosheets as stable anodes for lithium-ion batteries. *Sci. Rep.* **2012**, *2*, 986. [CrossRef]
15. Zhang, Y.; Gao, W.; Ji, S.; Jia, N. A unique nanosheet assembled $CoMn_2O_4$ hollow nanospheres as superior cyclability anode materials for lithium-ion batteries. *J. Alloys Compd.* **2019**, *786*, 428–433. [CrossRef]

16. Zhang, L.-X.; Wang, Y.-L.; Jiu, H.-F.; Zheng, W.-H.; Chang, J.-X.; He, G.-F. Controllable synthesis of spinel nano-$CoMn_2O_4$ via a solvothermal carbon templating method and its application in lithium ion batteries. *Electrochim. Acta* **2015**, *182*, 550–558. [CrossRef]
17. Yang, G.; Xu, X.; Yan, W.; Yang, H.; Ding, S. Single-spinneret electrospinning fabrication of $CoMn_2O_4$ hollow nanofibers with excellent performance in lithium-ion batteries. *Electrochimica Acta* **2014**, *137*, 462–469. [CrossRef]
18. Sanad, M.M.S.; Yousef, A.K.; Rashad, M.M.; Naggar, A.H.; El-Sayed, A.Y. Robust and facile strategy for tailoring $CoMn_2O_4$ and $MnCo_2O_4$ structures as high capacity anodes for Li-ions batteries. *Phys. B Condens. Matter* **2020**, *579*, 411889. [CrossRef]
19. Zhou, L.; Zhao, D.; Lou, X.W. Double-shelled $CoMn_2O_4$ hollow microcubes as high-capacity anodes for lithium-ion batteries. *Adv. Mater.* **2012**, *24*, 745–748. [CrossRef]
20. Zhang, C.; Zheng, B.; Huang, C.; Li, Y.; Wang, J.; Tang, S.; Deng, M.; Du, Y. Heterostructural Three-Dimensional Reduced Graphene Oxide/$CoMn_2O_4$ Nanosheets toward a Wide-Potential Window for High-Performance Supercapacitors. *ACS Appl. Energy Mater.* **2019**, *2*, 5219–5230. [CrossRef]
21. Alagar, S.; Kumari, S.; Upreti, D.; Aashi; Bagchi, V. High-Performance Mg-Ion Supercapacitor Designed with a N-Doped Graphene Wrapped $CoMn_2O_4$ and Porous Carbon Spheres. *Energy Fuels* **2022**, *36*, 14442–14452. [CrossRef]
22. Ma, Y.; Zhang, L.; Yan, Z.; Cheng, B.; Yu, J.; Liu, T. Sandwich-Shell Structured $CoMn_2O_4$/C Hollow Nanospheres for Performance-Enhanced Sodium-Ion Hybrid Supercapacitor. *Adv. Energy Mater.* **2022**, *12*, 2103820. [CrossRef]
23. Cai, D.; Qu, B.; Li, Q.; Zhan, H.; Wang, T. Reduced graphene oxide uniformly anchored with ultrafine $CoMn_2O_4$ nanoparticles as advance anode materials for lithium and sodium storage. *J. Alloys Compd.* **2017**, *716*, 30–36. [CrossRef]
24. Ren, F.; Wang, T.; Liu, H.; Liu, D.; Zhong, R.; You, C.; Zhang, W.; Lv, S.; Liu, S.; Zhu, H.; et al. $CoMn_2O_4$ nanoparticles embed in graphene oxide aerogel with three-dimensional network for practical application prospects of oxytetracycline degradation. *Sep. Purif. Technol.* **2021**, *259*, 118179. [CrossRef]
25. Zhang, J.; Yu, A. Nanostructured transition metal oxides as advanced anodes for lithium-ion batteries. *Sci. Bull.* **2015**, *60*, 823–838. [CrossRef]
26. Luo, S.; Yao, M.; Lei, S.; Yan, P.; Wei, X.; Wang, X.; Liu, L.; Niu, Z. Freestanding reduced graphene oxide-sulfur composite films for highly stable lithium-sulfur batteries. *Nanoscale* **2017**, *9*, 4646–4651. [CrossRef]
27. Zhang, L.; He, G.; Lei, S.; Qi, G.; Jiu, H.; Wang, J. Hierarchical hollow microflowers constructed from mesoporous single crystalline $CoMn_2O_4$ nanosheets for high performance anode of lithium ion battery. *J. Power Source* **2016**, *326*, 505–513. [CrossRef]
28. Tong, X.; Wang, H.; Wang, G.; Wan, L.; Ren, Z.; Bai, J.; Bai, J. Controllable synthesis of graphene sheets with different numbers of layers and effect of the number of graphene layers on the specific capacity of anode material in lithium-ion batteries. *J. Solid State Chem.* **2011**, *184*, 982–989. [CrossRef]

Article

Hydrogen Storage Properties of Economical Graphene Materials Modified by Non-Precious Metal Nickel and Low-Content Palladium

Yiwen Chen [1,2,†], Habibullah [3,†], Guanghui Xia [3], Chaonan Jin [3], Yao Wang [4,5], Yigang Yan [4,5,6], Yungui Chen [4,5,6], Xiufang Gong [1,2,*], Yuqiu Lai [1,2] and Chaoling Wu [3,4,5,*]

[1] State Key Laboratory of Clean and Efficient Turbomachinery Power Equipment, Deyang 618000, China; chenyiwenac@163.com (Y.C.); laiyuqiu2020@163.com (Y.L.)

[2] Dongfang Electric Corporation Dongfang Turbine Co., Ltd., Deyang 618000, China

[3] College of Materials Science and Engineering, Sichuan University, Chengdu 610064, China; habibullah@stu.scu.edu.cn (H.); xiagh1994@163.com (G.X.); jinchaonnan@163.com (C.J.)

[4] Engineering Research Center of Alternative Energy Materials & Devices, Ministry of Education, Chengdu 610064, China; wangyao516@scu.edu.cn (Y.W.); yigang.yan@scu.edu.cn (Y.Y.); chenyungui@scu.edu.cn (Y.C.)

[5] Institute of New Energy and Low-Carbon Technology, Sichuan University, Chengdu 610065, China

[6] Technology Innovation Center of Hydrogen Storage-Transportation and Fueling Equipments for State Market Regulation, Chengdu 610100, China

* Correspondence: gongxiufang@dongfang.com (X.G.); wuchaoling@scu.edu.cn (C.W.)

† These authors contributed equally to this work.

Abstract: Ni/Pd co-modified graphene hydrogen storage materials were successfully prepared by a solvothermal method using $NiCl_2 \cdot 6H_2O$ and $Pd(OAc)_2$ and reduced graphene oxide (rGO). By adjusting the hydrothermal temperature, Pd–Ni is successfully alloyed, and the size of the obtained nanoparticles is uniform. The electronic structure of Pd was changed by alloying, and the center of the D-band moved down, which promoted the adsorption of hydrogen. The NiPd-rGO-180 sample, in which 180 represents the solvothermal temperature in centigrade (°C), has the highest hydrogen storage capacity of 2.65 wt% at a moderate condition (RT/4MPa). The excellent hydrogen storage performance benefits from the synergistic hydrogen spillover effect of Pd–Ni bimetal. The calculated hydrogen adsorption energies of Ni_2Pd_2-rGO are within the ideal range (-0.20 to -0.60 eV) of hydrogen ads/desorption; however, the introduction of substrate defects and the cluster orientation alter the hydrogen adsorption energy. This work provides an effective reference for the design and optimization of carbon-based hydrogen storage materials.

Keywords: NiPd; graphene; carbon-based hydrogen storage; hydrogen adsorption

Citation: Chen, Y.; Habibullah; Xia, G.; Jin, C.; Wang, Y.; Yan, Y.; Chen, Y.; Gong, X.; Lai, Y.; Wu, C. Hydrogen Storage Properties of Economical Graphene Materials Modified by Non-Precious Metal Nickel and Low-Content Palladium. *Inorganics* **2023**, *11*, 251. https://doi.org/10.3390/inorganics11060251

Academic Editor: Hicham Idriss

Received: 18 May 2023
Revised: 2 June 2023
Accepted: 6 June 2023
Published: 8 June 2023

1. Introduction

Energy plays a crucial role in driving economic prosperity for nations worldwide. With the rapid growth of the population and evolving lifestyles, global energy demands continue to expand. Simultaneously, environmental concerns and the rapid depletion of resources have become alarming challenges. Fossil fuel consumption has notably contributed to environmental degradation, intensifying the need for clean and sustainable energy sources. In this context, hydrogen emerges as a promising solution to meet the growing energy requirements [1–3]. As an ecofriendly energy carrier, hydrogen offers a range of advantages, including a high gravimetric energy density, zero emissions, and a renewable nature. Its environmentally benign characteristics position it as a clean energy transporter [4]. With its immense potential, hydrogen is expected to serve as a highly efficient alternative fuel with diverse applications in the future. However, the development of highly efficient and safe hydrogen transportation and storage technologies remains a subject of discussion and demonstration. Currently, high-pressure compressed hydrogen is the most commonly employed storage technology, although concerns regarding safety persist. Among the three primary techniques for hydrogen storage, namely compressed gas, cryogenic liquid, and solid-state storage, only the former two methods have commercialized. Nevertheless, the cost-effectiveness and safety considerations of hydrogen storage have led to significant research efforts focusing on solid-state storage through the adsorption and/or absorption of various materials and alloys [5,6]. The growing development of the new energy automotive sector has spurred numerous studies on secure hydrogen storage. Porous materials, including zeolites, metal-organic frameworks (MOFs), covalent organic frameworks (COFs), and carbon-based materials (such as fullerenes, nanotubes and graphene), have received considerable attention [7]. However, the low hydrogen storage capacity of zeolites and porous silica, along with the low operating temperature (77K) of MOFs and COFs, pose disadvantages for hydrogen storage [8,9]. As a result, 2D carbonaceous materials [10], particularly graphene [11,12], have emerged as a topic of significant interest in the field of hydrogen storage materials. Graphene offers exceptional specific surface area, structural stability, lightweight properties and rapid adsorption kinetics [13–15]. Despite the relatively low hydrogen storage capacity of pure graphene (<1 wt% at room temperature), which falls short of the U.S. Department of Energy (DOE) target of 5.5 wt% for onboard applications by 2025 [16], it remains a subject of exploration and optimization for hydrogen storage purposes. Some researchers confirmed that there is an increment in the hydrogen storage capacity of graphene after being modified by metal atoms, clusters, or nanoparticles. Palladium, platinum, and other precious metals have been widely studied and used to enhance the hydrogen storage capacity of carbon-based materials owing to the hydrogen spillover effect [17–19]. However, the high cost has inhibited the large-scale industrial application of these precious metals. The spillover process in hydrogen storage involves the dissociation of hydrogen molecules on the catalyst, migration of hydrogen atoms from the catalyst to the substrate, and diffusion of hydrogen atoms on the substrate surface. However, certain pure transition metal-decorated graphene materials face limitations due to a high energy barrier. For instance, Psofogiannakis et al. [20] reported a migration barrier of 2.6 eV from the Pt_4 cluster to graphene. This high migration barrier hampers the efficient and thermodynamically favorable transfer of hydrogen atoms from the catalyst to the carbon receptor, thereby impeding the spillover rate of hydrogen.

To reduce the reliance on precious metals such as Pd, the formation of Pd-based alloys has emerged as an effective strategy. The introduction of alloying elements can alter the hydrogen spillover effect and consequently modify hydrogen storage capacity. Guo et al. conducted research on the highest occupied molecular orbital (HOMO) of Penta-Graphene (PG) decorated with Ni_4, Pd_4, and Ni_2Pd_2. Their findings revealed that the orbit of Ni overlaps with that of Pd, resulting in a decrease in the electric field of the Ni_2Pd_2 cluster compared to Ni_4 and Pd_4. This decrease leads to a reduction in hydrogen adsorption energy [21]. In another study by Wu et al., a systematic comparison of the properties of Pd_6Ni_4 and Pd_4Ni_6 was conducted [22]. It was observed that the migration of electrons between Pd and Ni caused the D-band center of Pd to shift downward. Our previous research demonstrated that P doping improves the electronic structure of Pd, resulting in a downward shift of the D-band center. This, in turn, weakens the adsorption energy of materials for H_2, lowers the diffusion energy barrier of H atoms, and promotes hydrogen spillover, thereby enhancing hydrogen absorption and desorption capacity [23]. Based on this understanding, the design and synthesis of Pd–Ni alloys hold promise for improving hydrogen storage performance by adjusting the D-band center.

Moreover, despite extensive research efforts in the field of hydrogen storage, the development of efficient onboard hydrogen storage systems still poses significant scientific and technological challenges. To address this issue, we have conducted a comprehensive investigation, taking into account previous experimental and computational studies towards the development of an economically viable graphene-based hydrogen storage material. In our study, we successfully synthesized the material using a solvothermal method, wherein palladium (Pd) was incorporated to partially replace nickel (Ni). Notably, the utilization of ethylene glycol as the hydrothermal solvent resulted in the coexistence of Ni and Pd in the final material. Concurrently, some Pd atoms infiltrated the Ni lattice, forming a NiPd alloy. Our findings reveal that the NiPd-rGO material exhibits a remarkable maximum hydrogen storage capacity of 2.65 wt% under mild conditions, specifically at 4 MPa and 298 K.

2. Material Preparation

2.1. Obtaining NiPd-rGO at Different Hydrothermal Temperatures

The experimental procedure commenced by dispersing 100 mg of graphene oxide (GO) powder in a mixed solution comprising 30 mL of ethylene glycol and 30 mL of de-ionized water. To this mixture, 78 mg of $Pd(OAc)_2$ and 150 mg of $NiCl_2 \cdot 6H_2O$ were added. The pH of the solution was adjusted to 12~13 using an appropriate amount of KOH. The resulting solution, containing the dispersed components, was transferred to a 100 mL Teflon-lined reaction kettle, sealed, and placed in an oven. The temperature of the oven was sequentially raised to 160 °C, 180 °C and 200 °C, allowing the hydrothermal reaction to proceed for 12 h at each temperature. After the hydrothermal reaction was completed, the resulting black solution was left undisturbed for 6 h. The supernatant was carefully decanted, and the remaining material was subjected to repeated washing with deionized water using suction filtration until the pH reached a neutral level. Subsequently, the material was freeze-dried for 12 h, resulting in the formation of a black powder. To further enhance the material, the obtained black powders underwent a heat treatment process at 500 °C for 1 h under a 10% H_2/Ar atmosphere, employing a heating rate of 5 °C/min. This final treatment led to the formation of NiPd-rGO, the desired product of the synthesis process.

2.2. Computational Methodology

For this study, a Ni_2Pd_2 cluster and a graphene slab consisting of sixty carbon atoms, along with a single attached oxygen atom (rGO), were employed. To prevent interactions between neighboring layers, a vacuum layer with a thickness of 15 Å was added in the Z-direction. The formation of single-vacancy-reduced graphene oxide (SVrGO) was achieved by removing a carbon atom from the center of the graphene sheet and attaching an oxygen atom to the resulting vacancy, resulting in a graphene sheet with fifty-nine atoms. Subsequently, the structure was optimized to attain a stable configuration. Density Functional Theory (DFT) was employed as the computational method in this study. It allowed for the calculation of important parameters such as the binding energy between the Ni_2Pd_2 cluster and SVrGO, as well as the adsorption energy of hydrogen (H_2) on the composite material. Vienna Ab initio Simulation Package (VASP) was used to perform the calculations. The exchange and correlation interactions were described using generalized gradient approximation (GGA) in the form of the Perdew–Burke–Ernzerhof (PBE) functional. The calculations used a Brillouin region of $5 \times 5 \times 1$ as the k-point value, and the geometry optimization structure was obtained by relaxation until the force on each atom was less than 0.02 eV/Å, and the convergence criterion for energy was chosen as 1×10^{-4} eV. The cutoff energy was 450 eV, and all calculations considered spin polarization. The converged bond length for pristine graphene was determined to be 1.42 Å which shows that all the parameters used in this calculation are valid. The study employed an equation to determine the adsorption energy of the cluster and slab:

$$E_{(B)} = E_{(Cluster+substarte)} - E_{(substrate)} - E_{(cluster)} \tag{1}$$

where $E_{(B)}$ is the binding energy, $E_{(cluster+substarte)}$ is the total energy of a system, $E_{(substarte)}$ is the total energy of slab and $E_{(cluster)}$ is the total energy of cluster. The following equation has been utilized for the adsorption energy of the H_2 molecule, taking into account the vibration effect for a gas phase only, as it is nearly negligible for a solid system.

$$E_{(ads)} = \left(E_{(H2+Cluster+substarte)} + \triangle ZPE \right) - E_{(cluster+substrate)} - \left(E_{(H2)} + \triangle ZPE \right) \tag{2}$$

where $E_{(ads)}$ is the H_2 adsorption energy, $E_{(H2+cluster+substarte)}$ is the total energy of a system, $E_{(cluster+substarte)}$ is the total energy of the cluster-decorated substrate, $E_{(H2)}$ is the total energy of a single H_2 molecule and $\triangle ZPE$ is the zero-point energy correction.

3. Results and Discussion

Figure 1 illustrates the X-ray diffraction (XRD) patterns obtained from the NiPd-rGO-X (X = 160, 180, 200 °C) samples, synthesized at different hydrothermal temperatures. A comparison with standard PDF cards allowed for the identification of a biphasic structure within the material, indicating the coexistence of Pd and Ni phases. The peaks observed at 40.2°, 46.8°, 68.3°, 82.4° and 86.9° correspond to Pd (111), Pd (200), Pd (220), Pd (311) and Pd (222), respectively, as per the PDF database. Of particular interest, the peaks associated with Ni (PDF#70-1849) exhibited a noticeable shift towards lower angles, implying the incorporation of Pd atoms into the Ni lattice, thereby forming a NiPd alloy [24]. This observation is evident from the shift of the Ni (111) diffraction peak towards smaller angles, indicating an expansion of the nickel unit cell due to alloying with palladium. The XRD peaks change with the variation in temperature. At 160 °C, the material predominantly exhibits a split phase composed of palladium and nickel. The XRD analysis indicates that the nickel content within the palladium lattice is relatively low, and correspondingly, the palladium content within the nickel lattice is also relatively small. As the temperature is increased to 180 °C, there is an augmentation in the quantity of palladium integrated into the nickel lattice, leading to the formation of a greater number of palladium–nickel alloys with nickel as the primary phase. Subsequently, at 200 °C, the amount of nickel incorporated into the palladium lattice increases, resulting in the transformation of the material into a palladium–nickel alloy with palladium as the principal phase. Moreover, the shift in the position of the XRD peaks can serve as evidence. At 160 °C, the phase exhibits closer proximity to our palladium and nickel composition. At 180 °C, the palladium content within the nickel lattice notably increases, leading to a greater shift in the peak position. Finally, at 200 °C, the position of the nickel peak shifts in the opposite direction, while the peak shift of palladium becomes more pronounced. This finding suggests a higher proportion of Ni alloyed with Pd in the sample synthesized at 180 °C, indicating that 180 °C may be the optimum hydrothermal temperature for the synthesis of NiPd. The higher proportion of Ni in the NiPd alloy synthesized at 180 °C is particularly significant as Ni is more cost-effective than Pd. Therefore, the increased proportion of Ni offers a promising avenue for achieving a cost-effective hydrogen storage material in the NiPd alloy synthesized at 180 °C.

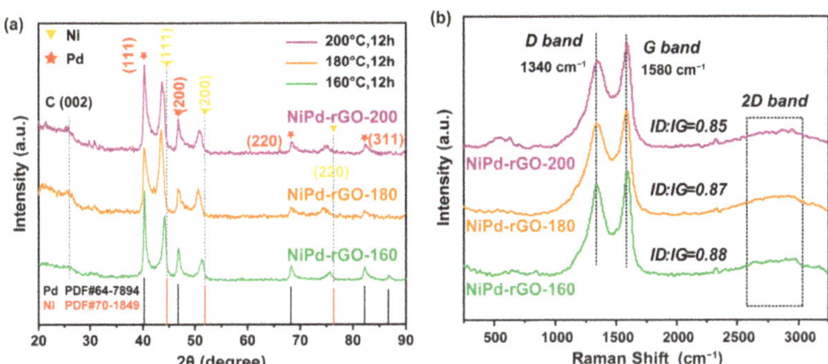

Figure 1. (**a**) XRD patterns, (**b**) Raman spectra of NiPd-rGO-X (X = 160, 180, 200 in °C) prepared at different hydrothermal temperatures.

In order to investigate the presence of defects within the materials, Raman spectroscopy was employed to analyze the NiPd-rGO-X samples (where X represents the hydrothermal temperature of 160, 180, and 200 °C), as illustrated in Figure 1b. The obtained spectra for all three samples exhibited the presence of the D-band and G-band. The D-band, positioned at 1340 cm^{-1}, corresponds to the disordered hybridization of sp^2 [25], while the G-band, observed at 1580 cm^{-1}, is associated with the stretching of the C-C bond within the graphene material. Furthermore, upon conducting calculations, it was discovered that the NiPd-rGO-160 sample exhibited the highest I_D/I_G ratio, measuring 0.88. It is worth noting that the I_D/I_G ratio tends to decrease as the hydrothermal temperature increases. Consequently, the NiPd-rGO-200 sample displayed the lowest I_D/I_G ratio, amounting to 0.85.

To examine the distribution of nanoparticles on the materials, field emission scanning microscopy (FESEM) was utilized to investigate the morphologies of the NiPd-rGO-X samples. The findings, depicted in Figure 2, demonstrate a uniform distribution of NiPd nanoparticles on the graphene surface. The grain size of these nanoparticles ranges from 5 to 45 nm, exhibiting a normal distribution pattern. A detailed analysis of Figure 2 indicates that as the hydrothermal temperature increases, the size of the nanoparticles loaded on the graphene surface tends to decrease. However, it is worth noting that beyond a temperature of 180 °C, the size variations become insignificant, thereby further corroborating that 180 °C represents the optimal hydrothermal temperature. Moreover, an interesting observation is made with the increase in hydrothermal temperature, wherein some larger particles present in the NiPd-rGO-160 sample disappear. In the NiPd-rGO-200 sample, the particle size distribution falls within the range of 5–40 nm, exhibiting a narrower distribution compared to that observed in NiPd-rGO-160.

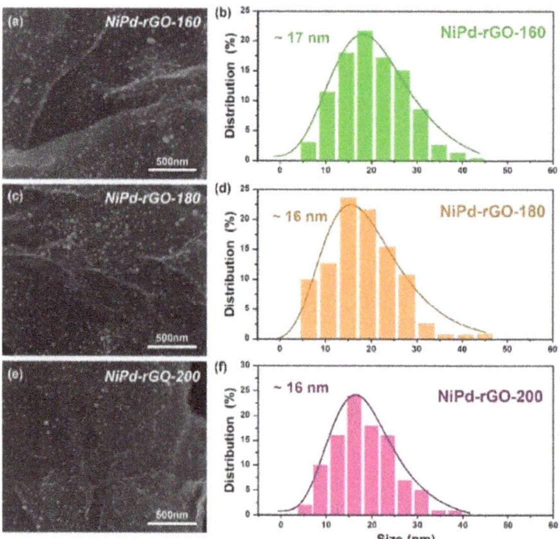

Figure 2. SEM images ((**a**) = NiPd-rGO-160, (**c**) = NiPd-rGO-180, (**e**) = NiPd-rGO-200) and particle size distribution ((**b**) = NiPd-rGO-160, (**d**) = NiPd-rGO-180, (**f**) = NiPd-rGO-200) of NiPd-rGO-X (X = 160, 180, 200 in °C) prepared at different hydrothermal temperatures.

The specific surface area and pore size distribution of the NiPd-rGO-X material were determined through N$_2$ isothermal adsorption–desorption tests conducted at a temperature of 77K. Figure 3a,b display the N$_2$ adsorption–desorption isotherms and pore size distribution curves of NiPd-rGO-X, respectively. It is evident that the N$_2$ adsorption–desorption isotherms of all three samples exhibit the characteristic features of type-IV adsorption isotherms, suggesting the presence of a mesoporous structure within the material. The observed hysteresis in the isotherms further supports the existence of mesopores. Among the samples, NiPd-rGO-200 exhibited the highest specific surface area, measuring 137.89 m^2/g.

It is noteworthy that the specific surface area demonstrated a decreasing trend as the hydrothermal temperature was reduced, which could be attributed to the exfoliation of graphene layers at higher temperatures. The average pore diameter of NiPd-rGO-X was determined to be approximately 2.5 nm, indicating a similarity in the pore structure among the three samples. Nevertheless, the pore volume diminished with decreasing hydrothermal temperature, consistent with the variation in specific surface area, thus suggesting a correlation between pore volume and surface area.

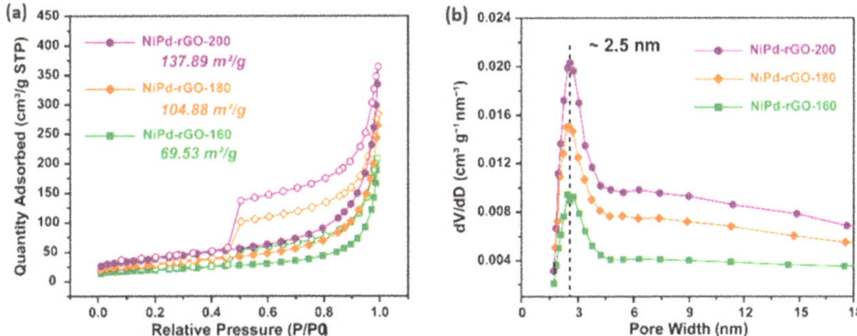

Figure 3. (a) N_2 isotherm adsorption–desorption curves, (**b**) pore size distribution of NiPd-rGO-X (X = 160, 180, 200 in °C) prepared at different hydrothermal temperatures.

The X-ray photoelectron spectroscopy (XPS) technique was utilized to investigate the surface state of the samples in detail. Figure 4 showcases the XPS spectrum of NiPd-rGO-X. Upon analyzing the survey spectra, distinct peaks corresponding to C, O, Pd and Ni are observed. In the 1s spectrogram of C, the peaks at 284 eV, 284.8 eV and 286.3 eV correspond to the C=C bond, C-O bond and C=O bond, respectively. These peaks indicate the presence of incompletely reduced oxygen-containing groups in rGO, thus reflecting its surface composition [26]. Moving on to Figure 4c, it is observed that the peaks at 335.1 eV and 340.2 eV correspond to the $3d_{5/2}$ orbital and $3d_{3/2}$ orbital of $Pd^{(0)}$, respectively. Additionally, the peaks at 336 eV and 341.5 eV correspond to the $3d_{5/2}$ orbital and $3d_{3/2}$ orbital of $Pd^{(II)}$, respectively. This indicates the coexistence of Pd in different oxidation states, suggesting the formation of PdO [27]. Notably, in comparison to Pd-rGO, the peak of Ni-Pd-rGO shifts towards a higher binding energy, indicating a change in the chemical state of the Pd atom. This forward movement suggests electron transfer from Pd to Ni after alloying, leading to a decrease in the center of the D-band [28]. The downward shift of the D-band center corresponds to a reduction in the adsorption of H by metals, thereby promoting hydrogen spillover and improving the hydrogen storage performance of the materials. Lastly, as depicted in Figure 4d, the peaks at 853.0 eV and 871.1 eV are associated with Ni, while the peaks at 855.5 eV and 873.1 eV are attributed to Ni^{2+}, indicating the oxidation of Ni on the surface of the sample.

The hydrogen adsorption kinetic curves of NiPd-rGO-X at a pressure of 4 MPa and temperature of 298 K are illustrated in Figure 4a. The measured hydrogen storage capacities of the NiPd-rGO-160, NiPd-rGO-180, and NiPd-rGO-200 samples are 2.60 wt%, 2.65 wt% and 2.63 wt%, respectively, indicating their close proximity as shown in Figure 5. The marginally higher hydrogen storage capacity observed in the NiPd-rGO-180 sample can be attributed to the enhanced alloying degree of NiPd, resulting in a more efficient hydrogen spillover effect compared to the Pd and Ni single phases. Notably, the NiPd-rGO-180 material exhibits the highest hydrogen storage capacity (2.65 wt%) among the Ni-rGO, Pd-rGO and NiPd-rGO materials. This remarkable improvement is ascribed to the synergistic effect of rGO and PdNi, which alters the diffusion behavior of hydrogen molecules within the composites. Specifically, the PdNi alloy particles act as activation centers, facilitating the decomposition of hydrogen molecules into hydrogen atoms. These atoms are then stored within the surface defects of graphene, thereby enhancing the overall hydrogen storage capacity of the material. This phenomenon is consistent with the findings of other researchers, who have reported on the higher affinity of defects for hydrogen adsorption from other surface sites [29,30].

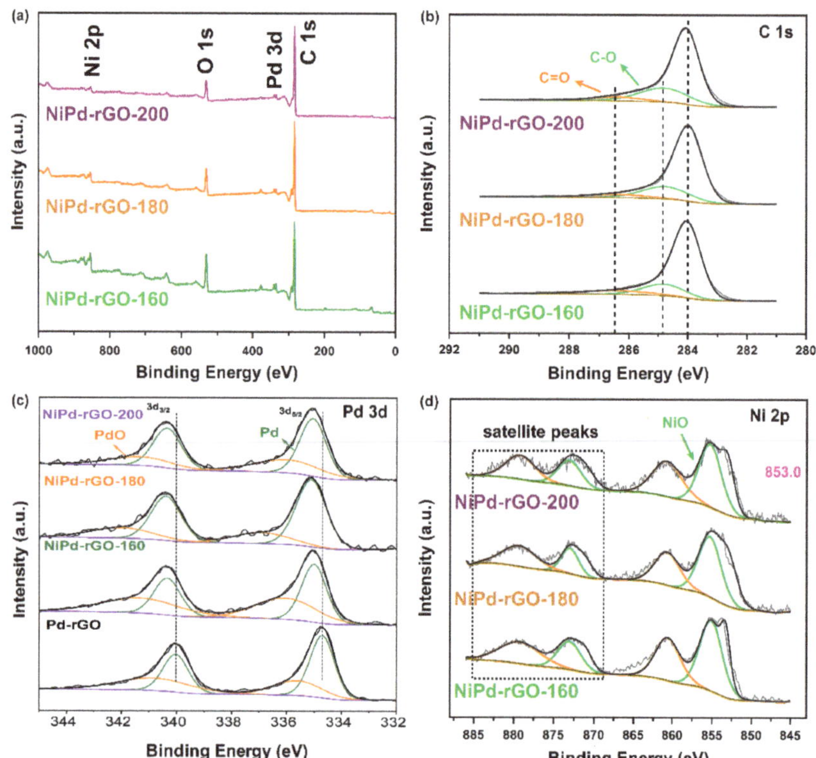

Figure 4. XPS spectra of NiPd-rGO-X (X = 160, 180, 200 in °C) prepared at different hydrothermal temperatures: (**a**) survey spectra, (**b**) C 1s, (**c**) Pd 3d and (**d**) Ni 2p orbitals.

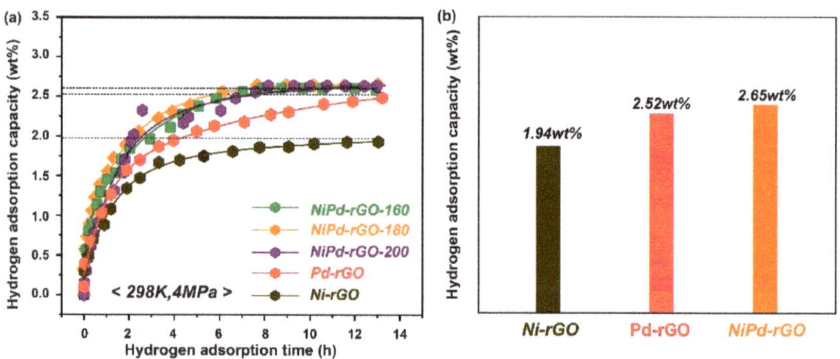

Figure 5. (**a**) Hydrogen adsorption kinetic curves and (**b**) hydrogen storage capacity at 4 MPa/298 K of different samples.

Following the optimization of the graphene slabs, as well as Ni_2Pd_2 clusters, the cluster was placed onto rGO and SVrGO, where the cluster of the Ni_2Pd_2 was tested in different orientations and is presented in Figure 6. To evaluate the stability of Ni_2Pd_2 on different graphene surfaces, the binding energies were determined using Equation (1), and the findings of our study suggest that the cluster is supported on rGO (with no defects) with three different orientations as shown in the following: Figure 6a, Pd atoms attached to carbon having a binding energy of -1.03 eV; Figure 6b, Ni atoms attached to carbon having a binding energy of -2.60 eV, which is higher than the former case due to bonding of the oxygen atom with the Ni atom; and Figure 6c, Pd and Ni atoms attached to carbon having a binding energy of -1.1 eV, which do not seem to be suitable materials for hydrogen storage. This is attributed to lower binding energy which can cause metal atoms agglomeration which reduces the hydrogen storage capacity. As we know, during metal–metal interaction, metal atoms tend to cluster together, resulting in larger clusters rather than being evenly dispersed on the substrate. This clustering phenomenon is much stronger than the cluster–substrate interaction, which minimizes the efficiency of hydrogen molecule interaction.

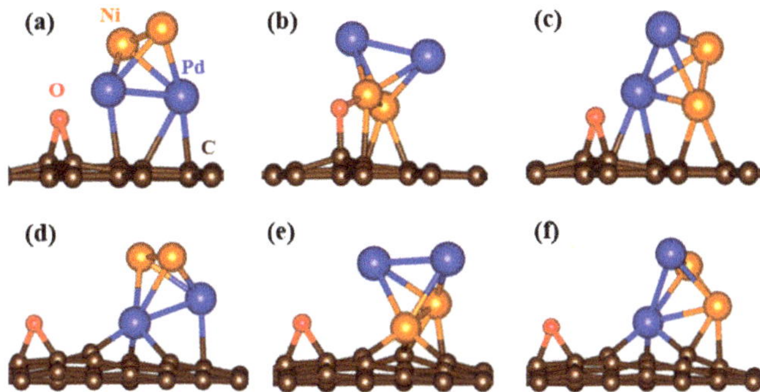

Figure 6. The side view of optimized structures of Ni_2Pd_2 cluster supported on rGO (**a**–**c**) and SVrGO (**d**–**f**) substrates.

Moreover, the desorption of metal–hydrogen complexes competes with the desorption of H_2 from pure graphene, further hindering the overall hydrogen storage process [31]. To address these issues, defect modifications have been employed as techniques to enhance the interaction between graphene and hydrogen molecules. Conversely, the binding energies of Ni_2Pd_2 supported on SVrGO with three different orientations are shown in the following: Figure 6d, Pd atoms attached to carbon vacancy having a binding energy of -4.72 eV; Figure 6e, Ni atoms attached to carbon having a binding energy of -6.61 eV; and, finally, Figure 6f, Pd and Ni atoms attached to carbon having a binding energy of -5.17 eV. Therefore, these substrates are not conducive to clustering. The results of our study also highlight the significance of defects in the graphene slab, as they significantly increase the binding energy between the transition metal cluster and the substrate. Our calculations indicate that the binding energy is approximately four times higher with the introduction of defects, which is consistent with previous findings reported by Kim et al. [32]. This increase in binding energy not only enhances the hydrogen storage capacity of the system but also addresses the challenge of complex hydride desorption during hydrogen molecule release [31]. By effectively increasing the binding energy, the presence of defects in the graphene structure offers a promising solution for improving hydrogen storage capabilities and overcoming the issue of complex hydride desorption.

In order to calculate the adsorption energy of the H_2 molecule, Equation (2) was applied which considers the vibration effect only for the gas phase since it is almost negligible for a solid system. The optimized structures and corresponding calculated adsorption energies of single H_2 molecules adsorbed on Ni_2Pd_2-rGO and Ni_2Pd_2-SVrGO at different positions within the orientation of the cluster are illustrated in Figure 7. The adsorption energies with negative values in Figure 7 indicate favorable adsorption, as they signify the attraction of H_2 molecules toward the cluster. Upon structural optimization, the H-H bond underwent relaxation, resulting in bond lengths ranging from 0.83 Å to 0.88 Å, and adsorbed H_2 molecules exhibited an increase in bond length on all materials as compared to the free H_2 bond length of 0.75 Å. However, when compared to an isolated H_2 molecule, the stretching of the H-H bond length was found to be minimal, with a difference of less than 0.3 Å [33], indicating non-dissociative chemisorption or "molecular" adsorption of H_2, with the association between the H_2 molecule and the Ni and Pd atoms resulting from Kubas-type interaction. Additionally, the bond length between Ni and H varies between 1.56 Å and 1.64 Å, and the bond length of Pd and H varies between 1.72 Å and 1.79 Å. As shown in Figure 7, it is observed that the adsorption energies are different from each other, which means that the orientation of the cluster and defects in the slabs alter the adsorption energy of the H_2 molecule. Successful hydrogen storage under ambient conditions requires an adsorption mechanism that falls between physical and chemical adsorption, with previous studies recommending an adsorption energy range of -0.20 to -0.60 eV [34], which is consistent with the adsorption energies observed in this study. All materials fall within this range except Figure 7(e, g), indicating their potential for hydrogen storage.

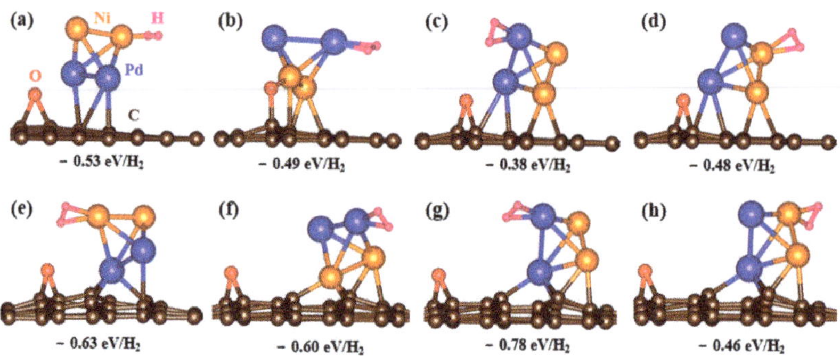

Figure 7. The side view of optimized structures of H_2 adsorbed on different sites on Ni_2Pd_2 cluster supported on rGO (**a–d**) and SVrGO (**e–h**) substrates.

4. Conclusions

In this study, we employed the hydrothermal method to synthesize NiPd-rGO, aiming to develop a novel hydrogen storage material with improved capacity at moderate operating conditions and reduced cost. The obtained material exhibited a higher hydrogen storage capacity of 2.65 wt% (at 4 MPa and 298 K) compared to its Ni-rGO and Pd-rGO counterparts. This enhanced capacity can be attributed to the synergistic effect between the alloy particles and rGO. The process of alloying induced changes in the electronic structure of Pd, leading to a decrease in the center of the D-band. Consequently, the adsorption of hydrogen on the metal surface weakened, promoting the hydrogen spillover effect. Additionally, we investigated the binding energies with different orientations. The range of binding energies was found to be -1.03 to -2.60 eV for Ni_2Pd_2-rGO and -4.72 to -6.61 eV for Ni_2Pd_2-SVrGO. These results indicate that the introduction of defects into the graphene substrate enhances the interaction between the supported cluster and the substrate, thereby improving the stability of the material and reducing the likelihood of transition metal atom agglomeration. The calculated adsorption energies fell within the ideal range of -0.20 to -0.60 eV, which is conducive to efficient hydrogen ad/desorption processes. Furthermore, the introduction of defects in the substrate and the orientation of the supported cluster were found to influence the hydrogen adsorption energy. These findings contribute to the advancement of hydrogen storage technologies and pave the way for the design and optimization of graphene-based materials with superior hydrogen storage properties.

Author Contributions: Conceptualization, Y.C. (Yungui Chen), X.G. and C.W.; Validation, G.X., C.J., Y.W., Y.Y., X.G., Y.L. and C.W.; Formal analysis, Y.C. (Yiwen Chen) and H.; Investigation, H., Y.W., X.G. and C.W.; Writing—original draft, Y.C. (Yiwen Chen) and H.; Writing—review & editing, Y.C. (Yiwen Chen), H., G.X., C.J., Y.Y. and Y.L.; Visualization, Y.W., X.G. and C.W.; Supervision, G.X. and C.J. All authors have read and agreed to the published version of the manuscript.

Funding: This research was funded by National Natural Science Foundation of China (U2030208, U2130208); National Key R&D Program of China (2022YFB3803700); State Key Laboratory of Clean and Efficient Turbomachinery Power Equipment (DECSKL202110); Fundamental Research Funds for Central Universities, China.

Data Availability Statement: No new data were created or analyzed in this study. Data sharing is not applicable to this article.

Conflicts of Interest: The authors declare no conflict of interest.

References

1. Shafiee, S.; Topal, E. When will fossil fuel reserves be diminished? *Energy Policy* **2009**, *37*, 181–189. [CrossRef]
2. Chakraborty, B.; Ray, P.; Garg, N.; Banerjee, S. High capacity reversible hydrogen storage in titanium doped 2D carbon allotrope Ψ-graphene: Density Functional Theory investigations. *Int. J. Hydrogen Energy* **2021**, *46*, 4154–4167. [CrossRef]
3. Li, H.-M.; Li, G.-X.; Li, L. Comparative investigation on combustion characteristics of ADN-based liquid propellants in inert gas and oxidizing gas atmospheres with resistive ignition method. *Fuel* **2023**, *334*, 126742. [CrossRef]
4. Midilli, A.; Ay, M.; Dincer, I.; Rosen, M.A. On hydrogen and hydrogen energy strategies I: Current status and needs. *Renew. Sustain. Energy Rev.* **2005**, *9*, 255–271. [CrossRef]
5. Liu, W.; Sun, L.; Li, Z.; Fujii, M.; Geng, Y.; Dong, L.; Fujita, T. Trends and future challenges in hydrogen production and storage research. *Environ. Sci. Pollut. Res.* **2020**, *27*, 31092–31104. [CrossRef] [PubMed]
6. Lijun, J. Expediting the Innovation and Application of Solid Hydrogen Storage Technology. *Engineering* **2021**, *7*, 731–733.
7. Sun, Y.; DeJaco, R.F.; Li, Z.; Tang, D.; Glante, S.; Sholl, D.S.; Colina, C.M.; Snurr, R.Q.; Thommes, M.; Hartmann, M.; et al. Fingerprinting diverse nanoporous materials for optimal hydrogen storage conditions using meta-learning. *Sci. Adv.* **2021**, *7*, eabg3983. [CrossRef]
8. Chen, Z.; Kirlikovali, K.O.; Idrees, K.B.; Wasson, M.C.; Farha, O.K. Porous materials for hydrogen storage. *Chem* **2022**, *8*, 693–716. [CrossRef]
9. Gor Bolen, M. Hydrogen Storage in Porous Silicon—A Review. *Silicon* **2023**. [CrossRef]
10. Guo, J.-H.; Li, S.-J.; Su, Y.; Chen, G. Theoretical study of hydrogen storage by spillover on porous carbon materials. *Int. J. Hydrogen Energy* **2020**, *45*, 25900–25911. [CrossRef]
11. Shiraz, H.G.; Tavakoli, O. Investigation of graphene-based systems for hydrogen storage. *Renew. Sustain. Energy Rev.* **2017**, *74*, 104–109. [CrossRef]
12. Wang, Y.; Meng, Z.; Liu, Y.; You, D.; Wu, K.; Lv, J.; Wang, X.; Deng, K.; Rao, D.; Lu, R. Lithium decoration of three dimensional boron-doped graphene frameworks for high-capacity hydrogen storage. *Appl. Phys. Lett.* **2015**, *106*, 063901. [CrossRef]
13. Luo, Z.; Fan, X.; Pan, R.; An, Y. A first-principles study of Sc-decorated graphene with pyridinic-N defects for hydrogen storage. *Int. J. Hydrogen Energy* **2017**, *42*, 3106–3113. [CrossRef]
14. Patchkovskii, S.; Tse, J.S.; Yurchenko, S.N.; Zhechkov, L.; Heine, T.; Seifert, G. Graphene nanostructures as tunable storage media for molecular hydrogen. *Proc. Natl. Acad. Sci. USA* **2005**, *102*, 10439–10444. [CrossRef] [PubMed]
15. Weizhi, T.; Ying, Z.; Yazhou, W.; Tong, L.; Hong, C. A study on the hydrogen storage performance of graphene-Pd(T)-graphene structure. *Int. J. Hydrogen Energy* **2020**, *45*, 12376–12383.
16. DOE Technical Targets for Onboard Hydrogen Storage for Light-Duty Vehicles. Available online: https://www.energy.gov/eere/fuelcells/doe-technical-targets-onboard-hydrogen-storage-light-duty-vehicles (accessed on 31 May 2023).
17. Xu, B.; Li, X.; Chen, Z.; Zhang, T.; Li, C. Pd@MIL-100(Fe) composite nanoparticles as efficient catalyst for reduction of 2/3/4-nitrophenol: Synergistic effect between Pd and MIL-100(Fe). *Microporous Mesoporous Mater.* **2018**, *255*, 1–6. [CrossRef]
18. Wan, W.C.; Zhang, R.Y.; Ma, M.Z.; Zhou, Y. Monolithic aerogel photocatalysts: A review. *J. Mater. Chem. A* **2018**, *6*, 754–775. [CrossRef]
19. Liu, D.; Xie, M.; Wang, C.; Liao, L.; Qiu, L.; Ma, J.; Huang, H.; Long, R.; Jiang, J.; Xiong, Y. Pd-Ag alloy hollow nanostructures with interatomic charge polarization for enhanced electrocatalytic formic acid oxidation. *Nano Res.* **2016**, *9*, 1590–1599. [CrossRef]
20. Psofogiannakis, G.M.; Froudakis, G.E. DFT Study of the Hydrogen Spillover Mechanism on Pt-Doped Graphite. *J. Phys. Chem. C* **2009**, *113*, 14908–14915. [CrossRef]
21. Guo, J.-H.; Li, X.-D.; Cheng, X.-L.; Liu, H.-Y.; Li, S.-J.; Chen, G. The theoretical study of the bimetallic Ni/Pd, Ni/Pt and Pt/Pd catalysts for hydrogen spillover on penta-graphene. *Int. J. Hydrogen Energy* **2018**, *43*, 19121–19129. [CrossRef]
22. Wu, S.-Y.; Chen, H.-T. Structure, Bonding, and Catalytic Properties of Defect Graphene Coordinated Pd-Ni Nanoparticles. *J. Phys. Chem. C* **2017**, *121*, 14668–14677. [CrossRef]

23. Jin, C.; Li, J.; Zhang, K.; Habibullah; Xia, G.; Wu, C.; Wang, Y.; Cen, W.; Chen, Y.; Yan, Y.; et al. Pd3P nanoparticles decorated P-doped graphene for high hydrogen storage capacity and stable hydrogen adsorption-desorption performance. *Nano Energy* **2022**, *99*, 107360. [CrossRef]
24. Chen, Z.; Zhang, J.; Zhang, Y.; Liu, Y.; Han, X.; Zhong, C.; Hu, W.; Deng, Y. NiO-induced synthesis of PdNi bimetallic hollow nanocrystals with enhanced electrocatalytic activities toward ethanol and formic acid oxidation. *Nano Energy* **2017**, *42*, 353–362. [CrossRef]
25. Reich, S.; Thomsen, C. Raman spectroscopy of graphite. *Philos. Trans. R. Soc. A Math. Phys. Eng. Sci.* **2004**, *362*, 2271–2288. [CrossRef]
26. Yang, G.; Chen, Y.; Zhou, Y.; Tang, Y.; Lu, T. Preparation of carbon-supported Pd-P catalyst with high content of element phosphorus and its electrocatalytic performance for formic acid oxidation. *Electrochem. Commun.* **2010**, *12*, 492–495. [CrossRef]
27. Qiu, B.; Li, Q.; Shen, B.; Xing, M.; Zhang, J. Stober-like method to synthesize ultradispersed Fe3O4 nanoparticles on graphene with excellent Photo-Fenton reaction and high-performance lithium storage. *Appl. Catal. B-Environ.* **2016**, *183*, 216–223. [CrossRef]
28. Belykh, L.B.; Sterenchuk, T.P.; Skripov, N.I.; Akimov, V.V.; Tauson, V.L.; Romanchenko, A.S.; Gvozdovskaya, K.L.; Sanzhieva, S.B.; Shmidt, F.K. Effect of the State of a Surface Layer on the Properties of Pd-P Catalysts in the Hydrogenation of Alkylanthraquinones. *Kinet. Catal.* **2019**, *60*, 808–817. [CrossRef]
29. Sunnardianto, G.K.; Bokas, G.; Hussein, A.; Walters, C.; Moultos, O.A.; Dey, P. Efficient hydrogen storage in defective graphene and its mechanical stability: A combined density functional theory and molecular dynamics simulation study. *Int. J. Hydrogen Energy* **2021**, *46*, 5485–5494. [CrossRef]
30. Kag, D.; Luhadiya, N.; Patil, N.D.; Kundalwal, S.I. Strain and defect engineering of graphene for hydrogen storage via atomistic modeling. *Int. J. Hydrogen Energy* **2021**, *46*, 22599–22610. [CrossRef]
31. López, M.J.; Cabria, I.; Alonso, J.A. Palladium Clusters Anchored on Graphene Vacancies and Their Effect on the Reversible Adsorption of Hydrogen. *J. Phys. Chem. C* **2014**, *118*, 5081–5090. [CrossRef]
32. Kim, G.; Jhi, S.-H.; Lim, S.; Park, N. Effect of vacancy defects in graphene on metal anchoring and hydrogen adsorption. *Appl. Phys. Lett.* **2009**, *94*, 173102. [CrossRef]
33. Kubas, J. Dihydrogen Complexes as Prototypes for the Coordination Chemistry of Saturated Molecules. *Proc. Natl. Acad. Sci. USA* **2007**, *104*, 6901–6907. [CrossRef] [PubMed]
34. Yoon, M.; Yang, S.; Wang, E.; Zhang, Z. Charged Fullerenes as High-Capacity Hydrogen Storage Media. *Nano Lett.* **2007**, *7*, 2578–2583. [CrossRef] [PubMed]

Review

Integration of CO$_2$ Capture and Conversion by Employing Metal Oxides as Dual Function Materials: Recent Development and Future Outlook

Wei Jie Tan and Poernomo Gunawan *

School of Chemistry, Chemical Engineering and Biotechnology, Nanyang Technological University, 62 Nanyang Drive, Singapore 637459, Singapore; c210023@e.ntu.edu.sg
* Correspondence: pgunawan@ntu.edu.sg

Abstract: To mitigate the effect of CO$_2$ on climate change, significant efforts have been made in the past few decades to capture CO$_2$, which can then be further sequestered or converted into value-added compounds, such as methanol and hydrocarbons, by using thermochemical or electrocatalytic processes. However, CO$_2$ capture and conversion have primarily been studied independently, resulting in individual processes that are highly energy-intensive and less economically viable due to high capital and operation costs. To enhance the overall process efficiency, integrating CO$_2$ capture and conversion into a single system offers an opportunity for a more streamlined process that can reduce energy and capital costs. This strategy can be achieved by employing dual function materials (DFMs), which possess the unique capability to simultaneously adsorb and convert CO$_2$. These materials combine basic metal oxides with active metal catalytic sites that enable both sorption and conversion functions. In this review paper, we focus on the recent strategies that utilize mixed metal oxides as DFMs. Their material design and characteristics, reaction mechanisms, as well as performance and limitations will be discussed. We will also address the challenges associated with this integrated system and attempt to provide insights for future research endeavors.

Keywords: CO$_2$ capture; CO$_2$ conversion; dual function material; materials design

Citation: Tan, W.J.; Gunawan, P. Integration of CO$_2$ Capture and Conversion by Employing Metal Oxides as Dual Function Materials: Recent Development and Future Outlook. *Inorganics* **2023**, *11*, 464. https://doi.org/10.3390/inorganics 11120464

Academic Editors: Roberto Nisticò, Torben R. Jensen, Luciano Carlos, Hicham Idriss, Eleonora Aneggi and Francis Verpoort

Received: 27 September 2023
Revised: 10 November 2023
Accepted: 22 November 2023
Published: 30 November 2023

1. Introduction

With the concentration of carbon dioxide (CO$_2$) in the earth's atmosphere increasing at an unprecedented rate, climate change has become one of the most pressing challenges of our time due to its adverse impact on environmental and ecological welfare, as reported by the Intergovernmental Panel on Climate Change (IPCC) [1]. Addressing this issue requires a comprehensive approach that encompasses CO$_2$ emission reductions, effective CO$_2$ storage, and the sustainable utilization of CO$_2$ as chemical feedstock to close the carbon cycle. Respectively, CO$_2$ capture and CO$_2$ conversion have garnered considerable attention as potential solutions to mitigate its impact and, concurrently, create other value-added products. These approaches are not only necessary to slow down global warming, but also to offer economic opportunities across various industries [2,3]. Naims shows that the economic feasibility for carbon capture and utilization (CCU) is specific to the technology; that it depends on the process efficiency in using CO$_2$ as a cheaper alternative feedstock to more expensive raw materials produced from fossil fuel [4].

Being primarily emitted from combustion processes, the strategies for CO$_2$ capture are determined by several factors, such as its partial pressure, operating conditions, and flue gas compositions. In general, there are typically three types of capture categories, as illustrated in Figure 1, namely (1) pre-combustion, (2) oxyfuel combustion, and (3) post-combustion, depending on the installation location [5]. A general comparison of their advantages and disadvantages in terms of their energy requirement, costs, CO$_2$ recovery, and limitations is presented in Table 1. Pre-combustion capture removes CO$_2$ before combustion is completed. It is typically integrated with a gasifier or reformer where syngas is produced. Through the intermediate removal of CO_2, H_2-rich gas is obtained, which can be further used for heat/power generation. Oxyfuel combustion employs either nearly-pure oxygen environments for combustion, or chemical looping combustion (CLC), in which solid oxides are used as oxygen carriers, resulting in a high concentration of CO_2 that allows for more efficient and inexpensive separation. However, the combustion itself would be relatively costly due to high oxygen purity demands. Lastly, post-combustion capture removes CO_2 from the flue gas. It is considered the most established technology [5], where it typically employs liquid amine absorption or solid adsorption. Due to the low CO_2 partial pressure in the flue gas (*ca.* 10–15%), large volumes of gas, as well as large amounts of energy and equipment size are required, making post-combustion capture processes relatively expensive. For comparison, the mean capture cost from high CO_2 purity sources via oxyfuel combustion ranges from USD 10 to USD 50 per ton of CO_2 mitigated, while that from post-combustion flue gas amine scrubbing ranges from USD 50 to USD 120 [3]. Following its capture, high-purity CO_2 can be recovered via temperature or pressure swing regeneration, after which it is compressed and transported for storage (e.g., geological injection, mineralization) or further utilized or converted.

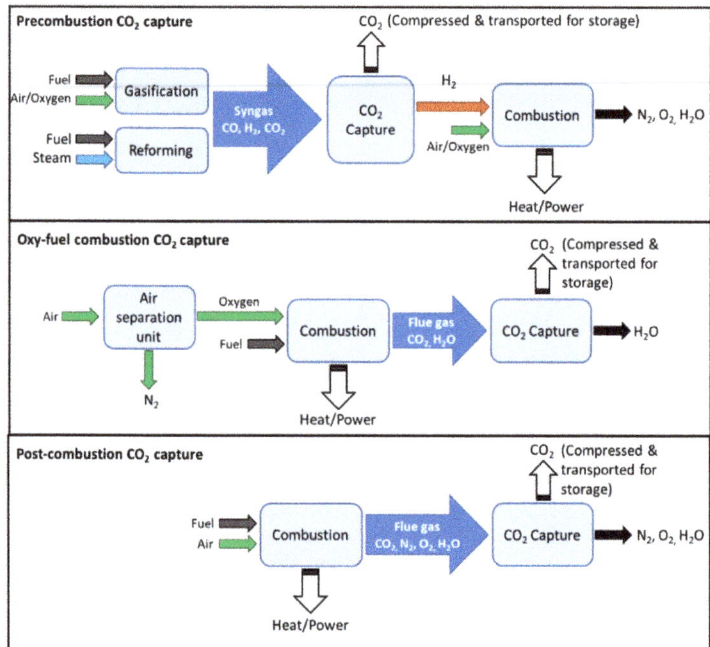

Figure 1. Schematic diagram of CO_2 capture processes from different sources of combustion.

Table 1. Comparison of different CO_2 capture categories. Reproduced from ref. [6] with permission from the American Chemical Society.

Category	Pre-Combustion	Oxyfuel Combustion	Post-Combustion	
			Amine Absorption	Solid Adsorption
CO_2 recovery	92–93%	90–94%	90–98%	80–95%
Energy requirement	low energy	low energy	high regeneration energy	high regeneration energy
Costs	less expensive	moderately expensive	expensive	expensive

Table 1. *Cont.*

Category	Pre-Combustion	Oxyfuel Combustion	Post-Combustion	
			Amine Absorption	Solid Adsorption
Challenges	increased process complexity high CO_2 concentration high pressure process	requires air separation unit high CO_2 concentration high temperature process	corrosion solvent degradation low CO_2 concentration	solid attrition high pressure drop easily poisoned by impurities (NO_x, SO_x) low CO_2 concentration

CO_2 conversion technologies seek to transform the captured CO_2 into useful chemicals, fuels, and materials. Although conversion processes may not directly lead to a net carbon reduction, it enables a circular utilization of CO_2 wastes as feedstock, which may reduce the overall carbon footprint of the corresponding chemical and fuel production, thus leading to more environmentally friendly production processes, as shown in a case study of polyol production for polyurethane [7]. A comprehensive review of sustainable CO_2 conversion by incorporating a life cycle assessment (LCA) in various CO_2 conversion pathways has been published by Artz et al. [8], who concluded that replacing energy-intensive feedstock with CO_2 seems highly promising, particularly when being integrated with renewable energy. In addition, as a C1 building block, the use of CO_2 may also create new, greener synthetic pathways [9].

Despite significant progress having been made for CO_2 capture and CO_2 conversion individually, numerous challenges remain for each technology, most notably the cost effectiveness and the process scalability. Generating high purity of CO_2 intermediates for subsequent utilization requires a large amount of energy for CO_2 recovery, the regeneration of the capture media, and the subsequent gas compression. The purified CO_2 stream then needs to be transported to the location where it will be converted to the desired product, thus further increasing energy demands. To address these issues, combining CO_2 capture and utilization in an integrated system, where in situ conversion will also serve to regenerate the capture media in the same reactor, would be a potential solution to eliminate both compression and transportation costs, thus making the whole process more economical and efficient. For example, the total energy required to produce 1 metric ton of methanol from separate direct air capture and hydrogenation reactions is about 49.4 GJ, out of which 27 GJ is consumed in the CO_2 capture, recovery, and distribution steps. For comparison, to achieve the same product yield, the integrated CO_2 capture and conversion (ICCC) system only requires less than 50% of the total energy requirement, as reported by Freyman et al. [10]. Additionally, ICCC could also reduce capital costs by approximately 38%, as fewer unit operations are required. Freyman et al. coined a term "Reactive Capture" to describe this ICCC system, the concept of which is depicted in Figure 2. To make this system work effectively, the optimization of the reaction conditions, the tuning of the material properties, as well as designing an appropriate reactor system configuration are necessary to ensure a good ICCC performance with good product selectivity and catalyst stability.

In this review paper, we will focus on the development of mixed metal oxides (MMO) as dual function materials (DFMs). To perform both CO_2 capture and CO_2 conversion, DFMs typically are synthesized as a combination of a CO_2 sorbent and a catalytic component. In the presence of CO_2 and other secondary reactants (e.g., H_2, for the methanation of CO_2), the sorbent component first captures CO_2, which then reacts with secondary reactants on the active catalytic sites, converting it into final products [10]. A schematic diagram of ICCC with DFMs is illustrated in Figure 3. The efficiency of the DFM would therefore depend on the CO_2 capture capacity and selectivity, as well as its catalytic activity. Among various solid sorbents, the choice of metal oxides, particularly alkali and alkaline earth oxides, as the sorbent component of a DFM, is convenient, in that the catalytic component would typically also be a transition or noble metal, which can be facilely incorporated into the DFM during its synthesis, such as via co-precipitation or incipient wetness impregnation.

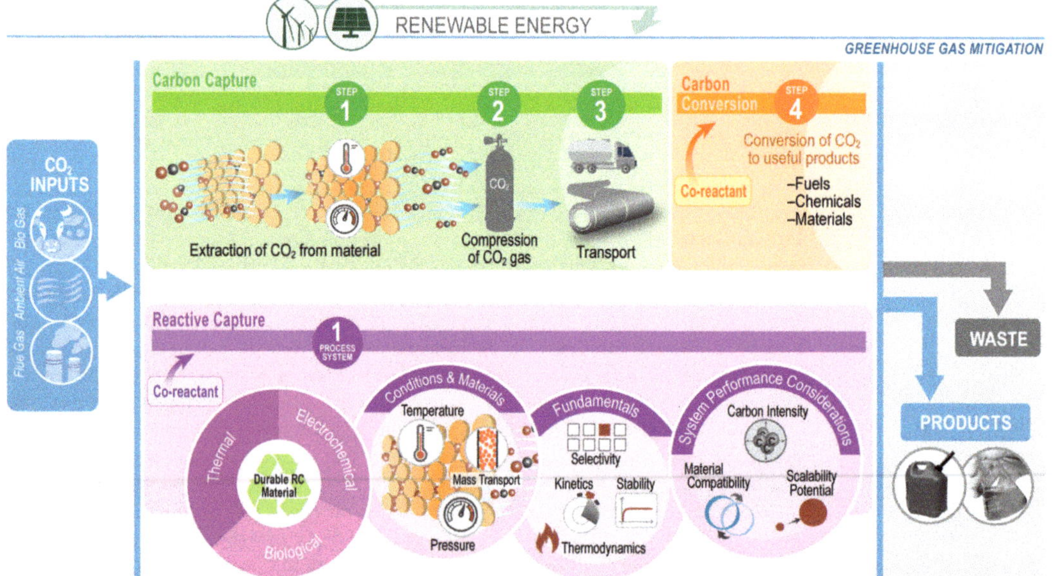

Figure 2. Integrated CO_2 capture and conversion system (Reactive Capture, bottom) in comparison with separate processes. Reprinted from ref. [10] with permission from Elsevier.

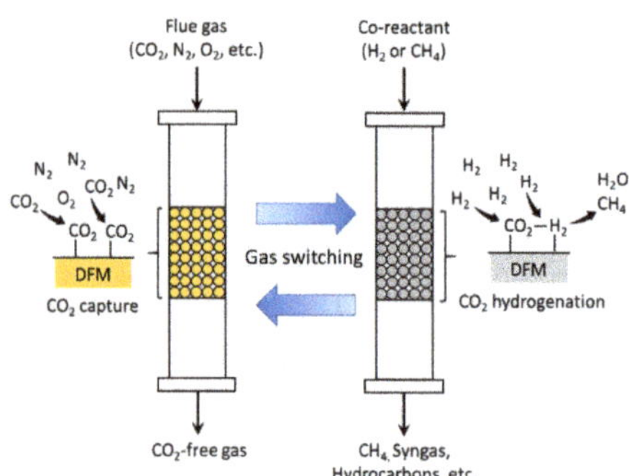

Figure 3. Schematic diagram of integrated CO_2 capture and conversion using DFMs.

To give an overview of the recent trends in this field, we have carried out a literature search on the Web of Science database with the following keywords:

"dual function material" AND "CO_2 capture" AND "CO_2 conversion"
"dual function material" AND "CO_2 capture" AND "CO_2 utilization"

It was found that about 60 relevant papers were published in the span of 10 years, between 2012 and 2022. Farrauto et al. were among the pioneering research groups who first published their work on the development of a DFM in 2015, which comprised ruthenium as the methanation catalyst and nano-dispersed CaO as the CO_2 sorbent, which are both supported on a porous γ-Al_2O_3 carrier [11]. The synthesized material was able to adsorb CO_2 and convert it in the same reactor into synthetic natural gas by using H_2 gas as the co-reactant at 320 °C. Upon the introduction of steam into the reactor, the material showed a stable methanation performance with more than 99% CH_4 purity after 20 cycles. The initial success of this experiment has since gathered increasing interests to explore DFM systems, as indicated by the increasing number of publications in recent years (Figure 4). We have seen efforts being made to improve the CO_2 capture capacity, conversion, and selectivity while reducing the precious metal content. The first review paper on DFMs was published in 2019 by Melo Bravo and Debecker [12], which focused on combined CO_2 capture and methanation, followed by several others in 2020 [6] and 2021 [13–16]. The highlights of these review papers are summarized in Table 2.

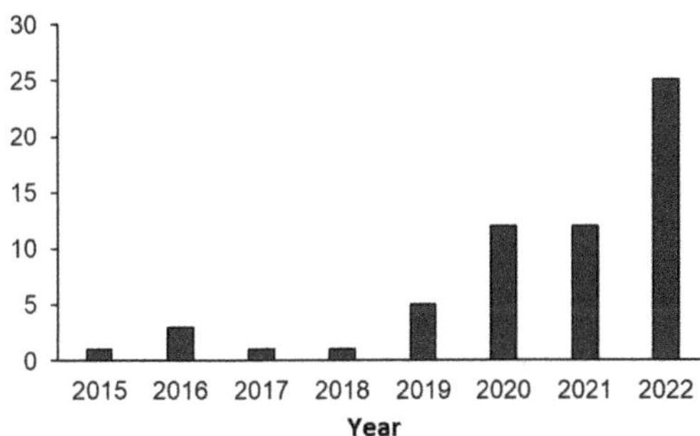

Figure 4. Annual number of publications in the field of dual function materials for integrated CO_2 capture and conversion (2015–2022).

Table 2. Summary of the published reviews on dual function materials (DFMs) for integrated CO_2 capture and conversion.

Authors	Year	Brief Description	Reference
Melo Bravo and Debecker	2019	Early review of DFMs with focus on CO_2 methanation reaction and different types of reactor configurations	[12]
Omodolor et al.	2020	Overview of oxide- and carbonate-based CO_2 adsorbents	[6]
		Investigation of the active metals, material characteristics, and reaction conditions for Ni-, Ru-, and Rh-based DFMs in CO_2 methanation, DRM, and RWGS reactions	
		Brief discussion on hydrotalcite-supported Fe-Cr-Cu catalyst as well as promoted Cu supported on alumina for syngas production	
Merkouri et al.	2021	Chronological review of advances in DRM, RWGS, and CO_2 methanation, as well as evaluation of the reaction mechanism of DFMs, by relating their performances with their physicochemical properties	[13]
Sun et al.	2021	Investigation of the process parameters and adsorbent–catalyst interaction on high-temperature CO_2 capture combined with in situ DRM, RWGS, and methanation reactions	[14]
Sabri et al.	2021	Environmental and economic evaluation of CO_2 capture and utilization process integration, as well as catalyst development for CO_2 conversion in various systems, such as photocatalysis, electrocatalysis, and thermocatalysis	[15]
Li et al.	2021	Discussion on the performances of dual-function oxide particles for CO_2 capture integrated with DRM, CO_2 hydrogenation, and chemical looping	[16]

DRM: dry reforming of methane; RWGS: reverse water–gas shift.

To continue building upon the previous reviews, herein we aim to cover the latest developments on the application of DFMs that employ solid metal oxides for ICCC, particularly from 2021–2023. We will begin our discussion by first giving an overview of the principles of CO_2 capture with metal oxides as solid sorbents, as well as the catalytic conversion of CO_2 from thermodynamic and material perspectives, followed by the typical synthetic methods of DFMs. Thereafter, the performance of various DFMs in ICCC systems will be further assessed for key reactions, such as methanation, RWGS, and DRM, by evaluating key metrics such as the CO_2 capture capacity, selectivity, reaction conversion, and stability. We will correlate these metrics with material design parameters, such as the types of metal oxides used as adsorbents, metallic active sites as catalysts, and the synthetic methods. Mechanistic studies on the molecular interaction between CO_2 and the adsorption and catalytic active sites will also be presented to shed some light on the relationship between the materials' physicochemical properties and their performance. As a conclusion, we will present our perspectives, as well as an outlook on the future opportunities with the emerging technologies of ICCC with DFMs.

2. Metal Oxides as Sorbents for CO_2 Capture

Various CO_2 capture methods have been extensively studied and explored, and several excellent reviews have been recently published [17–28]. An overview of these technologies is presented in Figure 5, among which, liquid absorption in organic amines, such as methanolamine (MEA) and n-methyldiethanolamine (MDEA), have seen the most implementation in large demonstration plants worldwide [29] due to its high absorption capacity. However, this system requires a high energy consumption for amine regeneration, which is prone to degradation. It is also highly corrosive, thus leading to high operating and capital costs.

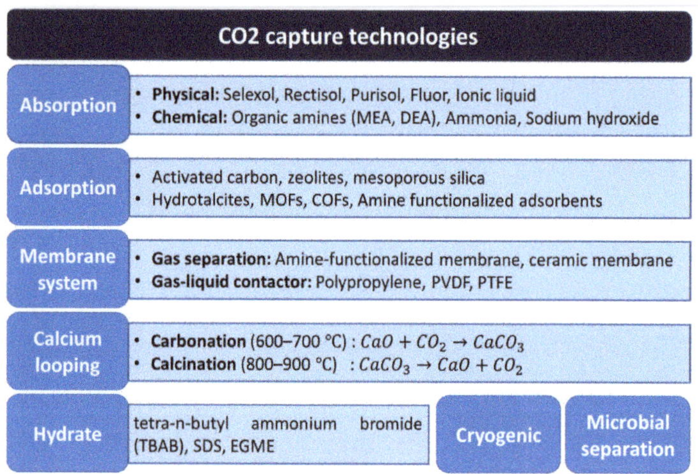

Figure 5. Overview of CO_2 capture technologies.

As potential alternatives to liquid absorption, solid sorbents, such as zeolites [30], hydrotalcites [17,20,31,32], magnesium oxides [33–36], metal organic frameworks (MOFs) [23,37], and porous organic polymers (POPs) [38–40], have been studied extensively due to their easy regeneration and tunable pore size and surface properties. Given their wide variety, the suitability of solid adsorbents depends on their application temperatures, typically ranging from low temperatures (<200 °C) for MOFs, POPs, and zeolites to medium temperatures (200–400 °C) for hydrotalcites and MgO-based sorbents. An overview of commonly used solid sorbents and their typical application temperatures is presented in Figure 6.

In principle, their CO_2 capture capability primarily relies on the physical adsorption of CO_2 without forming a chemical bond. Physisorption is typically mildly exothermic (−25 to −40 kJ/mol) and requires less energy for regeneration as compared to chemisorption processes. Owing to its exothermic nature, the adsorption of CO_2 decreases at a higher temperature, making it more suitable for low-to-medium operating temperatures (<400 °C). However, physisorption processes are usually less selective, leading to the adsorption of other gases, such as N_2 and water vapor, thus lowering the adsorption capacity of CO_2, which is the primary challenge that needs to be solved in order to scale up their application. A common strategy to improve the CO_2 adsorption capacity is by increasing the basicity of the materials via the incorporation of alkaline metals, such as K and Na, though their amounts should be carefully studied as high metal loading may decrease the surface area of the material. For example, for hydrotalcite-based materials, many attempts have been made to modify Mg/Al-based hydrotalcites with K [41–45], Na [43,46], and Cs [42,43,47], with typical optimum K loading of about 20 wt% to achieve the maximum sorption capacity at temperatures between 300 and 400 °C.

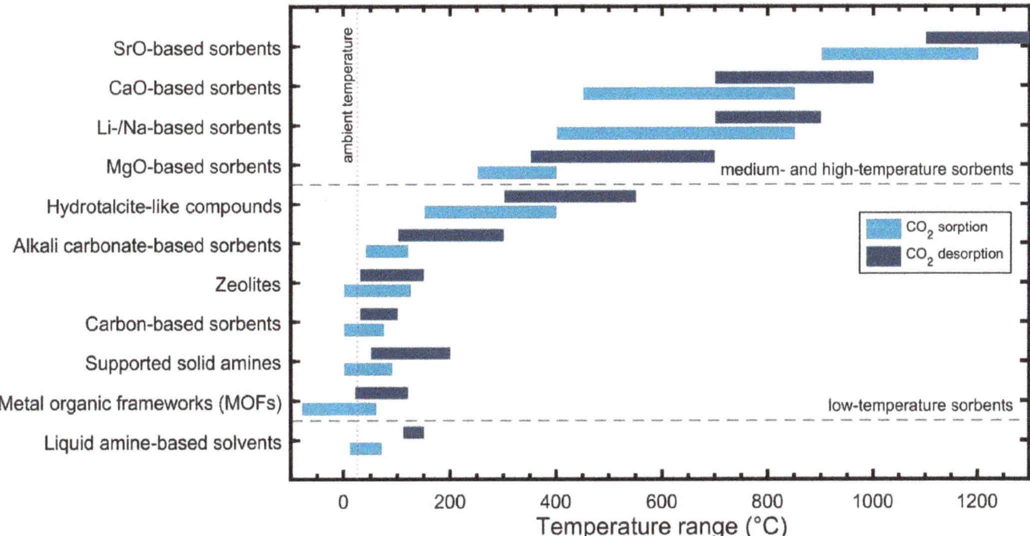

Figure 6. Overview of commonly used solid sorbents and their typical application temperatures. Liquid amine system is shown as the basis for comparison. Reprinted from ref. [25] with permission from the American Chemical Society.

On the other hand, for high-temperature CO_2 adsorption (see Figure 6), calcium looping has been found to be a promising technology [25,48]. It utilizes CaO to react with CO_2 to form calcium carbonate (carbonation), which can be calcined back to CaO for reuse, while recovering high-purity CO_2, as shown in Equation (1).

$$CaO_{(s)} + CO_{2(g)} \leftrightarrow CaCO_{3(s)} \Delta H^o_{298K} = -178\frac{kJ}{mol} \tag{1}$$

Owing to its low cost and high theoretical sorption capacity of up to 78.6 wt% (17.8 mmol CO_2/g sorbent), CaO has been extensively studied, and a pilot scale of this technology has been implemented in cement production [49,50]. An economic feasibility study by MacKenzie et al. shows that calcium looping is highly competitive with liquid amine-based systems [51]. However, to fully realize the industrial applications of the CaO sorbent, its material stability upon prolonged use at a high temperature needs to be improved as it is susceptible to sintering and surface carbonate poisoning [52], which may significantly decrease its CO_2 sorption capacity and limit its recyclability over time.

Extensive amounts of work have been devoted to the study and development of CaO-based sorbents for CO_2 capture, the kinetic and thermodynamic perspectives on the mechanisms of CO_2 sorption, as well as methods to improve the performance of the sorbents by mitigating their drawbacks [52–55]. The forward, highly exothermic carbonation of CaO with CO_2 and the backward, endothermic calcination of $CaCO_3$ as described in

Equation (1) occur at temperature ranges of 600–900 °C [53,55]. The calcination reaction typically takes place at a higher temperature than the carbonation reaction. However, at such temperature ranges, CaO-based CO_2 adsorbents may suffer from sintering, a phenomenon of the agglomeration of adsorbent particles when operating temperatures exceed the Tammann temperature (T_T), at which the mobility of the grains/crystallites in the solid material becomes appreciable, leading to major changes in the particle morphology. Given that T_T of $CaCO_3$ is typically about 533 °C [56], CaO crystallite growth would primarily occur during the calcination step of $CaCO_3$. As a result, particle sintering is inevitable under repeated carbonation–calcination cycles [57].

Lysikov et al. studied the behavior of CO_2 uptake by different CaO precursors in a series of carbonation–calcination cycles in the temperature range of 750–850 °C, where they observed that post-calcination CaO does not fully re-carbonate, and that the amount of unreacted CaO increases from cycle to cycle until it forms a rigid interconnected CaO skeleton after multiple cycles, as shown in Figure 7 [58]. They, therefore, proposed a model in which, upon repeated cycles, small unreacted CaO grains agglomerate, grow into larger, denser clusters, and ultimately form a stable, non-porous recalcitrant CaO skeleton that is resistant to further sintering, as illustrated in Figure 8.

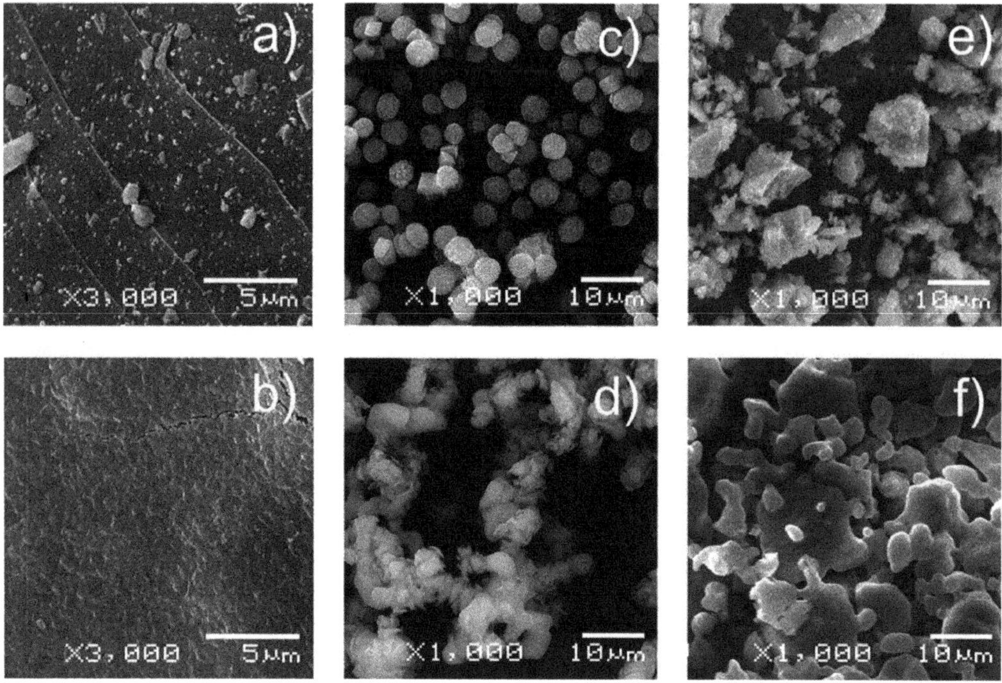

Figure 7. SEM images of the fresh CaO precursors: (**a**) MC, (**c**) MD, and (**e**) CP; and of the sorbents after carbonation–calcination cycles: (**b**) MC (211 cycles), (**d**) MD (86 cycles), and (**f**) CP (136 cycles). MC: monocrystal-derived sample, MD: monodispersed particles, CP: commercial powder. Reprinted from ref. [58] with permission from the American Chemical Society.

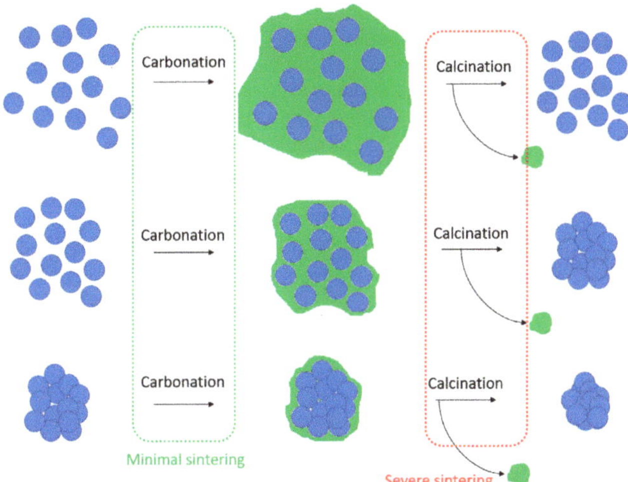

Figure 8. An illustration of the sintering process—blue spheres represent CaO-based adsorbent particles, and the green layer represents the formation of $CaCO_3$. Adapted from Lysikov et al. [58].

The sintering phenomenon shows to be destructive to the internal structure of the CaO-based sorbent, as it significantly reduces the particle surface area and pore volume due to the large, non-porous aggregate formation, which will eventually adversely affect the CO_2 sorption capacity with each carbonation–calcination cycle [59].

The common method of investigating the CO_2 sorption capacity of an adsorbent is thermogravimetric analysis (TGA), as depicted in Figure 9, which records the adsorbent's mass increase during carbonation and mass decrease during calcination. It can be seen that CO_2 uptake drastically decreases in the first few cycles, as reflected by the rapid decline of the sample weight after repeated carbonation–calcination reactions. CaO conversion to $CaCO_3$ decreased from an initial conversion of about 60% to a residual conversion of about 7–8%. This observation is in agreement with the model proposed by Lysikov et al. where the unreacted fraction of CaO increases with every cycle and only the outer layer of the non-porous recalcitrant CaO skeleton is carbonated, leading to a relatively constant residual conversion after sufficient cycles.

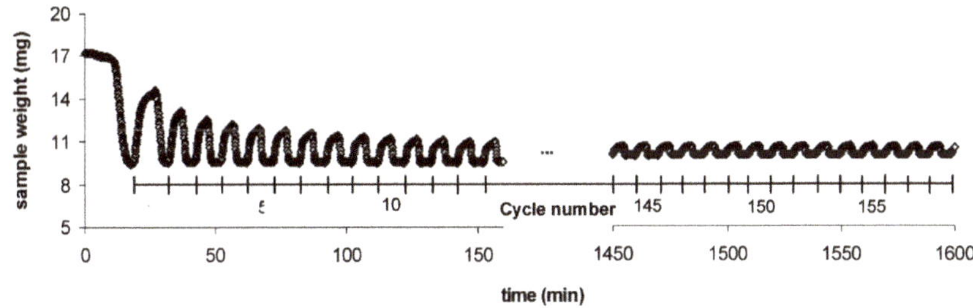

Figure 9. The profile of weight change vs. time for repeated calcination/carbonation cycles of limestone ($CaCO_3$). Calcination temperature is 850 °C; carbonation temperature is 650 °C. Reprinted from ref. [59] with permission from the American Chemical Society.

Meanwhile, kinetic analysis reveals that CO_2 adsorption characteristically takes place in two stages [52,54,55]—a fast, reaction-controlled stage, reflected by a spike in Figure 10 (first stage), followed by a slow, diffusion-controlled stage, reflected by a plateau in Figure 10 (second stage). As illustrated in Figure 11, in the reaction-controlled stage, carbonation may occur wherever CO_2 molecules come into contact with the exposed CaO. The rate of carbonation is then dependent on the rate at which CO_2 molecules reach the CaO surface, which in turn might be affected by internal characteristics such as pore structures and porosity. As the carbonation proceeds, a layer of $CaCO_3$ forms on the unreacted CaO particle. Following this, the reaction then transitions into the diffusion-controlled stage where incoming CO_2 molecules must diffuse through the $CaCO_3$ layer to be in contact with any unreacted CaO. Consequently, the rate of carbonation decreases drastically as it now becomes limited by mass transfer.

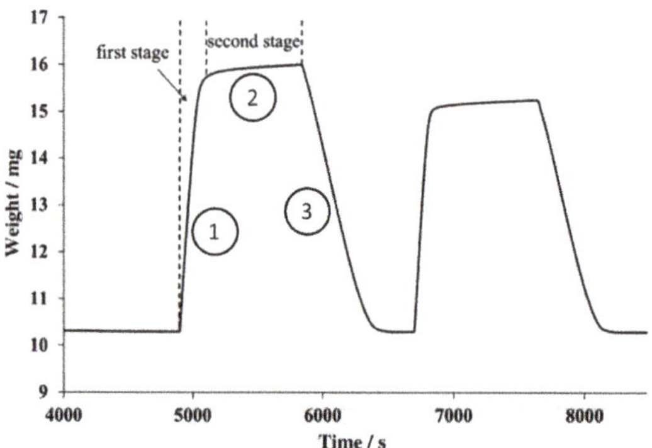

Figure 10. Typical carbonation–calcination cycle of CaO: (1) Fast carbonation stage, (2) Slow carbonation stage, and (3) Calcination step, as observed through thermogravimetric analysis (TGA). Reprinted from ref. [60] with permission from Wiley-VCH.

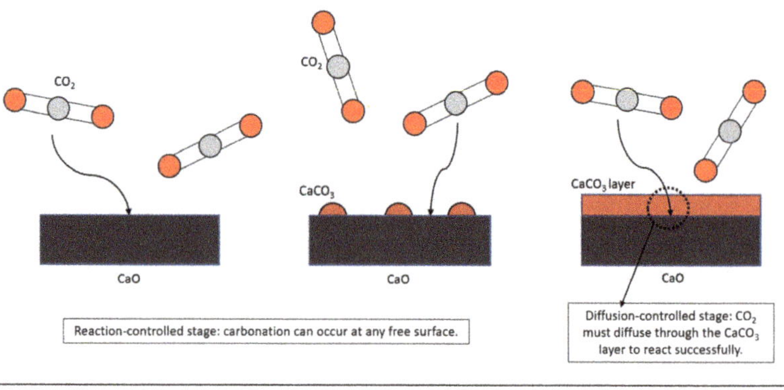

Figure 11. Illustration of the carbonation reaction mechanism of CaO. Adapted from Sun et al. [55] with permission from Elsevier.

The observed phenomena above allow us to glean important governing principles in the design of novel CO_2 sorbents, which can be incorporated into the synthesis of DFMs. Firstly, an ideal sorbent should be resistant to sintering at the required sorption temperatures to ensure its stability and prolonged use. In the case of CaO, metal supports such as the oxides of Al [61–64], Mg [65–67], and Yb [68,69], among many others, are commonly employed. These supports enhance an adsorbent's performance in numerous ways. For instance, certain metal oxides with high T_T can be used to decorate CaO particles [67], as in the case of MgO ($T_T = 1276\ ^\circ C$), which helps enhance the stability of the material. Others form stable oxides that disperse CaO particles and thus physically retard the sintering process, as in the case of Yb_2O_3 [68]. Some other metals aid in dispersion by forming composite metal oxides with CaO, as in the case of $Ca_{12}Al_{14}O_{33}$ and $Ca_3Al_2O_6$ [64]. These improvements allow supported sorbents to withstand severe reaction conditions while retaining CO_2 sorption capacities.

Secondly, an ideal sorbent should have a high internal surface area to allow for large sorption sites. The specific surface area measured via the Brunauer–Emmett–Teller (BET) method and the pore size measured with the Barrett–Joyner–Halenda (BJH) model are two common characteristics used to provide some insight into the internal morphology and microstructure of a sorbent. Primarily, the choice of synthetic method strongly influences the internal structures, and consequently, the specific surface area, of a sorbent [64]. Several studies evaluate the effect of various preparation methods on the sorbent's performance and show that sol–gel methods and flame spray pyrolysis, for example, produce sorbents with higher surface areas compared to those synthesized via simpler methods, such as the co-precipitation or mixing of metal precursor solutions (e.g., metal nitrate solutions). The choice of precursor chemicals seems also to influence the surface areas of a sorbent. It is expected that different chemical species will have varying interactions with CaO, giving rise to a variety of internal structures by the end of the synthesis depending on the chemical species present [55,70–72].

3. Overview of CO_2 Catalytic Conversion

Owing to its molecular structure, CO_2 is thermodynamically stable, as reflected in Figure 12, which compares the Gibbs free energy of formation (ΔG_f^0) for different carbon-containing compounds. Having the highest carbon oxidation state and lowest Gibbs free energy state, the chemical transformation of CO_2 will require considerable amounts of energy and, therefore, can only be achieved under harsh reaction conditions, such as a high temperature and pressure, leading to large energy penalties.

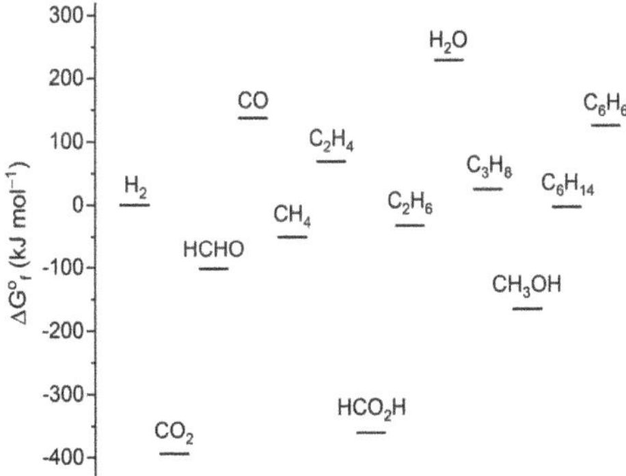

Figure 12. Gibbs free energy of formation (ΔG_f^0) for different carbon-containing compounds, including that of H_2 and H_2O as common co-reactants. Reprinted from Ref. [73] with permission from the American Chemical Society.

A strategy to make CO_2 transformation more thermodynamically favorable is by reacting it with a co-reactant that has a higher Gibbs free energy. For instance, the direct production of CO from CO_2 as a single reactant requires a large positive Gibbs free energy change (Equation (2)), whereas reacting it with H_2 significantly reduces the Gibbs free energy change of the reaction (Equation (3)). Consequently, most CO_2 transformations have been carried out by using a co-reactant, usually H_2, to produce a wide range of products, from small oxygenates to long-chain hydrocarbons.

$$CO_{2(g)} \rightleftharpoons CO_{(g)} + \frac{1}{2}O_{2(g)} \quad \Delta G^0_{298K} = 257.2\frac{kJ}{mol} \tag{2}$$

$$CO_{2(g)} + H_{2(g)} \rightleftharpoons CO_{(g)} + H_2O_{(g)} \Delta G^{\circ}_{298K} = 28.6\frac{kJ}{mol} \tag{3}$$

Another approach for a feasible process is through the addition of a solvent. For example, the gaseous phase conversion of CO_2 to formic acid is thermodynamically unfavorable (Equation (3a)), while conducting it in an aqueous phase in the presence of water makes the reaction slightly exergonic (Equation (3b)) [73].

$$CO_{2(g)} + H_{2(g)} \rightleftharpoons HCOOH_{(g)} \Delta G^{\circ}_{298K} = 32.9\frac{kJ}{mol} \tag{3a}$$

$$CO_{2(aq)} + H_{2(aq)} \rightleftharpoons HCOOH_{(aq)} \Delta G^{\circ}_{298K} = -4\frac{kJ}{mol} \tag{3b}$$

Given its inertness, extensive research efforts have been focused on tackling the high reaction energy barrier by designing more effective catalysts that convert CO_2 with good activity, selectivity, and stability. These metrics are important in designing the catalytic component of a DFM, which is usually in the form of supported noble metal or metal oxide nanoparticles that serve as the active sites for CO_2 activation. The typical reactions include the methanation of CO_2 (also known as the Sabatier reaction; Equation (4)) [74–76], a reverse water–gas shift reaction (RWGS, Equation (5)) [77–79], and the dry reforming of methane (DRM, Equation (6)) [80–82]. The resulting CO produced in the RWGS and DRM reactions can be further hydrogenated to produce methanol (Equation (7)) or further converted into long-chain paraffinic and olefinic hydrocarbons via Fischer–Tropsch synthesis (FTS, Equation (8a–c)) [83–85].

$$CO_2 \text{ methanation}: CO_2 + 4H_2 \rightarrow CH_4 + 2H_2O; \Delta H^{\circ}_R = -165\frac{kJ}{mol} \tag{4}$$

$$\text{Reverse water} - \text{gas shift (RWGS)}: CO_2 + H_2 \rightarrow CO + H_2O; \Delta H^{\circ}_R = 41.2\frac{kJ}{mol} \tag{5}$$

$$\text{Dry reforming of methane (DRM)}: CO_2 + CH_4 \rightarrow 2CO + 2H_2; \Delta H^{\circ}_R = 247.3\frac{kJ}{mol} \tag{6}$$

$$\text{Methanol formation}: CO + 2H_2 \rightarrow CH_3OH; \Delta H^{\circ}_R = -90.6\frac{kJ}{mol} \tag{7}$$

Fischer–Tropsch synthesis (FTS):

$$CO + (2n+1)H_2 \rightarrow C_nH_{2n+2} + nH_2O \tag{8a}$$

$$nCO + 2nH_2 \rightarrow C_nH_{2n} + nH_2O \tag{8b}$$

$$nCO + 2nH_2 \rightarrow C_nH_{2n+2}O + (n-1)H_2O \tag{8c}$$

Different types of transition metals are found to catalyze these reactions. For instance, Ni is commonly used for methanation and DRM reactions, Cu for the RWGS and subsequent hydrogenation to methanol, and Fe or Co for FTS. Owing to their low costs and high abundance, they are usually favored over their noble metal counterparts. However, many of these catalysts are prone to severe deactivation due to carbon deposition and sintering. Additionally, product selectivity would be an issue as the above reactions may all occur in parallel, which brings its complications, especially in industrial settings, where a certain reaction may be preferred to another. For example, the production of methanol via the RWGS pathway might be compromised by methanation occurring as an unwanted side reaction. Therefore, noble metals such as Ru, Rh, Pd, and Pt, though being more expensive, have received considerable attention because of their resistance to deactivation as well as high selectivity toward the desired products, which allow for more efficient downstream processes [86–89].

Despite major progress having been made in the development of various catalysts, significant challenges still exist in optimizing robust and economical catalysts with a good catalytic performance that can be utilized on an industrial scale. Through the better understanding of the reaction mechanisms, we can identify key descriptors that influence the performance of the catalysts [90], which eventually determine the effectiveness of DFMs in capturing CO_2 and catalyzing reactions.

4. Dual Function Materials: Synthetic Methods

A typical investigation into the sorption and catalytic performances of a metal oxide-based DFM starts with the choice of the desired precursors. This depends largely on the independent variables and the goals of the research, where some works aim to compare the efficacies of various adsorbent components [91–93], some seek to elucidate the mechanisms underlying CO_2 sorption and conversion [94,95], and others seek to evaluate the real-world applicability of DFMs by subjecting them to experiments that mimic realistic conditions.

The choice of adsorbent, catalytic components, and the support are known to influence the quality and performance of the as-synthesized DFM. For example, in an investigation by Bermejo-López et al. [96], supported CaO and Na_2CO_3 on Al_2O_3 were used as the adsorbent components of DFMs, with Ru as the catalytic metal. Various such DFMs were synthesized with varying CaO or Na_2CO_3 loadings. In general, higher adsorbent loadings promote CO_2 adsorption and hydrogenation. However, since CaO exhibits stronger basicity than Na_2CO_3, the former exhibits higher CO_2 capture, as shown in Figure 13. Consequently, the former requires higher calcination and hydrogenation temperatures, as compared to the latter.

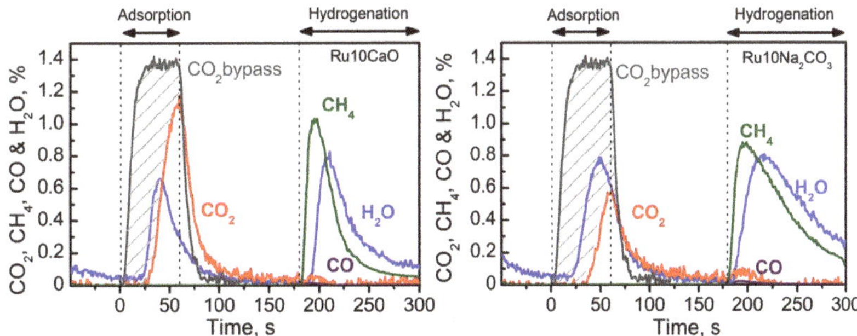

Figure 13. Concentration distribution profiles during a CO_2 adsorption and hydrogenation cycle at 370 °C with Ru10CaO and Ru10Na$_2$CO$_3$. Reprinted from ref. [96] with permission from Elsevier.

Upon deciding on the suitable precursors, the next step would be to employ an appropriate method to synthesize the DFM. Several synthetic methods are already well-known, which produce DFMs with certain desired characteristics, such as a high surface area and porosity, homogeneous particle dispersion, and unique morphology, with each of them having their respective benefits and disadvantages. A host of other factors also needs to be considered in selecting a synthetic method, including time, material, and labor costs, as well as ease of implementation, in addition to the typical structural characteristics. In this section, we will explore four commonly used methods, presented in an approximate order of their prevalence in the literature.

4.1. Incipient Wetness Impregnation

Incipient wetness impregnation (IWI) is a well-known and highly popular synthetic method for producing heterogeneous catalysts, and is, unsurprisingly, the method that is employed in most of the works on DFMs [11,91–101]. IWI is favored for its ease of implementation and relatively low costs [102]. This procedure aims to impregnate adsorbent and/or catalyst species onto a support by adding small amounts of adsorbent/catalyst precursor solution to a powdered support. The wet powder is then dried and may further be subjected to thermal activation treatment, e.g., calcination or reduction. The process is then repeated until the total volume of precursor solutions added equals the total pore volume of the support [103]. The final product is ideally a porous material with adsorbent and catalyst particles distributed throughout the surface area of the pores. Sietsma et al. found that the pore size distribution and drying rate affect the redistribution of the particles on the support, which eventually determines the final crystal size [102]. For supports with large pore size distribution, a fast drying rate leads to a larger crystal size due to a higher degree of particle redistribution. On the contrary, for supports with narrow pore size distribution, the effect of the drying rate is minimal.

Since surface area is one of the determinants in designing an effective DFM, the careful selection of the support materials, preferably ones with a larger surface area and well-defined pore structures, is an important consideration in optimizing the properties of DFMs after IWI post-treatment. Additionally, optimum loadings of adsorbent/catalyst components should be considered as well.

4.2. Sol–Gel Method

In order to gain greater control over the particles' morphology and porosity, sol–gel synthesis is an alternative method that can reliably ensure the formation of microstructures favorable to CO_2 capture, as well as uniform particulate dispersion and the minimal granulation of adsorbent/catalyst particles, owing to its ability to produce a solid-state material from a chemically homogeneous liquid precursor [104]. Sol–gel chemistry conventionally involves the hydrolysis and condensation of metal alkoxides, with reaction rates depending on the carbon length of the alkoxides (e.g., methoxide, ethoxide, propoxide, etc.) and the solvent/alkoxide ratio. A major limitation of the alkoxide-based synthetic method is the instability of the metal alkoxides itself, especially in the presence of water and moisture, which makes it difficult to handle. As a result, alternative methods have been developed to employ metal salts instead of alkoxides in aqueous solution, together with small organic molecules as chelating agents, such as hydroxycarboxylic acids, to modify the hydrolysis chemistry of metal ions in aqueous solution.

One of the most commonly used organic acids is citric acid. Given its low price and wide availability, the sol–gel method with citric acid has been widely used to synthesize various metal oxides, including binary and tertiary oxides [105]. The key feature of this method is the formation of a metal-citrate complex, which ensures the homogeneity of the starting liquid solution. The presence of a citrate organic matrix is believed to help maintain the uniform mixing and dispersion of the metal ions on the atomic scale, resulting in small crystallite sizes.

A modification to the citrate sol–gel method was developed by Pechini in 1967 [106], where polyhydroxy alcohol, such as ethylene glycol, was added into the aqueous citrate solution. This results in a polyesterification reaction, which forms a covalent polymeric network that entraps metal ions [104], allowing two or more metals to be dispersed homogeneously throughout the network, as shown in Figure 14. The intermediate product here is a brittle solid, which is subsequently subjected to thermal treatment, e.g., calcination, to drive off organic components and obtain the final metal oxide powder [107].

Further variations and modifications have been made since Pechini's initial proposition. For example, Pechini suggested metal oxides, hydroxides, and carbonates as possible cation precursor choices, but due to a need to control the amount of organic material in the final product, as well as economic reasons, nitrate salts have been a popular choice. Additionally, a modified pathway that substitutes the polyhydroxy alcohol for water subverts the esterification step and forms a gel-like substance instead.

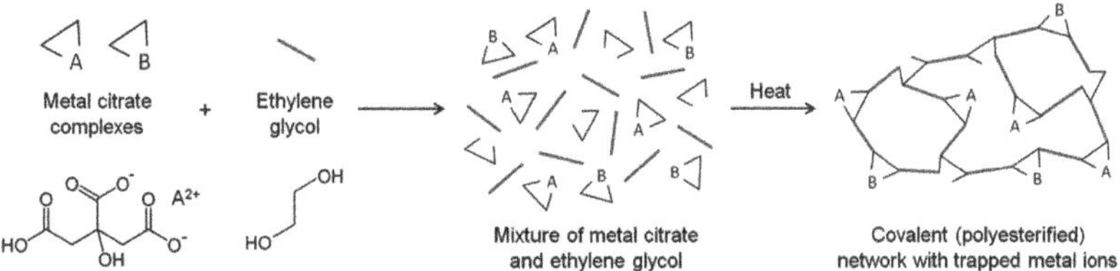

Figure 14. Formation of metal–organic gel via Pechini sol–gel synthetic method. Reprinted from ref. [104] with permission from the Royal Society of Chemistry.

Representative works on DFMs synthesized via a sol–gel method exhibit an expected degree of variability in terms of the synthetic pathways used. Nitrate precursors were common, though we note a study by Radfarnia et al. that used calcium acetate and aluminum isopropoxide as the metal cation precursors [108]. A study by Naeem et al. used ethylene glycol for polyesterification [69], as in the original Pechini method, while the aqueous citric acid variation is more common [109–111].

Notably, a study by Zhang et al. demonstrated that a sol–gel synthetic method produces DFMs with superior CO_2 capture capacities in comparison with other methods like co-precipitation and mixing [112]. Their study investigated the usage of CaO-based DFMs in the oxidative dehydrogenation of ethane. The sorbent component was comprised of ceria-doped CaO, which was synthesized via four methods—dry and wet mixing, co-precipitation, as well as the modified Pechini method with nitrate precursors of the calcium and cerium cations, and aqueous citric acid serving as the chelating agent. It was found that the sol–gel method produced a sample with an initial CO_2 capture capacity of 0.58 g/g, almost five times higher than that prepared by the dry mixing method, which exhibited an initial CO_2 capture capacity of 0.12 g/g. It was concluded that multiple factors influenced the CO_2 capturing abilities of a particular sorbent. While the available surface area is important, larger crystallite sizes severely impede the carbonation reaction when a surface layer of $CaCO_3$ forms around the sorbent material, causing the carbonation reaction to transit into the diffusion-controlled phase [52,54,55]. Therefore, a balance between maximizing the surface area and optimizing the crystallite sizes is necessary for ensuring excellent CO_2 capturing capacities, which can be obtained by fine-tuning the sol–gel synthesis parameters.

4.3. Co-Precipitation Method

Co-precipitation is a widely employed method in heterogeneous catalysis to synthesize materials such as mixed metal oxides [113,114]. The general method involves dissolving the precursor chemicals for the adsorbent, catalyst, and/or support/promoter (typically nitrate salts) in either deionized water or polar solvent under rigorous stirring to ensure homogeneity. A precipitating agent, e.g., NaOH, NH_4OH, etc., is then added dropwise to induce precipitation. Subsequently, aging to enhance crystallinity is practiced, though not necessary. The precipitate is collected via filtration followed by washing cycles in deionized water or another solvent. Lastly, the precipitate is subjected to thermal treatment, e.g., calcination, to obtain the final product.

As a synthetic method, co-precipitation is straightforward and inexpensive. The material's stoichiometry can be easily controlled by using appropriate amounts of precursors during the precipitation reaction. Depending on the aging conditions, adsorbents produced via this method typically exhibit a homogeneous particulate distribution with good crystallite sizes. However, disadvantages of this method include generally lower surface areas compared to adsorbents synthesized via IWI or sol–gel methods [105]. The process is also fairly laborious, with repeated washing being commonly necessary. Further, Li et al.

note that the structural properties of the resulting precipitate are sensitive and heavily dependent on reaction conditions and equipment setup, which makes reproducibility an issue [115].

Like the sol–gel pathways, there are several variations to the co-precipitation procedure. These include the types of precipitating agent, the solvent used to dissolve the precursors and wash the precipitate, any mechanical assistance to the reaction (e.g., stirring, sonication, microwave, etc.), as well as aging temperature and time, to enhance the crystallinity via Ostwald ripening [114,115] at a high temperature, e.g., hydrothermal treatment. With the careful selection and optimization of the reaction conditions, the wide variation of methods under the family of co-precipitation procedures could be used to overcome certain shortcomings or to reinforce certain desirable qualities in a DFM. For example, Molina-Ramírez et al. synthesized a DFM comprising of a Ni-Ba unsupported catalyst via a co-precipitation method assisted by sonication [116], where nitrate precursors were used and ammonia was employed as the precipitant. The as-synthesized DFM showed an exceptionally high BET surface area of 112 m^2/g, though we note that colloidal silica was used as a surface area promoter. Meanwhile, Karami and Mahinpey investigated the effect of synthetic methods on several Ca-alumina sorbents [117], such as the precipitation of a Ca salt solution onto colloidal gelled alumina, the co-precipitation of Ca and Al salt solutions with Na_2CO_3 or sodium aluminate as the precipitants, and the physical mixing of a Ca precipitate with a peptized alumina gel. It was found that co-precipitation produced a sorbent with the poorest CO_2 uptake performance, possibly due to the low surface area and porosity.

4.4. Mixing Method

The mixing of adsorbent/catalytic/support precursors under varying conditions to produce a DFM precursor is arguably the simplest and most straightforward synthetic method. However, it clearly lacks precision in terms of producing the most favorable microstructures required for the ICCC process. Slightly varied methods of mixing exist depending on the relative solubilities of the precursors used [53]. For example, the wide usage of nitrate precursors as described in the preceding sections lends itself to facile dissolution in water. The precursor solutions are then simply mixed under stirring for a few hours to ensure homogeneity, the resultant solution is then dried to obtain a powder, followed by calcination to obtain the DFM.

Mixing is more commonly used in conjunction with other synthetic methods to produce a DFM. For example, the adsorbent component might be synthesized by mixing, while the catalytic component is synthesized via another method, e.g., co-precipitation or impregnation. For example, in a work by Huang et al., who investigated the performance of $NaNO_3$-promoted Ni/MgO DFMs for integrated CO_2 capture and methanation, the unpromoted Ni/MgO catalysts were synthesized via a facile one-pot wet mixing method [118], whereas the alkali metal salt promoter $NaNO_3$ was subsequently impregnated upon the bare catalyst. When mixing is employed as the sole synthetic method for a DFM, its characteristics are decidedly inferior. Wu et al. demonstrated this in a comparative study investigating the use of Ni-CaO-CeO_2 DFMs for ICCC via the RWGS reaction [105]. Different DFMs were prepared using acetate/nitrate precursors of Ni, Ca, and Ce via wet mixing, co-precipitation, and sol–gel methods. It was found that the wet mixing method produced a sample with the lowest CO_2 capture capacity and product yield as compared to the other synthetic methods. The poor performance by wet mixing is mainly due to the low BET surface area, inferior pore structure, large Ni crystallite sizes, low surface basicity, and low reducibility.

5. Performance Evaluation of Dual Function Materials

With a general understanding of the kinetics and thermodynamics of CO_2 capture and conversion, as well as the typical synthetic methods of DFMs that enable the ICCC process, we proceed to evaluate the performance of DFMs across the literature. This section catalogues recent works according to the key ICCC reactions that the DFM catalyzes, namely

ICCC with a methanation reaction, a reverse water–gas shift (RWGS) reaction, and the dry reforming of methane (DRM). With each case study, we present examples of the synthetic, operating, and performance parameters of the DFMs, as well as the characteristics and mechanistic studies of the interaction between CO_2 and the adsorption and catalytic sites of the DFMs. Given the dual function characteristic of the materials, the efficacy of DFMs first depends on their ability to efficiently capture CO_2 with a high capacity, which becomes the reactant for the subsequent reactions. Therefore, the material selection for a DFM has to satisfy the specific operating conditions required for both CO_2 capture and reaction. As DFMs are subjected to repeated cycles of CO_2 capture and reaction, the robustness of the materials is also an important factor to be considered in the design and evaluation of DFMs.

5.1. ICCC with Methanation Reaction

The majority of studies on ICCC using DFMs focus on the methanation of CO_2, which takes place primarily via the Sabatier reaction (Equation (9)):

$$CO_2 + 4H_2 \rightleftharpoons CH_4 + 2H_2O \, \Delta H_R = -165\frac{kJ}{mol} \tag{9}$$

The reaction is typically carried out at temperatures around 300–350 °C and pressures of 5–20 MPa [119]. Given its highly exothermic nature, the reaction releases copious amounts of heat, which has to be managed accordingly to prevent the thermal degradation (e.g., sintering) of the DFMs. This can be achieved via the engineering of more thermally resistant DFMs, or by harnessing the heat of the reaction via efficient heat transfer and heat integration, e.g., powering endothermic reactions. In this view, CO_2 methanation offers an isothermal solution to the DFMs system because the exothermic reaction can supply the required heat for the decarbonation reaction. Additionally, the usage of real flue gas in industrial settings might lead to the poisoning and deactivation of the Ni-based DFMs typically used for methanation [119,120]. Practical solutions include the pre-scrubbing of the flue gas or considering the usage of alternative catalysts like Ru, which can be activated via pre-reduction at lower temperatures in a H_2 environment. The vast body of work studying the ICCC–methanation process is summarized in Table 3. It can be seen that Ru- and Ni-based DFMs supported on alkaline metal oxides, such as Na_2O, MgO, and CaO, are the most commonly studied materials. In fact, as shown in Figure 15, Na_2O, MgO, and CaO are among the most common sorbents for ICCC with methanation [14], due to their high CO_2 sorption capacity and methanation activity at moderate temperatures, given the thermodynamic limitation of the reaction.

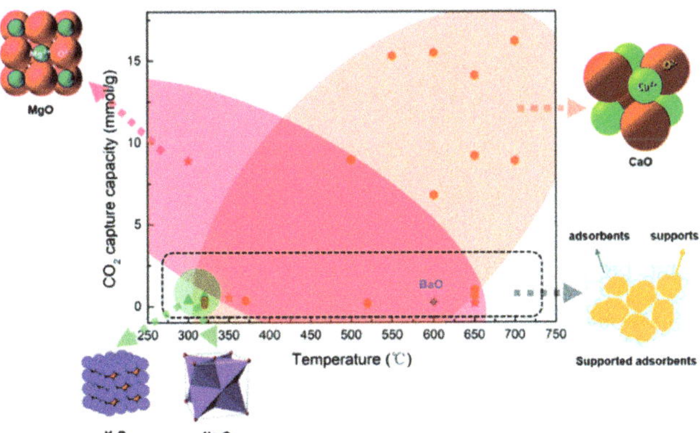

Figure 15. Typical CO_2 adsorption capacity of DFMs for ICCC–methanation. Orange zone: CaO; Red zone: MgO; Green zone: K_2O and Na_2O. Reprinted from ref. [14] with permission from the Royal Society of Chemistry.

Table 3. Literature summary of DFMs that catalyze the methanation reaction.

DFM	Synthesis Method	Carbonation Conditions	Methanation Conditions	CO$_2$ Sorption Capacity (mmol/g)	CO$_2$ Conversion (%)	Methanation Capacity (mmol/g)	Ref.
5% Ru 10% CaO/Al$_2$O$_3$	IWI	320 °C, 10%CO$_2$/air	320 °C, 5%H$_2$/N$_2$	0.41	76.17%	0.31	[11]
5% Ru 10% CaO/Al$_2$O$_3$	IWI	320 °C, 10%CO$_2$/N$_2$	320 °C, 4%H$_2$/N$_2$	-	-	0.5	[93]
5% Ru 10% K$_2$CO$_3$/Al$_2$O$_3$	IWI	320 °C, 10%CO$_2$/N$_2$	320 °C, 4%H$_2$/N$_2$	-	-	0.91	[93]
5% Ru 10% Na$_2$CO$_3$/Al$_2$O$_3$	IWI	320 °C, 10%CO$_2$/N$_2$	320 °C, 4%H$_2$/N$_2$	-	-	1.05	[93]
5% Ru 6.1% Na$_2$O/γ-Al$_2$O$_3$	IWI	300 °C, simulated flue gas	300 °C, 15%H$_2$/N$_2$	0.4 (50-cycle average)	77%	0.32	[98]
5% Ru-6.1% "Na$_2$O"/Al$_2$O$_3$	IWI	320 °C, 10%CO$_2$/N$_2$	320 °C, 10%H$_2$/N$_2$	0.651	96%	0.614	[92]
0.5% Rh-6.1% "Na$_2$O"/Al$_2$O$_3$	IWI	320 °C, 10%CO$_2$/N$_2$	320 °C, 10%H$_2$/N$_2$	0.626	69%	0.422	[92]
15% Ni 15% CaO/Al$_2$O$_3$	IWI	280–520 °C, 10%CO$_2$/Ar	280–520 °C, 10%H$_2$/Ar	-	-	0.142	[101]
15% Ni 10% Na$_2$CO$_3$/Al$_2$O$_3$	IWI	280–520 °C, 10%CO$_2$/Ar	280–520 °C, 10%H$_2$/Ar	-	-	0.186	[101]
4% Ru 10% CaO/Al$_2$O$_3$	IWI	370 °C, 1.4%CO$_2$/Ar	370 °C, 10%H$_2$/Ar	0.253	-	0.272	[96]
4% Ru 10% Na$_2$CO$_3$/Al$_2$O$_3$	IWI	370 °C, 1.4%CO$_2$/Ar	370 °C, 10%H$_2$/Ar	0.391	-	0.398	[96]
1% Ru, 10% Ni, 6.1% "Na$_2$O"/Al$_2$O$_3$	IWI	320 °C, 7.5%CO$_2$-4.5%O$_2$-15%H$_2$O-balance N$_2$	320 °C, 15%H$_2$/N$_2$	0.52 (3-cycle average)	81% (3-cycle average)	0.38 (3-cycle average)	[100]
1% Ru, 10% Ni, 6.1% "Na$_2$O"/Al$_2$O$_3$	IWI	320 °C, 7.5%CO$_2$-4.5%O$_2$-15%H$_2$O-balance N$_2$	320 °C, 15%H$_2$/N$_2$	0.52 (20-cycle average)	100% (20-cycle average)	0.38 (20-cycle average)	[100]
1% Pt 10% Ni 6.1% "Na$_2$O"/Al$_2$O$_3$	IWI	320 °C, 7.5%CO$_2$-4.5%O$_2$-15%H$_2$O-balance N$_2$	320 °C, 15%H$_2$/N$_2$	0.35 (3-cycle average)	87% (3-cycle average)	0.25 (3-cycle average)	[100]
2.5Ru/CeO$_2$-MgO	Impregnation/physical mixing	300 °C, 65%CO$_2$/N$_2$	300 °C, 5%H$_2$/N$_2$	-	60% (1st cycle); 39% (10th cycle)	5.73 (1st cycle); 1.13 (10th cycle)	[121]
5Ru/CeO$_2$-MgO	Impregnation/physical mixing	300 °C, 65%CO$_2$/N$_2$	300 °C, 5%H$_2$/N$_2$	-	74% (1st cycle); 79% (10th cycle)	6.6 (1st cycle); 3.36 (10th cycle)	[121]

Table 3. *Cont.*

DFM	Synthesis Method	Carbonation Conditions	Methanation Conditions	CO$_2$ Sorption Capacity (mmol/g)	CO$_2$ Conversion (%)	Methanation Capacity (mmol/g)	Ref.
10Ru/CeO$_2$-MgO	Impregnation/physical mixing	300 °C, 65%CO$_2$/N$_2$	300 °C, 5%H$_2$/N$_2$	-	89% (1st cycle); 69% (10th cycle)	7.07 (1st cycle); 2.31 (10th cycle)	[121]
Ni/Na-γ-Al$_2$O$_3$	IWI	450 °C, 5%CO$_2$/N$_2$	450 °C, H$_2$	0.209	96%	0.188	[122]
Ni/Na-γ-Al$_2$O$_3$	IWI	450 °C, 5%CO$_2$/N$_2$	450 °C, H$_2$	0.299	92%	0.266	[122]
0.95% Ru-5% K/Al$_2$O$_3$	IWI	350 °C, 2.5%H$_2$O/3%O$_2$/1%CO$_2$/He	350 °C, 4%H$_2$/He	0.028 (3rd cycle)	-	0.028 (3rd cycle)	[123]
0.95% Ru-5.1% Ca/Al$_2$O$_3$	IWI	350 °C, 2.5%H$_2$O/3%O$_2$/1%CO$_2$/He	350 °C, 4%H$_2$/He	0.116 (3rd cycle)	-	0.036 (3rd cycle)	[123]
0.84% Ru-16% Ba/Al$_2$O$_3$	IWI	350 °C, 2.5%H$_2$O/3%O$_2$/1%CO$_2$/He	350 °C, 4%H$_2$/He	0.165 (3rd cycle)	-	0.080 (3rd cycle)	[123]
0.5% Ru, 6.1% "Na$_2$O"/Al$_2$O$_3$	IWI	320 °C, 400 ppm CO$_2$/air	320 °C, 15% H$_2$/N$_2$	0.2	-	0.15	[124]
1%Ni/CeCaO-imp	One-pot citric acid chelation/wet impregnation	550 °C, 15%CO$_2$/N$_2$	550 °C, H$_2$	10.6	42%	3.3	[125]
1%Ni/CeCaCO$_3$-imp	One-pot citric acid chelation/carbonation/wet impregnation	550 °C, 15%CO$_2$/N$_2$	550 °C, H$_2$	14.1	52%	6	[125]
1%Ni/CeO$_2$-CaO-phy	Hydrothermal/one-pot citric acid chelation/physical mixing	550 °C, 15%CO$_2$/N$_2$	550 °C, H$_2$	15.3	62%	8	[125]
Ru-BaO/Al$_2$O$_3$ (intimate mixture)	IWI	350 °C, 1%CO$_2$/He	350 °C, 4%H$_2$/He	0.025	20% at 291 °C	0.151	[126]
BaO/Al$_2$O$_3$ + Ru/Al$_2$O$_3$ (mechanical mixture)	IWI	350 °C, 1%CO$_2$/He	350 °C, 4%H$_2$/He	0.055	20% at 362 °C	0.097	[126]
5Li-Ru/A	IWI	263 °C, 10%CO$_2$/N$_2$	263 °C, 10%H$_2$/N$_2$	-	98%	0.32	[127]
5Li-Ru/A	IWI	293 °C, 10%CO$_2$/N$_2$	293 °C, 10%H$_2$/N$_2$	-	97.4%	0.34	[127]
5Li-Ru/A	IWI	318 °C, 10%CO$_2$/N$_2$	318 °C, 10%H$_2$/N$_2$	-	95.4%	0.29	[127]
30% LaNiO$_3$/CeO$_2$	Citric acid/impregnation	400 °C, 1.4%CO$_2$/Ar,	400 °C, 10%H$_2$/Ar	0.089 (3-cycle average)	-	0.08 (3-cycle average)	[128]

Table 3. *Cont.*

DFM	Synthesis Method	Carbonation Conditions	Methanation Conditions	CO_2 Sorption Capacity (mmol/g)	CO_2 Conversion (%)	Methanation Capacity (mmol/g)	Ref.
AMS/CaMgO‖Ni1-Co3	Sol–gel/solvent evaporation	350 °C, 95%CO_2/N_2	350 °C, 10%H_2/N_2	14.7 (sorbent mass basis)	76.4%	5.46 (catalyst mass basis)	[129]
AMS/CaMgO‖Ni1-Co1	Sol–gel/solvent evaporation	350 °C, 95%CO_2/N_2	350 °C, 10%H_2/N_2	14.4 (sorbent mass basis)	90%	9.97 (catalyst mass basis)	[129]
AMS/CaMgO‖Ni3-Co1	Sol–gel/solvent evaporation	350 °C, 95%CO_2/N_2	350 °C, 10%H_2/N_2	14.1 (sorbent mass basis)	75%	6.66 (catalyst mass basis)	[129]
cDFM-2-0.4Ni-10Cs	Co-precipitation/IWI	350 °C, 15%CO_2/N_2	350 °C, H_2	0.48	-	0.33	[130]
20% $LaNiO_3$/CeO_2	Citric acid/impregnation	480 °C, 10%CO_2/Ar	480 °C, 10% H_2/Ar	0.113	-	0.075	[131]
20% La0.5Ca0.5NiO_3/CeO_2	Citric acid/impregnation	480 °C, 10%CO_2/Ar	480 °C, 10% H_2/Ar	0.175	-	0.140	[131]
Ni/CaZrO	Sol–gel	600 °C, 15%CO_2/N_2	600 °C, 66.7%H_2/N_2	9.9–9 (20 cycles)	86.3–78% (20 cycles)	6.7 (6th to 20th cycle)	[110]
NiO-MgO	Co-precipitation	320 °C 10%CO_2/5%O_2/He	320 °C, 20%H_2/He	~0.3 (10 cycles)	~96% (10 cycles)	~0.26 (10 cycles)	[132]
Cs-impregnated Ni-MgO-Al_2O_3 extrudate (EI)	Co-precipitation/impregnation	350 °C, 15%CO_2/N_2	350 °C, H_2	0.24	75% (250h on-stream)	0.18 (250h on-stream)	[133]

IWI: incipient wetness impregnation.

With a wide range of variables that are at play in the synthesis of a DFM, an extensive amount of work is naturally devoted to finding the best synthetic routes that produce DFMs with optimized makeups [11,95,101,110,117,121–123,125–131,134–136]. In a pioneering work by Duyar et al., CaO-based DFMs with Ru as the methanation catalyst supported on alumina were synthesized via incipient wetness impregnation (IWI) with varying adsorbent/catalyst loadings [11]. It was observed that the order of impregnation (i.e., Ru impregnated on CaO-alumina vs. CaO impregnated on Ru-alumina) affected the CO_2 capture and methanation performances at 320 °C. In particular, a DFM comprising 5 wt% Ru impregnated on 10 wt% CaO-γAl$_2$O$_3$ showed a single-pass methanation capacity of 0.5 mmol/g. Upon subjecting to cyclic testing, an average of 0.41 mmol/g of CO_2 capture could be obtained, with an average CO_2 conversion of 76.17% and a methanation capacity of 0.31 mmol CH_4/g over 19 cycles. The authors concluded that the order of impregnation was significant where the impregnation of Ru was found to be better than that of CaO due to the good dispersion of Ru. Should the order be reversed, the active Ru sites could be blocked by CaO particles, thus decreasing the catalytic activity of the DFM. Upon varying the CaO:Ru ratio, a high methane turnover (CH_4 yield/Ru) was observed at a high CaO:Ru ratio, indicating CO_2 spillover from the CaO to Ru sites where the methanation reaction takes place. It also suggests that the close proximity of CaO and Ru was an important factor in the DFM performance. However, low Ru loadings might not generate enough heat from the exothermic methanation reaction to liberate CO_2 chemisorbed to CaO, since methanation begins primarily where CO_2 is chemisorbed to Ru, thus compromising the overall CH_4 yield. Lastly, testing under simulated flue gas conditions showed that, in the presence of steam, the methanation capacity decreased to 0.27 mmol CH_4/g over 20 cycles, although the purity of the effluent gas improved to 99.9% CH_4 by volume, as compared to 83.6% CH_4 (balance CO_2) in the absence of steam, suggesting the competing adsorption of H_2O and CO_2 on the DFM surface. A follow-up study by Duyar et al. presented two new DFMs with similar adsorbent/catalyst loadings, but with K_2CO_3 or Na_2CO_3 substituting CaO as the adsorbent component [93]. These DFMs showed improved methanation capacities of 0.91 and 1.05 mmol CH_4/g for K and Na, respectively, owing to the increasing CO_2 adsorption capacity in the same order: $Na_2CO_3 > K_2CO_3 > CaO$. The results, therefore, show the importance of optimizing the adsorbent component of DFMs in future attempts to engineer more efficient materials.

In another study, Porta et al. assessed the CO_2 capture and methanation performances at 350 °C of Ru-based DFMs with six different adsorbent components (Li/Na/K/Mg/Ca/Ba) supported on alumina [123]. They found that metals capable of forming carbonates with high thermal stability led to the production of the highest amounts of CH_4, namely, DFMs promoted with K, Ca, and Ba. On the other hand, Arellano-Treviño et al. evaluated Ru-based DFMs supported on γ-Al$_2$O$_3$ containing different alkali and alkaline earth oxides, such as Na_2O, K_2O, CaO, and MgO, and found that Na_2O-promoted DFM (5%Ru-6.1%Na_2O/Al$_2$O$_3$) exhibited a high CO_2 capture capacity (0.651 mmol/g) and methanation rate (0.614 mmol CH_4/g) when tested in 10%CO_2/N_2 atmosphere [92]. Although the CH_4 yield significantly decreased upon exposure to a simulated flue gas condition (7.5%CO_2, 4.5%O_2, 15%H_2O, balance N_2), the DFM remained active, producing 0.291 mmol CH_4/g at 320 °C. In another experiment using a Rh-Na_2O-based DFM, the CO_2 capture capacity and catalytic activity of 0.5%Rh was compared with those of 5%Ru in order to obtain a similar material price. While both DFMs exhibited a similar CO_2 capture capacity (0.625 vs. 0.651 mmol/g, respectively), 5%Ru yielded higher methanation (0.614 mmol/g) than the 0.5%Rh counterpart (0.422 mmol/g), although one could argue that the methane yield per unit of metal is higher for Rh than that for Ru. As an alternative to noble metals, Ni-Na_2O-based DFM was also synthesized and tested, which produced 0.276 mmol CH4/g under a 10%CO_2/N_2 atmosphere. However, the material was deactivated when exposed to flue gas containing H_2O, with no methane being produced, as shown in Figure 16. Further analysis suggested that Ni atoms could not be completely reduced to the active metallic state in the presence of O_2 and H_2O, thus losing its catalytic activity.

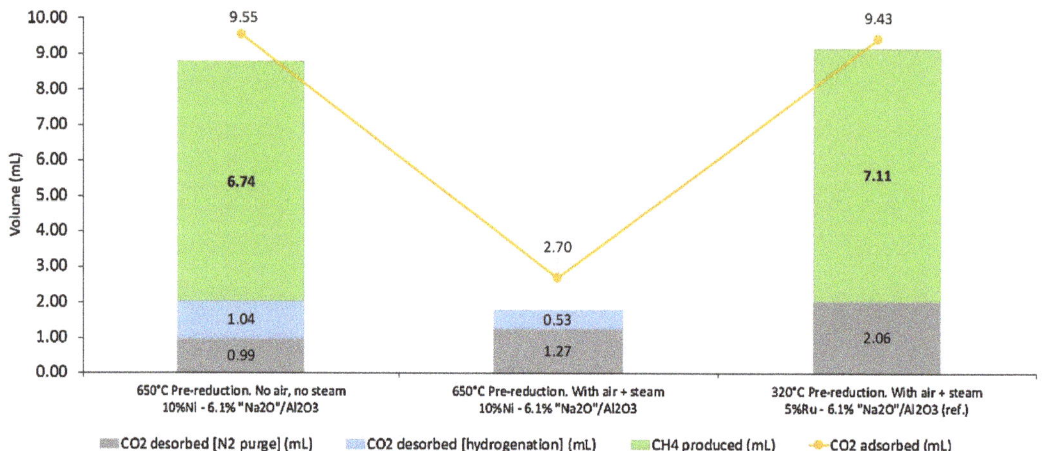

Figure 16. CO_2 adsorption, desorption, and CH_4 produced on 10%Ni-6.1%"Na_2O"/Al_2O_3 with and without O_2 and steam present in the CO_2 feed. CO_2 capture and methanation were performed at 320 °C, 1 atm. Reprinted from ref. [92] with permission from Elsevier.

Further possible variables during the synthesis of a DFM include the choice of adding promoters that can improve the CO_2 capture capacity and CH_4 selectivity, or the choice of precursor chemicals, which might affect the interactions between different components in the synthesized DFM. One study by Cimino et al. [135] found that promoting Ru/Al DFM with Li resulted in a four-to-five-fold increase in the CO_2 capture capacity compared to the unpromoted material. In addition, stable cyclic carbonation/methanation was possible at temperatures around 230 °C, instead of the commonly utilized temperatures of 300 to 320 °C. In another study, the incorporation of Cs into Ni-hydrotalcite-based DFMs was found to increase the CO_2 capture capacity up to 0.48 mmol/g and a CH_4 yield of 0.33 mmol/g at 350 °C [130]. These examples show the promotional effect of alkali metal as it increases the material's basicity, in particular, the number of medium and strong basic sites. Besides enhancing the basicity of the material, the addition of a promoter can also improve the stability of the material. For example, Ma et al. investigated the effect of doping Ni-CaO-based DFMs with various metal oxides, such as Mg, Al, Mn, Y, Zr, La, and Ce [110]. Among the synthesized DFMs, Zr-doped Ni/CaO exhibited the best performance by maintaining a stable CO_2 capture capacity of 9 mmol/g and 74% CH_4 selectivity after 20 cycles. It was attributed to the formation of $CaZrO_3$, which helped improve the thermal resistance to sintering.

To investigate the effect of catalyst loading, Bermejo-López et al. varied the Ni content on CaO- or Na_2CO_3-based DFMs with alumina support through the impregnation method [101]. Notably, it was found that increasing Ni loading would increase the particulate sizes, catalyst reducibility, and consequently, methanation rates, with 0.142 mmol CH_4/g produced by a DFM comprising 15 wt% Ni on 15% CaO/Al_2O_3 at 520 °C.

Lastly, the choice of the synthetic method has also been known to significantly influence the final microstructures and other physicochemical characteristics of a DFM, and a range of works have been devoted to comparatively evaluating the various possible synthetic routes [105,112,117,125]. For example, Zhang et al. noted that sol–gel synthetic methods tend to produce DFMs with physicochemical characteristics more favorable for ICCC processes, including an optimal balance of the BET surface area and crystallite sizes [112]. Sun et al. noted that the dispersion between catalyst and sorbent particles could even be tuned via different synthetic methods, thus giving rise to varying carbonation and methanation capabilities [125].

In addition to optimizing the materials design and synthesis, careful attention should also be given to the experimental parameters of the carbonation and methanation reaction, as well as the simulated flue gas conditions, as they can help uncover the possibility of translating ideal laboratorial conditions to a more real-life (e.g., industrial) setting, and provide plausible avenues for future work [98,100,122,124,137]. In one such study by Wang et al., a DFM consisting of Ru-Na_2O impregnated on an alumina support was subjected to parametric and cyclic tests [98]. Specifically, the effect of three parameters—gas space velocity, reaction temperature, and exposure to oxygen during adsorption—on performance indicators such as CO_2 uptake and CH_4 generation, were investigated. From the cyclic studies, a CO_2 uptake and CH_4 generation capacities of 0.40 mmol/g and 0.32 mmol/g, respectively, were reported after 50 cycles at 300 °C. Through parametric studies, it was found that gas space velocities and temperature would affect the rates and the extents of carbonation and methanation in accordance with kinetic and thermodynamic principles. For instance, increasing the gas space velocity would increase the rates of CO_2 uptake and CH_4 production due to a more efficient mass transfer, whereas increasing the temperature would decrease CO_2 adsorption and CH_4 generation due to the exothermic nature of the reactions. The oxygen exposure tests also showed that, minimally, 15% H_2/N_2 for methanation was necessary for the reduction of oxidized Ru due to exposure to oxygen-containing flue gas during adsorption. The presence of precious metals such as Ru in overcoming the limitations of Ni as a catalytic metal is further displayed by studies such as one by Arellano-Treviño et al., which suggests that Ru doping aids in the reduction and re-activation of Ni after oxygen exposure, leading to the possibility of developing DFMs which are more resistant to industrial reaction conditions at cheaper costs [100].

Meanwhile, other authors have studied reaction parameters such as optimal carbonation/methanation durations, feed gas concentrations, and process pressures. Kosaka et al., for example, investigated Ni-based DFMs promoted with alkali metals including Na/K/Ca supported on alumina for their carbonation and methanation at 450 °C [122]. It was found that increasing operational pressures (a range of 0.1 MPa to 0.9 MPa was used) could increase the CO_2 capture capacity of a Ni/Na-Al_2O_3 DFM from 0.209 mmol CO_2/g to 0.299 mmol CO_2/g, and its methanation capacity from 0.188 mmol CH_4/g to 0.266 mmol CH_4/g, in agreement with thermodynamics as the methanation reaction involves a decrease in the number of moles. For these tests, a gas feed containing 5% CO_2/N_2 was used, but it was additionally reported that increasing pressures enhanced carbonation and methanation at lower CO_2 concentrations of 100 and 400 ppm CO_2/N_2 as well. Jeong-Potter and Farrauto also attempted an investigation of the effectiveness of a Ru/Na-Al_2O_3 DFM at a CO_2 concentration of 400 ppm in air to assess the feasibility of utilizing the DFM for direct air capture purposes [124].

Considering that ICCC and methanation have been extensively studied, a good understanding of the kinetics and mechanisms of carbonation and methanation is equally important [94–96,138–140]. The mechanistic study is typically performed via investigative techniques including Fourier transform–infrared (FT-IR) spectroscopy, which is capable of measuring the gas composition at the reactor outlet and its changes with time, or by in situ diffuse reflectance infrared Fourier transform spectroscopy (DRIFTS), which detects the presence of chemical bonds that suggest the occurrence of possible reaction intermediates.

For instance, a study by Bermejo-López et al. uses FT-IR to study the temporal evolution of the gas composition within a reactor [96]. Their work involved the usage of Ru-based CaO and Na_2CO_3 DFMs supported on alumina. They proposed a mechanism where $CaCO_3$ formed through the carbonation of $Ca(OH)_2$ (Equations (10) and (11)) is decomposed upon the addition of H_2 during the methanation step (Equation (12)), releasing CO_2 which is subsequently converted on Ru sites according to the Sabatier reaction (Equation (13)). Additional observations include the delayed evolution of CO_2 itself during the carbonation step, suggesting the saturation of the CO_2 uptake sites, and the delayed evolution of H_2O during methanation (even though the Sabatier reaction produces H_2O), suggesting the hydration of CaO to form $Ca(OH)_2$ (Equation (14)). A similar reaction scheme is suggested for Na_2CO_3-based DFMs.

$$CaO + CO_2 \leftrightarrow CaCO_3 \tag{10}$$

$$Ca(OH)_2 + CO_2 \leftrightarrow CaCO_3 + H_2O \tag{11}$$

$$CaCO_3 \leftrightarrow CaO + CO_2 \tag{12}$$

$$CO_2 + 4H_2 \leftrightarrow CH_4 + 2H_2O \tag{13}$$

$$CaO + H_2O \leftrightarrow Ca(OH)_2 \tag{14}$$

Conversely, other studies utilize DRIFTS to detect the presence of reaction intermediates, and suggest possible reaction mechanisms [94,95,140]. One such study by Proaño et al. with Ru-based DFM [94,95] suggests the formation of bidentate carbonate species during carbonation and formate species during methanation (see Figure 17 for their detailed mechanism), but there exists a range of works that find corroborating [139] and disputing [140] results. Proaño's group has since conducted further studies into DFMs with a similar makeup (10% Ni, 6.1% Na_2O/Al_2O_3 DFMs enhanced with 1% Pt/Ru), utilizing in situ DRIFTS to study carbonation and methanation under oxidizing and non-oxidizing conditions, yielding largely corroborative results.

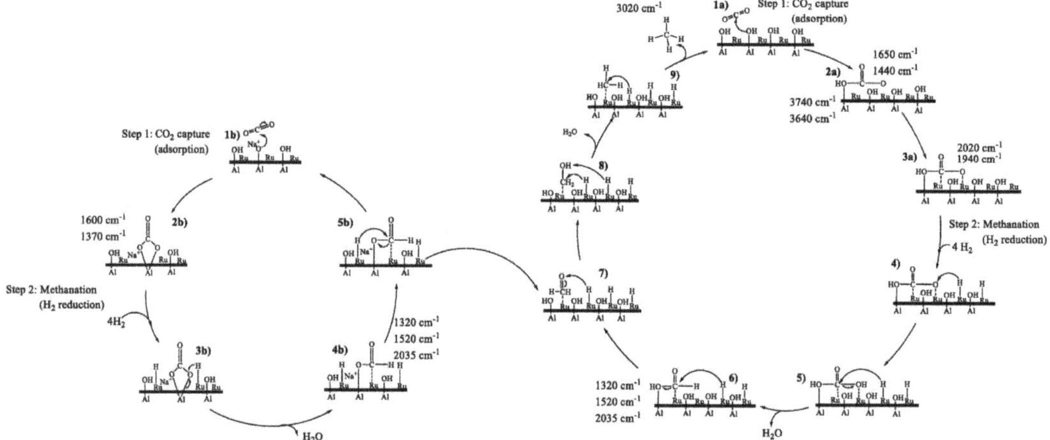

Figure 17. Proposed carbonation and methanation mechanisms for 5%Ru/Al_2O_3 (right) and 5%Ru-6.1%"Na_2O"/Al_2O_3 DFMs (left to right). Reprinted from ref. [95] with permission from Elsevier.

5.2. ICCC with Reverse Water–Gas Shift (RWGS) Reaction

RWGS is another key reaction in C1 chemistry that takes place according to the following reversible reaction (Equation (15)):

$$CO_2 + H_2 \rightleftarrows CO + H_2O; \Delta H_R = 41.2\,\frac{kJ}{mol} \tag{15}$$

RWGS is an important reaction for producing syngas with desired H_2/CO ratios, where CO can be further valorized into various chemicals, such as methanol via the CAMERE (carbon dioxide hydrogenation to form methanol via a reverse-water–gas-shift reaction) process or hydrocarbons via Fischer–Tropsch synthesis. However, although CO is industrially more useful than methane, the development of an effective DFM system to generate CO has been much less explored than that for methane generation. This is partly because the ability of DFMs to achieve a practical H_2/CO ratio remains a major challenge. The accurate tuning of syngas composition can be tricky because of other secondary reactions that may occur in parallel with the RWGS reaction, such as the Sabatier reaction (Equation (10)) and the methanation of CO (Equation (16)):

$$CO + 3H_2 \rightleftharpoons CH_4 + H_2O; \Delta H_R = -206\frac{kJ}{mol} \tag{16}$$

Considering the endothermic nature of RWGS, increasing operating temperatures might reduce the extent of the exothermic side reactions and favor the RWGS reaction. Indeed, in a review by González-Castaño et al. [141], where the equilibrium concentrations of chemical species involved in the RWGS, Sabatier, and CO methanation reactions were investigated, it was found that the equilibrium concentration of CH_4 decreases to near zero above 700 °C, as shown in Figure 18. In addition, the CO_2/H_2 molar ratio also plays a critical role in adjusting the CO composition, with a low CO_2/H_2 ratio being thermodynamically favorable for the RWGS reaction. These considerations are therefore important in maximizing the selectivity of CO, which turns out to be a key metric in evaluating DFMs used in ICCC processes involving RWGS.

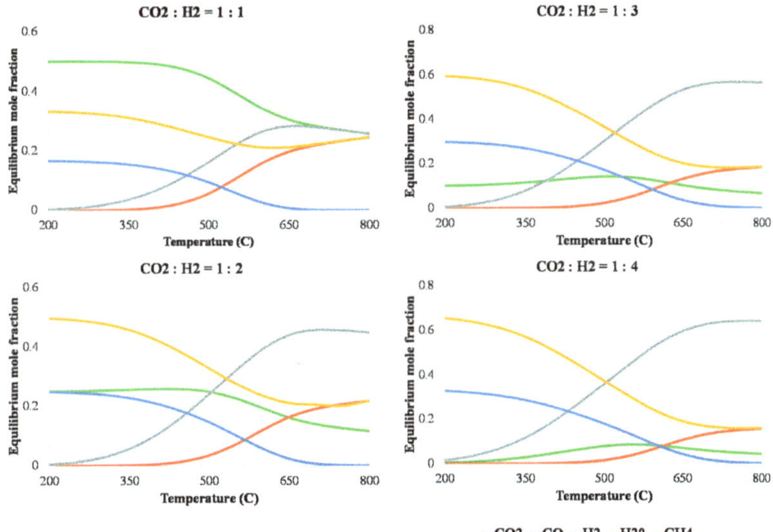

Figure 18. Equilibrium mole fractions of chemical species involved in the RWGS and its side reactions. Reprinted from ref. [141] with permission from the Royal Society of Chemistry.

Apart from thermodynamic considerations, ensuring a high selectivity of CO is also dependent on a prudent catalyst design, which is aided by the understanding of the kinetics and mechanisms of RWGS. Conventionally, Cu is favored for its CO selectivity and low methanation rates, while Pt is preferred for higher CO_2 conversions [78]. With regards to DFM selection for the RWGS reaction, a variety of materials have been developed, including those containing transition metals, such as Fe, Co, Ni, or combinations thereof [142,143]. Recently, there have also been studies that do not involve any transition metals, such as supported Na on CaO and Al_2O_3 [144,145]. Table 4 summarizes the DFMs that catalyze the RWGS reaction.

Table 4. Literature summary of DFMs that catalyze the RWGS reaction.

DFM	Synthesis Method	Carbonation Conditions	RWGS Conditions	CO_2 Sorption Capacity (mmol/g)	CO_2 Conversion (%)	Recycle Study and CO Yield Loss	Ref.
Ni-CaO-CeO$_2$ (CP)	Co-precipitation with acetate precursors	650 °C, 10% CO_2/N_2	650 °C, 10% H_2/N_2	9.81	75	10 cycles, ~10% loss	[105]
Ni-CaO-CeO$_2$ (WM)	Wet mixing with acetate precursors	650 °C, 10% CO_2/N_2	650 °C, 10% H_2/N_2	8.54	84	10 cycles, stable	[105]
Ni-CaO-CeO$_2$ (SG-A)	Citric acid-based sol–gel with acetate precursors	650 °C, 10% CO_2/N_2	650 °C, 10% H_2/N_2	15.34	92	10 cycles, 11.6% loss	[105]
Ni-CaO-CeO$_2$ (SG-N)	Citric acid-based sol–gel with nitrate precursors	650 °C, 10% CO_2/N_2	650 °C, 10% H_2/N_2	13.15	96	10 cycles, 3.5% loss	[105]
Ce-doped Ni/CaO (Ca$_1$Ni$_{0.1}$Ce$_{0.033}$)	Citric acid-based sol–gel with nitrate precursors	650 °C, 15% CO_2/N_2	650 °C, 5% H_2/N_2	14.1	51.8	20 cycles, stable	[111]
Ni-CaO-ZrO$_2$-12	Citric acid-based sol–gel with nitrate precursors	650 °C, 10% CO_2/N_2	650 °C, 45% H_2/N_2	16.72	63.2	12 cycles, 18.6% loss	[146]
Ni-CaO-6ZrO$_2$-6CeO$_2$	Citric acid-based sol–gel with nitrate precursors	650 °C, 10% CO_2/N_2	650 °C, 45% H_2/N_2	11.88	72.1	12 cycles, 12% loss	[146]
Ni/CeO$_2$-CaO	Physical mixing	650 °C, 20% CO_2/N_2	650 °C, 5% H_2/N_2	9.58	56.1	20 cycles, <5% loss	[147]
Na/Al$_2$O$_3$	Wetness impregnation	500 °C, 5% CO_2/N_2	500 °C, H_2	0.135	76.9	50 cycles at 450 °C, stable	[144]
Na/Al$_2$O$_3$	Wetness impregnation	500 °C, 400 ppm CO_2/N_2	500 °C, H_2	0.0907	77.7	-	[144]
Na-Pt/Al$_2$O$_3$	Wetness impregnation	350 °C, 1% $CO_2/10\% O_2/N_2$	350 °C, 5% H_2/N_2	0.19	89	6000 cycles, stable	[145]
Ni$_1$Fe$_9$-CaO	Citric acid-based sol–gel with nitrate precursors	650 °C, 10% CO_2/N_2	650 °C, H_2	14.78	82.5	10 cycles, 20.9% loss	[143]
Fe$_5$Co$_5$Mg$_{10}$CaO	Citric acid-based sol–gel with nitrate precursors	650 °C, 10% CO_2/N_2	650 °C, H_2	9.2	90	10 cycles, stable	[142]

In a recent work by Wu et al., Ni-CaO-CeO_2 DFMs were synthesized via their acetate/nitrate precursors, using several methods including wet mixing, co-precipitation, and a citric acid-based sol–gel pathway [105]. Among the synthesized samples, DFMs prepared by sol–gel method exhibited superior physicochemical properties and hence displayed the highest CO_2 capture capacities and conversion, as well as CO yields and selectivity. In particular, the sol–gel-derived material with acetate precursors showed a CO_2 uptake of 15.34 mmol/g and CO_2 conversion of 92.4%, with CO selectivity of 89.1% at 650 °C. Meanwhile, the material prepared with nitrate precursors displayed a lower CO_2 uptake of 13.15%, but higher CO_2 conversion of 96% and higher CO selectivity (87–96%). It was concluded that, firstly, the sol–gel synthesis resulted in an optimal morphology and internal structure for the CO_2 capture and conversion. For instance, large BET surface areas and pore volumes facilitated CO_2 diffusion and adsorption, whereas an enhanced surface basicity promoted CO_2 affinity. Meanwhile, a smaller NiO grain size, uniform Ni dispersion, and higher reducibility could have led to improved catalytic activity. Secondly, the choice of acetate/nitrate precursors could affect catalyst–support interactions, which in turn result in variations in CO_2 uptake and conversion, CO selectivity, as well as the material's stability upon recycle.

Similar DFMs consisting of Ni over Ce-modified CaO were reported earlier by Sun et al., who employed a sol–gel method with citric acid [111]. The incorporation of CeO_2 was found to enhance the material's stability due to the formation of well-dispersed CeO_2 that can effectively prevent the growth and agglomeration of CaO and NiO crystallites, resulting in a stable performance after 20 cycles. This is reflected by the synthesized material with Ca/Ni/Ce molar ratio of 1:0.1:0.033, which exhibited a CO_2 uptake of 14.1 mmol/g, 51.8% CO_2 conversion, and almost 100% CO selectivity at 650 °C. The authors proposed possible reaction mechanisms which are based on a redox (Scheme 1) or associative formate (Scheme 2) mechanism, as illustrated in Figure 19. According to the redox mechanism, upon switching to the hydrogenation step, the adsorbed CO_2 could spill over to the Ni active sites, which may result in the formation of CO and NiO. In the presence of CeO_2, the oxygen vacancy on the CeO_2 surface may also interact with CO_2, leading to the formation of CO and the oxidation of Ce^{3+} species. The introduction of H_2 into the reactor would subsequently reduce NiO back to Ni and release H_2O molecules. Likewise, the adsorbed H_2 species on the CeO_2 surface extracts the lattice oxygen from ceria, releasing a H_2O molecule along with the reduction of the catalyst. Alternatively, the associative formate mechanism suggests that the adsorbed CO_2 on the CeO_2 surface may form bidentate formate as a reaction intermediate, which decomposes to form CO and terminal hydroxyl groups. Subsequently, the adsorbed H_2 reacts with the hydroxyl group, releasing a H_2O molecule.

In another study by Guo et al., ZrO_2 was used as a dopant for Ni/CaO, where it enhanced the surface basicity and reducibility of the DFM, resulting in an increased CO_2 uptake and catalytic activity [146]. At the optimum 12 wt% ZrO_2 loading, the material exhibited a high CO_2 uptake of 16.7 mmol/g with 63.2% CO_2 conversion and a 10.5 mmol/g CO yield at 650 °C. The formation of $CaZrO_3$ was also found to reduce the sintering of CaO and NiO crystals. The further incorporation of CeO_2 into Ni/CaO-ZrO_2 resulted in increased CO_2 conversion to 72.1% due to the increase of an oxygen vacancy, although the CO_2 uptake capacity decreased to 11.9 mmol/g, which could be attributed to a lower surface area. A similar strategy was also implemented by Sun et al. who incorporated Fe into a Ni/CaO DFM. An optimum CO yield of 11.3 mmol/g was obtained by using Ni_1Fe_9-CaO at 650 °C. It was observed that the formation of $Ca_2Fe_2O_5$ acted as an oxygen carrier that promoted CO production and helped improve the stability of the DFM by acting as a physical barrier that slowed down CaO sintering [143].

As alternatives to transition-metal-based materials, Sasayama et al. prepared DFMs that only contain alkali/alkaline earth metals (e.g., Na, K, and Ca) supported on γ-Al_2O_3 via the impregnation method [144]. They were then tested with 5 vol% and 400 ppm CO_2 in N_2 to investigate the production of syngas under model flue gas or direct air capture modes, respectively. Upon testing with 5 vol% CO_2 at 500 °C, Na/Al_2O_3 exhibited a 0.135 mmol/g

CO_2 uptake with 77% CO_2 conversion and 99.8% CO selectivity. On the other hand, with 400 ppm CO_2, its CO_2 uptake was about 0.09 mmol/g with 77.7% CO_2 conversion and 94.3% CO selectivity. The results therefore indicate the potential use of transition-metal-free DFMs for ICCC process, although further works are needed to improve the materials' performance and to study the reaction mechanism as well as to understand the catalytic active sites.

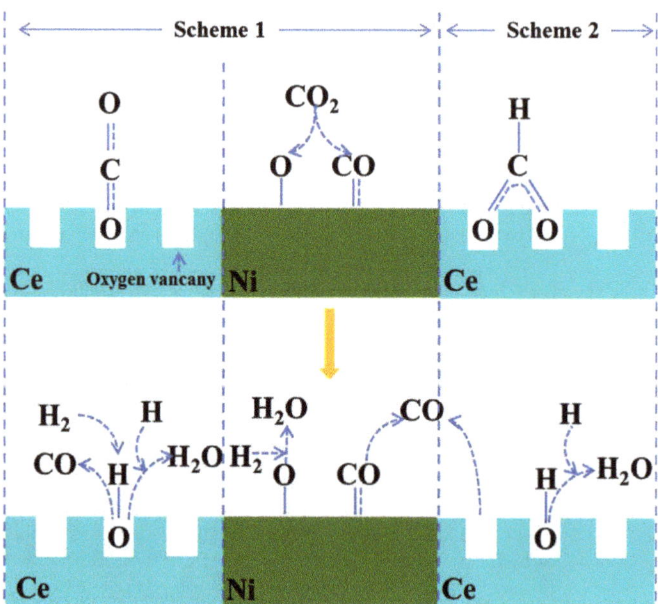

Figure 19. Reaction mechanism of RWGS over CeO_2-incorporated Ni/CaO DFM. Reprinted from ref. [111] with permission from the Elsevier.

In another study, the further incorporation of Pt on Na/Al_2O_3 was found to be selective for CO formation at lower temperatures of 350 °C [145], which is advantageous in reducing the typically high temperature requirement for the RWGS reaction. The DFM was prepared via the sequential impregnation method of Al_2O_3, first with Na, followed by Pt. The CO_2 capture was evaluated with 1% CO_2/10% O_2 in N_2, and its uptake capacity was found to be 0.19 mmol/g, comparable to that of Na/Al_2O_3 (0.21 mmol/g), and much higher than that of Pt/Al_2O_3 (0.08 mmol/g), indicating that Na serves as the CO_2 capture site. Upon hydrogenation, CO_2 was converted to CO with 89% conversion and 93% CO selectivity. In a control experiment, a physical mixture of Na/Al_2O_3 and Pt/Al_2O_3 was tested, and it was found that CO_2 conversion and CO selectivity decreased significantly to only 57% and 41%, respectively. The EDX spectroscopy (Figure 20) suggests that a core–shell structure of Pt-Na nanoparticles was formed on $Pt-Na/Al_2O_3$, with Pt as the core. As a result of the close interaction between the two metals, the Na species not only serves as the CO_2 capture site, but also as a promoter to enhance CO formation. Through in situ FTIR measurements, it was found that the adsorbed CO species was hardly observed over the material's surface, indicating that the presence of the Na species inhibited the adsorption of the generated CO, leading to high selectivity for CO.

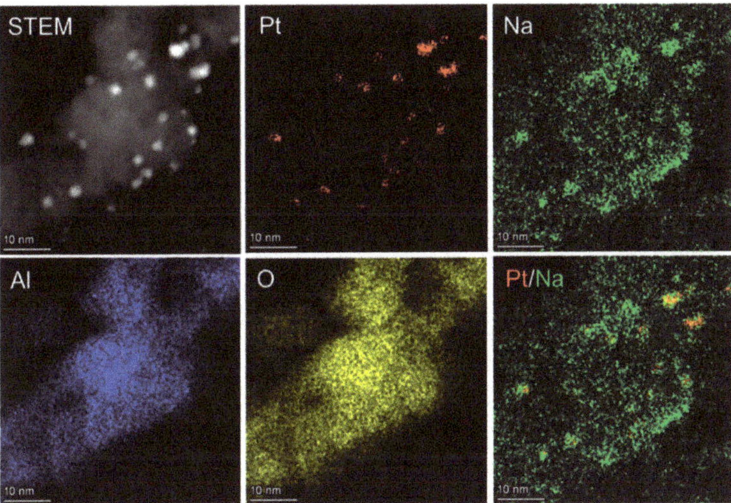

Figure 20. STEM image and EDX elemental mapping of Pt–Na/Al$_2$O$_3$. Reprinted from ref. [145] with permission from the American Chemical Society.

5.3. ICCC with Dry Reforming of Methane (DRM)

Besides RWGS, syngas can also be produced via the dry reforming of methane (DRM), where two potent greenhouse gases, CH$_4$ and CO$_2$, react with each other according to the following equation (Equation (17)):

$$CO_2 + CH_4 \rightarrow 2CO + 2H_2 \, \Delta H_R = 247.3 \frac{kJ}{mol} \tag{17}$$

Despite being favored for its "greening" capability, ICCC with DRM faces several disadvantages at industrial scales. First, the highly endothermic reaction requires a large energy input (Equation (17)). Second, the DFMs that catalyze the DRM reaction typically contain Ni in the catalytic component, which is susceptible to poisoning and deactivation via coking in the absence of steam. It has, therefore, been proposed that continued research in this field would focus on developing Ni-based DFMs that exhibit stronger coking resistance, e.g., by including alkaline promoters like Mg, or considering the use of Rh- or Ru-based DFMs to catalyze the reaction, in addition to uncovering further insights into the mechanisms of DRM so as to strengthen and optimize its industrial applicability [134,148]. The summary of DFMs that catalyze the DRM reaction is presented in Tables 5 and 6. The two tables categorize works according to their presentation of the ICCC performances of DFMs. In the former, the conversions of CO$_2$ and CH$_4$ are presented, whereas the yields of H$_2$ and CO are presented in the latter.

Table 5. Literature summary of DFMs that catalyze the dry reforming reaction (with conversions of CO_2 and CH_4).

DFM	Synthesis Method	Carbonation Conditions	Dry Reforming Conditions	CO_2 Sorption Capacity	CO_2 Conversion	CH_4 Conversion	Ref.
Ni_{10}-(K-Mg)$_{25}$/(γ-Al_2O_3)$_{75}$	Sol–gel/wet impregnation	650 °C, 10%CO_2/N_2	650 °C, 5%C_2H_6/N_2	0.22 mmol/g DFM	14%	16% (ethane)	[91]
Ni_{10}-(Na-Mg)$_{50}$/(γ-Al_2O_3)$_{50}$	Sol–gel/wet impregnation	650 °C, 10%CO_2/N_2	650 °C, 5%C_2H_6/N_2	0.16 mmol/g DFM	60%	47% (ethane)	[91]
Ni_{10}-(K-Ca)$_{50}$/(γ-Al_2O_3)$_{50}$	Sol–gel/wet impregnation	650 °C, 10%CO_2/N_2	650 °C, 5%C_2H_6/N_2	0.99 mmol/g DFM	65%	100% (ethane)	[91]
Ni_{10}-(Na-Ca)$_{50}$/(γ-Al_2O_3)$_{50}$	Sol–gel/wet impregnation	650 °C, 10%CO_2/N_2	650 °C, 5%C_2H_6/N_2	0.63 mmol/g DFM	75%	100% (ethane)	[91]
Ni/Ca-Zr	Precipitation/ammoniacal sol/impregnation	720 °C, 5%CO_2/Ar	720 °C, 8%CH_4/Ar	~3.64 mmol/g CaO (25 cycles)	-	~32%	[149]
NiCe/Ca-Zr	Precipitation/ammoniacal sol/impregnation	720 °C, 5%CO_2/Ar	720 °C, 8%CH_4/Ar	~5 mmol/g CaO (25 cycles)	-	~45%	[149]
Ni/MgO-Al_2O_3 (mixed with CaO)	Co-precipitation	720 °C, 20%CO_2/N_2	720 °C, 2.4%CH_4/N_2	14.1 mmol/g DFM (initial)	99.92%	99.94%	[150]
Ca-Fe-Mg oxide	Co-precipitation	600 °C, 100%CO_2	900 °C, 4.5%CH_4/N_2	0.64 mmol/g DFM	-	~50%	[151]

Table 6. Literature summary of DFMs that catalyze the dry reforming reaction (with yields of H_2 and CO).

DFM	Synthesis Method	Carbonation Conditions	Dry Reforming Conditions	CO_2 Sorption Capacity	H_2 Yield	CO Yield	Ref.
CaO/Ni_4	Sol–gel	600 °C, 20%CO_2/N_2	800 °C, 20%CH_4/N_2	~5.7 mmol/g CaO (initial)	5 mmol/g DFM	4.3 mmol/g DFM	[152]
CaO/Ni_9	Sol–gel	600 °C, 20%CO_2/N_2	800 °C, 20%CH_4/N_2	~5.1 mmol/g CaO (initial)	5 mmol/g DFM	4.5 mmol/g DFM	[152]
CaO/Ni_{19}	Sol–gel	600 °C, 20%CO_2/N_2	800 °C, 20%CH_4/N_2	~4.5 mmol/g CaO (initial)	3.5 mmol/g DFM	3.5 mmol/g DFM	[152]
Ni-CaO catal-sorbents	Sol–gel	700 °C, 10%CO_2/10%H_2O/N_2	700 °C, 10%CH_4/N_2	14.8 mmol/g DFM	131.7 mmol/g DFM	20.2 mmol/g DFM	[153]

Table 6. *Cont.*

DFM	Synthesis Method	Carbonation Conditions	Dry Reforming Conditions	CO_2 Sorption Capacity	H_2 Yield	CO Yield	Ref.
$Ni/Ca_{85}Ce_{15}$	Hydrothermal (sorbent); impregnation (DFM)	650 °C, 10%CO_2/Ar	650 °C, 6%CH_4/Ar	~13 mmol/g CaO (initial)	-	-	[154]
15% Ni-1% Ru, 10% Na_2O/CeO_2–Al_2O_3	Sequential impregnation	650 °C, 10%CO_2/N_2	650 °C, 10%CH_4/N_2	0.16 mmol/g DFM	25.774 mmol/g DFM	0.153 mmol/g DFM	[155]
15% Ni-1% Ru, 10% K_2O/CeO_2–Al_2O_3	Sequential impregnation	650 °C, 10%CO_2/N_2	650 °C, 10%CH_4/N_2	0.18 mmol/g DFM	22.512 mmol/g DFM	0.239 mmol/g DFM	[155]
15% Ni-1% Ru, 10% CaO/CeO_2–Al_2O_3	Sequential impregnation	650 °C, 10%CO_2/N_2	650 °C, 10%CH_4/N_2	0.26 mmol/g DFM	32.639 mmol/g DFM	0.338 mmol/g DFM	[155]
CaO-0.05Ni-0.05CeO_2	Co-precipitation	600 °C, 2%CO_2/N_2	800 °C, 2%CH_4/N_2	10.3 mmol/g CaO (initial); 7.9 mmol/g CaO (10th cycle)	754.4 mmol/g Ni (max, 2nd cycle)	454.6 mmol/g Ni (max, 2nd cycle)	[156]

In an earlier work, Kim et al. first demonstrated the calcium looping process combined with a catalytic system by using a physical mixture of CaO and Ni/MgO-Al$_2$O$_3$ as the CO$_2$ sorbent and DRM catalyst, respectively, in a single fluidized bed reactor [150]. The two-step process was operated in a cyclic manner where CO$_2$ was first captured on CaO during the carbonation step. Upon gas switching to CH$_4$, the CaO sorbent was regenerated while releasing CO$_2$, which instantaneously reacted with CH$_4$ on the Ni/MgO-Al$_2$O$_3$ catalyst, as depicted in Figure 21. CaO was derived from limestone via calcination at 800 °C whereas the Ni catalyst was derived from a hydrotalcite precursor that was synthesized via the co-precipitation method. At 720 °C, the CO$_2$ capture capacity was 14.1 mmol/g in the first cycle with an almost 100% conversion of CO$_2$ and CH$_4$. The resulting syngas thus had a H$_2$:CO ratio of 1.06:1. Due to the sintering of CaO, there was a significant loss of CO$_2$ uptake, which decreased to 9 mmol/g after 10 cycles of the reaction. Although, initially, Ni catalyst deactivation was observed due to coke formation, the carbon deposition amount was decreasing with increasing cycle numbers. It was found that the deposited carbon was removed during the carbonation step while the Ni metallic state was preserved, which could be due to the reverse Boudouard reaction C + CO$_2$ $\rightleftarrows$ 2CO that is favorable at high temperatures [152]. Therefore, CO$_2$ and CH$_4$ conversion can be maintained above 95% throughout the 10 cycles.

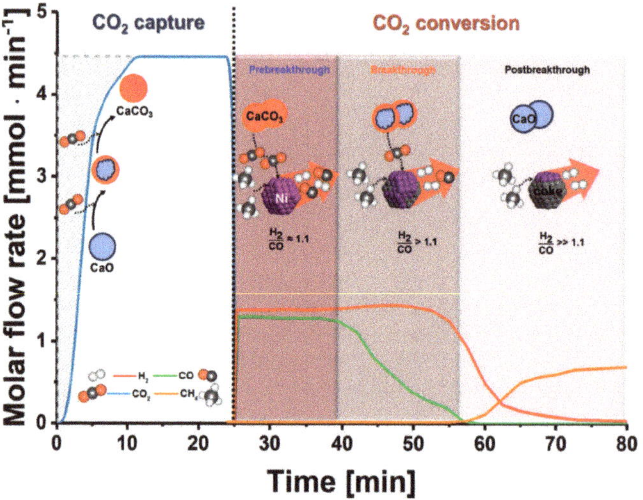

Figure 21. Cyclic process of ICCC with DRM over physical mixture of CaO and Ni/MgO-Al$_2$O$_3$ as CO$_2$ sorbent and catalyst, respectively. Reprinted from ref. [150] with permission from the American Chemical Society.

In subsequent works by Tian et al., a Ni/CaO-based DFM was prepared via citrate-based sol–gel synthesis which produced an average syngas yield of around 9 mmol/g throughout 10 cycles [152] of CO$_2$ capture (at 600 °C) and conversion with CH$_4$ (at 800 °C). The experimental H$_2$:CO ratio was about 1.1, higher than the theoretical ratio of 1. As the average CH4 conversion (86%) was higher than that of CO$_2$ (65%), it could be suggested that the Ni-CaO interface in the DFM was more active in dissociating CH$_4$ to yield H$_2$ than reducing CO$_2$ to yield CO. On the other hand, despite using a Ni/CaO DFM prepared with a similar method, Jo et al. obtained syngas with a high H$_2$:CO ratio of 6.52 at 700 °C, indicating that the CO$_2$ reduction reaction was suppressed [153]. It was observed that a large amount of carbon deposit was formed on the DFM, as well as the hydration of CaO to form Ca(OH)$_2$.

A further improvement of the Ni/CaO DFM was reported by Hu et al. where porous CeO$_2$-modified CaO microparticles were used as the support for Ni impregnation [154].

The introduction of CeO_2 into the material enhanced the DFM performance in two aspects, in that it acted as (i) a promoter that enhanced the CO_2 affinity towards CaO through the increase in lattice oxygen, thus enabling a high CO_2 uptake, and (ii) a physical stabilizer that enhanced the sintering resistance of CaO, improved Ni particle dispersion through the formation of small Ni crystallites, as well as increased Ni reducibility, thus enabling high and stable catalytic activity. Among the investigated samples, the material with $85\%CaO:15\%CeO_2$ ($Ni/Ca_{85}Ce_{15}$) showed a constant CO_2 uptake and retained about 80–90% of its initial syngas time-averaged space time yield (STY) over nine cycles of CO_2 sorption and conversion at 650 °C, as shown in Figure 22. In another experiment, a layer of ZrO_2 was coated on $CaCO_3$ nanoparticles, followed by the co-impregnation of Ni and Ce [149]. During the CO_2 capture–DRM cycle tests at 720 °C, 5% CO_2 was used as the gas feed to mimic flue gas composition where over 40% conversion of CO_2 and CH_4 can be achieved. The presence of ZrO_2 was also found to enhance material stability by preventing sintering. Through the in situ DRIFTS analysis (Figure 23), monodentate carbonate was observed on the CaO surface during the carbonation step, whereas polydentate carbonate was observed on ZrO_2 and bidentate carbonate on the CeO_2 surface. CO_2 dissociation to CO was also detected on NiCe/Ca-Zr oxide due to the presence of abundant oxygen vacancies on the reduced CeO_2. Upon gas switching to CH_4, the characteristic IR peaks of formate species and gaseous CO were observed, while carbonate peak intensities were decreased. These suggested that the occurrence of $CaCO_3$ decarbonation was coupled with the DRM reaction, during which the adsorbed H species produced from CH_4 decomposition reacted with the CO_3^{2-} species to form formate species, which further decomposed to CO [149].

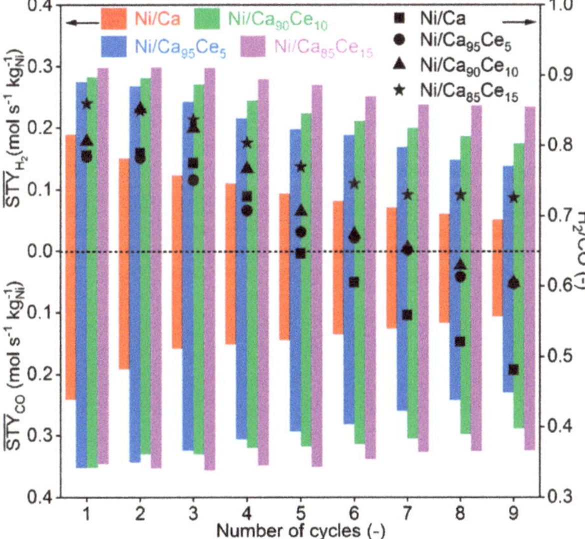

Figure 22. Time-averaged space time yield (STY) of H_2 (top) and CO (bottom) of the prepared Ni/CeO_2-CaO DFMs with various $CaO:CeO_2$ molar ratio throughout 9 cycles of CO_2 capture and conversion at 650 °C. Ni/CaO was used as reference material. Reprinted from ref. [154] with permission from Elsevier.

The combination of Ni with Ru supported on CeO_2/Al_2O_3 containing Na_2O, K_2O, or CaO was reported by Merkouri et al. [155]. The DFMs contained 15%Ni and 1%Ru, and were mostly active for ICCC with DRM at 650 °C, where the CaO-containing material produced 0.338 mmol/g CO and 32.6 mmol/g H_2, yielding H_2-rich syngas. The large formation of H_2 during DRM was mainly due to methane cracking and its decomposition into carbon, evidenced by CH_4-TPSR experiments which showed that methane cracking peaked at around 635 °C. By using an operando DRIFTS-MS method, the authors elucidated the CO_2 capture and reaction mechanism over the $Ni-Ru-Na_2O/CeO_2-Al_2O_3$ DFM through alternating successive CO_2 and CH_4 cycles at 550 °C. In agreement with Tian et al.'s previous work [152], CH_4 reduction on the metallic sites produced a H_2 and graphitic carbon layer, which further gasified with CO_2 during the capture step to yield CO through a reverse Bourdouard reaction. However, it was noted that the coke gasification kinetics were slower than the coke formation, resulting in the rapid deactivation of the catalyst. This mechanistic study thus highlights the importance of controlling CH_4 cracking during the DRM reaction via optimizing the material's design and reaction engineering in order to slow down the catalyst deactivation, as well as to obtain syngas with a favorable H_2:CO ratio.

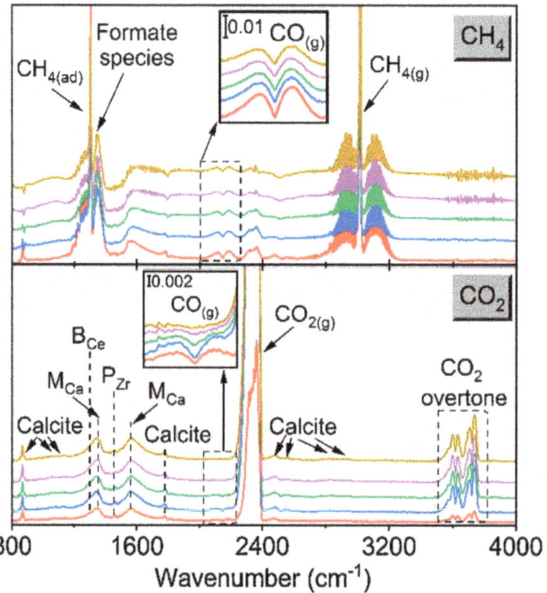

Figure 23. In situ DRIFTS spectra of the reduced NiCe/Ca@Zr in CO_2 and CH_4 environment at 550 °C and 1 bar, with different color to represent spectrum obtained at different time during exposure in different gas environment. Abbreviation—(g): gas phase; (ad): adsorbed; MCa: Monodentate carbonate on CaO surface; PZr: Polydentate carbonate on ZrO_2 surface; BCe: Bidentate carbonate on CeO_2 surface. Reprinted from ref. [149] with permission from Elsevier.

Further alternatives that include the dry reforming of other hydrocarbons, such as ethane (DRE), to produce syngas (Equation (18)) was reported by Al-Mamoori et al. [91], who investigated the use of CaO- and MgO-based double salts promoted by K/Ca, supported on alumina, and impregnated with Ni as the catalyst at 650 °C.

$$CO_2 + C_2H_6 \rightarrow 4CO + 3H_2; \Delta H_R = 428.1\,\frac{kJ}{mol} \tag{18}$$

The material's synthetic methods consisted of a sol–gel method for the alumina support, while the adsorbent and catalyst components were incorporated via wet impregnation. The CaO-based DFMs comprising 20 wt% Ni and a 1:1 ratio of adsorbent to support showed the best ICCC performance, with CO_2 adsorption capacities of 0.99 mmol/g and 0.63 mmol/g for the K-Ca-based and Na-Ca-based DFMs, respectively, a 65% and 75% conversion of CO_2 for the K-Ca-based and Na-Ca-based DFMs, respectively, and a 100% C_2H_6 conversion for both DFMs. Additionally, yields of approximately 45% and 37.5% for CO and H_2, respectively, were reported for the K-Ca-based DFM when subjected to a sustained DRE process (up to 10 h) at 650 °C. Despite a relatively stable performance, coke formation (9 wt%) was observed due to a high Ni content. From their investigations, it was concluded that the adsorbent/catalyst contents directly influenced CO_2 capture and conversion performances, and that reaction conditions strongly influenced the selectivity toward DRE or the occurrence of side reactions, such as oxidative dehydrogenation, ethane cracking, RWGS, and coking. Therefore, it is imperative to optimize their chemical and physical properties.

6. Perspectives and Future Outlook

It is evident that DFMs have increasingly played an important role in decarbonization efforts through CO_2 recycling and reutilization. Through numerous studies in the past decade, DFMs have shown to offer feasibility and versatility in capturing CO_2 from both model flue gas and air (direct air capture), and subsequently converting it to other C1 species, such as methane and syngas. The recent literature has reported extensive DFM formulations and configurations for ICCC that primarily include Ni, Ru, and Rh as the catalytic metals, basic oxides such as Na_2O, K_2O, CaO, and MgO as the CO_2 sorbents, and Al_2O_3 as a common support. From the various given examples, the combinations of these materials are shown to form flexible DFMs that catalyze CO_2 methanation, RWGS, and DRM, each of which has its own optimum operating conditions. For example, CO_2 methanation is favored at an intermediate temperature range of 300–350 °C, whereas RWGS and DRM are favored at high temperatures (600–750 °C).

The effectiveness of DFMs is determined by both their CO_2 capture capacity and catalytic activity. The CO_2 uptake principally depends on (i) the basicity of the materials, with medium to strong basic sites favorable for CO_2 adsorption, and (ii) the surface area and porosity. On the other hand, the catalytic activity depends on several factors, such as (i) the crystallite size and dispersion of the active metals, with a small size and high metal dispersion attributed to high catalytic activity, (ii) the reducibility of the active metals, which is determined by the interaction between the active metal particles and the oxide support, (iii) the basicity of the material, where weak basic sites are beneficial for CO_2 methanation and medium basic sites for the RWGS and DRM [155], and (iv) the availability of oxygen vacancies, which can often be enhanced by incorporating a dopant, such as CeO_2. In fact, there is a recent review by Hussain et al. that discusses the importance of oxygen vacancies for CO_2 methanation and its tuning strategies [157].

Given the complexity in developing and tuning DFMs with the desired properties, major progress has been achieved at the laboratory scale, leading to several promising candidates with good CO_2 conversion and product selectivity. However, there are still many challenges to implement ICCC with DFMs on a larger scale. For instance, the stability and recyclability of the materials need to be improved as DFMs are susceptible to sintering, coking, and poisoning. It, therefore, requires careful material engineering to synthesize robust DFMs that can withstand harsh conditions upon prolonged and repeated cycles. In view of this, lifetime studies of DFMs need to be carried out under real flue gas conditions that typically contain moisture, O_2, and other impurities, such as NO_x, to guarantee their industrial viability. To accelerate the screening process, the use of computational DFT modeling would be highly encouraged to aid in the fine-tuning of the material's properties while studying the reaction mechanisms to control the desired product selectivity as well as to minimize catalyst deactivation.

Considering that ICCC with DFMs is an emerging field, there are plenty of research opportunities in the future to develop versatile and robust DFMs that enable the production of other chemicals, such as alcohols and longer-chain hydrocarbons, which could offer the greater flexibility of producing different products within the same production facility. We believe that the advancement of this technology will continue promoting sustainable chemical manufacturing and a low carbon economy.

Author Contributions: Conceptualization, W.J.T. and P.G.; Methodology, W.J.T. and P.G.; Writing—original draft preparation, W.J.T. and P.G.; Writing—review and editing, W.J.T. and P.G.; Project administration, P.G.; Funding acquisition, P.G. All authors have read and agreed to the published version of the manuscript.

Funding: This research was funded by the C. N. Yang Scholars Program at Nanyang Technological University AY2022-2023.

Acknowledgments: This research was made possible by the C. N. Yang Scholars Program at Nanyang Technological University.

Conflicts of Interest: The authors declare no conflict of interest.

References

1. *An IPCC Special Report on the Impacts of Global Warming of 1.5 °C above Pre-Industrial Levels and Related Global Greenhouse Gas Emission Pathways, in the Context of Strengthening the Global Response to the Threat of Climate Change, Sustainable Development, and Efforts to Eradicate Poverty*; Intergovernmental Panel on Climate Change: Geneva, Switzerland, 2018.
2. Koytsoumpa, E.I.; Bergins, C.; Kakaras, E. The CO_2 economy: Review of CO_2 capture and reuse technologies. *J. Supercrit. Fluids* **2018**, *132*, 3–16. [CrossRef]
3. Leeson, D.; Mac Dowell, N.; Shah, N.; Petit, C.; Fennell, P.S. A Techno-economic analysis and systematic review of carbon capture and storage (CCS) applied to the iron and steel, cement, oil refining and pulp and paper industries, as well as other high purity sources. *Int. J. Greenh. Gas Control* **2017**, *61*, 71–84. [CrossRef]
4. Naims, H. Economics of carbon dioxide capture and utilization—A supply and demand perspective. *Environ. Sci. Pollut. Res.* **2016**, *23*, 22226–22241. [CrossRef] [PubMed]
5. Wilberforce, T.; Baroutaji, A.; Soudan, B.; Al-Alami, A.H.; Olabi, A.G. Outlook of carbon capture technology and challenges. *Sci. Total Environ.* **2019**, *657*, 56–72. [CrossRef]
6. Omodolor, I.S.; Otor, H.O.; Andonegui, J.A.; Allen, B.J.; Alba-Rubio, A.C. Dual-Function Materials for CO_2 Capture and Conversion: A Review. *Ind. Eng. Chem. Res.* **2020**, *59*, 17612–17631. [CrossRef]
7. von der Assen, N.; Bardow, A. Life cycle assessment of polyols for polyurethane production using CO_2 as feedstock: Insights from an industrial case study. *Green Chem.* **2014**, *16*, 3272–3280. [CrossRef]
8. Artz, J.; Müller, T.E.; Thenert, K.; Kleinekorte, J.; Meys, R.; Sternberg, A.; Bardow, A.; Leitner, W. Sustainable Conversion of Carbon Dioxide: An Integrated Review of Catalysis and Life Cycle Assessment. *Chem. Rev.* **2018**, *118*, 434–504. [CrossRef]
9. Huang, C.-H.; Tan, C.-S. A Review: CO_2 Utilization. *Aerosol Air Qual. Res.* **2014**, *14*, 480–499. [CrossRef]
10. Freyman, M.C.; Huang, Z.; Ravikumar, D.; Duoss, E.B.; Li, Y.; Baker, S.E.; Pang, S.H.; Schaidle, J.A. Reactive CO_2 capture: A path forward for process integration in carbon management. *Joule* **2023**, *7*, 631–651. [CrossRef]
11. Duyar, M.S.; Treviño, M.A.A.; Farrauto, R.J. Dual function materials for CO_2 capture and conversion using renewable H_2. *Appl. Catal. B Environ.* **2015**, *168–169*, 370–376. [CrossRef]
12. Melo Bravo, P.; Debecker, D.P. Combining CO_2 capture and catalytic conversion to methane. *Waste Dispos. Sustain. Energy* **2019**, *1*, 53–65. [CrossRef]
13. Merkouri, L.-P.; Reina, T.R.; Duyar, M.S. Closing the Carbon Cycle with Dual Function Materials. *Energy Fuels* **2021**, *35*, 19859–19880. [CrossRef]
14. Sun, S.; Sun, H.; Williams, P.T.; Wu, C. Recent advances in integrated CO_2 capture and utilization: A review. *Sustain. Energy Fuels* **2021**, *5*, 4546–4559. [CrossRef]
15. Sabri, M.A.; Al Jitan, S.; Bahamon, D.; Vega, L.F.; Palmisano, G. Current and future perspectives on catalytic-based integrated carbon capture and utilization. *Sci. Total Environ.* **2021**, *790*, 148081. [CrossRef] [PubMed]
16. Li, H.; Dang, C.; Yang, G.; Cao, Y.; Wang, H.; Peng, F.; Yu, H. Bi-functional particles for integrated thermo-chemical processes: Catalysis and beyond. *Particuology* **2021**, *56*, 10–32. [CrossRef]
17. Bhatta, L.K.G.; Subramanyam, S.; Chengala, M.D.; Olivera, S.; Venkatesh, K. Progress in hydrotalcite like compounds and metal-based oxides for CO_2 capture: A review. *J. Clean. Prod.* **2015**, *103*, 171–196. [CrossRef]
18. Khalilpour, R.; Mumford, K.; Zhai, H.; Abbas, A.; Stevens, G.; Rubin, E.S. Membrane-based carbon capture from flue gas: A review. *J. Clean. Prod.* **2015**, *103*, 286–300. [CrossRef]
19. Al-Mamoori, A.; Krishnamurthy, A.; Rownaghi, A.A.; Rezaei, F. Carbon Capture and Utilization Update. *Energy Technol.* **2017**, *5*, 834–849. [CrossRef]
20. Yang, Z.-Z.; Wei, J.-J.; Zeng, G.-M.; Zhang, H.-Q.; Tan, X.-F.; Ma, C.; Li, X.-C.; Li, Z.-H.; Zhang, C. A review on strategies to LDH-based materials to improve adsorption capacity and photoreduction efficiency for CO_2. *Coord. Chem. Rev.* **2019**, *386*, 154–182. [CrossRef]
21. Xie, K.; Fu, Q.; Qiao, G.G.; Webley, P.A. Recent progress on fabrication methods of polymeric thin film gas separation membranes for CO_2 capture. *J. Membr. Sci.* **2019**, *572*, 38–60. [CrossRef]
22. Gao, W.; Liang, S.; Wang, R.; Jiang, Q.; Zhang, Y.; Zheng, Q.; Xie, B.; Toe, C.Y.; Zhu, X.; Wang, J.; et al. Industrial carbon dioxide capture and utilization: State of the art and future challenges. *Chem. Soc. Rev.* **2020**, *49*, 8584–8686. [CrossRef] [PubMed]

23. Younas, M.; Rezakazemi, M.; Daud, M.; Wazir, M.B.; Ahmad, S.; Ullah, N.; Inamuddin; Ramakrishna, S. Recent progress and remaining challenges in post-combustion CO_2 capture using metal-organic frameworks (MOFs). *Prog. Energy Combust. Sci.* **2020**, *80*, 100849. [CrossRef]

24. Zheng, J.; Chong, Z.R.; Qureshi, M.F.; Linga, P. Carbon Dioxide Sequestration via Gas Hydrates: A Potential Pathway toward Decarbonization. *Energy Fuels* **2020**, *34*, 10529–10546. [CrossRef]

25. Dunstan, M.T.; Donat, F.; Bork, A.H.; Grey, C.P.; Müller, C.R. CO_2 Capture at Medium to High Temperature Using Solid Oxide-Based Sorbents: Fundamental Aspects, Mechanistic Insights, and Recent Advances. *Chem. Rev.* **2021**, *121*, 12681–12745. [CrossRef] [PubMed]

26. Fu, L.; Ren, Z.; Si, W.; Ma, Q.; Huang, W.; Liao, K.; Huang, Z.; Wang, Y.; Li, J.; Xu, P. Research progress on CO_2 capture and utilization technology. *J. CO_2 Util.* **2022**, *66*, 102260. [CrossRef]

27. Dubey, A.; Arora, A. Advancements in carbon capture technologies: A review. *J. Clean. Prod.* **2022**, *373*, 133932. [CrossRef]

28. Peres, C.B.; Resende, P.M.R.; Nunes, L.J.R.; Morais, L.C.d. Advances in Carbon Capture and Use (CCU) Technologies: A Comprehensive Review and CO_2 Mitigation Potential Analysis. *Clean Technol.* **2022**, *4*, 1193–1207. [CrossRef]

29. Idem, R.; Supap, T.; Shi, H.; Gelowitz, D.; Ball, M.; Campbell, C.; Tontiwachwuthikul, P. Practical experience in post-combustion CO_2 capture using reactive solvents in large pilot and demonstration plants. *Int. J. Greenh. Gas Control* **2015**, *40*, 6–25. [CrossRef]

30. Li, Y.; Li, L.; Yu, J. Applications of Zeolites in Sustainable Chemistry. *Chem* **2017**, *3*, 928–949. [CrossRef]

31. Suescum-Morales, D.; Jiménez, J.R.; Fernández-Rodríguez, J.M. Review of the Application of Hydrotalcite as CO_2 Sinks for Climate Change Mitigation. *ChemEngineering* **2022**, *6*, 50. [CrossRef]

32. Veerabhadrappa, M.G.; Maroto-Valer, M.M.; Chen, Y.; Garcia, S. Layered Double Hydroxides-Based Mixed Metal Oxides: Development of Novel Structured Sorbents for CO_2 Capture Applications. *ACS Appl. Mater. Interfaces* **2021**, *13*, 11805–11813. [CrossRef] [PubMed]

33. Gao, W.; Zhou, T.; Gao, Y.; Louis, B.; O'Hare, D.; Wang, Q. Molten salts-modified MgO-based adsorbents for intermediate-temperature CO_2 capture: A review. *J. Energy Chem.* **2017**, *26*, 830–838. [CrossRef]

34. Kwak, J.-S.; Kim, K.-Y.; Oh, K.-R.; Kwon, Y.-U. Performance enhancement of all-solid CO_2 absorbent based on Na_2CO_3-promoted MgO by using ZrO_2 dispersant. *Int. J. Greenh. Gas Control* **2019**, *81*, 38–43. [CrossRef]

35. Joo, H.; Cho, S.J.; Na, K. Control of CO_2 absorption capacity and kinetics by MgO-based dry sorbents promoted with carbonate and nitrate salts. *J. CO_2 Util.* **2017**, *19*, 194–201. [CrossRef]

36. Jeon, H.; Triviño, M.L.T.; Hwang, S.; Moon, J.H.; Yoo, J.; Seo, J.G. Unveiling the carbonation mechanism in molten salt-promoted MgO-Al_2O_3 sorbents. *J. CO_2 Util.* **2020**, *39*, 101153. [CrossRef]

37. Hu, Z.; Wang, Y.; Shah, B.B.; Zhao, D. CO_2 Capture in Metal–Organic Framework Adsorbents: An Engineering Perspective. *Adv. Sustain. Syst.* **2019**, *3*, 1800080. [CrossRef]

38. Song, K.S.; Fritz, P.W.; Coskun, A. Porous organic polymers for CO_2 capture, separation and conversion. *Chem. Soc. Rev.* **2022**, *51*, 9831–9852. [CrossRef]

39. Wang, J.; Wang, L.; Wang, Y.; Zhang, D.; Xiao, Q.; Huang, J.; Liu, Y.-N. Recent progress in porous organic polymers and their application for CO_2 capture. *Chin. J. Chem. Eng.* **2022**, *42*, 91–103. [CrossRef]

40. Luo, R.; Chen, M.; Liu, X.; Xu, W.; Li, J.; Liu, B.; Fang, Y. Recent advances in CO_2 capture and simultaneous conversion into cyclic carbonates over porous organic polymers having accessible metal sites. *J. Mater. Chem. A* **2020**, *8*, 18408–18424. [CrossRef]

41. Silva, J.M.; Trujillano, R.; Rives, V.; Soria, M.A.; Madeira, L.M. High temperature CO_2 sorption over modified hydrotalcites. *Chem. Eng. J.* **2017**, *325*, 25–34. [CrossRef]

42. Miguel, C.V.; Trujillano, R.; Rives, V.; Vicente, M.A.; Ferreira, A.F.P.; Rodrigues, A.E.; Mendes, A.; Madeira, L.M. High temperature CO_2 sorption with gallium-substituted and promoted hydrotalcites. *Sep. Purif. Technol.* **2014**, *127*, 202–211. [CrossRef]

43. Faria, A.C.; Trujillano, R.; Rives, V.; Miguel, C.V.; Rodrigues, A.E.; Madeira, L.M. Alkali metal (Na, Cs and K) promoted hydrotalcites for high temperature CO_2 capture from flue gas in cyclic adsorption processes. *Chem. Eng. J.* **2022**, *427*, 131502. [CrossRef]

44. Lee, J.M.; Min, Y.J.; Lee, K.B.; Jeon, S.G.; Na, J.G.; Ryu, H.J. Enhancement of CO_2 Sorption Uptake on Hydrotalcite by Impregnation with K_2CO_3. *Langmuir* **2010**, *26*, 18788–18797. [CrossRef]

45. Halabi, M.H.; de Croon, M.H.J.M.; van der Schaaf, J.; Cobden, P.D.; Schouten, J.C. High capacity potassium-promoted hydrotalcite for CO_2 capture in H_2 production. *Int. J. Hydrogen Energy* **2012**, *37*, 4516–4525. [CrossRef]

46. Kim, S.; Lee, K.B. Impregnation of hydrotalcite with $NaNO_3$ for enhanced high-temperature CO_2 sorption uptake. *Chem. Eng. J.* **2019**, *356*, 964–972. [CrossRef]

47. Oliveira, E.L.G.; Grande, C.A.; Rodrigues, A.E. CO_2 sorption on hydrotalcite and alkali-modified (K and Cs) hydrotalcites at high temperatures. *Sep. Purif. Technol.* **2008**, *62*, 137–147. [CrossRef]

48. Perejón, A.; Romeo, L.M.; Lara, Y.; Lisbona, P.; Martínez, A.; Valverde, J.M. The Calcium-Looping technology for CO_2 capture: On the important roles of energy integration and sorbent behavior. *Appl. Energy* **2016**, *162*, 787–807. [CrossRef]

49. Benhelal, E.; Shamsaei, E.; Rashid, M.I. Challenges against CO_2 abatement strategies in cement industry: A review. *J. Environ. Sci.* **2021**, *104*, 84–101. [CrossRef]

50. Plaza, M.G.; Martínez, S.; Rubiera, F. CO_2 Capture, Use, and Storage in the Cement Industry: State of the Art and Expectations. *Energies* **2020**, *13*, 5692. [CrossRef]

51. MacKenzie, A.; Granatstein, D.L.; Anthony, E.J.; Abanades, J.C. Economics of CO_2 Capture Using the Calcium Cycle with a Pressurized Fluidized Bed Combustor. *Energy Fuels* **2007**, *21*, 920–926. [CrossRef]

52. Geng, Y.-q.; Guo, Y.-x.; Fan, B.; Cheng, F.-q.; Cheng, H.-g. Research progress of calcium-based adsorbents for CO_2 capture and anti-sintering modification. *J. Fuel Chem. Technol.* **2021**, *49*, 998–1013. [CrossRef]

53. Hu, Y.; Lu, H.; Liu, W.; Yang, Y.; Li, H. Incorporation of CaO into inert supports for enhanced CO_2 capture: A review. *Chem. Eng. J.* **2020**, *396*, 125253. [CrossRef]

54. Krödel, M.; Landuyt, A.; Abdala, P.M.; Müller, C.R. Mechanistic Understanding of CaO-Based Sorbents for High-Temperature CO_2 Capture: Advanced Characterization and Prospects. *ChemSusChem* **2020**, *13*, 6259–6272. [CrossRef] [PubMed]

55. Sun, H.; Wu, C.; Shen, B.; Zhang, X.; Zhang, Y.; Huang, J. Progress in the development and application of CaO-based adsorbents for CO_2 capture—A review. *Mater. Today Sustain.* **2018**; 1–2, 1–27. [CrossRef]

56. Lu, H.; Khan, A.; Pratsinis, S.E.; Smirniotis, P.G. Flame-Made Durable Doped-CaO Nanosorbents for CO_2 Capture. *Energy Fuels* **2009**, *23*, 1093–1100. [CrossRef]

57. Li, L.; King, D.L.; Nie, Z.; Howard, C. Magnesia-Stabilized Calcium Oxide Absorbents with Improved Durability for High Temperature CO_2 Capture. *Ind. Eng. Chem. Res.* **2009**, *48*, 10604–10613. [CrossRef]

58. Lysikov, A.I.; Salanov, A.N.; Okunev, A.G. Change of CO_2 Carrying Capacity of CaO in Isothermal Recarbonation—Decomposition Cycles. *Ind. Eng. Chem. Res.* **2007**, *46*, 4633–4638. [CrossRef]

59. Grasa, G.S.; Abanades, J.C. CO_2 Capture Capacity of CaO in Long Series of Carbonation/Calcination Cycles. *Ind. Eng. Chem. Res.* **2006**, *45*, 8846–8851. [CrossRef]

60. Kierzkowska, A.M.; Pacciani, R.; Müller, C.R. CaO-Based CO_2 Sorbents: From Fundamentals to the Development of New, Highly Effective Materials. *ChemSusChem* **2013**, *6*, 1130–1148. [CrossRef]

61. Jing, J.-Y.; Li, T.-Y.; Zhang, X.-W.; Wang, S.-D.; Feng, J.; Turmel, W.A.; Li, W.-Y. Enhanced CO_2 sorption performance of $CaO/Ca_3Al_2O_6$ sorbents and its sintering-resistance mechanism. *Appl. Energy* **2017**, *199*, 225–233. [CrossRef]

62. Zhou, Z.; Qi, Y.; Xie, M.; Cheng, Z.; Yuan, W. Synthesis of CaO-based sorbents through incorporation of alumina/aluminate and their CO_2 capture performance. *Chem. Eng. Sci.* **2012**, *74*, 172–180. [CrossRef]

63. Han, R.; Gao, J.; Wei, S.; Su, Y.; Qin, Y. Development of highly effective $CaO@Al_2O_3$ with hierarchical architecture CO_2 sorbents via a scalable limited-space chemical vapor deposition technique. *J. Mater. Chem. A* **2018**, *6*, 3462–3470. [CrossRef]

64. Wang, N.; Feng, Y.; Liu, L.; Guo, X. Effects of preparation methods on the structure and property of Al-stabilized CaO-based sorbents for CO_2 capture. *Fuel Process. Technol.* **2018**, *173*, 276–284. [CrossRef]

65. López, J.M.; Grasa, G.; Murillo, R. Evaluation of the effect of inert support on the carbonation reaction of synthetic CaO-based CO_2 sorbents. *Chem. Eng. J.* **2018**, *350*, 559–572. [CrossRef]

66. Yu, Y.S.; Liu, W.Q.; An, H.; Yang, F.S.; Wang, G.X.; Feng, B.; Zhang, Z.X.; Rudolph, V. Modeling of the carbonation behavior of a calcium based sorbent for CO_2 capture. *Int. J. Greenh. Gas Control* **2012**, *10*, 510–519. [CrossRef]

67. Kurlov, A.; Broda, M.; Hosseini, D.; Mitchell, S.J.; Pérez-Ramírez, J.; Müller, C.R. Mechanochemically Activated, Calcium Oxide-Based, Magnesium Oxide-Stabilized Carbon Dioxide Sorbents. *ChemSusChem* **2016**, *9*, 2380–2390. [CrossRef] [PubMed]

68. Zhang, X.; Li, Z.; Peng, Y.; Su, W.; Sun, X.; Li, J. Investigation on a novel $CaO–Y_2O_3$ sorbent for efficient CO_2 mitigation. *Chem. Eng. J.* **2014**, *243*, 297–304. [CrossRef]

69. Naeem, M.A.; Armutlulu, A.; Imtiaz, Q.; Müller, C.R. CaO-Based CO_2 Sorbents Effectively Stabilized by Metal Oxides. *ChemPhysChem* **2017**, *18*, 3280–3285. [CrossRef]

70. Kim, S.M.; Kierzkowska, A.M.; Broda, M.; Müller, C.R. Sol-gel synthesis of $MgAl_2O_4$-stabilized CaO for CO_2 capture. *Energy Procedia* **2017**, *114*, 220–229. [CrossRef]

71. Koirala, R.; Reddy, G.K.; Smirniotis, P.G. Single nozzle flame-made highly durable metal doped Ca-based sorbents for CO_2 capture at high temperature. *Energy Fuels* **2012**, *26*, 3103–3109. [CrossRef]

72. Lu, H.; Smirniotis, P.G. Calcium oxide doped sorbents for CO_2 uptake in the presence of SO_2 at high temperatures. *Ind. Eng. Chem. Res.* **2009**, *48*, 5454–5459. [CrossRef]

73. De, S.; Dokania, A.; Ramirez, A.; Gascon, J. Advances in the Design of Heterogeneous Catalysts and Thermocatalytic Processes for CO_2 Utilization. *ACS Catal.* **2020**, *10*, 14147–14185. [CrossRef]

74. Younas, M.; Loong Kong, L.; Bashir, M.J.K.; Nadeem, H.; Shehzad, A.; Sethupathi, S. Recent Advancements, Fundamental Challenges, and Opportunities in Catalytic Methanation of CO_2. *Energy Fuels* **2016**, *30*, 8815–8831. [CrossRef]

75. Aziz, M.A.A.; Jalil, A.A.; Triwahyono, S.; Ahmad, A. CO_2 methanation over heterogeneous catalysts: Recent progress and future prospects. *Green Chem.* **2015**, *17*, 2647–2663. [CrossRef]

76. Fan, W.K.; Tahir, M. Recent trends in developments of active metals and heterogenous materials for catalytic CO_2 hydrogenation to renewable methane: A review. *J. Environ. Chem. Eng.* **2021**, *9*, 105460. [CrossRef]

77. Chen, X.; Chen, Y.; Song, C.; Ji, P.; Wang, N.; Wang, W.; Cui, L. Recent Advances in Supported Metal Catalysts and Oxide Catalysts for the Reverse Water-Gas Shift Reaction. *Front. Chem.* **2020**, *8*, 709. [CrossRef]

78. Daza, Y.A.; Kuhn, J.N. CO_2 conversion by reverse water gas shift catalysis: Comparison of catalysts, mechanisms and their consequences for CO_2 conversion to liquid fuels. *RSC Adv.* **2016**, *6*, 49675–49691. [CrossRef]

79. Nielsen, D.U.; Hu, X.-M.; Daasbjerg, K.; Skrydstrup, T. Chemically and electrochemically catalysed conversion of CO_2 to CO with follow-up utilization to value-added chemicals. *Nat. Catal.* **2018**, *1*, 244–254. [CrossRef]

80. le Saché, E.; Reina, T.R. Analysis of Dry Reforming as direct route for gas phase CO_2 conversion. The past, the present and future of catalytic DRM technologies. *Prog. Energy Combust. Sci.* **2022**, *89*, 100970. [CrossRef]

81. Jang, W.-J.; Shim, J.-O.; Kim, H.-M.; Yoo, S.-Y.; Roh, H.-S. A review on dry reforming of methane in aspect of catalytic properties. *Catal. Today* **2019**, *324*, 15–26. [CrossRef]

82. Singh, R.; Dhir, A.; Mohapatra, S.K.; Mahla, S.K. Dry reforming of methane using various catalysts in the process: Review. *Biomass Convers. Biorefinery* **2020**, *10*, 567–587. [CrossRef]

83. Dieterich, V.; Buttler, A.; Hanel, A.; Spliethoff, H.; Fendt, S. Power-to-liquid via synthesis of methanol, DME or Fischer–Tropsch-fuels: A review. *Energy Environ. Sci.* **2020**, *13*, 3207–3252. [CrossRef]

84. Atsbha, T.A.; Yoon, T.; Seongho, P.; Lee, C.-J. A review on the catalytic conversion of CO_2 using H_2 for synthesis of CO, methanol, and hydrocarbons. *J. CO_2 Util.* **2021**, *44*, 101413. [CrossRef]

85. Yang, H.; Zhang, C.; Gao, P.; Wang, H.; Li, X.; Zhong, L.; Wei, W.; Sun, Y. A review of the catalytic hydrogenation of carbon dioxide into value-added hydrocarbons. *Catal. Sci. Technol.* **2017**, *7*, 4580–4598. [CrossRef]

86. Tang, R.; Zhu, Z.; Li, C.; Xiao, M.; Wu, Z.; Zhang, D.; Zhang, C.; Xiao, Y.; Chu, M.; Genest, A.; et al. Ru-Catalyzed Reverse Water Gas Shift Reaction with Near-Unity Selectivity and Superior Stability. *ACS Mater. Lett.* **2021**, *3*, 1652–1659. [CrossRef] [PubMed]

87. Tawalbeh, M.; Javed, R.M.N.; Al-Othman, A.; Almomani, F.; Ajith, S. Unlocking the potential of CO_2 hydrogenation into valuable products using noble metal catalysts: A comprehensive review. *Environ. Technol. Innov.* **2023**, *31*, 103217. [CrossRef]

88. Pakhare, D.; Spivey, J. A review of dry (CO_2) reforming of methane over noble metal catalysts. *Chem. Soc. Rev.* **2014**, *43*, 7813–7837. [CrossRef]

89. Ojelade, O.A.; Zaman, S.F. A Review on Pd Based Catalysts for CO_2 Hydrogenation to Methanol: In-Depth Activity and DRIFTS Mechanistic Study. *Catal. Surv. Asia* **2020**, *24*, 11–37. [CrossRef]

90. Kattel, S.; Liu, P.; Chen, J.G.G. Tuning Selectivity of CO_2 Hydrogenation Reactions at the Metal/Oxide Interface. *J. Am. Chem. Soc.* **2017**, *139*, 9739–9754. [CrossRef]

91. Al-Mamoori, A.; Rownaghi, A.A.; Rezaei, F. Combined capture and utilization of CO_2 for syngas production over dual-function materials. *ACS Sustain. Chem. Eng.* **2018**, *6*, 13551–13561. [CrossRef]

92. Arellano-Treviño, M.A.; He, Z.; Libby, M.C.; Farrauto, R.J. Catalysts and adsorbents for CO_2 capture and conversion with dual function materials: Limitations of Ni-containing DFMs for flue gas applications. *J. CO_2 Util.* **2019**, *31*, 143–151. [CrossRef]

93. Duyar, M.S.; Wang, S.; Arellano-Treviño, M.A.; Farrauto, R.J. CO_2 utilization with a novel dual function material (DFM) for capture and catalytic conversion to synthetic natural gas: An update. *J. CO_2 Util.* **2016**, *15*, 65–71. [CrossRef]

94. Proaño, L.; Arellano-Treviño, M.A.; Farrauto, R.J.; Figueredo, M.; Jeong-Potter, C.; Cobo, M. Mechanistic assessment of dual function materials, composed of Ru-Ni, Na_2O/Al_2O_3 and Pt-Ni, Na_2O/Al_2O_3, for CO_2 capture and methanation by in-situ DRIFTS. *Appl. Surf. Sci.* **2020**, *533*, 147469. [CrossRef]

95. Proaño, L.; Tello, E.; Arellano-Trevino, M.A.; Wang, S.; Farrauto, R.J.; Cobo, M. In-situ DRIFTS study of two-step CO_2 capture and catalytic methanation over Ru,"Na_2O"/Al_2O_3 Dual Functional Material. *Appl. Surf. Sci.* **2019**, *479*, 25–30. [CrossRef]

96. Bermejo-López, A.; Pereda-Ayo, B.; González-Marcos, J.A.; González-Velasco, J.R. Mechanism of the CO_2 storage and in situ hydrogenation to CH4. Temperature and adsorbent loading effects over Ru-CaO/Al_2O_3 and Ru-Na_2CO_3/Al_2O_3 catalysts. *Appl. Catal. B Environ.* **2019**, *256*, 117845. [CrossRef]

97. Bobadilla, L.F.; Riesco-García, J.M.; Penelás-Pérez, G.; Urakawa, A. Enabling continuous capture and catalytic conversion of flue gas CO_2 to syngas in one process. *J. CO_2 Util.* **2016**, *14*, 106–111. [CrossRef]

98. Wang, S.; Farrauto, R.J.; Karp, S.; Jeon, J.H.; Schrunk, E.T. Parametric, cyclic aging and characterization studies for CO_2 capture from flue gas and catalytic conversion to synthetic natural gas using a dual functional material (DFM). *J. CO_2 Util.* **2018**, *27*, 390–397. [CrossRef]

99. Hu, L.; Urakawa, A. Continuous CO_2 capture and reduction in one process: CO_2 methanation over unpromoted and promoted Ni/ZrO2. *J. CO_2 Util.* **2018**, *25*, 323–329. [CrossRef]

100. Arellano-Treviño, M.A.; Kanani, N.; Jeong-Potter, C.W.; Farrauto, R.J. Bimetallic catalysts for CO_2 capture and hydrogenation at simulated flue gas conditions. *Chem. Eng. J.* **2019**, *375*, 121953. [CrossRef]

101. Bermejo-López, A.; Pereda-Ayo, B.; González-Marcos, J.A.; González-Velasco, J.R. Ni loading effects on dual function materials for capture and in-situ conversion of CO_2 to CH_4 using CaO or Na_2CO_3. *J. CO_2 Util.* **2019**, *34*, 576–587. [CrossRef]

102. Sietsma, J.R.A.; Jos van Dillen, A.; de Jongh, P.E.; de Jong, K.P. Application of ordered mesoporous materials as model supports to study catalyst preparation by impregnation and drying. In *Studies in Surface Science and Catalysis*; Gaigneaux, E.M., Devillers, M., De Vos, D.E., Hermans, S., Jacobs, P.A., Martens, J.A., Ruiz, P., Eds.; Elsevier: Amsterdam, The Netherlands, 2006; Volume 162, pp. 95–102.

103. Lu, Z.; Kunisch, J.; Gan, Z.; Bunian, M.; Wu, T.; Lei, Y. Gold catalysts synthesized using a modified incipient wetness impregnation method for propylene epoxidation. *ChemCatChem* **2020**, *12*, 5993–5999. [CrossRef]

104. Danks, A.E.; Hall, S.R.; Schnepp, Z. The evolution of 'sol–gel' chemistry as a technique for materials synthesis. *Mater. Horiz.* **2016**, *3*, 91–112. [CrossRef]

105. Wu, J.; Zheng, Y.; Fu, J.; Guo, Y.; Yu, J.; Chu, J.; Huang, P.; Zhao, C. Synthetic Ni–CaO–CeO_2 dual function materials for integrated CO_2 capture and conversion via reverse water–gas shift reaction. *Sep. Purif. Technol.* **2023**, *317*, 123916. [CrossRef]

106. Pechini, M.P. Method of Preparing Lead and Alkaline Earth Titanates and Niobates and Coating Method Using the Same to Form a Capacitor. U.S. Patent US304434A, 11 July 1967.

107. Sunde, T.O.L.; Grande, T.; Einarsrud, M.-A. Modified pechini synthesis of oxide powders and thin films. In *Handbook of Sol-Gel Science and Technology*; Springer: Berlin/Heidelberg, Germany, 2016.

108. Radfarnia, H.R.; Sayari, A. A highly efficient CaO-based CO_2 sorbent prepared by a citrate-assisted sol–gel technique. *Chem. Eng. J.* **2015**, *262*, 913–920. [CrossRef]

109. Jin, S.; Bang, G.; Liu, L.; Lee, C.-H. Synthesis of mesoporous $MgO–CeO_2$ composites with enhanced CO_2 capture rate via controlled combustion. *Microporous Mesoporous Mater.* **2019**, *288*, 109587. [CrossRef]

110. Ma, X.; Li, X.; Cui, H.; Zhang, W.; Cheng, Z.; Zhou, Z. Metal oxide-doped Ni/CaO dual-function materials for integrated CO_2 capture and conversion: Performance and mechanism. *AIChE J.* **2023**, *69*, e17520. [CrossRef]

111. Sun, H.; Wang, J.; Zhao, J.; Shen, B.; Shi, J.; Huang, J.; Wu, C. Dual functional catalytic materials of Ni over Ce-modified CaO sorbents for integrated CO_2 capture and conversion. *Appl. Catal. B Environ.* **2019**, *244*, 63–75. [CrossRef]

112. Zhang, X.; Liu, W.; Peng, P.; Zhang, Z.; Du, Q.; Shi, J.; Deng, L. A dual functional sorbent/catalyst material for in-situ CO_2 capture and conversion to ethylene production. *Fuel* **2023**, *351*, 128701. [CrossRef]

113. Machida, M.; Uto, M.; Kijima, T. Preparation of large surface area $MnOx-ZrO_2$ for sorptive NOx removal. In *Studies in Surface Science and Catalysis*; Elsevier: Amsterdam, The Netherlands, 2000; Volume 143, pp. 855–862.

114. Virji, M.; Stefaniak, A. *A Review of Engineered Nanomaterial Manufacturing Processes and Associated Exposures*; Elsevier: Amsterdam, The Netherlands, 2014.

115. Li, C.; Li, M.; van Veen, A.C. Synthesis of Nano-Catalysts in Flow Conditions Using Millimixers. In *Advanced Nanomaterials for Catalysis and Energy*; Elsevier: Amsterdam, The Netherlands, 2019; pp. 1–28.

116. Molina-Ramírez, S.; Cortés-Reyes, M.; Herrera, C.; Larrubia, M.; Alemany, L. CO_2-SR Cyclic Technology: CO_2 Storage and in situ Regeneration with CH4 over a new dual function NiBa unsupported catalyst. *J. CO_2 Util.* **2020**, *40*, 101201. [CrossRef]

117. Karami, D.; Mahinpey, N. Study of Al_2O_3 addition to synthetic Ca-based sorbents for CO_2 sorption capacity and stability in cyclic operations. *Can. J. Chem. Eng.* **2015**, *93*, 102–110. [CrossRef]

118. Huang, P.; Chu, J.; Fu, J.; Yu, J.; Li, S.; Guo, Y.; Zhao, C.; Liu, J. Influence of reduction conditions on the structure-activity relationships of $NaNO_3$-promoted Ni/MgO dual function materials for integrated CO_2 capture and methanation. *Chem. Eng. J.* **2023**, *467*, 143431. [CrossRef]

119. Wegener Kofoed, M.V.; Jensen, M.B.; Mørck Ottosen, L.D. Chapter 12—Biological upgrading of biogas through CO_2 conversion to CH4. In *Emerging Technologies and Biological Systems for Biogas Upgrading*; Aryal, N., Mørck Ottosen, L.D., Wegener Kofoed, M.V., Pant, D., Eds.; Academic Press: Cambridge, MA, USA, 2021; pp. 321–362. [CrossRef]

120. Seemann, M.; Thunman, H. 9—Methane synthesis. In *Substitute Natural Gas from Waste*; Materazzi, M., Foscolo, P.U., Eds.; Academic Press: Cambridge, MA, USA, 2019; pp. 221–243. [CrossRef]

121. Sun, H.; Zhang, Y.; Guan, S.; Huang, J.; Wu, C. Direct and highly selective conversion of captured CO_2 into methane through integrated carbon capture and utilization over dual functional materials. *J. CO_2 Util.* **2020**, *38*, 262–272. [CrossRef]

122. Kosaka, F.; Liu, Y.; Chen, S.-Y.; Mochizuki, T.; Takagi, H.; Urakawa, A.; Kuramoto, K. Enhanced Activity of Integrated CO_2 Capture and Reduction to CH4 under Pressurized Conditions toward Atmospheric CO_2 Utilization. *ACS Sustain. Chem. Eng.* **2021**, *9*, 3452–3463. [CrossRef]

123. Porta, A.; Matarrese, R.; Visconti, C.G.; Castoldi, L.; Lietti, L. Storage Material Effects on the Performance of Ru-Based CO_2 Capture and Methanation Dual Functioning Materials. *Ind. Eng. Chem. Res.* **2021**, *60*, 6706–6718. [CrossRef]

124. Jeong-Potter, C.; Farrauto, R. Feasibility Study of Combining Direct Air Capture of CO_2 and Methanation at Isothermal Conditions with Dual Function Materials. *Appl. Catal. B Environ.* **2021**, *282*, 119416. [CrossRef]

125. Sun, H.; Wang, Y.; Xu, S.; Osman, A.I.; Stenning, G.; Han, J.; Sun, S.; Rooney, D.; Williams, P.T.; Wang, F.; et al. Understanding the interaction between active sites and sorbents during the integrated carbon capture and utilization process. *Fuel* **2021**, *286*, 119308. [CrossRef]

126. Porta, A.; Visconti, C.G.; Castoldi, L.; Matarrese, R.; Jeong-Potter, C.; Farrauto, R.; Lietti, L. Ru-Ba synergistic effect in dual functioning materials for cyclic CO_2 capture and methanation. *Appl. Catal. B Environ.* **2021**, *283*, 119654. [CrossRef]

127. Cimino, S.; Russo, R.; Lisi, L. Insights into the cyclic CO_2 capture and catalytic methanation over highly performing $Li-Ru/Al_2O_3$ dual function materials. *Chem. Eng. J.* **2022**, *428*, 131275. [CrossRef]

128. Onrubia-Calvo, J.A.; Bermejo-López, A.; Pérez-Vázquez, S.; Pereda-Ayo, B.; González-Marcos, J.A.; González-Velasco, J.R. Applicability of $LaNiO_3$-derived catalysts as dual function materials for CO_2 capture and in-situ conversion to methane. *Fuel* **2022**, *320*, 123842. [CrossRef]

129. Sun, Z.; Shao, B.; Zhang, Y.; Gao, Z.; Wang, M.; Liu, H.; Hu, J. Integrated CO_2 capture and methanation from the intermediate-temperature flue gas on dual functional hybrids of AMS/CaMgO||NixCoy. *Sep. Purif. Technol.* **2023**, *307*, 122680. [CrossRef]

130. Faria, A.C.; Trujillano, R.; Rives, V.; Miguel, C.V.; Rodrigues, A.E.; Madeira, L.M. Cyclic operation of CO_2 capture and conversion into methane on Ni-hydrotalcite based dual function materials (DFMs). *J. CO_2 Util.* **2023**, *72*, 102476. [CrossRef]

131. Onrubia-Calvo, J.A.; Bermejo-López, A.; Pereda-Ayo, B.; González-Marcos, J.A.; González-Velasco, J.R. Ca doping effect on the performance of $La1-xCaxNiO_3/CeO_2$-derived dual function materials for CO_2 capture and hydrogenation to methane. *Appl. Catal. B Environ.* **2023**, *321*, 122045. [CrossRef]

132. Sakai, M.; Imagawa, H.; Baba, N. Layered-double-hydroxide-based Ni catalyst for CO_2 capture and methanation. *Appl. Catal. A Gen.* **2022**, *647*, 118904. [CrossRef]

133. Catarina Faria, A.; Miguel, C.V.; Ferreira, A.F.P.; Rodrigues, A.E.; Madeira, L.M. CO_2 capture and conversion to methane with Ni-substituted hydrotalcite dual function extrudates. *Chem. Eng. J.* **2023**, *476*, 146539. [CrossRef]
134. Wang, X.; Economides, M. CHAPTER 7—Gas-To-Liquids (GTL). In *Advanced Natural Gas Engineering*; Wang, X., Economides, M., Eds.; Gulf Publishing Company: Houston, TX, USA, 2009; pp. 243–287. [CrossRef]
135. Cimino, S.; Boccia, F.; Lisi, L. Effect of alkali promoters (Li, Na, K) on the performance of Ru/Al_2O_3 catalysts for CO_2 capture and hydrogenation to methane. *J. CO_2 Util.* **2020**, *37*, 195–203. [CrossRef]
136. Jeong-Potter, C.; Zangiabadi, A.; Farrauto, R. Extended aging of Ru-Ni, Na_2O/Al_2O_3 dual function materials (DFM) for combined capture and subsequent catalytic methanation of CO_2 from power plant flue gas. *Fuel* **2022**, *328*, 125283. [CrossRef]
137. Bermejo-López, A.; Pereda-Ayo, B.; González-Marcos, J.A.; González-Velasco, J.R. Simulation-based optimization of cycle timing for CO_2 capture and hydrogenation with dual function catalyst. *Catal. Today* **2022**, *394–396*, 314–324. [CrossRef]
138. Bermejo-López, A.; Pereda-Ayo, B.; González-Marcos, J.A.; González-Velasco, J.R. Modeling the CO_2 capture and in situ conversion to CH4 on dual function $Ru-Na_2CO_3/Al_2O_3$ catalyst. *J. CO_2 Util.* **2020**, *42*, 101351. [CrossRef]
139. Garbarino, G.; Bellotti, D.; Finocchio, E.; Magistri, L.; Busca, G. Methanation of carbon dioxide on Ru/Al_2O_3: Catalytic activity and infrared study. *Catal. Today* **2016**, *277*, 21–28. [CrossRef]
140. Dreyer, J.A.H.; Li, P.; Zhang, L.; Beh, G.K.; Zhang, R.; Sit, P.H.L.; Teoh, W.Y. Influence of the oxide support reducibility on the CO_2 methanation over Ru-based catalysts. *Appl. Catal. B Environ.* **2017**, *219*, 715–726. [CrossRef]
141. González-Castaño, M.; Dorneanu, B.; Arellano-García, H. The reverse water gas shift reaction: A process systems engineering perspective. *React. Chem. Eng.* **2021**, *6*, 954–976. [CrossRef]
142. Shao, B.; Hu, G.; Alkebsi, K.A.M.; Ye, G.; Lin, X.; Du, W.; Hu, J.; Wang, M.; Liu, H.; Qian, F. Heterojunction-redox catalysts of FexCoyMg10CaO for high-temperature CO_2 capture and in situ conversion in the context of green manufacturing. *Energy Environ. Sci.* **2021**, *14*, 2291–2301. [CrossRef]
143. Sun, S.; He, S.; Wu, C. Ni promoted Fe-CaO dual functional materials for calcium chemical dual looping. *Chem. Eng. J.* **2022**, *441*, 135752. [CrossRef]
144. Sasayama, T.; Kosaka, F.; Liu, Y.Y.; Yamaguchi, T.; Chen, S.Y.; Mochizuki, T.; Urakawa, A.; Kuramoto, K. Integrated CO_2 capture and selective conversion to syngas using transition-metal-free Na/Al_2O_3 dual-function material. *J. CO_2 Util.* **2022**, *60*, 102049. [CrossRef]
145. Li, L.; Miyazaki, S.; Yasumura, S.; Ting, K.W.; Toyao, T.; Maeno, Z.; Shimizu, K.-I. Continuous CO_2 Capture and Selective Hydrogenation to CO over Na-Promoted Pt Nanoparticles on Al_2O_3. *ACS Catal.* **2022**, *12*, 2639–2650. [CrossRef]
146. Guo, Y.; Wang, G.; Yu, J.; Huang, P.; Sun, J.; Wang, R.; Wang, T.; Zhao, C. Tailoring the performance of Ni-CaO dual function materials for integrated CO_2 capture and conversion by doping transition metal oxides. *Sep. Purif. Technol.* **2023**, *305*, 122455. [CrossRef]
147. Sun, S.; Zhang, C.; Guan, S.; Xu, S.; Williams, P.T.; Wu, C. Ni/support-CaO bifunctional combined materials for integrated CO_2 capture and reverse water-gas shift reaction: Influence of different supports. *Sep. Purif. Technol.* **2022**, *298*, 121604. [CrossRef]
148. Bao, Z.; Yu, F. Chapter Two—Catalytic Conversion of Biogas to Syngas via Dry Reforming Process. In *Advances in Bioenergy*; Li, Y., Ge, X., Eds.; Elsevier: Amsterdam, The Netherlands, 2018; Volume 3, pp. 43–76.
149. Hu, J.; Hongmanorom, P.; Galvita, V.V.; Li, Z.; Kawi, S. Bifunctional Ni-Ca based material for integrated CO_2 capture and conversion via calcium-looping dry reforming. *Appl. Catal. B Environ.* **2021**, *284*, 119734. [CrossRef]
150. Kim, S.M.; Abdala, P.M.; Broda, M.; Hosseini, D.; Copéret, C.; Müller, C. Integrated CO_2 Capture and Conversion as an Efficient Process for Fuels from Greenhouse Gases. *ACS Catal.* **2018**, *8*, 2815–2823. [CrossRef]
151. Zhao, Y.; Li, Y.; Jin, B.; Liang, Z. Layered double hydroxide derived bifunctional Ca-Fe-Mg material for integrated CO_2 capture and utilization via chemical looping strategy. *Chem. Eng. J.* **2022**, *431*, 133826. [CrossRef]
152. Tian, S.; Yan, F.; Zhang, Z.; Jiang, J. Calcium-looping reforming of methane realizes in situ CO_2 utilization with improved energy efficiency. *Sci. Adv.* **2019**, *5*, eaav5077. [CrossRef]
153. Jo, S.B.; Woo, J.H.; Lee, J.H.; Kim, T.Y.; Kang, H.I.; Lee, S.C.; Kim, J.C. CO_2 green technologies in CO_2 capture and direct utilization processes: Methanation, reverse water-gas shift, and dry reforming of methane. *Sustain. Energy Fuels* **2020**, *4*, 5543–5549. [CrossRef]
154. Hu, J.; Hongmanorom, P.; Chirawatkul, P.; Kawi, S. Efficient integration of CO_2 capture and conversion over a Ni supported CeO_2-modified CaO microsphere at moderate temperature. *Chem. Eng. J.* **2021**, *426*, 130864. [CrossRef]
155. Merkouri, L.-P.; Ramirez Reina, T.; Duyar, M.S. Feasibility of switchable dual function materials as a flexible technology for CO_2 capture and utilisation and evidence of passive direct air capture. *Nanoscale* **2022**, *14*, 12620–12637. [CrossRef] [PubMed]
156. Law, Z.X.; Pan, Y.-T.; Tsai, D.-H. Calcium looping of CO_2 capture coupled to syngas production using Ni-CaO-based dual functional material. *Fuel* **2022**, *328*, 125202. [CrossRef]
157. Hussain, I.; Tanimu, G.; Ahmed, S.; Aniz, C.U.; Alasiri, H.; Alhooshani, K. A review of the indispensable role of oxygen vacancies for enhanced CO_2 methanation activity over CeO_2-based catalysts: Uncovering, influencing, and tuning strategies. *Int. J. Hydrogen Energy* **2023**, *48*, 24663–24696. [CrossRef]

 *inorganics*

Article

ZnO–Doped CaO Binary Core–Shell Catalysts for Biodiesel Production via Mexican Palm Oil Transesterification

M. G. Arenas-Quevedo [1], M. E. Manríquez [1], J. A. Wang [1,*], O. Elizalde-Solís [1], J. González-García [2], A. Zúñiga-Moreno [1] and L. F. Chen [1]

[1] Escuela Superior de Ingeniería Química e Industrias Extractivas, Instituto Politécnico Nacional, Col. Zacatenco, Ciudad de México 07738, Mexico; 5167chaloarenas@gmail.com (M.G.A.-Q.); mmanriquez@ipn.mx (M.E.M.); oelizalde@ipn.mx (O.E.-S.); azunigam@ipn.mx (A.Z.-M.); lchen@ipn.mx (L.F.C.)

[2] TecNM-Tecnológico de Estudios Superiores de Coacalco, 16 de Septiembre 54, Col. Cabecera municipal, Estado de México 55700, Mexico; julio_gonzalez.iqu@tesco.edu.mx

* Correspondence: jwang@ipn.mx

Abstract: This work investigates biodiesel production via transesterification of Mexican palm oil with methanol catalyzed by binary solid base core–shell catalysts with improved catalytic stability. A series of CaO–ZnO mixed solids were prepared using an inexpensive co–precipitation method by varying ZnO content from 5 to 20 mol%. Several factors, such as surface basicity, ZnO content, phase compositions, and thermal treatment of the catalysts, were all proven to be crucial for the production of biodiesel with good quality. Thermal treatment could effectively remove the surface adsorbed water and impurities and improved the catalytic activity. The addition of ZnO to CaO significantly enhanced the catalysts' stability; however, it led to lower surface basicity and slightly diminished catalytic activity. ZnO doping inhibited the formation of surface $Ca(OH)_2$ and promoted the formation of Ca–Zn–O or $CaZn_2(OH)_6$ phase as the core and a surface $CaCO_3$ shell, which effectively decreased Ca^{2+} leaching by approximately 74% in methanol and 65% in a methanol–glycerol (4:1) mixture. A combined method of separation and purification for obtaining clean biodiesel with high quality was proposed. The biodiesel obtained under the control conditions exhibited properties which satisfied the corresponding standards well.

Keywords: transesterification; CaO–ZnO; biodiesel; palm oil; core–shell heterostructure

Citation: Arenas-Quevedo, M.G.; Manríquez, M.E.; Wang, J.A.; Elizalde-Solís, O.; González-García, J.; Zúñiga-Moreno, A.; Chen, L.F. ZnO–Doped CaO Binary Core–Shell Catalysts for Biodiesel Production via Mexican Palm Oil Transesterification. *Inorganics* **2024**, *12*, 51. https://doi.org/10.3390/inorganics12020051

Academic Editors: Roberto Nisticò, Torben R. Jensen, Luciano Carlos, Hicham Idriss and Eleonora Aneggi

Received: 26 October 2023
Revised: 26 December 2023
Accepted: 8 January 2024
Published: 3 February 2024

1. Introduction

Biodiesels, defined as the monoalkyl esters of fatty acids, can be produced by transesterification of renewable sources such as vegetable oils containing triglycerides of fatty acids via catalysis techniques. Usually, biodiesels are produced using homogeneous base catalysts such as NaOH and KOH. However, purification of the ester phase and removal of the liquid base catalyst from the products are a major problem [1,2]. Utilization of heterogeneous catalysts has attracted great attention for biodiesel production as it could solve most of the drawbacks encountered in the conventional homogeneous process [3,4]. For example, a heterogeneous catalyst can be rapidly separated from the reaction mixture by filtration, easily regenerated, and has a less corrosive character. Therefore, biodiesel production with a heterogeneous catalyst is an economical and environmentally friendly process [5].

In the process of biodiesel production, the purity of feedstocks and an effective catalyst are two key factors for obtaining a high-quality product. Water in the oil feedstocks usually poisons the catalyst [6,7]. Ma et al. claimed that during the transesterification reaction, the presence of water generates an even greater negative effect than free fatty acids [8]. Therefore, the feedstocks should be pretreated to minimize the water amount prior to the transesterification process. Some researchers have focused on the use of earth oxides like

CaO [9,10] and SrO [11] as solid base catalysts for biodiesel synthesis. However, it was reported that when CaO was used, catalyst leaching in methanol took place [12].

There are two main methods used for improving the catalytic activity and stability of solid base catalysts. One is to disperse a solid base like CaO on a proper support such as mesoporous solids SBA–15 [13]; another method is to use binary oxides as catalysts by doping the base solid with another metal oxide [14–19]. This kind of catalyst has a long life time, suggesting an advantage of heterogeneous catalysts in the form of mixed oxides. Dias et al. claimed that calcium–manganese oxide was active and stable for biofuel production [14]. Dai et al. reported a series of alkali metal–doped M_2ZrO_2 and $MNbO_3$ (X = Li, Na, K) catalysts for biodiesel production; both the acid and basic sites in the catalysts were proposed to take part in the formation of biodiesel [15,16]. Lithium–doped ZnO catalysts showed a conversion higher than 90% in soybean oil transesterification where the surface basicity of the catalysts significantly influenced the catalytic activity [17]. One successful example of a commercialized process for biodiesel production via vegetable oil transesterification was reported using a spinel oxide of Zn and Al solid catalyst in a two-consecutive-fixed-bed reactor system, where methyl esters reached 94.1% in the first reactor and 98.3% in the second reactor [18]. Rare earth oxide CeO_2–doped CaO was also tested for biodiesel production through transesterification of Pistacia chinensis oil and the catalysts could retain acceptable activity after 18 cycles of reaction [19]. Chen and coauthors reported that using $CaO–SiO_2$ catalysts, the reusability of the catalysts was obviously improved in comparison with CaO oxide alone in biodiesel production. The Ca^{2+} leaching content in the reaction media decreased as the Si content increased in the catalyst (0SiCa > 1SiCa > 2SiCa) [20]. It was reported that well-mixed CaO and MgO particles with a heterojunction structure synthesized via a wet ball milling process were used for biofuel production and the catalytic performance was improved relative to CaO nanoparticles [21]. Calcium–magnesium mixed oxide and dolomite containing CaO and MgO components as catalysts for transesterification of sunflower and soybean oil exhibited good activity and stability [22,23]. Woranuch et al. reported when Al_2O_3 was used as a binder ($CaO:Al_2O_3$ mass ratio 3.5:1), a 93.15% biodiesel yield was obtained, similar to that achieved using a CaO catalyst [24]. In the Ca–Al mixed oxides, Ca^{2+} leaching chiefly took place in the phase of CaO particles but not the mixed oxide, indicating that Al doping may improve stability [25]. Clearly, the surface basicity and the catalytic stability of CaO could be modified by doping with a second metal oxide, affecting the transesterification performance. The mechanism of the improvement of catalytic stability by doping is still an interesting research topic.

It is noted that a simple and economic technique for preparing catalysts is always at-tractive from the viewpoint of industrial applications. Keeping this in mind, the present work aims at designing heterogeneous base catalysts with low cost and enhanced catalytic stability for biodiesel production via Mexican palm oil transesterification with methanol. For this purpose, a series of ZnO–doped CaO solid catalysts were synthesized by a co-precipitation method due to their availability, low cost, strong basicity, and high catalytic activity. The ZnO content varied from 5 to 10, 15, and 20 mol% in the CaO–ZnO solids. Effects of ZnO content, surface basicity, and thermal treatment on the catalytic behaviors of the catalysts were investigated. For investigating the stability of ZnO–CaO catalysts, Ca^{2+} leaching in methanol or a methanol and glycerol mixture was measured. A mechanism towards a deep insight into the stability enhancement of ZnO–CaO catalysts was proposed. Some key physicochemical properties of the obtained biodiesels, like molecular weight, refractive index, density, and cetane number, were measured and all of these parameters satisfied the standards established for biodiesel well.

2. Results and Discussion

2.1. Phase Composition and XRD Analysis

XRD patterns of the pure CaO and the ZnO–doped CaO samples are shown in Figure 1. For the pure CaO sample (Figure 1a), the XRD peaks at 2θ approximately 32.2°, 37.3°, and

53.8° were indexed to (111), (200) and (220) planes of cubic CaO phase (JCPDS card file No.04-003-7161). The presence of Ca(OH)$_2$ phase was observed as evidenced by diffraction peaks at 2θ = 18.0°, 28.6°, 34.1°, 47.0°, and 50.8°. In addition, a very small amount of Ca–CO$_3$ was formed in the solid, probably resulting from the reaction of CaO with CO$_2$ in air during sample calcination. It is noted that the dried CaO solid possesses high capacity for adsorbing water, so surface Ca(OH)$_2$ phase was formed by CaO reacting with surface adsorbed water [26]. Therefore, in the pure CaO solid, two major phases, CaO and Ca(OH)$_2$, coexisted with a very small amount of surface calcium carbonate.

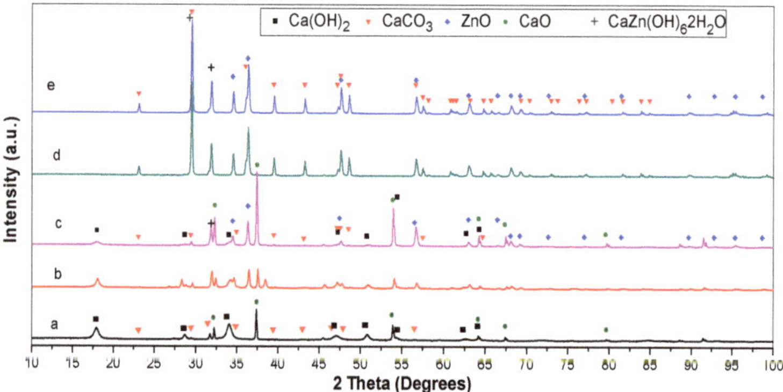

Figure 1. XRD patterns of CaO and CaO–ZnO samples with different ZnO content levels. (a): CaO; (b) CaO–ZnO–5%; (c) CaO–ZnO–10%; (d) CaO–ZnO–15%; (e) CaO–ZnO–20%.

In the ZnO–doped samples (Figure 1b–e), as the amount of ZnO in CaO–ZnO mixed oxide increased from 5 to 10 mol%, the intensity of the XRD peaks corresponding to ZnO with hexagonal structure (JCPDS card file No. 36-01-070-8072) increased, as evidenced by peaks at 2θ = 36.3°, 56.7°, 62.9°, and 68.0°. In these two samples, ZnO, CaO, and Ca(OH)$_2$, together with a trace amount of CaCO$_3$ and CaZn$_2$(OH)$_6$·2H$_2$O, were formed. When the ZnO content in the mixed sample was further increased to 15 mol% or greater, more CaO was converted into CaCO$_3$ as confirmed by the disappearance of the XRD peaks of CaO and a significant increase in intensity of the peaks related to CaCO$_3$ (Figure 1d,e). Moreover, Ca(OH)$_2$ phase concentration gradually reduced as the ZnO content increased from 5 to 10 wt%, and disappeared in the sample with higher ZnO content. In the meantime, CaZn$_2$(OH)$_6$ appeared in two samples: CaO–ZnO–15% and CaO–ZnO–20% (JCPDS card file No. 04-012-2175) [2,27,28]. Calcination may result in the departure of crystalline water from this phase, leading to the peak position slightly shifting to the low two–theta region. Formation of the phase CaZn$_2$(OH)$_6$ was due to the following reactions (1) and (2) [28]:

$$Ca^{2+} + 2Zn^{2+} + 8H_2O \rightarrow CaZn_2(OH)_6 \cdot 2H_2O + 6H^+ \tag{1}$$

$$CaZn_2(OH)_6 \cdot 2H_2O \rightarrow CaZn_2(OH)_6 + 2H_2O \tag{2}$$

Therefore, ZnO addition inhibited the formation of Ca(OH)$_2$ but generated ZnO and CaZn$_2$(OH)$_6$ phases. CaZn$_2$(OH)$_6$ may further dehydrate via two neighbored hydroxyls to release water, forming some oxygen defects (V_O^{2-}) in the structure. Moreover, a certain amount of CaCO$_3$ was formed due to the reaction between surface CO$_2$ and CaO during the calcination procedure. Therefore, the CaCO$_3$ phase serves as a shell of the catalyst.

2.2. FTIR Spectroscopy Analysis

The surface features of the catalysts were characterized with FTIR spectroscopy (Figure 2). For the pure CaO sample, a sharp IR band at 3625 cm^{-1} was assigned to stretching vibrations of hydroxyls ν(O–H), which is normally ascribed to the O–H bond in

Ca(OH)$_2$ [5]. The band at 420 cm^{-1} corresponded to the vibration of the Ca–O lattice bond. The characteristic bands of carbonate species were observed in the FTIR spectra. Usually, for CO$_3{}^{2-}$ species, four vibrational IR vibration modes can be observed with different band shapes and intensity: (i) the symmetric stretching, v_1[CO$_3$], at around 1080–1085 cm^{-1}; (ii) the out–Of–plane bend, v_2[CO$_3$], at around 700–720 cm^{-1}; (iii) the asymmetric stretching, v_3[CO$_3$], at around 1420–1580 cm^{-1}; and iv) the split in-plane bending vibrations, v_4[CO$_3$], at around 850–880 cm^{-1} [5,29,30]. The presence of carbonate species in all the samples was confirmed by the broad and strong doublet at 1420 and 1500 cm^{-1}. They should correspond to the asymmetric stretching vibrations of v_3[CO$_3$] species. A weak vibration of the v_1[CO$_3$] mode was observed at 1086 cm^{-1}, indicating that CaCO$_3$ was formed in the surface of the samples.

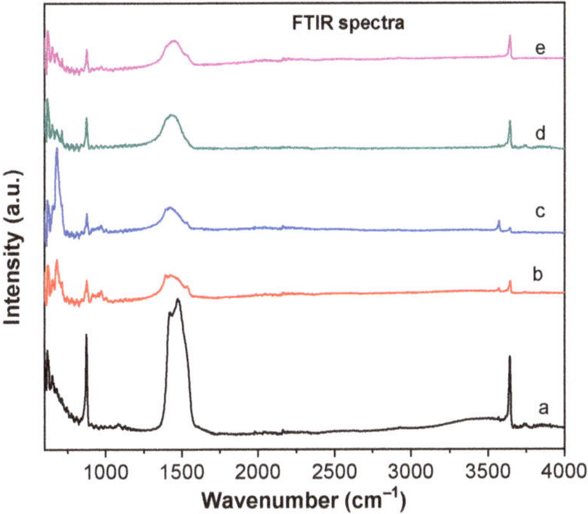

Figure 2. FTIR spectra of CaO and CaO–ZnO samples at different levels of ZnO content. (a) CaO–ZnO–0%; (b) CaO–ZnO–5%; (c) CaO–ZnO–10%; (d) CaO–ZnO–15%; (e) CaO–ZnO–20%.

In the ZnO–doped samples, the band at 3625 cm^{-1} was significantly diminished in intensity, indicating that the concentration of the Ca(OH)$_2$ phase was gradually decreased. In the meantime, one small band at 3610 cm^{-1} appeared, indicating the presence of hydroxyls in the CaZn$_2$(OH)$_6\cdot$2H$_2$O phase. Disappearance of the 3610 cm^{-1} band at greater ZnO content indicates the removal of molecular water from CaZn$_2$(OH)$_6\cdot$2H$_2$O and inhibition of calcium hydroxide formation.

2.3. Raman Spectroscopy

Micro Raman spectroscopy is generally more sensitive than FTIR in the characterization of surface species in oxides. Raman spectra of CaO and CaO–ZnO samples are shown in Figure 3. For all the catalyst samples, their surfaces contained some hydroxyl species and calcium carbonate phase. Similar to the FTIR characterization, the sharp band at 3622 cm^{-1} was assigned to stretching vibrations of hydroxyls v(O–H), which is normally ascribed to the O–H bond in Ca(OH)$_2$. Generally, there are three major Raman peaks, namely, at 263 and 1027 cm^{-1} and 3622 cm^{-1} for the Ca(OH)$_2$ phase [31]. The CO$_3{}^{2-}$ species are very clearly characterized by the major Raman bands at 1448 cm^{-1} and 1887 cm^{-1}, respectively. There were two major Raman peaks for ZnO at 347 and 1162 cm^{-1} [32].

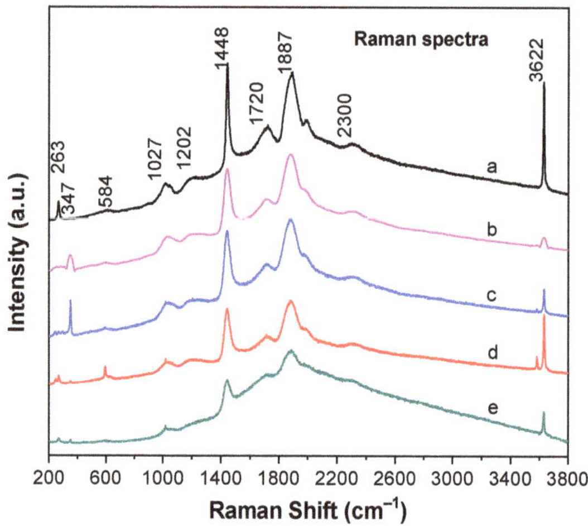

Figure 3. Raman spectra of CaO and CaO–ZnO samples at different levels of ZnO content. (a): CaO; (b) CaO–ZnO–5%; (c) CaO–ZnO–10%; (d) CaO–ZnO–15%; (e) CaO–ZnO–20%.

The Raman bands for CaO are difficult to characterize, as it is usually present in the surface as the $Ca(OH)_2$ phase and $CaCO_3$ in the samples. Some authors [32–34] claim that CaO does not possess any Raman-active modes, except for second-order effects involving the scattering of two phonons giving rise to two overlapping bands at 530 cm^{-1} and 660 cm^{-1}, as well as a broad signal at about 1000 cm^{-1} [35]. For our CaO sample, we did not observe Raman bands at the above-mentioned positions with obvious intensity, confirming that CaO indeed does not have a first–Order Raman spectrum. The second-order features of CaO might overlap with the broad band of $Ca(OH)_2$ and they were not able to be observed because they are much weaker than the first–Order Raman modes of calcium hydroxide.

For ZnO–doped samples, some Zn ions collaborated with Ca to form Ca–Zn–O phase in the bulk of the solid, so the Raman band at 3622 cm^{-1} gradually reduced in intensity in comparison with that shown in the CaO solid. However, amongst the ZnO–doped samples, this band intensity gradually increased on increasing the Zn content in the solid, reflecting the increment of –OH species in the $CaZn_2(OH)_6$ phase in the inner layer of the solid until 15mol% ZnO. At greater ZnO content, for example, in the CaO–ZnO–20% sample, more surface carbonate species may partially inhibit the presence of hydroxyls in the sample. In sum, both Raman and FTIR spectra confirmed the coexistence of hydroxyls and carbonate species in the samples, which are in good agreement with the XRD results.

2.4. XPS Analysis

The surface chemical valences and element oxidation states of the samples were characterized by the XPS technique. Figure 4 shows the core levels of Ca 2p, Zn 2p, O1s, and C1s XPS spectra of the different samples. The binding energy (BE) of the Ca 2p (Ca 2p1/2 and Ca 2p3/2) varied between 345.0 and 353.0 eV. These are two unsymmetrical peaks, indicating that Ca ions have different coordinates in different crystalline structures. For the pure CaO sample, its XPS signals can be deconvoluted into two main components corresponding to Ca 2p3/2 and Ca 2p1/2 in CaO and $Ca(OH)_2$ crystals. For the ZnO–doped CaO samples, XPS core levels of Ca 2p can be deconvoluted into four parts, corresponding to Ca 2p3/2 and Ca 2p1/2 in CaO (Ca 2p $\frac{1}{2}$ BE $\approx$ 346.0 eV and Ca 2p 3/2 BE $\approx$ 350.2 eV), $CaCO_3$ (Ca 2p $\frac{1}{2}$ BE $\approx$ 347 eV and Ca 2p 3/2 BE $\approx$ 351 eV), $Ca(OH)_2$ (Ca 2p $\frac{1}{2}$ BE $\approx$ 348 eV and Ca 2p 3/2 BE $\approx$ 352 eV), and $CaZn_2(OH)_6$ (Ca 2p $\frac{1}{2}$ BE $\approx$ 348.7 eV and Ca 2p 3/2 BE $\approx$ 352.5 eV), respectively [36]. These deconvolutions and assignments are in good agreement with the XRD patterns and FTIR and Raman spectroscopic characterizations [37].

For the samples with greater ZnO content, as ZnO content increased in the mixed oxide, the binding energy of Ca 2p slightly shifted toward the lower-value region due to the formation of a greater amount of the Ca–O–Zn phase.

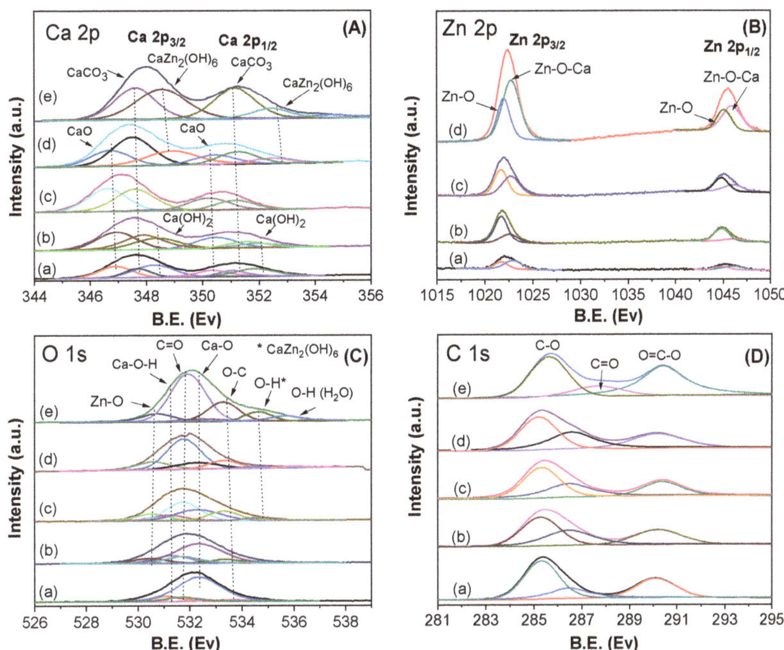

Figure 4. XPS spectra of the CaO and CaO–ZnO samples. (**A**): Ca 2p; (**B**): Zn 2p; (**C**): O 1s; (**D**): C 1s. (a): CaO; (b) CaO–ZnO–5%; (c) CaO–ZnO–10%; (d) CaO–ZnO–15%; (e) CaO–ZnO–20%. * The O–H species belong to $CaZn_2(OH)_6$ phase. * The O–H species belong to $CaZn_2(OH)_6$ phase.

For the core levels of Zn 2p, there were two peaks with binding energy around 1022.0 eV and 1045.3 eV, corresponding to Zn 2p3/2 and Zn 2p1/2, respectively [38]. The intensity of Zn 2p bands clearly increased as the ZnO content increased in the solids. Each of these two perks can also be deconvoluted into two parts, indicating Zn 2p bands in the ZnO and Ca–O–Zn phases, respectively.

The XPS O1s peak appeared in the BE range between 529 eV and 536 eV. These peaks were more pronounced in the samples containing ZnO. The asymmetrical and broad O1s spectrum peak indicated that more than one type of oxygen species was involved. Most of the oxygen was present in the form of Ca–O in CaO or Zn–O in ZnO crystals, along with oxygen species in carbonates and surface hydroxyls or adsorbed water. Therefore, the O1s peak can be deconvoluted into several peaks: the two at 532.17 eV and 530.2 eV may indicate the lattice O in CaO and ZnO oxides, respectively [38–40]. The band at 531.18 eV indicates the O–H bond in the $Ca(OH)_2$ phase [40]. The oxygen in surface CO_3^{2-} carbonate species is present at 531.76 eV (O=C) and at 533.5 eV (O–C) [38–41]. The peak at around 535.5 eV corresponds to OH in hydroxyls and water adsorbed in the surface.

Two peaks between 283.0~288.0 eV and 288.0~294.0 eV were well defined, corresponding to the C1s XPS spectra of the CaO–ZnO samples. These two peaks can be further deconvoluted into three components: the one at around 285.3 eV corresponds to a C–O bond; the one at around 286.6 eV corresponds to C=O in CO_3^{2-} [5], and the one at 290.1 eV is assigned to the vibration of the O–C=O species. These results suggest the formation of surface carbonate species in the samples, in agreement with FTIR characterization.

On the basis of the above analyses of XRD, FTIR, Raman, and XPS techniques, we may conclude that in the pure CaO and CaO–ZnO binary samples, there exist several phases. At least in the surface, hydroxyl species, $Ca(OH)_2$, and carbonate species are formed; further, CaO, $CaCO_3$, and $CaZn_2(OH)_6$ coexisted, depending on the phase composition and ZnO content. Without ZnO doping, both CaO and $Ca(OH)_2$ phases existed and surface carbonate species are also formed; when ZnO was added, the $CaZn_2(OH)_6$ phase dominated at a compensation of $Ca(OH)_2$ reduction or disappearance. In all the ZnO–doped samples, $CaCO_3$ was formed as a shell. ZnO doping promoted the formation of $CaCO_3$ as evidenced by an increasing amount of $CaCO_3$ in the samples with greater ZnO content.

2.5. Surface Basicity

The temperature-programmed desorption of CO_2 method was used to determine the surface basicity of the samples. The TPD–CO_2 profiles of the different catalysts are shown in Figure 5. Each TPD–CO_2 profile consists of two to three components: moderate strong basic sites (peak I, 325–400 °C; peak II, 375–475 °C or peak III, 450–475 °C) and strong basic sites (peak IV, 475–550 °C).

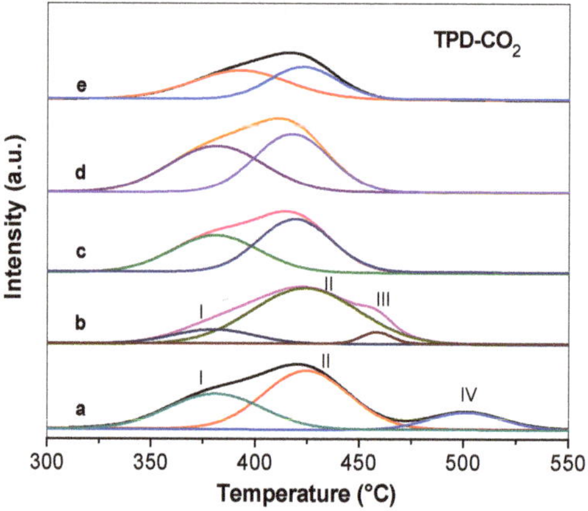

Figure 5. TPD–CO_2 profiles of CaO and CaO–ZnO samples. (a): CaO; (b) CaO–ZnO–5%; (c) CaO–ZnO–10%; (d) CaO–ZnO–15%; (e) CaO–ZnO–20%. Peak I (325–400 °C), peak II (375–475 °C) and peak III (450–475 °C): moderate strong basic sites; Peak IV (475–550 °C): strong basic sites.

The pure CaO (ZnO–0%) sample showed three peaks with peak maximums at approximately 375°, 425 °C, and 510 °C, corresponding to moderate–strong basic sites [42]. For the ZnO–doped samples, the peak near 510 °C disappeared. The number of total basic sites in each sample is reported in Table 1. The TPD–CO_2 data of each catalyst indicate that the basicity and base strength distributions were influenced by the presence of ZnO in the CaO–ZnO mixed oxides. The number of basic sites indicated by the peak area decreased on increasing the ZnO content [43]. For example, the catalyst containing 20% of ZnO showed a number of basic sites approximately 50% of that present in the pure CaO solid. Also, doping with ZnO content greater than 10 mol% eliminated the strong basic sites in the sample. Formation of $CaCO_3$ in the surface may also negatively affect the surface basicity.

The textural properties including the surface area, pore volume, and pore diameters are also measured. As shown in Table 1, a small amount of ZnO doping (5mol %) led to the surface area increasing; however, when ZnO content was greater than 5mol%, the surface area and pore volume decreased and pore diameter increased. These changes are related to the phase composition. Higher content of ZnO resulted in the formation of $CaCO_3$ and $CaZn_2(OH)_6$, altering the textural properties as confirmed by the XRD analysis. Therefore, ZnO doping not only modified the textural properties, but the surface basicity also.

Table 1. Textural data and basicity of the catalysts.

Catalysts	ZnO Concentration (mol%)	Surface Area (m²/g)	Pore Volume (cm³/g)	Pore Diameter (nm)	Surface Basicity (µmol/g)
CaO	0	85.5	0.378	17.7	10.8
CaO–ZnO–5%	5	108.0	0.465	17.2	10.4
CaO–ZnO–10%	10	44.8	0.366	14.5	8.6
CaO–ZnO–15%	15	13.5	0.068	20.1	7.4
CaO–ZnO–20%	20	12.6	0.056	21.3	5.4

2.6. Morphological Features and AFM and TEM Observations

The surface morphologies of the catalysts were studied using the AFM technique. Estimated values were obtained using ImageJ® software version 1.50I. The pure CaO sample had a relatively uniform surface morphology with many voids among particles. It was composed of particles with average size approximately 157 nm, and agglomeration of grains was absent (Figure 6). Topography images of ZnO–doped CaO catalysts (Figure 6b–d) showed the existence of the agglomeration of small particles with a triangular shape. For the sample with 20%ZnO, the large agglomerates with sizes between 350 and 390 nm consisted of many uniform particles with a smaller size.

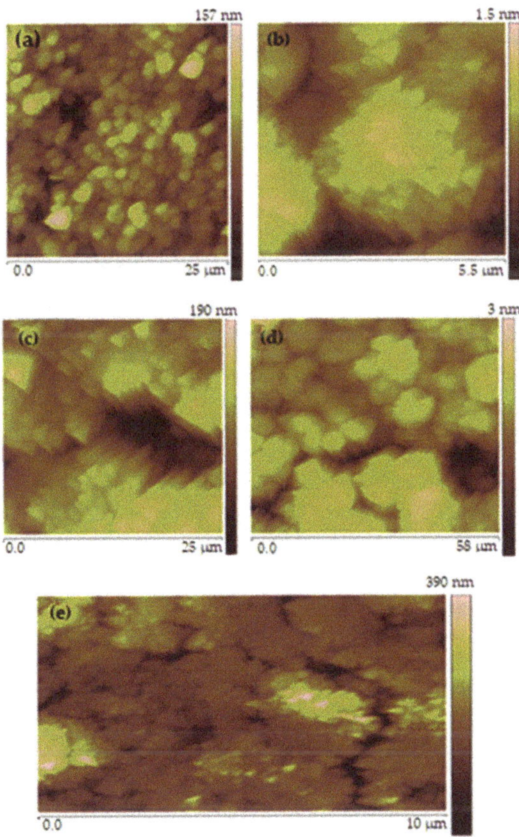

Figure 6. AFM micrographs of the CaO and CaO–ZnO samples. (**a**): CaO; (**b**) CaO–ZnO–5%; (**c**) CaO–ZnO–10%; (**d**) CaO–ZnO–15%; (**e**) CaO–ZnO–20%.

Figure 7 shows two TEM micrographs of CaO and CaO–ZnO–15 samples. For the pure CaO sample, all the particles are homogeneously distributed with an average size of approximately 30–40 nm. For the CaO–ZnO–15 sample, a close observation found many large aggregates consisting of small particles, forming core–shell structures with a shell thickness around 26 nm. ZnO doping led to the formation of bigger aggregates with respect to pure CaO, decreasing the surface area and surface basicity as confirmed by the N_2 physical adsorption and the TPD–CO_2 measurement (Table 1).

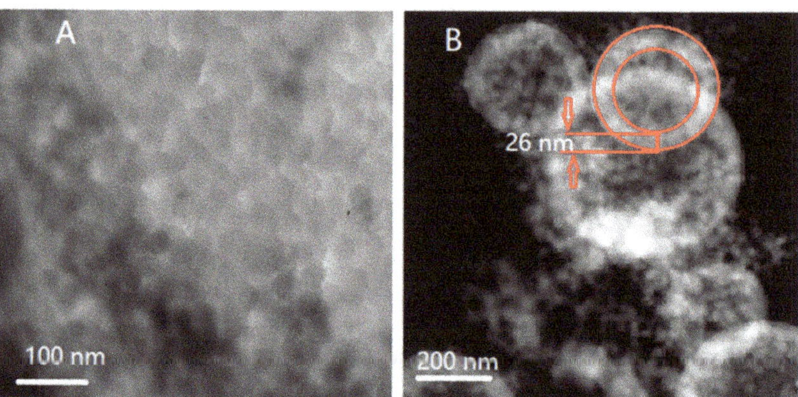

Figure 7. TEM micrographs of pure CaO (**A**) and CaO–ZnO–15 (**B**) samples.

2.7. Palm Oil Transesterification Reactions

2.7.1. Thermal Treatment Effect

In order to investigate the influence of thermal treatment of catalysts on biodiesel production, before the catalytic test, all the catalysts were thermally treated at 400 °C for 4 h to remove species adsorbed in the surface like adsorbed water. The catalytic activity of the catalysts with thermal treatment and without thermal treatment was comparatively evaluated. The results for biodiesels formation were reported in Table 2. The value of conversion of triacylglycerol obtained on different catalysts under conditions with thermal treatment or without thermal treatment are plotted in Figure 8. Using untreated catalysts, the maximum conversion achieved with the catalyst CaO catalyst was 97.5% and it decreased to around 70.3% when the ZnO content was 20%, showing palm oil conversion decreasing with the addition of ZnO. However, after the catalysts were thermally treated, palm oil conversion was rather stable, varying between 98.8% and 96.8%, which was higher than that obtained using the corresponding untreated catalysts. For example, the conversion achieved on the thermal-treated catalyst CaO–ZnO–20% was 96.8%, which is much greater than the data obtained from the same catalyst without thermal treatment, 70.3%.

Table 2. Comparison of Mexican palm oil conversions using the catalysts without and with thermal treatment.

Catalyst	Conversion (%) (without Thermal Treatment)	Conversion (%) (with Thermal Treatment)
CaO–ZnO–0%	97.5	98.8
CaO–ZnO–5%	97.1	97.0
CaO–ZnO–10%	94.1	96.8
CaO–ZnO–15%	79.3	96.9
CaO–ZnO–20%	70.2	96.8

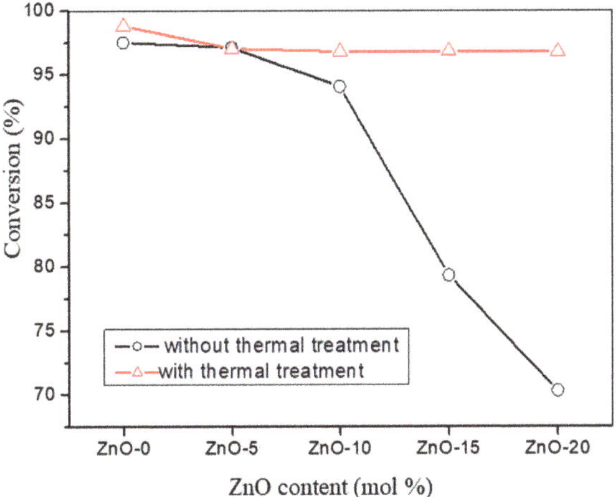

Figure 8. Effect of the thermal treatment of catalysts on palm oil conversion.

Thermally treated catalysts contained more basic centers compared with the catalysts without thermal treatment because the thermal treatment effectively removed the surface adsorbed water, which may poison the surface basic sites. Also, surface dehydration after thermal treatment promoted the generation of oxygen vacancies (O_v) that may serve as adsorption and surface reaction centers. These are responsible for the increase in the palm oil conversion via transesterification with methanol. However, the ZnO–doped catalysts exhibited lower activity in comparison with pure CaO, probably resulting from their lower surface basicity and smaller surface area.

2.7.2. Catalyst Stability

The pure CaO catalyst showed the highest initial catalytic activity (initial palm oil conversion was 98.8%); however, it decreased to around 78.0% after four runs of the reaction, which was lower than the conversion achieved with any one of the ZnO–doped catalysts under the same reaction conditions. The activity decrease in pure CaO probably resulted from its solubility in alcohol (herein, methanol or a mixture of methanol and glycerol). Ca^{2+} ions released from CaO into alcohols via the following reactions (3)~(9) [12]:

$$CaO \leftrightarrow Ca^{2+} + O^{2-} \tag{3}$$

$$Ca(OH)_2 \leftrightarrow Ca^{2+} + 2OH^- \tag{4}$$

$$CaO + 2CH_3OH \leftrightarrow Ca(OCH_3)_2 + H_2O \tag{5}$$

$$Ca(OH)_2 + 2CH_3\,OH \leftrightarrow Ca(OCH_3)_2 + 2H_2O \tag{6}$$

$$CaO + 2C_3H_8O_3 \leftrightarrow Ca(C_3H_7O_3)_2 + H_2O \tag{7}$$

$$Ca(OCH_3)_2 \leftrightarrow Ca^{2+} + 2OCH_3{}^- \tag{8}$$

$$Ca(C_3H_7O_3)_2 \leftrightarrow Ca^{2+} + 2C_3H_7O_3{}^- \tag{9}$$

Ca^{2+} concentrations in alcohols are reported in Table 3. For pure CaO, the Ca^{2+} concentration was 121 mg/L after 3h immersion in methanol at 60 °C. As CaO was combined with 5, 10, 15, and 20 mol% ZnO, Ca^{2+} concentration in methanol was 47.2, 34.6, 31.0, and 31.7 mg/L, respectively, after 3 h of catalyst immersion. Clearly, the Ca^{2+} solubility of ZnO–doped catalysts in alcohol was much lower than that of pure CaO. Similarly, in the mixture of methanol and glycerol, Ca^{2+} concentrations varied between 32 and 50 mg/L for the CaO–ZnO catalysts, lower than that for the value 101.9 mg/L obtained for the pure CaO sample.

Table 3. Ca^{2+} concentration in methanol or in mixture of methanol/glycerol (4:1) and palm oil after 3 h immersion.

Catalyst	Methanol Ca^{2+} (mg/L)	Methanol/Glycerol Mixture (4:1) Ca^{2+} (mg/L)	Palm Oil Ca^{2+} (mg/L)
CaO	121.7	101.9	7.2
CaO–ZnO–5%	47.2	49.6	5.5
CaO–ZnO–10%	34.6	32.6	0
CaO–ZnO–15%	31.0	34.0	0
CaO–ZnO–20%	31.7	35.3	0

Three factors may be responsible for the improvement of the CaO–ZnO stability: (i) elimination of $Ca(OH)_2$ on the catalyst surface; (ii) formation of $CaZn_2(OH)_6$ phase; and (iii) formation of surface $CaCO_3$ shell, as it is well known that calcium hydroxide is easily dissolved into water and alcohols. In the ZnO–doped CaO solids, surface calcium hydroxide formation was greatly inhibited, which contributes to the reduction in Ca^{2+} leaching. Because $CaCO_3$ was formed by CaO reacting with surface CO_2 or CO_2 in air, it mostly coated the particles as a shell of the catalyst. The formation of a $CaCO_3$ shell in the ZnO–doped CaO catalysts partially inhibits Ca^{2+} leaching in methanol or glycerol. This is similar to the results observed by Chen et al., who observed the inhibition of Ca^{2+} leaching from $CaO–SiO_2$ binary catalysts where a Si-related material shell on the CaO surface was formed [17]. On the other hand, as confirmed by the above characterization, Zn doping promotes the formation of $CaZn_2(OH)_6$. The crystalline structure of $CaZn_2(OH)_6$ consists of repeated $[CaO_6]$ octahedra and $[ZrO_4]$ tetrahedra. In one unit lattice cell, Ca atoms bonded to six oxygen atoms, forming $[CaO_6]$ clusters with octahedral coordination and Zn atoms are coordinated to four oxygen atoms in a tetrahedral $[ZrO_4]$ clusters. For charge balance, H atoms from H_2O molecules are bonded via hydrogen bridges to the hydroxyl groups [OH] [28]. In this crystalline structure, the Ca^{2+} bounded with Zn^{2+} via Ca–O–Zn bonds to from calcium zincate, which is more stable in alcohols in comparison with pure CaO. Formation of $Ca–Zn–O@CaCO_3$ core–shell nanoparticles with heterogeneous structure significantly inhibited Ca^{2+} leaching in the reaction mixture, which is demonstrated in Figure 9. It is noteworthy that $CaCO_3$ formation decreased the surface basicity of ZnO–CaO catalysts, and thus inhibited the catalytic activity to a certain degree. However, as $CaCO_3$ formation significantly enhanced the catalyst stability, from a practical viewpoint, the $Ca–Zn–O@CaCO_3$ binary base is an interesting and promising catalyst for biodiesel production.

2.7.3. FTIR Characterization of Biodiesel

FTIR spectroscopic technique was used to characterize the nature of the obtained biodiesel. A set of FTIR spectra of different biodiesel obtained with different catalysts is presented in Figure S1 in the Supplementary Materials. IR bands in three regions ($1000{\sim}1250$ cm^{-1}, $1700{\sim}1760$ cm^{-1}, and $2900{\sim}3050$ cm^{-1}) were related to the characteristics of biodiesel: strong bands at 1740 cm^{-1} and 1244 cm^{-1} were assigned to the vibrations of the methoxy group ($–O–CH_3$) and C=O in ester, respectively [44–48]. The IR bands of biodiesel in the region between 2950 cm^{-1} and 3050 cm^{-1} corresponded to the stretching vibration modes of the C–H bond in $–CH_2$ and $–CH_3$, respectively. The IR bands between 1170 cm^{-1} and 1183 cm^{-1} were assigned to the asymmetric vibration band C–O corresponding to the ester bonds [48]. There were doublet signals with moderate intensity at approximately 1485 cm^{-1}, which were attributed to the extension vibration of the $–CH_3$ group. FTIR characterization confirmed the presence of the typical bonds of methyl esters and evidenced the formation of biodiesel. It was noted that the IR spectra of all the biodiesel samples were quite similar because of the highly chemical similarities of methyl esters in the products.

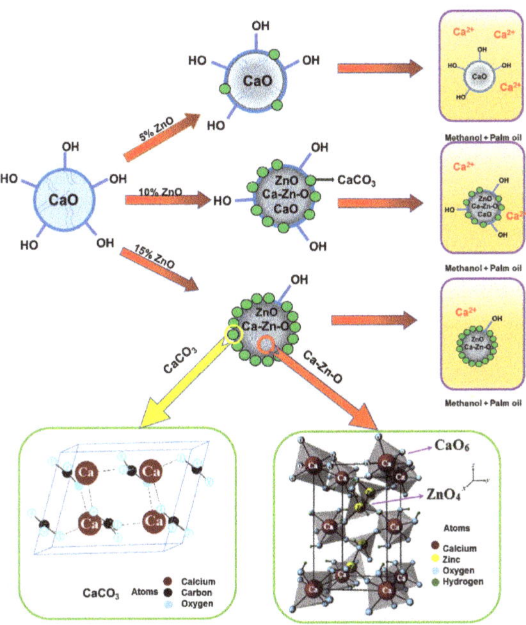

Figure 9. Ca–Zn–O@CaCO$_3$ core–shell heterostructures and Ca^{2+} solubility in methanol.

For comparison purposes, the result of the present work using the best catalyst CaO–ZnO–10% was compared with that reported in the literature. As shown in Table 4, under a soft reaction condition (less catalyst mass, lower methanol to oil ratio, shorter reaction time, and lower reaction temperature), our CaO–ZnO–10% core–shell catalyst shows a greater catalytic activity in comparison with the catalysts of CaO doped or modified with CeO$_2$ [19], SiO$_2$ [20], Al$_2$O$_3$ [24], MgO [49], and Fe$_3$O$_4$ [50].

Table 4. Comparison of catalytic activities obtained with different CaO–related catalysts under various experimental conditions.

Catalysts	Feedstocks	Alcohol/Oil Ratio	Catalyst/Oil (wt%)	Reaction Temperature (°C)	Reaction Time (h)	Yield or Conversion (%)	Refs.
CaO–ZnO (10mol% ZnO)	Palm oil + methanol	4:1	2.2	60	3	96.8	This work
CaO/γ-Al$_2$O$_3$	Palm oil + methanol	12:1	10	65	6	93.0	[24]
CaO–SiO$_2$ (2Si5Ca)	Palm oil + methanol	15:1	9	65	8	80.1	[20]
Na-CaO/MgO	Canola oil + methanol	12:1	5	65	6	95.4	[49]
CaO–CeO$_2$ (1Ca1Ce-650)	palm oil + methanol	20:1	5	85	10	>90	[19]
KF/CaO/Fe$_3$O$_4$ (Calcination at 600 °C)	Stillingia oil + methanol	12:1	4	65	3	95	[50]

2.8. Measurement of Biodiesel Indexes

The physicochemical parameters or indexes of the biodiesel products obtained by the present work such as molecular weight, density, cetane number, and refractive index were measured. Taking both the catalytic activity and stability into account, the two samples with 10 and 15 mol% ZnO samples are the best catalysts. All the parameters of biodiesel fuel obtained from the catalysts CaO–ZnO–10% and CaO–ZnO–15% are reported in Table 5. For comparison purposes, the parameters of the diesel obtained using CaO are also included in Table 5. Average molecular weight is dependent on the concentration of each compound in the feedstock (palm oil). As shown in Table 6, the components and molecular weight of palm oil principally contains around 45 mol% tripalmitin (molecular weight 807.5 g/mol) and 33 mol% triolein (molecular weight 885.6 g/mol). The palm oil feedstock has an average molecular weight of 774.4 g/mol. After the transesterification reaction, the average value of molecular weight of biodiesel is between 269.6 and 271.2 g/mol, which is reduced by approximately three times in comparison with that of the main component tripalmitin, indicating that the conversion of the triglycerides to the corresponding ester is rather complete [51–53].

Table 5. Data of density, cetane number, average molecular weight, and refractive index of the biodiesels obtained with different thermal-treated catalysts *.

Catalysts	Density (g/cm^3)	Cetane Number	Average Molecular Weight (g/mol)	Refractive Index.
CaO	0.8680	71.38	271.2	1.456
CaO–ZnO–10%	0.8753	67.43	269.6	1.442
CaO ZnO 15%	0.8789	62.16	270.3	1.458

* All the data were obtained at 20 °C.

Table 6. Components of Mexican palm oil.

Component	Mass Fraction (%)	Molar Fraction (%)	Molecular Mass (g/mol)
Tripalmitin	47.1	45.0	807.5
Triolein	37.6	33.2	885.6
Trilinolein	11.5	10.1	879.6
Palmitic acid	3.8	12.7	256.5

The value of the refractive index decreased as the temperature increased and varied in a narrow range between 1.452 and 1.463 in the temperature range between 20 and 70 °C (not all shown). The data in Table 5 were obtained at 20 °C. This parameter is acceptable as it meets the established norm for biodiesel well. The influence of ZnO content in the catalysts on the distribution of refractive index values is insignificant.

Another parameter related to the quality of biodiesel is fuel density, because it helps to directly define the mass of fuel injected into the combustion chamber. As expected, density value gradually decreases at higher temperatures; therefore, it is inversely proportional to the temperature. The obtained biodiesel had different densities, depending on the transesterification conversion of palm oil. A lower transesterification conversion of palm oil was achieved using the catalysts containing higher content of ZnO, and the obtained biodiesel showed an increased density. In our experiments, the fuel density remained within the standard region, indicating the good quality of the obtained fuel [54–56].

The cetane number in fuel represents the relative time delay between the injection and the autO–ignition. A high cetane number ensures the correct operation of the fuel during cold start and mild engine work. In contrast, fuels with a low cetane number tend to increase gas and particulate emissions due to incomplete combustion. The biodiesel obtained with the CaO–ZnO–10% catalyst shows a cetane number of 67.43, which is slightly smaller than the 71.38 value of the fuel produced with pure CaO. When the ZnO content in the CaO–ZnO binary catalysts was greater than 15 mol%, the cetane number of the biodiesel gradually decreased to 62.16. The values of the cetane number varied between 62 and 72. According to the standards of ASTM/EN ISO 5165, the minimum allowable value of the cetane number for Mexico is 51. Therefore, the quality of all biodiesels obtained with different catalysts was in accordance with the established standard, similar to the results reported by other authors [57].

3. Experimental Section

3.1. Reagents

$Ca(NO_3)_2 \cdot 4H_2O$ (>99.0%) and $Zn(NO_3)_2 \cdot 6H_2O$ (>98.0%) were of analytical grade and were obtained from Sigma-Aldrich. Double-distilled water produced in our own laboratory was used for catalyst preparation. NaOH was ACS-grade and was obtained from J.T. Baker. Methanol (99.9%) was purchased from Aldrich. Palm oil was obtained from the company Drotosa. The components of palm oil are reported in Table 5.

3.2. Catalyst Preparation

CaO–ZnO mixed oxides were prepared by the co–precipitation method using aqueous solutions of $Ca(NO_3)_2 \cdot 4H_2O$ and $Zn(NO_3)_2 \cdot 6H_2O$ as Ca and Zn precursors and 1M NaOH as the precipitation agent. The cO–precipitation was performed at pH around 10. Nominal ZnO content in the mixed oxides varied from 5 mol% to 10, 15, and 20 mol%. The precipitating solids were filtered and washed with deionized water, and then dried at 80 °C in air for 12 h. The dried samples were calcined at 600 °C for 4 h in air. For comparison purposes, pure CaO solid was also prepared by a similar method using $Ca(NO_3)_2 \cdot 4H_2O$ as the precursor. These samples were identified as CaO–ZnO–0% (pure CaO), CaO–ZnO–5%, CaO–ZnO–10%, CaO–ZnO–15%, and CaO–ZnO–20%, respectively.

3.3. Catalyst Characterization Techniques

X-ray diffraction (XRD) patterns of the catalysts were obtained in a two-theta region between 10 and 110° with Cu $K\alpha$ radiation (λ = 0.15418 nm) on an Empyrean Multi-Purpose Research X-ray Diffractometer.

Raman spectra of the catalysts were obtained with a micro Raman spectrophotometer (Confocal-Labram model HR800) at ambient temperature in wavenumbers ranging from 200 to 3800 cm^{-1}.

FTIR spectra were recorded at an ambient temperature in the wavenumber region between 400 and 4000 cm^{-1} on a 170-SX FTIR spectrometer.

Morphologies of catalysts were investigated using a NanoScope IV Atomic Force Microscope (Veeco Instruments Inc., Edina, MN, USA). Different n–doped silicon cantilevers (NanosensorsTM) with resonance frequencies f_0 = 190 kHz and 330 kHz, and the spring constants k = 48 N/m and 42 N/m, were applied. Morphology was also observed by transmission electron microscopy (JEOL 2100F) using a field emitter and 200 kV accelerator as illumination source.

X-ray photoelectronic spectroscopy (XPS) analyses were performed in a K-Alpha spectrophotometer (Thermo Scientific) equipped with a monochrome Al Kα (1478 eV) anode as the power supply of the X-ray. Analysis area was 400 μm^2 and a charge compensator was used. Mounting of the sample was performed as follows: the powder was set on the film and then was fixed in the sample holder with a double-face Cu tape.

The basicity of the samples was determined by means of temperature-programmed desorption of CO_2 (TPD–CO_2) employing a Micromeritics 2900 analyzer. A CO_2–Ar mixture (Praxair Inc., Norcross, GA 30093, USA) was used for the TPD–CO_2 experiments. First, 1 g catalyst was pretreated under argon gas flow (30 mL/min) at 300 °C for 2 h, and subsequently, it was cooled down to room temperature. Then, the sample was saturated under the CO_2–Ar mixture. After flushing the system with the carrier gas (N_2), desorption of carbon dioxide from the sample was performed from 200 to 600 °C at a temperature-increasing rate of 10 °C/min. The TPD signals were recorded by an on-line thermal conductivity detector (TCD).

3.4. Biodiesel Production, Purification, and Characterization

The transesterification reactions of palm oil with methanol were carried out in a 1000 mL stainless steel batch reactor under atmospheric pressure. Before the reaction, oil feedstock was passed through a column containing activated carbon to remove impurities and then passed through a column of calcium carbonate to remove the moisture. Reaction temperature was set at 60 °C, with a molar ratio of methanol to palm oil of 4. First, 1 g catalyst was dispersed in 200 mL methanol and stirred for 10 min at room temperature using a magnetic bar. Then, 50 mL palm oil was added into the reactor while stirring at 60 °C. The catalyst-to oil mass ratio was around 2.2 wt%. After 3h of reaction, the products were separated and purified for further analysis and characterization. In order to evaluate the reusability of the catalysts, after the first test run was finished, the CaO–ZnO catalysts were separated and tested again under the same experimental condition.

For product biodiesel purification, the liquid phase and solid catalyst were separated by centrifugation, and then biodiesel (upper layer) and glycerol (lower layer) were separated by a separation funnel; afterwards, excess methanol was evaporated at 60 °C under ambient pressure (in Mexico City, the ambient pressure is approximately 0.88 atm). Once alcohol was evaporated, biodiesel was passed through an activated carbon column to remove impurities and then passed through a column of calcium carbonate to remove the moisture.

The clean biodiesel was further characterized by both FTIR spectroscopy and ^{1}H-nuclear magnetic resonance (^{1}H-NMR) spectroscopy on a Varian Oxford 300MH3 apparatus. Usually, it is not necessary to quantify individual compounds in biodiesel oil but instead determine components by groups. For example, in the determination of total glycerol, it does not matter from which type of acylglycerol (mono-, di-, or tri-) the glycerol stems. In the present work, the conversion of palm oil was calculated by the ^{1}H NMR spectra according to the method developed by Gelbard and coworkers [58]. In this method, the protons of the methylene group adjacent to the ester moiety in triglycerols and the protons in the alcohol moiety of the methyl esters product were used to monitor the biodiesel production. The following Equation (10) was used to calculate the palm oil conversion:

$$C = 100 \left(\frac{2A_{ME}}{3A_{\alpha\text{-CH2}}} \right) \tag{10}$$

where C was the conversion of feedstock to the corresponding methyl ester (noted as ME); A_{ME} was defined as the integration value of the peak area of the protons of the methyl esters (the strong singlet peak); and $A_{\alpha\text{-CH2}}$ was the integration value of the peak area of the methylene protons (noted as α-CH$_2$). The factors 2 and 3 used in the above equation were due to the fact that the methylene carbon has two protons and the methanol-derived carbon links three protons. In the ^{1}H NMR spectra, the strong singlet at around 3.6 ppm indicated methyl ester (–(C=O)–O–CH$_3$) formation. These signals were used for the conversion calculation.

3.5. Ca^{2+} Solubility Measurement

To evaluate the stability of ZnO– doped CaO catalysts, Ca^{2+} solubility in the reaction mixture was measured. Because CaO dissolution in biodiesel is negligible, in the present experiments, Ca^{2+} leaching was measured only in the methanol and methanol–glycerol phases. The catalyst samples were immersed in methanol to test Ca^{2+} leaching in methanol. As an example, 100 mg CaO was added into 50 mL of methanol and stirred at 300 rpm for 3 h at 60 °C. After the experiment was finished, it was filtered and cooled, and then the equivalent volume of 0.1 M HCl was added to the methanol containing Ca^{2+} solution to keep calcium in a cationic state. It is noteworthy that methanol was consumed during the transesterification reaction, and another alcohol (glycerol) was produced in situ with increasing concentration at a compensation of methanol when the reaction proceeded continuously. Therefore, Ca^{2+} concentration was also measured in the mixture of glycerol–methanol. CaO was also added to a mixture of methanol and glycerol at a volumetric ratio = 4:1 and submerged for 3 h at 60 °C under stirring. Ca^{2+} concentration in methanol or the methanol–glycerol mixture was analyzed by inductively coupled plasma-mass spectrometry (ICP-MS, PerkinElmer/Sciex ELAN Dynamic Reaction Cell (DRCPlus)) coupled with a PerkinElmer AS-93 plus Autosampler with Elan v. 3.3 software for data collection.

3.6. Analyses of Biodiesel Properties

Biodiesel molecular weight was determined using MALDI ionization equipment (desorption/ionization by matrix-assisted laser) coupled to a TOF (time-of-flight) analyzer. Biodiesel density was measured using a densimeter (Anton Paar DMA 4500). This system allows fast density measurements with an accuracy of $\pm 5 \times 10^{-6}$ g/cm^3 in a wide temperature range (uncertainty of ± 0.03 °C and repeatability ± 0.01 °C).

The cetane number of biodiesel fuel was obtained from Equation (11) [59,60]:

$$\phi_I = -7.8 + 0.302 \times M_i - 20 \times N \tag{11}$$

where ϕ_I is the cetane number of the *i*th methyl ester, M_i represents the molecular mass of the *i*th methyl ester, and N is the number of double bonds.

The refractometer (Mode AR2008) was used for refractive index measurement. The refractive index of a substance or a transparent medium was defined as the ratio of the speed of light in vacuum to the speed of light in the substance or the transparent medium.

4. Conclusions

In the transesterification reaction of Mexican palm oil with methanol using CaO–ZnO base binary catalysts, factors such as thermal treatment, surface basicity, phase components, and ZnO content in the CaO–ZnO catalysts were all crucial for biodiesel production. Thermal treatment could remove the surface adsorbed water and impurities from catalysts and effectively enhanced the catalytic activity. Increasing the amount of ZnO in catalysts decreased the catalyst basicity to a certain extent, and slightly inhibited the catalytic activity; however, ZnO addition significantly improved the stability of the catalyst by inhibiting Ca^{2+} solving in alcohol or reaction mixture. The best catalytic activity was achieved on the catalysts containing 10 and 15 mol% ZnO in which a heterostructure consisting of Ca–Zn–O and ZnO as the core and $CaCO_3$ phase as the shell was formed, being responsible for the higher stability of the catalyst. The physicochemical parameters of the obtained clean biodiesel such as average molecular weight, density, refractive index, and cetane number of the biodiesel all satisfied the standards established for biodiesel well.

Supplementary Materials: The following supporting information can be downloaded at: https://www.mdpi.com/article/10.3390/inorganics12020051/s1, Figure S1: FTIR spectra for palm oil-base biodiesel obtained with thermal treated catalysts and after purification. (a): CaO; (b) CaO–ZnO–5%; (c) CaO–ZnO–10%; (d) CaO–ZnO–15%; (e) CaO–ZnO–20%.

Author Contributions: Conceptualization, M.E.M. and O.E.-S.; Methodology, M.G.A.-Q., M.E.M., O.E.-S. and J.A.W.; Data Curation, L.F.C., J.G.-G. and A.Z.-M.; Writing, M.G.A.-Q. and J.A.W.; Writing—Review and Editing, L.F.C.; Supervision, M.E.M. and O.E.-S. All authors have read and agreed to the published version of the manuscript.

Funding: The APC was funded by the Secretaría de Investigación y Posgrado del Instituto Politécnico Nacional.

Data Availability Statement: The data used to support the findings of this work are available from the corresponding author upon request.

Acknowledgments: M.G. Arenas-Quevedo thanks the scholarship from the CONHACyT-Mexico for supporting the investigation of his Master's thesis. The authors would like to acknowledge El Centro de Nanociencias y Micro y Nanotecnologías del Instituto Politécnico Nacional and the International Joint Research Center of Green Chemical Engineering at the East China University of Sciences and Technology for technical support.

Conflicts of Interest: The authors claim that there are no conflicts of interest.

References

1. Glîsić, S.; Lukic, I.; Skala, D. Biodiesel synthesis at high pressure and temperature: Analysis of energy consumption on industrial scale. *Bioresour. Technol.* **2009**, *100*, 6347–6354. [CrossRef]
2. Ngamcharussrivichai, C.; Totarat, P.; Bunyakiat, K. Ca and Zn mixed oxide as a heterogeneous base catalyst for transesterification of palm kernel oil. *Appl. Catal. A Gen.* **2008**, *341*, 77–85. [CrossRef]
3. Yan, S.; Di Maggio, C.; Mohan, S.; Kim, M.; Salley, S.O.; Ng, K.S. Advancements in Heterogeneous Catalysis for Biodiesel Synthesis. *Top. Catal.* **2010**, *53*, 721–736. [CrossRef]
4. Xie, W.; Yang, X.; Fan, M. Biodiesel production by transesterification using tetraalkylammonium hydroxides immobilized onto SBA–15 as a solid catalyst. *Renew. Energy* **2015**, *80*, 230–237. [CrossRef]
5. López–Granados, M.; Zafra–Poves, M.D.; Martín–Alonso, D.; Mariscal, R.; Cabello–Galisteo, F.; Moreno–Tost, R.; Santamaria, J.; Fierro, J.L.G. Biodiesel from sunflower oil by using activated calcium oxide. *Appl. Catal. B* **2007**, *73*, 317–326. [CrossRef]
6. Atadashi, I.M.; Aroua, M.K.; Abdul Aziz, A.R.; Sulaiman, N.M.N. The effects of water on biodiesel production and refining technologies: A review. *Renew. Sustain. Energy Rev.* **2012**, *16*, 3456–3470. [CrossRef]
7. Demirbas, A. Biodiesel production via non-catalytic SCF method and biodiesel fuel characteristics. *Energy Convers. Manag.* **2006**, *47*, 2271–2282. [CrossRef]
8. Ma, F.; Clements, I.D.; Hanna, M.A. The effects of catalyst, free fatty acids, and water on transesterification of beef tallow. *Trans. ASAE* **1998**, *41*, 1261–1264. [CrossRef]
9. Leung, D.Y.C.; Wu, X.; Leung, M.K.H. A review on biodiesel production using catalyzed transesterification. *Appl. Energy* **2010**, *87*, 1083–1095. [CrossRef]
10. Banković Ilić, B.; Miladinović, M.R.; Stamenković, O.S.; Veljković, V.B. Application of nano CaO–based catalysts in biodiesel synthesis. *Renew. Sustain. Energy Rev.* **2017**, *72*, 746–760. [CrossRef]
11. Liu, X.; He, H.; Wang, Y.; Zhu, S. Transesterification of soybean oil to biodiesel using SrO as a solid base catalyst. *Catal. Commun.* **2007**, *8*, 1107–1111. [CrossRef]
12. López Granados, M.; Martín Alonso, D.; Sadaba, I.; Mariscal, R.; Ocon, P. Leaching and homogeneous contribution in liquid phase reaction catalyzed by solids: The case of triglycerides methanolysis using CaO. *Appl. Catal. B* **2009**, *89*, 265–272. [CrossRef]
13. Albuquerque, M.C.G.; Jiménez Urbistondo, J.; Santamaría González, J.; Mérida Robles, J.M.; Monreno Tost, R.; Rodrígue Castellón, E. CaO supported on mesoporous silica SBA–15 as basic catalysts for transesterification reactions. *Appl. Catal. A* **2008**, *334*, 35–43. [CrossRef]
14. Dias, J.M.; Alvim Ferraz, M.C.M.; Almeida, M.F.; Díaz, J.D.M.; Pola, M.S.; Utrilla, J.R. Biodiesel production using calcium manganese oxide as catalyst and different raw materials. *Energy Convers. Manag.* **2013**, *65*, 647–653. [CrossRef]
15. Dai, Y.-M.; Li, Y.-Y.; Lin, J.-H.; Kao, I.-H.; Lin, Y.-J.; Chen, C.-C. Applications of M_2ZrO_2 (M = Li, Na, K) composite as a catalyst for biodiesel production. *Fuel* **2021**, *286*, 119392. [CrossRef]
16. Dai, Y.-M.; Li, Y.-Y.; Lin, J.-H.; Chen, B.Y.; Chen, C.-C. One-pot synthesis of acid-base bifunctional catalysts for biodiesel production. *J. Environ. Manag.* **2021**, *299*, 113592. [CrossRef] [PubMed]
17. Xie, W.; Yang, Z.; Chun, H. Catalysis properties of lithium–doped ZnO catalyst used for biofuel productions. *Ind. Eng. Chem. Res.* **2007**, *46*, 7942–7949. [CrossRef]
18. Bournay, L.; Casanave, D.; Delfort, B.; Hillion, G.; Chodorge, J.A. New heterogeneous process for biodiesel production: A way to improve the quality and the value of the crude glycerin produced by biodiesel plants. *Catal. Today* **2005**, *106*, 190–192. [CrossRef]
19. Thitsartarn, W.; Kawi, S. An active and stable CaO–CeO$_2$ catalyst for Transesterification of oil to biodiesel. *Green Chem.* **2011**, *13*, 3423–3430. [CrossRef]
20. Chen, G.; Shan, R.; Li, S.; Shi, J. A biomimetic silicification approach to synthesize CaO–SiO$_2$ catalyst for the transesterification of palm oil into biodiesel. *Fuel* **2015**, *153*, 48–55. [CrossRef]
21. Chansiriwat, W.; Wantala, K.; Khunphonoi, R.; Khemthong, P.; Suwannaruang, T.; Rood, S.C. Enhancing the catalytic performance of calcium-based catalyst derived from gypsum waste for renewable light fuel production through a pyrolysis process: A study on the effect of magnesium content. *Chemosphere* **2022**, *292*, 133516. [CrossRef]
22. Chansiriwat, W.; Chotwatcharanurak, L.; Khumta, W.; Suwannaruang, T.; Shahmoradi, B.; Kumsaen, T.; Wantala, K. Biofuel production from waste cooking oil by catalytic reaction over Thai dolomite under atmospheric pressure: Effect of calcination temperatures. *Eng. Appl. Sci. Res.* **2021**, *48*, 102–111. [CrossRef]
23. Albuquerque, M.C.G.; Azevedo, D.S.C.; Cavalgante, C.L.; Santamaría González, J.; Mérida Robles, J.M.; Moreno Tost, R.; Rodríguez Castellón, E.; Jiménez López, A.; Maireles Torres, P. Transesterification of ethyl butyrate with methanol using MgO/CaO catalysts. *J. Mol. Catal. A* **2009**, *300*, 19–24. [CrossRef]
24. Woranuch, W.; Ngaosuwan, K.; Kiatkittipong, W.; Wongsawaeng, D.; Appamana, W.; Powell, J.; Rokhum, S.L.; Assabumrungrat, S. Fine-tuned fabrication parameters of CaO catalyst pellets for transesterification of palm oil to biodiesel. *Fuel* **2022**, *323*, 124356. [CrossRef]
25. Kocík, J.; Hájek, M.; Troppová, I. The factors influencing stability of Ca–Al mixed oxides as a possible catalyst for biodiesel production. *Fuel Process. Technol.* **2015**, *134*, 297–302. [CrossRef]
26. Wu, X.; Kang, M.; Zhao, N.; Wei, W.; Sun, Y. Dimethyl carbonate synthesis over ZnO–CaO bi-functional catalysts. *Catal. Commun.* **2014**, *46*, 46–50. [CrossRef]

27. Margaretha, Y.Y.; Prastyo, S.H.; Ayucitra, A.; Ismadji, S. Calcium oxide from *Pomacea* sp. shell as a catalyst for biodiesel production. *J. Energy Environ. Eng.* **2012**, *3*, 33. [CrossRef]

28. Xavier, C.S.; Sczancoski, J.C.; Cavalcante, L.S.; Paiva Santos, C.O.; Li, M.S. A new processing method of $CaZn_2(OH)_6 \cdot 2H_2O$ powders: Photoluminescence and growth mechanism. *Solid State Sci.* **2009**, *11*, 2173–2179. [CrossRef]

29. Lavalley, J.C. Infrared spectrometric studies of the surface basicity of metal oxides and zeolites using adsorbed probe molecules. *Catal. Today* **1996**, *27*, 377–401. [CrossRef]

30. Fernández Carrasco, L.; Torrens Martín, D.; Morales, L.M.; Martínez Ramírez, S. *Infrared Spectroscopy in the Analysis of Building and Construction Materials*; Tech Open Publisher: London, UK, 2012. [CrossRef]

31. Genge, M.J.; Jones, A.P.; David Price, G. An Infrared and Raman study of carbonate glasses: Implications for the structure of carbonatite magmas. *Geochim. Cosmochim. Acta* **1995**, *59*, 927–937. [CrossRef]

32. Yang, C.C.; Chen, W.C.; Wang, C.L.; Wu, C.Y. Study the effect of conductive fillers on a secondary Zn electrode based on ball-milled ZnO and $Ca(OH)_2$ mixture powders. *J. Power Sources* **2007**, *172*, 435–445. [CrossRef]

33. Rieder, K.; Weinstein, B.; Cardona, M. Measurement and Comparative Analysis of the Second–Order Raman Spectra of the Alkaline-Earth Oxides with a NaCl Structure. *Phys. Rev. B* **1972**, *8*, 4780–4786. [CrossRef]

34. Schmida, T.; Dariz, P. Shedding light onto the spectra of lime: Raman and luminescence bands of CaO, $Ca(OH)_2$ and $CaCO_3$. *J. Raman Spectrosc.* **2015**, *46*, 141–146. [CrossRef]

35. Seehra, M. Comment on the Raman study of the thermal transformation of calcium hydroxide. *J. Solid State Chem.* **1986**, *63*, 344–345. [CrossRef]

36. Ni, M.; Ratner, B.D. Differentiation of Calcium Carbonate Polymorphs by Surface Analysis Techniques—An XPS and TOF-SIMS study. *Surf. Interface Anal.* **2008**, *40*, 1356–1361. [CrossRef]

37. Al-Gaashani, R.; Radiman, S.; Daud, A.R.; Tabet, N.; Al-Douri, Y. XPS and optical studies of different morphologies of ZnO nanostructures prepared by microwave methods. *Ceram. Int.* **2013**, *39*, 2283–2292. [CrossRef]

38. Alba Rubio, A.C.; Santamaría González, J.; Mérida Robles, J.M.; Moreno Tost, R.; Martín Alonso, D.; Jiménez López, A.; Maireles Torres, P. Heterogeneous transesterification processes by using CaO supported on zinc oxide as basic catalysts. *Catal. Today* **2010**, *149*, 281–287. [CrossRef]

39. Lee, J.; Chung, J.; Lim, S. Improvement of optical properties of postannealed ZnO nanorods. *Phys. E* **2010**, *42*, 2143–2146. [CrossRef]

40. Das, J.; Pradhan, S.K.; Sahu, S.R.; Mishra, D.K.; Sarangi, S.N.; Nayak, B.B.; Verma, S.; Roul, B.K. MicrO–Raman and XPS studies of pure ZnO ceramics. *Phys. B* **2010**, *405*, 2492–2497. [CrossRef]

41. Doyle, C.S.; Kendelewicz, T.; Carrier, X.; Brown, G.E. The interaction of carbon dioxide with single crystal CaO (100) surfaces. *Surf. Rev. Lett.* **1999**, *6*, 1247–1254. [CrossRef]

42. Weng, X.; Cui, Y.; Shaikhutdinov, S.; Freund, H.J. CO_2 Adsorption on CaO(001): Temperature-Programmed Desorption and Infrared Study. *J. Phys. Chem. C* **2019**, *123*, 1880–1887. [CrossRef]

43. Raouf, F.; Taghizadeh, M.; Yousef, M. Influence of CaO–ZnO supplementation as a secondary catalytic bed on the oxidative coupling of methane. *React. Kinet. Mech. Catal.* **2014**, *112*, 227–240. [CrossRef]

44. Xu, L.; Zhu, X.; Chen, X.; Sun, D.; Yu, X. Direct FTIR analysis of free fatty acids in edible oils using disposable polyethylene film. *Food Anal. Methods* **2015**, *8*, 857–863. [CrossRef]

45. Rabelo, S.N.; Ferraz, V.P.; Oliveira, L.S.; Franca, A.S. FTIR Analysis for quantification of fatty acid methyl esters in biodiesel produced by microwave-assisted transesterification. *Int. J. Environ. Sci. Dev.* **2015**, *6*, 964–969. [CrossRef]

46. Muik, B.; Lendl, B.; Molina, A.; Pérez, L.; Ayora, M.J. Determination of oil and water content in olive pomace using near infrared and Raman spectrometry. A comparative study. *Anal. Bioanal. Chem.* **2004**, *379*, 35–41. [CrossRef]

47. Shapaval, V.; Brandenburg, J.; Blomqvist, J.; Tafintseva, V.; Passoth, V.; Sandgren, M.; Kohler, A. Biochemical profiling, prediction of total lipid content and fatty acid profile in oleaginous yeasts by FTIR spectroscopy. *Biotechnol. Biofuels* **2019**, *12*, 140. [CrossRef]

48. Núñez, F.; Chen, L.; Wang, J.A.; Flores, S.O.; Salmones, J.; Arellano, U.; Noreña, L.E.; Tzompantzi, F. Bifunctional Co_3O_4/ZSM-5 Mesoporous Catalysts for Biodiesel Production via Esterification of Unsaturated Omega-9 Oleic Acid. *Catalysts* **2022**, *12*, 900. [CrossRef]

49. Murguía–Ortiz, D.; Cordova, I.; Manriquez, M.E.; Ortiz-Islas, E.; Cabrera-Sierra, R.; Contreras, J.L.; Alcántar-Vázquez, B.; TrejO–Rubio, M. Vázquez-Rodríguez, J.T.; Castro, L.V. Na-CaO/MgO dolomites used as heterogeneous catalysts in canola oil transesterification for biodiesel production. *Mater. Lett.* **2021**, *291*, 129587. [CrossRef]

50. Hu, S.; Guan, Y.; Wang, Y.; Han, H. NanO–magnetic catalyst $KF/CaO–Fe_3O_4$ for biodiesel production. *Appl. Energy* **2011**, *88*, 2685–2690. [CrossRef]

51. Knothe, G. Dependence of biodiesel fuel properties on the structure of fatty acid alkyl esters. *Fuel Process. Technol.* **2004**, *86*, 1059–1070. [CrossRef]

52. Knothe, G. "Designer" biodiesel: Optimizing fatty ester composition to improve fuel properties. *Energy Fuel* **2008**, *22*, 1358–1364. [CrossRef]

53. Hammond, E.G.; Lundberg, W.O. Molar refraction, molar volume and refractive index of fatty acid esters and related compounds in the liquid state. *J. Am. Oil Chem. Soc.* **1954**, *31*, 427–432. [CrossRef]

54. Tat, M.E.; Van Gerpen, J.H. The specific gravity of biodiesel and its blends with diesel fuel. *J. Am. Oil Chem. Soc.* **2000**, *77*, 115–119. [CrossRef]

55. Tekac, V.; Cibulka, I.; Holub, R. PVT properties of liquids and liquid mixtures: A review of the experimental methods and the literature data. *Fluid Phase Equilib.* **1985**, *19*, 133–149. [CrossRef]
56. Pratas, M.J.; Freitas, S.; Oliveira, M.B.; Monteiro, S.C.; Lima, A.S.; Coutinho, J.A.P. Densities and viscosities of fatty acid methyl and ethyl esters. *J. Chem. Eng. Data* **2010**, *55*, 3983–3990. [CrossRef]
57. Knothe, G.; Matheaus, A.C.; Ryan, T.W. Cetane numbers of branched and straight chain fatty esters determined in an ignition quality tester. *Fuel* **2003**, *82*, 971–975. [CrossRef]
58. Gelbard, G.; Brès, O.; Vargas, R.M.; Vielfaure, F.; Schuchardt, U.F. ^{1}H nuclear magnetic resonance determination of the yield of the transesterification of rapeseed oil with methanol. *J. Am. Oil Chem. Soc.* **1995**, *72*, 1239–1241. [CrossRef]
59. Ramírez Verduzco, L.F.; Rodríguez Rodríguez, J.E.; Jaramillo Jacob, A.R. Predicting cetane number, kinematic viscosity, density and higher heating value of biodiesel from its fatty acid methyl ester composition. *Fuel* **2012**, *91*, 102–111. [CrossRef]
60. Bamgboyel, A.I.; Hansen, A.C. Prediction of cetane number of biodiesel fuel from the fatty acid methyl ester (FAME) composition. *Int. Agrophysics* **2008**, *22*, 21–29.

 *inorganics*

Article

Basalt-Fiber-Reinforced Phosphorus Building Gypsum Composite Materials (BRPGCs): An Analysis on Their Working Performance and Mechanical Properties

Lei Wu [1,2], Zhong Tao [1,2,*], Ronggui Huang [1,2], Zhiqi Zhang [1,2], Jinjin Shen [1,2] and Weijie Xu [1,2]

[1] Faculty of Civil Engineering and Mechanics, Kunming University of Science and Technology, Kunming 650599, China; wulei0324@stu.kust.edu.cn (L.W.); 20212210030@stu.kust.edu.cn (R.H.)

[2] Yunnan Seismic Engineering Technology Research Center, Kunming University of Science and Technology, Kunming 650500, China

* Correspondence: taozhong@kust.edu.cn

Abstract: The preparation of fiber-reinforced phosphorus building gypsum composite materials (FRPGCs) is an important approach to enlarge the utilization of phosphogypsum resources. Through reinforcing phosphorus building gypsum (PBG) with basalt fiber (BF), this article probes into the effects of the length and fiber content of BF on the working performance and mechanical properties of basalt-fiber-reinforced phosphorus building gypsum composite materials (BRPGCs) and accesses the toughness of BRPGCs under bending loads using residual strength. The results showed that the addition of BF could significantly promote the mechanical properties of BRPGCs. However, due to the adverse effect of fibers on the working performance of BRPGCs, the fiber content was constrained. After adding 1.2% of 6 mm BF, the bending strength and compressive strength of FRPGCs reached maximum values of 10.98 MPa and 29.83 MPa, respectively. Under a bending load, BRPGCs exhibited an apparent ductile behavior. The P-δ curve presented five stages, with an evident phase of strength stability after cracking. A larger fiber content was conducive to the toughness of BRPGCs. When 1.6% of 6 mm BF was added, the residual strength of FRPGCs could reach 6.77 MPa.

Keywords: basalt fiber; phosphorus building gypsum; working performance; mechanical properties; load deformation curve

Citation: Wu, L.; Tao, Z.; Huang, R.; Zhang, Z.; Shen, J.; Xu, W. Basalt-Fiber-Reinforced Phosphorus Building Gypsum Composite Materials (BRPGCs): An Analysis on Their Working Performance and Mechanical Properties. *Inorganics* **2023**, *11*, 254. https://doi.org/10.3390/inorganics11060254

Academic Editors: Roberto Nisticò, Torben R. Jensen, Luciano Carlos, Hicham Idriss and Eleonora Aneggi

Received: 5 May 2023
Revised: 7 June 2023
Accepted: 8 June 2023
Published: 9 June 2023

1. Introduction

Phosphogypsum is a solid waste produced by the phosphate chemical industry that can cause soil and water pollution [1,2]. Disposing large amounts of newly generated and stockpiled phosphogypsum is a global challenge [3,4]. This problem is especially severe in China, where more than 500 million tons of phosphogypsum are currently stockpiled and more than 55 million tons of phosphogypsum are generated each year, with a recycling rate of only 40% [5,6]. One important way to recycle phosphogypsum is by preparing phosphorus building gypsum, but its use is limited due to inferior properties compared with cement or natural building gypsum [7,8]. Thus, improving the performance of PBG is a critical research task. The widespread use of PBG can promote the resource utilization of phosphogypsum and help reduce carbon emissions in the construction industry as a low-carbon building material [9].

In contrast with gypsum, fiber-reinforced gypsum composite materials (FRGCs) have better mechanical properties and can be used as paperless gypsum boards, gypsum planks, and molded gypsum [10,11]. Fibers such as polyvinyl alcohol fiber (PVA), polypropylene fiber (PP), and glass fiber are usually adopted to enhance the performance of gypsum [12]. Gonçalves et al. [13] evaluated the feasibility of utilizing recycled glass fibers for the preparation of FRGCs. After adding 6 wt% of recycled glass fibers, the bending strength of FRGCs could increase by 66%. Suarez [14] adopted three polymeric fibers to prepare

FRGCs and explored the influence law of fiber length on the fracture performance of FRGCs. Cong [15] prepared FRGCs using PP and PVA and made a thorough evaluation of the processability, bending properties, and toughness of FRGCs.

With the advantages of a low cost, excellent mechanical properties, and good fire resistance, basalt fiber (BF) has a great application potential in the field of building materials [16,17]. Li [18] treated BF with HCl, NaOH, and silane coupling agents, respectively, and analyzed the effects of the three treatment methods as well as fiber length and fiber content on the performance of BF-reinforced natural building gypsum composite materials (BFRGs). An appropriate fiber content can promote the bending strength of BFRGs by more than 50%. Meanwhile, Li [19] studied the effects of the dispersion mode of fibers on the working performance and mechanical properties of BFRGs. The bending strength of BFRGs prepared by the premix method increased by 40.6%. However, there was a decrease in compressive strength. Xie [20] used BF to reinforce desulfurized building gypsum and probed into the effects of BF on the working performance and mechanical properties of desulfurized building gypsum. Under the conditions of a fiber length of 9 mm and a fiber content of 0.7%, the bending strength of the composite material was significantly improved. Given the successful application of BF in reinforcing natural building gypsum and desulfurized building gypsum, BF may be used to improve the mechanical properties of phosphorus building gypsum. Wu [21] studied the effects of BF on the bending strength of phosphorus building gypsum. However, the effects of the fiber dosage and length on the working performance and bending toughness of basalt-fiber-reinforced phosphorus building gypsum composite materials (BRPGCs) were not involved.

To synthetically investigate the influence of BF on the working performance and mechanical properties of BRPGCs, two lengths of BF—6 mm and 12 mm—were adopted in this paper. The effects of fiber length and fiber addition on the fluidity and setting time of slurry from BRPGCs as well as the bending strength, bending toughness, and compressive strength of BRPGCs were explored through experiments. The fracture behavior of BRPGCs was analyzed through the P-δ curve. The mechanism of the influence of fibers on the performance of BRPGCs was preliminarily analyzed. This paper provides an essential research basis for the further application of BRPGCs.

2. Results and Discussion

2.1. Working Performance

2.1.1. Fluidity

The addition of BF caused a significant alteration in the fluidity of the slurry from BRPGCs, as shown in Figure 1. As the fiber content increased, the fluidity of the slurry from BRPGCs gradually decreased. For the 6 mm BF, when the fiber content reached 2%, the fluidity of the slurry from BRPGCs was lost and was remarked as fluidity of 60 mm. For the 12 mm BF, when the fiber content reached 1.6%, the slurry from BRPGCs lost its fluidity. With the same BF addition, the influence of the 6 mm BF on the fluidity of the slurry from BRPGCs was less than that of the 12 mm BF. This result was similar to the research on FRGCs, but the influence of BF on fluidity was greater than that of PP [15,18–20].

This phenomenon agreed with observations in other FRGCs [12,15]. When BF was added to the slurry of phosphorus building gypsum, the large surface area of the fibers partially adsorbed water, the water films on the surfaces of solid substances in the slurry became thinner, and resistance to particle movement increased, thus lowering the fluidity of the slurry. Meanwhile, fibers distributed in the slurry formed a grid structure, which impeded the flow of the slurry. Longer fibers and larger additions would strengthen this grid effect, leading to a greater decrease in the fluidity of the slurry.

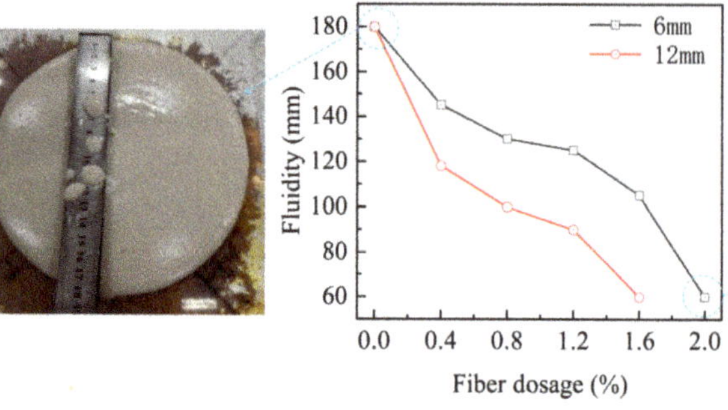

Figure 1. Influence of BF on the fluidity of the slurry from BRPGCs.

2.1.2. Setting Time

The addition of BF shortened the setting time of the slurry from BRPGCs, as shown in Figure 2. For the 6 mm BF, when the fiber content reached 1.6%, the initial setting time of the slurry from BRPGCs in the blank group decreased from 13 min and 20 s to 10 min and 10 s; the final setting time reduced from 30 min and 10 s to 25 min and 30 s. For the 12 mm BF, when the fiber content reached 1.2%, the initial setting time of the slurry from BRPGCs lessened to 10 min and 30 s and the final setting time decreased to 25 min and 10 s. On the whole, with the same addition, the fiber length hardly affected the setting time of the slurry from BRPGCs. This result was widely observed and was consistent with research on FRGCs [15,18–20].

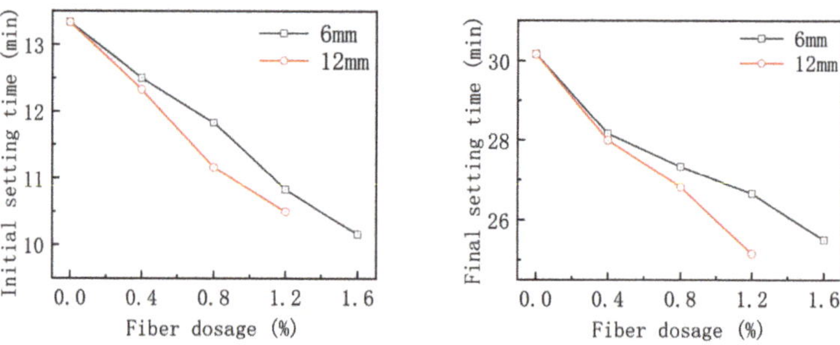

Figure 2. Influence of BF on the setting time of the slurry from BRPGCs.

Adding BF enlarged the total surface area of solid substances in the slurry from BRPGCs, thinned the water films on the surfaces of gypsum particles, narrowed the particle spacing, and accelerated the hydration reaction. Therefore, the setting time of the slurry from BRPGCs was shortened. The increase in the fiber surface area of composite materials was mainly relevant to the volume fraction of fibers, which were scarcely affected by fiber length. Therefore, the influence of fiber length on the setting time of the composite slurry was minimal.

2.2. Mechanical Properties

2.2.1. Analysis of P-δ Curve

During bending loading, the P-δ curve of BRPGCs had an apparent consistency. After cracking, BRPGCs possessed a certain residual bearing capacity, as shown in Figure 3. The load curve showed that the addition of BF greatly increased the peak load of BRPGCs. BRPGCs exhibited significant toughness after cracking and the residual strength also varied with changes in fiber length and fiber addition. The specific behavior is discussed in Sections 2.2.2 and 2.2.3.

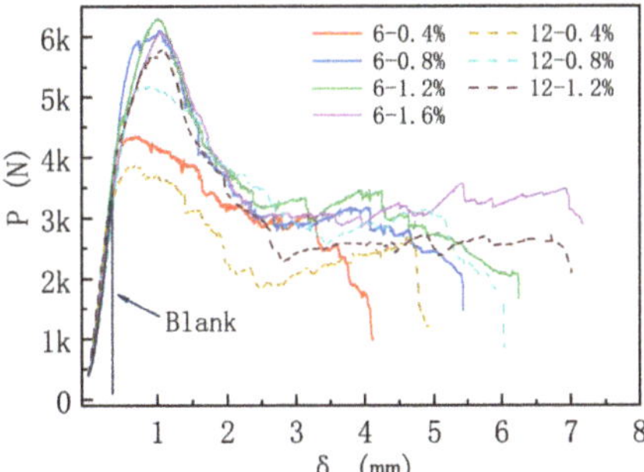

Figure 3. The P-δ curve of BRPGCs in the bending test.

The bending fracture behavior of BRPGCs can be summarized into five typical stages, as shown in Figure 4. The first stage (O–A) was the elastic phase, displaying elastic deformation. The phosphorus building gypsum matrix and BF were subjected to joint forces. The second stage (A–B) was the lifting phase. Due to the brittle feature of the phosphorus building gypsum matrix, the matrix cracked first. The crack did not rapidly expand, owing to the bridging effect of the fibers. The BF at the crack underwent elastic deformation under stresses. The bearing capacity of BRPGCs continued to increase until the peak point B. The third stage (B–C) was the declining phase. As the load continued to increase, the crack opening enlarged and BF at the crack began to pull out. The bearing capacity of BRPGCs decreased. The crack rapidly expanded, with more BF participating in bridging the crack. The fourth stage (C–D) was the stable phase. As the crack expanded, more and more BF bridged the crack. During pullout, the fibers at the opening of the crack dissipated energy through the interface as friction behavior between the fibers and the matrix. The fibers at the pointed ends of the crack showed elastic deformation and provided a bearing capacity. At this stage, the bearing capacity of BRPGCs was stable, with a slight decrease. The fifth stage (D–E) was the failure phase. With the continuous expansion of the crack, the fibers at the opening of the crack started to pull out and were no longer able to provide a bearing capacity. The bearing capacity of BRPGCs dropped sharply until complete failure.

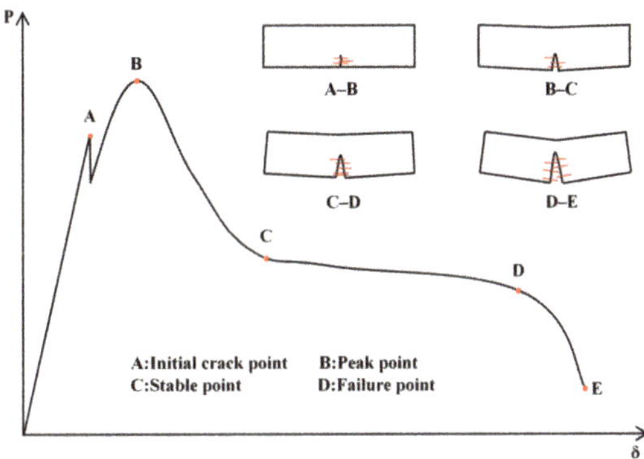

Figure 4. The typical P-δ curve of BRPGCs.

2.2.2. Bending Strength

The addition of BF had a significant impact on the bending strength of BRPGCs, as shown in Figure 5. The bending strength of BRPGCs increased with the fiber content. When the content of 6 mm BF reached 1.2%, the oven-dry bending strength of BRPGCs was up to the peak value of 12.14 MPa, 85.4% higher than the blank group. However, further increasing the fiber content caused a decrease in the bending strength of BRPGCs. This was mainly because when the fiber content was exceptionally high, the fluidity of the slurry from BRPGCs was quite low and a large number of pores appeared during the forming of BRPGCs, resulting in a decrease in the strength of the matrix. When the content of the 12 mm BF was added to 1.2%, the oven-dry bending strength of BRPGCs reached the maximum of 10.98 MPa, 67.7% higher than that of the blank group. With the same fiber addition, the reinforcement effect of the 6 mm BF was better than the 12 mm BF, which was also due to more defects generated by the 12 mm BF during the forming process of BRPGCs. Compared with other studies of FRGCs, the bending strength improvement of BRPGCs in this paper was more significant (see Table 1).

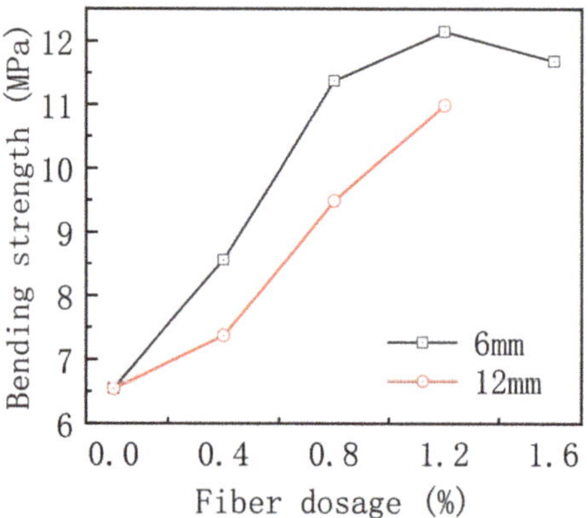

Figure 5. The effect of BF on the oven-dry bending strength of BRPGCs.

Table 1. Performance comparison with other research results.

	Fiber	Bending Strength/Flexural Strength	Compressive Strength
This paper	6 mm, 17.4 μm, and 1.2% BF	Increased by 85.4%	Increased by 69%
Gonçalves [13]	10 mm; 6 wt% E-glass fiber	Increased by 66%	-
Suarez [14]	12 mm, 31 μm, and 10 Kg/m^3	-	Decreased by 6.7%
Cong [15]	12 mm, 12 μm, and 1.2% PVA	Increased by 48%	Decreased by 6.6%
Li [18]	10 mm; 10% PVA	Increased by 78.3%	-
Li [19]	10 mm, 13 μm, and 0.5% BF	Increased by 40.6%	Decreased by 11.1%
Xie [20]	6 mm, 17 μm, and 0.7 wt% BF	Increased by 36.53%	Increased by 10.31%

2.2.3. Bending Toughness

The residual strength of BRPGCs was calculated according to the P-δ curve obtained in the bending test of BRPGCs, and the result is shown in Figure 6. It could be seen that the residual strength of BRPGCs increased with the fiber content, exhibiting a better toughness performance. The reason may be that when the fiber content was higher, after BRPGCs cracked, more fibers were distributed at the crack and provided a better bridging effect. However, the rise in fiber length had adverse effects on the residual bending strength of BRPGCs. It was mainly because the addition of 12 mm BF negatively affected the strength of BRPGCs and resulted in more defects in BRPGCs, which accelerated the expansion of the crack in the bending test. Expressing the contribution of the 12 mm BF to the toughness of BRPGCs solely through residual strength is still not comprehensive. From Figure 1, we could see that although the 12 mm BF had slight effects on peak load and residual strength, with the same addition, BRPGCs with 12 mm BF achieved greater ultimate deformation and exhibited a better deformation ability. This was because the 6 mm BF was easier to completely pull out from the matrix; in comparison, the 12 mm BF had a larger extraction length during the production process, resulting in a better energy dissipation capacity.

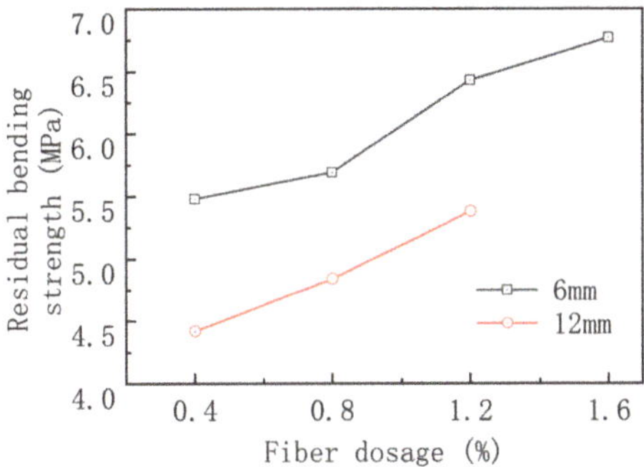

Figure 6. The influence of BF on the residual strength of BRPGCs.

When observing the micromorphology of BRPGCs after fracture through SEM, it could be seen that there was no significant plastic deformation of the BF during the fracture of BRPGCs, indicating that the BF was not a factor that consumed the energy of the fracture during the fracture. After being pulled out, BF left smooth marks and the interface between the hydration products of PBG and BF was neat. It could be inferred that BF and PBG hydration products did not form chemical coupling, but were combined through mechanical coupling. During the fracture process of BRPGCs, the energy of the fracture was mainly dissipated through sliding and friction between the BF and hydration products (see Figure 7).

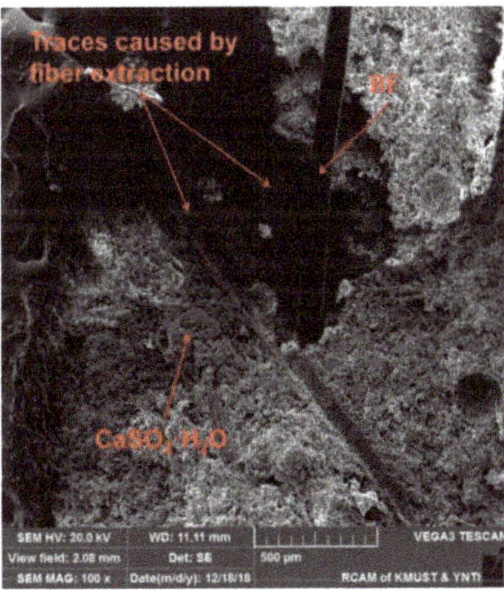

Figure 7. Microstructure of BRPGC.

2.2.4. Compressive Strength

The oven-dry compressive strength of BRPGCs remarkably increased with the addition of BF, as shown in Figure 8. Overall, the compressive strength of BRPGCs was enhanced with the fiber content. However, an excessive fiber content also caused a decrease in strength. When the content of 6 mm BF was 1.2%, the compressive strength of BRPGCs reached the maximum value of 29.83 MPa, 69.0% higher than that of the blank group. When the content of 12 mm BF was 1.2%, the compressive strength of BRPGCs reached a maximum value of 27.33 MPa, 54.8% higher than that of the blank group. With the same fiber content, the reinforcement effect of 6 mm BF was superior to that of 12 mm BF. Compared with other studies of FRGCs, the compressive strength improvement of BRPGCs in this paper was more significant (see Table 1).

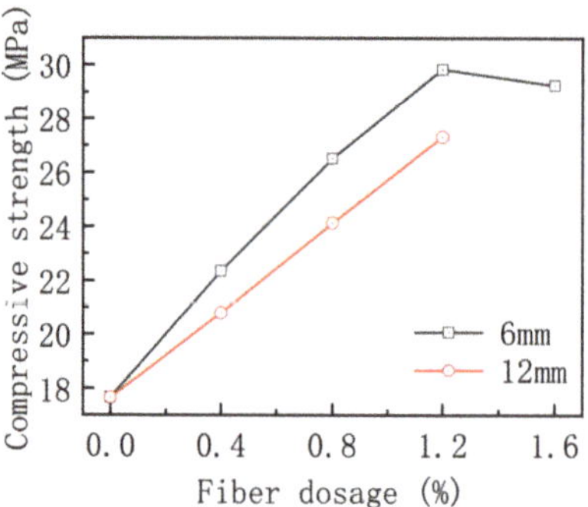

Figure 8. The impact of BF on the oven-dry compressive strength of BRPGCs.

As the elastic modulus of BF was much higher than that of phosphorus building gypsum, the addition of BF promoted the elastic modulus of BRPGCs. Under the action of a compressive load, a higher elastic modulus could lessen the transverse deformation of BRPGCs and delay material failure, thus enhancing the compressive strength of BRPGCs. The adverse effects of a larger fiber content and length on the compressive strength of composite materials were chiefly due to the defects in BRPGCs and the decrease in matrix strength during the forming process caused by the decline in the fluidity of the slurry from BRPGCs.

3. Materials and Methods

3.1. Raw Materials

The phosphogypsum used in this study was provided by Yunnan Yuntianhua Co., Ltd. (Kunming, China). After being washed in the laboratory to remove impurities, it was then calcined at 145 °C for 6 h to prepare phosphorus building gypsum (light yellow powders), as shown in Figure 9. Its chemical compositions are shown in Table 2. The XRD analysis of PBG is shown in Figure 10.

Figure 9. Phosphorus building gypsum.

Table 2. Chemical compositions of phosphogypsum.

Component	SiO$_2$	Al$_2$O$_3$	Fe$_2$O$_3$	MnO	MgO	CaO	K$_2$O	P$_2$O$_5$	SO$_3$	Organism
Content (%)	14.52	1.66	0.15	0.005	0.17	31.94	0.22	0.94	45.38	0.25

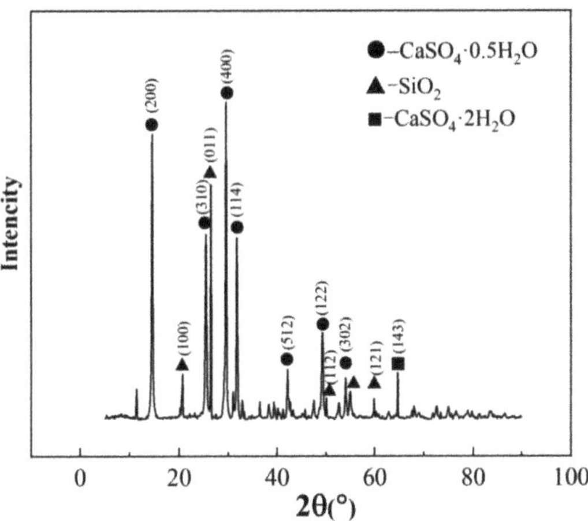

Figure 10. Results of XRD analysis.

BF was produced by Shanghai Chenqi Chemical Technology Co., Ltd., Shanghai, China, in the shape of bronze-colored coarse sand, as shown in Figure 11. The specifications adopted in this article were 6 mm- and 12 mm-long chopped sand. The physical properties are shown in Table 3.

(a) 6 mm   (b) 12 mm

Figure 11. Basalt fiber.

Table 3. Physical properties of BF.

Density	Tensile Strength	Elongation Rate	Maximum Working Temperature	Diameter	Tensile Modulus of Elasticity
2.65 g/cm	≥2000 MPa	≥2.5%	650 °C	17.4 μm	100 GPa

The water reducer used in this experiment was a polycarboxylic-acid-type water reducer prepared in the laboratory and was a light yellow oily liquid with a solid content of 40%.

3.2. Experiment Design

For the blank group in the experiment, the mixture ratio was 1200 g phosphorus building gypsum, 6 g water reducer, and 370 g water; the fluidity was 180 mm. The volume fractions for the BF of each length were 0.4%, 0.8%, 1.2%, 1.6%, and 2.0% (see Table 4). When the fluidity of the composite slurry could not be tested, the fiber content no longer increased. When the fluidity of the experimental group was unable to be tested, indicators such as the fluidity and mechanical properties stopped being tested.

Table 4. Mix proportions.

No.	PBG/g	Water Reducer/g	Water/g	Length of BF/mm	Volume Fractions for BF/%
1	1200	6	370	-	-
2	1200	6	370	6	0.4
3	1200	6	370	6	0.8
4	1200	6	370	6	1.2
5	1200	6	370	6	1.6
6	1200	6	370	6	2.0
7	1200	6	370	12	0.4
8	1200	6	370	12	0.8
9	1200	6	370	12	1.2
10	1200	6	370	12	1.6

3.3. Experiment Methods

3.3.1. Working Performance

The fluidity and setting time of the slurry from BRPGCs were primarily determined indicators and referred to the "Determination of Physical Properties of Building Gypsum Net Slurry". (GB/T 17669.4-1999) Chinese national standard [22].

The flow spread of different groups of PBG was tested using a cylinder with specifications of φ 50 × 50 mm. According to the mix proportions for the PBG plaster in Section 3.2, BF was added to the water and stirred until homogeneous. The PBG was then added and mixed for 60 s to obtain the plaster. The plaster was quickly poured into a clean, moist cylinder and excess plaster was scraped off. The cylinder was lifted to measure the vertical diameter of the disk formed by the expansion of the PBG plaster and calculate the average value as the flow spread of the test group.

The setting time was assessed using Vicat apparatus. The initial setting time was the time when the needle did not touch the glass floor and the final setting time was the time when the depth of the needle inserted into the slurry was not greater than 1 mm.

3.3.2. Mechanical Properties

The specification of the specimen of BRPGCs used for mechanical property testing was 40 mm × 40 mm × 160 mm. After forming, the specimen was naturally cured for 24 h and dried to a constant weight at 40 ± 4 °C. A microcomputer-controlled universal material testing machine was used for four-point bend loading, as shown in Figure 12. The P-δ curve of the specimen was collected. Displacement control was adopted in the loading process, with a loading rate of 0.5 mm/min and a sampling frequency of 10 Hz.

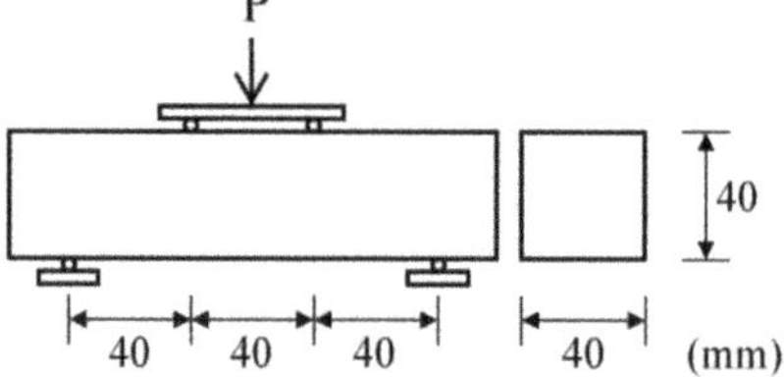

Figure 12. Schematic diagram of the test of four-point bend loading.

According to the obtained bending P-δ curve, the peak load of the curve was regarded as the maximum load (P) of the specimen. The flexural strength (R_f) of the specimen was obtained by the following equation:

$$R_f = PL/bh^2$$

where P represents the maximum load, L denotes the span of the specimen, and b and h are the width and height of the specimen, respectively.

The bending toughness was evaluated using the residual strength method [23], as shown in Figure 13. The deformation corresponding with the peak load of the P-δ curve was remarked as δ. Loads of P1, P2, P3, P4, and P5 corresponding with deformations of 1.5 δ, 2 δ, 3 δ, 4 δ, and 5 δ were found, respectively. Due to the remarkable fluctuation in the bearing capacity of BRPGCs after cracking, the average value of the load in the range of [−0.005 mm, 0.005 mm] before and after the corresponding deformation was taken as the load, as shown in Figure 13.

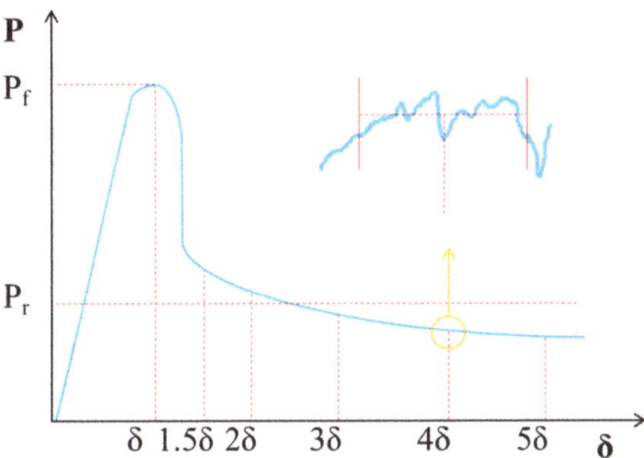

Figure 13. Schematic diagram of the calculation of bending toughness.

The residual load (P_r) was obtained by the following equation:

$$P_r = \frac{\alpha_1 P_1 + \alpha_2 P_2 + \alpha_3 P_3 + \alpha_4 P_4 + \alpha_5 P_5}{\alpha_1 + \alpha_2 + \alpha_3 + \alpha_4 + \alpha_5}$$

where α_1, α_2, α_3, α_4, and α_5 are weighted coefficients; the values were 0.25, 0.50, 0.75, 1, and 1, respectively.

The residual strength (R_r) could be obtained by the following equation:

$$R_r = P_r L/bh^2$$

Three half-cut parts from different specimens after the bending test were used for the test of compressive strength. A microcomputer-controlled universal material testing machine was adopted for loading through displacement control, with a loading rate of 0.5 mm/min and a sampling frequency of 10 Hz. We recorded the maximum compressive load (P_c) and calculated the flexural strength of the specimen as follows:

$$R_c = P_c/S$$

where P_c represents the maximum compressive load in units of N and S denotes the compression area of the specimen, with a value of 40 mm × 40 mm.

4. Conclusions

This article explored the influence law of the length and addition of BF on the working performance and mechanical properties of BRPGCs, emphasizing the fracture behavior of BRPGCs in a bending test and evaluating the bending toughness of BRPGCs through residual strength. The following conclusions were drawn:

(1)   The addition of BF had a negative impact on the working performance of the slurry from BRPGCs. As the fiber content and length increased, the fluidity of the slurry from BRPGCs decreased and the setting time shortened.

The effect of the recycled gypsum on the plastic shrinkage of the specimens (28 d) is shown in Figure 3. Gypsum did not affect the plastic shrinkage of the specimens whose shrinkage was already very small (cf. fly ash and biomass ash-based specimens **5** and **6**) and it increased the plastic shrinkage of OPC (cf. **11** and **12**). The plastic shrinkage of the other specimens was reduced by gypsum. The shrinkage of gypsum-free specimens (except **5**) was higher than that of OPC (**11**). However, the shrinkage-reducing effect of gypsum was significant, so that plastic shrinkage of gypsum-containing specimens was even smaller than the shrinkage in OPC (except the fly-ash-based specimen **4**). Based on previous studies, the shrinkage-reducing effect of gypsum is based on its tendency to form ettringite crystals with aluminium [29,30].

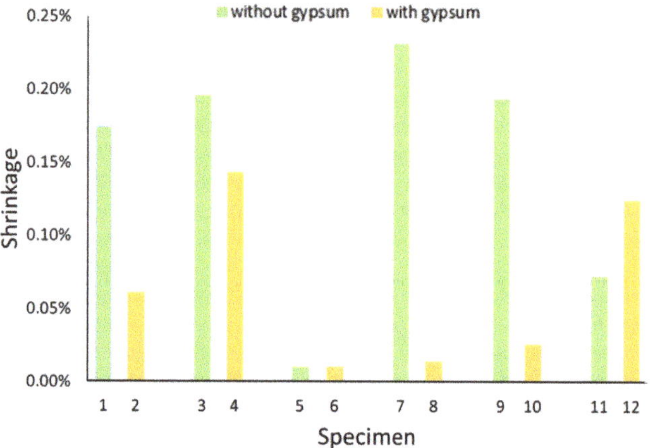

Figure 3. Effect of recycled gypsum on plastic shrinkage of specimens (28 d). Green columns indicate specimens without gypsum and yellow columns indicate specimens with gypsum.

2.2.2. Drying Shrinkage

Drying shrinkage occurs when the mass dries because of water evaporation. It causes large cracks that extend through the material. Drying shrinkage is usually the most significant cause of shrinkage [25,27]. Figure 4 shows the drying shrinkage of the specimens and the effect of the recycled gypsum on it. The drying shrinkage of fly ash and biomass ash-based specimens **5** and **6** could not be measured because they were so fragile that there were no intact pieces left. The smallest drying shrinkage was measured for the GGBS-containing specimens (**7**, **8**, **9**, and **10**). Their drying shrinkage was even smaller than the OPC-based specimen (**11**). The largest drying shrinkage was measured for fly-ash-based specimens (**3** and **4**). The recycled gypsum reduced the drying shrinkage of the GGBS-based specimen (cf. **7** and **8**) and increased the drying shrinkage of the metakaolin-based specimen (cf. **1** and **2**). Gypsum did not significantly affect the drying shrinkage of the other materials. The aggregates reduced drying shrinkage notably. The drying shrinkage of the specimen containing aggregates (**13**) was 0.07%, and the shrinkage of the same specimen without aggregates (**1**) was 0.17%.

2.2.3. Total Shrinkage

Figure 5 presents the total shrinkage of the specimens, and the proportions of plastic shrinkage and drying shrinkage. The smallest total shrinkage was measured for the gypsum-containing GGBS-based specimens (**8** and **10**). Their total shrinkages were even smaller than the OPC-based specimen (**11**). However, specimens **8** and **10** were cured at 70 °C and their initial length was measured later than the others, which may affect the results. The largest total shrinkage was measured for fly-ash-based specimens (**3** and **4**). The aggregates significantly reduced the shrinkage, which was expected based on a previous study [27]. The total shrinkage of aggregate-containing specimen (**13**: 0.10%) was significantly smaller than the shrinkage of the same specimen without aggregates (**1**: 0.35%).

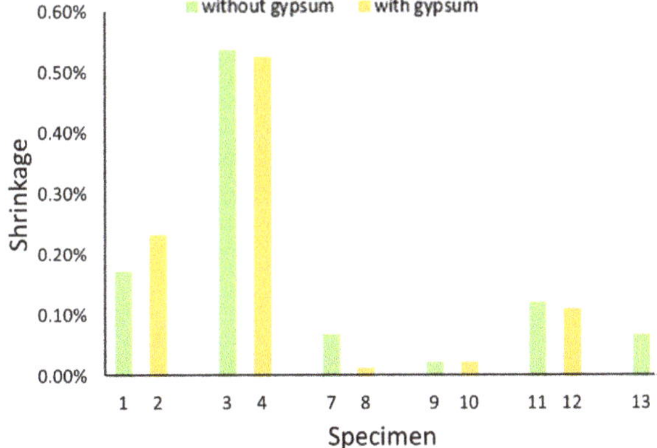

Figure 4. Drying shrinkage of specimens and effect of recycled gypsum on it. Green columns indicate specimens without gypsum and yellow columns indicate specimens with gypsum.

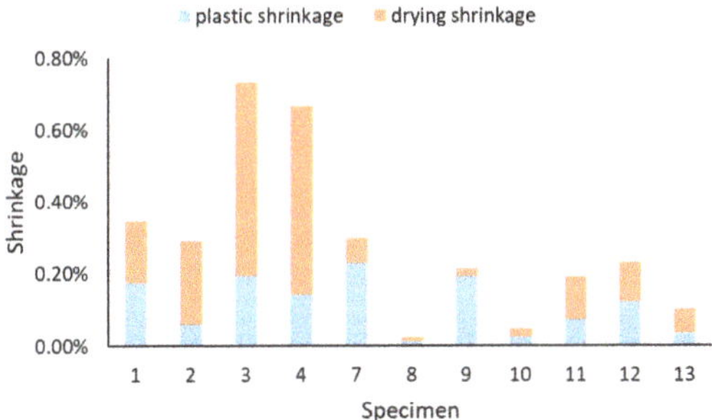

Figure 5. Total shrinkage of the specimens **1–13**, and proportions of plastic shrinkage (blue) and drying shrinkage (orange).

For nearly all specimens, at least half of the total shrinkage was caused by drying shrinkage, which was expected based on previous studies [25,27]. Although gypsum reduced the plastic shrinkage of the specimens, it did not affect their drying shrinkage. Therefore, the recycled gypsum had no significant effect on the total shrinkage of most specimens. For instance, gypsum reduced the plastic shrinkage of the metakaolin specimen, but it increased its drying shrinkage. Therefore, the total shrinkage of metakaolin-based specimens (**1** and **2**) remained almost the same (0.29% and 0.35%). Gypsum seemed to reduce the total shrinkage of the GGBS-containing specimens (cf. **7**, **8**, **9**, and **10**), but the effect could also be a result of the curing conditions.

In addition to shrinkage measurements, the cracks caused by the plastic shrinkage and the drying shrinkage in the specimens were observed (Figure 6). Plastic shrinkage caused cracks in the gypsum-free specimens based on metakaolin (**1**), GGBS (**7**), and fly ash (**3**), as well as in the fly-ash-based specimen with gypsum (**4**). Drying shrinkage caused cracks in the gypsum-containing specimens based on metakaolin (**2**), GGBS (**8**), and GGBS and biomass ash (**10**). Although specimens **8** and **10** had the lowest total shrinkage, they still showed shrinkage cracks. Therefore, it is reasonable to assume that their small shrinkage is due to the curing conditions. There were no cracks in the gypsum-free specimens based on fly ash and biomass ash (**5**) and OPC (**11**), and the gypsum-containing specimens based on fly ash and biomass ash (**6**), GGBS and biomass ash (**9**), and OPC (**12**).

Figure 6. Cracks of specimens caused by plastic shrinkage (**above**) and drying shrinkage (**below**).

2.3. Strength

2.3.1. Flexural Strength

The flexural strength of the specimens (7 d and 28 d) is presented in Figure 7. The flexural strength of the specimens developed mainly during the first week and, as expected from previous studies, the specimens were all quite weak [27,39]. In the 7 d measurement, the weakest specimens were gypsum-free and based on fly ash (**3**) and fly ash and biomass ash (**5**). The strongest specimens were based on GGBS without gypsum (**7**) and OPC with gypsum (**12**). Their flexural strength was even higher than that of the gypsum-free OPC (**11**). In the 28 d measurement, the weakest specimens were based on fly ash and biomass ash without gypsum (**5**) and with gypsum (**6**). The strongest was the gypsum-free GGBS-based specimen (**7**), which was even stronger than OPC with gypsum (**12**) and OPC (**11**). In addition, the gypsum-free specimens based on metakaolin (**1**) and GGBS and biomass ash (**9**) were also stronger than OPC (**11**). This has been observed in a previous study with a fly-ash-based specimen, which also had higher flexural strength than OPC [36].

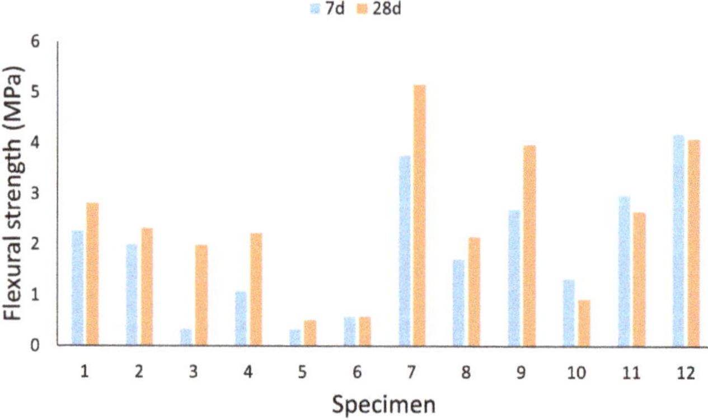

Figure 7. Flexural strength of the specimens at 7 d (blue) and 28 d (orange).

The effect of the recycled gypsum on the flexural strength of the specimens (28 d) is shown in Figure 8. Gypsum reduced the flexural strength of the GGBS-containing specimens (cf. **7** and **8**; **9** and **10**). The flexural strength of OPC was increased by gypsum (cf. **11** and **12**). There was no effect on the flexural strength of the other specimens.

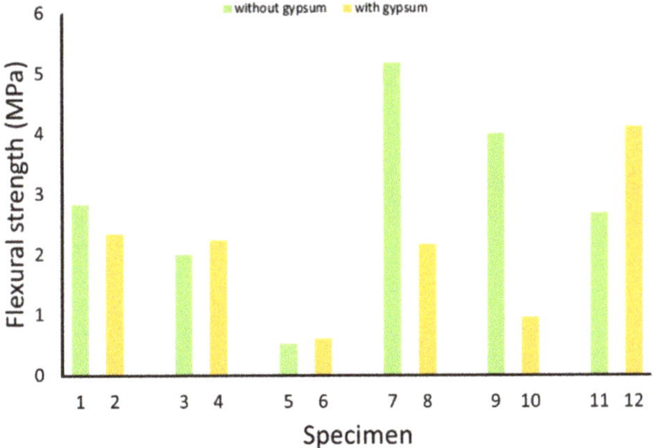

Figure 8. Effect of the recycled gypsum on flexural strength of specimens (28 d). Green columns indicate specimens without gypsum and yellow columns indicate specimens with gypsum.

2.3.2. Compressive Strength

Figure 9 shows the compressive strength of the specimens (7 d and 28 d). The compressive strength of the OPC-based specimens **11** (28 d) and **12** (7 d and 28 d) exceeded the limit of the device, therefore their true strength could not be determined. In the 7 d compressive strength measurement, the weakest were fly ash and biomass ash-based specimens without gypsum (**5**) and with gypsum (**6**), and the strongest were the specimens based on metakaolin (**1**) and OPC with gypsum (**12**), which were stronger than OPC without gypsum (**11**). In the 28 d compressive strength measurement, the weakest specimens were, again, **5** and **6**, and the strongest specimens were the ones based on OPC (**11**) and OPC with gypsum (**12**). The strongest of the geopolymer specimens was based on metakaolin (**1**). The strength of the other materials was below 50 MPa. Based on previous studies, the GGBS-based specimens are generally the strongest and, therefore, GGBS is often added with other raw materials to improve strength properties [8,32]. According to this study, the GGBS-based specimen (**7**) was as strong as the fly-ash-based specimen (**3**), but weaker than the metakaolin-based specimen (**1**).

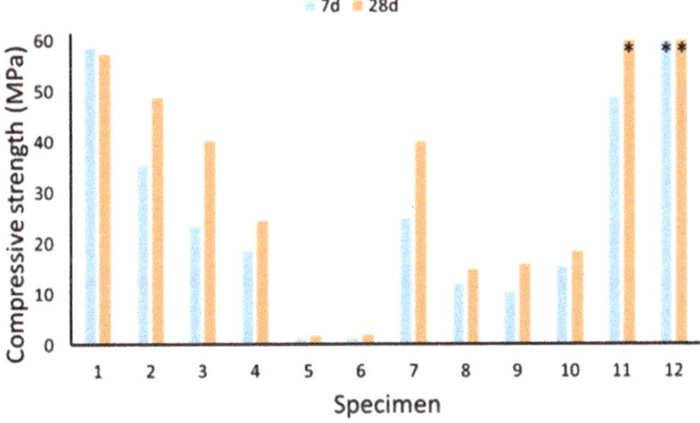

Figure 9. Compressive strength of the specimens at 7 d (blue) and 28 d (orange). * Compressive strength of the specimen exceeded the limit of the device.

The effect of recycled gypsum on the compressive strength of the specimens (28 d) is presented in Figure 10. It had no effect on the compressive strength of the already weak specimens (cf. fly ash and biomass ash-based specimens **5** and **6**, and biomass ash and GGBS-based specimens **9** and **10**), but it reduced the strength of the others. The effect of gypsum on the compressive strength of OPC could not be measured.

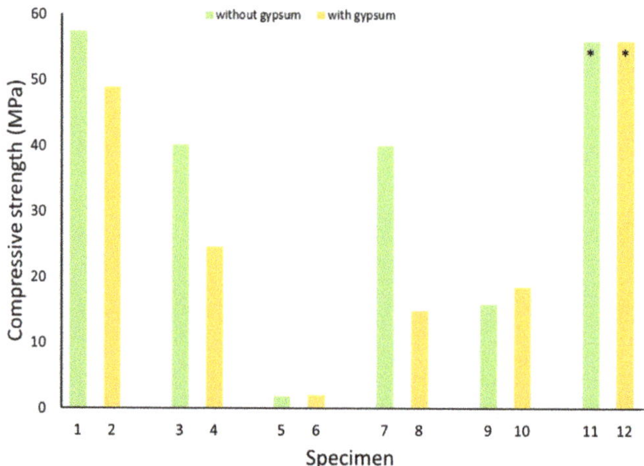

Figure 10. Effect of the recycled gypsum on compressive strength of the specimens (28 d). Green columns indicate specimens without gypsum and yellow columns indicate specimens with gypsum. * Compressive strength of the specimen exceeded the limit of the device.

2.4. Leaching of Si, Al, and Ca

Figure 11 presents the amount of aluminium leached from the specimens (7 d and 28 d). In general, less aluminium leached from the specimens at 28 d than at 7 d. For those specimens, from which aluminium leaching was already low (**6**, **8**, and **10–12**), there was no time dependence. In the 7 d measurement, leaching of aluminium was lower from the gypsum-containing specimens than from the other materials. In the 28 d measurement, the recycled gypsum reduced leaching of aluminium from the GGBS-based specimens (cf. **7**, **8**, **9**, and **10**) and increased it from the metakaolin-based specimens (cf. **1** and **2**). Gypsum had no significant effect on the other specimens.

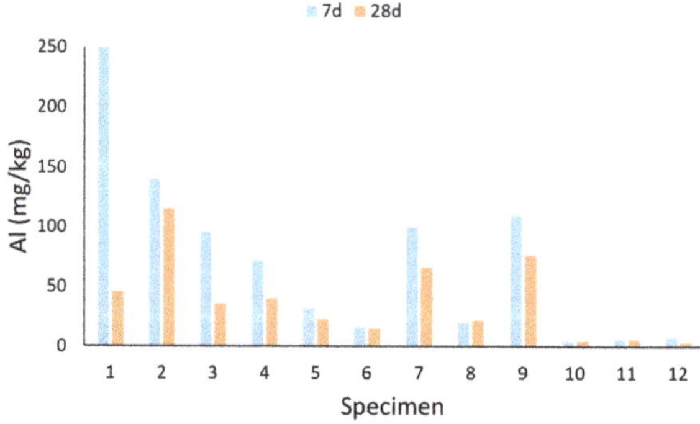

Figure 11. Amount of leached aluminium from the specimens at 7 d (blue) and 28 d (orange).

The amount of silicon leached from the specimens (7 d and 28 d) is shown in Figure 12. The amount of leached silicon was lower from specimens **1–6**, **8**, and **9** at 28 d than at the

7 d measurement. For the other materials, there was no difference in the amount of leached silicon from the specimens of different ages. Gypsum reduced leaching of silicon from the fly-ash-based specimens (cf. **3** and **4**), and increased it from the GGBS- and biomass ash-based specimens (cf. **9** and **10**). For the other specimens, it reduced slightly or had no effect on the amount of leached silicon.

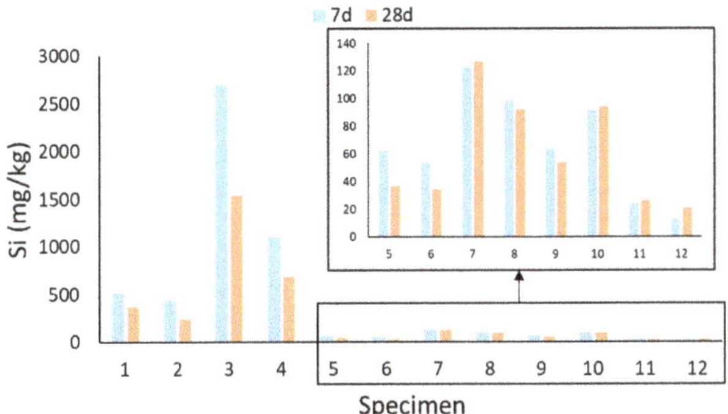

Figure 12. Amount of leached silicon from the specimens at 7 d (blue) and 28 d (orange).

Figure 13 shows the amount of leached calcium in specimens **5–12** (7 d and 28 d). Specimens **1–4** are missing because the amount of calcium leached from them was negligible. The amount of leached calcium was mainly influenced by the calcium content of the raw materials in the specimens. More calcium leached from the specimens made from calcium-rich raw materials (OPC, GGBS, and biomass ash) than from the specimens made from low-calcium raw materials (metakaolin and fly ash). The OPC-based specimens (**11** and **12**) clearly showed the largest amount of calcium leached. In general, the amount of leached calcium was lower from the specimens at 28 d than at 7 d. The exception was the OPC-based specimen (**11**), from which slightly more calcium was leached at 28 d than at the 7 d measurement. Gypsum increased the amount of leached calcium, which is explained by the fact that it brings more calcium into the specimen. The results measured by AAS supported the photometrically measured results presented in Figure 13.

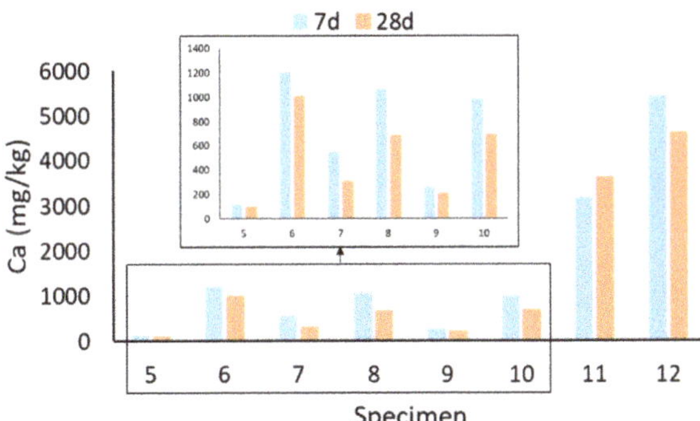

Figure 13. Photometrically measured amount of leached calcium from the specimens at 7 d (blue) and 28 d (orange). Specimens **1–4** are missing because the amount of calcium leached from them was negligible.

The amount of leached aluminium, silicon, and calcium depended on the age and the raw materials of the specimens, as well as the leaching of other substances. In general, less substance leached from the specimens at 28 d than at 7 d, indicating the maturation of the structure and the progress of the geopolymerization reaction. A notable leaching of a certain substance meant less leaching of the others. This was also observed in the gypsum-containing specimens. Gypsum increased calcium leaching but decreased the leaching of silicon and aluminium.

2.5. Crystalline Phase and SEM Measurements

Figures 14 and 15 present diffractograms of the metakaolin-based specimens (**1** and **2**) and fly ash-based specimens (**3** and **4**). The diffractograms for the other specimens are given in the supplementary material (Figures S1–S4). The structure of geopolymers is mostly amorphous, which caused a broad signal in diffractograms ($2\theta \approx 10$–$40°$) (Figures 14, 15 and S1–S3). The most abundant impurities in the specimens were quartz and mullite, which are weakly reactive compounds originating from the raw materials [14,59]. They were observed mostly in specimens made from biomass ash and fly ash (**3–6**, **9**, and **10**). The specimens made from metakaolin (**1** and **2**) and OPC (**11** and **12**) had only traces of quartz and mullite impurities. Specimens **1–4** contained CAC, which caused calcium aluminate signal ($27°$) in the diffractograms (Figures 14 and 15). Biomass ash and GGBS contain plenty of calcium, which caused calcium silicate hydrate signals (27–$29°$) in the diffractograms of specimens **5–10** (Figures S1–S3). As expected, based on previous studies, ettringite was observed in all specimens containing gypsum [29,30]. Although XRD measurement was not quantitative, conclusions could be drawn about the rate of ettringite formation and decomposition in the specimens. The formation is affected by the rate of diffusion and the amount of reactants [30]. Therefore, the rate varied a lot between the specimens and depended on the raw materials. The diffractogram of specimen **2** (Figure 14) shows that there was less ettringite in the specimen at 28 d than at 7 d, which means that the ettringite had already started to decompose. On the other hand, in some samples, the ettringite crystals persisted better, as can be seen in the diffractogram of specimen **4** (Figure 15).

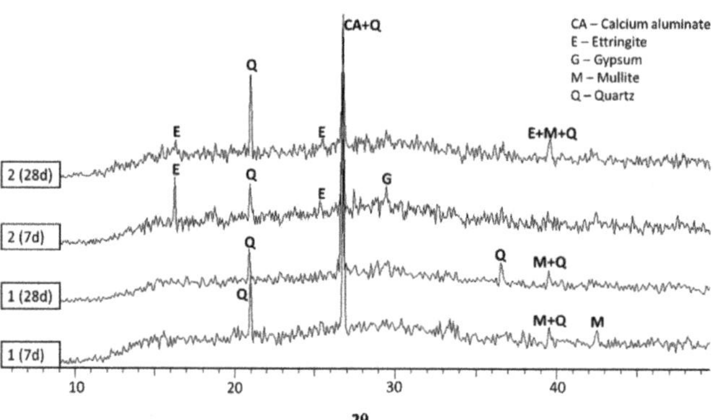

Figure 14. XRD of specimens **1** and **2**.

SEM measurements were performed to obtain further support of the relationship between the material fine structure and the strength of the specimens (Figure S6).

For example, specimen **8**, from blast furnace slag and biomass ash with added gypsum, showed low shrinkage. Its SEM figure shows needle-like crystals supporting the larger structural units (Figure 16a). It can be suggested that this prevents shrinkage when the residual water begins to leave the structure. By combining the EDS data of the observed crystals with the XRD analysis, it can be concluded that the crystals are formed from ettringite.

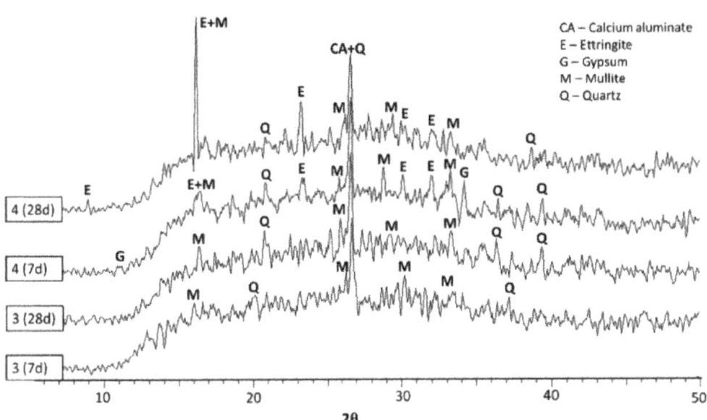

Figure 15. XRD of specimens **3** and **4**.

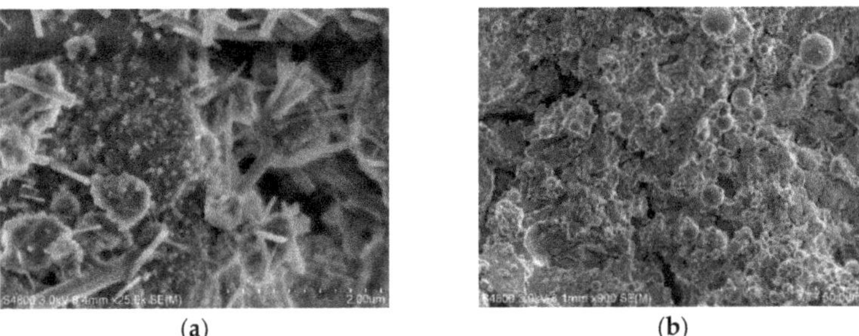

(a)   (b)

Figure 16. SEM figure of specimen **8** (a) and specimen **4** (b).

On the other hand, in specimen **4**, which is based on fly ash and gypsum, similar crystals are not observed; instead, the material is mainly a fusion of spherical fly ash particles (Figure 16b). The shrinkage of the specimen was high. The structure also had remarkable cracks caused by the plastic shrinkage, as also seen visually in the specimen.

The EDS of **4** shows surprisingly low sulfur content compared to specimen **8**. One possibility for this is that the reaction route or the kinetics are different in the formation of **4** than in the formation of **8**. This could lead to the flotation of gypsum to the surface of the specimen instead of it remaining in the material. This idea is supported by the formation of efflorescence on the surface of specimen **4** (Figure S5).

It is possible that, in the preparation of **4**, the solid substances were dissolved in the activator solution and the material was afforded strength so slowly in the geopolymerization reaction that the added gypsum did not form ettringite crystals. The progress of strength was slow even though the setting time was short. This indicates that the timely formed ettringite crystals support the structure units and hence prevent the collapse of the material when water starts to escape the structure.

Furthermore, the fragility of specimens **5** and **6** can be linked to the high crystallinity with poor interconnectivity between crystals.

3. Materials and Methods

3.1. Materials

In this study, 12 binder specimens with replicates were prepared from different raw materials (Tables 1 and 2). The main raw materials were metakaolin, fly ash (class F), biomass ash, and ground-granulated blast furnace slag (GGBS). Other materials than metakaolin were obtained from industrial side streams. In addition, a metakaolin-based specimen (**13**) containing aggregates was prepared (water to binder ratio of 0.61) and used as a control in the shrinkage measurements. The aggregate mixture (grain size ≤ 2 mm) consisted of mine tailings and bottom ash from fluidized bed waste incineration plant. Six of the specimens contained 5% recycled gypsum ($CaSO_4 \cdot 2H_2O$). Calcium aluminate cement (CAC) was used as an additive in specimens **1–4** and **13**. Sodium hydroxide (specimens **5–10**) or a mixture of sodium hydroxide and sodium silicate (specimens **1–4** and **13**) were used as an alkaline activator. Specimens **11** and **12** were made from ordinary Portland cement (OPC) and they were used as controls in this study.

Table 1. Elemental composition of raw materials as mass% of main cations.

Raw Material	Na + K	Ca	Si	Al
Metakaolin	1	-	25.7	21.2
CAC	-	27.2	1.7	22.0
Fly ash	1.7	14.3	21.0	10.0
Biomass ash	5.3	31.0	6.1	6.5
GGBS	0.9	27.9	17.8	4.8
OPC	0.4	36.3	7.2	7.3

Table 2. Numbering, solid raw materials, and water content (WC) of the specimens.

Specimen	Solid Raw Materials (Mass% of Total)	WC (%)	Specimen	Solid Raw Materials (Mass% of Total)	WC (%)
1	Metakaolin (34) CAC (12)	32.7	2	Metakaolin (32) CAC (11.4) $CaSO_4 \cdot 2H_2O$ (5)	31.0
3	Fly ash (54) CAC (9.1)	22.3	4	Fly ash (51) CAC (8.7) $CaSO_4 \cdot 2H_2O$ (5)	21.3
5	Fly ash (52) Biomass ash (24)	27.9	6	Fly ash (49) Biomass ash (21) $CaSO_4 \cdot 2H_2O$ (5)	26.8
7	GGBS (74)	25.1	8	GGBS (70) $CaSO_4 \cdot 2H_2O$ (5)	23.9
9	GGBS (48) Biomass ash (21)	25.8	10	GGBS (45.7) Biomass ash (20) $CaSO_4 \cdot 2H_2O$ (5)	25.3
11	OPC (75.8)	24.2	12	OPC (72.8) $CaSO_4 \cdot 2H_2O$ (5)	22.2
13	Metakaolin (10.3) CAC (3.6) Aggregates	16.2			

CAC = calcium aluminate cement; GGBS = ground-granulated blast furnace slag; OPC = ordinary Portland cement. NaOH was used as alkaline activator in specimens **1–4** and **13**. Mixture of NaOH and Na_2SiO_3 was used as alkaline activator in specimens **5–10**.

3.2. Preparation Process

Alkaline activator was added to solid substances (graded, d < 1 mm) and stirred for 3 min. The paste was poured into moulds ($40 \times 40 \times 160$ mm). Most of the specimens were cured sealed at room temperature for 1 d. However, specimens **5, 6, 8,** and **10** cured slowly. Therefore, they were cured sealed at 70 °C. After curing, the specimens were stored at room temperature in sealed bags for 28 d. During this time, measurements of the plastic phase were made (7 d and 28 d). The specimens were then taken from the bags and allowed to dry at room temperature for 28 d, after which the drying shrinkage (56 d) and SEM images were measured (100 d).

3.3. Testing Methods

3.3.1. Setting Time

Setting time of the specimens was measured by using Vicatronic Matest apparatus, which uses penetrating method. In this method, the setting time is defined when the penetration of the Vicat needle is 40 mm. Measurements were performed under ambient conditions.

3.3.2. Shrinkage

The shrinkage of the specimens was measured by comparing their length with a wooden standard piece. A Testing Bluhm and Feuerherdt shrinkage analyzator (accuracy 1 μm) was used in this study. Specimens **5** and **6** broke during the demoulding. Therefore, their shrinkage was measured with a calliper (accuracy 10 μm). The initial length of the specimens was measured immediately after demoulding (mostly 1 d, except for specimens **5** and **6**, for which the length was measured after 2 d and, for specimens **8** and **10**, after 4 d). The plastic shrinkage of the specimens was measured at 7 d and 28 d, and the drying shrinkage was measured at 56 d. The last measurement could not be conducted for specimens **5** and **6**, because they were so fragile that there were no intact pieces left.

3.3.3. Strength

Strengths of the specimens were measured with Instron 5581 (limit 50 kN). Flexural strength was measured from a prism (40 × 40 × 160 mm) by three-point flexing (press rate 25 MPa/s). Compressive strength was measured from a cube (40 × 40 × 40 mm) or rectangle (30 × 30 × 40 mm) if cube's compressive strength was above the limit (press rate 10 mm/min). The strengths of the specimens were measured at 7 d and 28 d.

3.3.4. Leaching of Si, Al, and Ca

The amounts of leached silicon, aluminium, and calcium from the specimens were measured at 7 d and 28 d. The measurement followed the standard EN 12457-2:2002 [60]. The amount of leached aluminium and silicon was measured photometrically (model YSI 9500, YSI Incorporated, Yellow Springs, OH, USA), and the amount of leached calcium was measured photometrically and by atomic absorption spectrometer (AAS, model SpecrAA 220, Varian Medical Systems, Inc., Palo Alto, CA, USA). Both methods have specific interferences, therefore several measurements were performed to obtain reliable results.

3.3.5. XRD

XRD was used to qualitatively identify crystalline phases of the specimens at 28 d. XRD test was performed via Bruker Advance D8 (Bruker Corporation, Billerica, MA, USA) powder diffractometer with the scanning range 5–50° (2θ) at a step 0.05° and a rate 5°/min. A database and literature were used to help interpret the diffractograms [61–67].

3.3.6. SEM

The surface structure of the samples was studied with a Hitachi S4800 FE-SEM (Hitachi, Ltd., Tokyo, Japan). Elemental identification of the surface structures was obtained with energy dispersive X-ray spectroscopy (EDS, Thermo Fisher Scientific, Waltham, MA, USA). Acceleration voltages of 3 and 10 kV were applied. The specimens (over 100 d) were cut into smaller species (approximately $1 \times 1 \times 1$ cm^3), dried in an oven at 110 °C for 1 h, and cooled to room temperature in desiccator before the measurements. Additional drying of the specimens was necessary to obtain the required vacuum for the measurement.

4. Conclusions

The combination of different raw materials of a geopolymer has a great impact on its properties. Hence, it is difficult to produce a geopolymer with the same properties as ordinary Portland cement (OPC). In this study, we compared the mechanical properties of specimens of geopolymers with and without recycled gypsum to establish trends in the setting time, shrinkage, and strength of the materials. In addition, the leaching of Al, Si, and Ca was studied. The results were supported by further analysis of the materials by XRD and SEM measurements. The main results can be summarized as follows:

- The metakaolin-based specimen (**1**) had the same setting time as OPC and its compressive strength was the closest to OPC, but the shrinkage was significantly greater, and the specimen had cracks caused by shrinkage. The GBSS and biomass ash-based specimen (**9**) was the closest to OPC in shrinkage without cracking, but had a much lower compressive strength and a much longer setting time.
- The leaching of Al, Si, and Ca from the specimens was mostly affected by the age and the raw materials of the specimen, as well as the leaching of other substances. However, the leaching of the measured substances did not seem to be related to the mechanical properties of the specimen.
- Recycled gypsum was found to have a notable effect on the setting time. Depending on the calcium content of the raw materials, it either accelerated or decreased the setting time.
- Gypsum effectively reduced the plastic shrinkage of the geopolymer specimens but had no effect on, or even increased, their drying shrinkage. This indicates that the shrinkage-reducing effect of gypsum is based on ettringite, which is not a stable compound. It decomposes over time and thus its shrinkage-reducing effect is also gradually lost.
- Gypsum had almost no effect on the total shrinkage of the geopolymers, as most of this consisted of drying shrinkage.
- It could be concluded from the diffractograms that the formation and decomposition of ettringite depends on the raw materials of the geopolymer.
- In this study, recycled gypsum reduced the compressive strength of the specimens, but this could perhaps be avoided by using a finer (<1 mm) gypsum powder. On the other hand, ettringite can also cause the material to become brittle.

Supplementary Materials: The supporting information can be downloaded at https://www.mdpi.com/article/10.3390/inorganics11070298/s1, and includes XRD data for specimens **5–13**, Figure of specimens **1–13** and SEM-images for **1–12**. Figure S1. XRD of specimens **5** and **6**; Figure S2. XRD of specimens **7** and **8**; Figure S3. XRD of specimens **9** and **10**; Figure S4. XRD of specimens **11** and **12**; Figure S5. Prepared specimens **1–13**; Figure S6. SEM-images of species **1–12**.

Author Contributions: H.K.: Investigation, writing—original draft. J.T.: Conceptualization, writing—review and editing, supervision. S.S.: Investigation, writing—review and editing. P.H.: Writing—review and editing, supervision. K.M.: Conceptualization, writing—review and editing. S.J.: Writing—review and editing, supervision. All authors have read and agreed to the published version of the manuscript.

Funding: This research received no external funding.

Data Availability Statement: The data presented in this study are available on request from the corresponding author.

Acknowledgments: We would like to thank Eetu Pietarinen for their assistance in the specimen preparation and measurements.

Conflicts of Interest: The authors declare no conflict of interest.

References

1. Siyal, A.A.; Shamsuddin, M.R.; Khan, M.I.; Rabat, N.E.; Zulfiqar, M.; Man, Z.; Siame, J.; Azizli, K.A. A Review on Geopolymers as Emerging Materials for the Adsorption of Heavy Metals and Dyes. *J. Environ. Manag.* **2018**, *224*, 327–339. [CrossRef] [PubMed]
2. Nazari, A.; Sanjayan, J.G. Synthesis of Geopolymer from Industrial Wastes. *J. Clean. Prod.* **2015**, *99*, 297–304. [CrossRef]
3. Duxson, P.; Fernández-Jiménez, A.; Provis, J.L.; Lukey, G.C.; Palomo, A.; van Deventer, J.S.J. Geopolymer Technology: The Current State of the Art. *J. Mater. Sci.* **2007**, *42*, 2917–2933. [CrossRef]
4. Hou, L.; Li, J.; Lu, Z.-Y. Effect of Na/Al on Formation, Structures and Properties of Metakaolin Based Na-Geopolymer. *Constr. Build. Mater.* **2019**, *226*, 250–258. [CrossRef]
5. Yu, G.; Jia, Y. Preparation of Geopolymer Composites Based on Alkali Excitation. *Arab. J. Geosci.* **2021**, *14*, 600. [CrossRef]
6. Kuenzel, C.; Vandeperre, L.J.; Donatello, S.; Boccaccini, A.R.; Cheeseman, C.; Torroja, E. Ambient Temperature Drying Shrinkage and Cracking in Metakaolin-Based Geopolymers. *J. Am. Ceram. Soc.* **2012**, *95*, 3270–3277. [CrossRef]
7. Kolezynski, A.; Król, M.; Zychowicz, M. The Structure of Geopolymers—Theoretical Studies. *J. Mol. Struct.* **2018**, *1163*, 465–471. [CrossRef]
8. Almutairi, A.L.; Tayeh, B.A.; Adesina, A.; Isleem, H.F.; Zeyad, A.M. Potential Applications of Geopolymer Concrete in Construction: A Review. *Case Stud. Constr. Mater.* **2021**, *15*, e00733. [CrossRef]
9. Provis, J.L.; van Deventer, J.S.J. *Geopolymers: Structure, Processing, Properties and Industrial Applications*; Provis, J.L., van Deventer, J.S.J., Eds.; Woodhead: Boca Raton, FL, USA; CRC Press: Oxford, UK, 2009; pp. 1–11.
10. van Deventer, J.S.J.; Provis, J.L.; Duxson, P. Technical and Commercial Progress in the Adoption of Geopolymer Cement. *Miner. Eng.* **2012**, *29*, 89–104. [CrossRef]
11. Kumar Mehta, P. Reducing the Environmental Impact of Concrete. *Concr. Int.* **2001**, *23*, 61–66.
12. Mclellan, B.C.; Williams, R.P.; Lay, J.; van Riessen, A.; Corder, G.D. Costs and Carbon Emissions for Geopolymer Pastes in Comparison to Ordinary Portland Cement. *J. Clean. Prod.* **2011**, *19*, 1080–1090. [CrossRef]
13. Ridtirud, C.; Chindaprasirt, P.; Pimraksa, K. Factors Affecting the Shrinkage of Fly Ash Geopolymers. *Int. J. Miner. Metall. Mater.* **2011**, *18*, 100–104. [CrossRef]
14. Davidovits, J. *Geopolymer Chemistry and Applications*, 4th ed.; Davidovits, J., Ed.; Institut Géopolymère: Saint-Quentin, France, 2015.
15. Wang, R.; Wang, J.; Dong, T.; Ouyang, G. Structural and Mechanical Properties of Geopolymers Made of Aluminosilicate Powder with Different SiO_2/Al_2O_3 Ratio: Molecular Dynamics Simulation and Microstructural Experimental Study. *Constr. Build. Mater.* **2020**, *240*, 117935. [CrossRef]
16. Kenne Diffo, B.B.; Elimbi, A.; Cyr, M.; Dika Manga, J.; Tchakoute Kouamo, H. Effect of the Rate of Calcination of Kaolin on the Properties of Metakaolin-Based Geopolymers. *J. Asian Ceram. Soc.* **2015**, *3*, 130–138. [CrossRef]
17. Singh, B.; Rahman, M.R.; Paswan, R.; Bhattacharyya, S.K. Effect of Activator Concentration on the Strength, ITZ and Drying Shrinkage of Fly Ash/Slag Geopolymer Concrete. *Constr. Build. Mater.* **2016**, *118*, 171–179. [CrossRef]
18. Duxson, P.; Lukey, G.C.; van Deventer, J.S.J. Thermal Evolution of Metakaolin Geopolymers: Part 1-Physical Evolution. *J. Non-Cryst. Solids* **2006**, *352*, 5541–5555. [CrossRef]
19. Barbosa, V.F.F.; MacKenzie, K.J.D. Synthesis and Thermal Behaviour of Potassium Sialate Geopolymers. *Mater. Lett.* **2003**, *57*, 1477–1482. [CrossRef]
20. Zhao, J.; Tong, L.; Li, B.; Chen, T.; Wang, C.; Yang, G.; Zheng, Y. Eco-Friendly Geopolymer Materials: A Review of Performance Improvement, Potential Application and Sustainability Assessment. *J. Clean. Prod.* **2021**, *307*, 127085. [CrossRef]
21. Wang, D.; Wang, Q.; Huang, Z. New Insights into the Early Reaction of NaOH-Activated Slag in the Presence of $CaSO_4$. *Compos. B Eng.* **2020**, *198*, 108207. [CrossRef]
22. El Alouani, M.; Saufi, H.; Moutaoukil, G.; Alehyen, S.; Nematollahi, B.; Belmaghraoui, W.; Taibi, M. Application of Geopolymers for Treatment of Water Contaminated with Organic and Inorganic Pollutants: State-of-the-Art Review. *J. Environ. Chem. Eng.* **2021**, *9*, 105095. [CrossRef]
23. Pol Segura, I.; Jensen, P.A.; Damø, A.J.; Ranjbar, N.; Jensen, L.S.; Canut, M. Influence of Sodium-Based Activators and Water Content on the Fresh and Hardened Properties of Metakaolin Geopolymers. *SN Appl. Sci.* **2022**, *4*, 283. [CrossRef]
24. Hassan, A.; Arif, M.; Shariq, M. Effect of Curing Condition on the Mechanical Properties of Fly Ash-Based Geopolymer Concrete. *SN Appl. Sci.* **2019**, *1*, 1694. [CrossRef]
25. Fang, G.; Bahrami, H.; Zhang, M. Mechanisms of Autogenous Shrinkage of Alkali-Activated Fly Ash-Slag Pastes Cured at Ambient Temperature within 24 h. *Constr. Build. Mater.* **2018**, *171*, 377–387. [CrossRef]
26. Mastali, M.; Kinnunen, P.; Dalvand, A.; Mohammadi Firouz, R.; Illikainen, M. Drying Shrinkage in Alkali-Activated Binders—A Critical Review. *Constr. Build. Mater.* **2018**, *190*, 533–550. [CrossRef]
27. Ranjbar, N.; Zhang, M. Fiber-Reinforced Geopolymer Composites: A Review. *Cem. Concr. Compos.* **2020**, *107*, 103498. [CrossRef]
28. Kotrla, J.; Soukal, F.; Bilek, V.; Alexa, M. Effects of Shrinkage-Reducing Admixtures on Autogenous Shrinkage in Alkali-Activated Materials. *IOP Conf. Ser. Mater. Sci. Eng.* **2019**, *583*, 012023. [CrossRef]
29. Bakharev, T.; Sanjayan, J.G.; Cheng, Y.-B. Effect of Admixtures on Properties of Alkali-Activated Slag Concrete. *Cem. Concr. Res.* **2000**, *30*, 1367–1374. [CrossRef]
30. Tao, S.; Yumei, Y. Quantitative Analysis of Ettringite Formed in the Hydration Products of High-Alite Cements. *Adv. Cem. Res.* **2015**, *27*, 497–505. [CrossRef]

31. Liu, J.; Hu, L.; Tang, L.; Zhang, E.Q.; Ren, J. Shrinkage Behaviour, Early Hydration and Hardened Properties of Sodium Silicate Activated Slag Incorporated with Gypsum and Cement. *Constr. Build. Mater.* **2020**, *248*, 118687. [CrossRef]
32. Divvala, S.; Rani, M.S. Early Strength Properties of Geopolymer Concrete Composites: An Experimental Study. *Mater. Today Proc.* **2021**, *47*, 3770–3777. [CrossRef]
33. Zaheer, M.; Khan, N.; Uddin, F.; Shaikh, A.; Hao, Y.; Hao, H. Synthesis of High Strength Ambient Cured Geopolymer Composite by Using Low Calcium Fly Ash. *Constr. Build. Mater.* **2016**, *125*, 809–820. [CrossRef]
34. Rattanasak, U.; Chindaprasirt, P. Influence of NaOH Solution on the Synthesis of Fly Ash Geopolymer. *Miner. Eng.* **2009**, *22*, 1073–1078. [CrossRef]
35. Pane, I.; Imran, I.; Budiono, B. Compressive Strength of Fly Ash-Based Geopolymer Concrete with a Variable of Sodium Hydroxide (NaOH) Solution Molarity. *MATEC Web Conf.* **2018**, *147*, 01004. [CrossRef]
36. Hardjasaputra, H.; Cornelia, M.; Gunawan, Y.; Surjaputra, I.V.; Lie, H.A.; Pranata Ng, G. Study of Mechanical Properties of Fly Ash-Based Geopolymer Concrete. *IOP Conf. Ser. Mater. Sci. Eng.* **2019**, *615*, 012009. [CrossRef]
37. Talha Ghafoor, M.; Khan, Q.S.; Qazi, A.U.; Sheikh, M.N.; Hadi, M.N.S. Influence of Alkaline Activators on the Mechanical Properties of Fly Ash Based Geopolymer Concrete Cured at Ambient Temperature. *Constr. Build. Mater.* **2021**, *273*, 121752. [CrossRef]
38. Gómez-Casero, M.A.; de Dios-Arana, C.; Bueno-Rodríguez, J.S.; Pérez-Villarejo, L.; Eliche-Quesada, D. Physical, Mechanical and Thermal Properties of Metakaolin-Fly Ash Geopolymers. *Sustain. Chem. Pharm.* **2022**, *26*, 100620. [CrossRef]
39. Zhang, H.Y.; Kodur, V.; Cao, L.; Qi, S.L. Fiber Reinforced Geopolymers for Fire Resistance Applications. *Procedia Eng.* **2014**, *71*, 153–158. [CrossRef]
40. Cao, Y.F.; Tao, Z.; Pan, Z.; Wuhrer, R. Effect of Calcium Aluminate Cement on Geopolymer Concrete Cured at Ambient Temperature. *Constr. Build. Mater.* **2018**, *191*, 242–252. [CrossRef]
41. Liew, Y.-M.; Cheng-Yong, H.; Al Bakri, M.; Hussin, K. Structure and Properties of Clay-Based Geopolymer Cements: A Review. *Prog. Mater. Sci.* **2016**, *83*, 595–629. [CrossRef]
42. Juengsuwattananona, K.; Winnefeldb, F.; Chindaprasirtc, P.; Pimraksa, K. Correlation between Initial SiO_2/Al_2O_3, Na_2O/Al_2O_3, Na_2O/SiO_2 and H_2O/Na_2O Ratios on Phase and Microstructure of Reaction Products of Metakaolin-Rice Husk Ash Geopolymer. *Constr. Build. Mater.* **2019**, *226*, 406–417. [CrossRef]
43. Qu, F.; Li, W.; Wang, K.; Zhang, S.; Sheng, D. Performance Deterioration of Fly Ash/Slag-Based Geopolymer Composites Subjected to Coupled Cyclic Preloading and Sulfuric Acid Attack. *J. Clean. Prod.* **2021**, *321*, 128942. [CrossRef]
44. Kozub, B.; Castro-Gomes, J. An Investigation of the Ground Walnut Shells' Addition Effect on the Properties of the Fly Ash-Based Geopolymer. *Materials* **2022**, *15*, 3936. [CrossRef] [PubMed]
45. Rojo-López, G.; González-Fonteboa, B.; Martínez-Abella, F.; González-Taboada, I. Rheology, Durability, and Mechanical Performance of Sustainable Self-Compacting Concrete with Metakaolin and Limestone Filler. *Case Stud. Constr. Mater.* **2022**, *17*, e01143. [CrossRef]
46. Weimann, K.; Adam, C.; Buchert, M.; Sutter, J. Environmental Evaluation of Gypsum Plasterboard Recycling. *Minerals* **2021**, *11*, 101. [CrossRef]
47. Ibrahim, Y.E.; Adamu, M.; Marouf, M.L.; Ahmed, O.S.; Drmosh, Q.A.; Malik, M.A. Mechanical Performance of Date-Palm-Fiber-Reinforced Concrete Containing Silica Fume. *Buildings* **2022**, *12*, 1642. [CrossRef]
48. Golewski, G.L. Combined Effect of Coal Fly Ash (CFA) and Nanosilica (NS) on the Strength Parameters and Microstructural Properties of Eco-Friendly Concrete. *Energies* **2023**, *16*, 452. [CrossRef]
49. Alfimova, N.; Pirieva, S.; Levickaya, K.; Kozhukhova, N.; Elistratkin, M. The Production of Gypsum Materials with Recycled Citrogypsum Using Semi-Dry Pressing Technology. *Recycling* **2023**, *8*, 34. [CrossRef]
50. Kozhukhova, N.I.; Alfimova, N.I.; Kozhukhova, M.I.; Nikulin, I.S.; Glazkov, R.A.; Kolomytceva, A.I. The Effect of Recycled Citrogypsum as a Supplementary Mineral Additive on the Physical and Mechanical Performance of Granulated Blast Furnace Slag-Based Alkali-Activated Binders. *Recycling* **2023**, *8*, 22. [CrossRef]
51. Nawaukkaratharnant, N. Utilization of Gypsum-Bonded Investment Mold Waste from Jewelry and Accessory Industry as Raw Material for Construction Materials Using Geopolymer Technology. *Rep. Grant-Support. Res. Asahi Glass Found.* **2021**, *89*, 1–12.
52. An, Q.; Pan, H.; Zhao, Q.; Du, S.; Wang, D. Strength Development and Microstructure of Recycled Gypsum-Soda Residue-GGBS Based Geopolymer. *Constr. Build. Mater.* **2022**, *331*, 127312. [CrossRef]
53. Abadel, A.A.; Alghamdi, H.; Alharbi, Y.R.; Alamri, M.; Khawaji, M.; Abdulaziz, M.A.M.; Nehdi, M.L. Investigation of Alkali-Activated Slag-Based Composite Incorporating Dehydrated Cement Powder and Red Mud. *Materials* **2023**, *16*, 1551. [CrossRef]
54. Ricciotti, L.; Occhicone, A.; Ferone, C.; Cioffi, R.; Tarallo, O.; Roviello, G. Development of Geopolymer-Based Materials with Ceramic Waste for Artistic and Restoration Applications. *Materials* **2022**, *15*, 8600. [CrossRef]
55. Mohammed, A.A.; Ahmed, H.U.; Mosavi, A. Survey of Mechanical Properties of Geopolymer Concrete: A Comprehensive Review and Data Analysis. *Materials* **2021**, *14*, 4690. [CrossRef]
56. Deb, P.S.; Nath, P.; Sarker, P.K. Drying Shrinkage of Slag Blended Fly Ash Geopolymer Concrete Cured at Room Temperature. *Procedia Eng.* **2015**, *125*, 594–600. [CrossRef]
57. Alnkaa, A.; Yaprak, H.; Selcuk, M.; Kaplan, G. Effect of Different Cure Conditions on the Shrinkage of Geopolymer Mortar. *Int. J. Eng. Res. Dev.* **2018**, *14*, 51–55.

58. Lu, T.; Li, Z.; van Breugel, K. Modelling of Autogenous Shrinkage of Hardening Cement Paste. *Constr. Build. Mater.* **2020**, *264*, 120708. [CrossRef]

59. Katare, V.D.; Madurwar, M.V. Experimental Characterization of Sugarcane Biomass Ash—A Review. *Constr. Build. Mater.* **2017**, *152*, 1–15. [CrossRef]

60. *EN 12457-2:2002*; Characterisation of Waste—Leaching—Compliance Test for Leaching of Granular Waste Materials and Sludges. Comite Europeen de Normalisation: Brussels, Belgium, 2002.

61. Levien, L.; Prewitt, C.T.; Weidner, D.J. Structure and Elastic Properties of Quartz at Pressure P = 1 Atm. *Am. Mineral.* **1980**, *65*, 920–930.

62. Balzar, D.; Ledbetter, H. Crystal Structure and Compressibility of 3:2 Mullite. *Am. Mineral.* **1993**, *78*, 1192–1196.

63. Goetz-Neunhoeffer, F.; Neubauer, J. Refined Ettringite ($Ca_6Al_2(SO_4)_3(OH)_{12}*26H_2O$) Structure for Quantitative X-Ray Diffraction Analysis. *Powder Diffr.* **2006**, *21*, 4–11. [CrossRef]

64. Schofield, P.F.; Knight, K.S.; Stretton, I.C. Thermal Expansion of Gypsum Investigated by Neutron Powder Diffraction T = 4.2 K. *Am. Mineral.* **1996**, *81*, 847–851. [CrossRef]

65. Abolhasani, A.; Aslani, F.; Samali, B.; Ghaffar, S.H.; Fallahnejad, H.; Banihashemi, S. Silicate Impurities Incorporation in Calcium Aluminate Cement Concrete: Mechanical and Microstructural Assessment. *Adv. Appl. Ceram.* **2021**, *120*, 104–116. [CrossRef]

66. Pacewska, B.; Nowacka, M.; Aleknevičius, M.; Antonovič, V. Early Hydration of Calcium Aluminate Cement Blended with Spent FCC Catalyst at Two Temperatures. *Procedia Eng.* **2013**, *57*, 844–850. [CrossRef]

67. Chen, B.; Zhang, Y.; Chen, Q.; Yang, F.; Liu, X.; Wu, J.; Wang, P. Effect of Mineral Composition and w/c Ratios to the Growth of AFt during Cement Hydration by In-Situ Powder X-Ray Diffraction Analysis. *Materials* **2020**, *13*, 4963. [CrossRef]

Review

How to Address Flame-Retardant Technology on Cotton Fabrics by Using Functional Inorganic Sol–Gel Precursors and Nanofillers: Flammability Insights, Research Advances, and Sustainability Challenges

Valentina Trovato [1,*], Silvia Sfameni [2], Rim Ben Debabis [1], Giulia Rando [3], Giuseppe Rosace [1], Giulio Malucelli [4,*] and Maria Rosaria Plutino [2,*]

[1] Department of Engineering and Applied Sciences, University of Bergamo, Viale Marconi 5, 24044 Dalmine, Italy; rim.bendebabis@unibg.it (R.B.D.); giuseppe.rosace@unibg.it (G.R.)

[2] Institute for the Study of Nanostructured Materials (ISMN–CNR), Palermo, c/o Department of ChiBioFarAm, University of Messina, Viale F. Stagno d'Alcontres 31, Vill. S. Agata, 98166 Messina, Italy; silvia.sfameni@ismn.cnr.it

[3] Department of ChiBioFarAm, University of Messina, Viale F. Stagno d'Alcontres 31, Vill. S. Agata, 98166 Messina, Italy; giulia.rando@unime.it

[4] Department of Applied Science and Technology, Politecnico di Torino, Viale T. Michel 5, 15121 Alessandria, Italy

* Correspondence: valentina.trovato@unibg.it (V.T.); giulio.malucelli@polito.it (G.M.); mariarosaria.plutino@cnr.it (M.R.P.)

Citation: Trovato, V.; Sfameni, S.; Ben Debabis, R.; Rando, G.; Rosace, G.; Malucelli, G.; Plutino, M.R. How to Address Flame-Retardant Technology on Cotton Fabrics by Using Functional Inorganic Sol–Gel Precursors and Nanofillers: Flammability Insights, Research Advances, and Sustainability Challenges. *Inorganics* **2023**, *11*, 306. https://doi.org/10.3390/inorganics11070306

Academic Editors: Roberto Nisticò, Torben R. Jensen, Luciano Carlos, Hicham Idriss and Eleonora Aneggi

Received: 27 June 2023
Revised: 14 July 2023
Accepted: 16 July 2023
Published: 18 July 2023

Abstract: Over the past decade, inorganic fillers and sol–gel-based flame-retardant technologies for textile treatments have gained increasing research interest as useful alternatives to hazardous chemicals previously employed in textile coating and finishing. This review presents the current state of the art of inorganic flame-retardant technology for cotton fabrics to scientists and researchers. Combustion mechanism and flammability, as well as the thermal behavior of neat cotton samples, are first introduced. The main section is focused on assessing the effect of inorganic and sol–gel-based systems on the final flame-retardant properties of cotton fabrics, emphasizing their fire safety characteristics. When compared to organic flame-retardant solutions, inorganic functional fillers have been shown to be more environmentally friendly and pollution-free since they do not emit compounds that are hazardous to ecosystems and humans when burned. Finally, some perspectives and recent advanced research addressing the potential synergism derived from the use of inorganic flame retardants with other environmentally suitable molecules toward a sustainable flame-retardant technological approach are reviewed.

Keywords: inorganic flame retardants; sol–gel technology; nanoclays; functional nanofillers; functional coatings; cotton fabrics

1. Introduction

In the last decades, continuous improvements in people's living standards, together with environmental problems and population growth, have generated a demand for advanced materials, thus resulting in ever-accelerating scientific progress and research. In this regard, a special place is covered by textiles, which are present in everyday human life, not only for conventional clothes and accessories, but also as decoration and comfort elements in homes, and public and private buildings, as components in transportation, and as structural elements for buildings. As a matter of fact, besides the conventional textiles sector, more complex fabric manufacturing is growing, aiming to a high level of product innovation and cutting-edge process technologies, continuously searching for specific textiles aesthetics and high-quality characteristics, as well as new functionalities (i.e., resulting from hybrid material additives capable of providing protection, comfort, and performance to the final textile products). At this stage, it appears very clear that textile fabrics must be treated not only as artistic surfaces for fashion but as materials between the wearers and the surroundings for all intents and purposes, with their tunable intrinsic structures and performances.

Today, the so-called technical textiles, designed for their functional properties rather than their aesthetic features, have become a revolutionary product category, providing the foundation for an entirely new variety of applications [1]. They can be utilized for various industrial sectors, also thanks to the possibility of introducing many functionalizations during the finishing step, such as water-repellent, antibacterial, or flame-retardant properties, among others [1,2]. Keeping an eye on nature and human health protection, stringent environmental restrictions and greater interest in natural resource usage have recently led to an increasing trend toward the exploitation of natural fibers in technological applications as eco-friendly options to synthetic counterparts [3].

Among others, cotton fabric represents one of the most widely employed natural fibers in the textile industry. Indeed, the average global production of cotton fibers reached about 25 million tonnes during recent years, representing 28% of all fibers' global production [4]. Cotton fiber structure consists of a polysaccharide component of β-d-glucopyranose units linked together by β-1,4 bonds (Figure 1); non-reducing and reducing sugar units stabilize the end terminal of cellulose polymer chains [5]. Chemical modifications and fiber behavior are mainly attributable to the -OH groups on positions C-2, C-3, and C-6 of the D-glucopyranosyl units [5,6].

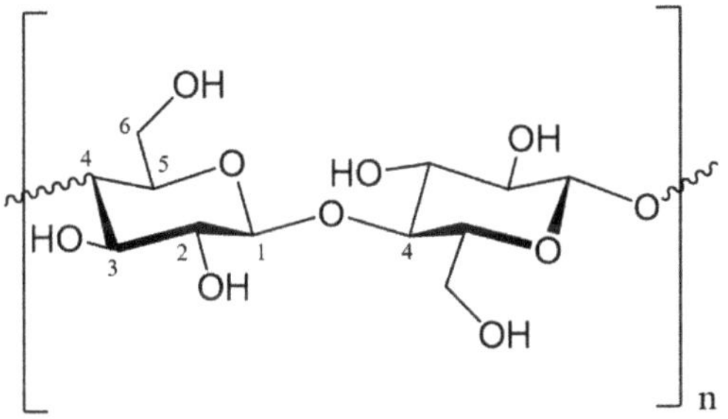

Figure 1. Cellulose structure with carbon atoms numbering. Numbers (1 and 4) were used in the text to explain the link between monomers (*Cotton fiber structure consists of a polysaccharide component of β-d-glucopyranose units linked together by β-1,4 bonds*).

Due to these superior characteristics (i.e., fast moisture absorption, mechanical properties, breathability, softness, comfort, biodegradability, and good thermal conductivity), cotton is used extensively in clothing, bedding, furniture, and wall hangings, as well as in apparel manufacturing, home furnishings, medical textiles, and other industrial products [7]. Moreover, cotton-based textiles have been used as protective clothing for workers and, more generally, as uniforms employed in workplaces where there is a chance of accidental contact with flames [8]. However, its disadvantage is represented by the easy ignition and aptitude to burn in the air (its limiting oxygen index, LOI, is around 18%) [9] as well as the quick flammability propensity, which have restricted its application in particular fields requiring textiles with enhanced flame retardancy [10]. As a significant component in a range of everyday textiles, most of the research efforts have been addressed to investigate the flammability of cotton fabrics [11].

Currently, the research on flame-retardant treatments for fabrics has become a necessary pathway, and many countries have created relevant testing standards and legislation to standardize the market [12]. As reported in the scientific literature [13], the estimated costs from fire losses are approximately 1% of the global gross domestic product. Only in the United States, home fires are the most frequent type of fire accidents [14]: the single most significant cause of civilian deaths refers to home fires (about 24%), due to the presence of residential upholstered furniture, with a yearly estimated average of 8900 fires, 610 deaths, and 1120 injuries, resulting in $566 million in direct damage [15]. Furthermore, the annual United Kingdom fire statistics demonstrate that most of the fire accidents that occur in houses involve upholstering furniture, bedding, and nightwear [16]. According to statistics from the International Association of Fire and Rescue Services [17], in the period 2016–2020, several fires involving textiles as a major fuel element have driven the research toward the development of new fire precautionary and preventative procedures.

The most common flame-retardant (hereafter, FR) finishes for cotton fabrics were developed from 1950 to 1980, and were based on halogen derivatives, phosphorous, and/or nitrogen [18]. In particular, during the combustion of halogen-containing flame-retardant fabrics, toxic gases, such as hydrogen halide, are produced, causing harm to human health and environmental pollution. Because of these concerns, halogenated finishes have been rigorously restricted in textile finishing [19].

In this regard, the development of flame-retardant textile finishing able to provide self-extinguishing properties and delay or inhibit the flame spread is relevant. Accordingly, the flame-retardant effect is attained when at least one or more factors among fuel, heat, and oxygen are reduced or eliminated, thus further enhancing the thermal stability of the polymer fabric and quenching the formed high-energy free radicals. Generally, flame-retardant mechanisms involve chemical, physical processes, or a combination of both. Mainly, the action mechanism of flame retardants can be traced back as follows [20]:

a. Gas phase. FRs following this mechanism act by diluting the gas phase and/or by chemical quenching of active radicals. The former effect is based on releasing non-combustible gases (e.g., H_2O and CO_2) that can dilute the oxygen or the fuel concentrations by lowering them under the flammability limit. Metal hydroxides and carbonates are generally believed to follow this mode of action due to their endothermic thermal decomposition and production of non-combustible gases. On the other side, because of the occurrence of radical reactions during combustion, the flame retardants decompose in radical species able to quench the high-energy free radicals formed during cellulose combustion (e.g., H• and •OH) by decreasing the burning rate of the combustible materials and, finally, interrupting the exothermic reactions of the combustion. However, the mechanism in the gas phase may be slightly different, depending on the used chemicals [21];

b. Condensed phase. The thermal cracking reaction process of cellulose can be modified by flame retardants. Indeed, many reactions (e.g., dehydration, condensation, cross-linking, and cyclization) take place at lower temperatures by producing coherent carbon layers on the fabric surface, thus lowering both the evolution of combustible gases and the decomposition rate of the fabric. For more in detail, the depolymerization of fiber materials is observable under the action of the flame retardant, as well as a decrease in the melting temperature that leads to a higher temperature difference between the melting and ignition point [22]. According to the chemical structure of the employed FR, an intumescent effect can also be observable. Moreover, a certain amount of heat is absorbed by interrupting the feedback of the heat to the fibers and, finally, the combustion process. Inorganic finishes containing phosphorus, boron, sol–gel precursors, nanoclay, and metal-based finishes, as well as carbon nanotubes and graphene, are believed to follow this mode of action.

However, combustion is a complex process, and the action of flame retardants may occur across both gas and condensed phases or in just one of the two. Indeed, synergistic flame-retardant actions can be obtained by combining different mechanisms that, conversely, are barely assignable to a single flame-retardant system [23].

As an alternative to most common FR treatments, whose flame-retardant mechanism mainly occurs in the gas phase, formulations acting in the condensed phase and containing nitrogen and phosphorus have been developed, promoting the formation of a char during thermal degradation, which provides an insulating layer to the underlying polymer. Indeed, organophosphate flame retardants containing synergistically active nitrogen may be more effective than pure phosphate flame retardants. According to [24], in systems based on the P–N synergism, the phosphoric acid undergoes the nucleophilic attack of the nitrogen, resulting in the formation of P–N bonded polymers. This bond is more polar than the already present P–O bonds, and the improved electrophilicity of the phosphorus atom boosts its capacity to phosphorylate the C(6) primary hydroxyl group of cellulose [25]. This, in turn, prevents the intramolecular C(6)–C(1) rearrangement process that produces levoglucosan. At the same time, the char formation derived by the action of the same flame retardants is promoted and consolidated by the auto-cross-linking of cellulose. Although several research works on P–N synergy in cotton have been published, to the best of our knowledge, only qualitative observations of this phenomenon are often discussed. Lewin [26] and Horrocks [27] have shown that the actual synergy between two species (i.e., phosphorus and nitrogen) can only be determined by calculating synergy parameters, according to which the effects of these two species may be additive or even antagonistic.

In this scenario, among others, organophosphorus compounds, such as tetrakis (hydroxymethyl) phosphonium chloride (THPC), hydroxyl functional organophosphorus oligomer (HFPO), as well as N-alkyl-substituted phosphono-propionamide derivatives, have been largely used on the market, dramatically increasing the number of applications for cellulose-based flame-retardant textiles [8,28].

A common FR process for cotton fabrics exhibiting washing fastness is provided by Proban® (Rhodia, La Défense, France), consisting of tetrakis-hydroxymethyl-phosphonium chloride (THPC) cross-linked with gaseous ammonia. The durability of this FR is explained by its resistance on textiles even after 70 washing cycles performed at temperatures beyond 70 °C in contrast with some drawbacks such as the formaldehyde release and the stiffness of treated fabrics. On the other side, Pyrovatex® (Huntsman, Woodlands, TX, USA), based on dimethyl-N-dydroxymethylcarbamoyl-ethylphosphonate, is another commercial FR with slightly lower durability than Proban® (Rhodia). This FR treatment is based on the acid-catalyzed N-methylol moiety chemical bonding of phosphorous-based compounds to cotton surfaces, but it has the problem of formaldehyde release both during the process and the use of the treated fabrics.

However, all these finishes need high loading of chemicals on the textile surface, requiring more synthetic substances in the treatments [29], which can alter the physical qualities and "hand" (i.e., soft touch) of the treated textiles. Moreover, according to recent research, some phosphorus-containing substances emit much smoke, may be toxic or potentially mutagenic, or pose other risks to the environment and human health [30]. Furthermore, as previously stated, most of them have an adverse environmental profile for producing free formaldehyde during application or use due to active hydroxymethyl units in their chemical structure [23].

The need to replace the above-mentioned fire-retardant finishes with eco-friendly, halogen- and formaldehyde-free formulations is an environmental progression according to the European directives and REACH regulation [31,32]. Moreover, internationally active private programs, such as Bluesign, Global Organic Textile Standard (GOTS), and Zero Discharge of Hazardous Chemicals (ZDHC), and major textile trading companies have defined guidelines limiting hazardous substances, which are usually even more restrictive than the government laws [33]. The design of alternative FRs to the conventional ones that are often banned due to their health and environmental concerns must fulfill several characteristics, including ease of application, maintenance of the main textile properties (e.g., comfort, appearance, aesthetics, tensile properties, and air permeability), durability, relatively low cost/performance value, no release of toxic substances (e.g., formaldehyde), no toxicity, and low environmental impact during application or use. Furthermore, innovative approaches focus not only on the selection of non-hazardous molecules but also on processes and methods exhibiting low environmental impact [16].

Recent advances in flame-retardant strategies have proposed the treatment of cotton textiles with insulating inorganic materials to mimic the formation of a char layer onto cellulose-based fabrics due to the presence of finishes containing zinc sulfide or oxide [19], Lewis acids, and aluminum sulphate, among others. Compared with organic counterparts, inorganic flame retardants are relatively green and pollution-free because they do not generate harmful chemicals to ecosystems and humans when they are burnt [34]. At low temperatures, they work by reducing the decomposition temperature by dehydration, decarboxylation, gradual depolymerization, and recombination of the breakdown products to form carbonaceous char. This prevents reactions that occur at high temperatures through the breakdown of the macromolecule by intramolecular transglycosylation to anhydroglucose units and their subsequent degradation to lower molecular weight flammable volatiles.

Following this overview of FR conventional solutions, this review aims to discuss the recent developments in the inorganic flame retardation of cotton fabrics. According to Ling et al. [23], several classifications are available for flame retardants: among them, the one based on the chemical element is often exploited. Furthermore, taking into consideration other relevant aspects, such as sustainability and performance, the flame retardants discussed in this review have been classified as follows: inorganic FRs (including phosphorous- or boron-based), sol–gel-derived FRs, and inorganic FR nanoparticles (including nanoclays, carbon nanotubes, graphene, and metal-based structures). First of all, a special emphasis is placed on the flammability and thermal behavior of untreated cotton fabrics with the aim of providing a detailed overview of the finishing processes, their optimization, and the related mechanism to researchers interested in the study of the influence of sol–gel- and inorganic-based flame-retardant finishes and coatings on cellulose-based textiles. Then, the comprehensive properties of inorganic sol–gel precursors and fillers as flame-retardant coatings and finishes for cotton are presented. Finally, an outline of the mentioned flame-retardant methods and materials, together with future challenges toward sustainability, is provided.

2. Mechanism of Cotton Combustion

As a cellulose-based polymer, the combustion of cotton is an exothermic oxidation process that takes place upon heating, consuming flammable gases, liquids, and solid residues produced during the pyrolysis of the textile material, thus generating heat. The process is considered to involve mainly four stages: heating, pyrolysis, ignition, and flame spread. During the heating step, the temperature of the cotton is raised by external ignition to a level that depends on the intensity of the ignition source and the thermal properties of the textile. When cellulose is heated at temperatures not exceeding 150 °C, water desorption occurs.

Once the cellulosic material exceeds 250 °C, pyrolysis occurs as an endothermic process [35,36], resulting in the irreversible polymer decomposition by producing tar (of which levoglucosan is the main component), flammable gases (methane, ethane, and carbon monoxide), non-flammable gases (carbon dioxide and formaldehyde), other by-products (water, alcohols, organic acids, aldehydes, and ketones), and char. If the combustion process generates sufficient heat, the heat is further transferred to the textile substrate, further accelerating the degradation processes and leading to a "self-sustaining" combustion cycle (Figure 2). The thermal decomposition of cotton fabrics generates approximately 51% water and gases, 47% tar, and 2% char [37].

More specifically, the depolymerization of cellulose can be mainly divided into three stages as a function of the temperature range:
- a first stage (between 300 and 400 °C), corresponding to the pyrolysis that takes place through two competing decomposition reactions: dehydration and depolymerization [38]. The former produces anhydro-cellulose (dehydro-cellulose), which further decomposes at higher temperatures and produces various volatile products such as alcohols, alkanes, aldehydes, fuel gases, carbon monoxide, methane, ethylene, and non-flammable gases (carbon dioxide and water vapor), and an aliphatic char (char I). Otherwise, depolymerization by breaking of glycosidic linkages results in the for-

mation of tar (condensed phase), which is mostly composed of levoglucosan and glycolaldehyde (hydrogen acetaldehyde);

- a second stage (between 400 and 600 °C), corresponding to the competitive conversion of aliphatic char to aromatic and char oxidation [38]. In this stage, tar is further decomposed into small volatile flammable molecules and aromatic char (char II, stable up to 800 °C), while volatile compounds from Stage 1 are also oxidized to produce similar oxidized char and aromatic molecules;
- a third stage (between 600 and 800 °C), corresponding to the further char decomposition to acetylene, CO, and CO_2 (beyond 800 °C), as well as other cellulose pyrolytic products [39–42].

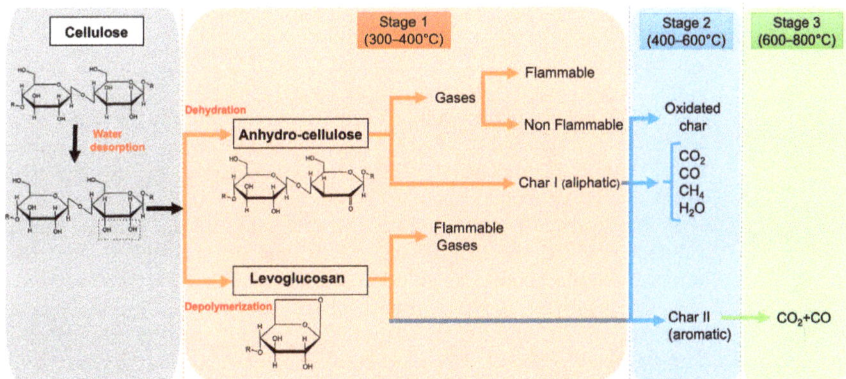

Figure 2. Schematic process of thermal degradation of cellulose.

The heating rate has been proven to influence the concentration of fuel products during the overall combustion steps: in particular, low heating rates promote dehydration and subsequent char formation, whereas high heating rates lead to depolymerization and fast volatilization by forming levoglucosan (whose yield can be reduced by dehydration, thus resulting in the formation of fewer volatile species), as well as higher gaseous flammable products. The formed char acts as an insulating layer on the surface of the cellulosic material, reducing the heat transfer, and as a diffusion barrier for combustible gases. Moreover, the level of material protection is strictly affected by its structure: a char, featuring an open channel structure, promotes the transfer of combustible gases to the flame, while a char with a closed structure hinders volatile species to reach the flame. Furthermore, the formation of char leads to condensed water that dilutes the flammable gases [17].

Generally, neat cellulose fibers lead to the formation of around 13% char, an amount that can be significantly enhanced by using FRs, thus reaching 30–60% [43]. Moreover, the presence of FRs on textile surfaces slows down the combustion due to the low heat involved, which is required to maintain pyrolysis. As a result of the highly flammable nature of cellulose-based materials, as a consequence of their chemical composition, a fire-hindering mechanism integrated into textiles is needed to decrease or eliminate possible fire dangers.

3. Flame Retardancy of Cotton Fabrics

Among the wide panorama of textile fibers, most of them are highly flammable, making exceptions only for protein fibers (e.g., wool and silk) and leather. Therefore, the term "combustible fibers" is intended to include readily and freely combustible fibers commonly found on the market, which are stored in relatively large quantities and pose a considerable fire hazard when stored. The leading causes of fires can be ascribed to faulty electrical equipment, friction between fibers, presence of foreign materials in the stored fibers (e.g., metals), fiber fermentation in the presence of high humidity that raises the temperature, possible spontaneous heating (oxidation reactions occur), as well as smoking [44].

Different fibers have varying flammability levels depending on their chemical composition and structure. Table 1 shows the thermal properties in terms of glass transition (T_g), melting temperature (T_m), degradation or pyrolysis temperature (T_d or T_p), and ignition combustion (T_c) values, as well as the limiting oxygen index (LOI), which are crucial parameters for evaluating the thermal stability and flame retardancy of some natural and synthetic fibers.

Table 1. Significant temperature values and LOI (%) of the most commonly used fibers.

Fiber	T_g (°C)	T_m (°C)	T_d (°C)	T_c (°C)	LOI (%)
Cotton	-	-	350	350	18.4
Viscose	-	-	350	420	18.9
Wool	-	-	245	600	25
Polyamide 6	50	215	431	450	20–21.5
Polyester	80–90	255	420–447	480	20–21.5

In general, for natural cellulosic fibers, the lower the T_c the hotter the flame, and the more flammable the fibers. As known, the primary component of mature cotton fibers is cellulose (from 88% to 96.5%). All the rest consists of non-cellulosic materials found in the outer layer and lumen.

These components include protein (1–1.9%), wax (0.4–1.2%), pectin (0.4–1.2%), and other substances [45]. Moreover, as evident from data reported in Table 1, cotton fabrics exhibit the lowest LOI value concerning other fibers, hence confirming high inherent flammability.

The expression "flame retardancy of textiles" concerns the possibility of making these materials less likely to ignite and, once they are ignited, to burn much less efficiently. Finishes used with this aim act to break the self-sustaining polymer combustion cycle and, thus, extinguish the flame or reduce the burning rate and flame propagation. Flame retardancy in cellulose-based textiles may be achieved in the followings ways and means, either by individual or combined mode (from a to f) as follows:

(a) by using fire-retardant materials that thermally decompose through strongly endothermic reactions to not easily achieve the heat required for thermal decomposition (temperature of pyrolysis (T_p) of the fiber should not be reached);

(b) by applying a material that forms an insulating layer around a temperature below the fiber T_p;

(c) by using phosphorous-containing fire retardants that modify the pyrolysis reaction ('condensed phase' mechanism) according to two mechanisms: (i) production of phosphoric acid through thermal decomposition, and cross-link with hydroxyl-containing polymers altering the pyrolysis pattern and yielding less-flammable byproducts; (ii) blocking of the primary hydroxyl group in the C-6 position of the cellulose units, preventing the formation of flammable byproducts (levoglucosan), and catalyzing the dehydration and char formation [46];

(d) by preventing combustion through scavenging the generated free radicals (e.g., $Br^{\bullet}/Cl^{\bullet}$ halogen-containing fire-retardant compounds), thus reducing the available heat ('gas phase' mechanism);

(e) by enhancing the T_c (combustion temperature) [47];

(f) by raising the initial decomposition temperature (i.e., T_p) for preventing/reducing the formation of flammable volatile species and increasing the formation of char and non-flammable gas.

For more in detail, the mechanism of FRs through the breaking the self-sustaining combustion cycle of cellulose is schematized in Figure 3. As a result, FRs should be able to (i) lower the heat, hence interrupting the combustion sustainability, (ii) modify the pyrolysis by lowering the formation of flammable volatile species and increasing the amount of char or intumescent coating (that acts as a barrier between the flame and the polymer), (iii) isolate the flame from oxygen, and (iv) decrease the heat flow back to the polymer to prevent further pyrolysis.

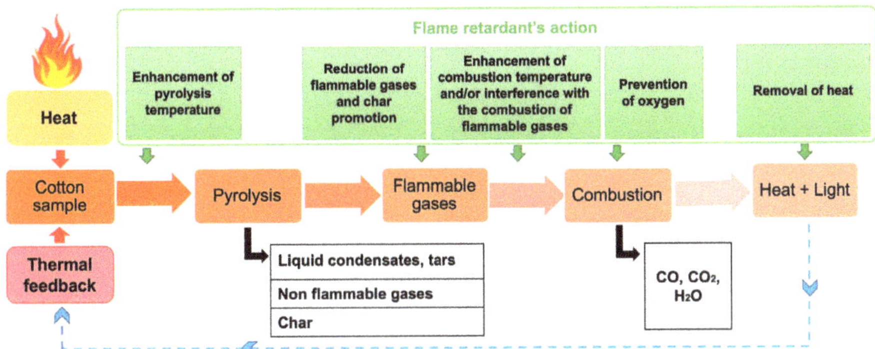

Figure 3. Combustion mechanism of cotton, together with the flame-retardant effect.

Therefore, all species that can acid-catalyze the dehydration and, subsequently, act as char formers in the condensed phase, are the best candidates for the flame retardation of cotton.

Thus, when textile fabrics are exposed to a heat source, a temperature increase occurs, and, if the temperature of the heat source (radiation or gas flame) is high enough and the net rate of heat transfer to the cellulosic substrate is high, only pyrolytic degradation of the fibrous matrix will occur.

The same reasoning applies to manufacturing flame-retardant fabric items, by employing a flame-retardant coating or laminate to any textile substrate to limit direct contact with free oxygen in the air, which actually helps any substance to burn. Such coatings or laminations of various agents, such as metal oxides, inorganic halides, organic compounds, and nanomaterials, are suitable for flame retardancy. Flame propagation rate, flame-retardant effectiveness, and combustion behavior depend on different materials and process parameters [2].

Although FRs can be classified by several methods, a simple method is based on their elemental compositions, according to which FRs are halogen-, phosphorous-, nitrogen-, and silicon-based, as well as metal hydroxides [23,48]. Furthermore, they can be classified as a function of their action mechanism, thus resulting in nano-FRs, co-effective FRs, or biological macromolecular FRs [23,49]. Another classification is based on the durability characteristics of the finishing applied to the fabrics, which represents another challenge. Persistent flame retardancy is, generally, obtained through chemical modification, which involves a chemical reaction of cellulose hydroxyl groups with reactive flame retardants. Accordingly, FRs can be classified as non-durable, semi-durable, and durable. Since non-durable FRs are easily removed by water, they are very often used for disposable goods (e.g., medical gowns) or, when used for work clothes, their application should be periodically performed.

Common examples of non-durable FRs are represented by water-soluble inorganic salts of phosphoric acid, zinc chloride, boric acid and borates, bases such as sodium hydroxide and potassium carbonate, as well as mono- or di-ammonium phosphates or water-soluble short-chain ammonium polyphosphate. Semidurable FRs are often used for textiles, which do not require frequent washings and can withstand a limited number of cleaning cycles, i.e., between 1–20. Some examples are provided by phosphates or borates of tin, aluminum, zinc, aluminates, stannates, and tungstates.

Moreover, metal oxides and hydroxides that are easily reducible can also catalytically alter cellulose's thermal decomposition path by combining water-insolubility with flame-retardant properties. Generally, FRs that withstand more than 50 laundry cycles are defined as durable and mainly represented by phosphorous- or halogen-based molecules (e.g., Tetrakis-(hydroxymethyl)phosphonium chloride).

Moreover, the difference between additive and reactive flame retardants is to be pointed out in the framework of FR classifications. The main difference is based on the means of their incorporation into the textile polymer chain. Usually, additives are represented by mineral fillers, organic or hybrid molecules physically deposited or incorporated into the textile polymer.

On the other side, reactive FRs are involved in the chemical modification of polymers by coating, chemical grafting, or copolymerization. As already introduced, since many available fire-retardant additives (e.g., halogen-based) are considered environmentally unacceptable [50], during the second half of the last century, several efforts have been made to develop novel flame-retardant treatments for textiles, meeting fire safety regulations.

4. Assessment of the Thermal Behavior and Flammability of Cotton Fabrics

With the development of innovative and performing flame-retardant chemical coatings and finishings, great attention has also been recently focused on the characterization techniques available to describe a realistic fire scenario. In this regard, to correctly interpret the results of flame-retardant treatments on cotton fabrics, it is essential to understand the flammability and thermal behavior of untreated (pristine) cotton.

However, the comprehensive characterization of the flammability properties of a polymeric material is not straightforward since the material behavior in a real fire scenario can be quite different from experimental conditions. Therefore, several analytical techniques and other prescriptive tests (small-, medium-, and large-scale) were developed over the years.

With the aim of providing fundamental knowledge for understanding and adequately comparing combustion data of untreated cotton with results obtained after flame-retardant treatments, the main techniques for characterizing flammability properties and thermal behavior of cotton samples are reported in this paragraph, indicating the measurable parameters and the most significant values reported in the literature for pure cotton, used as a reference for research in flame-retardant finishes.

Accordingly, common tests for determining the flame behavior of polymers such as the flammability tests (i.e., UL-94 and limiting oxygen index (LOI)), cone calorimetry, microscale combustion calorimetry (MCC), thermogravimetric analysis (TGA) and differential scanning calorimetry (DSC) are briefly described.

4.1. Flammability Tests

With the aim of depicting realistic fire scenarios, it is crucial to test the flammability of a sample under fire-propagation conditions, in terms of flammability by vertical/horizontal flame-spread tests (according to ASTM D-6413-11, ASTM D1230-22, and ISO 3795 standards).

Following UL-94 procedure, similar to those of the standard for fabrics (ISO 6940 (2003). Textile fabrics—Burning behaviour—Determination of ease of ignition of vertically oriented specimens), a textile sample with dimensions 50 mm × 150 mm is tested in vertical configuration by applying a methane flame for 5 s at the bottom of the specimen. As an index of flame-propagation resistance and combustion behavior of the material, some parameters can be evaluated, such as the total burning time(s), kinetics after the flame applications (total burning rate, mm/s), the final residue (%), as well as the time for flaming combustion (after-flame) and flameless combustion (after-glow) [51]. In particular, the flaming combustion corresponds to cellulose pyrolysis (production of volatile species and combustible gases). Otherwise, the after-glow combustion is due to the interaction between oxygen and residual carbon. Moreover, flameless combustion requires higher temperatures than those for flaming combustion [23,52]. To test the flammability of textile fabrics, a modified version of ASTM D3801 is performed by taking into account the actual flame height used, according to which the test method can be classified as being between the UL-V0 (ASTM D3801) and UL-94 5 V (ASTM D5048) tests in severity [53]. Depending on the dripping behavior and the self-extinguishing time, according to these test methods, materials could be rated as V-0, V-1, or V-2. However, since the method was developed for plastic materials, the above ratings should be used carefully when applied to textile fabrics.

For the 45° configuration, a special apparatus is used, in which a strip of fabric is held in a frame. A standard flame is applied to the surface near the lower edge for 10 s. The time is taken for the flame to penetrate the fabric and break the trigger thread and the physical reaction of the fabric at the ignition point is recorded. The data obtained from the 45° test in various laboratories show that the flame-spread rate is inversely proportional to the fabric weight per unit area for a given fabric [54]. This suggests that the rate of heat release from a given type of fabric, for example, cotton, is independent of the fabric weight. However, the total amount of heat released per unit area for any given fabric will be dependent on fabric weight, showing that the total heat transferred to a vertical surface from a burning fabric is proportional to the fabric weight [55].

The flammability performance index (FlaPI, %/s) is another parameter for the evaluation of flame-retardancy performance, and it is defined as the ratio of the final residue to the total burning time. Accordingly, better flame-retardancy performance is expected for higher FlaPI [56].

Using the horizontal configuration, according to which the flame is applied on the short side of the specimen, Rosace et al. [57] measured the time (t_1 and t_2) necessary for the flame to reach two horizontal lines drawn on the specimens (at 25 and 75 mm from the side on which the flame was applied, respectively). The findings confirmed that, immediately after ignition, on the untreated cotton a vigorous flame appears, the duration of which is about 23 s, followed by a prolonged after-glow (139 s); no residue was found at the end of the test.

In Table 2, and in the following ones in this section, data on the flammability properties of cotton fabrics when used as technical textiles (mass per unit area in the range of 110–300 g/m^2) are reported.

The presence of an FR treatment on the cotton surfaces results in a significant variation in its flammability. Indeed, FR-treated fabrics highlight a noticeable reduction in the total burning time, as well as a high final residue as a consequence of the formation of narrow char. In this regard, in [58] the authors measured an enhanced total burning time of up to 35 s, a total burning rate decreased to 4.2 mm/s and a final residue increased up to 46% for cotton treated with a silica phase containing nano-alumina particles.

On the other side, LOI is widely used to evaluate the flame retardancy of materials and is defined as the minimum oxygen percentage required to maintain the flame state when the sample is burned in oxygen and nitrogen atmosphere [59], by an oxygen index apparatus according to the ASTM D2863-10 standard method. Although it is not relevant for precisely determining mutual differences in FR properties, this parameter is a valuable method for the rapid characterization of samples with respect to flame resistance.

Table 2. Flammability tests performed on untreated cotton fabrics.

Mass Per Unit Area (g/m^2)	Configuration of Flammability Test	Total Burning Time (s)	Total Burning Rate (mm/s)	After-Flame Time (s)	After-Glow Time (s)	Residue (%)	FlaPI (%/s)	Ref.
145	horizontal	34	-	-	-	3.1	-	[51]
237	horizontal	149	0.67	-	-	0	-	[60]
331	horizontal	23	-	-	139	0	0	[57]
122	Vertical	-	-	15.8	20.7	-	-	[12]
129	Vertical	-	-	15	21	-	-	[61]
150	Vertical	-	-	15	26	0	-	[62]
184	Vertical	-	-	14.3	31.4	0	-	[63]
200	Vertical	45	2.2	-	27	<2	-	[64]
210	Vertical	38	-	-	-	14	0.37	[56]
220	Vertical	-	-	7	21	0	-	[65]
237	Vertical	35	7.50	-	-	-	-	[38]

LOI value is calculated as the maximum percentage of oxygen ($[O_2]$) in an oxygen-nitrogen gas mixture ($[O_2] + [N_2]$), able to sustain burning a standard sample for a specific time, according to the following equation:

$$LOI = \frac{[O_2]}{[O_2] + [N_2]} \times 100 \ [\%]$$ (1)

Untreated cotton fabrics show an LOI value in the range of 17–19.8% [62,66–69], exhibiting high flammability as materials with an LOI value below 21.0%. When LOI is between 21.0 and 25.0%, cotton samples turn to be moderately flammable and show flame retardancy when LOI is beyond 25.0%. However, to consider as complete and safe fire-retardant materials, it is better to have values of LOI nearly 35–36% at least [46]. In general, heavier fabrics experienced slower flame-spread rates [70].

The differences in the results given in the publications can be attributable to the amount of non-cellulosic components and crystallinity of the untreated cotton samples, as well as their mass per unit area values and fabric structural properties. Indeed, due to less air presence and less oxygen among fibers, heavier-weight and structured materials ignite less easily and burn more slowly than lighter-weight fabrics. Thus, the burning rate decreases as the fabric weight increases. At the same time, increasing the fabric weight, the flame temperature increases, indicating that heavy fabrics provide more fuel, sustaining the burning process. This results in a slower burn of heavier fabrics than light fabrics due to the high flame temperature during burning [71]. When the cotton sample is heated, the thermal decomposition occurs in multiple stages with different transition temperatures, which can affect the final burnability.

However, flammability is not entirely reliable and, as already noted in the literature [72,73], can only measure the ability of materials to burn when directly exposed to a small flame. For this purpose, more sophisticated instruments, such as the conical oxygen consumption calorimeter, are considered very useful tools [74,75]. These flammability tests can give surprising results, as products marketed as being constructed of flame-retardant materials are ineffective in inhibiting or preventing the combustion of polymers under a given heat flux [76,77]. Indeed, LOI values and combustion data collected by cone calorimetry often do not match. Although high LOI values have been accepted as an indication of the flame resistance of a product, this criterion is not sufficient to guarantee a high level of safety and it has never been used as an official test criterion for textiles. Indeed, new standards for transport, automotive, furniture, and protective clothing are required, in addition to high LOI values, high performance of fabrics against radiation, and cone thermometry [78–80].

4.2. Cone Calorimetry Test

Cone calorimetry is used for investigating the dynamic combustion behavior of polymeric materials in a forced-combustion scenario, providing information on time to ignition (TTI), peak of heat-release rate (PHRR) and total heat release (THR) as crucial parameters for assessing the flammability of polymers because they are reliable indicators of the size and fire-growth rate [81–83]. Moreover, reaction parameters, such as mass loss, smoke production, and generated gas composition (CO and CO_2 yields), are measured for the quantitative assessment of combustion phenomena. In addition, cone calorimetry allows assessing the flame-retardant behavior of polymer composites in the presence of different FRs, as well as the development of burning and pyrolysis models in gas and condensed phases, and the estimation of the potential hazards in fire scenarios [84]. The test is standardized in North America as ASTM E1354, "Standard Test Method for Heat and Visible Smoke Release Rates for Materials and Products Using an Oxygen Consumption Calorimeter", and internationally as ISO 5660-1:2015, "Reaction-to-fire tests—Heat release, smoke production and mass loss rate—Part 1: Heat release rate (cone calorimeter method) and smoke production rate (dynamic measurement)". The test is performed by applying an external irradiative heat flux on the textile surface able to generate a rapid increase in the surface temperature that, in a steady state before ignition, reaches a fixed value. In the cone calorimetry simulated fire scenario, the fabric surface temperature is higher than its decomposition temperature. Then, the increased temperature is responsible for the endothermic decomposition of the material and diffusion of reactive components toward the gas phase. Finally, pyrolysis and combustion processes are initiated within the cone calorimeter when it is reached the critical energy required for ignition [84].

Time to ignition (TTI) defines how quickly flaming combustion of material will occur when exposed to a heat source: accordingly, the shorter TTI, the easier the fabric ignition and spread as a threat to the surrounding materials. Peak of heat-release rate for the tested sample occurred after ignition and just before the flame went out. PHRR is the point where the material is burning most intensely and is, therefore, also crucial for estimating the fire cascading effect.

Among several parameters, heat-release rate (HRR), which is based on the oxygen consumption in the gas phase, is the most relevant for describing material flammability and related fire hazard [84]. In the HRR curve of cellulose-based samples, four significant points of interest can be observed [38]. The first point is the initial peak of heat-release rate (PHRR) that occurs when the sample surface ignites, causing great heat production and increasing HRR (heat-release rate). The second point of interest is the great decrease in HRR soon after the first PHRR. This is due to char formation, which acts as a protective barrier, preventing the exchange of volatile gases and oxygen. The third point of interest is a second PHRR close to the end of the combustion that occurs as a response to sample burn-through, which means that the heat gradient reaches the rear side of the sample. The second PHRR is highly dependent on the boundary condition defined by the rear material, which determines the heat losses at the rear side of the specimen, and, consequently, the temperature of the sample. The pyrolysis rate and, therefore, the second PHRR, increase in relation to the specimen temperature. High moisture content further prolongs the period until the second PHRR occurs because more water must go through a phase transition, requiring a high amount of energy. The fourth point of interest is the final decrease in HRR due to fuel depletion, leading to the sample smoldering or the fire being extinguished. The heat-release rate decreases down to the flame out (FO).

Furthermore, through the HRR curve generated by cone calorimetry is possible to calculate the slope of the FIGRA (fire-growth rate) line, extending from the origin to the earliest, highest peak, providing an estimation of both the spread rate and the size of the fire, thus estimating the contribution to fire growth of materials. Accordingly, the higher the FIGRA, the greater the fire hazard, suggesting a very low PHRR value with a high time to ignition, thus representing a thermal acceleration parameter.

Moreover, since fire deaths are caused more by toxic fumes than burns [85], the analysis of smoke and toxic fumes is very useful for predicting the fire toxicity of polymeric materials. As reported by Horrocks and co-workers, the evaluation of CO and CO_2 species in conjunction with smoke is significant for two reasons: first of all, CO and CO_2 are the main constituents of fire gases and high CO concentrations can be lethal; second, the analysis of these species can provide helpful information on the mechanism of decomposition of such polymer, as cotton [86]. Low CO_2/CO ratios suggest the inefficiency of combustion, inhibiting the conversion of CO to CO_2.

Alongi et al. [38] investigated the combustion behavior of square cotton samples ($50 \times 50 \times 0.5$ mm^3) using cone calorimetry under a 35 kW/m^2 irradiative heat flow in horizontal configuration. Such parameters as time to ignition (TTI, s), flame out time (FO, s), total heat release (THR, kW/m^2), peak of heat-release rate (PHRR, kW/m^2) were measured, as well as total smoke release (TSR, m^2/m^2), peak of rate of smoke release (PRSR, 1/s), smoke factor (SF, calculated as PHRR x TSR, MW/m^2), and CO, as well as CO_2 release (ppm and %, respectively). Time of production (s) of both CO and CO_2 gases generated during the oxidation of the char are also reported.

Some of these combustion data by cone calorimetry of pure cotton fabrics (under 35 kW/m^2 heat flow), selected in a range of mass per unit area 110–330 g/m^2, reported in the literature are summarized in Table 3. According to [87], the areal density values of samples significantly affect some flammability parameters, such as TTI, PHRR, or FIGRA, after ignition.

Table 3. Combustion data by cone calorimetry of neat cotton samples, selected with an areal density in the range 110–330 g/m^2.

Mass per Unit Area of the Sample (g/m^2)	TTI (s)	THR (MJ/m^2)	HRR		FIGRA (kW/m^2 s)	TSR (m^2/m^2)	CO$_2$/CO Peak	Residue (%)	Ref
			Peak (kW/m^2)	Time (s)					
110	-	2.35	179.2	15	11.94	-	71.5	2.6	[88]
115	6	2.84	196	20	10	-	77	7.5	[89]
118	7	2.68	181.55	25	7.26	-	77.05	0.93	[90]
122	2	6.3	232.1	35	6.6	-	22.1	4.3	[12]
129	7	2.8	181.5	22	8.3	5.3	88.7	0.9	[61]
145	48	2.1	147	-	-	-	-	2	[51]
150	8	1.2	97	23	4.2	-	-	3.4	[62]
200	22	2.0	83	-	-	4.3	-	0.01	[91]
220	4	4.6	269	7.7	34.9	-	-	0	[92]
237	18	4.6	143	58	-	26	0.024	4	[38]
280	45	19.2	223	55	4.05	-	-	0.6	[93]
290	18	3.8	131	-	-	24	-	<1	[94]
331	43	3.9	154.2	-	-	-	142.86	1	[57]

As observable from data reported in Table 3, the mass per unit area of cotton fabrics significantly affects the flame rate: the heavier the fabric, the slower the flame-spread rates. Moreover, no residue of cotton fabrics was observed at the end of the experiments. As reported in literature [95], cone calorimetry parameters are affected by the cotton grammage regardless of the heat flux set. According to data reported in Table 3, fabric with low mass per unit area provide high TTI and low THR, while just PHRR should be taken into consideration as an intrinsic characteristic of the material.

The presence of an inorganic FR coating on the cotton surface should affect the cone calorimetry parameters listed in Table 3. Generally, a delay is observed in TTI values, together with a decrease in THR and PHRR. Additionally, the coating has a significant impact on average TSR, which is a measure of how much smoke is produced during a full-scale fire. This is because the coating encourages the formation of char by preventing the generation of volatile species that can fuel thermal degradation, which results in a reduction in smoke production. Accordingly, Alongi et al. measured lower TSR peaks for Si and Si-ZnO treated cotton (11 m^2/m^2 and 9 m^2/m^2, respectively) with respect to untreated cotton (24 m^2/m^2).

4.3. Microscale Combustion Calorimetry (MCC)

The combustion behavior of a small amount of cotton samples can be monitored by microscale combustion calorimetry (MCC), also known as pyrolysis combustion flow calorimetry (PCFC), obtaining flammability parameters such as heat-release combustion (HRC), temperature at the maximum pyrolysis rate (T_{max}), heat-release rate (HRR), peak of heat-release rate (PHRR), total heat release (THR) values and temperature at peak release (TPHRR) [96].

The PCFC measurement is performed using an MCC-2 micro-scale combustion calorimeter, according to ASTM D7309-2007. The sample (about 5 mg) is heated to a selected temperature using a linear heating rate of 1 °C/s in a stream of nitrogen, with a flow rate of 80 cm^3/min. The thermal degradation products are mixed with a 20 cm^3/min stream of oxygen. The HRR vs. temperature curve indicates that the untreated cotton starts to decompose and to form fuel gases at less than 300 °C and presents the peak of heat-release rate (PHRR = 235 W/g) at about 384 °C (T_{max}), with an estimated value of about 12 kJ/g in THR. Moreover, Yang et al. [97] demonstrated that the MCC technique can be utilized as a trustworthy analytical methodology in determining the flammability of textile materials by providing good agreement between calculated and experimental LOI data.

Table 4 collects some significant MCC data of untreated cotton reported in the literature.

Table 4. MCC data of untreated cotton samples, selected with an areal density in the range 110–350 g/m^2.

Mass per Unit Area (g/m^2)	T_{PHRR} (or T_{max}) (°C)	PHRR (or Q_{max}) (W/g)	HRC (or ηc) (J/g·K)	THR (kJ/g)	Ref.
119	381	285	-	12.8	[98]
121	379.2	343.5	341.1	14.9	[99]
156	371	183.0	181.0	8.8	[100]
180	377.92	241.8	520.7	15.2	[101]
184	-	224.3	235.6	14.1	[63]
240	383.9	253.7	250.3	12.0	[96]
258	390.0	269.4	270.0	12.0	[102]
347	314.1	62.5	61.0	-	[103]

Contrary to the parameters of cone calorimetry, it was found that $T_{initial}$, THR, and PHRR decreased with increasing the cotton mass per unit area [95]. These differences can be explained considering the nature of the cellulose that forms char in cotton fabrics. In particular, by thermogravimetry, heavy cotton fabrics (around 400 g/m^2) lead to more char (35% and 23% in O_2 and N_2, respectively) compared to cotton at half or quarter weight (25% and 17%), thus indicating that a lower number of volatile molecules are released from heavier than lighter cotton fabrics. Accordingly, the higher volatile species generated from the lightest cotton (100 g/m^2) lead to more pyrolysis products that are further oxidized in PCFC, thus providing higher THR (11.7 kJ/g) than that of heavier cotton (8.8 kJ/g).

PHRR values assed with PCFC also differ from those obtained from cone calorimetry. Indeed, these values are a function of cotton weight and are directly related to the char formation characteristics.

In conclusion, PCFC provided the lowest THR and PHRR values for the largest cotton weights, which led to the maximum thermogravimetric char production [95].

On the other hand, the presence of FR coating on cotton surfaces accounts for a significant decrease in the main MCC parameters listed in Table 4 (THR, PHRR and T_{max}) as well as an increase in the final residue.

4.4. Pyrolysis Behavior of Cotton Fibers by Thermogravimetric Analysis

As already described, the generation of volatile fuel gas when a polymer or textile is heated may be roughly represented by the scission of the macromolecular chains, which results in a reduction of the molecular mass and the removal of monomers or oligomers, as well as the formation of non-saturated species.

The thermal stability of a sample is investigated by thermogravimetric analysis (TGA), which allows for measuring the changes in the sample weight, either as a function of time (isothermally) or with increasing temperature, in an atmosphere of air, nitrogen, helium, or other gases [104]. The thermogravimetric parameters are strictly related to the type of fibers and, for this reason, the thermal stability of natural-based fibers is measured by the decomposition of their constituents (i.e., cellulose, hemicellulose, lignin, and others) [105].

Generally, the oxidation resistance and the thermal stability of samples are investigated, respectively, under air and nitrogen atmosphere, in the temperature range of 30–700 °C, setting a heating rate of 10 °C/min. Similarly, the derivative thermogravimetric curve (DTG) shows the maximum rate of thermal decomposition, and fiber constituents are indicated by the appearance of peaks in each degradation range.

In nitrogen atmosphere, cotton samples follow a one-step thermal degradation, during which the maximum weight loss is observed [11]. According to literature [106,107], the pyrolysis of cotton in N_2 follows two alternative pathways involving the dehydration and polymerization of cellulose polymeric chain by forming stable aromatic char and volatile products. As demonstrated by Wang et al. [12], cotton samples rapidly degrade in the range of 331–391 °C, with a mass loss of about 85% due to flammable substances (e.g., L-glucose) produced by cellulose pyrolysis. As the temperature rises, the weight-loss rate is lowered, achieving a final residual weight of only 4% of the original sample, measured at 700 °C.

Since these two pathways concurrently occur, the two different weight losses as a function of temperature cannot be distinguished using thermogravimetric analysis.

Otherwise, the thermal degradation process of cotton textiles in air is more complicated, and proceeds according to three steps. The loss of bound water occurs in the first stage, resulting in a modest drop in the weight of cotton fabrics. During the first step of cellulose degradation in O_2 ($T_{max1} \approx 300$–$400\,^{\circ}C$), the two competitive dehydration and depolymerization pathways take place by providing the main and faster weight loss [51]. Further increasing the temperature, a second peak of maximum weight-loss rate appears ($T_{max2} \approx 400$–$500\,^{\circ}C$): it can be ascribed to the conversion of some aliphatic char in aromatic, and the simultaneous carbonization and char oxidation [108]. In particular, in the temperature range of 500–600 $^{\circ}C$, the mass retention rate of treated cotton tends to be stable, aromatic carbon decomposes into ethylene, and a competitive relationship between aliphatic carbon and volatile substances produced by cellulose dehydration and depolymerization process is established [65]. As the temperature continues to rise, the thermal degradation rate decreases, thus resulting in a slight mass decrease, while further char decomposition leads to CO, CO_2, H_2O, and CH_4 [65]. Another relevant parameter concerning thermal degradation is related to the maximum degradation rate of a sample, T_{deg} [109].

According to this mechanism, in [38], two decomposition peaks at 340 and 470 $^{\circ}C$ were found for cotton. Moreover, Wang et al. [110] observed during the first degradation stage in O_2 that neat cotton fabrics start dehydration at 268 $^{\circ}C$ (T_{onset}: the initial decomposing temperature), thus reaching the maximum rate degradation temperature at 346 $^{\circ}C$ (T_{max}). At T_{max}, almost 62.3% carbon layer can be measured, which is the result of the mass loss due to the volatiles produced from the cellulose decomposition, while a final residual weight of only 1.6% of the original sample was measured at 700 $^{\circ}C$. It was demonstrated that materials having a high T_{onset} do not feed the combustion reaction and, as a result, possess better flame-retardant features [111]. In another work, Simkovic et al. [112] found for bleached cotton one endotherm at 259 $^{\circ}C$, probably ascribed to the dehydration of hydroxyls and carboxyls introduced by bleaching. Moreover, the three exotherms observed at 327, 351, and 482 $^{\circ}C$ might be related to random oxidation on the surface, ignition temperature [113], and residual glowing [114], respectively. No residue was observed at 500 $^{\circ}C$, thus indicating that the process is completed before this temperature.

Some thermal degradation parameters by TGA of neat cotton fabrics, selected in a range of mass per unit area 110–330 g/m^2, reported in the literature are summarized in Table 5.

Table 5. TGA data of untreated cotton samples, selected with a mass per unit area between 110 and 300 g/m^2.

Mass per Unit Area (g/m^2)	Atm.	$T_{onset10\%}$ (°C)	$T_{50\%}$ (°C)	T_{max1} (°C)	R@T_{max1} (%)	T_{max2} (°C)	R@T_{max2} (%)	R > 600 °C (%)	Ref.
115	N_2	331	372	368	-	-	-	5.9	[89]
	O_2	327	378	352	-	-	-	1.1	
122	N_2	338	-	374	40	-	-	4	[12]
	O_2	327	-	353	42	-	-	1	
129	N_2	358.4	-	376.1	31.5	-	-	4.7	[61]
	O_2	345.6	-	349.2	43.1	489.5	4.5	0	
150	N_2	311	-	346	23.3	-	-	0.9	[62]
	O_2	297	-	327	49.7	-	-	0.6	
200	N_2	-	-	366	42	-	-	13	[58]
	O_2	-	-	341	52.5	482	6.0	<2.0	
220	N_2	-	-	365.7	-	-	-	-	[92]
	O_2	-	-	346.6	-	463.9	-	1.8	
237	N_2	336	360	362	50.3	-	-	8.3	[60]
	O_2	330	349	351	41.7	471	5.6	1.6	
290	N_2	316	-	366	42.3	-	-	13.4	[94]
	O_2	309	-	344	50.0	485	5	1.4	

In principle, a flame-retardant coating acts by strongly anticipating the cellulose decomposition, as observable from the temperature at which 10% weight loss occurs ($T_{onset\ 10\%}$) and maximum weight-loss rate temperature T_{max} [115].

In another study, the main thermal oxidative degradation process of control cotton was found between 250 and 440 °C, with the T_{max} at 327.5 °C and a weight loss of 90.4%. Finally, cotton fabrics led to a carbon residue of only 1.2% at 600 °C, ascribed to the instability of the aliphatic carbon residue in the O_2 atmosphere undergoing further oxidation at high temperatures. Conversely, FR-treated cotton fabrics started decomposing at low temperatures since $T_{5\%}$ and $T_{10\%}$ were found at 73.1 and 122.4 °C, respectively. Moreover, treated samples led to a T_{max} of 265.5 °C and a carbon residue of 12.5% at 600 °C, which are lower and higher than those of control cotton, respectively [116]. The decomposition of FR generates phosphoric acid by promoting the formation of carbon residue, which can effectively protect the underlying cellulose polymer from the action of heat flow and O_2, thus inhibiting the decomposition of treated cotton and highlighting its high thermal oxidation stability.

4.5. Differential Scanning Calorimetry

The thermal behavior of untreated cotton is also studied by differential scanning calorimetry (DSC). In DSC, analyses are performed in a temperature range of 30–600 °C with a flow rate range of nitrogen of 20–50 mL/min; the heating rate is 10 °C/min. Generally, due to the low heat conduction properties of cotton, the sample is cut into small pieces (in the range of 2–20 mg) to achieve uniform heat transfer from one point to another [70].

A typical DSC curve of untreated cotton is characterized by two endothermic peaks, corresponding to the water evaporation (100 °C) [117] and to the heat absorbed by the thermal decomposition of cellulose (350 °C) [118]. It was demonstrated [70] that the thermal degradation of hemicellulose starts above 200 °C with the breakage of bonds at 250 °C and the formation of volatile species.

At high temperatures, cellulose decomposes into L-glucose that further originates CO, other small molecules and char able to absorb a great amount of heat [119]. As the thermal degradation process goes on and more volatile species are generated, more energy is produced, which favors the breaking of the more resistant cellulosic chains and the formation of further volatile molecules. The volatilization of these molecules can be observed through the corresponding endothermic peaks above 360 °C. Overall, the thermal reaction from 220 °C and up to 370 °C can be considered a continuous reaction involving the decomposition of hemicellulosic and cellulosic components and volatilization of the resulting degradation products. Since the main endothermic peak is at 360–370 °C, it can be stated that the predominant reaction over 300 °C is related to cellulosic degradation that leads to the formation of less char and more tar.

Furthermore, since the dehydration rate of untreated cotton to char is significantly small, the exothermic peak at around 400 °C is very low and not immediately noticeable in the thermogram [61].

However, for untreated cotton, the decomposition of hemicellulose and cellulose, as well as the volatilization of formed compounds, are complete at 370 °C.

For more in detail, Teli et al. studied the peak temperature and corresponding heat release of cellulose and its components. The authors found a dehydration peak temperature at 92.2 °C that corresponds to a heat release of 127.3 J/g. Furthermore, the peak of cellulose and hemicellulose combustions were measured at 339.1 and 368.1 °C corresponding to 189.96 and 9.9 J/g heat release, respectively [120].

In another study [121], both untreated and FR-treated cotton samples provided an endothermic peak related to the dehydration phenomenon in the range of 100–160 °C. Moreover, for untreated cotton samples, ΔH of 155 J/g was measured. This value decreased to 125.48 and 113.08 J/g for FR-treated cotton samples resulting in higher intensity with respect to untreated cotton, thus highlighting the action of the FR applied. Moreover, the exothermic peak occurred for treated cotton at 250 °C, due to the high temperature of the cross-linking reaction, while at 380 °C the formation of levoglucosan was observed [121].

4.6. Study of Pyrolysis Products of Cellulose

The flammability of cellulose polymer is significantly affected by the pyrolysis; thus, the resulting products of FR and control fabric provide relevant information about the FR mechanism. In this regard, the pyrolysis and pyrolysis product of FR-cotton materials can be investigated by pyrolysis–gas chromatography–mass spectroscopy (PY–GC–MS) [122].

As reported in the literature [123–125], several flammable fragments are generated from the thermal decomposition of cellulose, such as levoglucosan, furfural, furans, ketones, and aldehydes. In this regard, the characterization of cotton fabrics by PY–GC–MS revealed more than 40 chromatographic peaks and more than 20 kinds of thermally decomposed products [122]. Similarly, in [125], fewer gas products as a consequence of a gas scavenging effect were detected for flame-retardant cotton fiber.

Accordingly, it was demonstrated that more peaks or more products are produced by the pyrolysis of the untreated cotton fibers than that of the flame-retardant cotton. Moreover, PY–GC–MS spectra of cotton fibers revealed a significant reduction in the numbers of peaks of the flame-retarded fiber, due to a reduction in the thermal degradation products. Indeed, the intensity of the peaks is strictly related to the amount of pyrolyzed products. Among the pyrolysis products, the incombustible ones (e.g., water and CO_2) play a key role in reducing the material flammability, contrary to the combustible ones (e.g., alcohol, aldehyde, ketone, furan, benzene ring, ester, and ether), whose amount is significantly different when they are released from thermal decomposition of FR-cotton or control cotton.

Furthermore, the pyrolyzed pure cotton can be characterized by Fourier transform infrared spectroscopy (FTIR) to assess the chemical composition of char. Spectra of cotton fabric char residue revealed some absorption bands at 3400, 2900, 1740, 1700, 1600, and 1444 cm^{-1}, assigned to OH, CH, CO stretching of aldehyde and unsaturated aldehyde, C-C stretching of alkenes, and aromatic C=C stretching vibration, respectively, as result of dehydration and loss of hydroxyl groups [126,127]. Moreover, another peak at 1032 cm^{-1} was assigned to the C-O-C bond, thus revealing the formation of ether molecules in the char [127].

5. Inorganic Chemicals Currently Used as Flame Retardants for Cotton

The use of flame retardants for developing protective cotton-based clothing has attracted increasing interest in recent years, owing to the unique properties of obtained materials that find numerous applications in many technical fields. According to their chemical composition, the broad field of finishes can be divided into two main categories: inorganic and organic compounds. Typically, the former may also be of natural origin but generally do not contain any carbon atoms, while the latter specifically includes carbon atoms. The exception to this rule refers to carbon nanotubes and graphene, which are arbitrarily categorized as inorganic fillers despite only being made of carbon atoms. Although many papers have been published on this topic, to the best of our knowledge, a detailed review of the effectiveness of inorganic-based formulations applied to cotton fabrics has not been published yet.

In the following sections, the most widely used inorganic products to enhance flame-retardant properties of cotton fabrics will be summarized. For each example, selected materials and their application to cellulose-based samples will be illustrated, emphasizing the structure–property relationships that enhance thermal behavior and fire-resistant properties of textiles.

The use and performance of phosphorus-containing inorganic molecules and salts, as well as the use of clays with other layered minerals of interest, will be discussed. Furthermore, FR finishing treatments on cotton fabrics by inorganic sol–gel precursors, as well as the use of metallic nanoparticles, will also be discussed. At the same time, a section will be dedicated to the rise of carbon-based finishes, such as carbon nanotubes and graphene, as promising candidates for developing new advanced materials with flame-retardant properties. A summary of the action mechanism and effect of each flame retardant discussed in the next paragraphs is reported in Table 6.

Table 6. Summary of the action mechanism and effect of the different flame retardants.

Inorganic Flame Retardant	Action Mechanism	Mode of Action	Effect
Red Phosphorous	Condensed Phase Gas Phase	- Char-forming agent - Reduce heat combustion	- Decrease the formation of combustible gases/mass loss - Production of H_3PO_4 (dehydration agent) enhancing the char formation - Radical quenching
Boron-based FRs	Condensed Phase	- Char-forming agent	- Enhance thermal insulation - Inhibit pyrolysis - Prevent heat transfer to the textile - Reduce the generation of smoke and toxic gases
Sol–gel-based FRs	Condensed Phase	- Char-forming agent	- Block O_2 transfer - Block heat exchange - Block oxidative decompositions
Nanoclays	Condensed Phase	- Char-forming agent	- Slow the evolution of combustible gases - Block the entry of O_2 - Prevent heat transfer to the textile - Reduce polymer degradation
CNTs	Condensed Phase	- Char-forming agent	- Act as a heat barrier - Act as a thermal insulator - Prevent heat transfer to the textile - Reduce polymer degradation
Graphene	Condensed Phase	- Char-forming agent	- Retard the mass- and heat-transfer processes
Metal-based NPs	Condensed Phase Gas Phase	- Char-forming agent - Catalytic role in the redox reactions	- Promote the dehydration of cotton - Reduce heat release - Absorb active species (e.g., free radicals) - Reduce O_2 concentration through redox mechanisms

5.1. Inorganic Phosphorus-Based Flame Retardants

Over the past 50 years, phosphates, phosphonates, phosphoramides, and phosphonium salts, as phosphorus-based flame retardants, have been used in flame-retardant finishing for cellulosic fabrics [118]. Among these phosphorus compounds, red phosphorus is one of the ecologically and physically most harmless alternative fire retardants, although its use was deterred by problems of handling safety, stability, and color. It is one of the primary allotropes of phosphorus, which includes white phosphorus and black phosphorus, and is the most concentrated source of phosphorus-containing flame retardant. A red phosphorus emits highly toxic PH_3 gas when exposed to air for an extended period of time, much progress has been achieved in its encapsulation, thanks to the commercial availability of masterbatches in a wide range of polymer precursors [128]. Therefore, the most significant disadvantage for its widespread application as a flame retardant is the reddish–brown color of treated polymeric materials. Red phosphorus can act in both the condensed and the gas phases [129]. In the former case, favored by the presence of water in the polymer, it acts as a char-forming agent, decreasing the formation of combustible gases and the mass loss. When red phosphorus burns, it produces phosphorus (V) oxide, which can absorb water vapor from the air, producing phosphoric acids that act as a dehydration agent, enhancing the formation of char. In the second case, in the gas phase, red phosphorus reduces the heat of combustion.

The flame-retardant effect of red phosphorus applied to a pure cotton fabric was confirmed by Mostashari et al. [130]. The optimum concentration of red phosphorus to impart flame retardancy onto cotton was around 4% (w/w). At this concentration, the comparative TG/DTG curves of the untreated and treated samples in air displayed the weight loss for the treated cotton around 290 °C, about 60 °C lower than that of the untreated fabric. At this temperature, the presence of red phosphorus promoted the thermal dehydration of the substrate and the formation of carbon residue.

Similarly, the combined effect between red phosphorus and calcium chloride on the flame-retardant behavior of cotton samples has also been studied [131]. According to the results of thermogravimetric analyses, a synergistic effect to promote the formation of non-volatile char residues and less flammable gases in the finished samples was observed. The authors proposed that the presence of phosphorous, with a mechanism similar to halogenated flame retardants but only in the condensed phase, could scavenge the free radicals produced during the thermal degradation of the polymer, delaying the non-oxidative pyrolysis of cotton samples. Furthermore, red phosphorous was found to show a synergistic effect in combination with zinc chloride due to their respective abilities to increase char formation and reduce the formation of combustible gases during the thermal decomposition of cotton [132]. After the deposition of 5.6% (w/w) anhydrous additives, vertical flame-spread tests on treated samples showed a flame-retardant effect, with a char length of 2 cm. Unfortunately, the authors admitted no uniformity in the observation during the burning process due to uneven fabric impregnation.

Furthermore, several inorganic phosphorus salts, such as urea phosphate, diammonium and ammonium phosphates, and ammonium sulfamate were studied to confer flame-retardant properties to cellulose-containing fabrics. When heated, these salts break down into acids, catalysing the dehydration and char-forming reactions that occur during the cellulose breakdown process.

A typical example of inorganic phosphate salt is ammonium polyphosphate (hereafter, APP), which is obtained by heating ammonium phosphate with urea. It is a branched or linear polymer compound with varying chain lengths (n). APP with short linear chains (where n is less than 100, crystal form I) is more sensitive to water and less thermally stable, while APP with long chains (n > 1000, crystal form II) has very low water solubility (<0.1 g/100 mL). APP is a stable and non-volatile compound. The long chains start to decompose at temperatures above 300 °C, producing polyphosphates and ammonia, while the short chains decompose at 150 °C. Therefore, it is important to change the crystal form of APP to match the polymer decomposition temperature. Incorporating APP into polymers containing oxygen and/or nitrogen atoms results in the charring of the polymer. The thermal decomposition of APP produces free acidic hydroxyl groups, which condense through thermal dehydration and produce highly cross-linked structures. Polyphosphoric acid reacts with polymers containing oxygen and nitrogen, catalyzing dehydration and charring formation. However, the effect of APP depends on its loading [133]. A simple treatment with an inorganic phosphorus aqueous solution also improves the flame resistance of cotton fabrics [134]. Cotton fabric treated with H_3PO_3 was not burnable and the residue increased by almost 50% at 800 °C. After treatment, the fibers became brittle with some disconnection, but the cellulose crystal structure suffered little damage. Some of H_3PO_3 reacts with the hydroxyl groups and is phosphorylated, which can promote dehydration and carbonization of cellulose and enhance its flame-retardant performance.

Recent findings by Lin et al. [135] confirmed the flame-retardancy properties of APP-coated cotton. Indeed, when exposed to heat, APP pyrolyzed earlier than cellulose to generate polyphosphoric acid and some inert gases, such as CO_2 and H_2O, boosting the pyrolysis of cellulose mostly toward the generation of intumescent char layers. Accordingly, the char residue of the APP-treated samples showed a significant rising trend, reaching 32.1 wt%. Moreover, due to the lower degradation temperature of APP, the temperature at 5 wt% mass loss ($T_{5\%}$) of the treated samples decreased accordingly.

Additionally, Wang et al. [136] reported an interlayer-confined approach for intercalating two-dimensional zirconium phosphate (ZrP) within reduced graphene oxide (RGO) interlayers, yielding a hierarchical ZrP-RGO product. This latter was combined with APP particles to create an intumescent flame-retardant coating on cotton fabrics via automatic screen printing. The treated samples exhibited excellent self-extinguishing performance after the withdrawal of flame. Moreover, compared to untreated cotton, its PHRR and THR values were significantly reduced by 92.1% and 61.8%, respectively.

5.2. Boron-Based Flame Retardant

Boron-based inorganic flame retardants have been used for more than 200 years [137]. When undergoing endothermic decomposition, the boron-based flame retardant shows a mechanism in the condensed phase forming a protective layer that enhances the thermal insulation barrier of treated samples, thus inhibiting pyrolysis processes and preventing heat transfer to the textile. At the same time, this layer prevents the exchange of organic volatiles in a fire and reduces the generation of smoke and toxic gases [138]. However, since boron-based FRs easily hydrolyzed, their washing fastness performance has been challenging for application on cotton and other cellulose fibers [139]. Some authors reported the use of hydrated sodium metaborate (SMB), as an example of an inexpensive and non-toxic flame retardant for cellulose-based materials, also exploiting its ability to release water molecules during combustion, in addition to the glassy charring effect of boron-hydroxyl compounds [140]. First, Mostashari et al. [141] demonstrated that impregnating sodium borate decahydrate onto cotton fabric resulted in better flame retardancy. The comparative TG curves of untreated and treated cotton fabric, with the salt at its optimum loading, displayed a main mass loss for the treated fabric around 275–325 °C, corresponding to a characteristic range of cotton thermal degradation. TG curves showed that the pure salt does not break down into oxide at the decomposition range of the pyrolyzing material.

Due to the removal of its crystallization water, the mass loss occurs at around 100 °C, thus permitting the remaining products to act as a barrier to absorb and dissipate the heat from the combustion zone. Then, Tawiah et al. [142] treated cotton fabrics with hydrated sodium metaborate (SMB) crystallized in situ in the pore spaces and on the surface of the impregnated cotton fabric. TGA results showed that SMB treatment improved the thermal stability of cotton fabric and enhanced the char yield. The treated cotton also had an LOI value of 28.5% with an after-glow time below 1 s in the UL-94 test (V-0 rating). Considerable reductions in peak heat-release rate (PHRR ~91.8%), total heat release (THR ~47.2%), peak carbon monoxide, and carbon dioxide produced (PCOP ~28.6, PCO$_2$P ~85.5%) were observed. The post-burn residues examined by SEM and Raman spectroscopy demonstrated a maintained fabric structure with high graphite content. SMB-treated cotton fabrics showed negligible changes in tensile strength and elongation at break values.

Furthermore, Akarslan [143] investigated the application of boric acid (BA) to cotton fabrics for flame-retardant purposes. Different concentrations of boric acid nanoparticles were tested to optimize the flame-retardant effect. The results showed increased flame-spreading times of the coated fabrics with increasing boric acid content. Compared to neat cotton (2.5 s and 2.4 s in weft and warp direction, respectively), the best result was observed by adding 30 g/L of BA, increasing the flame-spreading time by about 40% in the two directions. Unfortunately, an inevitable decrease in tensile strength was demonstrated for the treated samples.

Moreover, Liu et al. [144] used boron nitride nanosheets modified with hexachlorocyclotriphosphazene to develop a flame-retardant finishing by impregnation-drying method for cotton. Compared to the untreated fabric, the combustion rate of the treated counterpart decreased in vertical and horizontal flame-spread tests, and the limiting oxygen-index value increased to 24.1%, thus becoming less flammable than the reference.

On the other hand, zinc borate plays a leading role in delaying the thermal oxidation of the fabric. Indeed, over 300 °C, it loses crystal water diluting the oxygen in the air, significantly lowering the combustion surface temperature, and absorbing a lot of heat.

High temperatures cause some of the zinc borate to break down into B_2O_3 and hydrolyze to produce boric acid, which is then chemically and physically adsorbed onto the surface of the fabrics to form a coating.

Diboron trioxide inhibits the oxidation reaction of carbon compounds, reducing the production of free radicals and promoting the termination of chain reactions. In addition, when zinc borate is hydrolyzed, a tiny amount of boric acid will be created and utilized as an acid source to promote the creation of char layers, giving rise to synergistic effects with silica sol, and lowering the amount of smoke produced during combustion. Additionally, under the same conditions, roughly 38.0% of the zinc in zinc borate transforms into zinc oxide or zinc hydroxide, which dilutes the combustible gas and slows down its combustion rate. Microscale combustion calorimetry showed THR and PHRR values of zinc borate-treated fabrics decreased by 34% and 3.1% (9.2 kJ/g and 300.8 W/g) when compared to the pristine textile, respectively [66].

In another study, Durrani et al. [29] investigated the flame-retardant effect of nano-zinc borate (ZnB) on cotton fabrics. Compared to untreated cotton, the fabrics treated with 12 wt% of ZnB exhibited increased LOI values (+12%), accelerated char formation acceleration, decreased PHRR (21 vs. 90 kW/m^2 for treated and untreated cotton, respectively), and THE values (1.6 vs. 3.8 MJ/m^2 for treated and untreated cotton, respectively). Moreover, the TG curve of the synthesized zinc borate showed thermal stability up to 700 °C.

Further, cotton fabrics were treated with boric acid doped with TiO_2 obtained by titanium (IV) butoxide precursor [145], showing good flame resistance.

Compared to untreated cotton, as assessed by thermogravimetric analysis, the treated samples showed lower degradation temperature (250 °C), the formation of a thermally stable char at high temperatures, and a residue that was significantly more important than that of the reference. In particular, the best results in terms of thermal stability and flame retardancy were obtained for textiles finished with the sol doped with a 2.5 molar ratio of boric acid.

Moreover, the flammability behavior of the treated sample in vertical configuration, after 10 s of flame application, showed no full ignition after the flame was removed and a residue of more than 90%. The burned area was approximately 16 cm^2, and a higher FPI value of 45.46% demonstrated increased flame resistance of cellulose-based samples, due to the protective char layer and heat absorption introduced by the finishing treatment.

However, although boron-containing compounds are cited in several scientific papers as less toxic and cheaper alternatives to traditional flame retardants [137,139,146], many of them do not meet GOTS (global organic textile standard) compliance [147]. In particular, in the FR formulations, GOTS has kept the limit for boric acid, biboron trioxide, disodium octaborate, disodium tetraborate anhydrous, and tetraboron disodium heptaoxide hydrate individually under 250 mg/kg, while the use of disodium octaborate is forbidden.

5.3. Flame-Retardant Finishing by Sol–Gel Technique

Over time, the meaning of the sol–gel technique has evolved differently in the definitions of authors [148–152]. However, it can be summarized as a two-step reaction of hydrolysis and condensation, involving inorganic salts, (semi-) metal alkoxides and organosilanes as the most common components to form inorganic or hybrid organic–inorganic products [153–156]. The nature of precursors, pH, temperature, and reaction time values can affect the hydrolysis and condensation rates and, thus, the nature of new materials developed that show high homogeneity at the molecular level, and excellent physical and chemical properties [157]. Due to its tunable properties, the sol-gel technique has been extensively studied for textile applications to develop functional coatings on both natural and synthetic fabrics [158,159]. In textile finishing, colloidal solutions in aqueous media are usually used to impregnate fabrics at room temperature by padding. Then, the treated samples are cured at 130–160 °C to produce porous, fully inorganic, or hybrid organic–inorganic coatings with structures depending on the nature of precursors, water/precursor ratio, acidic or alkaline conditions of the medium, curing temperature, time of reaction, and presence of solvents [160]. In the literature, several properties developed by the sol–gel technique on textile materials are reported, such as water repellence [161,162], stimuli-responsiveness [163], dyeability enhancement [164], anti-wrinkle finishing [165], drug delivery [166], UV radiation protection [167], self-cleaning [168], stain-resistance [169,170], antimicrobial [171], and flame retardancy [172], also assessing the phosphorus–nitrogen synergism [25]. Due to the focus of this review, the thermal behavior and flammability of cotton textiles exclusively treated with sol–gel and fully inorganic precursors and dopants are described. The common inorganic precursors used are metal alkoxides, such as tetraethoxysilane (TEOS), tetramethoxysilane (TMOS), tetrabuthylorthosilicate (TBOS), tetraethylortho-titanate, zirconate, and aluminium isopropylate [173–176].

The first example of inorganic sol–gel-based coating to enhance flame-retardant properties on cotton was reported by Cireli et al. in 2007 [177]. The authors used TEOS as a sol–gel precursor in combination with phosphoric acid, the latter used as the agent to dehydrate cellulose samples. The TGA showed that P-doped silica-based powder has a weight-loss percentage equal to 37%, while the mass-loss value for silica powder only was 23%. After exposing fabrics to a flame for 15 s, the phosphoric acid-containing sol–gel-based finishing provided the production of nonflammable cotton fabrics. Moreover, the flame-retardant property was not completely lost after 10 washing cycles. The proposed finishing introduced a new approach to avoid additional post-treatment to improve washing fastness on treated cotton samples.

After that, another research investigated the influence of different $TEOS/H_2O$ ratios to explore the effect of the presence of inorganic silica-containing films on the thermal and fire stability of cotton fabrics [178]. At a $TEOS/H_2O$ ratio of 3:1, these films decreased up to 35% the peak of the heat-release rate (PHRR), and increased the time to ignition by up to 28%, as assessed by cone calorimetry tests. This behavior was attributed to the presence of the silica-based coating able to form a protective layer on treated cotton samples, also influencing their thermal stability in both nitrogen and air atmospheres. In the latter, the presence of coatings was responsible for slightly anticipating the decomposition step in the range between 300 and 400 °C and for an increase of about 20% of the final residue at 700 °C. These findings showed that, compared with pure cotton, the presence of the silica coatings did not improve in a remarkable way LOI values, though it changed the burning kinetics of treated samples. Indeed, the reference textile burned very quickly with a sparkling flame, while treated samples exhibited a very slow propagation of an incandescent front.

In 2012, Alongi et al. [94] developed sol–gel-based finishing of cotton fabrics using tetramethylorthosilicate precursor in the presence of various smoke suppressants, such as zinc oxide (ZnO), zinc acetate dihydrate (ZnAc), and zinc borate (ZnB), or flame retardants, such as ammonium pentaborate octahydrate (APB), boron phosphate (BP), and ammonium polyphosphate (APP), or in the presence of barium sulphate (BaS), which possesses both characteristics. According to TG studies, carried out both in nitrogen and air, the silica phase and the smoke suppressants showed combined effects that resulted in a remarkable increase in the residues at the various temperatures examined. As assessed by cone calorimetry, the presence of zinc in the silica coatings reduced the release of CO and CO_2 significantly. In particular, the joint effect of ZnO and silica promoted the most significant decrease in CO and CO_2 yields. This finding was ascribed to ZnO complete solubility in the sol solution that allows for the formation of a homogeneous coating, unlike ZnB, which formed macro aggregates. At variance, the highest residues at the maximum weight loss were found for BP and APP (69.3 and 69.0%, respectively).

In 2013, Colleoni et al. [179] focused on the sol–gel synthesis of silica thin films derived from TEOS on cotton fabrics. For this purpose, a new multi-step process consisting of one to six subsequent depositions was set to obtain samples with different numbers of silica layers, further testing the presence of dibutyltin diacetate (DBTA) as a condensation catalyst. Vertical flame-spread tests were carried out, also after several laundry cycles, to assess the washing fastness of the designed coatings. After removing the fire source, the untreated cotton burns quickly in a few seconds, leaving a very poor residue. In contrast, in 0.3 M TEOS six-layer finished fabrics, the flame spreads to the upper end of samples, then extinguishes, leaving slightly shrunken burned specimens, with 17% final residue. Again, the silica-containing coatings on cellulose-based fibers favored the char formation rather than the evolution of volatile species that could lead to further burning, regardless of the use of DBTA in the formulation. In addition, the authors highlighted the laundering durability of applied inorganic coatings that exhibit the same residue (about 13%) after combustion both after one and five laundry cycles.

However, it is interesting to consider one piece of evidence that emerges in the research reported in the literature: TEOS has low solubility in water, and it is highly reactive, which should be taken into account when used in sol–gel technology. The low solubility can be easily addressed by using such co-solvents as ethanol or other alcohol. Conversely, the high reactivity strongly affects the composition of the resulting coating [175]. In particular, the higher the hydrolysis/condensation rate, the lower is the conversion of TEOS on the fabric surface.

More recently, Zhang et al. used tetraethoxysilane precursor in combination with ammonium pentaborate in order to enhance the flame retardancy of cotton fabrics [180]. Thermogravimetric analyses of treated fabrics showed increased $T_{10\%}$, T_{max} values up to 288.3 °C and 357.6 °C, namely, 18.4 °C and 12.0 °C, higher than the untreated sample (269.9 °C and 345.6 °C). Finally, the treated sample weight loss in primary pyrolysis stage was 54.8%, lower than the reference (64.0%). Furthermore, the LOI value was almost unchanged with respect to the untreated fabric (i.e., 20.2% vs. 18.3%, respectively), while the residue at 700 °C was 31.3%. These results were attributed to the formation of an inorganic coating that acts as a physical barrier, blocking oxygen transfer, heat exchange, and oxidative decomposition, thus improving the overall thermal stability and flame retardancy. According to TG results, the silica coating evolved during pyrolysis, producing high-temperature-resistant Si-C compounds, which were further transformed into a char layer on cotton samples, resulting in high final char residue. In particular, compared with carbon layers formed by the thermal decomposition of other classes of FRs, the Si-based stable adiabatic char is denser, thicker, and stronger, thus resulting in a higher isolating effect toward heat and oxygen and further able to block the diffusion of flammable gases and inhibit the combustion [23].

Furthermore, TEOS was combined with zinc borate, investigating its flame-retardant properties on cotton fabrics [66]. It was confirmed that during the combustion the silica-containing film acts as a barrier forming a carbon layer on the fibers' surface due to the poly-condensation of the silica network at high temperature, promoting a decrease in THR and PHRR values of treated samples by 32.1% and 37.1% (9.5 kJ/g and 195.4 W/g), respectively, when compared to reference sample, as assessed by microscale combustion calorimetry.

In addition to TEOS, other silane sol–gel precursors were studied, investigating the water molar ratios and the alkoxide chain length. To this aim, TEOS was replaced with tetramethylorthosilicate (TMOS), an analogous silane bearing four methoxy groups [181]. Cone calorimetry tests showed that the best flame-retardant performance was achieved by the sol–gel-treated cotton samples (performing the treatment at 80 °C for 15 h, and using a 1:1 molar ratio of TMOS: H_2O, without HCl as condensation catalyst). Compared to the reference, TTI increased from 18 s to 28 s and PHRR decreased significantly (−20%), according to the formation of a silica coating on cotton fibers, which acts as a barrier to heat and oxygen diffusion on cellulose-based surfaces. The authors highlighted that the obtained results strictly depended on the degree of distribution and dispersion of silica on and within the fiber interstices. In fact, when the morphology of the coating is more homogeneous, flame-retardant performance is better, as observed from the correlation between cone calorimeter tests and ^{29}Si solid-state nuclear magnetic resonance results [181]. As confirmed by forced-combustion tests, the presence of the thermal barrier promoted by silica coatings effectively reduced the maximum heat-release rate (by about 15%) allowing for the formation of stable aromatic carbon [182]. Furthermore, the flammability and combustion behavior of textile samples were significantly affected by the presence of different alkoxide chain-length in the precursors [183]. To this aim, the flame retardancy of cotton fabrics separately treated with tetramethyl orthosilicate (TMOS), tetraethyl orthosilicate (TEOS), and tetrabutyl orthosilicate (TBOS), bearing four methoxy, ethoxy, and butoxy groups, respectively, was investigated and compared. Vertical flame-spread tests showed that even a low amount of silica was enough to slow down the burning rate. The final residue of untreated cotton was 10 wt%, while those of samples treated with TMOS, TEOS and TBOS were 48, 35, and 33 wt%, respectively: the shorter the chain length of the precursor, the lower the flammability of cotton. These findings were in agreement with the cone calorimetry results, carried out with an irradiating heat flux of 35 kW/m^2: the lowest PHRR and THR values were observed in the case of TMOS-treated cotton samples.

In another research, TMOS was combined with α-Zirconium dihydrogen phosphate (hereafter, ZrP) in order to enhance synergically the flame retardancy of cotton fabrics. ZrP was added to the sol solution prepared in a hydroalcoholic solution (TMOS:H$_2$O molar ratio = 1:1), in the presence of dibutyltin diacetate as condensation catalyst (0.9 wt%) [56]. The results collected in Table 7 highlight the enhanced FR properties of TMOS with respect to the untreated cotton and the synergistic effect of Si and P, that act, particularly, on the kinetics of the burning process and thermo-oxidative stability of the fabric. Instead of silica-based precursors, other oxidic phases that can improve the fire resistance of cotton through the sol–gel technique were investigated [176]. In particular, the fire performance of cotton fabrics treated with silica was compared with those with alumina, titania, and zirconia derived from aluminium isopropylate, tetraethylorthotitanate, and tetraethylorthozirconate, respectively. The results from vertical flame-spread tests showed a significant reduction of the burning rate and an enhancement of the final residue after combustion for the treated samples compared to the untreated cotton, regardless of the nature of the precursor. Generally, the findings confirm once again the role of sol–gel-based coatings as thermal insulators for protecting textiles from exposure to fire. The better performance of TEOS compared to the other metal-containing precursors in producing a flame-retardant coating on cotton samples was pointed out by cone calorimetry tests. Indeed, compared to the reference, the silica-containing finishing increased TTI by 56% and decreased PHRR by 20%, significantly better than those containing zirconia, titania, and alumina. The authors concluded that the thermal protection on cotton exerted by the sol–gel derived oxidic species can be referred to two main reasons: the thermal insulating effect of the ceramic layer and the presence of metal cations in the precursor. The results form TGA, and the forced-combustion and flammability tests are summarized in Table 7.

In the latter scenario, the thermo-oxidation of cotton is influenced by metal cations, also in combination with phosphorus-based flame retardants, which promote cellulose dehydration and, as a result, the development of significant amounts of char.

These sol–gel-based coatings developed from different inorganic precursors were responsible for an overall improvement of the thermal and fire stability of the treated cotton samples. Based on these findings, in further research, silicon and aluminum as two different ceramic film precursors were investigated to reach an optimal formulation for conferring flame retardancy to cotton fabrics, exploiting the respective characteristics [58]. In detail, tetramethylorthosilicate sol–gel precursors were exploited for depositing coatings consisting of silica added of micro- or nano-sized particles of alumina on cotton fabrics, thus investigating their thermal stability and flame retardancy by thermogravimetry, flammability and combustion tests. As a result, compared to untreated cotton, 1% and 0.5% of micro- and nano-sized alumina particles within the silica coating were able to decrease the total burning rate by at least 70%, increasing the final residue of treated samples by more than 40%.

Table 7. Forced-combustion, flammability, and thermal behavior of cotton fabrics treated by inorganic sol–gel precursors.

Weight per Unit Area (g/m^2)	Precursor (Molar Ratio of Precursor:H$_2$O)	Cone Calorimeter Tests			Thermal Behavior					Vertical Flame-Spread Tests			Ref.
					TGA in Air			TGA in Nitrogen		Total Burning Time (s)	Total Burning Rate (mm/s)	Residue (%)	
		TTI (s)	THR (MJ/m^2)	PHRR (kW/m^2)	T_{max1} ($^\circ$C)	T_{max2} ($^\circ$C)	Residue at $T > 600\,^\circ$C (%)	T_{max} ($^\circ$C)	Residue at >600 $^\circ$C (%)				
-	TEOS1 (1:1)	16	2	39	338	507 *	24	361	37	-	-	-	[178]
-	TEOS2 (2:1)	9	3	38	338	507 *	24	361	32	-	-	-	[178]
-	TEOS3 (3:1)	18	3	37	338	507 *	18	361	32	-	-	-	[178]
200	TMOS	30	1.8	82	338	496	20	-	-	70	1.43	48	[183]
200	TEOS	30	2.3	85	331	485	10	-	-	57	1.75	35	[183]
200	TBOS	28	2.2	90	338	495	12	-	-	57	1.75	33	[183]
200	TEOS	28	2.0	70	346	507	10	-	-	25	11	30	[176]
200	Tetraethylortho titanate	22	2.3	84	326	445	9	-	-	29	9	31	[176]
200	Tetraethylortho zirconate	22	2.7	82	340	498	7	-	-	24	12	21	[176]
200	Aluminium isopropylate	20	1.9	106	339	511	9	-	-	23	12	32	[176]
200	TMOS	24	3.5	114	346	484	24.0	362	34	26	5.8	23	[58]
200	TMOS + Micro-alumina	36	3.2	110	346	492	22.5	360	37	34	4.4	46	[58]
200	TMOS + Nano-alumina	22	3.4	101	346	486	27.0	361	38	35	4.2	46	[58]
210	TMOS	28	2.5	81	349	487	24	-	-	41	-	34	[56]
210	TMOS (1:1) + ZrP	22	2.9	107	344	498	31	-	-	110	-	62	[56]
290	TEOS	18	3.3	118	343	499	19.8	362	34.0	-	-	-	[94]
290	TEOS + ZnO	18	3.6	124	346	492	26.0	356	38.7	-	-	-	[94]
290	TEOS + ZnAc	18	3.6	135	345	498	18.5	356	30.7	-	-	-	[94]
290	TEOS + ZnB	18	3.6	133	346	503	23.3	359	34.8	-	-	-	[94]
290	TEOS + APB	22	3.4	116	341	501	26.0	349	35.5	-	-	-	[94]
290	TEOS + BP	20	3.6	132	349	505	24.0	358	40.8	-	-	-	[94]
290	TEOS + APP	20	3.6	130	347	498	24.1	358	41.8	-	-	-	[94]
290	TEOS + BaS	22	3.3	118	346	505	24.0	361	37.0	-	-	-	[94]

* T_{max3}.

5.4. Flame-Retardant Finishing by Inorganic Nanoparticles

Nanotechnology has been largely investigated since it represents a tool for obtaining non-toxic and high-performing flame-retardant systems of great scientific and industrial interest [184–186]. These particles have a size of 1–100 nm at least in one dimension, and they can be embedded in the fibers or coated on fabric surfaces to impart flame-retardant properties and improve the mechanical strength or the thermal stability of treated textiles. When uniformly dispersed, a small amount of inorganic filler (typically less than 5 wt%) can slow down the diffusion of gas molecules through the polymer matrix and the burning rate of materials [187].

According to the literature [75,175,188], nanoparticles can be applied to textiles by layer-by-layer (LbL) technique or nanoparticle adsorption. During combustion, these coatings give rise to intumescent barriers, the primary action of which is to form expanded char structures on the surface of the burning fabrics, thus reducing both the heat transferred by the flame and released flammable volatiles [189,190]. This mimics the FR of conventional intumescent coatings, which are macroscopically bigger, as they reach hundreds of microns in thickness [191].

The most common inorganic nanoparticles used to confer flame retardancy to cotton fabrics are reviewed in the following paragraphs.

5.4.1. Nanoclays

Nanoclay composites have attracted great attention from researchers and manufacturers due to their outstanding properties, such as excellent mechanical and thermal stability [192–195]. The term "clay" generally refers to a class of materials composed of layered silicates or clay minerals with trace amounts of metal oxides and organics [196].

Clay minerals are usually crystalline, hydrous aluminum phyllosilicates and may contain variable amounts of iron, magnesium, alkali metals, alkaline earth, and other cations. As low-cost inorganic materials, they are commonly used as catalysts, decolorizers, and adsorbents in different scientific fields. In contrast, for industrial uses, they are employed in oil drilling, as well as in ceramics and paper productions [197–199].

Nanoclays occur naturally, but they can also be synthesized. Individual nanolayers in nanoclays comprise SiO_4 tetrahedra or $[AlO_3(OH)_3]_6$ octahedra [200–202]. In general, clay particles have lateral dimensions in centimeters size, in-plane dimensions in the micron size of individual clay layers, and the thickness of a single clay plate in nanometers size. These layered clays are characterized by strong intralaminar covalent bonds within the individual sheets that form the clay [6].

In nanoclays, the thickness of the layered silicate sheets is about 0.7 nm, whereas the thickness of the bilayer is about 1.0 nm. The nanometric size of dispersed clays and the extremely high surface area induces unique properties, such as high tensile strength, low gas permeability, and a low coefficient of thermal expansion, without changing the optical homogeneity of the material [203,204]. The interaction between nanoclays and polymer chains is generally difficult because layered silicates are hydrophilic [205]. Therefore, to improve the organophilicity of nanoclays, the interlayer spacing is typically modified by organic molecules, surfactants, or exchangeable ions, such as Na^+, K^+, Mg^{2+}, and Ca^{2+} [193,194].

As shown in Table 8, clays can be classified according to the charge of their layers, exhibiting neutral lattice structures, low charge structures (negatively charged layers), and high charge structures (positively charged layers) [206,207].

Table 8. Classification of clays by layers' electrical charges.

	Type of Layers		
	Neutral	**Negatively Charged**	**Positively Charged**
Type of clay	Pyrophyllite, kaolinite, talc	Phyllosilicates: e.g., bentonites (main component: montmorillonite)	Hydrotalcite (HT): layered double hydroxides (HT-like family)
Main characteristic	Neutral clays (layers joined together by van der Waals interactions and/or hydrogen bonds)	Cationic clays (the negative layer charge is compensated by cations located in the interlayer space)	Anionic clays (the positive layer charge is compensated by anions located in the interlayer space)

Clays are also classified as phyllosilicates due to their layered structure, including tetrahedral silicon (Si) sheet(s) and octahedral aluminum (Al) sheet(s) [208]. Moreover, according to the literature [209], variable amounts of cations, such as Fe^{2+}, Mg^{2+},and other alkali metals, can be present in the space between layers or the lattice framework.

The layered silicates can be classified into three main groups: 1:1 (a), 2:1 (b), and 2:1:1 (c) (Figure 4).

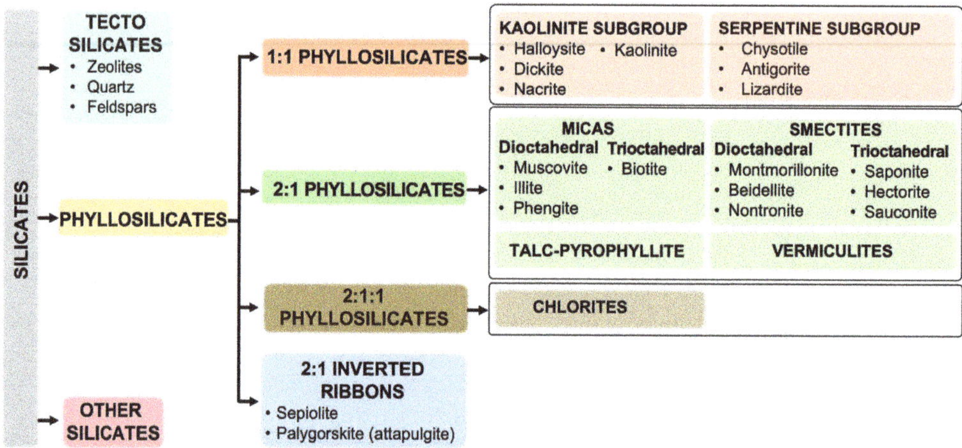

Figure 4. Silicates classified according to their layered structure.

This classification is based on the number of building block sheets (silica tetrahedral and alumina octahedral) involved in their structure. Type (a) 1:1 consists of one tetrahedral silica sheet and one octahedral alumina sheet (including kaolinite, halloysite, and serpentine). On the other hand, the 2:1 type consists of an octahedral alumina sheet sandwiched between two silica tetrahedral sheets (examples include bentonite, hectorite, illite, laponite, micas, montmorillonite, saponite, sepiolite, talc, and vermiculite). Finally, type (c) clay minerals (2:1:1) are formed with a unique structure consisting of two layers of tetrahedral silica, an octahedral alumina layer, and an octahedral magnesium hydroxide (brucite) layer (e.g., chlorites) [210].

The clays display a hydrophilic nature in contrast to carbonaceous nanostructures. The most-used clays in flame retardancy are layered materials [211], usually modified through exchangeable ions, organic molecules, or surfactants [212]. The reason is that well-dispersed clay tiles in the polymer matrix form a barrier layer during the combustion process, slowing the evolution of combustible gases and blocking the entry of oxygen, thus preventing heat transfer and reducing degradation [213,214].

The following sections review the flame-retardant properties induced in cotton fabrics treated by nanoclays without organic components. The most significant results are collected in Figure 4.

Montmorillonite

Montmorillonite (MMT) is a natural clay mineral that has been used in polymer nanocomposites for nearly three decades [215]. MMT nanoclay materials are a very soft phyllosilicate mineral belonging to the smectite family with an enlarged 2:1 crystal lattice [216]. The chemical structure of MMT is defined as $M_x(Al_{4-x}Mg_x)Si_8O_{20}(OH)_4$, where M is a univalent cation and x is an isomorphous substitution degree ranging from 0.5 to 1.3 [217]. It is a kind of smectite silicate, with a 2:1 layered structure about 1 nm thick and several microns in lateral dimensions, consisting of two tetrahedral silica layers (hereafter, T) separated by an octahedral alumina layer (hereafter, O), resulting in T:O:T structural unit [37,216]. MMT is the most significant species employed in creating commercial clay–polymer nanocomposites. It is important due to its swelling capabilities in water and other polar molecules. Idealized MMT has a negative charge of 0.67 units, so it behaves like a weak acid. MMT is usually characterized by a specific surface area of around 750–800 m^2/g (theoretical value 834 m^2/g) and a cation exchange capacity (CEC) of around 0.915 meq/g (corresponding to one ion per 1.36 nm^2 with anionic groups spaced about 1.2 nm) [218]. Moreover, the particle size of the commercially available MMT powder is around 8 μm, each containing approximately 3000 platelets with an aspect ratio of about 10–300. The toxicity profile of MMT appears to be safe and unlikely to pass the biological barriers, such as tissues and organs [218].

The required characteristics include a high aspect ratio, a large surface area, exceptional modulus, and nano-scale dispersion, dramatically encouraging improvements in mechanical, thermal, flammability, and barrier properties of polymers [219,220]. Figure 5 schematically illustrates the behavior of the polymer/MMT nanocomposites during combustion.

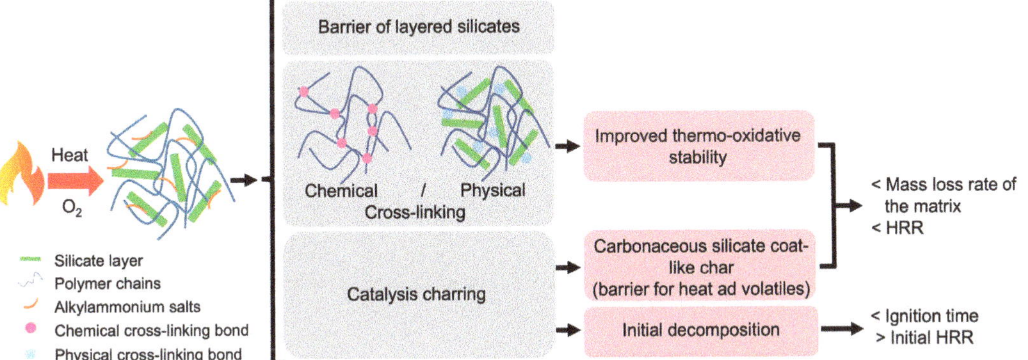

Figure 5. Schematic diagram of the flame-retardant mechanism of polymer/clay nanocomposites during combustion.

Due to the purpose of this review, only inorganic combinations were examined. In this regard, Alongi et al. studied a new concept of a fully inorganic intumescent flame-retardant nanocoating based on an inorganic expandable structure formed by montmorillonite (MMT) nanoparticles and ammonium polyphosphate (APP) using cotton as a model substrate [186]. A simple multi-step adsorption procedure was exploited to deposit APP/MMT coatings with the aim of improving the FR properties of cotton. Thermogravimetric analysis in nitrogen and air showed that the coating effectively promoted char formation from the cotton fibers and achieved self-extinction in horizontal flame-spread tests.

Furthermore, the results from the cone calorimetry tests, carried out under an irradiative heat flux of 35 kW/m^2, showed that the unmodified cotton ignites rapidly after 36 s, with an average PHRR of 61 kW/m^2 and burns without leaving residue at the end. Using 2.5% APP/MMT lowers the TTI to 22 s, indicating early ignition, whereas with the addition of 5%, no ignition of the treated sample was observed. The analysis of the final residues showed the formation of a swollen inorganic coating that could significantly enhance char formation by providing a barrier function against volatile emissions and heat transfer.

Furthermore, Oliveira and co-workers investigated the combination of a polymeric formulation containing different ratios of sodium montmorillonite (Na-MMT) and ammonium polyphosphate for improving the thermal degradation resistance of cotton fabrics [221]. A polymer paste containing the additive was mixed up and applied to one side of the fabric using a direct coating method. The thermal stability and heat release of the treated samples improved as a function of the Na-MMT/APP ratio loaded onto the fabric. At 800 °C the peak of heat-release rate (PHRR) decreased by around 51.6% and the carbonaceous residue increased.

In a further research effort, Furtana et al. assessed the possibility of using calcium-hypophosphite (CaHP) as flame retardant for cotton fabrics and evaluated its interaction with Na-MMT clay [222]. The TGA curves show that the Na-montmorillonite decomposes in two steps at 94 °C (water evaporation) and 638 °C (ascribed to dehydroxylation of the clay and the presence of CaHP), leaving 88% residue [223,224]. Indeed, at this temperature, the acidic compound phosphorylates the primary hydroxyl groups of cellulose through an esterification reaction that promotes the degradation of cellulose toward dehydration [225]. Accordingly, CaHP decomposes mainly in two steps corresponding to its decomposition into $Ca_2(HPO_4)_2$ and PH_3 at 417 °C, and into calcium pyrophosphate and water at 650 °C, thus yielding an inorganic residue of 87.4% [226]. When CaHP and clay were combined, no significant change in the $T_{5\%}$ value was observed. However, when the amount of clay applied reached 10% mass of CaHP, a shoulder at about 380 °C appeared [222].

Bronsted and Lewis acid sites present in the clay lattice are believed to provide a catalytic effect during the decomposition of CaHP [187]. The auxiliary effect of the nanoparticles was observed at different concentrations. For instance, when the coating CaHP:Clay was applied at the ratios of 1:99 and 10:90, the PHRR values increase from 72 kw/m^2 to 77 kw/m^2, respectively, and THE (total heat evolved) values remaining almost unchanged. Conversely, when the CaHP:Clay was applied at the ratio of 5:95, a stronger reduction in the PHRR (57 kW/m^2) and THE (0.85 kW/m^2) with respect to the untreated cotton (PHRR of 101 kW/m^2, and THE of 1.28 kW/m^2) was observed.

This finding demonstrated that when CaHP and nanoclay were used together, the fire performance of CaHP was further improved, with lower PHRR and THE values. Furthermore, the effect of clays at low concentrations was considered inadequate to form a continuous protective layer. At optimal concentration, the clay platelike structure acted as a barrier on the burning surface, reducing heat–mass transfer between the condensed and gas phases.

Hydrotalcite

Hydrotalcite (HT) is a white hydrous mineral of rhombohedral structure, with a chemical formula given as $Mg_6Al_2(CO_3)(OH_{16})_4(H_2O)$, showing low hardness (2.00) and low density (2.06 g/cm^3) [227]. HT is a naturally occurring, but rare, anionic mineral that is found on the earth's surface and is often associated with serpentine and calcite [228]. It gets its name from its similarity to talc and its high-water content. Indeed, hydrotalcite-type anionic clays and layered double hydroxides (LDH) containing exchangeable anions are materials that have attracted much attention in the last decade [228]. HT consists of stacked layers of brucite type built up of cationic octahedral units with common edges. Each octahedron consists of M(II) cations, some of which are replaced by M(III) cations. The M(III) cation is surrounded by six OH ions [227].

Alongi et al. [75] showed that a delay in TTI of the so-treated cotton fabrics might be caused by the high water content in the structure of hydrotalcite. These nanoparticles can partially protect the polymer from heat and oxygen since they form a ceramic layer during combustion and release large amounts of water upon heating. This way, the decomposition products released from the polymer are diluted, causing a significant delay in ignition.

The same research group investigated the influence of two impregnation times (namely, 30 and 60 min) on the thermal behavior of HT-treated cotton samples [229]. Comparing untreated cotton and the treated samples, it is possible to observe that the TTI increased from 14 s to 22 and 34 s and the PHRR decreased from 124 kW/m^2 to 87 and 94 kW/m^2, respectively. As a result, the FPI value of HT-treated fabric was higher than that of untreated cotton.

Furthermore, additional effects on thermal stability and combustion behavior may result from the combination of HT lamellar nanoparticles characterized by layers arranged parallel to each other [230] with spherical SiO_2 nanoparticles [229]. At both immersion times, corresponding to 30 and 60 min, the combination of the two nanofillers improved the fire resistance of cotton and significantly increased TTI (30 s and 37 s vs. 14 s) and the corresponding FPI (0.32 and 0.43 vs. 0.11). This means that using a combination of SiO_2 and HT nanoparticles allows for achieving better fire resistance than that observed after performing treatments with single nanoparticles.

Vermiculite

Vermiculite (VMT) is a layered magnesia aluminosilicate clay mineral, with a 2:1 structure, composed of two Al for Si-substituted tetrahedral silicate sheets, which are separated by Al-Fe for Mg substituted octahedral sheet [231], corresponding to the chemical formula $(Mg^{2+}, Fe^{2+}, Fe^{3+})_3[(SiAl)_4O_{10}]OH_2 \cdot 4H_2O$ [232].

When heated to a temperature higher than 200 °C, the VMT structure exfoliates as a result of water loss in the interlayer sheets and adaptation to the applied heat [233], inhibiting thermal transfer and imparting intumescent property to the treated materials [234]. Moreover, beyond 650 °C, vermiculite expands rapidly, showing excellent fire-retardant properties [235]. It is often used as packaging material in cardboard boxes when transporting chemical reagents [233].

Recent studies indicate that the concurrent use of VMT and nanoparticle structures (TiO_2/VMT) in LBL-based fire-retardant coatings plays an important role [236–240]. In this regard, a system of vermiculite clay and TiO_2 nanoparticles (VMT/TiO_2) in an aqueous solution was exploited for creating flame-retardant nanocomposite thin layers on cotton fabric as anionic species and cationized starch as cationic species [101]. The flame-retardant properties of LBL assemblies up to 15 BL were studied using thermogravimetric analysis, vertical flame-spread tests, and micro-combustion calorimetry. The after-burn surface and chemical analyses were performed to investigate the weave structure of the fabric. This is the first comprehensive investigation of the cationized starch (St)-VMT/TiO_2-based LBL-deposited flame retardant.

The (St)-VMT-7 sample (i.e., coated with seven bilayers) showed the highest pyrolysis inhibition, as assessed by microscale combustion calorimetry (~30% pyrolysis inhibition) and thermogravimetric analysis (~21% inhibition). Furthermore, the use of MCC data profiles confirmed that the thermal properties of StVMT-7 were significantly improved compared to other samples (PHRR ~193 W/g, THR ~10 kJ/g, HRC ~390 J/gK, and a final LOI value of ~22.2%).

In addition, the after-glow time and flame spread were significantly reduced for (St)-VMT-7 but increased as the number of layers augmented. As a result of the lower intake of the VMT/TiO_2 components and the selective deposition process, it was demonstrated that layering the coatings with more than seven bi-layers might reduce their ability to provide thermal protection (Table 9).

Table 9. Combustion and pyrolysis parameters obtained from MMC analysis (0, 7, 10, and 15 refer to the number of deposited bi-layers).

Sample	THR (kJ/g)	T_{PHRR} (°C)	PHRR (w/g)	FIGRA (PHHR/t_{max})	Hc (kJ g^{-1})	HRC (J/gK)	LOI (%)	Char Yield (%)
(St)-VMT-0	15.24	377.92	241.81	0.61	20.12	520.67	19.68	17.48
(St)-VMT-7	10.77	376.59	192.90	0.42	15.52	390.24	22.24	30.60
(St)-VMT-10	12.60	375.24	206.29	0.54	15.92	381.70	22.47	20.8
(St)-VMT-15	12.77	373.48	210.16	0.55	14.73	383.80	22.42	7.4

FIGRA: fire-growth rates, Hc: heat of combustion, HRC: values for each coated sample.

5.4.2. Carbon Nanotubes

Carbon nanotubes (CNTs) are characterized by tubular structure consisting of one (single-wall carbon nanotubes, SWCNTs), or more (multi-wall carbon nanotubes, MWCNTs) graphene sheets usually characterized by nanometric diameter. Each carbon of graphene is fully bonded to three adjacent carbon atoms through sp^2 hybridization, forming a seamless shell [241].

Since their discovery in 1991, carbon nanotubes have had a great impact on most fields of science and engineering due to their distinctive physical and chemical properties [242]. There is currently great interest in using CNTs as nanoscale modifiers to improve mechanical and electronic properties of conventional polymeric materials as well as to add new functionalities to these materials [243]. Due to their exceptional mechanical and electrical properties, as well as excellent thermal conductivity, CNTs have found many exciting applications and they have been used as additives to various polymer-based nanocomposites [244]. Given the unique physical properties of carbon nanotubes, the functionalization of carbon nanotube fiber materials is of great importance for fundamental research and practical applications [245]. Moreover, CNTs have been researched as a potential flame retardant for cotton by developing a new generation of flame-retardant additives [245,246]. In polymer/CNT systems, the main mechanisms of flame retardation rely on the formation of char layers that serve as a heat barrier and thermal insulator, re-emitting the radiation back to the gas phase, which, in turn, delays the degradation of the polymer. Additionally, CNTs enhance the thermal conductivity of the polymer nanocomposites and increase the polymer melt viscosity increasing the amount of carbon nanotubes [247]. The wide use of carbon-based nanomaterials, such as graphene, carbon black, and carbon nanotubes, as fillers for various polymeric materials is due to their excellent mechanical, thermal, and barrier properties [248–251]. This is mainly ascribed to the formation of carbon layers by continuously and three-dimensionally bonded carbon nanofillers in the polymer matrix [252,253]. It was reported that because of the fibrous morphology of CNTs, their effect as a barrier requires a high concentration compared to nanoclays. The majority of studies has been focused on the application of CNTs embedded in a polymer matrix, while others have focused on the addition of other fillers, such as clay, graphite, or intumescent compositions, into CNT/polymer compounds [254]. In this regard, Goncalves et al. [255] oxidized and incorporated multiwalled carbon nanotubes (MWCNTs) onto cotton fibers using a dyeing-like methodology. The authors evaluated the hydrophobicity and flame retardancy of the novel functionalized textiles, as well as the washing resistance of the MWCNTs-based coating. The incorporation of MWCNTs heat treated under N$_2$ flow at 400 °C (i.e., MWCNTs featuring a strong acidic character) in the cotton substrate slightly reduced the burning rate from 243 mm min^{-1} (untreated cotton) to 229 mm min^{-1} (treated cotton).

In this context, it appears that functionalizing textiles with oxidized MWCNTs increases the flame-retardant properties. In a further study [256], carboxylated single-walled carbon nanotubes (CSWCNT) were immobilized on cotton fabrics using citric acid as cross-linker and sodium hypophosphite (SHP) as catalyst. As a dispersing agent, sodium dodecyl sulfonate was employed. The flammability of cotton fabric modified by CSWCNT, CA, and SHP was evaluated by measuring the percentage of char yield. The results were discussed in relation to the experimental parameters adopted for treating the cotton fabrics (namely, different loading procedures of CSWCNTs, and the addition of citric acid and sodium hypophosphate).

The higher the concentration of the CSWCNT, the greater the flame retardancy observed on the treated samples. Indeed, considering the carbonization yield, compared to the untreated sample (1.72%), some treated samples increased significantly. Meanwhile, the cotton samples pretreated with CA and SHP achieved 6.12%, which can be a measure to evaluate the self-extinguish ability of the fabric. Recently, Xu et al. [257] investigated the potential of a novel carbonaceous nanomaterials-single-walled carbon nanohorns (SWCNHs) coating for improving the flame retardancy of cotton. The unique combination of SWCNHs and ammonium polyphosphate (APP) in a single halogen-free nanocoating showed a synergistic effect that conferred significant fire resistance to cotton fabric. By analyzing the impact of different SWCNHs concentrations on fabric flame retardancy, it was found that the optimum SWCNHs concentration was 0.15%, which provided cotton with self-extinction.

Furthermore, its LOI value reached 27.5% and the damage length was reduced to 6.5 cm (from 30 cm for cotton), preserving its original morphology apart from a partial carbonization. As assessed by cone calorimetry tests, APP/SWCNHs (0.15%)-treated cotton samples show a remarkable decrease in PHRR and THR, by 92.22% and 58.44%, respectively. The char residues of cotton, cotton-SWCNHs, and cotton-APP/SWCNHs (0.15) following vertical flammability tests were studied by FTIR and XPS spectroscopy in order to investigate the flame-retardant mechanism. The char layers of cotton-APP/SWCNHs (0.15) revealed the formation of complex cross-link structures spanned by C-C, C-O/C-O, C-N, and C-O-P bands [257]. Analyzing the gas products and char residues suggested the flame-retardant mechanism of APP/SWCNHs on cotton fabrics. Table 10 summarizes different examples of combustion and flammability tests for clay-based treated cotton, also previously described in montmorillonite and hydrotalcite sub-paragraphs.

The unique combination of SWCNHs and APP resulted in a thicker film and a more protective layer, showing a synergistic flame-retardant effect that promoted the creation of a char layer, reducing heat and oxygen diffusion. Indeed, the char layer containing carbon–nitrogen heterocycles acted as a barrier, preventing the release of volatile combustibles, such as hydrocarbons, ethers, and carbonyls.

5.4.3. Graphene

The unique graphene lamellar structure provides interesting flame-retardant properties to treated materials. Indeed, it forms protective char layers, which act as an efficient shield toward the polymer degradation, retarding the mass- and heat-transfer process. Recently, Liu et al. [258] fabricated phosphorus-(PGO) and nitrogen-(NGO) containing graphene, with which they finished cotton samples. The LOI values of PGO-treated cotton fabrics increased from 18.7% to 21.7%, exhibiting some smoke suppression and flame-retardant effects. Cotton fabrics treated with NGO showed a 20.4% LOI value, with significant smoke-suppression performance.

5.4.4. Metal-Based Nanoparticles

Metal-based nanoparticles have been used in textiles to improve flame resistance as they can act as heat and mass transfer barriers, altering the degradation pathways of polymers, limiting the mobility of polymer chains, and absorbing active species, such as free radicals [259]. In addition, they can enhance heat transfer within the material, slow the migration of air bubbles, and reduce heat release and local oxygen concentration through redox mechanisms [259–261].

Accordingly, in nanocomposites containing metal-based nanoparticles, improvements in many flame-retardancy parameters are evident. The following section presents more details on metal-based nanoparticles without organic components used as flame retardants in cotton fabrics.

Table 10. Combustion tests, flammability, and thermal behavior of cotton treated with clay additives.

| Weight per Unit Area (g/m^2) | Additive | Cone Calorimeter Tests | | | Thermal Behavior | | | | | | Vertical Flame-Spread Tests | | Ref. |
| | | | | | TGA in Air | | | TGA in Nitrogen | | | | | |
		TTI (s)	THR (MJ/m^2)	PHRR (kW/m^2)	$T_{onset\,n\%}$ (°C)	T_{max} (°C)	Residue at $T > 600$ °C (%)	$T_{n\%}$ (°C)	T_{max} (°C)	Residue >600 °C (%)	Burning Rate (mm/s)	Residue (%)	
100	2.5%APP/MMT	22	0.38	38	263 [n1]	508	4	266 [n1]	301	31.0	1.6	78	[186]
100	5%APP/MMT	NA *	NA *	NA *	280 [n1]	607	5	276 [n1]	310	34.0	1.3	85	[186]
-	CaHP/1 MMT	15	0.9	72	-	-	-	276 [n1]	344	28.0	-	-	[222]
-	CaHP/5 MMT	13	0.85	57	-	-	-	277 [n1]	343	26.6	-	-	[222]
-	CaHP/10 MMT	12	0.88	77	-	-	-	278 [n1]	341	30.0	-	-	[222]
170	SWCNHs	8	9.17	248.95	309 [n2]	547	0.85	313 [n2]	647	1.4	-	-	[257]
170	APP/SWCNHs (0.1)	-	3.59	26.17	266 [n2]	533	16.11	267 [n2]	599	29.3	-	-	[257]
170	APP/SWCNHs (0.15)	-	3.35	22.14	264 [n2]	542	18.51	264 [n2]	598	30.2	-	-	[257]
170	APP/SWCNHs (0.2)	-	3.52	26.66	264 [n2]	549	15.36	268 [n2]	608	28.4			[257]
214	SiO$_2$+ HT (30 + 30 min)	30	-	93	-	-	-	-	-	-	-	-	[229]
214	SiO$_2$+ HT (60 + 60 min)	37	-	86	-	-	-	-	-	-	-	-	[229]
214	HT (30 min)	34	-	87	300	478	23	-	-	-	-	-	[229]
214	HT (60 min)	22	-	94	302	480	23	-	-	-	-	-	[229]

* Parameters related to combustion are not available, as samples did not ignite during the test. n1: $T_{5\%}$ (°C), n2: $T_{10\%}$ (°C).

Metal Oxides

Instead of nanotechnology, a new approach to nanomaterials has attracted researchers and worldwide textile finishers [262,263]. Due to cost considerations, nanometal oxide-coated woven fabrics play an important role in developing functional properties that are more desirable than metal nanoparticles [264]. Various metal oxide nanoparticles can produce fire-resistant nanoscale networks, surrounded by char structures and surface defensive materials in the fibers. Metal oxides (for example, TiO_2, MgO, SiO_2, CuO, ZrO_2, and ZnO) have been used in a variety of ways to functionalize textile fabrics such as cotton [176,265]. The inclusion of nanoparticles in such fabric materials improves functional properties such as UV protection, antibacterial activity, flame retardancy, thermal stability, and physicochemical properties [266–269]. Nanoparticles can be easily incorporated into fabrics using a sol–gel technique followed by the pad-dry-cure method [270].

The optimum conditions for achieving high flame retardancy have been determined, and numerous mechanisms have been proposed [259]. The degradation pathway of the polymer can be changed, the mobility of the polymer chains is restricted, and active species, such as free radicals, can be absorbed by metal-based nanoparticles in addition to their barrier effect toward heat and mass transfer. Additionally, they may enhance the heat transfer within the material, which delays bubble migration and lowers the heat release and the concentration of local oxygen because of the oxidation-reduction mechanisms of the oxides [260,261]. The enhancement of significant flame-retardant parameters is evident in nanocomposites containing metallic nanoparticles. Some reports on the application of metal nanoparticles with only inorganic combinations are presented below. Silica spherical nanoparticles (SiO_2) have been widely used to improve the flame retardation of cotton. For instance, Alongi et al. [229] impregnated cotton fabrics in nanometric silica suspensions at different times to enhance the thermal stability and flame retardancy of the cotton. Cone calorimetry test shows that the silica accounted for effectively decreasing the PHRR of cotton (124 kW/m^2 for the untreated cotton vs. 95 and 99 kW/m^2 for the silica-treated cotton).

Titania (anatase crystalline form) is another globular form of metal oxide widely used as a flame retardant for cotton. In this regard, nano-sized titanium dioxide (TiO_2) particles were successfully synthesized and deposited on cellulose fibers using the sol–gel process by Moafi et al. [271]. The authors investigated the effect of these nanoparticles as flame retardants and as a semiconductor photocatalyst for self-cleaning, in order to create a flame-retardant cellulose photoreactive fabric. Vertical flame-spread tests showed that the effective amount of titanium dioxide as a flame retardant for cellulosic fabrics is about 4.8%, expressed in g per 100 g of dry fabric. In addition, this added value is an effective amount for providing flame retardancy to cellulose fabrics. It should be noted from the TGA curve that the untreated fibers lost about 70% of their mass at 350 °C. However, the treated sample lost only 50% of its mass at the same temperature. Therefore, it can be concluded that the application of titanium dioxide as a flame-retardant delays the formation of volatile products when the polymer undergoes pyrolysis. As assessed by TG analyses carried out in nitrogen atmosphere, the lower decomposition temperature of TiO_2-treated cotton with respect to the untreated fabric is due to catalytic dehydration of the fabric exerted by these nanoparticles. However, TiO_2 in the remaining residue appears to act as dust or a wall for heat absorption and dissipation in the combustion zone, as stated in Jolles' wall effect theory [272], according to which no flame propagation can occur in the presence of a sufficient concentration of dust in the air. Recently, Shen et al. [273] investigated a new biomineralization technique to produce thin and homogenous TiO_2 coatings on the surface of cotton. The fire behavior of the TiO_2-treated fibers was thoroughly evaluated using various methods ranging from micro-scale to macro-scale. Based on the surface morphology and chemical composition analysis, a uniform TiO_2 coating was successfully created. The results from TGA and PCFC (pyrolysis combustion flow calorimetry) tests show that this TiO_2 coating acts as a protective barrier in the condensed phase, directly slowing down the decomposition of cotton fibers and the release of flammable volatile pyrolysis products, protecting the char from thermal decomposition at low temperatures and reducing the intensity of combustion in the gas phase. Conversely, as assessed by cone calorimetry tests, the deposited protective coatings showed a limited protective effect on the underlying cotton fibers exposed to the irradiative heat flux. However, due to incomplete combustion and the flame-retardant mechanisms occurring in the condensed phase, the combustion propensity of the cotton fabric was significantly reduced by the TiO_2 coating. After one, three, and seven treatment cycles, the PHRR for TiO_2-coated cotton decreased by 5.5%, 27.1%, and 32.6%, respectively. It also showed the potential to slow down the rate of fire spread and the propensity of fire development. It can be concluded that under these experimental conditions, the flame-retardant performance is limited because a uniform TiO_2 coating cannot be formed on the cotton fiber surface after a single treatment cycle. Once a uniform TiO_2 coating is formed by biomineralization, the flammability of the cotton fabric is significantly reduced: after seven treatment cycles at a mass loading of 14.6 wt%, the LOI reaches 21.0% (LOI of pure cotton = 18.4%).

Additionally, in [274], the authors successfully prepared photoactive flame-retarded fibers by synthesizing nano-sized ZnO on cellulosic fabric through a sol–gel technique performed at low temperatures. From the experimental results of vertical flame-spread tests, it can be concluded that the effective amount of zinc oxide added as a flame retardant to cellulosic fabrics, expressed in grams per 100 g of dry fabric, is about 15.24%. Furthermore, the flame-retardant outcomes of the ZnO coating are consistent with the Wall effect theory and the coating theory [272]. Moreover, Prilla et al. [275] investigated the effects of copper (II) oxide nanoparticles (CuO-NPs) on the flame-retardant properties of cotton fabrics. CuO-NPs were prepared from Copper (II) chloride through a standard procedure carried out under alkaline conditions. The resulting powder was annealed separately at 200 °C and 600 °C to obtain two products with different sizes, increasing, consequently, the functionality and adhesiveness of CuO-NPs. Indeed, the higher the temperature, the smaller the size of copper (II) oxide nanoparticles. Consequently, the smaller the size, the higher their attachment to the surface of the cotton samples. Tetraethyl orthosilicate was used to immobilize CuO-NPs on textile fabrics by the pad-dry curing method. These experimental findings showed that the smallest CuO-NPs, as obtained at the highest temperature, increased the thermal stability of cotton fabrics. Moreover, compared to the untreated sample, an increase in the burning time of the treated fabrics was observed (27 s or 54 s, depending on the sample sizes).

Most researchers have investigated the synergistic effect of the blending of metal oxides to improve the flame-retardant attributes of nanocomposites and textiles. To this aim, Alongi et al. [94] studied the effect of the combination of ZnO and silica in cotton fabrics. Indeed, in the presence of zinc-based smoke retardants, the release of smoke, and CO_2 emissions were significantly reduced compared to fabrics treated with silica-coated fabrics alone. In particular, the combined effect of ZnO and silica led to the largest decrease in TSR and CO_2 emissions (by 62% and 35%, respectively). The recent scientific literature [276] also reports that SiO_2 and ZnO nanoparticles can provide fire resistance to cotton fabrics by the sol–gel method. The flame spread and ignition times of the untreated samples were much shorter than those of the treated samples. The cotton fabric samples (COT) treated with SiO_2 and ZnO nanosols showed flame resistance. The overall burn time as well as the ignition time were significantly improved. In addition, samples COT-1 (0.25% SiO_2, 0.25% ZnO), COT-2 (0.5% SiO_2, 0.25% ZnO), and COT-3 (0.25% SiO_2, 0.5% ZnO) effectively withstood the application of a flame and took, respectively, 17, 21, and 14 s to burn completely, unlike untreated cotton that showed a total burning time of just 8 s. The TGA curves showed that the presence of SiO_2 and ZnO in the nanosols causes thermal degradation of the COT samples. Silica and zinc oxide (acting as a physical barrier) protected the cotton from heat and oxygen, and promoted the formation of carbonaceous residues. As a result, silica degraded more efficiently than the silica/zinc combination; COT-2 provided the highest thermal stability to the cellulosic substrate. It can be concluded that the SiO_2 nanoparticles had a greater effect on the thermal properties of cotton fabrics than ZnO and the combination of the two metal oxides shows better fire resistance than when used separately. Furthermore, Dhineshbabu and co-workers [277] investigated the influence of ZrO_2/SiO_2 coating on the flammability of cotton fabrics. The finishes were prepared by the sol–gel technique and coated by the pad-dry method. Before and after washing, the burning performance of coated fabrics was in the order: ZrO_2/SiO_2 (19.5 s) > SiO_2 (11.3 s). According to this finding, the fire resistance of the ZrO_2/SiO_2-coated fabric was better than that of the SiO_2-coated fabric.

Furthermore Wang et al. [19] used seeding and secondary growth to generate ZnO and ZnS microparticles on cotton fabric. A dip coating approach was used to deposit a seed layer of ZnO, ZnS, or both, on cotton. After that, the seeded cotton fabric was immersed in a secondary growth solution to promote and direct crystal growth in a certain crystallographic direction, resulting in rod-like shaped particles. ZnO and ZnO/ZnS combinations on cotton fabric reduced HRR and TSR by 41% and 68%, respectively. Despite a considerable decrease in HRR, the coating did not achieve self-extinction, as evidenced by small changes in after-flame and after-glow time. In addition, the authors found that burning ZnO or ZnS-coated fabric produced higher CO and SO_2 amounts, both harmful to human health. It should also be stated that the attachment of ZnO and ZnS microparticles to the cotton surface was solely physical (i.e., physisorption), resulting in weakly bound particles. After repeated uses, the surface-finished cotton fabric may lose its flame-retardant efficiency.

Pursuing this research, Alongi et al. [58] also show the importance of combining alumina and silica regardless of the size of the employed Al_2O_3 particles by comparing pure silica, alumina, and alumina-doped silica coatings on cotton, and alumina-doped silica coatings on cotton. In particular, the best results in terms of decreased combustion rate and increased final residue (about 46%) were achieved by the concurrent presence of both ceramic particles, compared with the coating containing only alumina (about 32% final residue) or silica (about 23%). In another study [91], silica bilayer-based coatings were deposited on cotton fibers by layer-by-layer assembly using three different deposition methods (dipping, vertical, and horizontal spray). SEM observations showed that only horizontal spraying achieved a very homogeneous deposition of silica coating compared to vertical spraying or dipping. As a result, the horizontal spray proved to provide the best flame-retardant properties, with a significant increase in total burning time and final residue as evaluated in flammability tests. In addition, cone calorimetry measurements showed that fabrics treated with the horizontal spray had a 40% increase in time to ignition (TTI), a 30% decrease in heat-release rate (PHRR) and a significant 20% decrease in total smoke release (TSR).

Another possibility to exploit the benefit of the combination of metal oxides has been demonstrated by Rajendran et al. [278]. In this investigation, various metal oxide nanoparticles (namely, ZrO_2, MgO, and TiO_2) were prepared by hot air spray pyrolysis. The composite mixture of nanoparticles and sol (TEOS) was prepared by the sol–gel method and used to impregnate cotton fabrics. The addition of metal oxide nanoparticles significantly improved the thermal stability and flame resistance of the uncoated cotton fabric. Indeed, as far as the thermal stability is considered, the metal oxide composites obtained from the combination of silica hydrosol and metal oxide applied onto textile samples (hereafter, Zr-, Ti-, and Mg-based finishes) led to a significant decrease in initial decomposition temperature ($T_{onset\,5\%}$: 293, 284, and 299 °C, respectively) compared to the silica-coated fabric and the uncoated cotton fabric (319 °C and 315 °C, respectively). In addition, the synergistic effect of silica and metal oxide also led to a remarkable increase in the final residue at (T_{max}) 400 °C (52% for the Zr- based finish, 45% for the Mg-based finish and 44% for the Ti-based finish) with respect to silica-coated fabric 40%) and uncoated cotton fabric 38%. Therefore, according to these results, Zr-based sol-coated fabrics provide a larger amount of char than Si-based sol, Ti-based sol, and Mg-based sol-coated fabrics. Further, the flame-spread tests revealed that the char length of Zr-based sol-coated fabrics was less than 45% of those of untreated cotton and Ti-based sol, Mg-based sol, and Si-based sol-coated fabrics; besides, metal oxalate salts have been frequently utilized as precursors in producing high-purity oxides [261,279–286]. They decompose with the loss of carbon oxides at temperatures similar to many polymers (e.g., 200–400 °C) [261,280,282–286] and are water-insoluble [279].

Recently, they have received attention in this field [261,287], making them an interesting starting point for an investigation into potential new flame-retardant compounds either alone or as synergists. To this aim, Holdsworth et al. [288] reported the synthesis of six divalent metal oxalates, evaluating their potential flame-retardant properties on cotton fabrics, in combination with phosphorus- and bromine-containing flame retardants. In the presence of oxalates alone, none of the metal oxalates promoted the disappearance of the cotton substrate; only manganese (MnO_x) and iron (FeO_x) oxalates reduced the burning rate at 2.49 and 2.46 mm/s, compared to pure cotton (2.66 mm/s), while stannous oxalate (SnO_x) increased the burning rate to 3.04 mm/s. In contrast, in the presence of ammonium bromide (AB), all oxalates decreased the burning rate compared to untreated cotton, especially with calcium oxalate (CaO_x) and iron (FeO_x), which showed the lowest value (1.59 and 2.08 mm/s, respectively).

The effects of the application of AB or diammonium phosphate (DAP) alone on cotton samples were also found to be very different at relatively low concentrations (about 2.5 wt%).

Due to highly varied loadings achieved after impregnation compared to the pure cotton, only two oxalates (CaO_x and MnO_x) were evaluated with DAP: they showed a reduction in the burning rate (2.18 and 2.78 mm/s, respectively) compared to untreated cotton (3.02 mm/s). However, adding ammonium bromide decreased the burning rate for all oxalates. Although all combinations of flame retardants and oxalates were char-forming, oxalates alone did not promote the char formation, and samples containing SnO_x were observed to produce a qualitatively higher degree of smoke than other oxalates (Table 11). According to Horrocks et al. [27], the synergistic properties of flame retardants can be compared by calculating the synergistic effectiveness (E_S). In particular, $E_S > 1$ indicates a synergistic effect, while $E_S < 0$ is ascribed to antagonistic systems.

Table 11. Burning rates (mm/s) of cotton fabrics treated with different metal oxalates alone or in combination with ammonium bromide (AB) or diammonium phosphate (DAP) and calculated E_S of AB/metal oxalate systems.

Cotton Samples	Unfinished Sample	Metal Oxalates	Metal Oxalate + AB	Metal Oxalate + DAP	E_S (AB/Metal Oxalates)
Cotton	2.66	-	2.50	3.02	-
COT_FeO$_x$	-	2.46	2.08	-	0.801
COT_MnO$_x$	-	2.49	2.22	2.78	0.741
COT_CuO$_x$	-	2.73	2.56	-	0.360
COT_ZnO$_x$	-	2.83	2.15	-	0.774
COT_CaO$_x$	-	2.90	1.59	2.18	0.907
COT_SnO$_x$	-	3.04	2.41	-	0.600

Moreover, E_S values equal to 1 are referred to as additive systems, while values below 1 to less-than-additive systems. Following this classification, E_S calculated values for ammonium bromide combined with different oxalates (Table 11) highlight their less-than-additive effect ($E_S < 1$).

Metal Hydroxides

Metal hydroxides have the advantages of good thermal stability, non-toxicity, low emission of smoke during processing and burning, and low cost, among others. For their use as flame retardants, metal hydroxides must ensure their endothermic decomposition and water release at temperatures higher than the polymer processing temperature range and around the polymer decomposition temperature. Among the metal hydroxides, magnesium di-hydroxide ($Mg(OH)_2$) and aluminum tri-hydroxide Al(OH)$_3$) are the most common FR fillers [40]. They have a high specific heat capacity, which allows for absorbing a significant amount of heat prior to decomposition; besides the crystallized water produced during decomposition can also absorb heat when evaporating, at the same time diluting the concentration of combustible gases. The metal oxides formed by thermal decomposition have a high melting point, and they cover the surface of cotton fibers to form an effective physical barrier, promoting the dehydration of cotton fibers into char, while separating the cotton fiber from the outside and preventing the spread of flame. The produced metal oxides also act as catalysts in the redox and cross-linking reactions of cotton fibers, which can promote the conversion of carbon monoxide (CO) to carbon dioxide (CO_2) and reduce the generation of hazardous CO [289]. Aluminum hydroxide is only employed in polymers with low processing temperatures, such as polypropylene (PP) [290] and linear low-density polyethylene (LDPE) [291], due to the fact that it breaks down at about 200 °C. Magnesium hydroxide thermally decomposes at about

300 °C, making it suitable for usage in polymers that require higher processing temperatures. However, due to the low flame-retardant activity of inorganic hydroxides and the large amount added, they exert various effects on the physical performance of substrates, such as cotton fabrics, and are often used in combination with other FRs [23]. Ji et al. [289] investigated a novel and simple method for the preparation of flame-retarded cotton fabrics by using magnesium hydroxide ($Mg(OH)_2$) nanoparticles. A pad-dry-cure method was used to deposit $Mg(OH)_2$ nanoparticles on the surface of cotton fabrics. Thermogravimetric analysis in N_2 atmosphere revealed an increased final residue at 800 °C for the treated fabrics (28%) with respect to the untreated counterparts (13.3%). These results clearly show that the addition of this metal hydroxide effectively suppresses cracking and dramatically reduces the weight loss of the fabric. The reasons for the high thermal stability of $Mg(OH)_2$-coated cotton fabrics can be ascribed to the decomposition of magnesium hydroxide crystals into magnesium oxide (MgO) and water vapor. In addition, the precipitation of MgO acts as an insulator and prevents the transfer of heat. Vertical flame-spread tests revealed that the pristine cotton fabric ignited immediately and burned completely without any residue. Conversely, the cotton fabric treated with $Mg(OH)_2$ can self-ignite after 48 s, followed by glowing for 52 s, indicating that it has excellent flame-retardant properties.

Metal (Oxide/Hydroxide)

Bohemite nanoparticles (AlOOH) are aluminum oxide-hydroxides; they can be thought of as a two-dimensional nanomaterial and a mineral with a lamellar structure with {010}b (side pinacoid) on the particles [292]. Theoretically, these nanoparticles are inherently flame retardant, as they are hydroxide of aluminum oxide: γ-AlO(OH), dehydrates in the range of 100–300 °C to release water and then transforms into the crystalline γ-Al_2O_3 phase at about 420 °C [293]. This allows the volatile products generated from the polymer under irradiation to be diluted during the initial decomposition step, leading to ignition after a long time compared to unfilled polymer materials. In addition, the ceramic barrier due to the presence of freshly generated alumina inhibits further combustion. Bohemite nanoparticles are thought to behave like efficient flame retardants because of the cooling and dilution effects caused by the endothermic decomposition associated with the release of water [294]. The most important and also previously described experimental results, related to metallic nanoparticle-based flame-retardant cotton finishings, are summarized in Table 12.

Aiming to develop a new environmentally friendly flame retardant, a comparison of the performance of halogen-free finishing agents using sulfonate-modified boehmite nanoparticles (OS1) was reported [295]. The thermal stability and flame retardancy of the finished fabrics were compared with those of untreated cotton using thermogravimetric analysis (TGA) and cone calorimetry. Moreover, nanoparticles were found to exert a protective role in the thermal oxidation of cotton, modifying its degradation profile. Nanoparticles increased the thermal stability of cotton in air, facilitated the char formation, increased the final residue at high temperatures, and decreased the overall thermal oxidation rate.

The main outcome in the use of Bohemite nanoparticles is the development of a carbonaceous surface char that serves as a physical barrier to heat and oxygen transfer from the flaming zone to the polymer and vice versa. As revealed by cone calorimetry tests, compared to untreated cotton, the treated fabric showed increased TTI (22 s vs. 14 s, respectively) and decreased PHRR (57 vs. 50 kW/m^2 g, respectively, Table 12).

Zinc Carbonate

In comparison to nanoparticles, needle-like zinc carbonate has a high surface area and aspect ratio, allowing it to provide flame retardancy at low loadings. For the first time, Sharma et al. [296] synthesized and integrated zinc carbonate ($ZnCO_3$) into cotton fabrics. The treated cotton fabrics had self-extinguishing properties with a specific char length of 40 mm and a LOI value as high as 30%. As assessed by thermogravimetric analyses, the treated fabrics showed a significant mass-loss peak at low temperatures and an increased residue at high temperatures (500 °C) compared to untreated cotton.

Table 12. Results from forced-combustion, flammability, and thermogravimetric analyses for cotton treated with metal-based nanoparticle additives.

Weight per Unit Area (g/m^2)	Additive	Cone Calorimetry Tests			Thermal Behavior						Vertical Flame-Spread Tests		Ref.
					TGA in Air			TGA in Nitrogen					
		TTI (s)	FPI (sm^2/kW)	PHRR (kW/m^2)	$T_{onset5\%}$ (°C)	T_{max} (°C)	Residue (%)	$T_{onset5\%}$ (°C)	T_{max} (°C)	Residue (%)	Time to Ignite (s)	Flame-Spread Time (s)	
98	TiO$_2$ (1 cycle)	23	-	171	-	-	-	304	346	2.3	-	-	[273]
98	TiO$_2$ (3 cycles)	20	-	132	-	-	-	292	339	8.0	-	-	[273]
98	TiO$_2$ (7 cycles)	19	-	122	-	-	-	271	336	14.4	-	-	[273]
117	MgO	-	-	-	276	343	18.9 at 800 °C	324	422	28.0 at 800 °C	-	-	[289]
200	SiO$_2$ (by dipping)	20	-	75	-	-	-	-	-	-	22	-	[91]
200	SiO$_2$ (by vertical spray)	20	-	73	-	-	-	-	-	-	30	-	[91]
200	SiO$_2$ (by horizontal spray)	28	-	66	-	-	-	-	-	-	30	-	[91]
139	SiO$_2$	-	-	-	224	541	40 at 400 °C	-	-	-	-	13.6	[278]
139	ZrO$_2$	-	-	-	235	537	52 at 400 °C	-	-	-	-	19.4	[278]
139	MgO	-	-	-	222	562	45 at 400 °C	-	-	-	-	18.4	[278]
139	TiO$_2$	-	-	-	239	549	44 at 400 °C	-	-	-	-	18.9	[278]
155	0.25% SiO$_2$ + 0.25% ZnO	-	-	-	-	-	-	330 [b]	410 [b]	8.9 b, at 600 °C	4 [b]	17 [b]	[276]
		-	-	-	-	-	-	333 [a]	413 [a]	7.6 a, at 600 °C	3 [a]	15 [a]	
155	0.5% SiO$_2$ + 0.25% ZnO	-	-	-	-	-	-	350 [b]	428 [b]	8.5 b, at 600 °C	5 [b]	21 [b]	[276]
		-	-	-	-	-	-	341 [a]	418 [a]	7.9 a, at 600 °C	4 [a]	19 [a]	
155	0.25% SiO$_2$ + 0.5% ZnO	-	-	-	-	-	-	340 [b]	420 [b]	8.6 b, at 600 °C	5 [b]	14 [b]	[276]
		-	-	-	-	-	-	336 [a]	415 [a]	7.8 a, at 600 °C	4 [a]	13 [a]	
210	OS1	22	0.44	50	278	239	1 at 800 °C	-	-	-	-	-	[295]
214	SiO$_2$ (30 min)	17	0.17	99	296	500	25 at 400 °C	-	-	-	-	-	[229]
214	SiO$_2$ (60 min)	20	0.21	95	299	479	25 at 400 °C	-	-	-	-	-	[229]
-	TiO$_2$	-	-	-	-	-	-	314	-	26.36	-	296	[297]
400	ZnO microparticles	13.0	-	-	340	-	-	350	-	-	-	-	[19]
400	ZnO + ZnS microparticles	14.7	-	-	290	-	-	-	-	-	-	-	[19]

[a] Nanosol-treated sample after washing. [b] Nanosol-treated sample before washing.

6. Conclusions and Future Challenges

The unique properties of cotton, such as biocompatibility, breathability, hydrophilicity, softness, and comfortability, rendered this renewable resource to be largely used in daily life and different application fields. However, this cellulosic material is characterized by a limiting oxygen index (LOI) of about 18%, thus resulting in a high propensity to burn as demonstrated by the elevated proportion of fire accidents caused by cotton textiles with respect to other fibers. This drawback has limited the range of applications of cotton and required the need to improve its flame resistance. To this aim, several attempts were performed using flame-retardant finishes that have been identified as hazardous materials, and whose use was recently banned or restricted. In this regard, recent health and environmental awareness has prompted scientific research to develop environmentally friendly flame retardants without toxic components or byproducts, able to replace the hazardous conventional flame-retardant molecules.

Replacing flame retardants requires a variety of approaches, from fire chemistry to physics, and that must be non-toxic and environmentally friendly, i.e., sustainable. Over the last two decades, nanotechnology has attracted the interest of both academic and industrial researchers since the use of nano-sized coatings can improve the fire resistance of cotton. The surface modification of cellulose structures was performed through several implemented green approaches involving inorganic molecules (e.g., silanization, esterification, oxidation, polymer grafting, LbL, and sol–gel) for improving flame retardancy and thermal resistance of cotton.

Due to the increasing demand for technological innovation in the field of composite materials with improved flame retardancy, this review aimed to provide an overview of the inorganic coatings prepared by the use of sol–gel, LbL, and nanoparticle absorption as an effective strategy toward the development of sustainable and environmentally friendly flame-retardant treatments.

To better understand the relevance and the efficacy of the proposed technologies to impart flame retardancy to cotton fabrics, a throughout overview of the flammability and thermal behavior of neat cotton as a function of its weight per unit area was first proposed. Then, the most relevant scientific research conducted in the field of inorganic flame-retardant finishes for cotton fabrics was described.

According to the literature, the efficacy of inorganic- and sol–gel-based coatings on textile surfaces is ascribed to the protection action exerted by these inorganic coatings, which limits the heat- and mass-transfer phenomena involved during a fire occasion. In the wide panorama of inorganic molecules employed to impart flame retardancy to cotton fabrics, the most relevant study on the use and performance of phosphorous, boron-based chemicals, sol–gel precursors and inorganic nanoparticles (e.g., nanoclays, carbon nanotubes, and metal-based nanoparticles) were detailed, describing their physical barrier effect as well as fire-resistance mechanism and optimization conditions. The performance of each reported research was defined by experimental findings from the most relevant characterization techniques employed for assessing flame retardancy, such as vertical or horizontal flame-spread tests, cone calorimetry, and thermogravimetric analyses among others. Moreover, some studies reported the chemical analysis of the formed char residue after combustion performed by FTIR as an interesting tool for investigating the combustion behavior of the developed flame-retardant finishes and the fire-retardant mechanisms behind.

For a better understanding, Table 13 provides a summary of the most recent and relevant findings concerning the use of inorganic molecules for improving sustainable flame retardancy for cotton fabrics.

Table 13. Relevant inorganic finishings and their performances in providing flame retardancy to cotton fabrics.

Classification	Textile Finishing Process	Inorganic Chemicals	Main Outcomes	Ref.
Inorganic Phosphorus	Impregnation	Red phosphorus	- Lowering of the temperature of weight loss	[130]
		Red phosphorus + $CaCl_2$	- Synergistic effect that promotes the formation of non-volatile char residue and less flammable gases - Delay of the non-oxidative pyrolysis of cotton	[131]
		Red phosphorus + $ZnCl_2$	- Synergistic ability to increase char formation and - combustible gas during thermal decomposition of cotton	[132]
		APP + TEOS	- TGA, char formation increased by 40% compared to untreated cotton - Damage length decreased as assessed by vertical flame-spread tests	[135]
		APP + zirconium phosphate + graphene oxide	- Self-extinction following the withdrawal of the flame	[136]
Boron-based FRs	Impregnation	Hydrate sodium metaborate (SMB)	- Improved thermal stability - Enhanced char yield - Increased LOI (28.5%) - After-glow time < 1 s (UL-94) - Considerable reduction of PHRR (~−92%), THR (~−43%), CO peak, CO_2 peak	[142]
		Zinc borate	- Transformed in ZnO or ZnOH that dilutes the combustible gases and slow down the combustion rate - Decrease in THR (~34%) and PHRR (~3%)	[66]
		Boric acid + TiO_2	- Lower degradation onset - Formation of a thermally stable char at high temperatures - Residue > 90% (vertical flame-spread test) - Increased FPI (~46%)	[145]
		Nano-zinc borate (ZnB)	- Improvement in LOI value - Enhanced char formation - Reduced PHRR and THE values	[29]
		Boron nitride (BN) BN nanosheets BN modified with hexachlorocyclotriphosphazene	- The modified BN exhibited the highest flame retardancy with a LOI value of 24.1%	[144]
		Boric acid (BA)	- Reduction in the flame-spread rate - Decreased tensile strength	[143]
		Sodium borate decahydrate	- Reduction in the flame-spread rate	[141]
		TEOS Zinc oxide (ZnO) Zinc acetate dihydrate (ZnAc) Zinc borate (ZnB) Ammonium pentaborate octahydrate (APB) Boron phosphate (BP) Ammonium polyphosphate (APP) Barium sulphate (BaS)	- Maximum residues are obtained in the presence of APP and ZnO (30%) compared to sample coated with silica alone (20%) - The release of CO and CO_2 has been significantly reduced with respect to the fabric treated with the silica coating alone due to the presence of ZnO	[94]

Table 13. *Cont.*

Classification	Textile Finishing Process	Inorganic Chemicals	Main Outcomes	Ref.
Sol–gel	Impregnation	TEOS	- Significant decrease in PHRR - Increased TTI - Anticipation of the decomposition temperature - Increase in the final residue (~20%) - Laundering durability - Promoting char formation (regardless of the use of DBTA)	[178,179]
		TEOS Tetraethylortho titanate Tetraethylortho zirconate Aluminium isopropylate	Regardless of the nature of the precursor: - Significant reduction of the burning rate - Enhancement of the final residue after combustion (vertical flame-spread test) - Best performance observed for TEOS: - Increased TTI (~56%) - Decreased PHRR (~20%)	[176]
		TEOS + zinc borate	- Decrease in THR and PHRR (microscale combustion calorimetry)	[66]
		TEOS + H_3PO_4	- Synergistic effect of phosphoric acid and silica coating: cotton does not burn (vertical flame-spread test)	[177]
		TMOS	- Increased TTI - Significant decrease in PHRR (−20%)	[181]
		TMOS TEOS TBOS	- The shorter the chain length of the precursor the lower the flammability of cotton - Increased final residue (TMOS = 48%, TEOS = 35%, TBOS = 33%) - TMOS showed the lowest PHRR and THR values (cone calorimetry)	[183]
		TMOS + Al_2O_3 (micro and nano)	- Decreased total burning rate (~70%) - Increased final residue (>40%)	[58]
Nanoclay	Impregnation	Montmorillonite (MMT) + (ammonium polyphosphate) APP	- Promotion of char formation - Self-extinction on the horizontal flame-spread test - Lowering of TTI - Swollen inorganic coating - Thermal stability and heat release improved as a function of MMT/APP ratio - Reduction in PHRR (~51.6%) - Increase in char formation	[186,221]
		Calcium-hypophosphite (CaHP) + MMT clay	- Strong reduction in PHRR (~50%) and THE (~44%)	[222]
		Hydrotalcite (HT) + SiO_2	- Increase in TTI (almost doubled) - Decreased PHRR (~30%) - Increased FPI - Better performance when combined with SiO_2	[229]
	LbL	Vermiculite + TiO_2	Optimal conditions: seven layers - Improved thermal properties - Reduction of after-glow time and flame spread	[101]
Carbon nanotubes	Impregnation	MWCNTs	- Reduction of the burning rate	[255]
		SWCNTs + APP	- Synergistic flame-retardant effect - Increased LOI (27.5%) - Significant reduction of PHRR and THR	[257]

Table 13. *Cont.*

Classification	Textile Finishing Process	Inorganic Chemicals	Main Outcomes	Ref.
Graphene	Impregnation	Phosphorus-(PGO) Nitrogen-(NGO)	- The LOI values of PGO-treated cotton fabrics increased from 18.7% to 21.7% - The LOI values of NGO-treated cotton was 20.4%, with significant smoke-suppression performance	[258]
	LbL	Alumina-coated silica	- TTI increased by 40% - PHRR and TSR decreased by 30% and 20%, respectively	[91]
		SiO_2 nanoparticles	- Reduction of PHRR	[229]
		TiO_2 nanoparticles	- Effective FR concentration of 4.8% (g per 100 g of fabric) - Delay in the formation of volatile pyrolysis products - Lowering decomposition temperature (TGA in N_2)	[271]
		TiO_2 nanoparticles	- TGA and PCFC shows that TiO_2 coating provide a strong effect acting as the protective barrier in the condensed phase - Reduction of PHRR by 5.5%, 27.1%, and 32.6%, respectively, for the cotton coated with TiO_2 with one, three, and seven cycles of treatment - After seven processing cycles at a mass loading of 14.6 wt%, the LOI reaches 21.0% (LOI of pristine cotton = 18.4%)	[273]
		ZnO/SiO_2 nanoparticles	- The combined effect of ZnO/SiO_2 provided the largest reduction in TSR and CO_2 emissions - ZnO significantly decreased the release of smoke and CO_2 emissions than SiO_2	[94]
		ZnO/SiO_2 nanoparticles	- The combined effect of ZnO/SiO_2 led to flame spread and ignition time much shorter than those of untreated cotton and better fire resistance than when used separately - SiO_2 showed greater effect on thermal properties of cotton than ZnO alone	[276]
Metal-based nanoparticles	Impregnation	ZrO_2/SiO_2	- Flammability of ZrO_2/SiO_2 is better than SiO_2 providing the highest burning times	[277]
		ZnO and ZnS microparticles	- Reduction in HRR and TSR by ~41% and ~68%	[19]
		Al_2O_3 and SiO_2	- Best reduction of combustible rate and increase in the final residue (46%) than coating containing only Al_2O_3 (32%) or SiO_2 (23%)	[58]
		ZrO_2, MgO, TiO_2 combined with TEOS sol	- Each metal oxide combined with silica sol significantly decrease initial decomposition temperature compared to silica sol - Zr-based sol provided: • the large amount of char than Si-based sol, Ti-based sol, and Mg-based sol • char below 45% with respect to untreated cotton and Ti-based sol-, Mg-based sol-, and Si-based sol-coated fabrics	[278]
		$Mg(OH)_2$	- Increased char yields (TGA in N_2) - Increased ignition and glowing times	[289]
		Metal oxalates Ammonium bromide (AB) Diammonium phosphate (DAB)	- When applied alone only two metal oxalates (MnOx and FeOx) reduced the burning rate of cotton at.49 and 2.46 mm/s - When the metal oxalates combined with ammonium bromide, CaOx, and FeOx oxalates show an apparent possible additional effect in lowering the burning rate at 1.59 and 2.08 mm/s, respectively	[288]
		Sulfonate modified boehmite nanoparticles	- Increased thermal stability of cotton (TGA in air) - Enhanced char-forming character - Increased final residue at high temperature - Decreased overall thermal oxidation rate - Increased TTI - Decreased PHRR	[295]
		Zinc carbonate	- Cotton fabrics achieved self-extinction and a LOI value of 30%	[296]

As a result of the differences in the mutual interface contacts and multi-functional combined benefits, good synergistic effects may be obtained between inorganic, intumescent, and nanoparticle flame retardants. However, further investigations on the synergistic effect of interfacial compatibility between flame retardants, as well as methods for improving the interface compatibility between matrix and filler are important to better enhance the flame retardancy of materials. Indeed, most of the reported techniques are still in the stage of laboratory research and significant improvements in cost, process, and method are needed to ensure their continuous automatic production and commercialization. Furthermore, besides higher performance, an ideal flame retardant should ensure easy application, low cost, and environmentally friendliness. Indeed, although researchers are putting great efforts in the design of halogen-free, synergistic, multifunctional, and nanotechnology-based flame retardants, the topic has not yet been fully explored. Indeed, although many of the flame retardants described in this review present high performance and low environmental impact, the academic community is still working on solving some challenging issues. Indeed, designing new FRs requires large investments both for toxicity evaluation and production on at least a pre-industrial scale to maximize yields and reduce costs. In particular, the possibility of adopting green technology approaches at a pre-industrial scale is still a challenge, and whether it can be successfully tried will depend on the cost-effectiveness of the designed FR systems. A further challenge in implementing sustainable FRs concerns their durability to several washing cycles, often not at the level of conventional FRs yet, currently limiting their applications.

In conclusion, future efforts from both the academic and industrial communities should be focused on continuous implementation of even greener technologies for replacing hazardous conventional flame retardants, as well as on the design of more economical and environmentally friendly approaches for the surface modification of textiles, thus making possible the application of these methods also at the industrial scale.

Author Contributions: Conceptualization, V.T., S.S., R.B.D., G.R. (Giulia Rando), G.R. (Giuseppe Rosace), G.M. and M.R.P.; resources, G.R. (Giuseppe Rosace), G.M. and M.R.P.; data curation, V.T., S.S., R.B.D., G.R. (Giulia Rando), G.R. (Giuseppe Rosace), G.M. and M.R.P.; writing original draft preparation, V.T., S.S., R.B.D., G.R. (Giulia Rando), G.R. (Giuseppe Rosace), G.M. and M.R.P.; writing—review and editing, V.T., S.S., R.B.D., G.R. (Giulia Rando), G.R. (Giuseppe Rosace), G.M. and M.R.P.; supervision, V.T., G.R. (Giuseppe Rosace), G.M. and M.R.P. All authors have read and agreed to the published version of the manuscript.

Funding: This research received no external funding.

Data Availability Statement: Not applicable.

Acknowledgments: This review was carried out within the MICS (Made in Italy—Circular and Sustainable) Extended Partnership and received funding from the European Union Next-Generation EU (PIANO NAZIONALE DI RIPRESA E RESILIENZA (PNRR)—MISSIONE 4 COMPONENTE 2, INVESTIMENTO 1.3—D.D. 1551.11-10-2022, PE00000004). This manuscript reflects only the authors' views and opinions; neither the European Union nor the European Commission can be considered responsible for them. All authors wish to thank S. Romeo, G. Napoli, and F. Giordano for informatic and technical assistance.

Conflicts of Interest: The authors declare no conflict of interest.

References

1. Aldalbahi, A.; El-Naggar, M.; El-Newehy, M.; Rahaman, M.; Hatshan, M.; Khattab, T. Effects of Technical Textiles and Synthetic Nanofibers on Environmental Pollution. *Polymers* **2021**, *13*, 155. [CrossRef] [PubMed]
2. Pandit, P.; Singha, K.; Kumar, V.; Maity, S. Advanced flame-retardant agents for protective textiles and clothing. In *Advances in Functional and Protective Textiles*; Elsevier: Amsterdam, The Netherlands, 2020; pp. 397–414.
3. Sreenivasan, V.S.; Somasundaram, S.; Ravindran, D.; Manikandan, V.; Narayanasamy, R. Microstructural, physico-chemical and mechanical characterisation of Sansevieria cylindrica fibres—An exploratory investigation. *Mater. Des.* **2011**, *32*, 453–461. [CrossRef]

4. Pavlović, Ž.; Vrljičak, Z. Comparing double jersey knitted fabrics made of Tencel and modal yarns, spun by different spinning methods. *J. Eng. Fiber. Fabr.* **2020**, *15*, 155892502091985. [CrossRef]

5. Azman Mohammad Taib, M.N.; Hamidon, T.S.; Garba, Z.N.; Trache, D.; Uyama, H.; Hussin, M.H. Recent progress in cellulose-based composites towards flame retardancy applications. *Polymer* **2022**, *244*, 124677. [CrossRef]

6. Chakrabarty, A.; Teramoto, Y. Recent Advances in Nanocellulose Composites with Polymers: A Guide for Choosing Partners and How to Incorporate Them. *Polymers* **2018**, *10*, 517. [CrossRef]

7. Liu, Y.; Pan, Y.-T.; Wang, X.; Acuña, P.; Zhu, P.; Wagenknecht, U.; Heinrich, G.; Zhang, X.-Q.; Wang, R.; Wang, D.-Y. Effect of phosphorus-containing inorganic–organic hybrid coating on the flammability of cotton fabrics: Synthesis, characterization and flammability. *Chem. Eng. J.* **2016**, *294*, 167–175. [CrossRef]

8. Islam, M.S.; van de Ven, T.G.M. Cotton-based flame-retardant textiles: A review. *BioResources* **2021**, *16*, 4354–4381. [CrossRef]

9. Siriviriyanun, A.; O'Rear, E.A.; Yanumet, N. Self-extinguishing cotton fabric with minimal phosphorus deposition. *Cellulose* **2008**, *15*, 731–737. [CrossRef]

10. Lewin, M. *Handbook of Fiber Chemistry*; Wake, P.J., Ed.; CRC Press: Boca Raton, FL, USA, 2006; ISBN 9780824725655.

11. Alongi, J.; Colleoni, C.; Rosace, G.; Malucelli, G. Thermal and fire stability of cotton fabrics coated with hybrid phosphorus-doped silica films. *J. Therm. Anal. Calorim.* **2012**, *110*, 1207–1216. [CrossRef]

12. Wang, S.; Sun, L.; Li, Y.; Wang, H.; Liu, J.; Zhu, P.; Dong, C. Properties of flame-retardant cotton fabrics: Combustion behavior, thermal stability and mechanism of Si/P/N synergistic effect. *Ind. Crops Prod.* **2021**, *173*, 114157. [CrossRef]

13. Zammarano, M.; Cazzetta, V.; Nazaré, S.; Shields, J.R.; Kim, Y.S.; Hoffman, K.M.; Maffezzoli, A.; Davis, R.D. Smoldering and Flame Resistant Textiles via Conformal Barrier Formation. *Adv. Mater. Interfaces* **2016**, *3*, 1600617. [CrossRef] [PubMed]

14. Roberts, B.C.; Webber, M.E.; Ezekoye, O.A. Why and How the Sustainable Building Community Should Embrace Fire Safety. *Curr. Sustain. Energy Rep.* **2016**, *3*, 121–137. [CrossRef]

15. Hall, J.R. Estimating Fires When a Product is the Primary Fuel But Not the First Fuel, With an Application to Upholstered Furniture. *Fire Technol.* **2015**, *51*, 381–391. [CrossRef]

16. Salmeia, K.; Gaan, S.; Malucelli, G. Recent Advances for Flame Retardancy of Textiles Based on Phosphorus Chemistry. *Polymers* **2016**, *8*, 319. [CrossRef] [PubMed]

17. Available online: www.ctif.org (accessed on 25 May 2023).

18. Horrocks, A.R. Flame retardant challenges for textiles and fibres: New chemistry versus innovatory solutions. *Polym. Degrad. Stab.* **2011**, *96*, 377–392. [CrossRef]

19. Wang, Y.-W.; Shen, R.; Wang, Q.; Vasquez, Y. ZnO Microstructures as Flame-Retardant Coatings on Cotton Fabrics. *ACS Omega* **2018**, *3*, 6330–6338. [CrossRef]

20. Morgan, A.B.; Gilman, J.W. An overview of flame retardancy of polymeric materials: Application, technology, and future directions. *Fire Mater.* **2013**, *37*, 259–279. [CrossRef]

21. Salmeia, K.; Fage, J.; Liang, S.; Gaan, S. An Overview of Mode of Action and Analytical Methods for Evaluation of Gas Phase Activities of Flame Retardants. *Polymers* **2015**, *7*, 504–526. [CrossRef]

22. Shen, J.; Liang, J.; Lin, X.; Lin, H.; Yu, J.; Wang, S. The Flame-Retardant Mechanisms and Preparation of Polymer Composites and Their Potential Application in Construction Engineering. *Polymers* **2021**, *14*, 82. [CrossRef]

23. Ling, C.; Guo, L.; Wang, Z. A review on the state of flame-retardant cotton fabric: Mechanisms and applications. *Ind. Crops Prod.* **2023**, *194*, 116264. [CrossRef]

24. Hendrix, J.E.; Drake, G.L.; Barker, R.H. Pyrolysis and combustion of cellulose. III. Mechanistic basis for the synergism involving organic phosphates and nitrogenous bases. *J. Appl. Polym. Sci.* **1972**, *16*, 257–274. [CrossRef]

25. Alongi, J.; Colleoni, C.; Rosace, G.; Malucelli, G. Phosphorus- and nitrogen-doped silica coatings for enhancing the flame retardancy of cotton: Synergisms or additive effects? *Polym. Degrad. Stab.* **2013**, *98*, 579–589. [CrossRef]

26. Lewin, M. Synergism and catalysis in flame retardancy of polymers. *Polym. Adv. Technol.* **2001**, *12*, 215–222. [CrossRef]

27. Horrocks, A.R.; Smart, G.; Nazaré, S.; Kandola, B.; Price, D. Quantification of Zinc Hydroxystannate** and Stannate** Synergies in Halogen-containing Flame-retardant Polymeric Formulations. *J. Fire Sci.* **2010**, *28*, 217–248. [CrossRef]

28. Glogar, M.; Pušić, T.; Lovreškov, V.; Kaurin, T. Reactive Printing and Wash Fastness of Inherent Flame Retardant Fabrics for Dual Use. *Materials* **2022**, *15*, 4791. [CrossRef] [PubMed]

29. Durrani, H.; Sharma, V.; Bamboria, D.; Shukla, A.; Basak, S.; Ali, W. Exploration of flame retardant efficacy of cellulosic fabric using in-situ synthesized zinc borate particles. *Cellulose* **2020**, *27*, 9061–9073. [CrossRef]

30. van der Veen, I.; de Boer, J. Phosphorus flame retardants: Properties, production, environmental occurrence, toxicity and analysis. *Chemosphere* **2012**, *88*, 1119–1153. [CrossRef] [PubMed]

31. De Smet, D.; Weydts, D.; Vanneste, M. Environmentally friendly fabric finishes. In *Sustainable Apparel*; Elsevier: Amsterdam, The Netherlands, 2015; pp. 3–33.

32. Chivas, C.; Guillaume, E.; Sainrat, A.; Barbosa, V. Assessment of risks and benefits in the use of flame retardants in upholstered furniture in continental Europe. *Fire Saf. J.* **2009**, *44*, 801–807. [CrossRef]

33. Höhn, W. Textile Industry Effluent. In *Sustainable Textile and Fashion Value Chains*; Springer International Publishing: Cham, Switzerland, 2021; pp. 123–149.

34. Wang, X.; Hu, W.; Hu, Y. Polydopamine-Bridged Synthesis of Ternary h-BN@PDA@TiO$_2$ as Nanoenhancers for Thermal Conductivity and Flame Retardant of Polyvinyl Alcohol. *Front. Chem.* **2020**, *8*, 587474. [CrossRef]

35. Price, D. *Fire Retardant Materials*; Elsevier Science & Technology: Amsterdam, The Netherlands, 2001; ISBN 9781855734197.

36. Horrocks, A.R.; Price, D. *Fire Retardant Materials*, 1st ed.; Woodhead Publishing Limited: Cambridge, UK, 2001; ISBN 1855734192.

37. Little, R.W. Fundamentals of Flame Retardancy. *Text. Res. J.* **1951**, *21*, 901–908. [CrossRef]

38. Alongi, J.; Colleoni, C.; Malucelli, G.; Rosace, G. Hybrid phosphorus-doped silica architectures derived from a multistep sol–gel process for improving thermal stability and flame retardancy of cotton fabrics. *Polym. Degrad. Stab.* **2012**, *97*, 1334–1344. [CrossRef]

39. Horrocks, A.R. An Introduction to the Burning Behaviour of Cellulosic Fibres. *J. Soc. Dye Colour.* **1983**, *99*, 191–197. [CrossRef]

40. Laoutid, F.; Bonnaud, L.; Alexandre, M.; Lopez-Cuesta, J.-M.; Dubois, P. New prospects in flame retardant polymer materials: From fundamentals to nanocomposites. *Mater. Sci. Eng. R Rep.* **2009**, *63*, 100–125. [CrossRef]

41. Shafizadeh, F.; Fu, Y.L. Pyrolysis of cellulose. *Carbohydr. Res.* **1973**, *29*, 113–122. [CrossRef]

42. Visakh, P.M.; Arao, Y. (Eds.) *Flame Retardants*; Engineering Materials; Springer International Publishing: Cham, Switzerland, 2015; ISBN 978-3-319-03466-9.

43. Kaur, B.; Gur, I.S.; Bhatnagar, H.L. Thermal degradation studies of cellulose phosphates and cellulose thiophosphates. *Die Angew. Makromol. Chem.* **1987**, *147*, 157–183. [CrossRef]

44. Merritt, M.J. *Textile Fibres, Their Physical, Microscopic, and Chemical Properties*, 6th ed.; John Wiley & Sons, Inc.: New York, NY, USA, 1954.

45. Cabrales, L.; Abidi, N. Kinetics of Cellulose Deposition in Developing Cotton Fibers Studied by Thermogravimetric Analysis. *Fibers* **2019**, *7*, 78. [CrossRef]

46. Xu, F.; Zhong, L.; Xu, Y.; Zhang, C.; Wang, P.; Zhang, F.; Zhang, G. Synthesis of three novel amino acids-based flame retardants with multiple reactive groups for cotton fabrics. *Cellulose* **2019**, *26*, 7537–7552. [CrossRef]

47. Pal, A.; Samanta, A.K.; Bagchi, A.; Samanta, P.; Kar, T.R. A Review on Fire Protective Functional Finishing of Natural Fibre Based Textiles: Present Perspective. *Curr. Trends Fash. Technol. Text. Eng.* **2020**, *7*, 19–30. [CrossRef]

48. Chang, S.; Slopek, R.P.; Condon, B.; Grunlan, J.C. Surface Coating for Flame-Retardant Behavior of Cotton Fabric Using a Continuous Layer-by-Layer Process. *Ind. Eng. Chem. Res.* **2014**, *53*, 3805–3812. [CrossRef]

49. Lu, H.; Song, L.; Hu, Y. A review on flame retardant technology in China. Part II: Flame retardant polymeric nanocomposites and coatings. *Polym. Adv. Technol.* **2011**, *22*, 379–394. [CrossRef]

50. Lomakin, S.M.; Zaikov, G.E.; Artsis, M.I. Advances in Nylon 6,6 Flame Retardancy. *Int. J. Polym. Mater.* **1996**, *32*, 173–202. [CrossRef]

51. Faheem, S.; Baheti, V.; Tunak, M.; Wiener, J.; Militky, J. Flame resistance behavior of cotton fabrics coated with bilayer assemblies of ammonium polyphosphate and casein. *Cellulose* **2019**, *26*, 3557–3574. [CrossRef]

52. Emsley, A.M.; Stevens, G.C. Kinetics and mechanisms of the low-temperature degradation of cellulose. *Cellulose* **1994**, *1*, 26–56. [CrossRef]

53. *UL-94*; Standard for Tests for Flammability of Plastic Materials for Parts in Devices and Appliances. UL Solutions: Northbrook, IL, USA, 1996.

54. Birky, M.M.; Yeh, K.-N. Calorimetric study of flammable fabrics. I. Instrumentation and measurements. *J. Appl. Polym. Sci.* **1973**, *17*, 239–253. [CrossRef]

55. Webster, C.T.; Wraight, H.G.H.; Thomas, P.H. 3—Heat-Transfer from Burning Fabrics. *J. Text. Inst. Trans.* **1962**, *53*, T29–T37. [CrossRef]

56. Alongi, J.; Ciobanu, M.; Malucelli, G. Novel flame retardant finishing systems for cotton fabrics based on phosphorus-containing compounds and silica derived from sol–gel processes. *Carbohydr. Polym.* **2011**, *85*, 599–608. [CrossRef]

57. Rosace, G.; Castellano, A.; Trovato, V.; Iacono, G.; Malucelli, G. Thermal and flame retardant behaviour of cotton fabrics treated with a novel nitrogen-containing carboxyl-functionalized organophosphorus system. *Carbohydr. Polym.* **2018**, *196*, 348–358. [CrossRef]

58. Alongi, J.; Malucelli, G. Thermal stability, flame retardancy and abrasion resistance of cotton and cotton–linen blends treated by sol–gel silica coatings containing alumina micro- or nano-particles. *Polym. Degrad. Stab.* **2013**, *98*, 1428–1438. [CrossRef]

59. Chang, S.; Condon, B.; Graves, E.; Uchimiya, M.; Fortier, C.; Easson, M.; Wakelyn, P. Flame retardant properties of triazine phosphonates derivative with cotton fabric. *Fibers Polym.* **2011**, *12*, 334–339. [CrossRef]

60. Rosace, G.; Colleoni, C.; Trovato, V.; Iacono, G.; Malucelli, G. Vinylphosphonic acid/methacrylamide system as a durable intumescent flame retardant for cotton fabric. *Cellulose* **2017**, *24*, 3095–3108. [CrossRef]

61. Liao, Y.; Chen, Y.; Wan, C.; Zhang, G.; Zhang, F. An eco-friendly N P flame retardant for durable flame-retardant treatment of cotton fabric. *Int. J. Biol. Macromol.* **2021**, *187*, 251–261. [CrossRef]

62. Vishwakarma, A.; Singh, M.; Weclawski, B.; Reddy, V.J.; Kandola, B.K.; Manik, G.; Dasari, A.; Chattopadhyay, S. Construction of hydrophobic fire retardant coating on cotton fabric using a layer-by-layer spray coating method. *Int. J. Biol. Macromol.* **2022**, *223*, 1653–1666. [CrossRef]

63. Hu, M.-Y.; Xiong, K.-K.; Li, J.-R.; Jing, X.-B.; Zhao, P.-H. Novel P/N/Si/S-containing Mononickel Complex as a Metal-based Intumescent Flame Retardant for Cotton Fabrics. *Fibers Polym.* **2019**, *20*, 1794–1802. [CrossRef]

64. Manfredi, A.; Carosio, F.; Ferruti, P.; Alongi, J.; Ranucci, E. Disulfide-containing polyamidoamines with remarkable flame retardant activity for cotton fabrics. *Polym. Degrad. Stab.* **2018**, *156*, 1–13. [CrossRef]

65. Li, P.; Wang, B.; Xu, Y.-J.; Jiang, Z.; Dong, C.; Liu, Y.; Zhu, P. Ecofriendly Flame-Retardant Cotton Fabrics: Preparation, Flame Retardancy, Thermal Degradation Properties, and Mechanism. *ACS Sustain. Chem. Eng.* **2019**, *7*, 19246–19256. [CrossRef]

66. Li, D.; Wang, Z.; Zhu, Y.; You, F.; Zhou, S.; Li, G.; Zhang, X.; Zhou, C. Synergistically improved flame retardancy of the cotton fabric finished by silica-coupling agent-zinc borate hybrid sol. *J. Ind. Text.* **2022**, *51*, 8297S–8322S. [CrossRef]
67. Cheng, X.; Shi, L.; Fan, Z.; Yu, Y.; Liu, R. Bio-based coating of phytic acid, chitosan, and biochar for flame-retardant cotton fabrics. *Polym. Degrad. Stab.* **2022**, *199*, 109898. [CrossRef]
68. Gaan, S.; Sun, G. Effect of phosphorus flame retardants on thermo-oxidative decomposition of cotton. *Polym. Degrad. Stab.* **2007**, *92*, 968–974. [CrossRef]
69. Camlibel, N.O.; Avinc, O.; Arik, B.; Yavas, A.; Yakin, I. The effects of huntite–hydromagnesite inclusion in acrylate-based polymer paste coating process on some textile functional performance properties of cotton fabric. *Cellulose* **2019**, *26*, 1367–1381. [CrossRef]
70. Parmar, M.S.; Chakraborty, M. Thermal and Burning Behavior of Naturally Colored Cotton. *Text. Res. J.* **2001**, *71*, 1099–1102. [CrossRef]
71. Alaybeyoglu, E.; Duran, K.; Körlü, A. Flammability Behaviours of Knitted Fabrics Containing PLA, Cotton, Lyocell, Chitosan Fibers. *Mugla J. Sci. Technol.* **2022**, *8*, 1–8. [CrossRef]
72. Babrauskas, V. Upholstered Furniture Heat Release Rates: Measurements and Estimation. *J. Fire Sci.* **1983**, *1*, 9–32. [CrossRef]
73. Babrauskas, V.; Baroudi, D.; Myllymäki, J.; Kokkala, M. The Cone Calorimeter Used for Predictions of the Full-scale Burning Behaviour of Upholstered Furniture. *Fire Mater.* **1997**, *21*, 95–105. [CrossRef]
74. Hirschler, M.M. Polyurethane foam and fire safety. *Polym. Adv. Technol.* **2008**, *19*, 521–529. [CrossRef]
75. Alongi, J.; Tata, J.; Carosio, F.; Rosace, G.; Frache, A.; Camino, G. A Comparative Analysis of Nanoparticle Adsorption as Fire-Protection Approach for Fabrics. *Polymers* **2014**, *7*, 47–68. [CrossRef]
76. Tata, J.; Alongi, J.; Carosio, F.; Frache, A. Optimization of the procedure to burn textile fabrics by cone calorimeter: Part I. Combustion behavior of polyester. *Fire Mater.* **2011**, *35*, 397–409. [CrossRef]
77. Tata, J.; Alongi, J.; Frache, A. Optimization of the procedure to burn textile fabrics by cone calorimeter: Part II. Results on nanoparticle-finished polyester. *Fire Mater.* **2012**, *36*, 527–536. [CrossRef]
78. *ISO 5659*; Plastics—Smoke Generation—Part 1: Guidance on Optical-Density Testing 1996 + Part 2: Determination of Optical Density by a Single-Chamber Test 2002. International Organization for Standardization: Geneva, Switzerland, 2017.
79. *ISO 5660*; Fire Test, Reaction to Fire—Part 1: Rate of Heat Release (Cone Calorimeter Method) + Part 2: Smoke Production Rate (Dynamic Measurement) + Part 3: Guidance on Measurement 2002 and 2003. International Organization for Standardization: Geneva, Switzerland, 2015.
80. *ISO 9239*; Reaction to Fire Tests for Floorings—Part 1: Determination of the Burning Behaviour Using a Radiant Heat Source + Part 2: Determination of Flame Spread at a Heat Flux Level of 25 kW/m^2 2002. International Organization for Standardization: Geneva, Switzerland, 2010.
81. Nazaré, S.; Kandola, B.; Horrocks, A.R. Use of cone calorimetry to quantify the burning hazard of apparel fabrics. *Fire Mater.* **2002**, *26*, 191–199. [CrossRef]
82. Babrauskas, V.; Peacock, R.D. Heat release rate: The single most important variable in fire hazard. *Fire Saf. J.* **1992**, *18*, 255–272. [CrossRef]
83. Zhang, J.; Wang, X.; Zhang, F.; Richard Horrocks, A. Estimation of heat release rate for polymer–filler composites by cone calorimetry. *Polym. Test.* **2004**, *23*, 225–230. [CrossRef]
84. Quan, Y.; Zhang, Z.; Tanchak, R.N.; Wang, Q. A review on cone calorimeter for assessment of flame-retarded polymer composites. *J. Therm. Anal. Calorim.* **2022**, *147*, 10209–10234. [CrossRef]
85. Gann, R.G.; Babrauskas, V.; Peacock, R.D.; Hall, J.R. Fire conditions for smoke toxicity measurement. *Fire Mater.* **1994**, *18*, 193–199. [CrossRef]
86. Nazaré, S.; Kandola, B.K.; Horrocks, A.R. Smoke, CO, and CO_2 Measurements and Evaluation using Different Fire Testing Techniques for Flame Retardant Unsaturated Polyester Resin Formulations. *J. Fire Sci.* **2008**, *26*, 215–242. [CrossRef]
87. Hernandez, N.; Sonnier, R.; Giraud, S. Influence of grammage on heat release rate of polypropylene fabrics. *J. Fire Sci.* **2018**, *36*, 30–46. [CrossRef]
88. Gou, T.; Wu, X.; Zhao, Q.; Chang, S.; Wang, P. Novel phosphorus/nitrogen-rich oligomer with numerous reactive groups for durable flame-retardant cotton fabric. *Cellulose* **2021**, *28*, 7405–7419. [CrossRef]
89. Li, S.; Zhong, L.; Huang, S.; Wang, D.; Zhang, F.; Zhang, G. A novel flame retardant with reactive ammonium phosphate groups and polymerizing ability for preparing durable flame retardant and stiff cotton fabric. *Polym. Degrad. Stab.* **2019**, *164*, 145–156. [CrossRef]
90. Wan, C.; Liu, S.; Chen, Y.; Zhang, F. Facile, one–pot, formaldehyde-free synthesis of reactive N P flame retardant for a biomolecule of cotton. *Int. J. Biol. Macromol.* **2020**, *163*, 1659–1668. [CrossRef] [PubMed]
91. Alongi, J.; Carosio, F.; Frache, A.; Malucelli, G. Layer by Layer coatings assembled through dipping, vertical or horizontal spray for cotton flame retardancy. *Carbohydr. Polym.* **2013**, *92*, 114–119. [CrossRef]
92. Guo, W.; Wang, X.; Huang, J.; Zhou, Y.; Cai, W.; Wang, J.; Song, L.; Hu, Y. Construction of durable flame-retardant and robust superhydrophobic coatings on cotton fabrics for water-oil separation application. *Chem. Eng. J.* **2020**, *398*, 125661. [CrossRef]
93. Li, Z.-F.; Zhang, C.-J.; Cui, L.; Zhu, P.; Yan, C.; Liu, Y. Fire retardant and thermal degradation properties of cotton fabrics based on APTES and sodium phytate through layer-by-layer assembly. *J. Anal. Appl. Pyrolysis* **2017**, *123*, 216–223. [CrossRef]
94. Alongi, J.; Malucelli, G. Cotton fabrics treated with novel oxidic phases acting as effective smoke suppressants. *Carbohydr. Polym.* **2012**, *90*, 251–260. [CrossRef]

95. Alongi, J.; Cuttica, F.; Carosio, F.; Bourbigot, S. How much the fabric grammage may affect cotton combustion? *Cellulose* **2015**, *22*, 3477–3489. [CrossRef]
96. Grancaric, A.M.; Colleoni, C.; Guido, E.; Botteri, L.; Rosace, G. Thermal behaviour and flame retardancy of monoethanolamine-doped sol-gel coatings of cotton fabric. *Prog. Org. Coat.* **2017**, *103*, 174–181. [CrossRef]
97. Yang, C.Q.; He, Q.; Lyon, R.E.; Hu, Y. Investigation of the flammability of different textile fabrics using micro-scale combustion calorimetry. *Polym. Degrad. Stab.* **2010**, *95*, 108–115. [CrossRef]
98. Laufer, G.; Carosio, F.; Martinez, R.; Camino, G.; Grunlan, J.C. Growth and fire resistance of colloidal silica-polyelectrolyte thin film assemblies. *J. Colloid Interface Sci.* **2011**, *356*, 69–77. [CrossRef]
99. Ma, Z.; Zhang, Z.; Zhao, F.; Wang, Y. A multifunctional coating for cotton fabrics integrating superior performance of flame-retardant and self-cleaning. *Adv. Compos. Hybrid Mater.* **2022**, *5*, 2817–2833. [CrossRef]
100. Xing, W.; Jie, G.; Song, L.; Hu, S.; Lv, X.; Wang, X.; Hu, Y. Flame retardancy and thermal degradation of cotton textiles based on UV-curable flame retardant coatings. *Thermochim. Acta* **2011**, *513*, 75–82. [CrossRef]
101. Ur Rehman, Z.; Huh, S.-H.; Ullah, Z.; Pan, Y.-T.; Churchill, D.G.; Koo, B.H. LBL generated fire retardant nanocomposites on cotton fabric using cationized starch-clay-nanoparticles matrix. *Carbohydr. Polym.* **2021**, *274*, 118626. [CrossRef]
102. Chang, S.; Condon, B.; Nam, S. Development of Flame-Resistant Cotton Fabrics with Casein Using Pad-dry-cure and Supercritical Fluids Methods. *Int. J. Mater. Sci. Appl.* **2020**, *9*, 53. [CrossRef]
103. Kaurin, T.; Pušić, T.; Dekanić, T.; Flinčec Grgac, S. Impact of Washing Parameters on Thermal Characteristics and Appearance of Proban®—Flame Retardant Material. *Materials* **2022**, *15*, 5373. [CrossRef]
104. Krishnasamy, S.; Thiagamani, S.M.K.; Muthu Kumar, C.; Nagarajan, R.; Shahroze, R.M.; Siengchin, S.; Ismail, S.O.; Devi, I. Recent advances in thermal properties of hybrid cellulosic fiber reinforced polymer composites. *Int. J. Biol. Macromol.* **2019**, *141*, 1–13. [CrossRef]
105. Neto, J.; Queiroz, H.; Aguiar, R.; Lima, R.; Cavalcanti, D.; Doina Banea, M. A Review of Recent Advances in Hybrid Natural Fiber Reinforced Polymer Composites. *J. Renew. Mater.* **2022**, *10*, 561–589. [CrossRef]
106. Price, D.; Horrocks, A.R.; Akalin, M.; Faroq, A.A. Influence of flame retardants on the mechanism of pyrolysis of cotton (cellulose) fabrics in air. *J. Anal. Appl. Pyrolysis* **1997**, *40–41*, 511–524. [CrossRef]
107. Alongi, J.; Colleoni, C.; Rosace, G.; Malucelli, G. Sol–gel derived architectures for enhancing cotton flame retardancy: Effect of pure and phosphorus-doped silica phases. *Polym. Degrad. Stab.* **2014**, *99*, 92–98. [CrossRef]
108. Alongi, J.; Carosio, F.; Malucelli, G. Influence of ammonium polyphosphate-/poly(acrylic acid)-based layer by layer architectures on the char formation in cotton, polyester and their blends. *Polym. Degrad. Stab.* **2012**, *97*, 1644–1653. [CrossRef]
109. Montava-Jordà, S.; Torres-Giner, S.; Ferrandiz-Bou, S.; Quiles-Carrillo, L.; Montanes, N. Development of Sustainable and Cost-Competitive Injection-Molded Pieces of Partially Bio-Based Polyethylene Terephthalate through the Valorization of Cotton Textile Waste. *Int. J. Mol. Sci.* **2019**, *20*, 1378. [CrossRef] [PubMed]
110. Wang, S.; Liu, J.; Sun, L.; Wang, H.; Zhu, P.; Dong, C. Preparation of flame-retardant/dyed cotton fabrics: Flame retardancy, dyeing performance and flame retardant/dyed mechanism. *Cellulose* **2020**, *27*, 10425–10440. [CrossRef]
111. Sánchez-Jiménez, P.E.; Pérez-Maqueda, L.A.; Crespo-Amorós, J.E.; López, J.; Perejón, A.; Criado, J.M. Quantitative Characterization of Multicomponent Polymers by Sample-Controlled Thermal Analysis. *Anal. Chem.* **2010**, *82*, 8875–8880. [CrossRef]
112. Šimkovic, I. TG/DTG/DTA evaluation of flame retarded cotton fabrics and comparison to cone calorimeter data. *Carbohydr. Polym.* **2012**, *90*, 976–981. [CrossRef]
113. Davies, D.; Horrocks, A.R.; Greenhalgh, M. Ignition studies on cotton cellulose by DTA. *Thermochim. Acta* **1983**, *63*, 351–362. [CrossRef]
114. Shafizadeh, F.; Bradbury, A.G.W.; DeGroot, W.F.; Aanerud, T.W. Role of inorganic additives in the smoldering combustion of cotton cellulose. *Ind. Eng. Chem. Prod. Res. Dev.* **1982**, *21*, 97–101. [CrossRef]
115. Ghanadpour, M.; Carosio, F.; Wågberg, L. Ultrastrong and flame-resistant freestanding films from nanocelluloses, self-assembled using a layer-by-layer approach. *Appl. Mater. Today* **2017**, *9*, 229–239. [CrossRef]
116. Li, J.; Jiang, W. Synthesis of a novel P-N flame retardant for preparing flame retardant and durable cotton fabric. *Ind. Crops Prod.* **2021**, *174*, 114205. [CrossRef]
117. Li, Q.-L.; Huang, F.-Q.; Wei, Y.-J.; Wu, J.-Z.; Zhou, Z.; Liu, G. A Phosphorus-Nitrogen Flame-retardant: Synthesis and Application in Cotton Fabrics. *Mater. Sci.* **2018**, *24*, 448–452. [CrossRef]
118. Gaan, S.; Sun, G. Effect of phosphorus and nitrogen on flame retardant cellulose: A study of phosphorus compounds. *J. Anal. Appl. Pyrolysis* **2007**, *78*, 371–377. [CrossRef]
119. Lawler, T.E.; Drews, M.J.; Barker, R.H. Pyrolysis and combustion of cellulose. VIII. Thermally initiated reactions of phosphonomethyl amide flame retardants. *J. Appl. Polym. Sci.* **1985**, *30*, 2263–2277. [CrossRef]
120. Teli, M.; Pandit, P. Development of thermally stable and hygienic colored cotton fabric made by treatment with natural coconut shell extract. *J. Ind. Text.* **2018**, *48*, 87–118. [CrossRef]
121. El Messoudi, M.; Boukhriss, A.; Bentis, A.; El Bouchti, M.; Ait Chaoui, M.; El Kouali, M.; Gmouh, S. Flame retardant finishing of cotton fabric based on ionic liquid compounds containing boron prepared with the sol-gel method. *J. Coat. Technol. Res.* **2022**, *19*, 1609–1619. [CrossRef]
122. Zhu, P.; Sui, S.; Wang, B.; Sun, K.; Sun, G. A study of pyrolysis and pyrolysis products of flame-retardant cotton fabrics by DSC, TGA, and PY–GC–MS. *J. Anal. Appl. Pyrolysis* **2004**, *71*, 645–655. [CrossRef]

123. Bourbigot, S.; Chlebicki, S.; Mamleev, V. Thermal degradation of cotton under linear heating. *Polym. Degrad. Stab.* **2002**, *78*, 57–62. [CrossRef]

124. Tian, C.M.; Shi, Z.H.; Zhang, H.Y.; Xu, J.Z.; Shi, J.R.; Guo, H.Z. Thermal degradation of cotton cellulose. *J. Therm. Anal. Calorim.* **1999**, *55*, 93–98. [CrossRef]

125. Nakanishi, S.; Masuko, F.; Hori, K.; Hashimoto, T. Pyrolytic Gas Generation of Cotton Cellulose With and Without Flame Retardants at Different Stages of Thermal Degradation: Effects of Nitrogen, Phosphorus, and Halogens. *Text. Res. J.* **2000**, *70*, 574–583. [CrossRef]

126. Gaan, S.; Sun, G. Effect of nitrogen additives on thermal decomposition of cotton. *J. Anal. Appl. Pyrolysis* **2009**, *84*, 108–115. [CrossRef]

127. Kang, M.; Chen, S.; Yang, R.; Li, D.; Zhang, W. Fabrication of an Eco-Friendly Clay-Based Coating for Enhancing Flame Retardant and Mechanical Properties of Cotton Fabrics via LbL Assembly. *Polymers* **2022**, *14*, 4994. [CrossRef] [PubMed]

128. Weil, E.D.; Levchik, S. Current Practice and Recent Commercial Developments in Flame Retardancy of Polyamides. *J. Fire Sci.* **2004**, *22*, 251–264. [CrossRef]

129. Granzow, A. Flame retardation by phosphorus compounds. *Acc. Chem. Res.* **1978**, *11*, 177–183. [CrossRef]

130. MOSTASHARI, S.M.; FAYYAZ, F. A Thermogravimetric Study of Cotton Fabric Flame-Retardancy by Means of Impregnation with Red Phosphorus. *Chin. J. Chem.* **2008**, *26*, 1030–1034. [CrossRef]

131. Mostashari, S.M.; Baie, S. TG studies of synergism between red phosphorus (RP)–calcium chloride used in flame-retardancy for a cotton fabric favorable to green chemistry. *J. Therm. Anal. Calorim.* **2010**, *99*, 431–436. [CrossRef]

132. Mostashari, S.M.; Fayyaz, F. A Combination of Red Phosphorus-Zinc Chloride for Flame-Retardancy of a Cotton Fabric. *Int. J. Polym. Mater.* **2007**, *57*, 125–131. [CrossRef]

133. Mngomezulu, M.E.; John, M.J.; Jacobs, V.; Luyt, A.S. Review on flammability of biofibres and biocomposites. *Carbohydr. Polym.* **2014**, *111*, 149–182. [CrossRef]

134. Lee, H.C.; Lee, S. Flame retardancy for cotton cellulose treated with H_3PO_3. *J. Appl. Polym. Sci.* **2018**, *135*, 46497. [CrossRef]

135. Lin, D.; Zeng, X.; Li, H.; Lai, X.; Wu, T. One-pot fabrication of superhydrophobic and flame-retardant coatings on cotton fabrics via sol-gel reaction. *J. Colloid Interface Sci.* **2019**, *533*, 198–206. [CrossRef]

136. Wang, D.; Ma, J.; Liu, J.; Tian, A.; Fu, S. Intumescent flame-retardant and ultraviolet-blocking coating screen-printed on cotton fabric. *Cellulose* **2021**, *28*, 2495–2504. [CrossRef]

137. Lu, S.-Y.; Hamerton, I. Recent developments in the chemistry of halogen-free flame retardant polymers. *Prog. Polym. Sci.* **2002**, *27*, 1661–1712. [CrossRef]

138. Zhang, Q.-H.; Zhang, W.; Chen, G.-Q.; Xing, T.-L. Combustion properties of cotton fabric treated by boron doped silica sol. *Therm. Sci.* **2015**, *19*, 1345–1348. [CrossRef]

139. Zhu, W.; Yang, M.; Huang, H.; Dai, Z.; Cheng, B.; Hao, S. A phytic acid-based chelating coordination embedding structure of phosphorus–boron–nitride synergistic flame retardant to enhance durability and flame retardancy of cotton. *Cellulose* **2020**, *27*, 4817–4829. [CrossRef]

140. Nine, M.J.; Tran, D.N.H.; Tung, T.T.; Kabiri, S.; Losic, D. Graphene-Borate as an Efficient Fire Retardant for Cellulosic Materials with Multiple and Synergetic Modes of Action. *ACS Appl. Mater. Interfaces* **2017**, *9*, 10160–10168. [CrossRef]

141. Mostashari, S.M.; Fayyaz, F. TG of a cotton fabric impregnated by sodium borate decahydrate ($Na_2B_4O_7 \cdot 10H_2O$) as a flame-retardant. *J. Therm. Anal. Calorim.* **2008**, *93*, 933–936. [CrossRef]

142. Tawiah, B.; Yu, B.; Yang, W.; Yuen, R.K.K.; Fei, B. Facile flame retardant finishing of cotton fabric with hydrated sodium metaborate. *Cellulose* **2019**, *26*, 4629–4640. [CrossRef]

143. Akarslan, F. Investigation on Fire Retardancy Properties of Boric Acid Doped Textile Materials. *Acta Phys. Pol. A* **2015**, *128*, B-403–B-405. [CrossRef]

144. Liu, H.; Du, Y.; Lei, S.; Liu, Z. Flame-retardant activity of modified boron nitride nanosheets to cotton. *Text. Res. J.* **2020**, *90*, 512–522. [CrossRef]

145. Bentis, A.; Boukhriss, A.; Gmouh, S. Flame-retardant and water-repellent coating on cotton fabric by titania–boron sol–gel method. *J. Sol-Gel Sci. Technol.* **2020**, *94*, 719–730. [CrossRef]

146. Shen, K.K. Review of Recent Advances on the Use of Boron-based Flame Retardants. In *Polymer Green Flame Retardants*; Elsevier: Amsterdam, The Netherlands, 2014; pp. 367–388.

147. Global Organic Textile Standard (GOTS) 2020. Version 6.0. Available online: https://www.global-standard.org (accessed on 25 May 2023).

148. Dislich, H. Glassy and crystalline systems from gels, chemical basis and technical application. *J. Non. Cryst. Solids* **1984**, *63*, 237–241. [CrossRef]

149. Segal, D.L. Sol-gel processing: Routes to oxide ceramics using colloidal dispersions of hydrous oxides and alkoxide intermediates. *J. Non. Cryst. Solids* **1984**, *63*, 183–191. [CrossRef]

150. Brinker, C.J.; Scherer, G.W. *The Physics and Chemistry of Sol–Gel Processing*; Academic Press: New York, NY, USA, 1990.

151. Sfameni, S.; Hadhri, M.; Rando, G.; Drommi, D.; Rosace, G.; Trovato, V.; Plutino, M.R. Inorganic Finishing for Textile Fabrics: Recent Advances in Wear-Resistant, UV Protection and Antimicrobial Treatments. *Inorganics* **2023**, *11*, 19. [CrossRef]

152. Sfameni, S.; Del Tedesco, A.; Rando, G.; Truant, F.; Visco, A.; Plutino, M.R. Waterborne Eco-Sustainable Sol–Gel Coatings Based on Phytic Acid Intercalated Graphene Oxide for Corrosion Protection of Metallic Surfaces. *Int. J. Mol. Sci.* **2022**, *23*, 12021.

153. Innocenzi, P. The Precursors of the Sol-Gel Process. In *The Sol-to-Gel Transition. SpringerBriefs in Materials*; Springer: Cham, Switzerland, 2019; pp. 7–19.

154. Ielo, I.; Giacobello, F.; Castellano, A.; Sfameni, S.; Rando, G.; Plutino, M.R. Development of Antibacterial and Antifouling Innovative and Eco-Sustainable Sol–Gel Based Materials: From Marine Areas Protection to Healthcare Applications. *Gels* **2022**, *8*, 26.

155. Giacobello, F.; Ielo, I.; Belhamdi, H.; Plutino, M.R. Geopolymers and Functionalization Strategies for the Development of Sustainable Materials in Construction Industry and Cultural Heritage Applications: A Review. *Materials* **2022**, *15*, 1725. [CrossRef]

156. Sfameni, S.; Rando, G.; Marchetta, A.; Scolaro, C.; Cappello, S.; Urzì, C.; Visco, A.; Plutino, M.R. Development of Eco-Friendly Hydrophobic and Fouling-Release Coatings for Blue-Growth Environmental Applications: Synthesis, Mechanical Characterization and Biological Activity. *Gels* **2022**, *8*, 528. [CrossRef]

157. Figueira, R.B.; Silva, C.J.R.; Pereira, E.V. Organic–inorganic hybrid sol–gel coatings for metal corrosion protection: A review of recent progress. *J. Coat. Technol. Res.* **2014**, *12*, 1–35. [CrossRef]

158. Mahltig, B.; Textor, T. *Nanosols and Textiles*; World Scientific Publishing Co. Pte. Ltd.: Singapore, 2008.

159. Sfameni, S.; Rando, G.; Plutino, M.R. Sustainable Secondary-Raw Materials, Natural Substances and Eco-Friendly Nanomaterial-Based Approaches for Improved Surface Performances: An Overview of What They Are and How They Work. *Int. J. Mol. Sci.* **2023**, *24*, 5472. [CrossRef]

160. Sakka, S. (Ed.) *Sol-Gel Science and Technology: Topics and Fundamental Research and Applications*; Kluwer Academic Publishers: Norwell, Australia, 2003; ISBN 978-1402072918.

161. Colleoni, C.; Guido, E.; Migani, V.; Rosace, G. Hydrophobic behaviour of non-fluorinated sol-gel based cotton and polyester fabric coatings. *J. Ind. Text.* **2015**, *44*, 815–834. [CrossRef]

162. Sfameni, S.; Lawnick, T.; Rando, G.; Visco, A.; Textor, T.; Plutino, M.R. Super-Hydrophobicity of Polyester Fabrics Driven by Functional Sustainable Fluorine-Free Silane-Based Coatings. *Gels* **2023**, *9*, 109. [CrossRef] [PubMed]

163. Guido, E.; Colleoni, C.; De Clerck, K.; Plutino, M.R.; Rosace, G. Influence of catalyst in the synthesis of a cellulose-based sensor: Kinetic study of 3-glycidoxypropyltrimethoxysilane epoxy ring opening by Lewis acid. *Sens. Actuators B Chem.* **2014**, *203*, 213–222. [CrossRef]

164. Min, L.; Xiaoli, Z.; Shuilin, C. Enhancing the wash fastness of dyeings by a sol-gel process. Part 1; Direct dyes on cotton. *Color. Technol.* **2003**, *119*, 297–300. [CrossRef]

165. Schramm, C.; Binder, W.H.; Tessadri, R. Durable Press Finishing of Cotton Fabric with 1,2,3,4-Butanetetracarboxylic Acid and TEOS/GPTMS. *J. Sol-Gel Sci. Technol.* **2004**, *29*, 155–165. [CrossRef]

166. Haufe, H.; Muschter, K.; Siegert, J.; Böttcher, H. Bioactive textiles by sol–gel immobilised natural active agents. *J. Sol-Gel Sci. Technol.* **2008**, *45*, 97–101. [CrossRef]

167. El-Hady, M.M.A.; Farouk, A.; Sharaf, S. Flame retardancy and UV protection of cotton based fabrics using nano ZnO and polycarboxylic acids. *Carbohydr. Polym.* **2013**, *92*, 400–406. [CrossRef]

168. Colleoni, C.; Massafra, M.R.; Rosace, G. Photocatalytic properties and optical characterization of cotton fabric coated via sol–gel with non-crystalline TiO_2 modified with poly(ethylene glycol). *Surf. Coat. Technol.* **2012**, *207*, 79–88. [CrossRef]

169. Sfameni, S.; Lawnick, T.; Rando, G.; Visco, A.; Textor, T.; Plutino, M.R. Functional Silane-Based Nanohybrid Materials for the Development of Hydrophobic and Water-Based Stain Resistant Cotton Fabrics Coatings. *Nanomaterials* **2022**, *12*, 3404. [CrossRef] [PubMed]

170. Sfameni, S.; Rando, G.; Plutino, M.R. *Perspective Chapter: Functional Sol–Gel Based Coatings for Innovative and Sustainable Applications*; Singh, D.J.P., Acharya, D.S.S., Kumar, D.S., Dixit, D.S.K., Eds.; IntechOpen: Rijeka, Croatia, 2023; p. Ch. 3. ISBN 978-1-80355-415-0.

171. Poli, R.; Colleoni, C.; Calvimontes, A.; Polášková, H.; Dutschk, V.; Rosace, G. Innovative sol–gel route in neutral hydroalcoholic condition to obtain antibacterial cotton finishing by zinc precursor. *J. Sol-Gel Sci. Technol.* **2015**, *74*, 151–160. [CrossRef]

172. Alongi, J.; Malucelli, G. Cotton flame retardancy: State of the art and future perspectives. *RSC Adv.* **2015**, *5*, 24239–24263. [CrossRef]

173. Hench, L.L.; West, J.K. The sol-gel process. *Chem. Rev.* **1990**, *90*, 33–72. [CrossRef]

174. Pierre, A. *Introduction to Sol-Gel Processing*; The Kluwer International Series in Sol-Gel Processing: Technology and Applications; Springer: Boston, MA, USA, 1998; ISBN 978-0-7923-8121-1.

175. Malucelli, G.; Carosio, F.; Alongi, J.; Fina, A.; Frache, A.; Camino, G. Materials engineering for surface-confined flame retardancy. *Mater. Sci. Eng. R Rep.* **2014**, *84*, 1–20. [CrossRef]

176. Alongi, J.; Ciobanu, M.; Malucelli, G. Thermal stability, flame retardancy and mechanical properties of cotton fabrics treated with inorganic coatings synthesized through sol–gel processes. *Carbohydr. Polym.* **2012**, *87*, 2093–2099. [CrossRef]

177. Cireli, A.; Onar, N.; Ebeoglugil, M.F.; Kayatekin, I.; Kutlu, B.; Culha, O.; Celik, E. Development of flame retardancy properties of new halogen-free phosphorous doped SiO_2 thin films on fabrics. *J. Appl. Polym. Sci.* **2007**, *105*, 3748–3756. [CrossRef]

178. Alongi, J.; Ciobanu, M.; Tata, J.; Carosio, F.; Malucelli, G. Thermal stability and flame retardancy of polyester, cotton, and relative blend textile fabrics subjected to sol–gel treatments. *J. Appl. Polym. Sci.* **2011**, *119*, 1961–1969. [CrossRef]

179. Colleoni, C.; Donelli, I.; Freddi, G.; Guido, E.; Migani, V.; Rosace, G. A novel sol-gel multi-layer approach for cotton fabric finishing by tetraethoxysilane precursor. *Surf. Coat. Technol.* **2013**, *235*, 192–203. [CrossRef]

180. Zhang, X.; Wang, Z.; Zhou, S.; You, F.; Li, D.; Zhou, C.; Pan, Y.; Wang, J. Enhanced flame retardancy level of a cotton fabric treated by an ammonium pentaborate doped silica-KH570 sol. *J. Ind. Text.* **2022**, *52*, 152808372211165. [CrossRef]

181. Alongi, J.; Ciobanu, M.; Malucelli, G. Sol–gel treatments for enhancing flame retardancy and thermal stability of cotton fabrics: Optimisation of the process and evaluation of the durability. *Cellulose* **2011**, *18*, 167–177. [CrossRef]
182. Alongi, J.; Malucelli, G. *Thermal Degradation of Cellulose and Cellulosic Substrates*; Wiley: Hoboken, NJ, USA, 2015.
183. Alongi, J.; Ciobanu, M.; Malucelli, G. Sol–gel treatments on cotton fabrics for improving thermal and flame stability: Effect of the structure of the alkoxysilane precursor. *Carbohydr. Polym.* **2012**, *87*, 627–635. [CrossRef] [PubMed]
184. Birnbaum, L.S.; Staskal, D.F. Brominated flame retardants: Cause for concern? *Environ. Health Perspect.* **2004**, *112*, 9–17. [CrossRef] [PubMed]
185. Barcelo, D.; Kostianoy, A.G. Handbook environmental chemistry. In *Brominated Flame Retardants*; Heidelberg, S.-V.B., Ed.; Springer: New York, NY, USA, 2011; Volume 16.
186. Alongi, J.; Carosio, F. All-inorganic intumescent nanocoating containing montmorillonite nanoplatelets in ammonium polyphosphate matrix capable of preventing cotton ignition. *Polymers* **2016**, *8*, 430. [CrossRef]
187. Kiliaris, P.; Papaspyrides, C.D. Polymer/layered silicate (clay) nanocomposites: An overview of flame retardancy. *Prog. Polym. Sci.* **2010**, *35*, 902–958. [CrossRef]
188. Alongi, J.; Carosio, F.; Kiekens, P. Recent Advances in the Design of Water Based-Flame Retardant Coatings for Polyester and Polyester-Cotton Blends. *Polymers* **2016**, *8*, 357. [CrossRef]
189. Carosio, F.; Alongi, J. Ultra-Fast Layer-by-Layer Approach for Depositing Flame Retardant Coatings on Flexible PU Foams within Seconds. *ACS Appl. Mater. Interfaces* **2016**, *8*, 6315–6319. [CrossRef]
190. Carosio, F.; Alongi, J.; Malucelli, G. Flammability and combustion properties of ammonium polyphosphate-/poly(acrylic acid)-based layer by layer architectures deposited on cotton, polyester and their blends. *Polym. Degrad. Stab.* **2013**, *98*, 1626–1637. [CrossRef]
191. Alongi, J.; Han, Z.; Bourbigot, S. Intumescence: Tradition versus novelty. A comprehensive review. *Prog. Polym. Sci.* **2015**, *51*, 28–73. [CrossRef]
192. Hall, P.L. The application of electron spin resonance spectroscopy to studies of clay minerals: I. Isomorphous substitutions and external surface properties. *Clay Miner.* **1980**, *15*, 321–335. [CrossRef]
193. Murray, H.H. Chapter 2 Structure and Composition of the Clay Minerals and their Physical and Chemical Properties. In *Developments in Clay Science*; Elsevier: Amsterdam, The Netherlands, 2006; pp. 7–31.
194. Bergaya, F.; Lagaly, G. Chapter 1 General Introduction: Clays, Clay Minerals, and Clay Science. In *Developments in Clay Science*; Elsevier: Amsterdam, The Netherlands, 2006; pp. 1–18.
195. Rando, G.; Sfameni, S.; Galletta, M.; Drommi, D.; Cappello, S.; Plutino, M.R. Functional Nanohybrids and Nanocomposites Development for the Removal of Environmental Pollutants and Bioremediation. *Molecules* **2022**, *27*, 4856. [CrossRef]
196. Kotal, M.; Bhowmick, A.K. Polymer nanocomposites from modified clays: Recent advances and challenges. *Prog. Polym. Sci.* **2015**, *51*, 127–187. [CrossRef]
197. Murray, H.H. Overview—Clay mineral applications. *Appl. Clay Sci.* **1991**, *5*, 379–395. [CrossRef]
198. Kloprogge, J.T. Synthesis of Smectites and Porous Pillared Clay Catalysts: A Review. *J. Porous Mater.* **1998**, *5*, 5–41. [CrossRef]
199. Ielo, I.; Galletta, M.; Rando, G.; Sfameni, S.; Cardiano, P.; Sabatino, G.; Drommi, D.; Rosace, G.; Plutino, M.R. Design, synthesis and characterization of hybrid coatings suitable for geopolymeric-based supports for the restoration of cultural heritage. *IOP Conf. Ser. Mater. Sci. Eng.* **2020**, *777*, 012003. [CrossRef]
200. Masini, J.C.; Abate, G. Guidelines to Study the Adsorption of Pesticides onto Clay Minerals Aiming at a Straightforward Evaluation of Their Removal Performance. *Minerals* **2021**, *11*, 1282. [CrossRef]
201. Perelomov, L.; Mandzhieva, S.; Minkina, T.; Atroshchenko, Y.; Perelomova, I.; Bauer, T.; Pinsky, D.; Barakhov, A. The Synthesis of Organoclays Based on Clay Minerals with Different Structural Expansion Capacities. *Minerals* **2021**, *11*, 707. [CrossRef]
202. Undabeytia, T.; Shuali, U.; Nir, S.; Rubin, B. Applications of Chemically Modified Clay Minerals and Clays to Water Purification and Slow Release Formulations of Herbicides. *Minerals* **2020**, *11*, 9. [CrossRef]
203. Jacquet, A.; Geatches, D.; Clark, S.; Greenwell, H. Understanding Cationic Polymer Adsorption on Mineral Surfaces: Kaolinite in Cement Aggregates. *Minerals* **2018**, *8*, 130. [CrossRef]
204. Lazorenko, G.; Kasprzhitskii, A.; Yavna, V. Comparative Study of the Hydrophobicity of Organo-Montmorillonite Modified with Cationic, Amphoteric and Nonionic Surfactants. *Minerals* **2020**, *10*, 732. [CrossRef]
205. Li, A.; Wang, A.-Q.; Chen, J.-M. Preparation and Properties of Poly (acrylic acid-potassium acrylate)/Attapulgite Superabsorbent Composite. *J. Funct. Polym.* **2004**, *17*, 200–206.
206. Taylor, R.K.; Smith, T.J. The engineering geology of clay minerals: Swelling, shrinking and mudrock breakdown. *Clay Miner.* **1986**, *21*, 235–260. [CrossRef]
207. Chi, M. Cation Exchange Capacity of Kaolinite. *Clays Clay Miner.* **1999**, *47*, 174–180. [CrossRef]
208. Theng, B.K.G. *Formation and Properties of Clay-Polymer Complexes*; Elsevier: Amsterdam, The Netherlands, 2012; ISBN 9780080885278.
209. Rezaei, A.; Daeihamed, M.; Capanoglu, E.; Tomas, M.; Akbari-Alavijeh, S.; Shaddel, R.; Khoshnoudi-Nia, S.; Boostani, S.; Rostamabadi, H.; Falsafi, S.R.; et al. Possible health risks associated with nanostructures in food. In *Safety and Regulatory Issues of Nanoencapsulated Food Ingredients*; Elsevier: Amsterdam, The Netherlands, 2021; pp. 31–118.
210. Churchman, G.J.; Lowe, D.J. Alteration, formation, and occurrence of minerals in soils. In *Handbook of Soil Sciences: Properties and Processes*; CRC Press: Boca Raton, FL, USA, 2012.

211. Chalasani, R.; Gupta, A.; Vasudevan, S. Engineering New Layered Solids from Exfoliated Inorganics: A Periodically Alternating Hydrotalcite—Montmorillonite Layered Hybrid. *Sci. Rep.* **2013**, *3*, 3498. [CrossRef]
212. Kausar, A.; Ahmad, I.; Maaza, M.; Eisa, M.H. State-of-the-Art Nanoclay Reinforcement in Green Polymeric Nanocomposite: From Design to New Opportunities. *Minerals* **2022**, *12*, 1495. [CrossRef]
213. Shan, G.; Jin, W.; Chen, H.; Zhao, M.; Surampalli, R.; Ramakrishnan, A.; Zhang, T.; Tyagi, R.D. Flame-Retardant Polymer Nanocomposites and Their Heat-Release Rates. *J. Hazard. Toxic Radioact. Waste* **2015**, *19*, 04015006. [CrossRef]
214. Kang, D.J.; Park, G.U.; Park, H.Y.; Park, J.-U.; Im, H.-G. A high-performance transparent moisture barrier using surface-modified nanoclay composite for OLED encapsulation. *Prog. Org. Coat.* **2018**, *118*, 66–71. [CrossRef]
215. Król-Morkisz, K.; Pielichowska, K. Thermal Decomposition of Polymer Nanocomposites With Functionalized Nanoparticles. In *Polymer Composites with Functionalized Nanoparticles*; Elsevier: Amsterdam, The Netherlands, 2019; pp. 405–435.
216. Jayrajsinh, S.; Shankar, G.; Agrawal, Y.K.; Bakre, L. Montmorillonite nanoclay as a multifaceted drug-delivery carrier: A review. *J. Drug Deliv. Sci. Technol.* **2017**, *39*, 200–209. [CrossRef]
217. Shunmugasamy, V.C.; Xiang, C.; Gupta, N. Clay/Polymer Nanocomposites: Processing, Properties, and Applications. In *Hybrid and Hierarchical Composite Materials*; Springer International Publishing: Cham, Switzerland, 2015; pp. 161–200.
218. Khalid, M.; Walvekar, R.; Ketabchi, M.R.; Siddiqui, H.; Hoque, M.E. Rubber/Nanoclay Composites: Towards Advanced Functional Materials. In *Nanoclay Reinforced Polymer Composites*; Springer: Berlin/Heidelberg, Germany, 2016; pp. 209–224.
219. Zare, Y.; Garmabi, H.; Sharif, F. Optimization of mechanical properties of PP/Nanoclay/CaCO3 ternary nanocomposite using response surface methodology. *J. Appl. Polym. Sci.* **2011**, *122*, 3188–3200. [CrossRef]
220. Zare, Y.; Garmabi, H. Modeling of interfacial bonding between two nanofillers (montmorillonite and CaCO3) and a polymer matrix (PP) in a ternary polymer nanocomposite. *Appl. Surf. Sci.* **2014**, *321*, 219–225. [CrossRef]
221. de Oliveira, C.R.S.; Batistella, M.A.; Lourenço, L.A.; de Souza, S.M.d.A.G.U.; de Souza, A.A.U. Cotton fabric finishing based on phosphate/clay mineral by direct-coating technique and its influence on the thermal stability of the fibers. *Prog. Org. Coat.* **2021**, *150*, 105949. [CrossRef]
222. Furtana, S.; Mutlu, A.; Dogan, M. Thermal stability and flame retardant properties of calcium- and magnesium-hypophosphite-finished cotton fabrics and the evaluation of interaction with clay and POSS nanoparticles. *J. Therm. Anal. Calorim.* **2020**, *139*, 3415–3425. [CrossRef]
223. He, H.; Tao, Q.; Zhu, J.; Yuan, P.; Shen, W.; Yang, S. Silylation of clay mineral surfaces. *Appl. Clay Sci.* **2013**, *71*, 15–20. [CrossRef]
224. Romanzini, D.; Piroli, V.; Frache, A.; Zattera, A.J.; Amico, S.C. Sodium montmorillonite modified with methacryloxy and vinylsilanes: Influence of silylation on the morphology of clay/unsaturated polyester nanocomposites. *Appl. Clay Sci.* **2015**, *114*, 550–557. [CrossRef]
225. Illy, N.; Fache, M.; Ménard, R.; Negrell, C.; Caillol, S.; David, G. Phosphorylation of bio-based compounds: The state of the art. *Polym. Chem.* **2015**, *6*, 6257–6291. [CrossRef]
226. Tang, G.; Huang, X.; Ding, H.; Wang, X.; Jiang, S.; Zhou, K.; Wang, B.; Yang, W.; Hu, Y. Combustion properties and thermal degradation behaviors of biobased polylactide composites filled with calcium hypophosphite. *RSC Adv.* **2014**, *4*, 8985. [CrossRef]
227. Costantino, U.; Nocchetti, M.; Sisani, M.; Vivani, R. Recent progress in the synthesis and application of organically modified hydrotalcites. *Z. Für Krist.* **2009**, *224*, 273–281. [CrossRef]
228. Forano, C.; Hibino, T.; Leroux, F.; Taviot-Guého, C. Chapter 13.1 Layered Double Hydroxides. In *Developments in Clay Science*; Elsevier: Amsterdam, The Netherlands, 2006; pp. 1021–1095.
229. Alongi, J.; Tata, J.; Frache, A. Hydrotalcite and nanometric silica as finishing additives to enhance the thermal stability and flame retardancy of cotton. *Cellulose* **2011**, *18*, 179–190. [CrossRef]
230. Ishihara, S.; Sahoo, P.; Deguchi, K.; Ohki, S.; Tansho, M.; Shimizu, T.; Labuta, J.; Hill, J.P.; Ariga, K.; Watanabe, K.; et al. Dynamic Breathing of CO_2 by Hydrotalcite. *J. Am. Chem. Soc.* **2013**, *135*, 18040–18043. [CrossRef]
231. Wang, S.; Gainey, L.; Mackinnon, I.D.R.; Allen, C.; Gu, Y.; Xi, Y. Thermal behaviors of clay minerals as key components and additives for fired brick properties: A review. *J. Build. Eng.* **2023**, *66*, 105802. [CrossRef]
232. Shmuradko, V.T.; Panteleenko, F.I.; Reut, O.P.; Panteleenko, E.F.; Kirshina, N.V. Composition, structure, and property formation of heat insulation fire- and heat-reflecting materials based on vermiculite for industrial power generation. *Refract. Ind. Ceram.* **2012**, *53*, 254–258. [CrossRef]
233. Addison, J. Vermiculite: A Review of the Mineralogy and Health Effects of Vermiculite Exploitation. *Regul. Toxicol. Pharmacol.* **1995**, *21*, 397–405. [CrossRef] [PubMed]
234. Cain, A.A.; Plummer, M.G.B.; Murray, S.E.; Bolling, L.; Regev, O.; Grunlan, J.C. Iron-containing, high aspect ratio clay as nanoarmor that imparts substantial thermal/flame protection to polyurethane with a single electrostatically-deposited bilayer. *J. Mater. Chem. A* **2014**, *2*, 17609–17617. [CrossRef]
235. Suvorov, S.A.; Skurikhin, V.V. Vermiculite—A promising material for high-temperature heat insulators. *Refract. Ind. Ceram.* **2003**, *44*, 186–193. [CrossRef]
236. Ortelli, S.; Malucelli, G.; Cuttica, F.; Blosi, M.; Zanoni, I.; Costa, A.L. Coatings made of proteins adsorbed on TiO_2 nanoparticles: A new flame retardant approach for cotton fabrics. *Cellulose* **2018**, *25*, 2755–2765. [CrossRef]
237. Ortelli, S.; Malucelli, G.; Blosi, M.; Zanoni, I.; Costa, A.L. NanoTiO$_2$@DNA complex: A novel eco, durable, fire retardant design strategy for cotton textiles. *J. Colloid Interface Sci.* **2019**, *546*, 174–183. [CrossRef] [PubMed]

238. Apaydin, K.; Laachachi, A.; Ball, V.; Jimenez, M.; Bourbigot, S.; Ruch, D. Layer-by-layer deposition of a TiO2-filled intumescent coating and its effect on the flame retardancy of polyamide and polyester fabrics. *Colloids Surfaces A Physicochem. Eng. Asp.* **2015**, *469*, 1–10. [CrossRef]

239. Qin, S.; Pour, M.G.; Lazar, S.; Köklükaya, O.; Gerringer, J.; Song, Y.; Wågberg, L.; Grunlan, J.C. Super Gas Barrier and Fire Resistance of Nanoplatelet/Nanofibril Multilayer Thin Films. *Adv. Mater. Interfaces* **2019**, *6*, 1801424. [CrossRef]

240. Ali, Z.A. A Seq to Seq Machine Translation from Urdu to Chinese. *J. Auton. Intell.* **2021**, *4*, 1. [CrossRef]

241. Beyer, G. Short communication: Carbon nanotubes as flame retardants for polymers. *Fire Mater.* **2002**, *26*, 291–293. [CrossRef]

242. Coleman, J.N.; Khan, U.; Blau, W.J.; Gun'ko, Y.K. Small but strong: A review of the mechanical properties of carbon nanotube–polymer composites. *Carbon* **2006**, *44*, 1624–1652. [CrossRef]

243. Moniruzzaman, M.; Winey, K.I. Polymer Nanocomposites Containing Carbon Nanotubes. *Macromolecules* **2006**, *39*, 5194–5205. [CrossRef]

244. Zare, Y.; Garmabi, H. Attempts to Simulate the Modulus of Polymer/Carbon Nanotube Nanocomposites and Future Trends. *Polym. Rev.* **2014**, *54*, 377–400. [CrossRef]

245. Liu, Y.; Wang, X.; Qi, K.; Xin, J.H. Functionalization of cotton with carbon nanotubes. *J. Mater. Chem.* **2008**, *18*, 3454. [CrossRef]

246. Kashiwagi, T.; Du, F.; Douglas, J.F.; Winey, K.I.; Harris, R.H.; Shields, J.R. Nanoparticle networks reduce the flammability of polymer nanocomposites. *Nat. Mater.* **2005**, *4*, 928–933. [CrossRef]

247. Ye, L.; Wu, Q.; Qu, B. Synergistic effects and mechanism of multiwalled carbon nanotubes with magnesium hydroxide in halogen-free flame retardant EVA/MH/MWNT nanocomposites. *Polym. Degrad. Stab.* **2009**, *94*, 751–756. [CrossRef]

248. Janas, D.; Rdest, M.; Koziol, K.K.K. Flame-retardant carbon nanotube films. *Appl. Surf. Sci.* **2017**, *411*, 177–181. [CrossRef]

249. Araby, S.; Philips, B.; Meng, Q.; Ma, J.; Laoui, T.; Wang, C.H. Recent advances in carbon-based nanomaterials for flame retardant polymers and composites. *Compos. Part B Eng.* **2021**, *212*, 108675. [CrossRef]

250. Cho, C.; Song, Y.; Allen, R.; Wallace, K.L.; Grunlan, J.C. Stretchable electrically conductive and high gas barrier nanocomposites. *J. Mater. Chem. C* **2018**, *6*, 2095–2104. [CrossRef]

251. Rizkalla, S.; Dawood, M.; Schnerch, D. Development of a carbon fiber reinforced polymer system for strengthening steel structures. *Compos. Part A Appl. Sci. Manuf.* **2008**, *39*, 388–397. [CrossRef]

252. He, W.; Gao, J.; Liao, S.; Wang, X.; Qin, S.; Song, P. A facile method to improve thermal stability and flame retardancy of polyamide 6. *Compos. Commun.* **2019**, *13*, 143–150. [CrossRef]

253. Lazar, S.T.; Kolibaba, T.J.; Grunlan, J.C. Flame-retardant surface treatments. *Nat. Rev. Mater.* **2020**, *5*, 259–275. [CrossRef]

254. Montazer, M.; Harifi, T. Flame-retardant textile nanofinishes. In *Nanofinishing of Textile Materials*; Elsevier: Amsterdam, The Netherlands, 2018; pp. 163–181.

255. Gonçalves, A.G.; Jarrais, B.; Pereira, C.; Morgado, J.; Freire, C.; Pereira, M.F.R. Functionalization of textiles with multi-walled carbon nanotubes by a novel dyeing-like process. *J. Mater. Sci.* **2012**, *47*, 5263–5275. [CrossRef]

256. Motaghi, Z.; Shahidi, S. Improvement the Conductivity and Flame Retardant Properties of Carboxylated Single-Walled Carbon Nanotube/Cotton Fabrics Using Citric Acid and Sodium Hypophosphite. *J. Nat. Fibers* **2018**, *15*, 353–362. [CrossRef]

257. Xu, J.; Niu, Y.; Xie, Z.; Liang, F.; Guo, F.; Wu, J. Synergistic flame retardant effect of carbon nanohorns and ammonium polyphosphate as a novel flame retardant system for cotton fabrics. *Chem. Eng. J.* **2023**, *451*, 138566. [CrossRef]

258. Liu, H.; Du, Y.; Yang, G.; Zhu, G.; Gao, Y.; Ding, W. Flame retardance of modified graphene to pure cotton fabric. *J. Fire Sci.* **2018**, *36*, 111–128. [CrossRef]

259. Norouzi, M.; Zare, Y.; Kiany, P. Nanoparticles as Effective Flame Retardants for Natural and Synthetic Textile Polymers: Application, Mechanism, and Optimization. *Polym. Rev.* **2015**, *55*, 531–560. [CrossRef]

260. Cinausero, N.; Azema, N.; Lopez-Cuesta, J.-M.; Cochez, M.; Ferriol, M. Synergistic effect between hydrophobic oxide nanoparticles and ammonium polyphosphate on fire properties of poly(methyl methacrylate) and polystyrene. *Polym. Degrad. Stab.* **2011**, *96*, 1445–1454. [CrossRef]

261. Rault, F.; Pleyber, E.; Campagne, C.; Rochery, M.; Giraud, S.; Bourbigot, S.; Devaux, E. Effect of manganese nanoparticles on the mechanical, thermal and fire properties of polypropylene multifilament yarn. *Polym. Degrad. Stab.* **2009**, *94*, 955–964. [CrossRef]

262. Coyle, S.; Wu, Y.; Lau, K.-T.; De Rossi, D.; Wallace, G.; Diamond, D. Smart Nanotextiles: A Review of Materials and Applications. *MRS Bull.* **2007**, *32*, 434–442. [CrossRef]

263. Sundarrajan, S.; Chandrasekaran, A.R.; Ramakrishna, S. An Update on Nanomaterials-Based Textiles for Protection and Decontamination. *J. Am. Ceram. Soc.* **2010**, *93*, 3955–3975. [CrossRef]

264. Yadav, A.; Prasad, V.; Kathe, A.A.; Raj, S.; Yadav, D. Functional finishing in cotton fabrics using zinc oxide nanoparticles. *Bull. Mater. Sci.* **2006**, *29*, 641–645. [CrossRef]

265. Xue, C.-H.; Yin, W.; Jia, S.-T.; Ma, J.-Z. UV-durable superhydrophobic textiles with UV-shielding properties by coating fibers with ZnO/SiO$_2$ core/shell particles. *Nanotechnology* **2011**, *22*, 415603. [CrossRef] [PubMed]

266. Dastjerdi, R.; Montazer, M. A review on the application of inorganic nano-structured materials in the modification of textiles: Focus on anti-microbial properties. *Colloids Surf. B Biointerfaces* **2010**, *79*, 5–18. [CrossRef]

267. El-Nahhal, I.M.; Zourab, S.M.; Kodeh, F.S.; Selmane, M.; Genois, I.; Babonneau, F. Nanostructured copper oxide-cotton fibers: Synthesis, characterization, and applications. *Int. Nano Lett.* **2012**, *2*, 14. [CrossRef]

268. Selvam, S.; Rajiv Gandhi, R.; Suresh, J.; Gowri, S.; Ravikumar, S.; Sundrarajan, M. Antibacterial effect of novel synthesized sulfated β-cyclodextrin crosslinked cotton fabric and its improved antibacterial activities with ZnO, TiO$_2$ and Ag nanoparticles coating. *Int. J. Pharm.* **2012**, *434*, 366–374. [CrossRef] [PubMed]

269. Esmail, W.A.; Darwish, A.M.Y.; Ibrahim, O.A.; Abadir, M.F. The effect of magnesium chloride hydrates on the fire retardation of cellulosic fibers. *J. Therm. Anal Calorim.* **2001**, *63*, 831–838. [CrossRef]

270. Gulrajani, M.L.; Deepti, G. Emerging techniques for functional finishing of textile. *Indian J. Fibre Text. Res.* **2011**, *36*, 388–397.

271. Moafi, H.F.; Shojaie, A.F.; Zanjanchi, M.A. Flame-retardancy and photocatalytic properties of cellulosic fabric coated by nano-sized titanium dioxide. *J. Therm. Anal. Calorim.* **2011**, *104*, 717–724. [CrossRef]

272. Jolles, Z.E.; Jolles, G.I. Some notes on flame-retardant mechanisms in polymers. *Plast Polym.* **1972**, *40*, 319.

273. Shen, R.; Fan, T.; Quan, Y.; Ma, R.; Zhang, Z.; Li, Y.; Wang, Q. Thermal stability and flammability of cotton fabric with TiO$_2$ coatings based on biomineralization. *Mater. Chem. Phys.* **2022**, *282*, 125986. [CrossRef]

274. Fallah, M.H.; Fallah, S.A.; Zanjanchi, M.A. Synthesis and Characterization of Nano-sized Zinc Oxide Coating on Cellulosic Fibers: Photoactivity and Flame-retardancy Study. *Chin. J. Chem.* **2011**, *29*, 1239–1245. [CrossRef]

275. Prilla, K.A.V.; Jacinto, J.M.; Ricardo, L.J.O.; Box, J.T.S.; Lim, A.B.C.; Francisco, E.F.; De Vera, G.I.N.; Yaya, J.A.T.; Natividad, V.V.M.; Awi, E.N.; et al. Flame Retardant and Uv-Protective Cotton Fabrics Functionalized with Copper (II) Oxide Nanoparticles. *Antorcha* **2020**, *7*, 11–16.

276. Saleemi, S.; Naveed, T.; Riaz, T.; Memon, H.; Awan, J.A.; Siyal, M.I.; Xu, F.; Bae, J. Surface Functionalization of Cotton and PC Fabrics Using SiO$_2$ and ZnO Nanoparticles for Durable Flame Retardant Properties. *Coatings* **2020**, *10*, 124. [CrossRef]

277. Dhineshbabu, N.R.; Manivasakan, P.; Yuvakkumar, R.; Prabu, P.; Rajendran, V. Enhanced Functional Properties of ZrO$_2$/SiO$_2$ Hybrid Nanosol Coated Cotton Fabrics. *J. Nanosci. Nanotechnol.* **2013**, *13*, 4017–4024. [CrossRef] [PubMed]

278. Rajendran, V.; Dhineshbabu, N.R.; Kanna, R.R.; Kaler, K.V.I.S. Enhancement of Thermal Stability, Flame Retardancy, and Antimicrobial Properties of Cotton Fabrics Functionalized by Inorganic Nanocomposites. *Ind. Eng. Chem. Res.* **2014**, *53*, 19512–19524. [CrossRef]

279. Frąckowiak, A.; Skibiński, P.; Gaweł, W.; Zaczyńska, E.; Czarny, A.; Gancarz, R. Synthesis of glycoside derivatives of hydroxyanthraquinone with ability to dissolve and inhibit formation of crystals of calcium oxalate. Potential compounds in kidney stone therapy. *Eur. J. Med. Chem.* **2010**, *45*, 1001–1007. [CrossRef] [PubMed]

280. Małecka, B.; Drożdż-Cieśla, E.; Małecki, A. Mechanism and kinetics of thermal decomposition of zinc oxalate. *Thermochim. Acta* **2004**, *423*, 13–18. [CrossRef]

281. Echigo, T.; Kimata, M.; Kyono, A.; Shimizu, M.; Hatta, T. Re-investigation of the crystal structure of whewellite [Ca(C$_2$O$_4$)·H$_2$O] and the dehydration mechanism of caoxite [Ca(C$_2$O$_4$)·3H$_2$O]. *Mineral. Mag.* **2005**, *69*, 77–88. [CrossRef]

282. Majumdar, R.; Sarkar, P.; Ray, U.; Roy Mukhopadhyay, M. Secondary catalytic reactions during thermal decomposition of oxalates of zinc, nickel and iron(II). *Thermochim. Acta* **1999**, *335*, 43–53. [CrossRef]

283. Gabal, M.; El-Bellihi, A.; El-Bahnasawy, H. Non-isothermal decomposition of zinc oxalate–iron(II) oxalate mixture. *Mater. Chem. Phys.* **2003**, *81*, 174–182. [CrossRef]

284. Mohamed, M.A.; Galwey, A.K.; Halawy, S.A. A comparative study of the thermal reactivities of some transition metal oxalates in selected atmospheres. *Thermochim. Acta* **2005**, *429*, 57–72. [CrossRef]

285. Vlaev, L.; Nedelchev, N.; Gyurova, K.; Zagorcheva, M. A comparative study of non-isothermal kinetics of decomposition of calcium oxalate monohydrate. *J. Anal. Appl. Pyrolysis* **2008**, *81*, 253–262. [CrossRef]

286. Donkova, B.; Mehandjiev, D. Mechanism of decomposition of manganese(II) oxalate dihydrate and manganese(II) oxalate trihydrate. *Thermochim. Acta* **2004**, *421*, 141–149. [CrossRef]

287. Weil, E.D.; Levchik, S.; Moy, P. Flame and Smoke Retardants in Vinyl Chloride Polymers—Commercial Usage and Current Developments. *J. Fire Sci.* **2006**, *24*, 211–236. [CrossRef]

288. Holdsworth, A.F.; Horrocks, A.R.; Kandola, B.K.; Price, D. The potential of metal oxalates as novel flame retardants and synergists for engineering polymers. *Polym. Degrad. Stab.* **2014**, *110*, 290–297. [CrossRef]

289. Ji, W.; Wang, H.; Yao, Y.; Wang, R. Mg(OH)$_2$ and PDMS-coated cotton fabrics for excellent oil/water separation and flame retardancy. *Cellulose* **2019**, *26*, 6879–6890. [CrossRef]

290. Plentz, R.S.; Miotto, M.; Schneider, E.E.; Forte, M.M.C.; Mauler, R.S.; Nachtigall, S.M.B. Effect of a macromolecular coupling agent on the properties of aluminum hydroxide/PP composites. *J. Appl. Polym. Sci.* **2006**, *101*, 1799–1805. [CrossRef]

291. Sabet, M.; Hassan, A.; Ratnam, C.T. Flammability and Thermal Characterization of Aluminum Hydroxide Filled with LDPE. *Int. Polym. Process.* **2013**, *28*, 393–397. [CrossRef]

292. Lee, M.-Y.; Yen, F.-S.; Hsiang, H.-I. Generating Self-Shaped 2D Aluminum Oxide Nanopowders. *Nanomaterials* **2022**, *12*, 2955. [CrossRef]

293. Esposito Corcione, C.; Frigione, M.; Maffezzoli, A.; Malucelli, G. Photo—DSC and real time—FT-IR kinetic study of a UV curable epoxy resin containing o-Boehmites. *Eur. Polym. J.* **2008**, *44*, 2010–2023. [CrossRef]

294. Grand, A.F.; Wilkie, C.A. (Eds.) *Fire Retardancy of Polymeric Materials*; Marcel Dekker: New York, NY, USA, 2000; Chapter 9.

295. Alongi, J.; Brancatelli, G.; Rosace, G. Thermal properties and combustion behavior of POSS- and bohemite-finished cotton fabrics. *J. Appl. Polym. Sci.* **2012**, *123*, 426–436. [CrossRef]

296. Sharma, V.; Basak, S.; Rishabh, K.; Umaria, H.; Ali, S.W. Synthesis of zinc carbonate nanoneedles, a potential flame retardant for cotton textiles. *Cellulose* **2018**, *25*, 6191–6205. [CrossRef]
297. de Paiva Teixeira, M.H.; Lourenço, L.A.; Artifon, W.; de Castro Vieira, C.J.; Gómez González, S.Y.; Hotza, D. Eco-Friendly Manufacturing of Nano-TiO$_2$ Coated Cotton Textile with Multifunctional Properties. *Fibers Polym.* **2020**, *21*, 90–102. [CrossRef]

Review

Nanomaterials Used in the Preparation of Personal Protective Equipment (PPE) in the Fight against SARS-CoV-2

Pierantonio De Luca [1,*], Janos B.Nagy [1] and Anastasia Macario [2]

[1] Department of Mechanical and Energy Engineering, University of Calabria, 87036 Arcavacata, Italy; janos.bnagy1@gmail.com
[2] Department of Environmental Engineering, University of Calabria, 87036 Arcavacata, Italy; anastasia.macario@unical.it
* Correspondence: pierantonio.deluca@unical.it

Abstract: Following the well-known pandemic, declared on 30 January 2020 by the World Health Organization, the request for new global strategies for the prevention and mitigation of the spread of the infection has come to the attention of the scientific community. Nanotechnology has often managed to provide solutions, effective responses, and valid strategies to support the fight against SARS-CoV-2. This work reports a collection of information on nanomaterials that have been used to counter the spread of the SARS-CoV-2 virus. In particular, the objective of this work was to illustrate the strategies that have made it possible to use the particular properties of nanomaterials, for the production of personal protective equipment (DIP) for the defense against the SARS-CoV-2 virus.

Keywords: COVID-19; coronaviruses; personal protective equipment; nanomaterials; PPE; SARS-CoV-2

Citation: De Luca, P.; B.Nagy, J.; Macario, A. Nanomaterials Used in the Preparation of Personal Protective Equipment (PPE) in the Fight against SARS-CoV-2. *Inorganics* **2023**, *11*, 294. https://doi.org/10.3390/inorganics11070294

Academic Editor: Roberto Nisticò

Received: 29 May 2023
Revised: 10 July 2023
Accepted: 11 July 2023
Published: 12 July 2023

1. Introduction

Nanotechnology is one of the sectors of applied science and technology that concerns the control of matter on a dimensional scale in the order of the nanometer [1].

The Recommendation of the European Commission of 18 October 2011 clarifies that "nanomaterial" means a natural, derived, or manufactured material containing particles in the free state, aggregate, or agglomeration, and in which, for at least 50% of the particles in the numerical size distribution, one or more external dimensions are between 1 nm and 100 nm" [2].

By virtue of their very small size, nanomaterials show behaviors and properties that are very different from those of the same material on a macroscopic scale. First, they have a very high ratio between the atoms on the surface and those inside, since they have very large surfaces compared to the volume; moreover, the atoms on the surface, having unsaturated binding sites, are very reactive and excellent catalysts [3,4].

Nanomaterials include nanopowders with all three dimensions smaller than 100 nm (height, width, depth), nanotubes with two smaller dimensions, thin films with one smaller dimension, and finally nanostructured materials having dimensions greater than 100 nm but made from elements with at least one lower dimension [5].

The reduction of dimensions at the nanometric level constitutes an important step forward towards the miniaturization of matter, which no longer behaves as it can usually be observed at the macroscopic level but assumes a new behavior in which the force of gravity has no value, while the van der Waals forces, the surface tension forces and all those forces that concern the atom and the interaction between atoms become important. The explanation lies in the fact that structures with nanometric dimensions are characterized by a number of surface atoms comparable if not higher than the number of atoms in the rest of the structure. In materials of higher dimensions, on the other hand, the number of surface atoms is much lower than the number of internal atoms and this numerical prevalence determines the behavior of matter perfectly described by the scientific laws of classical physics and chemistry.

From the above, it is clear how nanostructured materials have aroused great interest as their properties differ significantly from all other materials due to the increase in specific area and the quantum effects present in them. These two factors cause a change or can increase properties such as reactivity, mechanical strength, electrical characteristics of the material, and optical and magnetic properties [5].

Nanotechnology is one of the sectors of applied science and technology that concerns the control of matter on a dimensional scale in the order of the nanometer, or one billionth of a meter, and the development of devices on this scale [6].

Nanotechnologies, since their debut in the last decades of the twentieth century, have provided a significant contribution in many fields of science, opening new scenarios in the textile sector [7,8], food [9,10], construction [11,12], electronic components [13,14] and in agriculture [15,16].

Interest was also directed to the use of nanomaterials in the field of environmental prevention and protection. Materials such as carbon nanotubes have proven to be excellent materials for the purification of contaminated water [17,18].

In recent years, there has been a significant and successful use of nanotechnology in the biomedical sciences due to the unique physicochemical properties of nanomaterials, which offer versatile chemical functionalization for the creation of advanced biomedical tools. Nanomedicine consists precisely of the application of nanotechnologies in the medical and pharmacological fields for the diagnosis, treatment, control, and prevention of diseases [19,20].

The field of study of nanomedicine has two related macro-areas: diagnostics [21] and therapy [22–24]. The value of nanotechnologies in medicine lies in their ability to act on a nanometric scale, therefore with dimensions much larger than that of the human cell, thus allowing nanoparticles to move at the same dimensional level as biological processes, paving the way for the so-called target medicine [25,26].

The recent pandemic, which hit our planet with fatal outcomes and a high rate of reproduction and transmission, was caused by the spread of COVID-19, an acronym for the English coronavirus disease, where 19 indicates the year in which the virus was identified for the first time [27,28].

COVID-19 is an acute respiratory disease caused by the virus called SARS-CoV-2 belonging to the coronavirus family [29].

Coronaviruses cause infections in humans and in various animals, they are capable of infecting different species and this "jump of species" takes place thanks to mutations in the genetic heritage of the virus which make it capable of infecting new animal species, including human beings [30,31].

SARS-CoV-2 would appear to have first appeared in the Wuhan seafood market, where live animals were also present, however, the steps regarding transmission to humans are still poorly understood. There are no definitive indications of whether the infection occurred through contact between bats and humans or whether the virus passed from bats to humans through an intermediate host. The intermediate host may have been the pangolin, also known as the "scaly anteater" [32–34].

Due to the high transmissibility of the virus and international travel, the worldwide spread of SARS-CoV-2 increased exponentially during the pandemic, thus posing a serious threat to global public health, so much so that on 11 March 2020 the WHO director- generally officially recognized the outbreak as a pandemic [35].

The following work generally concerns the applications of nanotechnology in the prevention and control of infection caused by SARS-CoV-2. In particular, the aim is to illustrate the strategies and methods by which the unique properties of nano-sized materials are used in the development of PPE. Following an ever-increasing sensitivity that has developed to the problem of the pandemic, a large research activity has led to many publications in the literature. In this work, the attention was focused only on some research which has proven to be more significant and which allows us to represent a general framework on the application of nanomaterials for the creation of personal protective equipment.

2. Nanotechnology in the Fight against SARS-CoV-2

Nanotechnology can provide promising solutions in the fight against the SARS-CoV-2 virus, thanks to the precious peculiarities of the nanoparticles involved such as a high surface-to-volume ratio, easy surface modification, high physic-chemical stability, optical properties specifications, and targeted and controlled release capabilities [36–38]. The potential of nanotechnologies toward SARS-CoV-2 is currently established in prevention, diagnosis, and therapy [39,40] (Figure 1).

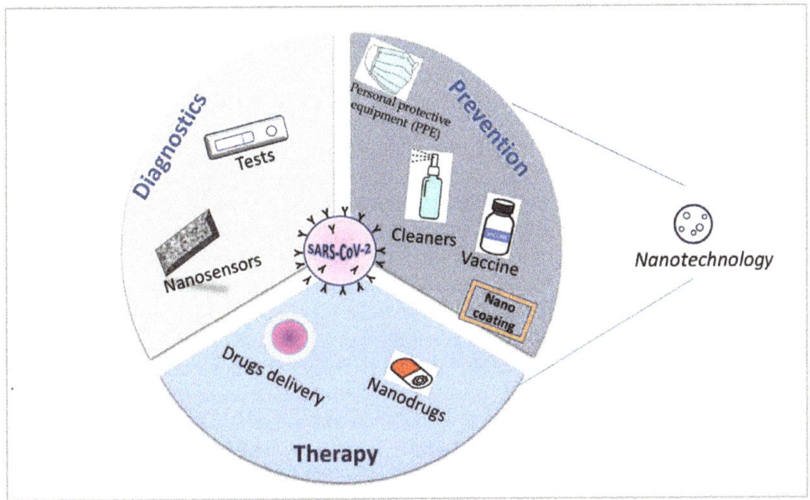

Figure 1. Main applications of nanotechnology for the fight against SARS-CoV-2.

Scientists around the world have worked hard to study and develop highly sensitive diagnostic tools to detect the virus quickly and accurately and to replace them with techniques that instead require more time, high costs, and expert users [41–43].

The new sensing tests, based on smart nanometer sensors, are fast, sensitive, and cost-effective, requiring a smaller sample volume and less laboratory equipment. Nanomaterials are used in the design of highly specific and sensitive nanometer sensors for rapid identification of infection. Nanobiosensors are analytical devices comprising a biomarker, a transducer, and a signal amplifier used for the detection of molecules. They have the advantage of selectively detecting all types of analytes by combining the considerable electrical and optical properties of nanomaterials with biological or synthetic molecules employed as receptors [44–46].

Graphene-containing electrochemical biosensors can be used advantageously for the detection of the SARS-CoV-2 virus. The detection mechanism involves the immobilization of specific biomolecules on the surface of the graphene, capable of interacting with the virus, causing a change in the electrical properties of the graphene [47]. To be used in these biosensors, graphene must first be functionalized, creating functional groups on its surface, such as carboxylic acids, amines, and thiols, etc., capable of reacting and therefore immobilizing biomolecules, such as antibodies or nucleic acids, sensitive to SARS-CoV-2 (Figure 2).

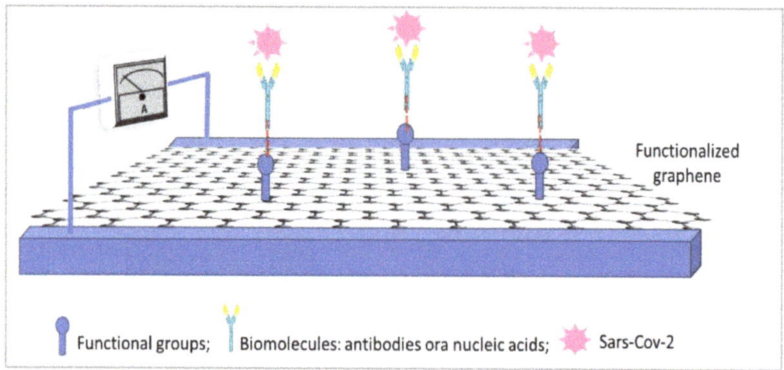

Figure 2. Biosensor with functionalized graphene containing biomolecules attracting SARS-CoV-2.

Graphene has shown itself to be very sensitive towards the detection of the virus thanks to its unique properties such as the high surface area and its high conductivity. The use of graphene in biosensors has proved to be more advantageous than other nanomaterials such as gold nanoparticles, silicon being cheaper, and offering greater sensitivity and therefore faster response times. Despite these evident advantages in the use of graphene for the realization of biosensors for the detection of the SARS-CoV-2 virus, an intense research phase is still needed to optimize, for example, the functionalization phases of graphene and more generally a development of more convenient methods of graphene production so that these biosensors can be increasingly cheaper and widely commercialized [48,49].

In general, nano-sized pharmaceuticals exhibit lower toxicities and higher therapeutic efficacy than conventional formulations in the prevention and treatment of viral diseases. The advantages of nanotechnology in antiviral research are many. In fact, in the biomedical field, its use promotes the administration of water-insoluble drugs; improves in vivo drug circulation time, and drug utilization efficiency, and reduces side effects [50,51].

Nanotechnology has recently shown considerable potential in the development of drugs to be used in the treatment of COVID-19 disease [52,53].

In the first months of 2020, numerous research centers in various countries launched experimental pathways aimed at creating safe and effective vaccines against SARS-CoV-2 [54–56].

Although the expected waiting times were longer, some pharmaceutical and biotech companies involved in the trial managed to develop vaccines in less than a year. These vaccines fall into the category of "mRNA drugs" and quickly obtained the approval of the European Medicines Agency (EMA). Vaccines developed by Moderna and Pfizer-BioNTech use messenger ribonucleic acid (mRNA) molecules. The latter is equipped with all the information necessary for the cells to be able to synthesize the Spike protein, present on the external surface of SARS-CoV-2, which is the one that allows the virus access into the host's cells [57].

The Spike protein, produced inside the ribosomes of the host cell, is then expelled and therefore recognized as an antigen by the immune system, which will be able to produce antibodies.

In fact, if the vaccinated person were to later meet SARS-CoV-2, their immune system would recognize it and be ready to fight it. The injected mRNA is encapsulated in lipid nanoparticles (LNPs) to prevent it from being degraded in the bloodstream before reaching the cytoplasm of cells. LNPs are lipids of the size of a few nanometers increasingly used in drug delivery [58].

3. PPE with Nanomaterials to Fight the Spread of the SARS-CoV-2 Virus

Nanotechnology is used in the fight against the SARS-CoV-2 pandemic, through the design of new materials capable of inactivating and preventing the spread of the virus. These nanomaterials have found application above all in the production of personal protective equipment (PPE) such as masks and fabrics.

The FFP1, FFP2, and FFP3 filtering masks are the so-called filtering facepieces (hence the acronym FFP), i.e., respirators with filtering facepieces, defined by current legislation as Personal Protective Equipment (PPE) [59].

They are classified in a very precise manner and subjected to tests that certify their adherence to certain parameters. Unlike surgical masks, they are also highly effective in blocking droplets that remain in the air; breathing, therefore, avoids breathing the droplets of humidity which are the main vector of the virus.

The FFP1, FFP2, and FFP3 masks are required to bear the CE mark; they are produced in compliance with the EN 149-2001 standard which sets the standards of efficiency, breathability, stability of the structure, as well as the technical tests of biocompatibility and the performances of the masks. The progressive numbering (FFP1, FPP2, FFP3) indicates the progressive air filtering capacity of the different types of devices [60].

In a viral epidemic, the protection of healthcare workers and individuals at risk is very important and it is here that nanomaterials can play a fundamental role through their incorporation into personal protective equipment [61,62].

The growing attention that COVID-19 has aroused in the last two years has encouraged and accelerated the design of masks by many researchers, who have worked to modify the surface of the PPE, not only to capture and inactivate the virus, but also to make them washable, reusable, and environmentally friendly, without compromising their effectiveness and safety. For this purpose, various types of nanoparticles have been studied and used, among which the most used are nanoparticles of silver, copper, copper iodide, copper oxide, graphene, graphene oxide, and nanofibres.

3.1. PPE with Graphene

Since the beginning of the pandemic, several companies have launched PPE enhanced with graphene microfiber fabrics on the market.

Graphene, as is known, is a nanomaterial consisting of a monoatomic layer of carbon atoms, arranged to form a hexagonal lattice and its structure can be defined as two-dimensional, as there are only two dimensions of the plane [63,64]. By exploiting the antimicrobial, antistatic, and electrically conductive properties of the nanomaterial, researchers have developed face masks with antiviral properties that can be sterilized and reused. We need to dwell on this type of mask.

The new graphene FFP2 produced by the C&S Italy company of Arezzo, called AV Mask Pro [65] would be able to fulfill several functions:

-   repel and block viruses and nanoparticles, ensuring antimicrobial protection, thanks to the inherent antistatic and waterproof capabilities of graphene;
-   reduce the risk of secondary contact infections, caused by the possible presence of bacteria and/or viruses on the external surface of the masks;
-   improve the comfort of those who wear them, guaranteeing lightness, breathability, softness, and resistance.

These masks were made using an original bioactivation process, which allows the integration of graphene within the polypropylene polymer, avoiding the possibility of detachment of nanoparticles, which could occur using more traditional processes such as coating. These masks have passed all tests, and have antiviral activity certified according to ISO 21702:2019 and ISO 18184:2019 standards (Figure 3).

The Abruzzo startup Hygraner managed to patent and create the first antimicrobial, filtering, and breathable non-woven fabric in graphene nanofiber, recyclable and sustainable, with the melt-blown technique [66].

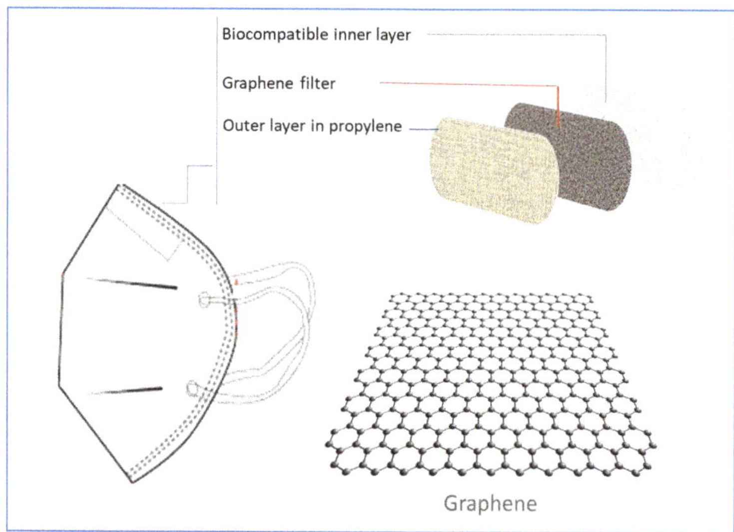

Figure 3. Mask with graphene filter.

The fabric is made up of 2 μm diameter polypropylene microfibers, used as an intermediate filtering layer in IR, FFP2, and FF3 surgical-type masks.

With an innovative patent, the company has developed a melt-blown fabric in which fibers are made entirely of a polymer and functionalized graphene, capable of blocking the proliferation of the virus and oxidizing its cellular components, inhibiting all activity.

The fabric, certified according to ISO 18184 which determines the requirements of the antiviral activity of textile products, is hydrophobic, hypoallergenic, antistatic, and non-toxic, soft to the touch, and sustainable for the environment.

Its antiviral property does not decline over time, not even after use, after repeated washing, or other permitted treatments, so the fabric can also be sterilized and reused.

Other research has been conducted for the development of new surgical masks capable of self-sterilization and can be reused or recycled. For these purposes, an original and unique method has been developed to functionalize surgical masks ensuring their self-cleaning and antimicrobial properties [67].

The keystone of the development of this innovative method was the use of the dual-mode laser, which allows depositing an ultra-thin graphene coating of a few layers on a non-woven mask at a low melting temperature (Figure 4).

Specifically, the laser uses a high-energy and precision collimated light beam capable of inducing specific morphological modifications on the surface of the target material. The laser coating allows the so-called graphene-like layers (GL) to be deposited on the support of the mask, which consists of stacked layers of graphene-like nanometric dimensions obtained from a controlled top-down demolition of a target graphite. Generally, to obtain an ultra-thin coating it is possible to use various techniques, many of which involve the use of solvents, which could damage or alter the surface to be coated. The laser technique offers advantageous aspects for the realization of coatings on biomedical materials such as facial masks, thanks to the possibility of controlling various process parameters that could damage the support on which the deposition takes place. In fact, it is not always easy for an implantable material to have all the required characteristics, for example, it can have the right mechanical characteristics but be sensitive to solvents. The laser deposition system can preserve the functionality of the material to be treated by not requiring the use of solvents in direct contact with the surface to be coated.

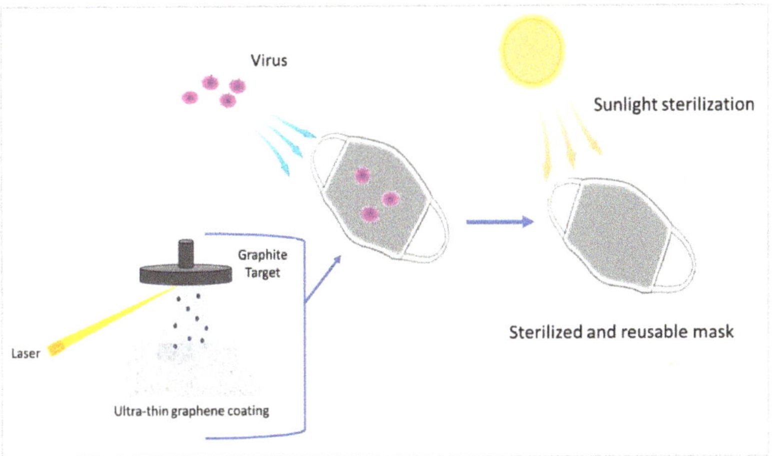

Figure 4. The mask, made using an ultra-thin graphene layer deposited by the Laser dual method, is reusable and sterilizable with the aid of solar radiation.

In laboratory tests, the surfaces of these masks were found to be highly hydrophobic and capable of causing incoming water droplets to bounce back. These masks have photothermal properties which give them two important prerogatives: the ability to self-sterilize and virucidal efficacy (inactivation of the virus). The presence of graphene, in fact, determines a photothermal effect, i.e., the production of heat following the absorption of light. The surface temperature of the functional mask, irradiated by the sun's rays, can rapidly rise to over 80 °C, making the masks reusable. As is known, most viruses cannot survive at high temperatures, therefore, the solar radiation to which the surface of the graphene layer is subjected causes it to reach temperatures of up to 80 °C which consequently allow the inactivation of the virus. In particular, the results of such studies found that the antimicrobial efficiency of graphene material, with sunlight sterilization, could improve by 99% within 15–20 min [68].

Recent studies have also tested the effectiveness of graphene oxide (GO). It is a stratified material produced by the oxidation of graphite, but unlike the latter, it is strongly oxygenated, and recent studies have tested its antiviral efficacy [69,70].

To improve the protective properties of the masks, functionalized fabrics have been created, i.e., normal fabrics (cotton and polyurethane), in whose fibers graphene and graphene oxide sheets have been inserted [71]. These studies demonstrated that when cotton or polyurethane fabrics are functionalized with graphene or graphene oxide the infectivity of the fabric towards the SARS-CoV-2 virus is inhibited.

The data reported in the literature show that both graphene and graphene oxide can be advantageously used as materials for the manufacture of personal protective equipment. However, the latter have different characteristics. Graphene oxide, in fact, thanks to the presence on the surface of functional groups such as hydroxyl, epoxy, and carboxylic groups, has a greater hydrophilicity compared to graphene (Figure 5a). Consequently, depending on the final product to be created, graphene can be used to create a more hydrophobic material. Graphene oxide can be more advantageously used for internal layers or more generally in molecular diagnostic devices involving applications in aqueous solutions [72].

Both graphene and graphene oxide can block the virus, preventing its diffusion, facilitated by the different dimensions between the carbon rings, present in both graphene and graphene oxide [73], and the dimensions of the virus [74] (Figure 5b).

Hu et al., reported the creation of a cotton fabric, with antimicrobial properties, on which graphene oxide was inserted and subsequently implanted with Fe^{3+} ions [75].

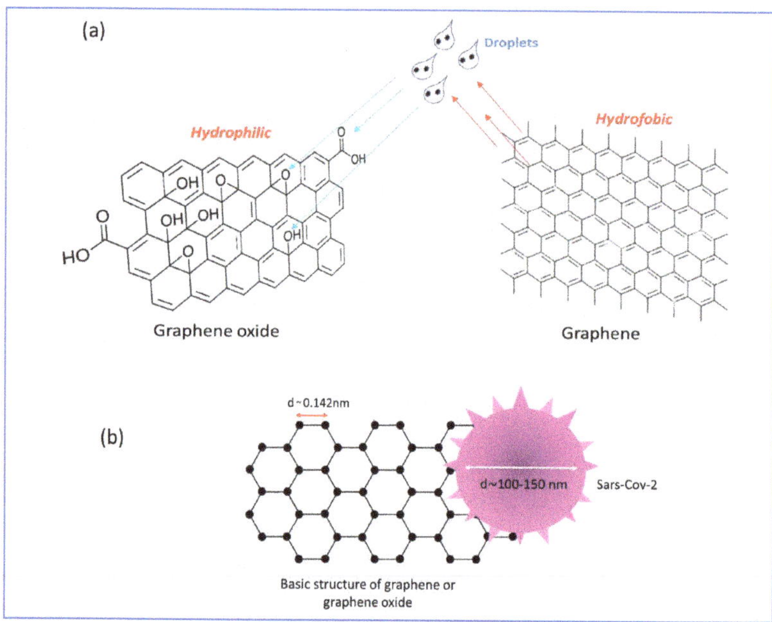

Figure 5. (**a**) Hydrophilic and hydrophobic action, respectively, of graphene oxide and graphene, with respect to droplets; (**b**) size of virus and carbon rings.

Ion implantation is a technique capable of modifying the properties, structure, and morphology of surfaces of carbon-based materials using ions with different energies and doses in a controlled way.

Specifically, cotton fabric was treated with an aqueous dispersion of graphene oxide and placed in a microwave oven for programmed times. The fabric containing graphene oxide, obtained after drying, was subjected to ion implantation of Fe^{3+} using an ion implanter, with programmed ion doses and acceleration voltage.

The antimicrobial tests showed that the cotton fabric, containing graphene oxide and implanted with Fe^{3+} ions had a greater activity than the untreated ones and that increasing the dose of Fe^{3+} showed a greater antimicrobial activity. Although the tests have not been performed strictly on the SARS-CoV-2 virus, this study is important as regards the technique performed and the results of antimicrobial activity that the treated tissues have shown. Given the results obtained, it can be expected that these fabrics can also be efficient for the SARS-CoV-2 virus in the same way and therefore it is to be hoped that there will be an in-depth analysis of the research activities on this study.

The set of information reported in the literature shows that both graphene and graphene oxide can inactivate the SARS-CoV-2 virus, although they use different inactivation mechanisms and therefore, they can be used to make protective devices against the SARS-CoV-2 virus.

Now it is not easy to make a precise comparison between the activity of the two materials as the data reported are carried out under different experimental conditions and therefore further studies are desirable.

3.2. Nanoparticles against SARS-CoV-2

As is known, viruses replicate only within a host cell, through different mechanisms that can be summarized as follows:

-   attack on the cell membrane;
-   penetration of the virus into the cytoplasm of the cell;

- replication exploiting enzymes and organelles of the host cell;
- escape of the virus from the cell.

In general, the antiviral action of nanoparticles mainly intervenes in these interaction mechanisms of the virus with the cell.

Lysenko et al. (2018) [76] proposed that nanoparticles adsorbed on the cell surface induce an alteration of the membrane with consequent contrast to the attachment and penetration of the virus. Most often, nanoparticles work by altering the structure of the capsid protein and reducing viral activity. The capsid of the SARS-CoV-2 virus consists of four structural proteins, known as Spike, Envelope, Membrane, and Nucleocapsid. The large Spike (S) protein, which forms a kind of crown on the surface of the viral particles, acts as a real "anchor" which allows the virus to dock to a receptor expressed on the membrane of the host cells, the receptor ACE2 (Angiotensin-Converting Enzyme 2), allowing the initiation of entry of the virus into the cell. Once penetrated inside, the virus exploits the functional mechanisms of the cell to multiply, which dies, releasing millions of new viruses (Figure 6a). The attack between the Spike protein and the AC2 cell receptor, in the presence of nanoparticles, is prevented with the consequent impossibility of the virus to enter the cell and multiply (Figure 6b).

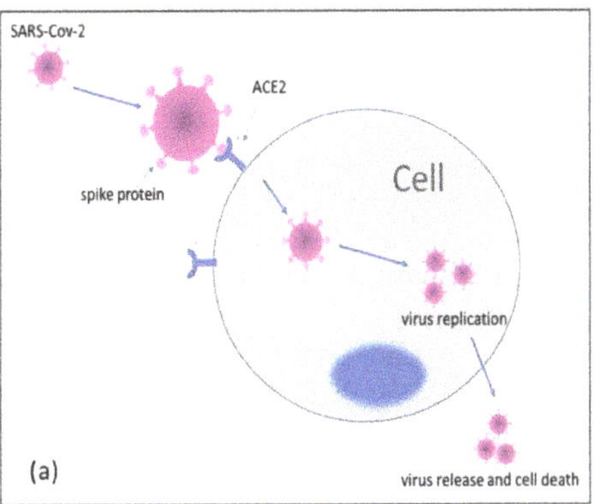

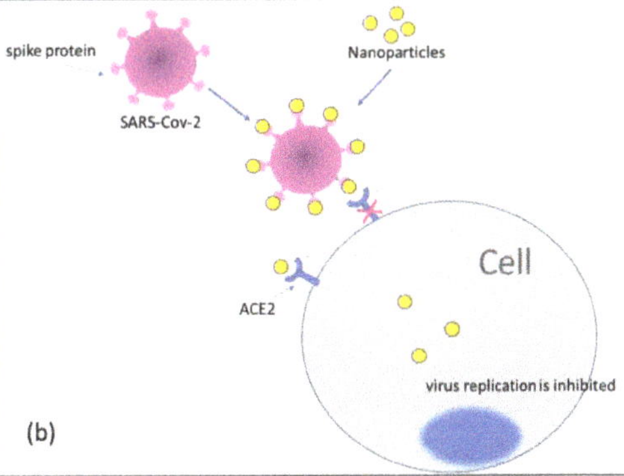

Figure 6. Synthetic scheme of the mechanism of attack of the virus to a cell: in the absence (**a**) and in the presence (**b**) of nanoparticles.

3.2.1. PPE with Silver Nanoparticles

Since ancient times, such as the time of the Greeks and Romans, products based on Ag^+ ions, known for their antimicrobial properties, were used as medicines and for water conservation [77].

The antibacterial action of silver depends on the biologically active silver ion (Ag^+), capable of irreversibly damaging the key system of enzymes in the membrane of pathogens.

Silver nanoparticles (AgNPs) have a size between 1 nm and 100 nm; although they are often described as silver, some of them are composed, in large percentages, of silver oxide, due to the high ratio between surface and volume.

AgNPs exhibit physicochemical characteristics, as their high surface/volume ratio allows a great synergy with microbial surfaces, which leads to high antimicrobial safety [78]. The size and shape of the AgNPs play a determining role; their size, in fact, allows a constant release of ions having antimicrobial activity, unlike salts which rapidly release the ions, quickly wearing off this effect.

With the use of antibiotics, silver-based disinfectants have become obsolete, surviving only in traditional ethnic medicine, but in recent years they have made a comeback because they are used in cases of antibiotic resistance, which leave the sick defenseless against numerous infections. Colloidal silver has proved to be a very valid remedy for the skin, both to ward off dangerous infections in burn victims and to reactivate the metabolism of damaged skin tissues. Colloidal silver, being made up of nanometric particles of silver suspended in totally purified water, is a liquid compound.

Today, silver is considered one of the most powerful natural disinfectants [79].

Several companies that are at the forefront of nanotechnology have produced innovative masks to fight the SARS-CoV-2 virus.

It is appropriate to dwell on some of the innovative masks made recently containing silver nanoparticles. A team of researchers from the National Autonomous University of Mexico (UNAM), led by researcher Sandra Rodil, has created a face mask called "SakCu" (in the Mayan language "Sak" means silver, while "Cu" indicates the symbol of copper) able not only to protect against COVID-19 but also to inactivate the virus [80,81].

The mask consists of three layers: the central one is composed of silver-copper nanolayers deposited in polypropylene, while the first and third are made of cotton. The researchers thus used a silver-copper mixture, which forms a nanolayer between 30 and 40 nanometers thick, which ensures double protection against viruses and bacteria.

To test the mask, the researchers used a few drops contaminated with the virus and placed them on the silver-copper film deposited on polypropylene.

The result was satisfactory, since when the viral concentration was high, the virus disappeared by more than 80% in about 8 h; however, when the viral load was low, no virus RNA was detected within 2 h. The researchers, illustrating their discovery, concluded that the virus RNA was damaged by contact with the Ag-Cu nanolayer.

Since the surface of the mask does not remain contaminated, unlike many others, its disposal would not create any problems, moreover, the mask is reusable and washable up to ten times without losing its properties. Based on this research, the Kolzer Company of Cologno Monzese has given its availability to use the technology it possesses in the production of masks with an antibacterial, water-repellent, waterproof, ionizing, and temperature-regulating device [82]. The NANOX biotechnology company, supported by the FAPESP (Program Innovative Research in Small Enterprises) of São Paulo of Brazil, has produced a particular fabric impregnated on the surface with silver nanoparticles, capable of inactivating the SARS-CoV-2 virus through the simple contact.

In tests carried out in the laboratory, it was found, in fact, that it took only two minutes, after contact, to eliminate 99.9% of the virus. This extraordinary tissue was developed by an international research team led by Brazilian researchers from the Institute of Biomedical Sciences of the University of São Paulo (ICB-USP).

The fabric is made from a blend of polyester and cotton (polycotton) containing silver nanoparticles soaked into the surface through a process of dipping, followed by drying and fixing, called "dry curing" [83].

The laboratory tests analyzed several tissue samples, with and without embedded silver nanoparticles, in test tubes with a solution containing large amounts of cellular SARS-CoV-2. All samples were kept in direct contact with the viruses at different time intervals to evaluate their antiviral activity.

The results of the analyzes were satisfactory as they established that the tissue samples, containing silver nanoparticles, inactivated 99.9% of the virus after a few minutes of contact.

3.2.2. PPE with Copper Nanoparticles

Copper is bacteriostatic, i.e., it fights the proliferation of bacteria on its surface. In the health sector, it has always been used for hospital door handles, IV poles, and all those accessories that are touched, and which can constitute a vehicle for contagion.

Copper works as a barrier for bacteria and viruses, as it contains positively charged ions that trap viruses, which are instead negatively charged and so the positive copper ions penetrate the viruses, preventing them from replicating. The characteristics of copper are comparable to those of other much more expensive metals with antibacterial activity, such as silver and gold [84].

Copper nanomaterials are an emerging class of nano-antimicrobials that provide complementary effects and characteristics, compared to other nano-sized metals, such as silver or zinc oxide nanoparticles.

It is established that soluble copper compounds provide excellent antimicrobial activity against bacteria, fungi, algae, and viruses. It appears that copper can generate reactive hydroxyl radicals, which can cause irreparable damage such as oxidation of proteins, cleavage of DNA and RNA molecules, and membrane damage due to lipid peroxidation [85].

The antibacterial and antiviral activity of copper has been known since ancient times. Its antimicrobial properties were known in ancient Egypt, as the metal was used for the sterilization of water and the treatment of wounds, as reported in medical texts dating back to around 2500 BC. As was the case with silver, also for copper the advent of antibiotics in 1932 determined its eclipse [53].

Some nanotechnology experts have invested their efforts and expertise in the development of fabrics treated with copper nanoparticles, to prevent the spread and replication of the virus.

Firmly convinced of the bacteriostatic and antiviral properties of copper, researchers have developed different types of hypoallergenic masks that offer a high level of protection combined with comfort and environmental sustainability.

A textile company in the Como area, Italtex SpA in partnership with the Ambrofibre company in Milan, has patented "Virkill", the registered trademark of the first fabric made in Italy with copper nanoparticles that are very effective against bacteria and viruses [86].

Colored between yellow and orange, Virkill has the consistency of a normal fabric, but the element that makes it innovative is the insertion, with an industrial process, of "fused" copper nanoparticles in the thread. This fabric is suitable for its use in the PPE sectors against the propagation of viruses such as SARS-CoV-2.

It should be emphasized that the copper nanoparticles are not deposited on the surface of the fabric but are inserted into the polymer of the thread before its formation.

The copper nanoparticles, inserted inside the thread, preserve the surface of the fabric from contamination by viruses and bacteria, as certified by the ISO 18184:2019 tests for the determination of its antiviral activity against the SARS-CoV-2 virus, and by UNI EN ISO 20743: 2013 for the determination of its antibacterial activity. In particular, the test found an antiviral activity that corresponds to a virus inactivation of more than 99.9%.

The presence of nanoparticles inside the polymer ensures a high resistance to washing, in water and dry, so the antiviral property of Virkill is long-lasting.

Other appreciable characteristics of the fabric are breathability and compatibility in contact with the wearer's skin.

The high breathability of Virkill has been confirmed by the results of tests compliant with the UNI EN ISO 11092:2014 standard.

The second peculiarity was tested with positive results by the patch test, an exam used for the diagnosis of allergic contact reactions, resulting in a hypoallergenic fabric.

3.2.3. PPE with Copper Iodide Nanoparticles

Copper iodide particles have proven to be potent microbicidal agents due to their high surface area-to-volume ratio. A recent study reports the deposition of copper iodide nanoparticles on cotton fabric with an ultrasound method [87].

This ultrasonic coating is free of toxic materials and is a process that has a significant practical advantage as it is fast, simple, economical, and environmentally friendly by involving the principles of "green chemistry".

In particular, the study concerned the antimicrobial activity of natural cotton coated with copper iodide and covered with Hibiscus rosa-sinensis flower extract (CuI-FE) rich in anthocyanin and cyanidin 3-soforoside.

As is known, copper iodide (CuI) is an inorganic chemical compound, while Hibiscus rosa sinensis is a shrub belonging to the Malvaceae family, originally from China.

Since CuI particles exhibit good antibacterial and antitumor activity, a "molecular docking" study was conducted with CuI-FE synthesized to act against the COVID-19 main protease protein. It should be remembered that "molecular docking" is a method capable of predicting the preferred orientation of one molecule towards another, favoring their bond and therefore the formation of a stable complex. The synthesized CuI-FE, which was ultrasonically deposited on natural cotton to improve its multifunctional properties, was well adsorbed on the cotton surface due to physical and chemical interactions. Cotton coated with CuI-FE, showing good antibacterial activity against infections caused by broad-spectrum bacteria and against the main protease of the coronavirus, answers the need for an efficient material to counter the SARS-CoV-2 virus.

3.2.4. PPE Containing Copper Oxide Nanoparticles

A study conducted in 2018 by Redwanul Hasan (Bangladesh University of Textiles) dealt with the development of antimicrobial cotton fabrics, using copper oxide nanoparticles applied directly to the fabric, using the pad-dry-cure method [88].

This method consists in preparing a suspension of NPs in an aqueous solution containing both stabilizing agents, essential to prevent the particles from settling, and fixing or binding agents, which bind the particles to the fibers. At this point the tissue is immersed in the suspension for a certain period, then it is extracted, sterilized, dried, and subjected to curing treatments and possible washings (Figure 7).

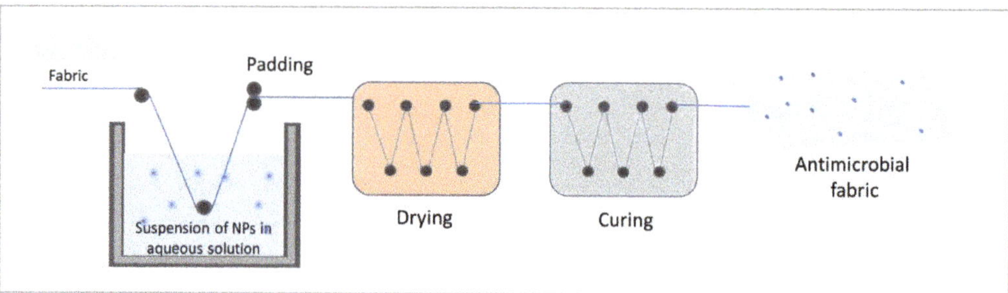

Figure 7. Pad–dry–cure method.

The method, which is quite simple, however, has some drawbacks because it requires various steps which involve manual work and the use of time, but above all heating steps with high temperatures imply considerable energy consumption.

In laboratory tests, treated and untreated fabrics were compared, and the results obtained found the incorporation of copper nanoparticles into the treated fabrics, as well as a significant antibacterial activity. Another test showed that the treated fabric can withstand up to 25 washing cycles This technology, in addition to cotton, can be extended to polyester, silk, and other fabrics to make PPE against SARS-CoV-2.

3.2.5. PPE with Zinc Oxide

Zinc oxide nanoparticles exhibit well-known antimicrobial properties [89,90].

They represent a possible alternative, albeit with a less powerful antimicrobial action, to other metals such as silver and copper, being cheaper and easier to find. The antiviral action is determined by the action of the zinc ions which prevent the binding of the virus with the host cell thus leading to the deactivation of the virus [91].

A recent study reports a method in which zinc oxide nanoparticles are synthesized directly into textile materials which can be used for the creation of PPE [92]. This method can potentially be applied to both natural and synthetic fibers and is called "Crescoating". It consists of the growth of zinc nanoparticles in situ starting from an initial solution in which the dissolved ionic precursor is present. The fabric is soaked in the initial solution and subsequently subjected to heat treatment to trigger the nucleation of the nanoparticles on the fabric. The results showed that the particles are strongly anchored to the fabric resulting in greater stability and durability compared to other surface-treated materials. The antiviral properties of the fabric show a reduction of the virus of 99.99% before washing, after 10 min of contact with the virus, and an efficacy of 99.8% after 50 washing cycles, demonstrating excellent adhesion of the nanoparticles and an important reuse of fabric. The "Crescoating" method is, therefore, to be considered valid for the creation of masks that are cheap, durable, reusable, and effective against the SARS-CoV-2 virus.

3.2.6. PPE with Titanium Oxide

Titanium oxide (TiO_2) is known to be a readily available photocatalytic material. In general, the mechanism of antimicrobial action of titanium oxide can be attributed to the production of reactive oxygen species (ROS) which lead to an alteration of the virus proteins or damage to the DNA [93] The use of titanium oxide nanoparticles represents an excellent strategy to increase contrast actions against SARS-CoV-2. These nanoparticles, therefore, through the combined action with light can be used to develop self-decontaminating materials capable of inactivating the virus. The researchers of the Coraero project, in collaboration with the University of Milan-Bicocca, have highlighted, through experimental tests and simulations, that the titanium oxide nanoparticles, deposited on the surfaces, can both adsorb the virus, but also make it inactive through a combined action of light or thermal treatments [94].

The adsorption phase of the virus on the surface of the titanium oxide nanoparticles is due to the interaction of the latter with the Spike protein of the virus. In particular, in the amino acids that make up the protein, there are nitrogen, sulfur, and oxygen atoms which, with their electronic doublets, can interact with the titanium atom which acts as a Lewis acid. The functional groups $-NH_2$, -SH, -OH, -COOH present in the spike protein, therefore interacting with the surface of the titanium oxide, favor the adsorption of the viral particles. As regards the deactivation of the SARS-CoV-2 virus following exposure to UV light, the mechanisms of action of titanium oxide nanoparticles are of three types: (I) denaturation of the viral protein; (II) damage to the viral genome by absorption of UV radiation by the nucleic acid; (III) photocatalytic oxidation of the virus. The deactivation of the virus through thermal treatments, on the other hand, is caused by the separation of the proteins, adsorbed by the TiO_2 nanoparticles, from the virus structure with consequent inactivation of the virus. The adsorption and inactivation activity of photosensitive titanium oxide nanoparticles is particularly interesting when talking about their possible use in the production of personal protective equipment such as masks, gowns, etc.

Most of the masks available on the market are made of polyester, polyamide, or synthetic non-woven fibers containing titanium oxide. However, it is important to specify that studies are still needed to identify reliable methods of analysis to evaluate the release of particles and estimate their exposure.

A 2021 study by scientists from the Belgian Institute of Public Health, Sciensano, highlighted the presence of titanium dioxide nanoparticles in the synthetic fibers of masks sold to the public with concentrations from a few milligrams to 150 mg per mask [95].

It should be noted that according to the regulation on classification, labeling, and packaging (CLP) titanium dioxide particles with a diameter less than or equal to 10 μm in a concentration greater than 0.1% are designated as suspected carcinogenic by inhalation. Considering prolonged and intensive use of the masks, the value of 3.6μg was evaluated as the limit fraction of particles that must be released from the fibers of the mask to exceed the acceptable exposure level [96].

3.3. PPE with Nanofibers

Nanofiber fabrics consist of filaments of polymer materials of nanometric dimensions [97]. Nanofibers are usually not used as a single strand but in the form of a messy weave with a non-woven structure. Nanofibers, thanks to a very small diameter compared to other natural fibers or microfibres (Table 1), have a high surface/volume ratio and a high specific surface area (lateral area/weight). These fabrics, being generally constituted by polymeric nanofibres, are hydrophobic and highly flexible [98,99].

Table 1. Diameter of different fibers.

Fiber	Diameter (μm)
Wool	30–120
Silk	20
Microfiber	2–5
Nanofiber	0.05–1

Nanofibers can be produced by electrospinning. The process is based on the application of high voltages to a flow of a polymer solution. The polymeric jet during its flight towards the manifold stretches and becomes thinner. Following the evaporation of the solvent, the nanofibers are deposited on a substrate (Figure 8).

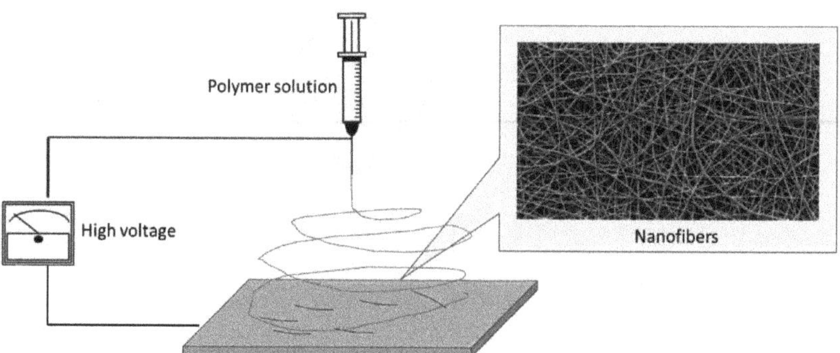

Figure 8. Electrospinning nanofibers at high voltage.

These nanofibers, therefore, have perfect characteristics for multiple areas of application, including the creation of high-performance filter masks [100].

Since SARS-CoV-2 spreads mainly through respiratory droplets produced and by close contact between infected people, the use of appropriate PPE, with novel functions such as hydrophobicity and antimicrobial activity, without compromising the consistency or breathability of the material used is a desirable expectation.

The hydrophobicity of PPE products alone can provide an effective barrier against airborne droplets emitted when you cough or sneeze.

Traditional face masks have a gap between the fibers, on average 10–30 μm, which is not adequate to avoid contact with the virus. A nanofiber fabric has a marked porosity and small pores.

In principle, a face mask should be comfortable and effective. Nanofibers for making masks respond well to these needs. The latter significantly reduces, compared to traditional filters, the resistance of the airflow, with a consequent improvement in breathability. Therefore, the use of nanofibers for the production of face masks or respiratory filters increases respiratory comfort as well as having high filtering properties.

The above-mentioned ensures far superior protection compared to traditional surgical masks which do not protect against particles of size between 10 and 80 nm.

Nanofiber fabrics can therefore be used advantageously for the creation of masks thanks to a double action: water repellency towards the droplets that can carry the virus and blocking the action of the virus thanks to a structure made up of pores of smaller dimensions than those of the virus (Figure 9)

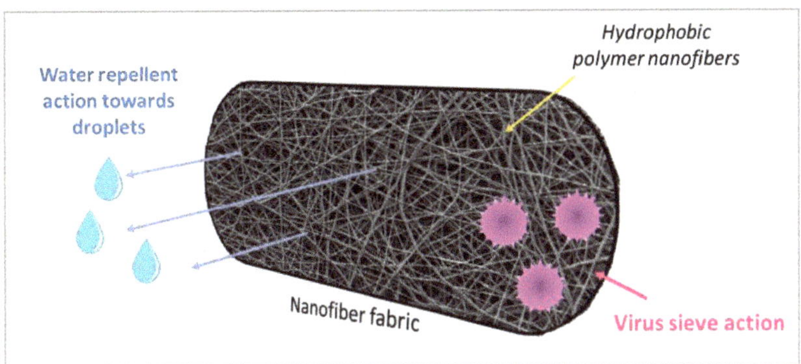

Figure 9. The double action of nanofiber fabrics: water repellency towards droplets and sieving and blocking of the virus.

As is known, air transport for COVID-19 consists of tiny droplets and particulate matter of infected people and environmental aerosols present in the air. These poly-dispersed aerosols are of different sizes and can interact with each other by interfering with the filtration process. Chinese scientists, Leung and Sun, for the first time studied the effects of the shape and size of environmental aerosols that simulate the bio-nano aerosols of the coronavirus, on the efficiency of the nanofibers filters [101]. Tests were performed on electrostatically charged PVDF nanofiber filters and manufactured as multiple modules of 2, 4, and 6, with each module having 0.765 g/m^2 of electrostatically charged PVDF nanofibers. The authors reported that a PVDF filter, either charged or uncharged, can capture ambient aerosols with sizes ranging from 10 to 400 nm. When the nanofibers are charged, dielectrophoresis improves the capture of aerosols larger than 80 nm. A nanofiber filter charged and made up of 6 modules can achieve an efficiency of 88%, 88%, and 96% for the ambiental aerosol size of 50, 100, and 300 nm respectively.

Nanomaterials such as nanofibres, used as components of high-performance filter masks (FFP2, FFP3), minimize the dispersion of respiratory droplets, released by infected subjects in the form of aerosols. General characteristics of anti-SARS-CoV-2 masks prepared with different nanomaterials are reported in Table 2.

Table 2. General characteristics of anti-SARS-CoV-2 masks prepared with different nanomaterials.

Nanomaterials	Classification	General Characteristics
Graphene [60–65]		• Ability to block the virus thanks to the pore size of the graphene structure smaller than the size of the SARS-CoV-2 virus. • Water-repellent properties towards droplets. • Photothermal effect. • Sunlight sterilization. • Maintains the filtering, antiviral capacity of the mask intact after repeated washing. • The graphene layer is inserted in layers of a polymeric and patibiocompatible nature.

Table 2. *Cont.*

Nanomaterials	Classification	General Characteristics
Graphene oxide (GO) [66–71]		• Ability to block the virus, thanks to the size of the pores of the graphene oxide structure smaller than the size of the SARS-CoV-2 virus. • Hydrophilic properties, cannot repel droplets. • Mainly used for the inner layers of masks. • The graphene oxide layer is inserted into layers of polymeric and bio-combustible nature.
Nanoparticles [73–85]	• Silver nanoparticles (AgNPs) • Copper nanoparticles (CuNPs) • Copper iodide nanoparticles (CuINPs) • Copper oxide (CuONPs) • Zinc oxide nanoparticles (ZnONPs)	• Effectiveness up to 99.9% in deactivating the SARS-CoV-2 virus. • Nanoparticles are deposited or inserted on tissue or polymeric supports. • The nanoparticles act by releasing ions that interact with the genetic material of the virus, preventing its replication.
[93,94]	• Titanium oxide (TiO$_2$NPs)	• Nanoparticles are deposited or inserted on tissue or polymeric supports. • High antiviral efficiency with adsorption and photo inactivation of the virus
Nanofibers [90–94]	• Polymeric nanofibers • PVDF (Polyvinylidene fluoride Polyvinylidene fluoride)	• High antiviral efficiency with 99.9% virus-blocking ability. • The high breathability of the masks. • Low cost of preparation.

4. Conclusions

The work of prevention and control of the recent pandemic has made use not only of previous experiences and knowledge acquired in the management of previous epidemics but also of the progress made by nanotechnologies in the design and implementation of feasible solutions to prevent and contain the infection.

Reusable, washable, and sustainable masks have been developed that are able not only to protect against the virus but also to inactivate it as some of them have nanolayers inside them that damage the viral RNA; others are covered in a water-repellent fabric that releases metal ions capable of repelling droplets in the air that deposit on the mask. These masks, even if they cost more than traditional ones, have the advantage of providing effective prevention of contagion and reducing the environmental impact caused by disposable masks.

In this period of greater sensitivity toward the problems created by the recent pandemic, this research sector is very active and highly prolific in publications.

Therefore, this work aimed to identify only some of the most significant results reported in the literature and on the market to allow a broad and general overview of the application of nanomaterials for the creation of PPE.

The various nanomaterials reported in this review first of all show that they are all capable of contributing in a positive and effective way to countering the spread of the SARS-CoV-2 virus, although their way of acting, their characteristics are often different.

Graphene, graphene oxide, and nanofibers mainly act as real filters capable of blocking the virus, thanks to the size of the pores present in their structures which are smaller than the size of the SARS-CoV-2 virus. So in this respect, it is more precise to talk about blocking the virus.

Metallic nanoparticles act differently, releasing ions that come into contact with the genetic material of the virus, thus leading to its inactivation. So in the case of metal nanoparticles, it is more correct to speak of inactivation of the virus. Beyond this substantial difference in the mechanism of action, there are differences even within these two categories of nanomaterials.

Graphene and polymeric nanofibres, in general, have a water-repellent behavior, and therefore exhibit an action of removing droplets, which, as is known, are small water droplets and are among the most common carriers of the virus. These materials, therefore, have a double action: filtering of the virus that manages to reach the surface and preventive action against droplets.

Graphene oxide, on the other hand, has a hydrophilic nature, therefore it has no action to contrast droplets but only a virus-blocking action. In fact, in the creation of masks, their use is mainly intended for internal layers.

Graphene has another important characteristic that of the photothermic effect, i.e., if subjected to sunlight it heats up leading to the death of the virus.

Graphene, therefore, presents itself, under this aspect, as one of the best materials exhibiting virus-blocking properties, preventive droplet contrast, and virus inactivation after exposure to sunlight. This means that the graphene masks could potentially be reused after sterilization in sunlight.

As far as nanoparticles are concerned, almost all of them have a high virus deactivation efficacy which can reach up to 99.9%. The data reported in the literature still do not allow for the creation of a precise scale of efficacy as the laboratory tests are not yet standardized. However, it is possible to say that the effectiveness of nanoparticles depends on their size, shape, nanoparticle/support ratio, and the construction characteristics of the nanomaterial.

Other important features of the face masks against the SARS-CoV-2 virus, in addition to effectiveness, are comfort, breathability, and safety.

Safety is a fundamental aspect, but to date with respect to nanomaterials there is a scarcity of information against a high level of pervasiveness on the market. These gaps are to be attributed to a certain fragmentation of the studies available so far in the literature, which does not find a real common line.

A common fear is that the nanomaterials inserted into the masks could be released and meet the human body [102,103].

In most cases, however, manufacturers ensure that the insertion of nanomaterials inside the masks has been carried out with technologies, often covered by patent secrets, capable of preventing any release harmful to human health.

Most of the products put on the market, according to the declarations of the producers, are tested and certified by independent bodies, while for the products in the research phase, the tests are carried out as a preliminary phase, in the laboratory.

Comparing the materials presented, with respect to the comfort and breathability of the masks, an important position is occupied by nanofibres. In fact, they themselves form a nanofibre membrane that allows high breathability, flexibility, and therefore high comfort for the wearer. For other materials, support is always necessary, such as synthetic or natural fabric, in which to insert the nanomaterials which, depending on their nature, can interfere with the breathability and comfort of the mask.

According to the European Chemicals Agency (ECHA), nanotechnologies will be increasingly present in daily life and this sector is expected to develop very rapidly which will make a significant contribution to industrial and economic growth in the coming years. According to widespread scientific opinion, nanotechnologies will go through 4 stages of development. The first two are already underway and extensively developed. The first concerns the use of nanomaterials in "passive nanostructures" which can be the creation of nanoreinforced nanomaterials or the application of nanocoatings that can be used for example for the creation of self-cleaning or antimicrobial surfaces.

The second phase, which has already begun, is aimed above all at what could be defined as "active nanostructures", i.e., capable, thanks to their bioactivity, of being used as target drugs, for example by coating the nanoparticles with specific proteins capable of interacting with the virus, the bacterium, the organ or with the diseased cell. The third and fourth phases will be more complex and highly sophisticated and will concern the creation of nanosystems to then move on to the generation of molecular nanorobotics capable, for example, of using energy and promoting a series of specific functions such as intercepting and virus elimination.

Current research has clearly shown the important role that nanomaterials, as well as nanoparticles, can play in the fight against viral diseases and especially against SARS-CoV-2. Nanotechnology paves the way for the research of new drugs and treatments against COVID-19 by addressing the problems of low biocompatibility, poor stability, and toxicity issues.

Nanotechnology applications, however, present some critical issues that need to be addressed to facilitate their wider implementation in the healthcare field. It is important to underline that particular attention must be paid to the behavior of nanomaterials when they come into contact with the human body, for example, inhaled or entering the bloodstream, where they can potentially change their behavior. Most of the studies reported in the literature have evaluated biocompatibility only through laboratory and in vitro tests and using generic protocols. Precise standardized protocols for characterizing the properties of nanomaterials, with particular reference to their behavior with respect to human health, are desirable.

Overcoming these challenges requires effective collaboration between experts in materials science, pharmacology, toxicology, and regulatory agencies.

Another obstacle to overcome is the production of nanomaterials on a large scale in order to make possible feasible and economic commercialization of nanoparticle-based products.

In conclusion, it can be said that nanotechnologies have already amply highlighted the potential ability to improve various aspects that are crucial for the fight against SARS-CoV-2, such as diagnostics, therapies, and protection.

Author Contributions: Conceptualization, P.D.L. and A.M.; methodology, P.D.L. and A.M.; writing—review and editing, P.D.L. and A.M.; supervision, P.D.L., A.M. and J.B.N. All authors have read and agreed to the published version of the manuscript.

Funding: This research received no external funding.

Conflicts of Interest: The authors declare no conflict of interest.

References

1. Poole, C.P.; Owens, F.J. *Introduction to Nanotechnology*; John Wiley & Sons: New York, NY, USA, 2003.
2. Available online: https://eur-lex.europa.eu/legal-content/IT/TXT/PDF/?uri=CELEX:32011H0696 (accessed on 10 May 2023).
3. Mohanraj, V.J.; Chen, Y. Nanoparticles—A review. *Trop. J. Pharm. Res.* **2006**, *5*, 561–573. [CrossRef]
4. Xu, L.; Liang, H.W.; Yang, Y.; Yu, S.H. Stability and Reactivity: Positive and Negative Aspects for Nanoparticle Processing. *Chem. Rev.* **2018**, *118*, 3209–3250. [CrossRef]
5. Kolahalam, L.A.; Viswanath, I.K.; Diwakar, B.S.; Govindh, B.; Reddy, V.; Murthy, Y.L.N. Review on nanomaterials: Synthesis and applications. *Mater. Today Proc.* **2019**, *18*, 2182–2190. [CrossRef]
6. Nasrollahzadeh, M.; Sajadi, S.M.; Sajjadi, M.; Issaabadi, Z. Chapter 1—An Introduction to Nanotechnology. *Interface Sci. Technol.* **2019**, *28*, 1–27. [CrossRef]
7. Mishra, R.; Militky, J. *Nanotechnology in Textiles Theory and Application*; Elsevier: Amsterdam, The Netherlands, 2019; ISBN 978-0-08-102609-0.
8. Ali, K.; Ye, T.; Hang, Q.; Amir, M.; Butt, H.; Mehmet, R.; Dokmeci, J.P.; Hinestroza, M.S.; Khademhosseini, A.; Yun, S.H. Nanotechnology in Textiles. *ACS Nano* **2016**, *10*, 3042–3068.
9. López, O.V.; Castillo, L.A.; Garcia, M.A.; Villar, M.A.; Barbosa, S.E. Food packaging bags based on thermoplastic corn starch reinforced with talc nanoparticles. *Food Hydrocoll.* **2015**, *43*, 18–24. [CrossRef]
10. Singh, P. Nanotechnology in food preservation. *Food Sci.* **2018**, *9*, 435–441. [CrossRef]
11. De Luca, P.; De Luca, P.; Candamano, S.; Macario, A.; Crea, F.; B.Nagy, J. Preparation and characterization of plasters with photodegradative action. *Buildings* **2018**, *8*, 122. [CrossRef]
12. Konsta-Gdoutos, M.S.; Metaxa, Z.S.; Shah, S.P. Highly dispersed carbon nanotube reinforced cement based materials. *Cem. Concr. Res.* **2010**, *40*, 1052–1059. [CrossRef]
13. Kalam, K.; Otsus, M.; Kozlova, J.; Tarre, A.; Kasikov, A.; Rammula, R.; Link, J.; Stern, R.; Vinuesa, G.; Lendinez, J.M.; et al. Memory Effects in Nanolaminates of Hafnium and Iron Oxide Films Structured by Atomic Layer Deposition. *Nanomaterials* **2022**, *12*, 2593. [CrossRef] [PubMed]
14. Hsu, C.S.; Chan, C.C.; Huang, H.T.; Peng, C.H.; Hsu, W.C. Electrochromic properties of nanocrystalline MoO_3 thin films. *Thin Solid Films* **2008**, *516*, 4839–4844. [CrossRef]
15. Shang, Y.; Hasan, M.K.; Ahammed, G.J.; Li, M.; Yin, H.; Zhou, J. Applications of Nanotechnology in Plant Growth and Crop Protection: A Review. *Molecules* **2019**, *24*, 2558. [CrossRef] [PubMed]

16. Singh, H.; Sharma, A.; Bhardwaj, S.K.; Arya, S.K.; Bhardwaj, N.; Khatri, M. Recent advances in the applications of nano-agrochemicals for sustainable agricultural development. *Environ. Sci. Process. Impacts* **2021**, *23*, 213–239. [CrossRef]

17. De Luca, P.; Macario, A.; Siciliano, C.; B.Nagy, J. Recovery of Biophenols from Olive Vegetation Waters by Carbon Nanotubes. *Materials* **2022**, *15*, 2893. [CrossRef] [PubMed]

18. De Luca, P.; Chiodo, A.; Macario, A.; Siciliano, C.; B.Nagy, J. Semi-continuous adsorption processes with multi-walled carbon nanotubes for the treatment of water contaminated by an organic textile dye. *Appl. Sci.* **2021**, *11*, 1687. [CrossRef]

19. Agrahari, P.A.J.R.A.; Bhatia, A.K.; Sexna, A.G.; Pant, G.; Singh, R.P.; Mishra, R.R.; Mishra, V.K. Advance Applications of Nanotechnology in Medicine. *Int. J. Sci. Res.* **2016**, *7*, 1284–1315.

20. Kargozar, S.; Mozafari, M. Nanotechnology and Nanomedicine: Start small, think big. *Mater. Today Proc.* **2018**, *5*, 15492–15500. [CrossRef]

21. Hou, S.; Zhang, A.; Su, M. Nanomaterials for Biosensing Applications. *Nanomaterials* **2016**, *6*, 58. [CrossRef]

22. Zhang, L.; Gu, F.X.; Chan, J.M.; Wang, A.Z.; Langer, R.S.; Farokhzad, O.C. Nanoparticles in medicine: Therapeutic applications and developments. *Clin. Pharmacol. Ther.* **2008**, *83*, 761–769. [CrossRef]

23. Davis, M.E.; Chen, Z.; Shin, D.M. Nanoparticle therapeutics: An emerging treatment modality for cancer. *Nat. Rev. Drug Discov.* **2008**, *7*, 771–782. [CrossRef]

24. Li, Z.; Shan, X.; Chen, Z.; Gao, N.; Zeng, W.; Zeng, X.; Mei, L. Applications of surface modification technologies in nanomedicine for deep tumor penetration. *Adv. Sci.* **2021**, *8*, 2002589. [CrossRef]

25. Goldberg, M.; Langer, R.; Jia, X. Nanostructured materials for applications in drug delivery and tissue engineering. *J. Biomater. Sci. Polym.* **2007**, *18*, 241–268. [CrossRef]

26. Wang, J.L.; Chen, G.H.; Jiang, H.; Li, Z.Y.; Wang, X.M. Advances in Nano-Scaled Biosensors for Biomedical Applications. *Analyst* **2013**, *138*, 4427–4435. [CrossRef]

27. Wu, Y.C.; Chen, C.S.; Chan, Y.J. The outbreak of COVID-19: An overview. *J. Chin. Med. Assoc.* **2020**, *83*, 217–220. [CrossRef] [PubMed]

28. Singhal, T. A Review of Coronavirus Disease-2019 (COVID-19). *Indian J. Pediatr.* **2020**, *87*, 281–286. [CrossRef]

29. Hu, B.; Guo, H.; Zhou, P.; Shi, Z.L. Characteristics of SARS-CoV-2 and COVID-19. *Nat. Rev. Microbiol.* **2021**, *19*, 141–154. [CrossRef] [PubMed]

30. Deng, S.Q.; Peng, H.J. Characteristics of and public health responses to the coronavirus disease 2019 outbreak in China. *J. Clin. Med.* **2020**, *9*, 575. [CrossRef]

31. Cui, J.; Li, F.; Shi, Z.L. Origin and evolution of pathogenic coronaviruses. *Nat. Rev. Microbiol.* **2019**, *17*, 181–192. [CrossRef]

32. Andersen, K.G.; Rambaut, A.; Lipkin, W.I.; Holmes, E.C.; Garry, R.F. The proximal origin of SARS-CoV-2. *Nat. Med.* **2020**, *26*, 450–452. [CrossRef]

33. Zhu, H.; Wei, L.; Niu, P. The novel coronavirus outbreak in Wuhan, China. *Glob. Health Res. Policy* **2020**, *5*, 6. [CrossRef] [PubMed]

34. Cheng, Z.J.; Shan, J. 2019 Novel coronavirus: Where we are and what we know. *Infection* **2020**, *48*, 155–163. [CrossRef]

35. Wang, C.; Horby, P.W.; Hayden, F.G.; Gao, G.F. A novel coronavirus outbreak of global health concern. *Lancet* **2020**, *395*, 470–473. [CrossRef]

36. Campos, E.V.R.; Pereira, A.E.S.; de Oliveira, J.L.; Carvalho, L.B.; Guilger-Casagrande, M.; de Lima, R.; Fraceto, L.F. How can nanotechnology help to combat COVID-19? Opportunities and urgent need. *J. Nanobiotechnol.* **2020**, *18*, 125. [CrossRef]

37. Zhou, J.; Krishnan, N.; Jiang, Y.; Fang, R.H.; Zhang, L. Nanotechnology for virus treatment. *Nano Today* **2021**, *36*, 101031. [CrossRef]

38. Kirtane, A.R.; Verma, M.; Karandikar, P.; Furin, J.; Langer, R.; Traverso, G. Nanotechnology approaches for global infectious diseases. *Nat. Nanotechnol.* **2021**, *16*, 369–384. [CrossRef]

39. Solanki, R.; Shankar, A.; Modi, U.; Patel, S. New insights from nanotechnology in SARS-CoV-2 detection, treatment strategy, and prevention. *Mater. Today Chem.* **2023**, *29*, 101478. [CrossRef] [PubMed]

40. Hasanzadeh, A.; Alamdaran, M.; Ahmadi, S.; Nourizadeh, H.; Bagherzadeh, M.A.; Mofazzal, J.M.; Simon, P.; Karimi, M.; Hamblin, M.R. Nanotechnology against COVID-19: Immunization, diagnostic and therapeutic studies. *J. Control. Release* **2021**, *336*, 355–365. [CrossRef] [PubMed]

41. Abdelhamid, H.N.; Badr, G. Nanobiotechnology as a Platform for the Diagnosis of COVID-19: A Review. *Nanotechnol. Environ. Eng.* **2021**, *6*, 19. [CrossRef]

42. Al-Hindawi, A.; AlDallal, U.; Waly, Y.M.; Hussain, M.H.; Shelig, M.; Saleh ElMitwalli, O.S.M.M.; Deen, G.R.; Henari, F.Z. An Exploration of Nanoparticle-Based Diagnostic Approaches for Coronaviruses: SARS-CoV-2, SARS-CoV and MERS-CoV. *Nanomaterials* **2022**, *12*, 3550. [CrossRef]

43. Krishnan, S.; Dusane, A.; Morajkar, R.; Venkat, A.; Vernekar, A.A. Deciphering the role of nanostructured materials in the point-of-care diagnostics for COVID-19: A comprehensive review. *J. Mater. Chem. B* **2021**, *9*, 5967–5981. [CrossRef] [PubMed]

44. Choi, J.R. Development of Point-of-Care Biosensors for COVID-19. *Front. Chem.* **2020**, *8*, 517. [CrossRef] [PubMed]

45. Shan, B.; Broza, Y.Y.; Li, W.; Wang, Y.; Wu, S.; Liu, Z.; Wang, J.; Gui, S.; Wang, L.; Zhang, Z.; et al. Multiplexed nanomaterial-based sensor array for detection of COVID-19 in exhaled breath. *ACS Nano* **2020**, *14*, 12125–12132. [CrossRef]

46. Bhalla, N.; Pan, Y.; Yang, Z.; Payam, A.F. Opportunities and Challenges for Biosensors and Nanoscale Analytical Tools for Pandemics: COVID-19. *ACS Nano* **2020**, *14*, 7783–7807. [CrossRef] [PubMed]

47. Sengupta, J.; Hussain, C.M. Graphene-Based Electrochemical Nano-Biosensors for Detection of SARS-CoV-2. *Inorganics* **2023**, *11*, 197. [CrossRef]

48. Lee, J.H.; Park, S.J.; Choi, J.W. Electrical Property of Graphene and Its Application to Electrochemical Biosensing. *Nanomaterials* **2019**, *9*, 297. [CrossRef] [PubMed]

49. Sengupta, J.; Hussain, C.M. Graphene-Based Field-Effect Transistor Biosensors for the Rapid Detection and Analysis of Viruses: A Perspective in View of COVID-19. *Carbon Trends* **2021**, *2*, 100011. [CrossRef]

50. Chakravarty, M.; Vora, A. Nanotechnology-based antiviral therapeutics. *Drug Deliv. Transl. Res.* **2021**, *11*, 748–787. [CrossRef]

51. Weiss, C.; Carriere, M.; Fusco, L.; Capua, I.; Regla-Nava, J.A.; Pasquali, M.; Scott, J.A.; Vitale, F.; Unal, M.A.; Mattevi, C.; et al. Toward Nanotechnology-Enabled Approaches against the COVID-19 Pandemic. *ACS Nano* **2020**, *14*, 6383–6406. [CrossRef]

52. Bwalya, A.W.; Pedzisai, A.M.; Larry, L.M.; Pascal, V.N.; Melissa, T.R.C.; Scott, K.M.; Chiluba, M.; Steward, M.; Jonathan, K.; Roderick, B.W. Nano-Biomimetic Drug Delivery Vehicles: Potential Approaches for COVID-19 Treatment. *Molecules* **2020**, *25*, 5952. [CrossRef]

53. Tang, Z.; Zhang, X.; Shu, Y.; Guo, M.; Zhang, H.; Tao, W. Insights from Nanotechnology in COVID-19 Treatment. *Nano Today* **2021**, *36*, 101019. [CrossRef]

54. Nanomedicine and the COVID-19 vaccines. *Nat. Nanotechnol.* **2020**, *15*, 963. [CrossRef]

55. Krammer, F. SARS-CoV-2 vaccines in development. *Nature* **2020**, *586*, 516–527. [CrossRef]

56. Chauhan, G.; Madou, M.J.; Kalra, S.; Chopra, V.; Ghosh, D.; Martinez-Chapa, S.O. Nanotechnology for COVID-19: Therapeutics and vaccine research. *ACS Nano* **2020**, *17*, 7760–7782. [CrossRef]

57. Meo, S.A.; Bukhari, I.A.; Akram, J.; Meo, A.S.; Klonoff, D.C. COVID-19 vaccines: Comparison of biological, pharmacological characteristics and adverse effects of Pfizer/BioNTech and Moderna vaccines. *Eur. Rev. Med. Pharmacol. Sci.* **2021**, *25*, 1663–1669. [PubMed]

58. Khurana, A.; Allawadhi, P.; Khurana, I.; Allwadhi, S.; Weiskirchen, R.; Banothu, A.K.; Chhabra, D.; Joshi, K.; Bharani, K.K. Role of nanotechnology behind the success of mRNA vaccines for COVID-19. *Nano Today* **2021**, *38*, 101142. [CrossRef]

59. Holland, M.; Zaloga, D.J.; Friderici, C.S. COVID-19 Personal Protective Equipment (PPE) for the emergency physician. *Vis. J. Emerg. Med.* **2020**, *19*, 100740. [CrossRef] [PubMed]

60. Lee, S.A.; Hwang, D.C.; Li, H.Y.; Tsai, C.F.; Chen, C.W.; Chen, J.K. Particle Size-Selective Assessment of Protection of European Standard FFP Respirators and Surgical Masks against Particles-Tested with Human Subjects. *J. Healthc. Eng.* **2016**, *2016*, 8572493. [CrossRef] [PubMed]

61. McCarthy, R.; Gino, B.; d'Entremont, P.; Barari, A.; Renouf, T.S. The Importance of Personal Protective Equipment Design and Donning and Doffing Technique in Mitigating Infectious Disease Spread: A Technical Report. *Cureus* **2020**, *12*, e12084. [CrossRef]

62. Abbasinia, M.; Karimie, S.; Haghighat, M.; Mohammadfam, I. Application of Nanomaterials in Personal Respiratory Protection Equipment: A Literature Review. *Safety* **2018**, *4*, 47. [CrossRef]

63. Nasir, S.; Hussein, M.Z.; Zainal, Z.; Yusof, N.A. Carbon-Based Nanomaterials/Allotropes: A Glimpse of Their Synthesis, Properties and Some Applications. *Materials* **2018**, *11*, 295. [CrossRef]

64. Urade, A.R.; Lahiri, I.; Suresh, K.S. Graphene Properties, Synthesis and Applications: A Review. *JOM* **2023**, *75*, 614–630. [CrossRef] [PubMed]

65. C&S ITALY S.R.L. Available online: https://avmaskpro.it/ (accessed on 2 April 2023).

66. Hygraner Srl. Available online: https://www.hygraner.it/ (accessed on 2 April 2023).

67. Pal, K.J.; Kyzas, G.Z.; Kralj, S.; Gomes de Souza, F. Sunlight sterilized, recyclable and super hydrophobic anti-COVID laser-induced graphene mask formulation for indelible usability. *J. Mol. Struct.* **2021**, *1233*, 130100. [CrossRef] [PubMed]

68. Zhong, H.; Zhu, Z.; Lin, J.; Cheung, C.F.; Lu, V.L.; Yan, F.; Chan, C.Y.; Li, G. Reusable and Recyclable Graphene Masks with Outstanding Superhydrophobic and Photothermal Performances. *ACS Nano* **2020**, *14*, 6213–6221. [CrossRef] [PubMed]

69. Fukuda, M.; Islam, M.S.; Shimizu, R.; Nasser, H.; Rabin, N.N.; Takahashi, Y.; Sekine, Y.; Lindoy, L.F.; Fukuda, T.; Ikeda, T.; et al. Lethal Interactions of SARS-CoV-2 with Graphene Oxide: Implications for COVID-19 Treatment. *ACS Appl. Nano Mater.* **2021**, *4*, 11881–11887. [CrossRef]

70. Rhazouani, A.; Aziz, K.; Gamrani, H.; Gebrati, L.; Uddin, M.S.; Faissal, A. Can the application of graphene oxide contribute to the fight against COVID-19? Antiviral activity, diagnosis and prevention. *Curr. Res. Pharmacol. Drug Discov.* **2021**, *2*, 100062. [CrossRef]

71. Palmieri, V.; Papi, M. Can graphene take part in the fight against COVID-19? *Nano Today* **2020**, *33*, 100883. [CrossRef]

72. De Maio, F.; Palmieri, V.; Babini, G.; Augello, A.; Palucci, I.; Perini, G.; Salustri, A.; Spilman, P.; De Spirito, M.; Sanguinetti, M.; et al. Graphene nanoplatelet and graphene oxide functionalization of face mask materials inhibits infectivity of trapped SARS-CoV-2. *iScience* **2021**, *24*, 7. [CrossRef]

73. Mbayachi, V.B.; Ndayiragije, E.; Sammani, T.; Taj, S.; Mbuta, E.R.; Khan, A. Graphene synthesis, characterization and its applications: A review. *Results Chem.* **2021**, *3*, 100163. [CrossRef]

74. Laue, M.; Kauter, A.; Hoffmann, T.; Möller, L.; Michel, J.; Nitsche, A. Morphometry of SARS-CoV and SARS-CoV-2 particles in ultrathin plastic sections of infected Vero cell cultures. *Sci. Rep.* **2021**, *11*, 3515. [CrossRef]

75. Hu, J.; Liu, J.; Gan, L.; Long, M. Surface-modified graphene oxide-based cotton fabric by ion implantation for enhancing antibacterial activity. *ACS Sustain. Chem. Eng.* **2019**, *7*, 7686–7692. [CrossRef]

76. Lysenko, V.; Lozovski, V.; Lokshyn, M.; Gomeniuk, Y.V.; Dorovskih, A.; Rusinchuk, N.; Pankivska, Y.; Povnitsa, O.; Zagorodnya, S.; Tertykh, V.; et al. Nanoparticles as antiviral agents against adenoviruses. *Adv. Nat. Sci. Nanosci. Nanotechnol.* **2018**, *9*, 025021. [CrossRef]

77. Rai, M.; Yadav, A.; Gade, A. Silver nanoparticles as a new generation of antimicrobials. *Biotechnol. Adv.* **2009**, *27*, 76–80. [CrossRef] [PubMed]

78. Khoshnevisan, K.; Maleki, H.; Baharifar, H. Nanobiocide Based-Silver Nanomaterials Upon Coronaviruses: Approaches for Preventing Viral Infections. *Nanoscale Res. Lett.* **2021**, *16*, 100. [CrossRef] [PubMed]

79. Fung, M.C.; Bowen, D.L. Silver products for medical indications: Risk-benefit assessment. *J. Toxicol. Clin. Toxicol.* **1996**, *34*, 119–124. [CrossRef]

80. New Silver-Copper Nanolayer-based Antimicrobial Mask Inactivates SARS-CoV-2. Industry News. Edited Special Chem. 2021. Available online: https://omnexus.specialchem.com/news/industry-news/silver-copper-mask-000225616 (accessed on 8 May 2023).

81. Bello-Lopez, M.; Silva-Bermudez, P.; Prado, G.; Martínez, A.; Ibáñez-Cervantes, G.; Cureño-Díaz, M.A.; Rocha-Zavaleta, L.; Manzo-Merino, J.; Almaguer-Flores, A.; Ramos-Vilchis, C.; et al. Biocide effect against SARS-CoV-2 and ESKAPE pathogens of a noncytotoxic silver–copper nanofilm. *Biomed. Mater.* **2021**, *17*, 015002. [CrossRef] [PubMed]

82. Kolzer Srl. Available online: https://www.kolzer.com/it/kolzer/news/418-nanoparticelle-argento-mascherine-coronavirus (accessed on 8 May 2023).

83. São Paulo-Based Company Develops Fabric That Eliminates Novel Coronavirus by Contact. Available online: https://agencia.fapesp.br/sao-paulo-based-company-develops-fabric-that-eliminates-novel-coronavirus-by-contact/33568/ (accessed on 8 May 2023).

84. Grass, G.; Rensing, C.; Solioz, M. Metallic copper as an antimicrobial surface. *Appl. Environ. Microbiol.* **2011**, *77*, 1541–1545. [CrossRef]

85. Govind, V.; Bharadwaj, S.; Sai Ganesh, M.R.; Vishnu, J.; Shankar, K.V.; Shankar, B.; Rajesh, R. Antiviral properties of copper and its alloys to inactivate COVID-19 virus: A review. *Biometals* **2021**, *34*, 1217–1235. [CrossRef]

86. Italtex S.P.A. Available online: https://virkill.it/ (accessed on 8 May 2023).

87. Archana, K.M.; Rajagopal, R.; Krishnaswamy, V.G.; Aishwarya, S. Application of green synthesised copper iodide particles on cotton fabric-protective face mask material against COVID-19 pandemic. *J. Mater. Res. Technol.* **2021**, *15*, 2102–2113. [CrossRef]

88. Hasan, R. Production of antimicrobial textiles by using copper oxide nanoparticles. *IJCRR* **2018**, *9*, 20195–20200. [CrossRef]

89. Pasquet, J.; Chevalier, Y.; Pelletier, J.; Couval, E.; Bouvier, D.; Bolzinger, M.A. The contribution of zinc ions to the antimicrobial activity of zinc oxide. *Colloids Surf. A* **2014**, *457*, 263–274. [CrossRef]

90. Moezzi, A.; McDonagh, A.M.; Cortie, M.B. Zinc oxide particles: Synthesis, properties and applications. *Chem. Eng. J.* **2012**, *185–186*, 1–22. [CrossRef]

91. Read, S.A.; Obeid, S.; Ahlenstiel, C.; Ahlenstiel, G. The role of zinc in antiviral immunity. *Adv. Nutr.* **2019**, *10*, 696–710. [CrossRef] [PubMed]

92. Gonzalez, A.; Aboubakr, H.A.; Brockgreitens, J.; Hao, W.; Wang, Y.; Goyal, S.M.; Abbas, A. Durable nanocomposite face masks with high particulate filtration and rapid inactivation of coronaviruses. *Sci. Rep.* **2021**, *11*, 24318. [CrossRef] [PubMed]

93. Forman, H.J.; Augustus, O.; Brigelius-Flohe, R.; Dennery, P.A.; Kalyanaraman, B.; Ischiropoulos, H.; Mann, G.E.; Giovanni, E.; Radi, R.; Roberts, L.J.; et al. Even free radicals should follow some rules: A guide to free radical re-search terminology and methodology. *Free Root. Biol. Med.* **2015**, *78*, 233–235. [CrossRef]

94. Kohantorabi, M.; Wagstaffe, M.; Creutzburg, M.; Ugolotti, A.; Kulkarni, S.; Jeromin, A.; Krekeler, T.; Feuerherd, M.; Herrmann, A.; Ebert, G.; et al. Adsorption and Inactivation of SARS-CoV-2 on the Surface of Anatase TiO$_2$(101). *ACS Appl. Mater. Interfaces* **2023**, *15*, 8770–8782. [CrossRef]

95. Available online: https://chemicalwatch.com/425786/scientists-urge-regulatory-action-on-titanium-dioxide-particles-in-face-masks (accessed on 5 July 2023).

96. Verleysen, E.; Ledecq, M.; Siciliani, L.; Cheyns, K.; Vleminckx, C.; Blaude, M.N.; De Vos, S.; Brassinne, F.; Van Steen, F.; Nkenda, R.; et al. Titanium dioxide particles frequently present in face masks intended for general use require regulatory control. *Sci. Rep.* **2022**, *12*, 2529. [CrossRef]

97. Essa, W.K.; Yasin, S.A.; Saeed, I.A.; Ali, G.A.M. Nanofiber-Based Face Masks and Respirators as COVID-19 Protection: A Review. *Membranes* **2021**, *11*, 250. [CrossRef]

98. Rasmi, Y.; Saloua, K.S.; Nemati, M.; Choi, J.R. Recent Progress in Nanotechnology for COVID-19 Prevention, Diagnostics and Treatment. *Nanomaterials* **2021**, *11*, 1788. [CrossRef]

99. Available online: https://www.inveniosolutions.it/electrospinning/nanofibre (accessed on 8 May 2023).

100. Hemmer, C.J.; Hufert, F.; Siewert, S.; Reisinger, E. Protection From COVID-19–The Efficacy of Face Masks. *Dtsch. Arztebl. Int.* **2021**, *118*, 59–65. [CrossRef]

101. Leung, W.W.F.; Sun, Q. Charged PVDF multilayer nanofiber filter in filtering simulated airborne novel coronavirus (COVID-19) using ambient nano-aerosols. *Sep. Purif. Technol.* **2020**, *245*, 116887. [CrossRef]

102. Mosselhy, D.A.; Virtanen, J.; Kant, R.; He, W.; Elbahri, M.; Sironen, T. COVID-19 Pandemic: What about the Safety of Anti-Coronavirus Nanoparticles? *Nanomaterials* **2021**, *11*, 796. [CrossRef] [PubMed]
103. Available online: https://www.lindipendente.online/2021/06/08/possibili-danni-da-inalazione-di-grafene-la-francia-sconsiglia-le-mascherine-ffp2/ (accessed on 10 June 2023).

Review

Emerging Nanomaterials Biosensors in Breathalyzers for Detection of COVID-19: Future Prospects

Saravanan Rajendrasozhan [1,2], Subuhi Sherwani [2,3,*], Faheem Ahmed [4,*], Nagih Shaalan [5,6], Abdulmohsen Alsukaibi [1,2], Khalid Al-Motair [2] and Mohd Wajid Ali Khan [1,2]

[1] Department of Chemistry, College of Science, University of Ha'il, Ha'il 55473, Saudi Arabia; s.rajendrasozhan@uoh.edu.sa (S.R.); a.alsukaibi@uoh.edu.sa (A.A.); mw.khan@uoh.edu.sa (M.W.A.K.)

[2] Medical and Diagnostic Research Center, University of Ha'il, Ha'il 55473, Saudi Arabia; k.almutier@uoh.edu.sa

[3] Department of Biology, College of Science, University of Ha'il, Ha'il 55473, Saudi Arabia

[4] Department of Applied Sciences & Humanities, Faculty of Engineering & Technology, Jamia Millia Islamia, New Delhi 110025, India

[5] Department of Physics, College of Science, King Faisal University, Al-Ahsa 31982, Saudi Arabia; nmohammed@kfu.edu.sa

[6] Physics Department, Faculty of Science, Assiut University, Assiut 71516, Egypt

* Correspondence: s.sherwani@uoh.edu.sa (S.S.); fahmed@jmi.ac.in (F.A.)

Abstract: In recent times, the global landscape of disease detection and monitoring has been profoundly influenced by the convergence of nanotechnology and biosensing techniques. Biosensors have enormous potential to monitor human health, with flexible or wearable variants, through monitoring of biomarkers in clinical and biological behaviors and applications related to health and disease, with increasing biorecognition, sensitivity, selectivity, and accuracy. The emergence of nanomaterial-based biosensors has ushered in a new era of rapid and sensitive diagnostic tools, offering unparalleled capabilities in the realm of disease identification. Even after the declaration of the end of the COVID-19 pandemic, the demand for efficient and accessible diagnostic methodologies has grown exponentially. In response, the integration of nanomaterial biosensors into breathalyzer devices has gained considerable attention as a promising avenue for low-cost, non-invasive, and early detection of COVID-19. This review delves into the forefront of scientific advancements, exploring the potential of emerging nanomaterial biosensors within breathalyzers to revolutionize the landscape of COVID-19 detection, providing a comprehensive overview of their principles, applications, and implications.

Keywords: COVID-19; exhaled breath analysis; CO; breathalyzer; biosensors

Citation: Rajendrasozhan, S.; Sherwani, S.; Ahmed, F.; Shaalan, N.; Alsukaibi, A.; Al-Motair, K.; Khan, M.W.A. Emerging Nanomaterials Biosensors in Breathalyzers for Detection of COVID-19: Future Prospects. *Inorganics* **2023**, *11*, 483. https://doi.org/10.3390/ inorganics11120483

Academic Editors: Roberto Nisticò, Torben R. Jensen, Luciano Carlos, Hicham Idriss and Eleonora Aneggi

Received: 16 September 2023
Revised: 4 December 2023
Accepted: 14 December 2023
Published: 18 December 2023

1. Introduction

Coronaviruses (CoVs) represent a class of enveloped viruses characterized by their single-stranded RNA genome [1]. These viruses possess the capability to infect a broad spectrum of hosts, including mammals and avian species, often exhibiting respiratory and enteric pathologies [2]. The process of urbanization, coupled with inadequate urban planning, suboptimal sanitation and water provisions, high population density, and encroachment upon formerly undisturbed ecosystems, collectively foster the proliferation of zoonotic infectious diseases [3]. Despite the presence of established protocols for detection and mitigation of established coronaviruses through various initiatives, the escalating occurrence of newly emergent diseases, such as the recent Coronavirus disease 2019 (COVID-19) pandemic, caused by the severe acute respiratory syndrome coronavirus 2 (SARS-CoV-2), serves as a reminder of the massive costs to the global healthcare systems. Severe cases of COVID-19 can lead to life-threatening situations and are typically linked with risk factors such as age, co-morbidities, and immune status [4].

Timely and early diagnosis of COVID-19 is of paramount importance to minimize the likelihood of encountering severe complications. Diagnosis relies heavily on expensive techniques such as real-time reverse-transcriptase polymerase chain reaction (rRT-PCR) positivity, lateral flow tests, rapid antigen tests, and enzyme-linked immunosorbent assays (ELISA), in addition to clinical symptoms and radiological diagnostics [5]. rRT-PCR, which detects viral nucleic acid from a nasopharyngeal swab, is regarded as the gold standard due to its specificity and accuracy [6]. However, accuracy often depends on the method of sample collection and preservation, prior to testing as well as at the time point of collection. Minor mistakes can lead to false positives. Additionally, these techniques cannot be readily employed in remote or resource-constrained environments. Hence, timely medical intervention assumes critical significance in curbing the progression of the disease and preventing fatalities. Even after the declaration of the end of the pandemic, fast and precise diagnostic testing of COVID-19 continues to be the main strategy in determining isolation or medical intervention.

In the quest for enhanced diagnostic tools to combat the ongoing COVID-19 pandemic, the amalgamation of nanomaterial-based biosensors with breathalyzer technology has emerged as a frontier approach [7]. The convergence of nanotechnology and biosensing has paved the way for innovative and sensitive detection methodologies, holding the promise of revolutionizing disease identification. With the imperative need for rapid, reproducible, non-invasive, and accurate diagnostic techniques, the incorporation of nanomaterial biosensors into breathalyzers offers a promising avenue for addressing these challenges in COVID-19 detection [8]. This scientific research review embarks on an exploration of the scientific progress in this field, elucidating the potential of emerging nanomaterial biosensors within breathalyzers to establish a paradigm shift in the landscape of COVID-19 diagnostics. By delving into the underlying principles, experimental methodologies, and anticipated implications, this study endeavors to contribute to the evolving discourse surrounding the application of nanomaterial biosensors for the early and efficient detection of COVID-19 through breath analysis.

2. Nano-Biosensors as Diagnostic Tools

Nano-biosensors have emerged as a groundbreaking class of diagnostic tools, offering remarkable potential for precise and timely disease detection. These sensors harness the unique properties of nanomaterials such as high surface area and the ability to be functionalized with specific molecules to enable highly sensitive and specific detection of biomolecular interactions, and thus have the potential to be used in a wide variety of diagnostic applications, including the detection of diseases, toxins, and pollutants [9–11]. Their miniaturized scale allows for efficient integration into various diagnostic platforms, offering advantages such as rapid response times, reduced sample volumes, and enhanced portability. Nano-biosensors exhibit exceptional versatility, and are capable of detecting a diverse range of analytes, including proteins, nucleic acids, and small molecules. The incorporation of nanomaterials, such as nanoparticles, nanowires, and nanotubes, into biosensor designs enhances surface area interactions and signal transduction mechanisms, thereby amplifying the detection signal and improving overall sensor performance [10].

In totality, an optimal biosensor should encompass several critical attributes, such as being amenable to mass production, autonomous operation, notable sensitivity and selectivity, rapid response kinetics, capacity for multiplexing, multiple sensing modalities, disposability, extended shelf-life, cost-effectiveness, and user-friendly operation.

Nano-biosensors have been used for the detection of cancer, infectious diseases, and toxins. Studies have shown significant potential in the early detection of cancer biomarkers. Wang et al. (2019) devised a methodology to detect the tumor marker carcinoembryonic antigen (CEA) through the application of a solid-state nanopore as an instrumental component [12]. This innovative approach capitalizes on the inherent binding affinity between CEA and aptamer-modified magnetic Fe_3O_4 nanoparticles, obviating the need for direct monitoring of CEA translocation through the nanopore. The aptamer modified magnetic Fe_3O_4 nanoparticles were intricately combined with tetrahedral DNA nanostructures (TDNs) to facilitate identification of a low concentration (0.1 nm) of CEA, affirming the specificity of the interaction between the aptamer and the target molecule. Thus, the nanopore sensing approach demonstrated its capability to detect CEA in actual samples. Another study evaluated human epidermal growth factor receptor-2 (HER2) status, which is routinely used in the screening, diagnosis, and surveillance of breast cancer (BC). They employed an innovative electrochemical biosensor specifically targeting the detection of HER2-positive circulating tumor cells. The hybrid nanocomposite, formed through the amalgamation of reduced graphene oxide nanosheets and rhodium nanoparticles onto the surface of a graphite electrode, yielded enhancements in surface area, electrochemical activity, and biocompatibility with outstanding performance in discerning HER2-overexpressing SKBR3 cancer cells. It exhibited a linear dynamic range, an impressively low analytical limit of detection at 1.0 cell/mL, and a limit of quantification of 3.0 cells/mL [13].

Nano-biosensors have also demonstrated their utility in point-of-care (POC) infectious disease detection. Chowdhary et al. developed a rapid and quantitative biosensor for the specific detection of Dengue virus (DENV) serotypes 1 to 4 through the optimization of a stable interaction between cadmium selenide tellurium sulfide fluorescent quantum dots and gold nanoparticles, and were able to successfully detect target virus DNA at standard dilutions ranging from 10^{-15} to 10^{-10} M. This versatile biosensor could detect and quantify serotypes, thereby exhibiting significant potential as a diagnostic tool for point-of-care DENV detection [14]. Pang et al. developed a fluorescent nano-biosensor employing magnetic nanoparticles and aptamers for the detection of avian influenza virus H5N1 [15]. The fluorescent aptasensor employed metal-enhanced fluorescence and core-shell Ag@SiO2 nanoparticles for the highly sensitive detection of recombinant hemagglutinin protein associated with the H5N1 influenza virus. This system capitalizes on the dual advantages of the rHA aptamer's exceptional selectivity and the substantial fluorescence enhancement capabilities inherent to core-shell Ag@SiO2 nanoparticles, with robust selectivity and remarkable sensitivity when applied in the presence of diverse proteins or within complex matrices, achieving a detection limit of 2 ng/mL in buffer and 3.5 ng/mL in the specified context [15].

Nano-biosensors are also powerful tools for the sensitive and selective detection of various toxins across diverse applications. They combine the high specificity of biological recognition elements with the enhanced properties of nanomaterials, resulting in improved sensitivity and rapid response times. For instance, He et al. developed a nano-biosensor utilizing gold nanoparticles and aptamers to detect ochratoxin A (OTA), a mycotoxin commonly found in food products, with exceptional sensitivity, achieving a remarkable detection limit of 0.05 $U \cdot L^{-1}$ [16]. The method was effectively used to develop a target-responsive aptasensor for colorimetric detection of OTA, associated with food poisoning and various adverse health effects in humans, and thus providing substantial potential for applications in the field of mycotoxin-related food safety monitoring.

Skotadis et al. developed an electrochemical hybrid biosensor using a combination of noble nanoparticles and DNAzymes for the simultaneous detection of Pb^{2+}, Cd^{2+}, and Cr^{3+} ions. The double-helix structure of DNAzymes undergoes disintegration into smaller fragments upon encountering specific heavy metal ions, with discernable alteration in device resistance due to collapse of conductive inter-nanoparticle DNAzymes. The biosensor allowed successful detection of the three target ions at concentrations significantly below their maximum allowable levels in tap water, with considerable promise for integration into autonomous and remote sensing systems to detect water pollution [17].

Nano-biosensors have also emerged as a promising class of diagnostic tools in the battle against COVID-19. Broadly, biosensor platforms designed for SARS-CoV-2 detection center on the following pivotal facets: specified target for identification (such as viral RNA, proteins, or human antibodies), the modality of identification (leveraging aptamers, immunoglobulins, nucleic acids, and receptors), and augmentation of the signal generation and transduction mechanisms (involving electrical, surface plasmon resonance, electrochemical, optical, mechanical systems, and fluorescence signals). Leveraging the unique properties of nanomaterials, these biosensors offer a powerful platform for the rapid and sensitive detection of the SARS-CoV-2 virus. Through strategic functionalization of nanomaterials with specific biomolecular recognition elements, such as antibodies or aptamers targeting viral antigens, nano-biosensors enable precise and early identification of COVID-19 infection [18]. The amplified surface area provided by nanomaterials enhances the interaction between the sensor and the target virus, resulting in enhanced sensitivity. Moreover, the integration of nano-biosensors into portable and point-of-care devices facilitates efficient and real-time diagnostics, thereby addressing the urgent need for rapid testing and screening strategies and contributing to effective pandemic management.

A study by Karakuş et al. reported the development of a colorimetric assay for the rapid detection of SARS-CoV-2 using gold nanoparticles functionalized with SARS-CoV-2 spike antigens. The SARS-CoV-2 spike antigens trigger rapid and irreversible aggregation of gold nanoparticles due to antibody–antigen interactions, resulting in a noticeable change in color, observable with the naked eye or measured using UV-Vis spectrometry. It exhibits the capability to identify SARS-CoV-2 spike antigens at an astonishingly low concentration and does not cross-react with other antigens, including influenza A (H1N1), MERS-CoV, and *Streptococcus pneumoniae*, even at high concentrations [19]. Another nano-biosensor array for SARS-CoV-2 detection involved the surface-enhanced Raman scattering-immune substrate, which was meticulously engineered using an innovative oil/water/oil three-phase liquid–liquid interface self-assembly technique. The method yielded two layers of compact and homogeneous gold nanoparticle films. The detection procedure entailed an immunoreaction involving the SARS-CoV-2 spike antibody-modified surface-enhanced Raman scattering (SERS) immune substrate, the spike antigen protein, and Raman reporter-labeled immuno-Ag nanoparticles. This process demonstrated the ability to identify the presence of the SARS-CoV-2 spike protein at concentrations as low as 0.77 fg mL^{-1} in phosphate-buffered saline and 6.07 fg mL^{-1} in untreated saliva. These findings underscore its potential utility as a viable option for the early diagnosis of COVID-19 [20].

These examples highlight the versatility and effectiveness of nano-biosensors in various diagnostic applications. As the field continues to evolve, nano-biosensors hold immense promise for revolutionizing disease diagnostics by providing earlier detection, more accurate results, and greater accessibility, thereby addressing critical challenges in effective disease management and healthcare, and fostering advancements in personalized medicine.

3. Uses of Breath Analysis

Breath analysis has emerged as a promising diagnostic strategy with broad applications across various medical fields. By analyzing the composition of exhaled breath, known as the breathome, this non-invasive approach offers valuable insights into an individual's lifestyle and physiological and pathological states, as well as microbiome metabolism, including the presence and/or abundance of pathogens [21]. The breathome contains a complex mixture of volatile organic compounds (VOCs) originating from metabolic processes, reflecting the underlying clinical state and overall health condition of the individual [8]. Volatolomics denotes the intricate chemical mechanisms associated with VOCs discharged from bodily fluids, including blood, urine, sweat, feces, nasal and skin secretions, and exhaled breath [22]. The respiratory system is a prominent substrate for the investigation of these VOCs which originate endogenously within the human body, undergo systemic circulation through the bloodstream, and ultimately traverse the alveolar interface, depending upon solubility to manifest in exhaled breath. Human breath contains up to 3500 VOCs. Nonpolar VOCs are characterized by limited solubility in blood (indicated by a blood–air partition coefficient, $\lambda_{b:a}$, of less than 10) and undergo exchange primarily in the alveoli. Conversely, blood-soluble VOCs (with $\lambda_{b:a}$ values exceeding 100) are predominantly exchanged within the lung airways. VOCs exhibiting intermediate solubility, with blood–air partition coefficients falling within the range of 10 to 100, participate in pulmonary gas exchange in both the alveoli and the lung airways [21]. These VOCs can be quantified at minute concentrations, typically within the parts-per-million by volume (ppmv) and parts-per-billion by volume (ppbv) thresholds or even lower levels. This strategy holds the potential to detect and monitor a range of diseases, including respiratory disorders, metabolic diseases, and infectious illnesses. High-humidified exhaled gas consists of more than one compound quantifiable as concentration of ppbv to ppmv [23]. Studies have assessed fluctuations in exhaled breath as vital indicators of a variety of disease conditions, including metabolic disorders, infections, and cancers [24–26]. Human breath is composed of VOCs such as acetone, isoprene, ethane, and pentane, as well as inorganic gases such as carbon dioxide (CO_2), carbon monoxide (CO), and nitric oxide (NO), in addition to non-volatile compounds and exhaled breath condensates such as peroxynitrite, cytokines, and isoprostanes [21]. Various exhaled gases, such as NO, ammonia (NH_3), hydrogen sulfide (H_2S), and volatile organic compounds such as acetone (CH_3COCH_3), methane (CH_4), toluene ($C_6H_5CH_3$), and pentane are recognized as vital indicators of disease [27,28].

A study by Shorter et al. reported that breath analysis is a useful technique for diagnosing and monitoring asthma and chronic obstructive pulmonary disease (COPD) [29]. Carbon monoxide (CO) and nitric oxide in exhaled breath are signs of airway inflammation and can indicate the development of respiratory diseases. The change in NO levels of exhaled air between exacerbations coincide with the collected spirometry data. Breath analysis is important to understand the dynamic of exchange of NO and other types in the lungs and airways. It was found that NO flows are higher for those who suffer from simple asthma and during development. The concentration of NO level was higher in adults with asthma and COPD, as well as in children [29].

Boots et al. investigated two hundred samples of the region of gas immediately surrounding a bacterial sample or headspace, originating from four distinct microorganisms: *Escherichia coli*, *Pseudomonas aeruginosa*, *Staphylococcus aureus*, and *Klebsiella pneumonia*, and found a significant level of disparity in the occurrence of VOCs among various bacterial cultures [30]. Furthermore, the study successfully delineated a demarcation between methicillin-resistant and methicillin-sensitive isolates of *Staphylococcus aureus*, potentially offering a valuable diagnostic tool within the field of bacterial resistance.

Neurodegenerative conditions such as Alzheimer's disease (AD) and Parkinson's disease (PD) exhibit pathophysiological processes that commonly initiate well in advance of any detectable cognitive impairment symptoms [21]. A study used an assortment of metal oxide nanosensors integrated onto an SnO2 thin film, layered with Au, Cr, or Wo, configuring these sensors in two distinct ways. Analysis of exhaled breath samples using GC-MS and sensor systems revealed distinct dissimilarities in the chemical constituents present in the AD and PD patient cohorts (alkanes, benzene, and acetamide) as compared to those in healthy individuals. The configuration encompassing a greater number of sensors demonstrated enhanced distribution patterns and a superior capacity for discrimination based on factors such as oxidative stress and corresponding metal homeostasis [31].

Electronic nose (e-nose/EOS) devices employ multisensor arrays to differentiate intricate VOC combinations present in clinical air samples. They generate distinct sensor-response patterns, known as "smellprint signatures", which are disease-specific. Specialized statistical models and pattern recognition algorithms are then utilized to aid in discriminating among various samples. A study utilized a commercial metaloxide semiconductor chemiresistive e-nose (aeoNose™), comprising three distinct non-specific micro-hotplate metal-oxide nanosensors. Exposure of the sensors to exhaled gas prompts a redox reaction, leading to a conductivity alteration. Through the quantification of these alterations, the breath print was acquired which effectively differentiated individuals with the neurological disease multiple sclerosis, not undergoing medication, from healthy subjects, with a sensitivity of 93%, a specificity of 74%, and an overall accuracy of 80% [32].

A customized nanoscale artificial olfactory system, NA-Nose, comprising an assemblage of cross-reactive gas sensors composed of spherical gold nanoparticles, was employed to discern various odors within the alveolar exhaled breath of both cancer patients and healthy counterparts [33]. The nanosensors exhibited the capability to differentiate between individuals with head and neck cancer (HNC) and healthy subjects (83%), lung cancer (LC) patients (95%), and LC patients versus healthy controls (87%), relying on detection of some common volatile biomarkers, thus positioning it as a promising prospect for future screening applications [33].

Chronic kidney disease (CKD) poses a significant public health challenge, with its global prevalence on the rise, and is often undetected by patients. Specific compounds found in exhaled breath are associated with biochemical processes linked to CKD or the buildup of toxins resulting from impaired kidney function. To address this, carbon-nanotube-based sensors and breath sample analysis were employed as non-invasive tools for diagnosing end-stage renal disease in an animal model [34]. In an experimental model involving bilateral nephrectomy in rats, researchers differentiated between exhaled breath from healthy individuals versus those with chronic renal failure (CRF) through 15 common volatile organic compounds present in both healthy and CRF states, and twenty-seven VOCs unique to the CRF condition. Utilizing an array of organically coated single-walled carbon chemiresistive nanotubes, online breath analysis demonstrated accurate discrimination between different breath states. This discrimination could be further improved by reducing breath humidity before sensor analysis. In a separate study, gold nanoparticle-based sensors in conjunction with support vector machine analysis successfully detected CKD in breath samples. The sensor effectively distinguished between the exhaled breath of CKD patients and that of healthy individuals. Combining 2–3 gold nanoparticle sensors yielded notable differentiation between early-stage CKD and healthy individuals (79% accuracy) as well as between stage 4 and 5 CKD (85% accuracy). Furthermore, single sensor exhibited the capability to diagnose accurately (76%) both early- and late-stage CKD [35].

Numerous published studies have established associations between specific elements present in exhaled breath and various diseases. However, the practical implementation of these findings for diagnostic purposes has been hindered by the limited application, largely attributable to the inconsistent and conflicting identification of breath biomarkers and lack of large-scale trials. However, the potential of nano-biosensors in the field of nanomedicine, for testing and determination of an individual's health and application for a varied range of disease diagnosis, based on VOC emissions in exhaled breath analysis, is ever expanding.

4. Breathomics in COVID-19 Diagnostics

Expiratory breathomics, an emerging field at the intersection of metabolomics and exhaled breath analysis, has also gained traction as a promising avenue for advancing COVID-19 diagnostics. It can be used to evaluate inflammatory and oxidative tension in the respiratory system [36]. In the context of COVID-19, breathomics explores the unique quantifiable VOCs profiles associated with SARS-CoV-2 infection, including those arising from viral replication, host immune responses, and potential co-infections [37].

Remy et al. identified that the majority of volatile organic compound concentrations (hydrogen sulfide, methanol, acetic acid, butyric acid, and crotonaldehyde) exhibited suppression in individuals with COVID-19 [38]. Distinctions in VOC concentrations were also noted between symptomatic and asymptomatic cases. The breath-based markers reflected the impact of infections on the host's cellular metabolism and microbiome, with reduced levels of specific VOCs attributed to the suppressive influence of SARS-CoV-2 on microbial metabolism within the gastrointestinal or pulmonary systems.

A research investigation identified a set of volatile organic compounds comprising ethanal, 2-butanone, octanal, acetone, and methanol that effectively distinguished COVID-19 cases from other disease states [39]. In addition to these VOC biomarkers, other noteworthy differentiators between COVID-19-positive and/or negative states encompassed propanal, propanol, heptanal, and isoprene. Remarkably, an unidentified VOC denoted as M7-VOC exhibited remarkably strong predictive capabilities for assessing the severity of the disease or the risk of mortality. Furthermore, heptanal was also deemed significant in this context. The study's findings collectively suggest that these nine biomarker VOCs provide clear indications of COVID-19-related alterations in breath biochemistry, reflecting processes associated with ketosis, gastrointestinal effects, and inflammatory responses [39]. The predominant and recurring impact of SARS-CoV-2 infection on human metabolic pathways primarily pertains to certain aspects of amino acid metabolism. Notably, tryptophan metabolism has strong correlation with COVID-19 through a metabolic pathway, in which the production of nicotinamide adenine dinucleotide (Kynurenine pathway) has been observed in recent investigations [40].

It is worth noting that in a retrospective cohort study carried out at a single tertiary care hospital in Northern Italy during the initial wave of the pandemic, dual-energy computed tomography was used to quantitatively evaluate pulmonary gas and blood distribution [41]. The findings of the study revealed a significant difference between patients who underwent invasive ventilation and those receiving non-invasive respiratory support. Specifically, invasively ventilated patients exhibited a lower percentage of normally aerated tissue, amounting to 33% of their lung volume. In contrast, patients receiving non-invasive respiratory support had a substantially higher percentage of normally aerated tissue, accounting for 63% of their lung volume. A disparity in gas-to-blood volume mismatch between the two groups was also observed. In patients subjected to invasive ventilation, the extent of this mismatch was notably higher, constituting 43% of their lung volume compared to non-invasively ventilated patients, at 25% of their lung volume. Interestingly, the poorly aerated tissue in both groups exhibited distinctive characteristics, with some regions displaying a high gas-to-blood volume ratio and others presenting a low gas-to-blood volume ratio [41].

Diffusing capacity (DLCO) is a vital parameter that quantifies the volume of carbon monoxide transferred from alveolar gas to blood per minute. It is calculated by dividing this volume by the difference between the mean pressure of carbon monoxide in alveolar-capillary spaces and the mean pressure in pulmonary capillaries. DLCO serves as a valuable measure to evaluate the lungs' ability to facilitate the transfer of gases from inspired air into the bloodstream. A study aimed at investigating the underlying pathophysiology of alveolar-to-capillary gas exchange in patients recovering from COVID-19 pneumonia utilized both decreased lung-diffusing capacity for nitric oxide (DLNO) and DLCO [42]. The findings from this study revealed that DLCO exhibited variable results among subjects during the initial 3 months of recovery from severe COVID-19. In contrast, DLNO was more frequently and persistently impaired when compared to the standard DLCO measurement. This disparity suggests a specific impairment in diffusive conductance, likely attributed to damage to the alveolar unit alveolar-capillary interface, while the capillary volume remains relatively preserved. Notably, alterations in gas transport were detected even in subjects who had experienced mild COVID-19 pneumonia and displayed no or minimal lingering abnormalities in computed tomography scans [42]. These findings underscore the potential of both DLCO and DLNO as valuable functional biomarkers for COVID-19 breathomics-based point-of-care diagnostics, even in follow-up post-acute care facilities.

Another interesting study explored the potential of exhaled volatile organic compounds to differentiate COVID-19 vaccine recipients from non-vaccinated individuals with respect to the metabolic adjustments prompted by vaccination and the underlying mechanisms of immune-driven metabolic reprogramming [43]. The study looked at breath samples from both COVID-19 vaccine recipients and non-vaccinated controls analyzed for VOCs using comprehensive two-dimensional gas chromatography coupled with time-of-flight mass spectrometry. Specific VOCs were identified as biomarkers, forming a distinct VOC fingerprint through alveolar gradients and machine learning. This signature effectively distinguished vaccine recipients from the control group with an impressive prediction accuracy of 94.42% [43]. Further analysis of the metabolic pathways associated with these biomarkers revealed that host–pathogen interactions during vaccination enhanced enzymatic activity and microbial metabolism in the liver, lung, and gut, representative of the mechanisms behind vaccine-induced metabolic regulation. In comparison to non-vaccinated individuals, the levels of biomarkers—6-methyl-5-hepten-2-one, CPTO, methanesulfonyl chloride, and benzothiazole—as well as benzene/3-MHeptane levels, increased by approximately 70% in vaccine recipients, notably affecting ketone synthesis pathways such as β-oxidation of fatty acids with branched chains. Five other individual biomarkers (hexanal, benzene, acetonitrile, 2-MOctane, and phenol) decreased to half the levels observed in the control group among vaccine recipients. The findings indicated that vaccination accelerated the metabolism of aromatic hydrocarbons and monomethylated alkanes [43]. This study supports the potential use of breathomics in aspects of vaccine-induced metabolic regulation and in advancing the development of early clinical diagnostics.

Nurputra et al. outlined the selection of acetone as a volatile organic compound model in their investigation. Acetone was chosen due to its presence in rebreathed breath, typically found within the range of 0.8 to 2.0 parts per million (ppm), alongside other VOCs such as alcohol and carbon monoxide. Additionally, acetone stands out as a significant COVID-19 biomarker carried in the breath. Furthermore, in clinical settings, the presence of acetone in exhaled breath has been extensively studied for diagnosing various other conditions, including lung cancer, diabetes mellitus, starvation, and adherence to ketogenic diets [44].

Although the translation of breathomics into clinical practice warrants standardization and validation, its potential for non-invasive, rapid, and scalable COVID-19 diagnosis positions it as a promising tool for complementing existing testing methodologies and advancing our capacity to combat the disease.

5. Nano-Biosensor Breathalyzers for COVID-19 Detection

The nanobody-centered electrochemical methodology has enabled accelerated virus detection owing to its independence from requirements of reagents or extensive processing stages.

A. Promising bio-functionalization techniques and nanomaterials used in bio marker detection in COVID-19 diagnostics:

Biofunctionalization refers to the process of modifying a biomaterial to confer a particular biological function. This is typically achieved by integrating biomimetic molecules, serving as mediators or imitators of cellular systems. These mediators can be introduced into the scaffolds through surface adsorption or via nanospheres, and scaffold-bound complexes, which are subsequently released during scaffold degradation. Surface biofunctionalization entails either the immobilization of bioactive molecules on scaffold surfaces or the introduction of biofunctional surface groups to induce specific cellular responses.

Owing to their distinctive attributes, including novel optical characteristics, biodegradability, low toxicity, biocompatibility, size, and highly catalytic properties, these materials are deemed superior [9–11]. Nonetheless, functionalized nanoparticles face challenges such as rapid aggregation, oxidation, and related issues that impede their effective utilization. Effective surface fabrication of nanoparticles through the application of organic and inorganic compounds leads to enhanced physicochemical properties, surmounting these obstacles and potentially allowing a wide range of diagnostic applications. The underlying principle of development of COVID-19 breathalyzers relies on evidence indicating that the virus microenvironment emits VOCs detectable in exhaled air (Table 1). The appearance of VOCs in exhaled breath occurs during the early stages of infection, enabling the immediate detection of COVID-19.

The study by Shan et al. described a sensor array based on nanomaterials with multiplexed detection and monitoring abilities for COVID-19 in expirated air [45]. These sensing elements consisted of diverse spherical gold nanoparticles associated with organic ligands, forming a versatile and expandable or contractable silicon wafer sensing layer subsequent to VOC exposure, leading to changes in electrical resistance. The inorganic nanomaterials within the layers exhibited electrical conductance, while the organic elements adsorbed VOCs. Upon exposure, VOCs diffused into the sensor or settled on the surface, reacting with either organic segments or functional groups (thermal silicon oxide) on the inorganic nanomaterials. These interactions resulted in volume changes in the nanomaterial film, and the contacts among the inorganic nanomaterials restrained the changes in conductivity.

When the nanomaterial layer was exposed to VOCs, rapid transfer of charge with the inorganic nanomaterial occurred, leading measurable conductivity variations, even in the absence of sensor structural changes. Assorted functional group(s) coating the nanoparticles creates sensor variability, which could be utilized as cross-reactive semi-selective sensor arrays, imitating the olfactory repertoire of human systems. The resultant adaptability and potential for training for identification of chemical patterns, or "fingerprints," associated with diverse health conditions using pattern realization and machine learning are the primary focus for diagnostics development, relevant to both different stages of COVID-19 as well as other disease states.

Table 1. Sensor arrays for biomarker detection in COVID-19 breathalyzers.

Sensing Element	Biomarkers Detected	Reference
Molecularly capped AuNPs (8 ligands)	NA	[45]
γ-phase WO_3 sensor	Nitrogen oxide	[46]
Micro hotplate MOS	Carbon monoxide, nitrogen oxide and VOCs.	[47]
10 MOS Sensor Array	Ammonia and aromatic molecules, nitrogen oxide, hydrogen, Methane, propane, and aliphatics, Sulfur and chlorine containing organics Sulfur-containing organics, broad alcohols, and carbon chains.	[48,49]
4 MOS sensor array	NA	[50]

Exline et al. introduced a specialized NO nanoprobe within an electronic COVID-19 breathalyzer, employing a solitary catalytically active, resistive sensor highly specific for NO [46]. Nitric oxide is an oxidizing gas known for its strong affinity towards metal oxides possessing the perovskite modified cubic ReO_3 structure. Among the various metal oxides that conform to this polymorphic structure, γ-phase WO_3 stands out as a widely utilized material for resistive gas sensors. Accordingly, the developed WO_3 sensor demonstrated a remarkable sensitivity to NO in human exhaled breath, even in the presence of various concentrations of NO and prevalent breath VOCs such as acetone, isoprene, and ammonia. The monoclinic phase nanostructure of WO_3 was prepared through the alcoholic hydrolysis of tungsten VI isopropoxide, a metal alkoxide precursor. This nanostructure was then deposited as a thin film on alumina substrates with Pt electrodes and heaters, which were coated using the sol–gel technique.

Wintjens et al. detailed a sampling apparatus designed to detect SARS-CoV-2 in the oropharyngeal environment. This device utilizes machine learning classifiers alongside an electronic portable nose, known as Aeonose (The Aeonose Company in Zutphen, the Netherlands) [47]. The Aeonose is equipped with three micro-hotplate metal-oxide sensors: carbon monoxide (AS-MLC), nitrogen dioxide (AS-MLN), and VOC (AS-MLX) sensors. Each measurement cycle involves the sensors undergoing a sinusoidal temperature variation ranging from 260 to 340 °C, with 32 intervals comprising a single breath analysis. Exhaled volatile organic compounds interact with these sensors through a redox reaction, eliciting changes in conductivity and producing distinctive numerical patterns. These numerical data are transmitted to a data center for storage and subsequent analysis utilizing Aethena software. The software employs pattern recognition techniques to assess and indicate the probability of infection based on the captured sensor data.

The PEN3 eNose used by Snitz et al. in their study differentiates between different types of organics based on the specific functionalization of each of the 10 sensors in the device, each functionalized to detect specific VOCs or gases at different concentration levels [48]. These sensors include W1C detecting up to 5 ppm aromatics; W5S detecting up to 1 ppm ammonia and aromatic molecules; W3C detecting up to 5 ppm broad-nitrogen oxide; W6S detecting up to 5 ppm hydrogen; W5C detecting up to 1 ppm methane, propane, and aliphatics; W1S detecting up to 5 ppm broad-methane; W1W detecting up to 0.1 ppm sulfur-containing organics; W2S detecting up to 5 ppm broad-alcohols and broad-carbon chains; W2W detecting up to 1 ppm aromatics, sulfur- and chlorine-containing organics; and W3S detecting up to 5 ppm methane and aliphatics. The response patterns of each sensor to a given intranasal sample are analyzed using pattern recognition algorithms. By identifying and differentiating between various odor profiles or chemical signatures, the eNose can recognize unique patterns associated with specific organic compounds or mixtures.

Bax et al. utilized the self-powered "electronic nose" developed by SACMI S.C known as EOS-AROMA, which is used for measuring odors in various industries. The instrument is equipped with metal oxide semiconductor (MOS) sensors that respond to the presence of odorous compounds in the analyzed air [50]. The MOS sensor chamber is assisted by an electrical signal processing system that recognizes the "olfactory footprint" of a particular odor and quantifies its concentration. The EOS system is characterized by outstanding stability and high sensitivity, which allows it to measure even very diluted odors at very low olfactory thresholds. The EOS version of the electronic nose used in this study is equipped with three MOS sensors with an active film consisting of a tin oxide thin layer and tin oxide catalyzed with palladium [50]. The specific types of MOS sensors used in the EOS device include SnO_2, $SnO_2 + SiO_2$, $SnO_2 + Mo$, and $SnO_2 + MoO_2$ [49]. These sensors are designed to detect and quantify odors in various industrial and environmental, and even clinical, applications.

An Indonesian study by Nurpurta et al. used the GeNose C19, which is an electronic nose that integrates an array of metal oxide semiconductor gas sensors, optimized feature extraction, and machine learning models to rapidly sniff out COVID-19 based on exhaled breath-print recognition [44]. These nano-biosensors are antibody-based or DNA-based and allow for optical, electrochemical, or field effect transistor (FET)-based transduction. The transducer in the nano-biosensors is modified to capture the target element, convert the biological response into electrical signals, and quickly detect it with high accuracy. They offer several benefits, including high sensitivity, fast response, miniaturization, portability, and accuracy, making detection highly effective, even at very low concentrations. GeNose C19 consists of gas sensors and an artificial intelligence-based pattern recognition system, consisting of four different machine learning algorithms, enabling it to be used as a rapid, non-invasive screening tool in less than two minutes.

B. Device architecture

COVID-19 nano-biosensor breathalyzers have emerged as innovative diagnostic tools in the fight against the pandemic, due to their versatility in the detection of a range of gases and VOCs emitted during the different stages of infection. Figure 1 illustrates the structural layout utilized in constructing a typical breathalyzer setup. This model comprises a compact handheld apparatus adaptable for a disposable tube mouthpiece connected to the sensor chamber via a HEPA filter. Alternatively, a cost-effective single-use device, potentially 3D printed, with an incorporated mouthpiece, has also been developed, eliminating the need for disinfection after each use. The procedure involves exhaling into the mouthpiece briefly and observing the outcome on a separate device.

Within the sensor chamber, expiratory breath condensation gathers on a dielectric cold mirror, usually coated with a hydrophobic film, and angled at 45 degrees, facilitating breath interaction. Exhaled breath transforms into minute droplets that envelop an electrochemical biosensor. The efficacy of the breathalyzer's detection hinges on the nano-biosensor type employed, featuring distinct electrodes for various targets. Developing a nanosensor tailored for COVID-19-associated gases or VOCs is pivotal.

Hybrid organic/inorganic configurations featuring 3D nano-architectures such as nanofibers, nanowires, and nanofins are instrumental in augmenting the active surface-area-to-volume ratios. In this context, the surfaces of semiconductor nanostructures frequently undergo functionalization with specific self-assembled monolayers or polymers. Organic materials exhibit limited durability. They have been widely recognized for their tendency to degrade relatively quickly during usage, leading to alterations in their chemical compositions and consequently compromising sensor performance. Consequently, sensor manufacturing prioritizes the production of pure inorganic materials, specifically metal oxide semiconductors, which remain extensively employed in gas-sensing applications due to their superior stability and reliability over time [44]. Electrodes are connected to contacts to fabricate the sensor. The crafted differentially sensitive and selective sensor layers and arrays facilitate precise detection of the targeted gas molecules and/or VOCs. Sensing assessments are often conducted within a quartz chamber comprising a regulated electric furnace and electrical terminals linked to computerized data acquisition. The resultant signal is converted into an electric circuit. Most prototypes integrate an electrical readout integrated circuit linked to a microsystem of sensor arrays capable of detecting exhaled breath and analyzing gas/VOC content. Test evaluation outcomes would be accessible to end-users through a smartphone application.

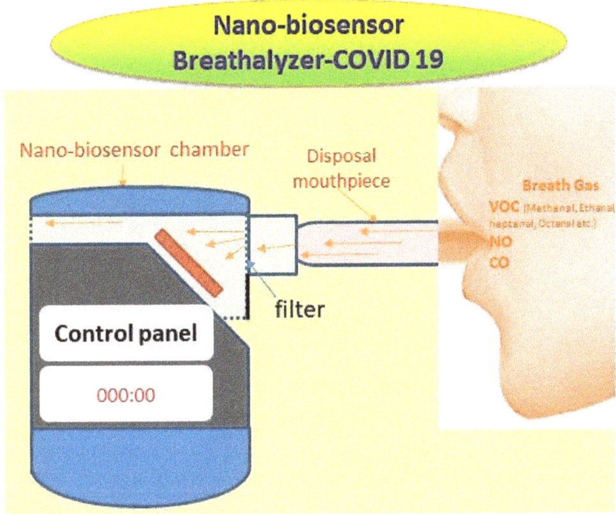

Figure 1. Potential nano-biosensor breath-analyzer for COVID-19.

C. Breath analysis studies for COVID-19

A study from the early months of the pandemic in Wuhan, China focused on a handheld nanomaterial-based sensor array-based breathalyzer called the NaNose electronic nose (e-nose) developed with wider functionality for detection and continuous monitoring of COVID-19 in expirated air [45]. Gold nanoparticles were used in the sensors, which were connected to organic ligands, forming a diverse sensing layer capable of expanding or contracting upon exposure to VOCs, resulting in varying electrical resistance. Within the sensor, inorganic nanomaterials conduct electricity, while the organic elements adsorp VOCs. The system was reliant on a training machine learning algorithm. The cohort consisted of 49 COVID-19 patients, 58 healthy controls, and 33 individuals with non-COVID lung infections. Utilizing the test data, the system achieved an accuracy of 94% and 76%, respectively, in distinguishing patients from controls, and it exhibited accuracy rates of 90% and 95% in distinguishing patients with COVID-19 from those with other lung infections.

Exline and colleagues introduced an innovative electronic breathalyzer technology featuring a solitary sensor composed of γ-phase tungsten trioxide (WO_3) as a catalytically active semiconductor sensing film [46]. This film was fabricated through sol–gel processing employing tungsten alkoxide precursors, and the sensor was designed for the rapid detection of NO and ammonia in exhaled breath, providing diagnostic results within a mere 15 s. The breathalyzer demonstrated the detection of elevated exhaled NO levels, exhibiting a unique pattern characteristic of individuals suffering from active COVID-19 pneumonia. Remarkably, it achieved an 88% accuracy rate in identifying COVID-19 pneumonia patients upon their admission to the intensive care unit (ICU). Furthermore, the sensitivity index associated with the breath print, which is closely linked to the concentration of the critical biomarker ammonia, appeared to exhibit a correlation with the duration of COVID-19 infection.

Among the many electronic nose device-based studies thus far, the majority are equipped with metal oxide semiconductor (MOS) sensors, typically containing 10 or fewer MOS sensors [47–49]. However, there are a few exceptions with sensor array type instruments that stand out. For instance, the handy EOS Cyranose 320 is noteworthy for its use of thirty-two sensors of carbon black polymer composite, while the NaNose utilizes eight sensors of gold nanoparticle [51,52]. It is worth noting that all these devices, except for one, were in existence and had undergone prior developmental stages before the emergence of the pandemic. Although not originally designed for COVID-19, however, evaluation for their potential utility in identifying other respiratory diseases and cancer was successfully carried out. Kwiatkowski and colleagues engineered an e-nose with innovative sensor array architecture and that is a cheap POC testing and screening device. This array comprises three micro-electro-mechanical system (MEMS) sensors and a solitary MOS sensor. MEMS are chip-sensors, constructed with a suspended mass between two capacitive plates. When sensors are tilted, this configuration generates a disparity in electrical potential, a phenomenon that is quantified as a variation in capacitance [53]. Thus, nano-breathalyzers have the potential to revolutionize the COVID-19 testing process with cheap, non-invasive, point-of-care detection devices with wide employability to service sectors such as schools, workplaces, hospitals, and large public gatherings/events (Figure 2).

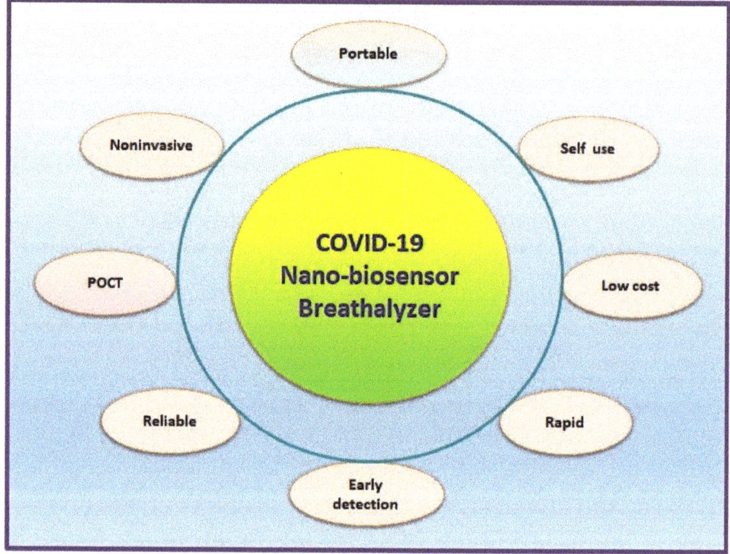

Figure 2. Advantages and uses of nano-biosensor breathalyzer for COVID-19.

D. Future prospects

Following the stabilization of the acute phase of the COVID-19 pandemic, scientists are now exploring ways to continuously monitor indoor spaces for airborne transmission of viruses in real-time without necessitating specialized expertise. Leveraging recent advancements in aerosol sampling technology and highly sensitive nano-biosensing techniques, researchers are directing efforts towards creating tools capable of promptly detecting the SARS-CoV-2 virus in indoor environments.

A compact, cost-effective, and real-time proof-of-concept air quality (pAQ) monitor with the capability to identify airborne SARS-CoV-2 virus variants within indoor settings within a 5 min timeframe has been engineered by Puthussery et al. [54] This innovative device holds potential applications in hospitals, healthcare facilities, schools, and public areas, aiding in the detection of CoV-2 and potentially monitoring other respiratory virus aerosols. The system integrates a high-volume, high-flow (approximately 1000 L per minute) wet cyclone air sampler with an ultrasensitive micro-immunoelectrode (MIE) previously utilized in Alzheimer's disease detection. The nanobody, derived from llamas and specifically targeting the SARS-CoV-2 spike antigen, is bonded covalently to the MIE biosensor. Employing screen-printed carbon electrodes, the biosensor utilizes the MIE technique to detect the presence of virus aerosols. Through square wave voltammetry, it identifies oxidation of tyrosine amino acids within the spike protein, offering a quantitative measure of the virus concentration in the sampled air. Comparative assessments demonstrate that the wet cyclone device exhibits similar/superior virus identification compared to commercial samplers, boasting a sensitivity ranging from 77% to 83% and a detection limit of 7–35 viral RNA copies per cubic meter of air [54]. This air quality monitor is tailored for immediate surveillance of virus variants that are detectable indoors and holds potential for adaptation to simultaneously detect various other respiratory pathogens of interest.

Artificial intelligence (AI) and machine learning (ML) play a significant role in enhancing the capabilities of nanomaterial-based biosensors. They have provided new avenues in biomedical monitoring, clinical diagnosis, and high-precision detection of diseases such as COVID-19 [55]. ML-enabled biosensors can offer real-time analysis of breath samples, leading to intelligent healthcare management and early disease detection. In the context of COVID-19, the combination of nano-biosensors with AI and ML can lead to the development of highly sensitive and specific breathalyzers for the rapid and accurate detection of the virus. These advancements hold promise for the future, offering the potential for widespread and accessible screening and early detection of COVID-19 and other respiratory viruses [56,57].

To address limitations associated with variations in test accuracy related to sample composition and storage in breath analysis, research efforts should prioritize standardizing sampling protocols and procedural methodologies [58]. Enhancements in systems should target mitigating technical, physiological, and pathophysiological variables that might affect the results. This approach aims to identify endogenous VOCs and establish reliable exhaled biomarker patterns [8].

Efforts in this direction involve establishing standardized correlations between concentrations of VOCs in blood and breath [8]. Additionally, development of handy and cost-effective nanomaterial sensors that resist humidity should be prioritized. These sensors should fulfill clinical requirements, displaying selectivity for VOCs and inorganic gases identified in specific diseases, and ensuring swift population screening. Moreover, optimizing sensor training and validation using diverse subjects becomes crucial for devising sensors applicable in clinical diagnosis [59]. A noteworthy aspect is the emerging capacity of such systems to differentiate between various diseases, providing diagnoses of conditions that manifest similar clinical presentations. This aspect, previously overlooked, now demands considerable attention in the pursuit of advancing diagnostic capabilities [59].

The ongoing advancements in innovative nanomaterials present a significant avenue for enhancing sensing components, particularly for the creation of sensors that exhibit both selectivity and cross-reactivity. This progress holds promise, particularly in the realms of POC diagnosis and treatment. Several fundamental challenges within this emerging area impede practical application of breathomics in medical settings, necessitating focused attention.

6. Conclusions

Rapid advancements in sensor technologies, coupled with sophisticated data analysis techniques, have paved the way for accurate and sensitive detection of gases as well as specific VOC patterns associated with different diseases. The utilization of large, costly, and intricate analytical devices for identifying exhaled volatile organic compounds faces limitations in hospital settings, and their integration into portable point-of-care systems remains unachievable. The simplicity of sample collection, coupled with the ability to provide real-time results, positions breath analysis as an attractive tool for early COVID-19 diagnosis, disease progression monitoring, and treatment response evaluation. Electronic-nose technology and related VOC detection methods hold the potential for early and noninvasive identification of COVID-19 individuals who are pre-symptomatic or asymptomatic carriers of the virus, displaying no overt symptoms following infection. The multitude of advantages associated with early COVID-19 detection through noninvasive means offers substantial improvements in precision medicine. These benefits encompass reduced transmission rates, enhanced effectiveness of regional epidemiological management, expanded choices for more efficacious treatments, improved prognostic capabilities, and potentially lower COVID-19-related mortality rates across various age groups, racial backgrounds, and ethnic populations. However, validation of breath analysis results is important due to the impact of trace levels of expired VOCs on accuracy. Concurrently, the absence of well-defined breath sampling protocols and procedures for breath collection and storage pose crucial challenges, potentially altering sample composition and consequently affecting analysis outcomes. Identifying the specific VOCs that consistently serve as reliable chemical biomarkers for SARS-CoV-2 infections in individuals with diverse ailment histories, health conditions, immune responses, and nutritional statuses greatly enhances the effectiveness of COVID-19 diagnoses through the utilization of nano-biosensor breathalyzer devices. Fulfilment of these objectives hinges upon interdisciplinary research and collaboration as fundamental prerequisites. Breath analysis emerges as a potent diagnostic tool that, upon surmounting prevailing challenges, holds the potential for adoption in clinical settings or within portable, compact health monitoring systems.

Author Contributions: Conceptualization, S.R., S.S., F.A. and N.S.; writing—original draft preparation, S.R., S.S. and F.A.; writing—review and editing, A.A., K.A.-M. and M.W.A.K.; supervision, S.S.; funding acquisition, S.S. All authors have read and agreed to the published version of the manuscript.

Funding: This research has been funded by Scientific Research Deanship at University of Ha'il-Saudi Arabia through project number MDR-22 016.

Data Availability Statement: Not applicable.

Conflicts of Interest: The authors declare no conflict of interest.

References

1. Guan, W.J.; Ni, Z.Y.; Hu, Y.; Liang, W.H.; Ou, C.Q.; He, J.X.; Liu, L.; Shan, H.; Lei, C.L.; Hui, D.S.C.; et al. Clinical characteristics of coronavirus disease 2019 in China. *N. Engl. J. Med.* **2020**, *382*, 1708–1720. [CrossRef] [PubMed]
2. Sherwani, S.; Khan, M.W.A.; Mallik, A.; Khan, M.; Saleem, M.; Raafat, M.; Shati, A.A.; Alam, N. Seroprevalence of Anti-S1-RBD Antibodies in Pre-pandemic and Pandemic Subjects From Hail Region, KSA. *Front. Public Health* **2022**, *10*, 874741. [CrossRef] [PubMed]
3. Alluwaimi, A.M.; Alshubaith, I.H.; Al-Ali, A.M.; Abohelaika, S. The coronaviruses of animals and birds: Their zoonosis, vaccines, and models for SARS-CoV and SARS-CoV-2. *Front. Vet. Sci.* **2020**, *7*, 582287. [CrossRef] [PubMed]
4. Sherwani, S.; Khan, M.W.A. Cytokine response in SARS-CoV-2 infection in the elderly. *J. Inflamm. Res.* **2020**, *13*, 737–747. [CrossRef] [PubMed]
5. Peaper, D.R.; Kerantzas, C.A.; Durant, T.J. Advances in molecular infectious diseases testing in the time of COVID-19. *Clin. Biochem.* **2023**, *117*, 94–101. [CrossRef] [PubMed]
6. Dutta, D.; Naiyer, S.; Mansuri, S.; Soni, N.; Singh, V.; Bhat, K.H.; Singh, N.; Arora, G.; Mansuri, M.S. COVID-19 diagnosis: A comprehensive review of the RT-qPCR method for detection of SARS-CoV-2. *Diagnostics* **2022**, *12*, 1503. [CrossRef] [PubMed]
7. Gowri, A.; Kumar, N.A.; Anand, B.S. Recent advances in nanomaterials based biosensors for point of care (PoC) diagnosis of COVID-19–a minireview. *TrAC Trends Anal. Chem.* **2021**, *137*, 116205. [CrossRef]
8. Das, S.; Pal, M. Non-invasive monitoring of human health by exhaled breath analysis: A comprehensive review. *J. Electrochem. Soc.* **2020**, *167*, 037562. [CrossRef]
9. Chamorro-Garcia, A.; Merkoçi, A. Nanobiosensors in diagnostics. *Nanobiomed* **2016**, *3*, 1849543516663574. [CrossRef]

10. Naresh, V.; Lee, N. A Review on Biosensors and Recent Development of Nanostructured Materials-Enabled Biosensors. *Sensors* **2021**, *21*, 1109. [CrossRef]

11. Willner, M.R.; Vikesland, P.J. Nanomaterial enabled sensors for environmental contaminants. *J. Nanobiotechnology* **2018**, *16*, 95. [CrossRef] [PubMed]

12. Tian, R.; Weng, T.; Chen, S.; Wu, J.; Yin, B.; Ma, W.; Liang, L.; Xie, W.; Wang, Y.; Zeng, X.; et al. DNA nanostructure-assisted detection of carcinoembryonic antigen with a solid-state nanopore. *Bioelectrochemistry* **2023**, *149*, 108284. [CrossRef] [PubMed]

13. Sadeghi, M.; Kashanian, S.; Naghib, S.M.; Askari, E.; Haghiralsadat, F.; Tofighi, D. A highly sensitive nanobiosensor based on aptamer-conjugated graphene-decorated rhodium nanoparticles for detection of HER2-positive circulating tumor cells. *Nanotechnol. Rev.* **2022**, *11*, 793–810. [CrossRef]

14. Chowdhury, A.D.; Takemura, K.; Khorish, I.M.; Nasrin, F.; Tun, M.M.N.; Morita, K.; Park, E.Y. The detection and identification of dengue virus serotypes with quantum dot and AuNP regulated localized surface plasmon resonance. *Nanoscale Adv.* **2020**, *2*, 699–709. [CrossRef] [PubMed]

15. Pang, Y.; Rong, Z.; Wang, J.; Xiao, R.; Wang, S. A fluorescent aptasensor for H5N1 influenza virus detection based-on the core–shell nanoparticles metal-enhanced fluorescence (MEF). *Biosens. Bioelectron.* **2015**, *66*, 527–532. [CrossRef]

16. He, Y.; Tian, F.; Zhou, J.; Zhao, Q.; Fu, R.; Jiao, B. Colorimetric aptasensor for ochratoxin A detection based on enzyme-induced gold nanoparticle aggregation. *J. Hazard. Mater.* **2020**, *388*, 121758. [CrossRef]

17. Skotadis, E.; Aslanidis, E.; Tsekenis, G.; Panagopoulou, C.; Rapesi, A.; Tzourmana, G.; Kennou, S.; Ladas, S.; Zeniou, A.; Tsoukalas, D. Hybrid Nanoparticle/DNAzyme Electrochemical Biosensor for the Detection of Divalent Heavy Metal Ions and Cr^{3+}. *Sensors* **2023**, *23*, 7818. [CrossRef]

18. Samson, R.; Navale, G.R.; Dharne, M.S. Biosensors: Frontiers in rapid detection of COVID-19. *3 Biotech* **2020**, *10*, 385. [CrossRef]

19. Karakuş, E.; Erdemir, E.; Demirbilek, N.; Liv, L. Colorimetric and electrochemical detection of SARS-CoV-2 spike antigen with a gold nanoparticle-based biosensor. *Anal. Chim. Acta* **2021**, *1182*, 338939. [CrossRef]

20. Zhang, M.; Li, X.; Pan, J.; Zhang, Y.; Zhang, L.; Wang, C.; Yan, X.; Liu, X.; Lu, G. Ultrasensitive detection of SARS-CoV-2 spike protein in untreated saliva using SERS-based biosensor. *Biosens. Bioelectron.* **2021**, *190*, 113421. [CrossRef]

21. Kaloumenou, M.; Skotadis, E.; Lagopati, N.; Efstathopoulos, E.; Tsoukalas, D. Breath Analysis: A Promising Tool for Disease Diagnosis—The Role of Sensors. *Sensors* **2022**, *22*, 1238. [CrossRef] [PubMed]

22. Vishinkin, R.; Haick, H. Nanoscale Sensor Technologies for Disease Detection via Volatolomics. *Small* **2015**, *11*, 6142–6164. [CrossRef] [PubMed]

23. Corradi, M.; Mutti, A. Exhaled breath analysis: From occupational to respiratory medicine. *Acta Bio-Medica Atenei Parm.* **2005**, *76* (Suppl. S2), 20.

24. Güntner, A.T.; Abegg, S.; Königstein, K.; Gerber, P.A.; Schmidt-Trucksäss, A.; Pratsinis, S.E. Breath Sensors for Health Monitoring. *ACS Sens.* **2019**, *4*, 268–280. [CrossRef] [PubMed]

25. de Lacy Costello, B.; Amann, A.; Al-Kateb, H.; Flynn, C.; Filipiak, W.; Khalid, T.; Osborne, D.; Ratcliffe, N.M. A review of the volatiles from the healthy human body. *J. Breath Res.* **2014**, *8*, 14001. [CrossRef]

26. Kabir, E.; Raza, N.; Kumar, V.; Singh, J.; Tsang, Y.F.; Lim, D.K.; Szulejko, J.E.; Kim, K.-H. Recent Advances in Nanomaterial-Based Human Breath Analytical Technology for Clinical Diagnosis and the Way Forward. *Chem* **2019**, *5*, 3020–3057. [CrossRef]

27. Wallace, M.A.G.; Pleil, J.D. Evolution of clinical and environmental health applications of exhaled breath research: Review of methods and instrumentation for gas-phase, condensate, and aerosols. *Anal. Chim. Acta* **2018**, *1024*, 18–38. [CrossRef]

28. Cikach, F.S., Jr.; Dweik, R.A. Cardiovascular Biomarkers in Exhaled Breath. *Prog. Cardiovasc. Dis.* **2012**, *55*, 34–43. [CrossRef]

29. Shorter, J.H.; Nelson, D.D.; McManus, J.B.; Zahniser, M.S.; Sama, S.R.; Milton, D.K. Clinical study of multiple breath biomarkers of asthma and COPD (NO, CO_2, CO and N_2O) by infrared laser spectroscopy. *J. Breath Res.* **2011**, *5*, 037108. [CrossRef]

30. Boots, A.W.; Smolinska, A.; van Berkel, J.J.B.N.; Fijten, R.R.R.; Stobberingh, E.E.; Boumans, M.L.L.; Moonen, E.J.; Wouters, E.F.M.; Dallinga, J.W.; Van Schooten, F.J. Identification of microorganisms based on headspace analysis of volatile organic compounds by gas chromatography–mass spectrometry. *J. Breath Res.* **2014**, *8*, 027106. [CrossRef]

31. Lau, H.-C.; Yu, J.-B.; Lee, H.-W.; Huh, J.-S.; Lim, J.-O. Investigation of Exhaled Breath Samples from Patients with Alzheimer's Disease Using Gas Chromatography-Mass Spectrometry and an Exhaled Breath Sensor System. *Sensors* **2017**, *17*, 1783. [CrossRef] [PubMed]

32. Ettema, A.R.; Lenders, M.W.P.M.; Vliegen, J.; Slettenaar, A.; Tjepkema-Cloostermans, M.C.; de Vos, C.C. Detecting multiple sclerosis via breath analysis using an eNose, a pilot study. *J. Breath Res.* **2020**, *15*, 027101. [CrossRef] [PubMed]

33. Hakim, M.; Billan, S.; Tisch, U.; Peng, G.; Dvrokind, I.; Marom, O.; Abdah-Bortnyak, R.; Kuten, A.; Haick, H. Diagnosis of head-and-neck cancer from exhaled breath. *Br. J. Cancer* **2011**, *104*, 1649–1655. [CrossRef] [PubMed]

34. Haick, H.; Hakim, M.; Patrascu, M.; Levenberg, C.; Shehada, N.; Nakhoul, F.; Abassi, Z. Sniffing Chronic Renal Failure in Rat Model by an Array of Random Networks of Single-Walled Carbon Nanotubes. *ACS Nano* **2009**, *3*, 1258–1266. [CrossRef] [PubMed]

35. Marom, O.; Nakhoul, F.; Tisch, U.; Shiban, A.; Abassi, Z.; Haick, H. Gold nanoparticle sensors for detecting chronic kidney disease and disease progression. *Nanomedicine* **2012**, *7*, 639–650. [CrossRef]

36. Ibrahim, W.; Carr, L.; Cordell, R.; Wilde, M.J.; Salman, D.; Monks, P.S.; Thomas, P.; Brightling, C.E.; Siddiqui, S.; Greening, N.J. Breathomics for the clinician: The use of volatile organic compounds in respiratory diseases. *Thorax* **2021**, *76*, 514–521. [CrossRef]

37. Lichtenstein, M.; Turjerman, S.; Pinto, J.M.; Barash, O.; Koren, O. Pathophysiology of SARS-CoV-2 Infection in the Upper Respiratory Tract and Its Relation to Breath Volatile Organic Compounds. *mSystems* **2021**, *6*, e00104-21. [CrossRef]

38. Remy, R.; Kemnitz, N.; Trefz, P.; Fuchs, P.; Bartels, J.; Klemenz, A.-C.; Rührmund, L.; Sukul, P.; Miekisch, W.; Schubert, J.K. Profiling of exhaled volatile organics in the screening scenario of a COVID-19 test center. *iScience* **2022**, *25*, 105195. [CrossRef]

39. Ruszkiewicz, D.M.; Sanders, D.; O'Brien, R.; Hempel, F.; Reed, M.J.; Riepe, A.C.; Bailie, K.; Brodrick, E.; Darnley, K.; Ellerkmann, R.; et al. Diagnosis of COVID-19 by analysis of breath with gas chromatography-ion mobility spectrometry—A feasibility study. *EClinicalMedicine* **2020**, *29–30*, 100609. [CrossRef]

40. Badawy, A.A.B. The kynurenine pathway of tryptophan metabolism: A neglected therapeutic target of COVID-19 pathophysiology and immunotherapy. *Biosci. Rep.* **2023**, *43*, BSR20230595. [CrossRef]

41. Ball, L.; Robba, C.; Herrmann, J.; Gerard, S.E.; Xin, Y.; Mandelli, M.; Battaglini, D.; Brunetti, I.; Minetti, G.; Seitun, S.; et al. Lung distribution of gas and blood volume in critically ill COVID-19 patients: A quantitative dual-energy computed tomography study. *Crit. Care* **2021**, *25*, 214. [CrossRef] [PubMed]

42. Fuschillo, S.; Ambrosino, P.; Motta, A.; Maniscalco, M. COVID-19 and diffusing capacity of the lungs for carbon monoxide: A clinical biomarker in postacute care settings. *Biomarkers Med.* **2021**, *15*, 537–539. [CrossRef] [PubMed]

43. Cen, Z.; Lu, B.; Ji, Y.; Chen, J.; Liu, Y.; Jiang, J.; Li, X.; Li, X. Virus-induced breath biomarkers: A new perspective to study the metabolic responses of COVID-19 vaccinees. *Talanta* **2023**, *260*, 124577. [CrossRef] [PubMed]

44. Nurputra, D.K.; Kusumaatmaja, A.; Hakim, M.S.; Hidayat, S.N.; Julian, T.; Sumanto, B.; Mahendradhata, Y.; Saktiawati, A.M.I.; Wasisto, H.S.; Triyana, K. Fast and noninvasive electronic nose for sniffing out COVID-19 based on exhaled breath-print recognition. *NPJ Digit. Med.* **2022**, *5*, 115. [CrossRef] [PubMed]

45. Shan, B.; Broza, Y.Y.; Li, W.; Wang, Y.; Wu, S.; Liu, Z.; Wang, J.; Gui, S.; Wang, L.; Zhang, Z.; et al. Multiplexed Nanomaterial-Based Sensor Array for Detection of COVID-19 in Exhaled Breath. *ACS Nano* **2020**, *14*, 12125–12132. [CrossRef]

46. Exline, M.C.; Stanacevic, M.; Bowman, A.S.; Gouma, P.-I. Exhaled nitric oxide detection for diagnosis of COVID-19 in critically ill patients. *PLoS ONE* **2021**, *16*, e0257644. [CrossRef]

47. Wintjens, A.G.W.E.; Hintzen, K.F.H.; Engelen, S.M.E.; Lubbers, T.; Savelkoul, P.H.M.; Wesseling, G.; van der Palen, J.A.M.; Bouvy, N.D. Applying the electronic nose for pre-operative SARS-CoV-2 screening. *Surg. Endosc.* **2021**, *35*, 6671–6678. [CrossRef]

48. Snitz, K.; Andelman-Gur, M.; Pinchover, L.; Weissgross, R.; Weissbrod, A.; Mishor, E.; Zoller, R.; Linetsky, V.; Medhanie, A.; Shushan, S.; et al. Proof of concept for real-time detection of SARS-CoV-2 infection with an electronic nose. *PLoS ONE* **2021**, *16*, e0252121. [CrossRef]

49. Eusebio, L.; Derudi, M.; Capelli, L.; Nano, G.; Sironi, S. Assessment of the Indoor Odour Impact in a Naturally Ventilated Room. *Sensors* **2017**, *17*, 778. [CrossRef]

50. Bax, C.; Robbiani, S.; Zannin, E.; Capelli, L.; Ratti, C.; Bonetti, S.; Novelli, L.; Raimondi, F.; Di Marco, F.; Dellacà, R.L. An Experimental Apparatus for E-Nose Breath Analysis in Respiratory Failure Patients. *Diagnostics* **2022**, *12*, 776. [CrossRef]

51. Zamora-Mendoza, B.N.; de León-Martínez, L.D.; Rodríguez-Aguilar, M.; Mizaikoff, B.; Flores-Ramírez, R. Chemometric analysis of the global pattern of volatile organic compounds in the exhaled breath of patients with COVID-19, post-COVID and healthy subjects. Proof of concept for post-COVID assessment. *Talanta* **2022**, *236*, 122832. [CrossRef] [PubMed]

52. VR, N.; Mohapatra, A.K.; VK, U.; Lukose, J.; Kartha, V.B.; Chidangil, S. Post-COVID syndrome screening through breath analysis using electronic nose technology. *Anal. Bioanal. Chem.* **2022**, *414*, 3617–3624. [CrossRef]

53. Kwiatkowski, A.; Borys, S.; Sikorska, K.; Drozdowska, K.; Smulko, J.M. Clinical studies of detecting COVID-19 from exhaled breath with electronic nose. *Sci. Rep.* **2022**, *12*, 15990. [CrossRef] [PubMed]

54. Puthussery, J.V.; Ghumra, D.P.; McBrearty, K.R.; Doherty, B.M.; Sumlin, B.J.; Sarabandi, A.; Mandal, A.G.; Shetty, N.J.; Gardiner, W.D.; Magrecki, J.P.; et al. Real-time environmental surveillance of SARS-CoV-2 aerosols. *Nat. Commun.* **2023**, *14*, 3692. [CrossRef] [PubMed]

55. Manickam, P.; Mariappan, S.A.; Murugesan, S.M.; Hansda, S.; Kaushik, A.; Shinde, R.; Thipperudraswamy, S.P. Artificial Intelligence (AI) and Internet of Medical Things (IoMT) Assisted Biomedical Systems for Intelligent Healthcare. *Biosensors* **2022**, *12*, 562. [CrossRef]

56. Velumani, M.; Prasanth, A.; Narasimman, S.; Chandrasekhar, A.; Sampson, A.; Meher, S.R.; Rajalingam, S.; Rufus, E.; Alex, Z.C. Nanomaterial-Based Sensors for Exhaled Breath Analysis: A Review. *Coatings* **2022**, *12*, 1989. [CrossRef]

57. Junaid, S.B.; Imam, A.A.; Abdulkarim, M.; Surakat, Y.A.; Balogun, A.O.; Kumar, G.; Shuaibu, A.N.; Garba, A.; Sahalu, Y.; Mohammed, A.; et al. Recent Advances in Artificial Intelligence and Wearable Sensors in Healthcare Delivery. *Appl. Sci.* **2022**, *12*, 10271. [CrossRef]

58. Bikov, A.; Lazar, Z.; Horvath, I. Established methodological issues in electronic nose research: How far are we from using these instruments in clinical settings of breath analysis? *J. Breath Res.* **2015**, *9*, 034001. [CrossRef]

59. Koureas, M.; Kirgou, P.; Amoutzias, G.; Hadjichristodoulou, C.; Gourgoulianis, K.; Tsakalof, A. Target Analysis of Volatile Organic Compounds in Exhaled Breath for Lung Cancer Discrimination from Other Pulmonary Diseases and Healthy Persons. *Metabolites* **2020**, *10*, 317. [CrossRef]

MDPI

St. Alban-Anlage 66

4052 Basel

Switzerland

www.mdpi.com

Inorganics Editorial Office

E-mail: inorganics@mdpi.com

www.mdpi.com/journal/inorganics